Cellular and Molecular Aspects of Implantation

Cellular and Molecular Aspects of Implantation

Edited by
Stanley R. Glasser
and
David W. Bullock
Baylor College of Medicine
Houston, Texas

PLENUM PRESS · NEW YORK AND LONDON

Library of Congress Cataloging in Publication Data

Main entry under title:

Cellular and molecular aspects of implantation.

Proceedings of a conference held in Houston, Tex., Sept. 17-19, 1979, which was sponsored by the Center for Population Research, National Institute of Child Health and Human Development, and other bodies.

Includes index.

1. Ovum implantation—Congresses. I. Glasser, Stanley. II. Bullock, David W. III. United States. National Institute of Child Health and Human Development. Center for Population Research.

QP275.C44 599.01′6 80-20471

ISBN-13: 978-1-4613-3182-7 e-ISBN-13: 978-1-4613-3180-3
DOI: 10.1007/978-1-4613-3180-3

Softcover reprint of the hardcover 1st edition 1981
A Division of Plenum Publishing Corporation
227 West 17th Street, New York, N.Y. 10011

"It is characteristic of eggs and early embryos of lower animals that they are prepared to develop without shelter and nutriment from the mother. When the mammals evolved the phenomenon of utero-gestation, the chosen place of shelter, the uterus, was developed from the part of the oviduct, a channel that had for its purpose the efficient transportation and discharge of eggs, not their retention and maintenance. To fit it for gestational functions, the endocrine mechanism of the corpus luteum was evolved. In the light of this thought it is not surprising that the uterine chamber is actually a less favorable place for early embryos than, say, the anterior chamber of the eye, except when the hormones of the ovary act upon it and change it to a place of superior efficiency for its new function."

George W. Corner
The Hormones in Human Reproduction, 1947

Preface

Long was I hugg'd close—long and long.
Immense have been the preparations for me,
Faithful and friendly the arms that have help'd me.
Cycles ferried my cradle, rowing and rowing like friendly boatmen.
For room to me stars kept aside in their own rings,
They sent influences to look after what was to hold me.
Before I was born out of my mother, generations guided me,
My embryo has never been torpid, nothing could overlay it.

—Walt Whitman, "Song of Myself"

The womb is the seat of all mammalian life. In pregnancy, the uterus acquires this importance with the arrival of the fertilized egg, which takes up residence for periods ranging from about 2 weeks in the opossum to about 2 years in the elephant. The arrival of the embryo signals a crucial time for the establishment of pregnancy. For several days the blastocyst remains free in the uterine lumen, where it depends on uterine secretions for its survival and differentiation. During this time, essential changes in the endometrium take place in preparation for attachment of the blastocyst and implantation. Early embryonic loss is an economic problem of global proportions in animal husbandry, where, in pigs and cattle for example, some 30% of all fertilizations fail to result in a pregnancy. In humans this figure may be even higher, and estimates of early spontaneous abortions range from 40 to 60% of all conceptions. Because the time of implantation is before the end of the menstrual cycle, failure of the embryo to implant could occur with little or no delay in a woman's menses. Thus many women engaging in unprotected intercourse may initiate a pregnancy without ever being aware of the fact. From the point of view of human fertility and of animal food production, therefore, knowledge of the mechanisms of implantation and the factors governing this process assumes considerable importance.

Research in the past has provided much information about the developmental steps related to implantation in many species. While the results have been of immense value, the potential of the techniques commonly employed has been exhausted, and research on implantation has stagnated as a consequence. Our understanding of the interaction between the embryo and the uterus and of its regulation remained primitive. In recent years powerful new tools have become available from advances in the fields of cellular and molecular biology. Few investigators in implantation had access to these new approaches and the people working in molecular biology were removed from reproductive and developmental biology.

We were fortunate that our concern about this state of affairs was sympathetically received by others. With the encouragement of Dr. Bert O'Malley, Chairman of the Department of Cell Biology and Director of the Center for Population Research and Reproductive Biology at Baylor College of Medicine, we sought to organize a conference on the cellular and molecular aspects of implantation. Dr. William A. Sadler, Chief, Population and Reproduction Branch, CPR, NICHD, was enthusiastic about the idea and invited us to plan such a meeting under the auspices of the National Institute of Child Health and Human Development. The intent was to provide a compendium of information on the new approaches and their application to research on implantation. Two immediate benefits of the meeting were expected to be the opportunity for interaction between scientists from different fields who had never previously been given a common forum, and the stimulation arising from such interaction that might rekindle interest in research on implantation among scientists and granting agencies.

With the generous support of additional sponsors and the aid of our scientific colleagues, the conference was convened in Houston in September 1979 and proved to be a great success, to the pleasant surprise of some who had argued against its premise and to the gratification of its organizers and sponsors. This volume contains the edited proceedings of the conference, at which some of the leading participants in the "new implantation" shared their findings and thoughts with an appreciative audience that engaged in lively discussions. We hope that this book will propagate this impetus to a wider sphere and will be of use not only to reproductive and developmental biologists but also to graduate and medical students and to practitioners in the fields of animal and human fertility.

We are grateful to The Rockefeller Foundation, The National Science Foundation, The Upjohn Company, Ortho Pharmaceutical Corporation, G.D. Searle & Co., Schering Corporation, Merck & Co., Inc., Abbott Laboratories, and to the Population Program and the Department of Cell Biology of Baylor College of Medicine for their supplementary financial support, to Ms. Joanne Julian, Ms. Linda Hall, Mr. Tony May, and Mr. James Hawkins for their help, to Mr. Richard Cunningham of Houston, Texas for designing and donating the implantation logo, and to Messrs. Kirk Jensen and Richard Jannaccio of Plenum Publishing Corporation for their help and guidance.

Stanley R. Glasser
David W. Bullock

Houston, Texas

Contents

Part I: Perspectives

Chapter 1
Viviparity
E. C. Amoroso

Chapter 2
My Life with Mammalian Eggs
M. C. Chang

Part II: Cell Biology of the Developing Egg

Chapter 5

Relationship between the Programs for Implantation and Trophoblast Differentiation

Michael I. Sherman, Martin H. Sellens, Sui Bi Atienza-Samols, Anna C. Pai, and Joel Schindler

Chapter 6

Cellular and Genetic Analysis of Mouse Blastocyst Development

Roger A. Pedersen and Akiko I. Spindle

Part III: Macromolecular Synthesis in the Developing Egg

Chapter 7
Mobilization of Genetic Information in the Early Rabbit Trophoblast
Cole Manes, Michael J. Byers, and Andrew S. Carver

Chapter 8
Activity of RNA and DNA Polymerases in Delayed-Implanting Mouse Embryos
H. M. Weitlauf and A. A. Kiessling

Chapter 9
A Reexamination of Messenger RNA Populations in the Preimplantation Mouse Embryo
Gilbert A. Schultz, Jeremy R. Clough, Peter R. Braude, Hugh R. B. Pelham, and M. H. Johnson

Chapter 17
Uterine DNA Polymerase: Acquisition of Responsiveness to Estrogen during Postnatal Development of the Rat
Alvin M. Kaye, Michael D. Walker, and Bertold R. Fridlender

Chapter 18
The Artificially Stimulated Decidual Cell Reaction in the Mouse Uterus: Studies of RNA Polymerases and Histone Modifications
Martin J. Serra, Billy Baggett, Judith C. Rankin, and Barry E. Ledford

Part VI: Blastocyst–Uterine Interactions

Chapter 19
Embryonic Signals and Maternal Recognition
R. B. Heap, A. P. F. Flint, and J. E. Gadsby

Perspectives

1

Viviparity

E. C. Amoroso

I. Introduction

Reproduction is one of the cardinal attributes of all living things and although the end result in all instances is the same, the ways by which it is accomplished may be quite different. Irrespective of their ordinal rank, living creatures face much the same problems in their efforts to survive as individuals and as species. They must adjust, in some way or the other, to the same physical influences, such as those caused by seasonal and environmental changes, and they have at their disposal the same choice of chemical elements from which to synthesize the humoral agencies that regulate their reproductive activities. The endocrine secretions of the reproductive organs are not, however, of vital importance for the well-being of the individual; any or all of the structures directly concerned with reproduction may fail to function or be surgically removed without affecting the general health or life expectancy of the individual. This relative independence of the reproductive processes, as true of vertebrates as it is of invertebrates, has made it possible for widely different adaptive mechanisms to occur, without marked interference with the general economy of the body; it is with a consideration of these, as well as the hormonal regulations in the adaptations of the different species to their established environmental niches, that we will be most concerned. Many of the facts are familiar; perhaps the relationships are too; but there are reasons to believe that they are worth pointing out again, even if they are not new.

II. Viviparity as a Reproductive Mechanism

Broadly speaking, viviparity embraces a wide range of adaptations of animals for the retention of young in the body of the parent during embryonic development, and most major groups have individuals that practice this method of caring for their young. As a reproductive mechanism it is, if we exclude birds, an essential part of the stock-in-trade of all living vertebrate classes from fish to man, and attains its highest expression in Primates with the development of a placental connection of a unique kind, and this through speciali-

E. C. Amoroso • A.R.C. Institute of Animal Physiology, Babraham, Cambridge, England.

zation of the embryonic membranes (chorion) as an organ of exchange and internal secretion.

This widespread occurrence of viviparity in animals suggests that those species that adopted it were to a large extent preadapted for it. One such adaptation—a prerequisite to internal gestation—was the perfection of internal fertilization, which is as common among animals that subsequently lay eggs as among those that retain their young until birth. The evolution of a variety of intromittent organs from a number of different structures provides clear evidence that viviparity itself must have arisen independently many times during the course of evolution.

A large number of animals with internal fertilization thus show some degree of viviparity; consequently, to a species that had attained this modest degree of internal gestation, the further development of viviparity involved a relatively moderate transition. Nevertheless, it is worth pointing out that before true viviparity could be attained, the problem of endocrine adaptation to the retention of fertilized eggs had to be solved. These evolutionary innovations have involved: the suppression of ovulation during pregnancy (e.g., tsetse flies and a majority of eutherian mammals); intrauterine care of a few young until relatively advanced stages of development; the adoption of the corpus luteum in the family of endocrine glands; the maternal recognition of pregnancy; the participation of the endocrine secretions of the conceptus in extending gestation beyond the limits of the normal sex cycle; the abrogation of the maternal immune response against the products of conception; and the adaptation of the endocrine control of lactation in preparation for the early nutrition of the newborn.

III. Adaptations for Ovoviviparity and Viviparity in Invertebrates

The acquirement of a viviparous relationship between parent and offspring may involve only simple harboring of young in a brood chamber under conditions in which the embryos subsist entirely on their own yolk reserves, or they may be nourished by the surrounding parental tissue, or take advantage of both possibilities. As examples of adaptations of this sort among invertebrates, one may cite such diverse relationships as that observed in certain molluscs, some annelids, and a few cockroaches, where embryonic development occurs entirely within the mother, yet no nutrients are obtained from her; or, that seen in the viviparous tsetse fly (*Glossina*), where the single egg develops in the uterus and special structures serve to nourish the larva (*adenotrophic viviparity*). In this instance, once the egg is hatched, the larva is nourished from a milk gland that secretes into the uterus through a nipple to which the mouth of the larva is closely applied, a relationship reminiscent of that between the early marsupial young and its mother. Another obvious feature of this relationship is that it provides an opportunity for the passive immunization of the larva, prior to its deposition by the mother, that is not unlike the postpartum transmission of antibodies by way of mammary secretions (colostrum) in many ungulates. Equally noteworthy is the fact that while the larva is developing in the uterus, another egg matures in the nongestational ovary and is ovulated only after the third instar larva is deposited, thus mimicking the condition in mammals where ovulation is suppressed by the presence of the products of conception.

Ordinarily in insects, the corpora allata are inactivated during gestational periods; as a consequence no eggs mature. This phenomenon suggests that certain control measures for egg maturation and ovulation do exist in *Glossina* species. As yet, however, no control mechanisms are known to account for the maturation of one egg in alternate succession in the ovaries of *Glossina*. In this connection attention should also be directed to the unusual sensitivity to vertebrate hormones shown by the tsetse fly: females fed on pregnant goats produced more offspring than those fed on nonpregnant hosts, suggesting that *Glossina* females obtain a factor in the blood of the pregnant goats that promotes fecundity. Whether the host's steroids directly cause the maturation of eggs in these insects by acting as gonadotropic hormones or whether the insect's own endocrine glands become active in these situations remains equivocal. However, as the gonadotropin in insects is almost certainly the same substance as the authentic juvenile hormone, the second explanation is probably correct (Hogarth, 1976). There is also evidence for the existence of a second juvenile hormone that, according to Englemann (1970), may be implicated in the synthesis of the female-specific protein in some cockroaches, thus recalling conditions in mammals.

A third type of viviparity occurs in invertebrates such as the earwig (*Hemimerus talpoides*), parasitic on African rats, where the thickened follicular epithelium functions as a pseudoplacenta, and the polyctenid bug (*Hesperoctenes fumarius*), parasitic on Jamaican bats (Amoroso, 1952, 1955a). In these insects the extra embryonic membranes give rise to a placenta in a system that is analogous to that of viviparous reptiles. Although the placenta never vascularizes, it successfully accommodates the nutritional and respiratory needs of the embryos, which are born at an advanced stage of development. As with *Hemimerus*, egg development in *Hesperoctenes* begins in the ovariole, but a noteworthy feature of its development is that as pregnancy advances, the egg, which has little yolk, moves down the reproductive tract so that embryos at various stages of development can be found in the gravid females at any time. Thus, in contrast to *Glossina* in which we find some kind of restraining control of egg maturation during certain periods of the animal's life, no such control of egg maturation is found in *Hesperoctenes*.

For a more advanced viviparity among the invertebrates one must turn to the Ascidians and to the Onychophora in particular, the phylum that contains the celebrated *Peripatus* from Trinidad. Both groups possess individuals that display many of the adaptations of viviparity seen in higher animals. In the Tunicate *Salpa democratica*, the egg, which is small and poor in yolk, is nourished within the brood pouch of the mother by means of a complex placenta. Here the tissues of the embryo are in immediate contact with maternal blood in a manner that has a superficial resemblance to that encountered in the hemochorial condition of some higher mammals: after a relatively long gestation period the embryo is born as a small but fully developed salp (Amoroso, 1955a).

In the primitive wormlike arthropod *Peripatus*, on the other hand, each fertilized egg

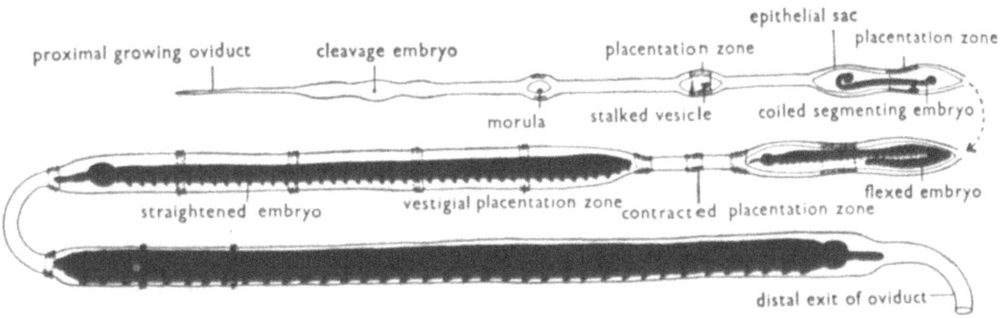

Figure 1. Diagrammatic reconstruction of the oviduct of a reproductively mature specimen of *Epiperipatus trinidadensis*. Seven embryo stages are represented. From Anderson and Manton (1972).

becomes implanted in a newly formed segment of the oviduct in a manner analogous in certain ways to that of the embryo and the endometrium of higher forms that develop a yolk sac placenta. As each embryo develops it remains attached to the parental tissues within its associated segment of oviduct, and is displaced distally (Fig. 1) while additional lengths of oviduct containing younger embryos are added proximally (Anderson and Manton, 1972). Curiously enough, this placental structure is a transient one, for subsequent to gastrulation the young animal sunders its connection with the parental organism and survives for a time by utilizing the secretions that accumulate in the cavity of the oviduct, foreshadowing conditions in elasmobranch fishes. Nothing is known about the nature of substances with the properties of hormones in the Onychophora, and the whole problem of the endocrinology of reproduction in these unique arthropods remains to be investigated.

IV. Adaptations for Viviparity in Fishes

Among the anamnia, the elasmobranch fishes (as well as the coelacanth *Latimeria* among bony fishes) show conditions of viviparity that in many respects more closely resemble those commonly present in amniotes (Amoroso, 1960). Although most elasmobranchs are oviparous, quite a few species are ovoviviparous and some show advanced conditions of viviparity in which the young embryos are retained for prolonged periods in the lower part of the oviduct. Their chief adaptations for viviparity are: perfection of internal fertilization; marked specialization of the Mullerian ducts as repositories for the developing young; the utilization of the fetal yolk sac as a placenta; and development of corpora lutea in the ovaries (Amoroso, 1960; Hisaw, 1961).

A. Elasmobranchs

In the spotted dogfish (*Scylliorhinus canicula*), an oviparous species, Dodd *et al.* (1960) have shown that the pituitary gland, and the pars distalis in particular, plays an important role in gonadotropin production and in maintaining the integrity of the gonads. However, when the pituitary gland was removed soon after fertilized eggs had entered the uterus in the smooth dogfish (*Mustelus canis*), a viviparous species, the embryos continued to develop for at least 3½ months of an estimated 11-month gestational period, suggesting that pituitary stimulation may not be involved in sustaining the corpora lutea and that these luteal bodies may not be essential for the maintenance of gestation in this fish, at least in its early stages (Hisaw and Abramowitz, 1938, 1939). It is assumed but not yet proven that the active principle elaborated by the elasmobranch pituitary gland is a gonadotropin that regulates steroid production by gonadal tissue. Large quantities of estradiol-17β, estrone, estriol, and progesterone have been found at different times in the whole ovary and testis, as well as in the circulating plasma, egg yolk, and embryos of several oviparous and viviparous species (see Amoroso and Perry, 1977). These facts, together with the ability of estradiol implants to stimulate the oviducts of females of *S. cuniculus,* seem to indicate that these structures are under steroid control (Wotiz *et al.*, 1958). As yet, however, and excepting the interesting discovery of Lupo di Prisco and his associates (1967) that total corticosteroid concentration was lower in the plasma of *T. marmorata* during pregnancy than at other times—the reverse of the situation in mammals—there is very little evidence of any hormonal involvement in gestation beyond those leading up to and including ovulation. By implication, the transition from oviparity to viviparity in this group of fishes cannot, as yet, be definitely linked with fundamental changes in endocrine regulation. But despite any clear indications of endocrine involvement, elasmobranch fishes have developed a capacity to provide nutrient fluids for the developing embryos and there is a parallelism between the degree of dependence of the embryos on uterine secretions and the degree to which the liver is reduced in size during development; this fluctuation in liver weight appears to be integrated in the endocrine cycle of reproduction (Ranzi, 1935).

It is a point of interest, as the ablation experiments of Dodd and his colleagues have shown, that pituitary–gonadal interactions are well developed in elasmobranchs and that this adaptation, already present in cyclostomes, may have been a cardinal feature at the very inception of the vertebrates (Barrington, 1964). Clearly, therefore, it would be of the greatest interest to discover the point in the evolutionary series at which viviparity came under the control of the pituitary gland, remembering of course that internal gestation was not a new discovery of the vertebrates; that reproductive processes were carried out successfully in a coordinated fashion long before the advent of pituitary gonadotropins; and that so common a phenomenon as ovulation, which is very generally thought of in terms of gonadotropic function, occurs in all animals and was likewise the rule for millions of years before the advent of a pituitary gland.

The fact that the essential features of pituitary organization are already fully outlined in the larva of the lamprey, the most primitive of living vertebrates, attests to its central importance in the economy of the vertebrate endocrine system. Thus, for clues as to its early evolutionary history, resort must be made to the protochordates, in many of which specialized cephalic sensory organs have been described. Highly suggestive is the neural gland of Tunicates, which has from time to time been held to be homologous with the vertebrate pituitary gland and may play some part in the coordination of reproductive activity. Such a cephalic sensory structure, by virtue of its sensitivity to environmental change or to the products of other individuals (pheromones), and capable of releasing signals, might have become sensitive to the products of its own body. In their turn, these secretions, even though they are initially available only in trace quantities, could then evoke in some way the release of germ cells from its own body, thus foreshadowing not only the far-reaching regulatory power that the pars distalis of the adenohypophysis now possesses, but also the reciprocal feedback relationships with other endocrine glands, especially the gonads (Danforth, 1939; Amoroso, 1969). Accordingly, the situation, as it applies to mammals and other vertebrates, must represent the culmination of a long series of adaptive mechanisms in which the pituitary gonadotropins may be regarded as a link in the informational chain between the central nervous system and the gonads, and the pituitary–gonadal interactions as representing a device for regulating gonadal activities, which at their inception were capable of going on in direct response to environmental stimuli (Hisaw, 1961).

B. Teleosts

The habit of viviparity has been developed independently in several families of teleost fishes and to different degrees. In the majority of these live-bearing fish (the Coelacanth *Latimeria* is a notable exception), the ovary not only is the source of ova and presumably ovarian hormones, but also provides housing and nourishment for the developing young; gestation may take place within either the ovarian or the follicular cavity, conditions that are strikingly different from other vertebrates. In many cases (Poeciliid and Anablepid fishes) the ovarian follicle plays the dominant role and its epithelium assumes an important function in the nourishment of the embryos (follicular gestation); in other instances (Goodeid and Jenynsiid fishes), the embryos are discharged into the ovarian cavity (ovarian gestation) where they continue to develop through the mediation of a series of temporary but unique pseudoplacental adaptations. It is clear, however, that fertilization within the follicle prevents true ovulation, and as ovulation is controlled by a pituitary hormone, it would appear that some endocrine change has been brought about by, or is concurrent with, the innovation of intrafollicular fertilization: In certain teleost fishes (e.g., the mosquito fish, *Heterandria*), follicular gestation is often accompanied by superfetation in which two or more broods of embryos, at different stages of development, are harbored within the follicles of the ovary at the same time, and, as the successive broods mature, each in turn is born. Presumably this is so because each follicle, to some extent at least, has some autonomy in regulating the cyclical changes attendant on intrafollicular fertilization and the timing of ovulation/birth. Effective superfetation, as far as is known, occurs only in the poeciliid fishes among the cyprinodonts and does not occur in those vertebrates that have uterine gestation.

While the female parent is most often the one to which the young have become associated, the male parent in some species of teleost fishes (pipe fishes and sea horses) has specialized pouches that serve as repositories for the developing young. Such embryos are provided with vascular placentalike structures and in some cases a nutrient secretion as well. However, although the development of the brood pouch relies on the presence of the male sex hormone testosterone, there is no evidence that gestation itself is immediately dependent on hormones from the gonads, but there are clear indications that a pituitary prolactin (PRL) may be necessary to meet the osmoregulatory demands of the pouch young. Hypophysectomy of a pregnant male sea horse is reported to cause high mortality of the young and their premature delivery (see Hogarth, 1976); also it has been suggested that the female sex hormone progesterone is required for gestation. If confirmed this would imply an extragonadal source of the hormone, for gonadectomy of pregnant males is without effect on the course of gestation; the adrenal glands may represent the alternative source.

Equally curious, of course, is the fact that the tissues of the male reproductive tract, though totally different from the female, have not only acquired responsiveness to female gonadal hormones and PRL but also have become adapted for the facilitation of viviparity. In fact, it appears that the mechanisms of viviparity are methods mostly for establishing an embryo in a particular locality rather than maintaining it over a prolonged period. Furthermore, this phenomenon, that male as well as female structures may provide nutrient fluids, which would seem the more remarkable if it were not so familiar to zoologists, suggests that in mammalian evolution, the functioning of the mammae has only recently become confined to the female, explaining their slight sexual dimorphism.

As with other vertebrate gonads, the poeciliid ovary is capable of synthesizing steroids, and it has been suggested that the so-called preovulatory corpora lutea, which are universally present in fish ovaries, form the main mass of endocrine tissue of the teleost ovary. As yet, however, nothing substantial is known of possible endocrine influences in these fishes, and the question whether the luteal bodies have a function in controlling gestation remains unanswered. Indeed there is a strong case for it being unanswerable until ovariectomy is performed on a pregnant coelacanth, for, with this sole exception, extirpation of the corpora lutea or removal of the ovaries is unlikely to be successfully performed on a pregnant teleost with follicular or ovarian gestation and would in any case certainly terminate pregnancy (Amoroso, 1968).

Like all vertebrate gonads, the poeciliid ovary is under pituitary control, as ablation of this organ in gestation results in high embryonic mortality. This result may be averted if hypophysectomized pregnant individuals are maintained on a mixture of salt and fresh water or are treated with PRL, suggesting that the role of the pituitary gland is in maintaining a suitable salt/water balance of the fetal environment during gestation. But although the gonadotropic hormones and estrogens do not seem to play any significant role once pregnancy has started, there is a strong suggestion that a substance very similar in structure to mammalian oxytocin (arginine vasotocin) may be responsible for expelling the young from the ovary at birth (Hogarth, 1976). A similar polypeptide hormone may be involved in parturition in male sea horses and pipe fishes, but delivery cannot take place until the animal finds suitable mechanical support in the environment around which its tail can be entwined.

C. Latimeria: The Living Coelacanth

The discovery of living coelacanths, which until 1939 were known only from fossils and supposed to have been long extinct, has special significance not only for students of paleontology, but also for us, concerned as we are with the evolution of viviparity. A little more than fifty years ago Watson (1927) described two small skeletons of the Jurassic coelacanth *Undina* (*Holophagus*) found inside the body cavity of a much larger specimen of the same taxon. He suggested that *Holophagus* was viviparous. Noting that recently captured coelacanths lack any fins in the form of intromittent organs, it was suggested that these fishes were oviparous and Watson's specimen was interpreted as a case of cannibalism. Subsequently, on the basis of a female with 19 apparently fully developed, shell-less eggs in the right oviduct (as in the elasmobranchs the left oviduct is nonfunctional), it was concluded that *Latimeria* was indeed oviparous, until the discovery of five advanced young with moderately large yolk sacs (yolk sac placentas) lying free within the right oviduct of a gravid female (Smith *et al.*, 1975). Accordingly, we can now state with considerable confidence that the coelacanth *Latimeria chalumnae*, with its well-developed yolk sac placenta, shows advanced conditions of viviparity similar to those of the smooth dogfish *Mustelus canis*, and which in many respects closely resemble those commonly present in amniotes. However, future research will have to show what precise endocrine controlling mechanisms exist and to what extent, if any, pituitary or ovarian hormones are essential for uterine development of young in this newly rediscovered but ancient fish.

V. Amphibian Viviparity

Unique innovations for viviparity are also found in amphibians, but the only ones to show true viviparity are the anurans. A few urodeles are ovoviviparous and use the oviducts for this purpose, but there is comparatively little tendency to the formation of a pseudoplacenta, so characteristic of teleosts (see Amoroso, 1952; Hogarth, 1976; Amoroso *et al.*, 1979). Among the urodeles, salamanders are of special interest, not only because of the apparent flexibility of the reproductive processes imposed by the environment (e.g., *S. atra*), but also because certain species, of which the cave salamander *Proteus anguinis* (the olm) is an example, produce both live young and eggs; the same may also be true of certain reptiles, e.g., *Lacerta vivipara*, which under exceptional circumstances lay eggs that eventually hatch.

In *S. atra* the eggs (up to 60) are retained in the oviduct, but under normal conditions few of them produce young; only the egg most distant from the ovary in the oviduct actually develops. If, however, the temperature rises above 120°C or so, all of the eggs may be laid although they will probably not develop normally. In its normal habitat in the Alps at altitudes of about 800–3000 m, *S. atra* produces two live young (one from each oviduct) after a gestation period of 1 year; but at higher altitudes, where hibernation is more prolonged, pregnancy may last 2–3 years, emphasizing yet again the importance of the environment on reproductive processes in these amphibians.

While the oviduct of *S. atra* appears to be under some form of control by the ovary, there is uncertainty about the function of the postovulatory corpora lutea in pregnancy. Although these luteal bodies persist throughout gestation and regress rapidly after the young are born, they do not seem to have any function in controlling the course of gestation, as castration of gravid females at the beginning of pregnancy did not interfere with the development of the larvae. There is evidence, nevertheless, that the corpora lutea secrete progesterone and during pregnancy may be involved in the inhibition of oocyte maturation.

In Anura, ovoviviparity is very rare, yet it is among these, if not in Amphibia as a whole, that the most advanced condition of viviparity is found. The two best known examples are the small tropical West African toad, *Nectophrynoides occidentalis*, which lives in an area of alternating wet and dry seasons at altitudes of around 1200 m, and a newly discovered species, *Eleuterodactylus jasperi*, living in Puerto Rico (Drewry and Jones, 1976). As an adaptation to viviparity *N. occidentalis* lives underground during the dry season and its reproductive cycle is so adjusted that fertilization occurs at the end of one wet season and birth in the middle of the next, thus avoiding the exposure of the young to dry conditions during development. In both *N. occidentalis* and *E. jasperi*, about ten eggs, poor in yolk, are retained in a distended, richly vascularized segment of the Mullerian ducts where, despite the lack of a complex placental relationship, they complete their development. Fully metamorphosed young toads are born after a pregnancy of 9 months, and, as with certain male pipe fishes and sea horses, an oxytocic principle similar to mammalian arginine vasotocin may be responsible for expelling the young from the oviducts.

In certain species of tree frogs (Gastrotheca), on the other hand, the fertilized eggs are carried in a pouch on the back of the female (del Pino *et al.*, 1975); in *G. marsupiata* the young are retained up to the tadpole stage (Amoroso, 1955a; Amoroso *et al.*, 1957) and in *G. ovifera* to the froglet stage (Mertens, 1957). In these anurans the ruptured follicles are organized into corpora lutea that persist throughout gestation in the various forms (9 months in *N. occidentalis*, 3–4 months in *G. marsupiata*, 3–4 years in *G. ovifera*), and appear to be capable of secreting progesterone, presumably under the influence of a PRL secreted by the pituitary gland (Xavier, 1970; Hogarth, 1976). Nevertheless, these luteal bodies (like those of *S. salamdra*) do not appear to be essential for the maintenance of pregnancy; castration caused abortion only when the operation was performed on *N. occidentalis* at the beginning of pregnancy in primigravid females (Xavier, 1970). In fact, a rather surprising result of ovariectomy in *N. occidentalis* is that pregnancy proceeds more rapidly than in intact females; gestation time is curtailed, and normal, fully developed toads are born 3 or 4 months early. It thus appears that within the two orders of modern Amphibia, the Urodeles and Anura, corpora lutea are formed with the capacity to produce progesterone and that the steroid may function in early pregnancy as an endocrine regulator (Lofts, 1974; Hogarth, 1976).

VI. Reptilian Viviparity

Reptiles like birds have evolved the cleidoic egg and with it the allantois, primarily a receptacle for waste products and secondarily a respiratory organ. Some reptiles (notably lizards, chameleons, and snakes) have advanced beyond this achievement and attained viviparity; in doing so, they have developed various forms of rudimentary placentas that are more complex than those of elasmobranchs and coelacanths because, being amniotes, the chorion and the allantois, as well as the yolk sac, are available to take part in their formation. The mammals have inherited these structures and, excepting the Monotremes (e.g., the spiny anteater and the duck-billed platypus), have abandoned oviparity in favor of true viviparity and the production of a chorioallantoic placenta, although in many, the yolk sac still plays an important role, at least in the early stages of pregnancy.

While the discharged follicle of reptiles gives rise to luteal bodies of essentially mammalian structure, the specific function of the reptilian corpus luteum remains a matter of controversy (see Callard and Lance, 1977; Amoroso *et al.*, 1979). Though its presence in the ovaries of both oviparous and viviparous species would seem to indicate that it is not primarily related to viviparity, there appears, nevertheless, to be a definite correlation between the longevity of the corpus luteum and the egg-laying or egg-retaining habit of the species. As might be expected, progesterone secretion in reptiles shares features in common with birds on the one hand and with certain mammals on the other. Thus preovulatory progesterone secretion has been described in certain oviparous species (*Chrysemys picta* and *Chelonia mydas*), and its production in these turtles is in all probability related to the luteinization of the follicles, for the corpus luteum is short-lived, as in birds, and persists only for the time eggs are in the oviduct. By contrast, in viviparous reptiles the secretory function of the corpus luteum persists for most if not the whole of gestation, while the concentration of progesterone in the peripheral plasma peaks at about the time of maximum development of the corpus luteum—endocrine adaptations that anticipate the evolution of the dominance of luteal control in more advanced forms of viviparity (Highfill and Mead, 1975; Colombo and Yaron, 1976; Callard and Lance, 1977).

It is a point of interest that in certain viviparous lizards (e.g., *Xantusia vigilis*), the removal of the ovaries or of the corpora lutea during pregnancy (even when performed as early as the first week) does not lead to abortion or affect the development of the embryos, but will interfere with parturition, causing the retention of young beyond term or their being born dead (Panigel, 1956; Yaron, 1977). Similarly, in the water snake *Natrix sipedon*, embryos also continue to develop after bilateral oophorectomy of pregnant females, while in the garter snake *Thamnophis elegans*, the only unusual effect noticed was an extension of the gestation interval and the concomitant postponement of parturition (Bragdon, 1951). Browning (1973) believes this to be due to the removal of an ovarian estrogen, though direct evidence to support this view is lacking. This interference with parturition was more complete after hypophysectomy than after ovariectomy. Such, however, might be expected, considering the debilitating effects resulting from the loss of the neurophypophysis and the consequent elimination of its oxytocic polypeptides because of pituitary ablation. These observations are in agreement with the instances cited previously in which corpora lutea do not function as endocrine glands of gestation, a condition probably true of all viviparous anamnia and all but a few notable exceptions among live-bearing reptiles (e.g., *Chameleo pumilus pumilus*) in which progesterone secretion by the corpus luteum is essential for the maintenance of gestation (Veith, 1974).

The evidence here summarized notwithstanding, it must be admitted that the involvement of progesterone in reptilian pregnancy remains equivocal (Callard and Lance, 1977). On the other hand, it should be remembered that even in those cases in which experimental observations seem to deny ovarian participation in the maintenance of gestation in reptiles, there is abundant evidence that the ovaries are capable of producing steroid hormones, but their action on target organs may be different from those so far studied in mammals (Amoroso and Perry, 1977). If this holds, then it would appear that progesterone took no direct part in the evolution of viviparity in any of the anamnia and it probably was not of importance in reptiles either, though it might have acquired endocrine status in the reptilian ancestors of mammals. It is thus clear, as Hisaw (1959) suggested, that the next stage in the evolution of viviparity leading to conditions found in mammals was the adoption of the corpus luteum into the family of endocrine glands. He suggested that synthesis of progesterone, probably on a small scale, was the first step in the specialization of the corpus luteum, and that this process was later augmented by the acquisition of competence to respond to the pituitary luteotropic hormone (PRL). However, although the exact steps through which the corpus luteum was incorporated as an endocrine gland may remain obscure, there can be little doubt that this adaptation was the most important contributing factor in the evolution of viviparity in mammals. In essence, this involved the addition of a luteal phase in sequence with a preceding follicular phase that terminated at ovulation; thus the two components of the mammalian estrous cycle were established (Hisaw, 1961).

VII. Adaptations for Viviparity in Mammals

The universal and almost unique feature of the evolutionary adaptations for viviparity in mammals, i.e., the establishment of a trophic and respiratory placental connection between parent and offspring has not been accomplished by the evolution of new fetal structures beyond those encountered in reptiles; both the yolk sac and the allantois may, with the chorion, form placental structures. Typically, the major placental organ in marsupials is the yolk sac; and the bandicoots, the koala, and the wombats use the allantois as well (Amoroso, 1952). In higher mammals the degree of development of the yolk sac placenta varies from one group to another, but in all higher mammals it is the chorioallantoic placenta that is the principal and in some the only instrument of placentation.

Another consideration that often has been overlooked is that the chorioallantoic placenta has become modified during its evolutionary development into an organ that exercises an endocrine influence on gestation (Amoroso, 1955b; Porter and Amoroso, 1977). It also is significant that in many mammalian species this specialization should involve the elaboration of steroids, such as estrogens, progesterone, and adrenal corticoids, as well as glycoprotein substances resembling pituitary gonadotropins in their actions. "The surprising thing," as Hisaw (1961) put it, "is that in certain instances it succeeded in accomplishing both."

A. Monotremes (*Prototheria*)

The Monotremes are the only mammals that lay eggs, and reproduction is accomplished by retaining eggs in the uterus and incubating them after laying until hatching. Compared to reptiles, the egg of present-day monotremes is small and the yolk content insufficient to support development beyond the earliest stages. There is good evidence, however, that endometrial gland secretions are absorbed by the egg for some time before the outer shell is applied to it so that intrauterine development can proceed to a stage beyond the compass of the initial stores of yolk within the egg (Hill, 1933, 1941), and, as with oviparous reptiles, egg laying is correlated with the involution of the corpora lutea (Griffiths, 1968, 1978).

As a functional source of progesterone, the montreme corpus luteum has long been credited with controlling the transport of the egg and with the development of the mammary glands (Allen *et al.*, 1939). It now seems that during gestation it also is responsible for suppressing ovulation and for maintaining the endometrium in a state equivalent to that in viviparous mammals during the luteal phase of the estrous cycle (Hill, 1941). This innovation of a luteal phase in the reproductive cycle of mammals, the dominant feature of which is the secretion of progesterone, would thus set them apart from other vertebrates, for although luteal bodies have been described in representatives of every other vertebrate class, there is no certain evidence that they are endocrine glands specialized for the secretion of progesterone (Veith, 1974; Amoroso and Perry, 1977). Of course, this does not signify that the same is true of all live-bearing reptiles, as is attested by the case of the ovoviviparous chameleon *C. pumilus pumilus* (previously alluded to), in which progesterone secretion by the corpus luteum is essential for the maintenance of pregnancy (Veith, 1974). In fact, it would appear quite in keeping with evolutionary expectations if more instances were found of hormonal interactions in viviparous reptiles similar in some fundamental way to the more specialized conditions in mammals.

With respect to the basic endocrine apparatus for controlling gestation, Monotremes share features in common with oviparous reptiles on the one hand and higher mammals on the other. Thus, besides the endocrine control over oocyte maturation, ovulation, and support of the uterus (Boyd, 1942), they have retained the old reptilian custom of carrying fully formed eggs in their reproductive tracts for some time prior to oviposition, and this has developed into a timing device for laying the egg in coordination with the involution of the corpus luteum. By contrast, the principal innovations that monotremes share with other mammals include the formation of a corpus luteum by the action of the pituitary luteinizing hormone (LH) on the ruptured follicle following ovulation, and the secretion of progesterone in response to the luteotropic hormone, or prolactin (PRL). Nevertheless, the whole problem of the endocrinology of reproduction in these unique mammals remains to be investigated.

The dependence of lactation in higher mammals upon the luteotropic hormone prolactin and the influence this hormone exerts as a potent activator of parental behavior during the period of lactation make these observations in Monotremes of particular interest (Riddle *et al.*, 1942; Eayrs *et al.*, 1977). In many Eutheria, this response of the ovaries to prolactin is dependent on the production of membrane receptors that provide luteal cells with the competence to respond to luteotropic stimuli (Richards *et al.*, 1978). Prolactin also increases the number of LH receptors in luteal cells and enhances the secretion of luteal progesterone, which is essential for mammary development (Holt *et al.*, 1976). In other words, the height of specialization of the corpus luteum beyond conditions found in reptiles has involved enhanced secretion of progesterone as well as its implication in the complex of endocrine factors that leads to full secretory activity of the mammary gland. Therefore it seems not an unlikely possibility that this one characteristically mammalian feature, lactation, and consequent nutritional dependence on the mother after birth, which is totally lacking in reptiles, may have appeared first in an egg-laying species and, if so, was not associated directly with viviparity. Nor should we overlook the fact that the anterior pituitary principle referred to as PRL is a very ancient molecule and was performing hormonal functions of an osmoregulatory kind long before the advent of progesterone-secreting corpora lutea, and certainly before mammary glands were invented (Riddle and Bates, 1939; Pickford and Atz, 1957).

Prolactins of great variety have been described; even in the primitive acorn worms (e.g., *Lineus ruber*), which do not possess a pituitary gland, the evidence points to the occurrence of certain cells in the pharyngeal membrane, the staining and ultrastructural properties of which have many points in common with the various classes of cells in the anterior pituitary glands of vertebrates (Willmer, 1974). Among the acidophils there are cells whose granules are very similar in size, shape, electron density, and staining reactions to the prolactin cells (mammotrophs) of the adenohypophysis, and much recent work shows that, in fish, prolactin is instrumental in maintaining the osmotic balance of the animal by raising the Na^+ concentration of the blood, which it does, at least partly, by altering the activity of the chloride cells. It is significant in this connection that it is the acidophil cells particularly that lose their granules when *Lineus* is placed in diluted seawater, because the prolactin cells of euryhaline fish do exactly the same (see Willmer, 1974). "Not even the possession of a prolactin, then," as Medawar (1953) put it, "can be said to be distinctive of mammals; what is distinctive of mammals is the evolution of a new mode of tissue response to hormones of a category already in being"—an idea perhaps first formulated by Danforth (1939). And although references may be cited that alternatively affirm and deny that this is true, the results of more recent studies (Amoroso *et al.*, 1979) suggest that it may indeed be the case, for to quote Danforth: "It is unlikely that protoplasm could have attained any degree of complexity without, at the same time, developing a capacity to react with some components of the internal environment more readily than with others." Put more succinctly, "It is the tissues and not really the hormones that do the *reacting*" (Danforth, 1939).

B. Marsupials (Metatheria)

This modest gesture toward viviparity in monotremes is reflected in certain aspects in marsupials, such as the American opossum (*Didelphis*), the Australian native cat (*Dasyurus*), and the brush-tailed possum (*Trichosurus*). In these species the young are born at the end of the luteal phase and parturition is coincident with the involution of the uterine endometrium, itself attributable to the functional demise of the corpus luteum. The chief difference between this and the situation in monotremes is that instead of laying eggs, embryonic development is accelerated and within a relatively brief period viable young are born. From the endocrinological standpoint, on the other hand, hormonal conditions in monotremes and marsupials are alike in that the progestational transformation of the uterus is dependent on progesterone produced by pituitary–corpus luteum interaction.

The fact that the duration of pregnancy, in the majority of species so far studied, does not exceed the length of the estrous cycle implies an endocrinological equivalence of the pregnant and nonpregnant states and suggests that no enhanced hormone secretion occurs during pregnancy for such is the inference to which these premises must lead. That this interpretation may not be entirely correct, however, is suggested by the continuance of pregnancy, in the American opossum and several Australian wallabies and kangaroos, following complete ablation of the ovaries or corpora lutea from pregnant females during the second half of gestation (Tyndale-Biscoe, 1963a, 1973; Renfree, 1974). Also, Heller (1973) found that in quokkas the myometrium of late pregnancy is less sensitive to the contractile effect of oxytocin than that of the nonpregnant animal at the same postovulatory stage. The finding that the marsupial placenta is capable of synthesizing progesterone from a natural precursor (Bradshaw *et al.*, 1975) may account for this difference. It would be precipitate, nevertheless, to assume that these results, as they stand, point to a sustaining action on gestation of endocrinologically active agents emanating from the products of conception; there are reasons for thinking this interpretation insufficient. Indeed, fetal or placental participation can probably be discounted, in some marsupials at least, as the uterine luteal phase persists in nonpregnant, ovariectomized quokkas, *Setonix brachyrus*, for the same time as in pregnant animals (Tyndale-Biscoe, 1963a). There is thus strong presumptive evidence that the maintenance of pregnancy in marsupials may not involve endocrine mechanisms other than those that participate in the regulation of the estrous cycle.

So much may be true, of course, but the empirical facts do suggest that requirements for the endocrinological specialization of the marsupial yolk sac placenta do exist (Renfree, 1977; see also Amoroso *et al.*, 1979); the *in vitro* conversion of [^{3}H]pregnenolone to progesterone by the choriovitelline placenta of the quokka (Bradshaw *et al.*, 1975), together with the discovery that steroid precursors are actively metabolized by the equivalent structure of the tammar wallaby, *M. eugenii* (Renfree, 1977), is evidence of this. Furthermore, these findings suggest that enzymatic activity associated with steroid synthesis and metabolism, so long denied the trophoblast (chorionic) cells of the marsupial choriovitelline placenta and of the chorioallantoic placenta of the bandicoot (*Perameles*), may be unwarranted. It would be as well, therefore, to keep an open mind.

In macropodid marsupials (wallabies and kangaroos) the gestation period, with but a few exceptions (e.g., the swamp kangaroo and the gray kangaroo), is as long as the luteal phase of the estrous cycle. In these species, the females usually conceive again shortly after birth of the single young, but the second embryo develops only to the blastocyst stage and does not implant until the suckling "joey" is weaned, or removed, or is lost accidentally (Sharman, 1963; Tyndale-Biscoe, 1963b; Merchant and Sharman, 1966); the phase of dormancy is known as *embryonic diapause* to distinguish it from delayed implantation in Eutheria. It seems likely, therefore, that diapause results from an inhibitory effect of lactation and/or suckling on luteal growth, and by analogy with higher mammals it has been suggested that suckling induces the release of oxytocin from the neurohypophysis.

Apart from stimulating milk ejection, this hormone is believed to inhibit the corpus luteum, and in consequence to suppress the development of the postpartum embryo (Berger and Sharman, 1969b). Perhaps, then, it is right to think that oxytocin may have an additional physiological role in the maintenance of embryonic diapause in macropodid marsupials, although the mechanism by which this may be achieved remains to be disclosed. A possible action of the neurohypophyseal hormones has been reported for some vertebrate groups and concerns the control of secretion of adenohypophyseal hormone (Wallis, 1975). Like the hypothalamic releasing factors, the neurohypophyseal hormones are synthesized in the hypothalamus and they have often been proposed as possible regulators of adenohypophyseal secretion in some vertebrate groups and this may have played a part in determining the evolution of these hormones.

For the resumption of growth of the quiescent blastocyst and for the endometrial activity necessary for implantation, either the ovary or injections of progesterone (and sometimes estrogen) are necessary (Tyndale-Biscoe, 1963a,b; Berger and Sharman, 1969a). On the other hand, although hypophysectomy of the tammar wallaby, *Macropus eugenii*, has no effect on the development of a pregnancy, it prevents successful parturition at term and abolishes lactation; but when the operation is performed during embryonic diapause, it causes both the blastocyst and the quiescent corpus luteum to resume development and the pregnancy then to continue to term. These observations have been interpreted to mean that in this macropodid marsupial, "the adenohypophysis is essential for the initiation and maintenance of follicular development and ovulation as in the eutheria, but is not essential for development of a corpus luteum after ovulation" (Hearn, 1974). Apart from this, there is no evidence to suggest that the functions of the neuro- and adenohypophysis of metatherian mammals are fundamentally different from those of Eutheria, although the observations could hardly be described as adequate.

Ealey (1963) has tentatively suggested that embryonic diapause is an advantage to those marsupials in which it occurs: he regards the presence of viable embryos ready to implant immediately if the pouch young should be lost as a stratagem that reduces the time that must elapse before replacement is possible after such an accident. Equally, however, the condition of arrested pregnancy in marsupials can be regarded as an evolutionary device whereby the postnatal development of one pouch young is safeguarded by its ability to delay subsequent pregnancies, thus avoiding the simultaneous presence in the pouch of several broods of young, all competing for the limited resources of the mother. This statement derives its significance from the fact that the "joey" does not always contrive to do so. The proof that this is the case is well illustrated by the red kangaroo (*Megaleia rufa*), in which the female may have as many as three successive broods depending on her at any time: one young in diapause, one on the teat, and the third and eldest returning periodically to suckle (Hogarth, 1976). A curious feature of this relationship, which ia analogous to superfetation, is that the young retained in the pouch are provided with milks of different composition.

Finally, in those marsupial species where the young are born before the end of the cycle, a different stratagem is employed. Under these conditions embryonic development is accelerated and within a relatively brief period, viable young are born and, becoming attached to the teats, effectively bring about postponement of estrous cycles for the duration of lactation (Sharman, 1963). It is only when the young animal relinquishes its hold on the teat and begins to leave the pouch periodically that this inhibition is lifted, thus permitting a return to estrus and probably conception. It is evident, therefore, that the period of lactation in marsupials is much longer than that of pregnancy and the delay in implantation it imposes (up to 204 days in the red kangaroo, *Megaleia rufa*) has many features in common with the facultative embryonic diapause of suckling rats and mice, where lactation delays the time of implantation by several days, according to the size of the litter. The proximate factor in both appears to be the physical stimulus of suckling young at the nipple (Tyndale-Biscoe, 1963c; Sharman, 1976), and in both, the delay in implantation is apparently associated with a reduced secretion of pituitary gonadotropins other than prolactin (see Wallis, 1975).

C. Eutheria

The endocrinological events associated with the establishment and maintenance of pregnancy in marsupials that we have so far been considering have involved the direct interaction between the pituitary gland, the gonads, and to a lesser extent the embryo. This predominantly maternal control over gestation implies an endocrinologically inactive placenta and is contingent on a luteal phase of the estrous cycle with a duration the same as, or in excess of, that of true pregnancy. It is accordingly supposed that the state of pregnancy is maintained entirely by the maternal endocrine system (pituitary, ovaries, and adrenals) and the conceptus plays a passive role, although it may participate in the immunological events of pregnancy and in the induction of labor as it does in higher mammals (Porter and Amoroso, 1977). For the majority of mammals, above marsupials, however, such a mechanism is not possible as the luteal phase of their respective estrous cycles is short relative to the period of gestation, and in some species (e.g., rats and mice) even to the ovulation–implantation interval. Therefore it would seem that the evolutionary trend of viviparity beyond conditions found in marsupials was one of extending gestation for periods outlasting the duration of a normal estrous cycle.

Although the evolutionary sequence of the adaptations involved is far from clear, there is general agreement that this lengthening of gestation is accomplished either by prolonging the functional life span of the corpus luteum or by the adoption of complete endocrinological control of pregnancy by the placenta or a combination of both; it has been suggested that the first adaptations were concerned with extending the life of the luteal bodies (Amoroso, 1969). As a consequence, there is, in many species of eutherians, an endocrinological initiative on the part of the conceptus (trophoblast) that, seemingly, may vary from a simple antiluteolytic signal to prevent the collapse of the cyclic corpus luteum, to an extensive participation in, or even the appropriation of the functions of a number of adult endocrine organs, including the pituitary, the ovaries, the adrenal cortices, and even the hypothalamus (Porter and Amoroso, 1977; Khodr and Siler-Khodr, 1978). The almost complete structural identity between human growth hormone and human placental lactogen is evidence of this (Sherwood *et al.*, 1972).

VIII. The Corpus Luteum and the Establishment of Pregnancy

Apart from structural modifications in the arrangement of the female genital ducts, mammals have uniformly adopted endocrine mechanisms to solve many of the problems posed by viviparity. Three organs or organ systems are known to be implicated in these changes: the corpora lutea, the hypophysis, and the placenta. From what has already been said, it is evident that the success of viviparity in mammals owes much to the adoption of the corpora lutea in the family of endocrine glands. In essence this involved the addition of a luteal phase in sequence to a previous follicular phase that terminated at ovulation. But whereas the endocrine control of the first few days of pregnancy is apparently the same in all mammals, as it is fundamentally that of the estrous cycle, the humoral events attendant on the retention and expulsion of the products of conception may be quite different.

The action of the natural secretion of the corpus luteum, progesterone, is believed to be essential for the earliest stages of pregnancy in all mammals. In the later stages of gestation, the persistence of the luteal bodies may or may not be essential. In mice, rats, rabbits, goats, and cows, the removal of the ovaries at any stage of pregnancy brings gestation to an end; in women, monkeys, mares, sheep, and guinea pigs, the ovaries are not essential for the completion of pregnancy (Amoroso and Perry, 1977; Porter and Amoroso, 1977). It is, therefore, presumed that in the latter animals the placenta takes over the endocrine functions of the corpus luteum and is capable of producing the same secretions (progesterone and estrogen) or secretion of equivalent effect. Nor is the integrity of the pituitary gland essential for the completion of pregnancy in all mammals. In women, monkeys, mice, and rats, for example, but not in pigs, rabbits, and dogs, pregnancy may be completed after hypophysectomy. It is accordingly supposed that the corpora lutea in these species are maintained by a luteotropic complex that deputizes for secretions otherwise produced by the anterior lobe of the pituitary gland. It also is significant that the time at which hypophysectomy does not interrupt pregnancy in mice and rats coincides with that at which the chorioallantoic placental circulation is established. However, not only are the maternal ovaries and pituitary gland dispensable during gestation in women and monkeys, but the primate placenta becomes so completely autonomous in function that even when fetuses are removed, leaving their placentae *in situ*, gestation of the placentae continues to term, and some aspects of a second phenomenon, the onset of labor, are likewise shown to be independent of the immediate presence of the fetus.

In the rat, in which the pituitary–ovarian relationship was first and most extensively studied, the corpus luteum formed at ovulation persists only if pregnancy or pseudopregnancy supervenes; otherwise it has little effect in regulating the ovarian cycle. In pregnant individuals luteal action is prolonged and the maintenance of the corpora lutea, through at least the earliest stages of gestation, is agreed to be under the influence of a pituitary luteotropic complex consisting of LH, FSH, and PRL (Everett, 1964). In a similar way the estrous cycle of a rat (4 to 5 days) is automatically prolonged into a pseudopregnant period of 12 to 14 days if the mating is sterile, as after copulation with a vasectomized male, or after electrical or mechanical stimulation of the *cervix uteri,* which replicates the effect of mating (Everett, 1968). In this species the hormone responsible for the induction of pseudopregnancy is prolactin, but it is noteworthy that LH is the only pituitary gonadotropin that is needed for implantation (Madwha Raj and Moudgal, 1970).

Stimulation of the *cervix uteri* during estrus is the most effective way of eliciting prolactin secretion, but the same stimuli during diestrus may result in a period of estrus that is then followed by pseudopregnancy (Everett, 1969). Thus it seems that a comparatively long-lasting sensitivity to circulating hormones is imparted to the neural structures concerned in prolactin release by cervical stimulation, a specialization that provides for the maximal release of the hormone and, therefore, for the optimal function of the luteal bodies. Perhaps, then, there are general reasons for thinking that the endocrines regulating pseudopregnancy in these murine rodents are the same as those of the luteal phase of the estrous cycle, that is, luteal function is maintained by prolactin.

IX. The Maternal Recognition of Pregnancy

The prolongation of the functional life span of the corpus luteum evoked by the presence of a viable embryo and which is necessary for the establishment of gestation in many eutherian mammals has been referred to as *the maternal recognition of pregnancy* (Short, 1969; Heap and Perry, 1977; Heap *et al.*, 1978). It is a mechanism whereby the embryo signals its presence by actively controlling maternal endocrine functions involved in maintaining the secretory activity of the luteal bodies, thus ensuring a high concentration of progesterone in maternal plasma; it depends on the production of membrane receptors, which provide luteal cells with the competence to respond to luteotropic stimuli (Richards *et al.*, 1978). These embryonic signals are of several kinds and vary from species to species; their role in several species that have been studied more thoroughly than others has been discussed recently by Heap and his collaborators (Amoroso *et al.*, 1979; Heap *et al.*, this volume). Therefore, only certain of the more salient features will be used for the present purpose.

In spontaneously ovulating animals, the corpora lutea remain functional for only a short period unless pregnancy or pseudopregnancy supervenes. In some species (sheep, pig, cow) this attrition of the luteal bodies is achieved by the release of a uterine luteolytic substance (probably prostaglandin $F_{2\alpha}$, $PGF_{2\alpha}$) that is transported to the ovary directly through a utero-ovarian venous shunt. Thus, in sheep, which have an estrous cycle of about 16 days, there is strong evidence that relatively small surges of uterine $PGF_{2\alpha}$ secretion are observed on Days 12–14 of the normal ovarian cycle and these are followed by a much larger surge of secretion just prior to ovulation. There is equally strong evidence that comparable initial surges are observed at equivalent stages of pregnancy, but the subsequent, enhanced release of the luteolytic agent that coincides with ovulation is either suppressed or abolished altogether (Thorburn *et al.*, 1973; Barcikowski *et al.*, 1974). Therefore it seems probable that from about Day 12, the presence of a viable embryo (blastocyst) within the uterus of the ewe, by inhibiting the production of $PGF_{2\alpha}$, prevents the extinction of the luteal bodies, thus ensuring the survival of the embryo (Moor and Rowson, 1966a,b; Caldwell *et al.*, 1969; Peterson *et al*, 1976). But whatever the mechanism involved, the maternal recognition of pregnancy in the sheep appears to be primarily a matter of interfering with a luteolytic action rather than by promoting a luteotropic one; the antiluteolytic action of trophoblastin, a protein present in early pregnancy in sheep, is evidence of this (Martal *et al.*, 1979).

Although the nature of this inhibiting substance in the sheep remains unresolved, the discovery that trophoblastic tissue from the pig, a species in which estrogen has a luteotropic action (Gardner *et al.*, 1963), possesses the enzymatic capacity to synthesize estrogens from neutral steroid precursors (Perry *et al.*, 1973) and that estrogens inhibit the uterine secretion of $PGF_{2\alpha}$ (Bazer *et al.*, 1979) all support the idea that estrogens of embryonic origin may play an important role in furthering luteal function. It may also be significant that large amounts of estrogens are secreted by the pig's placenta in late pregnancy (Lunaas, 1962; Rombauts, 1962; Raeside, 1963).

Similar studies in Primates reveal that there is a luteotropic mechanism in early pregnancy that overrides a luteolytic process normally active within the ovary itself. In both the rhesus monkey (Knobil, 1973) and human female (Ross, 1979) the trophoblast of the early conceptus produces a gonadotropic hormone similar in structure and activity to LH, and designated CG and hCG, respectively. There is now ample evidence that these hormones are responsible either alone or in combination with other agencies for rescuing the corpus luteum of the cycle by protecting it from the local luteolytic action of estrogens and prostaglandins, thereby prolonging the secretion of luteal progesterone.

This two-way interaction between the early conceptus and its mother, which is primarily concerned with the extension of the life span of the corpora lutea of the estrous cycle into the corpora lutea of pregnancy (i.e., the secretion of luteal progesterone) by direct or indirect processes, emphasizes the importance of chemical signals attributable to the blastocyst before and during implantation. But whether the local effect of these signals on the uterine endometrium plays a significant role in the events associated with subsequent implantation/placentation remains a moot point. However, as the maternal response to these signals is crucial for the maternal recognition of pregnancy, and hence embryo survival, it is obvious that their inadequate expression will contribute to the failure of pregnancy that commonly occurs around the time of implantation in man and in many polytocous species and is not directly attributable to genetic abnormalities (Bishop, 1964). It will be as well to be aware of this fact, for even if we set down all the known causes of antenatal mortality, such as genetic abnormalities (Short, 1979), aberrations of maternal nutrition (Lamming, 1966), or the stress of a high ambient temperature (Smith *et al.*, 1966), the unexplained residue is of sterling proportions.

X. Immunological Transactions in Early Pregnancy

The transformation of the cyclical corpus luteum is but one of the many expressions of the maternal recognition of pregnancy. Yet another is the implantation of the embryo on the wall of the hormonally prepared uterus, a process that ranges in complexity from the superficial nidation of marsupials and of pigs and horses, to the highly invasive hemochorial condition of the primates, where trophoblast is in contact with and bathed by circulating maternal blood as in conventional man-made grafts. This intimate intraspecific contact or fusion between tissues of different genotype thus constitutes a naturally occuring instance of tissue transplantation, and its continued development and survival as an intrauterine allograft represent an apparent infringement of one of the fundamental laws of transplantation immunology.

Although a number of protective specific and nonspecific immunoregulatory mechanisms may be at play to reduce the antigenicity of the conceptus, it should not be supposed that the unmolested development of the mammalian embryo during a gestation period considerably in excess of the time required for experimental allograft rejection can be explained entirely apart from the endocrine framework in which viviparity has evolved. Indeed, because implantation of the blastocyst is so wholly regulated by endocrine-controlling mechanisms to which it also contributes, it is not beyond conjecture that the trophoblast as the main fabric of the internal secretions of the pre- and postimplantation embryo contains some of the clues to the failure of the conceptus to provoke an effective immunological rejection response in the maternal organism (Amoroso and Perry, 1975).

Among eutherian mammals there is now overwhelming evidence for steroid synthesis and catabolism (e.g., progestagens, estrogens, and corticoids) by the trophoblast (chorionic) cells, properties which are acquired before implantation in several species including the rat, mouse, hamster, rabbit, pig, sheep, and cow (cf. Heap *et al.*, 1979). Because of the preeminent role of progesterone in the full differentiation of the reproductive tract and mammary glands in the female mammal, as well as its indispensability for maintaining the life of the conceptus *in utero*, this hormone is commonly described as the hormone of pregnancy and an immunosuppressive role has been claimed for it (Siiteri *et al.*, 1977; Beer and Billingham, 1979; see also O'Malley and Means, 1973).The same appears to be true also of estrogens and glucocorticoids, of which there is a steady increase in production and excretion during pregnancy in many mammals.

The versatility of the synthetic mechanisms attributable to the trophoblast becomes even more impressive in considering the production of glycoprotein hormones. The placenta of many, and perhaps most, viviparous mammals secretes at least one gonadotropic hormone (e.g., hCG, CG, PMSG, pCG, PRL), and the presence of a lactogenic hormone (hPL) and the luteinizing hormone-releasing factor (LRF) in the human placenta has now been demonstrated beyond cavil. Placental lactogens have also been detected by biological methods in several other species, including monkey, rat, goat, sheep, and cow, as well as in the bank vole and the field vole (Amoroso *et al.*, 1979), and the development of immunological techniques has made it possible to display an array of glycoproteins peculiar to human pregnancy but distinct from hCG or hPL (Klopper, 1980). It seems likely that these glycoprotein hormones may be signals directed by the conceptus at the mother, perhaps acting in concert with steroids (progesterone) to reduce the maternal immune response by suppressing lymphocyte transformation (Contractor and Davies, 1973).

But whatever other functions this complex of hormones may exercise during pregnancy, a crucial consideration is the status of the trophectoderm (trophoblast), the outermost fetal layer, which remains intact throughout gestation in all Eutheria so far examined. This tissue, which forms an encapsulating layer around the early developing embryo from the time of inner cell mass formation in the blastocyst and which also constitutes the main fabric of the definitive placenta, has the ability to synthesize both steroid and glycoprotein hormones. It is thus evident that throughout the entire course of development, the possibility exists for the presence of glycoprotein hormones in conjunction with steroids and other sialosubstances at the periphery of trophoblast (chorionic) cells; such materials we consider as being present in a "cell coat" covering the plasma membrane of the microvilli, rather than in the membrane itself. From this viewpoint the question is, do chorionic gonadotropins and placental lactogens in conjunction with steroids (estrogens and progestagens) provide a perfect immunological insulation to the trophoblast, rendering it both incapable of provoking maternal sensitization and insusceptible to an existing state of sensitization; and if this is so, is the placental trophoblast of mammals fully equipped to exercise these functions? The gist of the answer is that chorionic gonadotropins (hCG, hCS) will inhibit an immunological system dependent on *antigen recognition* and will synergize with steroids; furthermore, the implanting mammalian blastocyst has the ability to synthesize both estrogens and gonadotropins in sufficient quantities to be a factor in the local survival of the trophoblast (Amoroso and Perry, 1975).

The specific identification of glycoprotein and steroid hormones in the trophoblast of the early mammalian conceptus is avowedly imperfect, but placental glycoproteins in great variety contrive to be identified in an ever increasing number of species (Heap *et al.*, 1978; Klopper, 1980), and although it is unlikely that immunological protection of the fetus depends on an identical mechanism throughout the group, it is probable that it evolved by the exploitation of existing properties of the chorion. Indeed, the development of the chorion in mammals is all the more impressive when one considers that in origin it was a thin nonglandular membrane that served the requirements of the embryo within the confines of a *cleidoic* egg.

It is evident, nevertheless, that viviparity has evolved on numerous different occasions in widely separated groups of vertebrates and invertebrates alike. It takes a variety of forms, and it would not be surprising if the immunological problems it entails had been solved differently in different cases; conclusions about how the developing offspring escapes rejection as a homograft in one species may not apply to another species in which viviparity evolved independently. However, granted that the immunological events in the reproductive process leading to successful viviparity in eutherian mammals are influenced by the complex of hormones that have well-known effects upon the behavior of the reproductive system, it is at least possible that among live-bearing insects, primitive arthropods, and nonmammalian vertebrates in which rudimentary placentae have been described, kindred agencies, as yet undisclosed, may be involved in the maternal acceptance of the fetal graft.

References

Allen, E., Hisaw, F. L., and Gardner, W. U., 1939, The endocrine functions of the ovaries, in: *Sex and Internal Secretions* (E. Allen, ed.), pp. 452–629, Williams & Wilkins, Baltimore.

Amoroso, E. C., 1952, Placentation, in: *Marshall's Physiology of Reproduction,* Vol. 2 (A. S. Parkes, ed.), pp. 119–224, Longmans, London.

Amoroso, E. C., 1955a, The comparative anatomy and histology of the placental barrier, in: *Gestation* (L. B. Flexner, ed.), pp. 119–224, 1st Conference, Josiah Macy Foundation, New York.

Amoroso, E. C., 1955b, Endocrinology of pregnancy, *Br. Med. Bull.* **11:**117–125.

Amoroso, E. C., 1960, Viviparity in fishes, *Symp. Zool. Soc. London* **1:**153–181.

Amoroso, E. C., 1968, The evolution of viviparity, *Proc. R. Soc. Med.* **61:**1188–1199.

Amoroso, E. C., 1969, Physiological mechanisms in reproduction, *J. Reprod. Fertil. Suppl.* **6:**5–18.

Amoroso, E. C., and Perry, J. S., 1975, The existence during gestation of an immunological buffer zone at the interface between maternal and foetal tissues, *Philos. Trans. R. Soc. London B Ser.* **271:**343–361.

Amoroso, E. C., and Perry, J. S., 1977, Ovarian activity during gestation, in: *The Ovary*, Vol. 2 (Lord Zuckerman and B. J. Weir, eds.), pp. 315–398, Academic Press, London.

Amoroso, E. C., Austin, J., Goffin, A., and Langford, E., 1957, Breeding habits of an amphibian *Gastrotheca marsupiata, J. Physiol. London* **135:**38P.

Amoroso, E. C., Heap, R. B., and Renfree, M. B., 1979, Hormones and the evolution of viviparity, in: *Hormones and Evolution, Vol. 2 (E. J. Barrington, ed.), pp. 925*–989, Academic Press, London.

Anderson, D. T., and Manton, S. M., 1972, Studies on the Onychophora. VIII. The relationship between the embryos and the oviduct in the viviparous placental onychophorans *Epiperipatus trinidadensis* (Bouvier) and *Macroperipatus torquatus* (Kennel) from Trinidad, *Philos. Trans. R. Soc. London B Ser.* **264:**161–189.

Barcikowski, B., Carlson, J. C., Wilson, L., and McCracken, J. A., 1974, The effect of endogenous and exogenous estradiol-17β on the release of prostaglandin $F_{2\alpha}$ from the ovine uterus, *Endocrinology* **95:**1340–1349.

Barrington, E. J. W., 1964, *Hormones and Evolution*, English Univ. Press, London.
Bazer, F. W., Roberts, R. M., and Thatcher, W. W., 1979, Actions of hormones on the uterus and effect on conceptus development, *J. Anim. Sci. Suppl.* **11**:49–35.
Beer, A. E., and Billingham, R. E., 1979, Maternal immunological recognition mechanisms during pregnancy, in: *Maternal Recognition of Pregnancy* (J. Whelan, ed.), Ciba Foundation Symposium No. 64 (new series), Excerpta Medica, Amsterdam.
Berger, P. J., and Sharman, G. B., 1969a, Progesterone-induced development of dormant blastocysts in the tammar wallaby, *Macropus eugenii* (Desmarest); Marsupialia, *J. Reprod. Fertil.* **20**:201–210.
Berger, P. J., and Sharman, G. B., 1969b, Embryonic diapause initiated without the suckling stimulus in the wallaby, *Macropus eugenii, J. Mammal.* **50**:630–632.
Bishop, M. W. H., 1964, Paternal contribution to embryonic death, *J. Reprod. Fertil.* **7**:383–396.
Boyd, M. M., 1942, The oviduct, foetal membranes and placentation of *Hoplodactylus maculatus* (Gray), *Proc. Zool. Soc. London Ser. A* **112**:65–104. .
Bradshaw, S. D., McDonald, I. R., Hähnel, R., and Heller, H. 1975, Synthesis of progesterone by the placenta of a marsupial, *J. Endocrinol.* **65**:451–452.
Bragdon, D. E., 1951, The non-essentiality of the corpora lutea for the maintenance of gestation in certain livebearing snakes, *J. Exp. Biol.* **118**:419–435.
Browning, H. C., 1973, The evolutionary history of the corpus luteum, *Biol. Reprod.* **8**:128–157.
Caldwell, B. V., Rowson, L. E. A., Moor, R. M., and Hay, M. F., 1969, The utero-ovarian relationship and its possible role in infertility, *J. Reprod. Fertil. Suppl.* **8**:59–76.
Callard, I. P., and Lance, V., 1977, The control of reptilian follicular cycles, in: *Reproduction and Evolution* (J. H. Calaby and C. H. Tyndale-Biscoe, eds.), pp. 199–210, Australian Academy of Science, Canberra.
Colombo, L., and Yaron, Z., 1976, Steroid 21-hydroxylase activity in the ovary of the snake *Storeria dekayi* during pregnancy, *Gen. Comp. Endocrinol.* **28**:403–412.
Contractor, S. F., and Davis, H., 1973, Effect of human chorionic somatomammotrophin and human chorionic gonadotrophin on phytohaemagglutinin-induced lymphocyte transformation, *Nature (London) New Biol.* **243**:283–285.
Danforth, C. H., 1939, Relation of genic and endocrine factors in sex, in: *Sex and Internal Secretions* (E. Allen, ed.), pp. 328–350, Williams & Wilkins, Baltimore.
del Pino, E. M., Galarza, M. L., de Albuja, C. M., and Humphries, A. A., Jr., 1975, The maternal pouch and development in the marsupial frog *Gastrotheca rio Cambae* (Fowler), *Biol. Bull.* **149**:480–491.
Dodd, J. M., Evennett, P. J., and Godard, C. K., 1960, Reproductive endocrinology in Cyclostomes and Elasmobranchs, *Symp. Zool. Soc. London* **1**:77–103.
Drewry, G. E., and Jones, K. L., 1976, A new ovo-viviparous frog, *Eleutherodactylus jasperi* (Amphibia, Anura, Leptodactylidae), from Puerto Rico, *J. Herpetol.* **10**:161–165.
Ealey, E. H. M., 1963, The ecological significance of delayed implantation in a population of the hill kangaroo (*Macropus robustus*), in *Delayed Implantation* (A. C. Enders, ed.), pp. 33–48, Univ. of Chicago Press, Chicago.
Eayrs, J. T., Glass, A., and Swanson, H. H., 1977, The ovary and nervous system in relation to behaviour, in: *The Ovary* (Lord Zuckerman and B. J. Weir, eds.), Vol. 2, pp. 399–455, Academic Press, London.
Englemann, F., 1970, *The Physiology of Insect Reproduction*, Pergamon Press, Oxford.
Everett, J. W., 1964, Central neural control of reproductive functions of the adenohypophysis, *Physiol. Rev.* **44**:373–471.
Everett, J. W. 1968, Delayed pseudopregnancy in the rat: A tool for the study of central nervous mechanisms in reproduction, in: *Reproduction and Sexual Behaviour* (M. Diamond, ed.), pp. 25–31, Univ. of Indiana Press, Urbana, Ill.
Everett, J. W., 1969, Neuroendocrine aspects of mammalian reproduction, *Annu. Rev. Physiol.* **31**:383–416.
Gardner, M. L., First, N. L., and Casida, L. E., 1963, Effect of exogenous estrogens on corpus luteum maintenance in gilts, *J. Anim. Sci.* **22**:132–134.
Griffiths, M. E., 1968, *Echidnas*, Pergamon Press, Oxford.
Griffiths, M. E., 1978, *The Biology of Monotremes*, Academic Press, London.
Heap, R. B., and Perry, J. S., 1977, Maternal recognition of pregnancy, in: *Contemporary Obstetrics and Gynaecology* (G. V. P. Chamberlain, ed.), pp. 3–7, Northwood Publications, London.
Heap, R. B., Flint, A. P., and Jenkin, G., 1978, Control of ovarian function during the establishment of gestation, in: *Control of Ovulation* (D. B. Crighton, N. B. Haynes, G. R. Foxcroft, and G. E. Lamming, eds.), pp. 295–318, Butterworths, London.
Heap, R. B., Flint, A. P., and Gadsby, J. E., 1979, Role of embryonic signals in the establishment of pregnancy, *Br. Med. Bull.* **35**(2):129–135.

Hearn, J. P., 1974, The pituitary gland and implantation in the tammar wallaby, *Macropus eugenii, J. Reprod. Fertil.* **39:**235–241.

Heller, H., 1973, Effects of oxytocin and vasopressin during the oestrous cycle and pregnancy on the uterus of *Setonix brachyurus, J. Endocrinol.* **58:**657–671.

Highfill, D. R., and Mead, R. A., 1975, Function of corpora lutea of pregnancy in the viviparous garter snake, *Thamnophis elegans, Gen. Comp. Endocrinol.* **27:**401–407.

Hill, C. J., 1933, The development of the Monotremata. Part 1. The histology of the oviduct during gestation, *Trans. Zool. Soc. London* **21:**413–443.

Hill, C. J., 1941, The development of the Monotremata. Part V. Further observations on the histology and the secretory activities of the oviduct prior to and during gestation, *Trans. Zool. Soc. London* **25:**1–31.

Hisaw, R. L., 1959, Endocrine adaptations of the mammalian estrous cycle and gestation, in: *Comparative Endocrinology* (A. Gorbman, ed.), pp. 533–552, Wiley, New York.

Hisaw, R. L., 1961, Endocrines and the evolution of viviparity among the vertebrates, in: *Physiology of Reproduction*, Proc. 22nd Biol. Colloq. Oregon State University, 1961, Oregon State Univ. Press, Corvallis, Oreg.

Hisaw, F. L., and Abramowitz, A. A., 1938, The physiology of reproduction in the dogfish, *Mustelus canis, Rep. Woods, Hole Oceanogr. Inst.* **1937:**21–22.

Hisaw, F. L., and Abramowitz, A. A., 1939, Physiology of reproduction in the dogfishes, *Mustelus canis* and *Squalus acanthias, Rep. Woods Hole Oceanogr. Inst.* **1938:**22.

Hogarth, P. J., 1976, *Viviparity*, Studies in Biology No. 75, Arnold, London.

Holt, J. A., Richards, J. S., Midgley, A. R., and Reichert, L. E., 1976, Effect of prolactin on LH receptors in rat luteal cells, *Endocrinology* **98:**1005–1013.

Khodr, G. S., and Siler-Khodr, T., 1978, Localization of luteinizing hormone-releasing factor in the human placenta, *Fertil. Steril.* **29:**523–526.

Klopper, A., 1980, The new placental proteins, *Placenta* **1:**77–89.

Knobil, E., 1973, On the regulation of the primate corpus luteum, *Biol. Reprod.* **8:**246–268.

Lamming, G. E., 1966, Nutrition and the endocrine system, *Nutr. Abstr. Rev.* **36:**1–13.

Lofts, B., 1974, Reproduction, in: *Physiology of the Amphibia* (B. Lofts, ed.), Vol. 2, pp. 107–218, Academic Press, New York/London.

Lunaas, T., 1962, Urinary oestrogen levels in the sow during the oestrous cycle and pregnancy, *J. Reprod. Fertil.* **4:**13–20.

Lupo di Prisco, C., Vellano, C., and Chieffi, G., 1967, Steroid hormones in the plasma of the elasmobranch *Torpedo marmorata* at various stages of the sexual cycle, *Gen. Comp. Endocrinol.* **8:**325–331.

Madhwa Raj, H. G., and Moudgal, N. R., 1970, Hormonal control of gestation in the intact rat, *Endocrinology* **86:**847–889.

Martal, J., Lacroix, M. C., Loudes, C., Saunier, M., and Wintenberger-Torrès, S., 1979, Trophoblastin, an antiluteolytic protein present in early pregnancy in sheep, *J. Reprod. Fertil.* **56:**63–73.

Medawar, P. B., 1953, Some immunological and endocrinological problems raised by the evolution of viviparity in vertebrates, *Symp. Soc. Exp. Biol.* **7:**328–338.

Merchant, J. C., and Sharman, G. B., 1966, Observations on the attachment of marsupial pouch-young to the teats and on the rearing of pouch-young by foster mothers of the same or different species, *Aust. J. Zool.* **14:**593–609.

Mertens, R., 1957, Zür Naturgeschichte des venezolanischen Reisen-Beutel Frosches, *Gastrotheca ovifera, Zool. Garten* (N.F.).

Moor, R. M., and Rowson, L. E. A., 1966a, Local uterine mechanisms affecting luteal function in the sheep, *J. Reprod. Fertil.* **11:**307–310.

Moor, R. M., and Rowson, L. E. A., 1966b, Local maintenance of the corpus luteum in sheep with embryos transferred to various isolated portions of the uterus, *J. Reprod. Fertil.* **12:**539–550.

O'Malley, B. W., and Means, A. R. (eds.), 1973, *Receptors for Reproductive Hormones*, Plenum Press, New York.

Panigel, M., 1956, Contribution à l'étude de l'ovoviviparité chez les reptiles: Gestation et parturition chez le lézard vivipare, *Zootoca vivipara, Ann. Sci. Nat. Zool. Biol. Anim. XI* **18:**569–668.

Perry, J. S., Heap, R. B., and Amoroso, E. C., 1973, Steroid hormone production by pig blastocysts, *Nature (London)* **245:**45–47.

Peterson, A. J., Tervit, H. R., Fairclough, R. J., Hawik, P. G., and Smith, J. F., 1976, Jugular levels of 13,14-dihydro-15-keto-prostaglandin F and progesterone around luteolysis and early pregnancy in the ewe, *Prostaglandins* **12:**551–558.

Pickford, G. E., and Atz, J. W., 1957, *The Physiology of the Pituitary Gland of Fishes*, New York Zoological Society, New York.

Porter, D. G., and Amoroso, E. C., 1977, The endocrine function of the placenta, in: *Scientific Foundations of Obstetrics and Gynaecology* (E. E. Philipp, J. Barnes, and M. Newton, eds.), pp. 675–712, Heinemann, London.

Raeside, J. I., 1963, Urinary oestrogen excretion in the pig during pregnancy and parturition, *J. Reprod. Fertil.* **6:**427–431.

Ranzi, S., 1935, Risultati di ricerche su varie forme di gestazione, *Boll. Zool.* **6:**153.

Renfree, M. B. 1974, Ovariectomy during gestation in the American opossum, *Didelphis marsupialia virginiana, J. Reprod. Fertil.* **39:**127–130.

Renfree, M. B., 1977, Feto-placental influences in marsupial gestation, in: *Reproduction and Evolution* (J. H. Calaby and C. H. Tyndale-Biscoe, eds.), pp. 325–332, Australian Academy of Science, Canberra.

Richards, J. S., Rao, M. C., and Ireland, J. J., 1978, Actions of pituitary gonadotrophins on the ovary, in: *Control of Ovulation* (D. B. Crighton, G. R. Foxcroft, N. B. Haynes, and G. E. Lamming, eds.), pp. 197–216, Butterworths, London.

Riddle, O., and Bates, R. W., 1939, The preparation, assay and actions of lactogenic hormones, in: *Sex and Internal Secretions* (E. Allen, ed.), pp. 1088–1117, Williams & Wilkins, Baltimore.

Riddle, O., Lahr, E. L., and Bates, R. W., 1942, The role of hormones in the initiation of maternal behaviour in rats, *Am. J. Physiol.* **137:**200–217.

Rombauts, P., 1962, Excrétion urinaire d'oestrogènes chez la truie pendant la gestation, *Ann. Biol. Anim. Biochim. Biophys.* **2:**151–156.

Ross, E. T., 1979, Human chorionic gonadotrophin and the maternal recognition of pregnancy, in: *Maternal Recognition of Pregnancy* (J. whelan, ed.), Ciba Foundation Symposium No. 64 (new series), pp. 191–201, Excerpta Medica, Amsterdam.

Sharman, G. B., 1963, Delayed implantation in marsupials, in: *Delayed Implantation* (A. C. Enders, ed.), pp. 3–14, Univ. of Chicago Press, Chicago.

Sharman, G. B., 1976, Evolution of viviparity in mammals, in: *Reproduction in Mammals: The Evolution of Reproduction* (C. R. Austin and R. V. Short, eds.), pp. 32–70, Cambridge Univ. Press, London/New York.

Sherwood, L. M., Handwerger, S., and McLauren, W. D., 1972, Structure and function of human placental lactogen, in: *Lactogenic Hormones,* CIBA Foundation Symposium in Memory of Professor S. J. Folley (G. E. W. Wolstenholme and J. Knight, eds.), pp. 27–45, Churchill, London.

Short, R. V., 1969, Implantation and the maternal recognition of pregnancy, in: *Foetal Autonomy* (G. E. W. Wolstenholme and M. O'Connor, eds.), CIBA Foundation Symposium, pp. 2–26, Churchill, London.

Short, R. V., 1979, When a conception fails to become a pregnancy, in: *Maternal Recognition of Pregnancy* (J. Whelan, ed.), Ciba Foundation Symposium No. 64 (new series), pp. 377–387, Excerpta Medica, Amsterdam.

Siiteri, P. K., Febres, F., Clemens, L. E., Chang, J. R., Gondos, B., and Stites, D., 1977, Progesterone and maintenance of pregnancy: Is progesterone nature's immunosuppressant?, *Ann. N.Y. Acad. Sci.* **286:**384–397.

Smith, C. L., Rand, C. S., Schaeffer, B., and Atz, J. W., 1975, *Latimeria,* the living coelacanth, is oviparous, *Science* **190:**1105–1106.

Smith, I. D., Bell, G. H., and Chanlet, G., de, 1966, Embryonic mortality in merino ewes exposed to high ambient temperatures, *Aust. Vet. J.* **42:**468–470.

Thorburn, G. D., Cox, R. I., Currie, W. B., Restall, B. J., and Schneider, W., 1973, Prostaglandin F and progesterone concentrations in the uterovarian venous plasma of the ewe during the oestrous cycle and early pregnancy, *J. Reprod. Fertil. Suppl.* **18:**151–158.

Tyndale-Biscoe, C. H., 1963a, Effects of ovariectomy in the marsupial *Setonix brachyurus, J. Reprod. Fertil.* **6:**25–40.

Tyndale-Biscoe, C. H., 1963b, Blastocyst transfer in the marsupial *Setonix brachyurus, J. Reprod. Fertil.* **6:**41–48.

Tyndale-Biscoe, C. H., 1963c, The role of the corpus luteum in the delayed implantation of marsupials, in: *Delayed Implantation* (A. C. Enders, ed.), pp. 15–32, Univ. of Chicago Press, Chicago.

Tyndale-Biscoe, C. H., 1973, *Life of Marsupials,* Arnold, London.

Veith, W. J., 1974, Reproductive biology of *Chameleo pumilus pumilus* with special reference to the role of the corpus luteum and progesterone, *Zool. Afr.* **9:**161–183.

Wallis, M., 1975, The molecular evolution of pituitary hormones, *Biol. Rev.* **50:**35–98.

Watson, D. M. S., 1927, The reproduction of the coelacanth fish, *Undina, Proc. Zool. Soc. London* **Part I–11:**453–457.

Willmer, E. N., 1974, Nemertines as possible ancestors of the vertebrates, *Biol. Rev.* **49:**321–363.

Wotiz, H. H., Botticelli, C., Hisaw, F. L., Jr., and Ringler, I., 1958, Identification of estradiol-17β from dogfish ova (*Squalus suckleyi*), *J. Biol. Chem.* **231**:589–592.
Xavier, F., 1970, Analyse du rôle des 'corpora lutea' dans le maintien de la gestation chez *Nectophrynoides occidentalis* (Ang), *C. R. Acad. Sci. Ser. D.* **270**:2018–2020.
Yaron, Z., 1977, Embryo-maternal interrelations in the lizard *Xantusia vigilis*, in: *Reproduction and Evolution* (J. H. Calaby and C. H. Tyndale-Biscoe, eds.), pp. 271–278, Australian Academy of Science, Canberra.

2

My Life with Mammalian Eggs

M. C. Chang

I. Introduction

It is my belief that the late Sir John Hammond of Cambridge University accepted me as a research student in 1939 because I found two cow eggs under an old-fashioned compound microscope after removal of the objective lens. But I did very little work on mammalian eggs there. My research work in England was mainly concerned with the metabolism of ram spermatozoa and artificial insemination. I started my work on rabbit eggs in early 1945 when I came to the Worcester Foundation for Experimental Biology, Shrewsbury, Massachusetts. The Foundation was newly established by H. Hoagland, a neurophysiologist, and the late Gregory Pincus, who was at that time the only person in the world able to fertilize rabbit eggs *in vitro*.

When I went to Pincus' office the morning after my arrival, I realized that my projected work included the perfusion of cow ovaries, and the induction of superovulation. I was expected to find the eggs, to fertilize them *in vitro,* and then to transfer them to different cows. I was quite startled by the American way of research and remarked that this project could take my whole life but that I planned to stay only a year in his laboratory. During the past 34 years, my academic life has been mainly involved with mammalian eggs. I have had frustrations and failures but I have enjoyed my work tremendously. On this occasion, I shall summarize my research and experience in the hope that it may reveal a little bit about mammalian fertilization and early development, and may also shed some light on implantation.

II. In Vitro Fertilization of Mammalian Eggs

After several years of studying mammalian spermatozoa in England, I was quite convinced that in order to understand the basic function of spermatozoa, one had to learn the process and mechanisms of fertilization. After all, not only do the spermatozoa take up oxy-

M. C. Chang • Worcester Foundation for Experimental Biology, Shrewsbury, Massachusetts 01545. The author is a Research Career Awardee of NICHD (K6-HD-18, 334).

gen, utilize substrate, and move about, but their primary function is to fertilize eggs. So I decided first I would learn *in vitro* fertilization with G. Pincus. This would enable me to have an objective and concise procedure to examine the physiology of spermatozoa. In 1944 I wrote asking him for permission to come to his laboratory. I mentioned my work on artificial production of monstrosities in the rabbit by treating sperm with colchicine (Chang, 1944) because he had done some work on the induction of polyploidy by treating rabbit eggs with colchicine.

From 1945 to 1950, I tried everything that came to my mind attempting to fertilize rabbit eggs *in vitro*. I had spent a lot of time working by myself and occasionally with him but without success. I thought that there must be some tubal factor(s) for fertilization that we did not understand. When I deposited mixed sperm and eggs, or sperm alone, into the oviducts at various times before or after ovulation, fertilization occurred only when sperm were deposited into the oviducts more than 6 hr before ovulation. This indicated that sperm must have undergone physiological changes during these 6 hr after the deposition in the female tract. This observation led to the discovery of the development, in the female tract, of the fertilizing capacity of sperm (Chang, 1951a, 1955a). This process was termed "capacitation of spermatozoa" by Austin (1951, 1952). Although the cytological evidence for *in vitro* fertilization of rabbit eggs had been described by Thibault *et al.* (1954), the incontestable evidence was reported only after I obtained offspring, genetically true to their parents, following the transfer of *in vitro* fertilized rabbit eggs (Chang, 1959). The recent publicity about the test-tube baby girl in England reported by Steptoe and Edwards (1978) caused a great deal of public interest, but I joked to my friends that it took 20 years for the medical profession to catch up with the work of an experimental biologist.

In vitro fertilization of hamster eggs was achieved in my laboratory by Yanagimachi (Yanagimachi and Chang, 1964), who has done a great deal of important work on mammalian fertilization. To date, *in vitro* fertilization of mammalian eggs has been extended to Chinese hamsters (Pickworth and Chang, 1969), mice [first by uterine sperm (Whittingham, 1968), then by sperm capacitated *in vitro* (Iwamatsu and Chang, 1969)], deer mice (Fukuda *et al.*, 1979), rats (Miyamoto and Chang, 1973; Toyoda and Chang, 1974), cats (Hamner *et al.*, 1970), guinea pigs (Yanagimachi, 1972), squirrel monkeys (Gould *et al.*, (1973), cows (Iritani and Niwa, 1977), pigs (Iritani *et al.*, 1978), and humans (Edwards *et al.*, 1969, 1970; Steptoe and Edwards, 1978). We have also determined the fertilizability of mouse ovarian oocytes during maturation (Iwamatsu and Chang, 1972) and that of rat eggs before and after ovulation (Niwa and Chang, 1975) by *in vitro* techniques. It should be pointed out here that the important observations that ensured successful *in vitro* fertilization were the recognition of capacitation of sperm in the female tract, the induction of capacitation, and the acrosome reaction of spermatozoa, by incubation of sperm in a particular medium, and careful manipulation of sperm and eggs.

The detailed methods and procedures for *in vitro* fertilization of mammalian eggs were developed by my Japanese associates when they were working with me and after they left my laboratory. However, the technique of mounting a group of whole eggs between a cover slip and a slide (Chang, 1952a) and the technique of spreading a whole blastocyst on a slide (Chang, 1954) for the examination of histological details were also very important steps forward for the study of eggs. By studying the fertilization of zona-free eggs we have demonstrated that the major function of the zona pallucida is to block polyspermy and the penetration by foreign sperm. However, the vitelline membrane also plays a role (Hanada and Chang, 1972, 1978). The inhibition of mammalian fertilization *in vitro* and *in vivo* by active and passive immunization with anti-ovary, anti-egg, and anti-zona serum was also reported from my laboratory (Tsunoda and Chang, 1976a,b, 1977, 1978).

Our basic knowledge of fertilization was obtained mainly by the study of sea urchin eggs from the time of Lillie (1919) up to the present. During the past 20 years we have learned a great deal about fertilization in higher species using the procedures of *in vitro* fertilization. It should be pointed out, however, that, as a high proportion of polyspermy was always observed following *in vitro* fertilization (Fukuda and Chang, 1978), there may be some differences between *in vitro* and *in vivo* fertilization.

III. Storage and Transfer of Eggs

When I was at Cambridge University, my major work was the study of the storage of sperm for artificial insemination of farm animals. The purpose of artificial insemination is the utilization of the germ cells of genetically superior males. Likewise, the development of the technique of egg transfer was to allow the full utilization of the germ cells of genetically superior females. When I started my work in the United States, the first thing that came into my mind was to determine whether fertilized rabbit eggs can be stored in a refrigerator without loss of their development potential as is the case for mammalian sperm. As my technique of recovery, manipulation, and transfer of eggs improved, I was able to store the fertilized rabbit eggs in a refrigerator for several days and then transfer them to recipient rabbits. When I obtained normal young after the transfer of fertilized eggs stored at 5 or 10°C for several days (Chang, 1947), there was a great deal of publicity in the media because this was considered an achievement in those days. Further study revealed that unfertilized rabbit eggs (Chang 1952a, 1953, 1955b), as well as fertilized eggs at different stages of development (Chang, 1948, 1950a), could be stored at 5 or 10°C for 2 days without losing their ability to be fertilized or impairing subsequent development. Since the discovery of cryoprotective agents (Smith and Polge, 1950) and the successful transplantation of cow eggs (Rowson *et al.*, 1971), deep-freezing of mammalian embryos kept at −196°C is commonly performed in several laboratories (Leibo, 1977; Polge, 1977). I should like to point out here that successful deep-freezing of unfertilized mouse and hamster eggs without loss of fertilizability was reported from my laboratory (Tsunoda *et al.*, 1976; Parkening *et al.*, 1976) a year before the others.

I have mentioned before that one of my assignments in 1945 was the transfer of eggs from one cow to another. In those days very little had been done with mammalian eggs and egg transfer, so I concentrated my study on rabbit eggs. The adverse effect of heterologous sera (Chang, 1949) and seminal plasma (Chang, 1950b) on fertilized rabbit eggs was reported first. By transferring rabbit eggs at different stages of development to recipient rabbits at various times after ovulation, I was able to demonstrate the importance of synchronization between the age of the embryo and the developmental stage of the endometrium to ensure proper implantation (Chang, 1950c). This was confirmed later in various species by many scientists. I am very pleased with this work because this observation stimulated many investigators and attracted many scientists to the field of implantation. In passing, it should be pointed out that the effectiveness of oral contraceptives has been attributed to the fact that the endometrium is desynchronized by progestin treatment, thereby preventing proper implantation (Chang, 1967a, 1978).

Besides the practical applications of egg transfer for animal breeding, the technique of egg transfer can also be used for investigating many reproductive processes including implantation. For instance, I have transferred rabbit eggs to nonovulated rabbits treated with progesterone and obtained young (Chang, 1951b).

IV. Interspecific Fertilization and Egg Transfer

Another of my interests was mammalian hybridization (Chang *et al.*, 1969). The impossibility of crosses between domestic rabbits (*Oryctolagus cuniculus*) and hares (*Lepus europaeus*) was reported by my teachers, but the possibility that fertilization occurred and was followed by embryonic degeneration had never been examined. We inseminated female domestic rabbits with the sperm of cottontail rabbits (*Syvilagus floridanus*) and studied the details of fertilization and early development. We found that about 30–40% of the eggs were fertilized but degenerated before implantation (Chang and McDonough, 1955; Chang, 1960). However, after inseminating domestic rabbits with sperm of the snowshoe hare (*Lepus americanus*), we observed a fertilization rate of 97%. Although fertilized eggs again degenerated during cleavage and blastocyst formation, approximately 1 out of 64 eggs implanted and developed into a small embryo. On the other hand, only 1 of 10 eggs was fertilized following insemination of female snowshoe hares with sperm of domestic rabbits (Chang *et al.*, 1964). Development to the blastocyst stage was possible following transfer of 11 fertilized snowshoe hare eggs into the oviducts of domestic rabbits, but decidual formation and implantation failed (Chang, 1965a). In a study of reciprocal transfer of eggs between rabbit and ferret, it was found that ferret eggs at different stages of development can survive and develop in the oviduct but not in the uterus of rabbits, whereas rabbit eggs cannot survive in the oviduct or uterus of ferrets. The decidual reaction was not induced in the rabbit or ferret uterus by foreign embryos (Chang, 1966a). In fact we have obtained normal fetuses following the retransfer of ferret early blastocysts to the ferret uterus after they had been kept in the rabbit oviducts, stored at 1 or 5°C, or cultured at 38°C for 2 or 3 days (Chang *et al.*, 1971a).

Implantation of ferret (*Mustela putorius*) eggs fertilized by mink (*M. vision*) sperm was observed (Chang, 1965b), but fertilization of mink eggs by ferret sperm was not possible (Chang, 1968). By transferring fertilized ferret eggs to mink and vice versa, it was found that ferret eggs developed at a slow rate in the mink uterus but degenerated and failed to implant, while the mink eggs developed in the ferret uterus, implanted without delay, but degenerated after implantation (Chang, 1968). This indicated that delayed implantation is controlled mainly by the conditions of the endometrium rather than the inherent characteristics of the embryo. Reciprocal fertilization between ferrets and short-tailed weasels (*M. erminea*) was possible. Implantation of ferret eggs fertilized by weasel sperm occurred on Days 13–14 of pregnancy as is characteristic for ferrets, and was not delayed as is the case with weasels. A nearly full-grown hybrid fetus was obtained 42 days after insemination of ferret with weasel sperm (Wu and Chang, 1973a). Most work on hormonal requirements for implantation has been done on rats and mice, but we have also determined the hormonal requirements in the ferret (Wu and Chang 1972, 1973b) and Mongolian gerbil (Wu, 1974).

Rabbit sperm, unfertilized eggs, and fertilized eggs were subjected to radiocobalt irradiation. Following insemination with irradiated sperm, transfer of irradiated unfertilized eggs to mated rabbits, or transfer of irradiated fertilized eggs to recipients, it was found that the radiation was harmful to the subsequent development of the embryo: Sperm is more radioresistant than unfertilized or fertilized eggs (Chang *et al.*, 1957, 1958). Further study of *in vitro* radiocobalt irradiation of rabbit eggs at different stages of development and of recipient or pregnant rabbits has shown that before implantation, eggs irradiated *in vitro* or *in vivo* either died or developed into apparently normal fetuses. There was no evidence of differential radiosensitivity at various stages of development. Irradiation of recipient rabbits alone also affected the subsequent implantation and embryonic development.

The proportion of eggs implanting, however, was decreased more by irradiation of pregnant rabbits than by irradiation of eggs or recipients alone (Chang and Hunt, 1960). Irradiation of pregnant rabbits at 400 R before implantation reduced the proportion of implantation to 24%, with 19% living fetuses, while irradiation at the time of implantation (Day 7 or 8) caused a complete degeneration of embryos after implantation (Chang *et al.*, 1963). It appears that irradiation during implantation is more harmful to subsequent embryonic development than at any other time. Moreover, we have found that irradiation of exteriorized ovary or uterus also reduced the proportion of implantation and interfered with embryonic development in the rat (Gibbons and Chang, 1973a,b).

V. Effects of Steroids and Other Compounds on the Transportation and Development of Eggs

In the development of a postcoital pill, we found that feeding pregnant rabbits norethynodrel, clomiphene, or triethylamine 1 to 3 days after mating caused a disturbance of egg transport or expulsion of eggs from the uterus, and a possible disturbance of implantation. But feeding these compounds during the time of implantation, on Days 6 to 8, had no effect. Transplantation of 6-day normal blastocysts to the uterus of recipient rabbits treated with norethynodrel, but not with the other two compounds, also reduced the proportion of implantations significantly (Chang, 1964). Further study of the effect of estrogens, an A-nor steroid (H241) with estrogenic activity, or other compounds revealed that natural estrogens or other compounds that have estrogenic activities could be effective for contraceptive purposes when administered soon after mating but not during implantation (Chang and Yanagimachi, 1965). An A-nor steroid (Anordrin AF53) has been used in China (Ku *et al.*, 1975) as a postcoital pill. Treating rabbits with estrogen 1 to 3 days after insemination led us to conclude that estrogen treatment interferes with egg development by disturbing egg transport; thus the eggs enter the uterus too soon or are left in the oviduct too long. This was not due to a toxic effect of estrogen on the eggs because a similar proportion of fetuses was obtained after transfer of morulae from the oviducts of treated animals to untreated recipients or vice versa (Chang and Harper, 1966).

The disturbance of egg transport and interruption of pregnancy in animals by administration of estrogens after mating have been known for a long time. The accelerated egg transport and the degeneration of eggs caused by administration of progesterone before ovulation were first reported by us (Chang and Bedford, 1961; Chang, 1966b). As unfertilized rabbit eggs (Chang, 1955c) or newly fertilized rabbit eggs (Chang, 1950c) could not survive in the uterus (although newly fertilized ferret eggs could persist slightly better in the uterus; Chang, 1969a), it was thought that progestins administered before ovulation could also be used for antifertility purposes. By treating rabbits before ovulation with medroxyprogesterone acetate or after ovulation with estrogen, complete degeneration of blastocysts on Day 6 was observed (Chang, 1966b). Treatment of ferrets with medroxyprogesterone before ovulation, but not with estrogen after ovulation, also inhibited fertilization, hastened the transport of eggs from the oviduct to the uterus, and prevented pregnancy (Chang, 1967b). Further study revealed that treating female rabbits with progesterone or related compounds before ovulation caused an inhibition of fertilization, due to the disturbance of sperm capacitation and rapid egg transport, and degeneration of eggs in the uterus (Chang, 1967c). Fertilization and egg transport were also disturbed in pseudopregnant or progesterone-treated rabbits. Transfer of fertilized eggs from untreated animals to the pseudopregnant or progesterone-treated rabbits led to the degeneration of eggs and failure of implantation (Chang, 1969b). Obviously, preconditioning of the female tract by progestational compounds also disturbs the implantation process and embryonic development.

Subcutaneous implantation of a Silastic tube containing estrogen, but not various progestins, prevented pregnancy in the rat until its removal at 45 days (Casas and Chang, 1970). In the rabbit, implantation of estrogen, but not progesterone or chlormadinone acetate (a potent progestin), caused a decrease in the number of eggs recovered and the degeneration of eggs before implantation. The ineffectiveness of implanted progestins may be due to their lower rate of absorption and other undefined reasons (Chang *et al.*, 1970). Pregnancy in rabbits was suppressed by subcutaneous implantation of a Silastic tube containing various doses of estrogenic compounds because of the degeneration of eggs and failure of implantation. Transfer of early blastocysts recovered from these estrogen-treated rabbits to normal recipient rabbits did not diminish the rate of implantation or disturb normal development, but transfer of eggs from untreated rabbits to estrogen-treated rabbits led to a complete failure of implantation (Chang *et al.*, 1971b).

We have also studied the effects of progestins or estrogens (Chang and Hunt, 1970) and prostaglandins (Chang *et al.*, 1972) on gamete transport and fertilization. In contrast to the report of Dickmann (1973) that injections of medroxyprogesterone acetate on Day 1 followed by estrogen on Day 3 destroyed rat eggs, we found that such treatment had little effect on rabbit eggs (Chang, 1974). The induction of abortion in laboratory animals by administration of prostaglandins (Chang and Hunt, 1972; Lau *et al.*, 1974, 1975; Saksena *et al.*, 1975) or Trichosanthin (Chang *et al.*, 1979) was also reported from my laboratory.

I have also been interested in the process of implantation, especially the time of implantation in different species. I often wonder why rabbit and human eggs implant on Day 7 whereas monkey eggs implant on Day 9. On the other hand, a physical attachment of the embryo to the endometrium (implantation) occurs 40–50 days after mating in the cow (Chang, 1952b) and about 18 days in the sheep (Chang and Rawson, 1965). The process of implantation takes place when the blastocyst and endometrium are at the right and proper stage of development. Therefore, unless there is a healthy blastocyst in a properly prepared endometrium, any antifertility compound claimed to be an inhibitor of implantation should be seriously reevaluated. My work has been mainly concerned with general morphological and physiological aspects, and I am very much looking forward to this conference and to learning about the recent advances in the study of animal reproduction at the cellular and molecular levels.

VI. Summary

I have had a very pleasant and satisfying life earning a living by the study of the mammalian egg. In this article I have neither mentioned my frustrations, failures, or unpublished observations nor discussed the important contributions by other scientists. I have only summarized most of the work published by myself and my associates to whom I am very much indebted. I should like to claim that the techniques and procedures for successful *in vitro* fertilization of different species were developed mainly in my laboratory. The storage and transplantation of eggs were carried out not only to provide a better understanding of embryonic development but also for the improvement of animal breeding. Interspecific fertilization is possible between some species, but hybrid embryos either degenerate before implantation in some species or die at various times after implantation in other species. Following the transfer of eggs from one species to another, development of eggs in the oviduct and implantation are also possible in some species. The effects of irradiation of sperm, eggs, ovary, uterus, recipient pseudopregnant, or pregnant animals on embryonic development and implantation were also mentioned. Finally, the success and failure of our attempts to inhibit implantation or interrupt pregnancy by administration of estrogen, progestins, and other compounds were described in the hope of finding more and better methods for the control of human fertility.

References

Austin, C. R., 1951, Observations on the penetration of the sperm into the mammalian egg, *Aust. J. Sci. Res. Ser. B* **4**:581–596.

Austin C. R., 1952, The capacitation of mammalian sperm, *Nature (London)* **170**:326–327.

Casas, J. H., and Chang, M. C., 1970, Effects of subcutaneous implantation or intra-uterine insertion of Silastic tube containing steroids on the fertility of rats, *Biol. Reprod.* **2**:315–325.

Chang, M. C., 1944, Artificial production of monstrosities in the rabbit, *Nature (London)* **154**:150.

Chang, M. C., 1947, Normal development of fertilized rabbit ova stored at low temperature for several days, *Nature (London)* **159**:602–603.

Chang, M. C., 1948, Transplantation of fertilized rabbit ova: The effect of viability of age, *in vitro* storage period, and storage temperatures, *Nature (London)* **161**:978–979.

Chang, M. C., 1949, Effects of heterologous sera on fertilized rabbit ova, *J. Gen. Physiol.* **32**:291–300.

Chang, M. C., 1950a, Transplantation of rabbit blastocysts at late stage: Probability of normal development and viability at low temperature, *Science* **3**:544–545.

Chang, M. C., 1950b, The effect of seminal plasma on fertilized rabbit ova, *Proc. Nat. Acad. Sci. USA* **36**:188–191.

Chang, M. C., 1950c, Development and fate of transferred rabbit ova or blastocysts in relation to the ovulation time of recipients, *J. Exp. Zool.* **114**:197–216.

Chang, M. C., 1951a, Fertilizing capacity of spermatozoa deposited into the Fallopian tubes, *Nature (London)* **168**:697–698.

Chang, M. C., 1951b, Maintenance of pregnancy in intact rabbits in the absence of corpora lutea, *Endocrinology* **481**:17–24.

Chang, M. C., 1952a, Fertilizability of rabbit ova and effects of temperature *in vitro* on their subsequent fertilization and activation *in vivo*, *J. Exp. Zool.* **121**:351–382.

Chang, M. C., 1952b, Development of bovine blastocysts with a note on implantation, *Anat. Rec.* **113**:143–162.

Chang, M. C., 1953, Storage of unfertilized rabbit ova: Subsequent fertilization and the probability of normal development, *Nature (London)* **172**:353–354.

Chang, M. C., 1954, Development of parthenogenetic rabbit blastocysts induced by low temperature storage of unfertilized ova, *J. Exp. Zool.* **125**:127–149.

Chang, M. C., 1955a, Development of fertilizing capacity of rabbit spermatozoa in the uterus, *Nature (London)* **175**:1036–1037.

Chang, M. C., 1955b, Fertilization and normal development of follicular oocytes in the rabbit, *Science* **121**:867–869.

Chang, M. C., 1955c, Développement de la capacité fertilizatrice des spermatozoïdes du lapin a l'intérieur du tractus génital femelle et fécondabillé des oeufs de lapine, in: *La Fonction Tubaire et Ses Troubles* (R. Moricard, ed.), pp. 40–52, Masson, Paris.

Chang, M. C., 1959, Fertilization of rabbit ova *in vitro*, *Nature (London)* **184**:466–467.

Chang, M. C., 1960, Fertilization of domestic rabbit (*Oryctolagus cuniculus*) ova by cottontail rabbit (*Syvilagus transitionalis*) sperm, *J. Exp. Zool.* **144**:1–10.

Chang, M. C., 1964, Effects of certain antifertility agents on the development of rabbit ova, *Fertil. Steril.* **15**:97–106.

Chang, M. C., 1965a, Artificial insemination of snowshoe hares (*Lepus americanus*) and the transfer of their fertilized eggs to the rabbit (*Oryctolagus cuniculus*), *J. Reprod. Fertil.* **10**:447–449.

Chang, M. C., 1965b, Implantation of ferret ova fertilized by mink sperm, *J. Exp. Zool.* **160**:67–80.

Chang, M. C., 1966a, Reciprocal transplantation of eggs between rabbit and ferret, *J. Exp. Zool.* **161**:297–306.

Chang, M. C., 1966b, Effects of oral administration of medroxyprogesterone acetate and ethinyl estradiol on the transportation and development of rabbit eggs, *Endocrinology* **79**:939–948.

Chang, M. C., 1967a, Physiological mechanisms responsible for the effectiveness of oral contraceptives, in: *Proceedings, 8th International Congress on Planned Parenthood, Santiago, Chile*, pp. 386–392.
Chang, M. C., 1967b, Effects of medroxyprogesterone acetate and ethinyl oestradiol on the fertilization and transportation of ferret eggs, *J. Reprod. Fertil.* **13:**173–174.
Chang, M. C., 1967c, Effects of progesterone and related compounds on fertilization, transportation and development of rabbit eggs, *Endocrinology* **81:**1251–1260.
Chang, M. C., 1968, Reciprocal insemination and egg transfer between ferrets and mink, *J. Exp. Zool.* **168:**49–59.
Chang, M. C., 1969a, Development of transferred ferret eggs in relation to the age of corpora lutea, *J. Exp. Zool.* **171:**459–464.
Chang, M. C., 1969b, Fertilization, transportation and degeneration of eggs in pseudopregnant or progesterone-treated rabbits, *Endocrinology* **84:**356–361.
Chang, M. C., 1974, Effects of medroxyprogesterone acetate and estrogen on the development of the early rabbit embryos, *Contraception* **10:**405–409.
Chang, M. C., 1978, Development of the oral contraceptives, *Amer. J. Obstet. Gynecol.* **132:**217–219.
Chang, M. C., and Bedford, J. M., 1961. Effects of various hormones on the transportation of gametes and fertilization in the rabbit, in: *4th International Congress on Animal Reproduction, The Hague, Netherlands*, Vol. 1, pp. 367–370.
Chang, M. C., and Harper, M. J. K., 1966, Effects of ethinyl estradiol on egg transport and development in the rabbit, *Endocrinology* **78:**860–872.
Chang, M. C., and Hunt, D. M., 1960, Effects of *in vitro* radiocobalt irradiation of rabbit ova on subsequent development *in vivo* with special reference to the irradiation of maternal organism, *Anat. Rec.* **137:**511–519.
Chang, M. C., and Hunt, D. M., 1970, Effects of various progestins and estrogens on the gamete transport and fertilization in the rabbit. *Fertil. Steril.* **21:**683–686.
Chang, M. C., and Hunt, D. M., 1972, Effect of prostaglandin $F_{2\alpha}$ on the early pregnancy of rabbits, *Nature (London)* **236:**120–121.
Chang, M. C., and McDonough, J. J., 1955, An experiment to cross the cottontail and domestic rabbit, *J. Hered.* **46:**41–44.
Chang, M. C., and Rowson, L. E. A., 1965, Fertilization and early development of Dorset Horn sheep in the spring and summer, *Anat. Rec.* **152:**303–316.
Chang, M. C., and Yanagimachi, R., 1965, Effect of estrogens and other compounds as oral antifertility agents on the development of rabbit ova and hamster embryos, *Fertil. Steril.* **16:**281–291.
Chang, M. C., Hunt, D. M., and Romanoff, E. B., 1957, Effects of radiocobalt irradiation of rabbit spermatozoa *in vitro* on fertilization and early development, *Anat. Rec.* **129:**211–230.
Chang, M. C., Hunt, D. M., and Romanoff, E. B., 1958, Effects of radiocobalt irradiation of unfertilized or fertilized rabbit ova *in vitro* on subsequent fertilization and development *in vivo*, *Anat. Rec.* **132:**161–180.
Chang, M. C., Hunt, D. M., and Harvey, E. B., 1963, Effects of radiocobalt irradiation of pregnant rabbits on the development of fetuses, *Anat. Rec.* **145:**455–466.
Chang, M. C., Marston, J. H., and Hunt, D. M., 1964, Reciprocal fertilization between the domesticated rabbit and the snowshoe hare with special reference to insemination of rabbits with equal number of hare and rabbit spermatozoa, *J. Esp. Zool.* **155:**437–446.
Chang, M. C., Pickworth, S., and McGaughey, R. W., 1969, Experimental hybridization, in: *Comparative Mammalian Cytogenetics* (K. Benirschke, ed.), pp. 132–143, Springer-Verlag, New York.
Chang, M. C., Casas, J. H., and Hunt, D. M., 1970, Prevention of pregnancy in the rabbit by subcutaneous implantation of Silastic tube containing oestrogen, *Nature (London)* **226:**1262–1265.
Chang, M. C., Casas, J. H., and Hunt, D. M., 1971a, Development of ferret eggs after 2 to 3 days in the rabbit Fallopian tube, *J. Reprod. Fertil.* **25:**129–131.
Chang, M. C., Casas, J. H., and Hunt, D. M., 1971b, Suppression of pregnancy in the rabbit by subcutaneous implantation of Silastic tubes containing various estrogenic compounds, *Fertil. Steril.* **22:**383–388.
Chang, M. C., Hunt, D. M., and Polge, C., 1972, Effects of prostaglandins (PGs) on sperm and egg transport in the rabbit, *Adv. Biosci.* **9:**805–809.
Chang, M. C., Saksena, S. K., Lau, I. F., and Wang, Y. H., 1979, Induction of mid-term abortion by Trichosanthin in laboratory animals, *Contraception* **19:**175–184.
Dickmann, Z., 1973, Postcoital contraceptive effects of medroxyprogesterone acetate and oestrogen in rats, *J. Reprod. Fertil.* **32:**65–69.
Edwards, R. G., Bavister, B. D., and Steptoe, P. C., 1969, Early stages of fertilization *in vitro* of human oocytes, *Nature (London)* **221:**632–635.
Edwards, R. G., Steptoe, P. C., and Purdy, J. M., 1970, Fertilization and cleavage *in vitro* of preovulation human oocytes, *Nature (London)* **227:**1307–1309.

Fukuda, Y., and Chang, M. C., 1978, Relationship between sperm concentration and polyspermy in intact and zona-free mouse egg inseminated *in vitro*, *Arch. Androl.* **1**:267–273.

Fukuda, Y., Maddock, M. B., and Chang, M. C., 1979. *In vitro* fertilization of two species of deer mouse eggs by homologous or heterologous sperm and penetration of laboratory mouse eggs by deer mouse sperm, *J. Exp. Zool.* **207**:481–490.

Gibbons, A. F. E., and Chang, M. C., 1973a, Indirect effects of X-irradiation on embryonic development: Irradiation of the exteriorized rat uterus, *Biol. Reprod.* **9**:133–141.

Gibbons, A. F. E., and Chang, M. C., 1973b, The effect of X-irradiation of the rat ovary on implantation and embryonic development, *Biol. Reprod.* **9**:343–349.

Gould, K. G., Cline, E. M., and Williams, W. L., 1973, Observations on the induction of ovulation and fertilization *in vitro* in the squirrel monkey (*Saimiri sciureus*), *Fertil. Steril.* **24**:260–268.

Hamner, C. E., Jennings, L. L., and Sojka, N. J., 1970, Cat (*Felis catus* L.) spermatozoa require capacitation, *J. Reprod. Fertil.* **23**:477–480.

Hanada, A., and Chang, M. C., 1972, Penetration of zona-free eggs by spermatozoa of different species, *Biol. Reprod.* **6**:300–309.

Hanada, A., and Chang, M. C., 1978, Penetration of the zona-free or intact eggs by foreign spermatozoa and the fertilization of deer mouse eggs *in vitro*, *J. Exp. Zool.* **203**:277–286.

Iritani, A., and Niwa, K., 1977, Capacitation of bull spermatozoa and fertilization *in vitro* of cattle follicular oocytes matured in culture, *J. Reprod. Fertil.* **50**:119–121.

Iritani, A., Niwa, K., and Imai, H., 1978, Sperm penetration *in vitro* of pig follicular oocytes matured in culture, *J. Reprod. Fertil.* **54**:379–383.

Iwamatsu, T., and Chang, M. C., 1969, *In vitro* fertilization of mouse eggs in the presence of bovine follicular fluid, *Nature (London)* **224**:919–920.

Iwamatsu, T., and Chang, M. C., 1972, Sperm penetration *in vitro* of mouse oocytes at various times during maturation, *J. Reprod. Fertil.* **31**:237–247.

Ku, C. P., Chu, M. K., Chiang, H. C., Chao, S. H., Pang, T. W., and Tsou, K., 1975, Pharmacological studies of a contraceptive drug—Anordrin, *Sci. Sin.* **18**:262–270.

Lau, I. F., Saksena, S. K., and Chang, M. C., 1974, Midterm abortion with Silastic-PVP implant containing prostaglandin $F_{2\alpha}$ in rabbits, rats, and hamsters, *Fertil. Steril.* **25**:839–844.

Lau, I. F., Saksena, S. K., and Chang, M. C., 1975, Prostaglandin $F_{2\alpha}$ for induction of midterm abortion: A comparative study, *Fertil. Steril.* **26**:74–79.

Leibo, S. P., 1977, Fundamental cryobiology of mouse ova and embryos, in: *Freezing of Mammalian Embryos* (K. Elliott and J. Whelan, eds.), pp. 69–96, Elsevier, Excerpta Medica, North-Holland, Amsterdam.

Lillie, F. R., 1919, *Problems of Fertilization*, Univ. of Chicago Press, Chicago.

Miyamoto, H., and Chang, M. C., 1973, Fertilization of rat eggs *in vitro, Biol. Reprod.* **9**:384–393.

Niwa, K., and Chang, M. C., 1975, Fertilization of rat eggs *in vitro* at various times before and after ovulation with special reference to fertilization of ovarian oocytes matured in culture, *J. Reprod. Fertil.* **43**:435–451.

Parkening, T. A., Tsunoda, Y., and Chang, M. C., 1976, Effects of various low temperatures, cryoprotective agents, and cooling rates on the survival, fertilizability, and development of frozen–thawed mouse eggs, *J. Exp. Zool.* **197**:369–374.

Pickworth, S., and Chang, M. C., 1969, Fertilization of Chinese hamster eggs *in vitro, J. Reprod. Fertil.* **19**:371–374.

Polge, C., 1977, The freezing of mammalian embryos: Perspectives and possibilities, in: *Freezing of Mammalian Embryos* (K. Elliott and J. Whelan, eds.) pp. 1–18, Elsevier, Excerpta Medica, North-Holland, Amsterdam.

Rowson, L. E. A., Lawson, R. A. S., and Moor, R. M., 1971, Production of twins in cattle by egg transfer, *J. Reprod. Fertil.* **25**:261–268.

Saksena, S. K., Lau, I. F., and Chang, M. C., 1975, Induction of midterm abortion in rabbits and hamsters with Silastic-PVP tubes containing prostaglandins E_2 and $F_{2\alpha}$, *Contraception* **11**:479–487.

Smith, A. U., and Polge, C., 1950, Survival of spermatozoa at low temperatures, *Nature (London)* **166**:668–669.

Steptoe, P. C., and Edwards, R. G., 1978, Birth after the reimplantation of a human embryo, *Lancet* **2**(No. 8085):366.

Thibault, C., Dauzier, L., and Wintenberger, S., 1954, Étude cytologique de la fécondation *in vitro* de l'oeuf de la lapine, *C.R. Séances Soc. Biol. Paris* **148**:789–790.

Toyoda, Y., and Chang, M. C., 1974, Fertilization of rat eggs *in vitro* by epididymal spermatozoa and the development of such eggs following transfer, *J. Reprod. Fertil.* **36**:9–22.

Tsunoda, Y., and Chang, M. C., 1976a, Reproduction in rats and mice isoimmunized with homogenates of ovary or testis with epididymis, or sperm suspensions, *J. Reprod. Fertil.* **46**:379–382.

Tsunoda, Y., and Chang, M. C., 1976b, The effect of passive immunization with hetero and isoimmune anti-ovary antiserum on fertilization of mouse, rat and hamster eggs, *Biol. Reprod.* **15**:361–365.

Tsunoda, Y., and Chang, M. C., 1977, Further studies of antisera on fertilization of mouse, rat and hamster eggs *in vivo* and *in vitro, Int. J. Fertil.* **22:**129–139.

Tsunoda, Y., and Chang, M. C., 1978, Effect of antisera against eggs and zonae pellucidae on fertilization and development of mouse eggs in vivo and in culture, *J. Reprod. Fertil.* **54:**233–237.

Tsunoda, Y., Parkening, T. A., and Chang, M. C., 1976, *In vitro* fertilization of mouse and hamster eggs after freezing and thawing, *Experientia* **32:**223–224.

Whittingham, D. G., 1968, Fertilization of mouse eggs *in vitro, Nature (London)* **220:**592–593.

Wu, J. T., 1974, Artificial insemination and induction of pregnancy in the Mongolian gerbil (*Meriones unquiculatus*), *J. Reprod. Fertil.* **37:**139–140.

Wu, J. T., and Chang, M. C., 1972, Effects of progesterone and estrogen on the fate of blastocysts in the ovariectomized pregnant ferrets: A preliminary study, *Biol. Reprod.* **7:**231–237.

Wu, J. T., and Chang, M. C., 1973a, Reciprocal fertilization between the ferret and short-tailed weasel with special reference to the development of ferret eggs fertilized by weasel sperm, *J. Exp. Zool.* **183:**281–290.

Wu, J. T., and Chang, M. C., 1973b, Hormonal requirement for implantation and embryonic development in the ferret, *Biol. Reprod.* **9:**350–355.

Yanagimachi, R., 1972, Fertilization of guinea pig eggs *in vitro, Anat. Rec.* **174:**9–20.

Yanagimachi, R., and Chang, M. C., 1964, *In vitro* fertilization of hamster ova, *J. Exp. Zool.* **156:**361–378.

II

Cell Biology of the Developing Egg

Introduction
Cell Biology of the Developing Egg

John D. Biggers

In mammals a new genotype is formed at fertilization, but pregnancy is not established until the mother recognizes physiologically the presence of an embryo and modifies her functions to permit its survival. In general, this physiological switch in maternal function is closely correlated with the process of implantation. The mechanisms involved in the establishment of pregnancy, therefore, embrace the preimplantation and implantation periods. A physiological analysis of these periods requires the study of independent events in the embryo and mother and also interactions between these two compartments. Consequently, the contributions to this volume involve the analysis of phenomena separated in space and time.

Figure 1 illustrates a theoretical model for a compartmental analysis of the events leading to the initiation of implantation. The model demonstrates five types of change: (a) the irreversible, one-way processes involved in the formation and development of the embryo, (b) the two-way exchanges that occur between an embryo and its microenvironment, which alter as the embryo travels down the oviduct to the uterus, (c) the two-way exchanges that occur between the mother and the microenvironment of the embryo, which also changes as far as the embryo is concerned because of the tubal journey, (d) the exchanges that occur between the secretions of the genital tract in the neighborhood of the embryo and the secretions in the proximal and distal segments of the genital tract, and (e) the specialized events that lead to implantation through the elimination of the fluid compartment around the blastocyst and the establishment of intimate contact between the cells of the mother and the cells of the embryo.

The chapters in Parts II and III are concerned with the irreversible developmental changes that occur in the establishment of pregnancy, represented by the left column in Figure 1. The problems involved are in the realms of developmental and cellular biology, i.e., the mechanisms that lead to the division of labor between cells involving the differentiation of new cell types that interact to form a multicellular organism. Part II deals with

John D. Biggers • Department of Physiology; Laboratory of Human Reproduction and Reproductive Biology, Harvard Medical School, Boston, Massachusetts 02115.

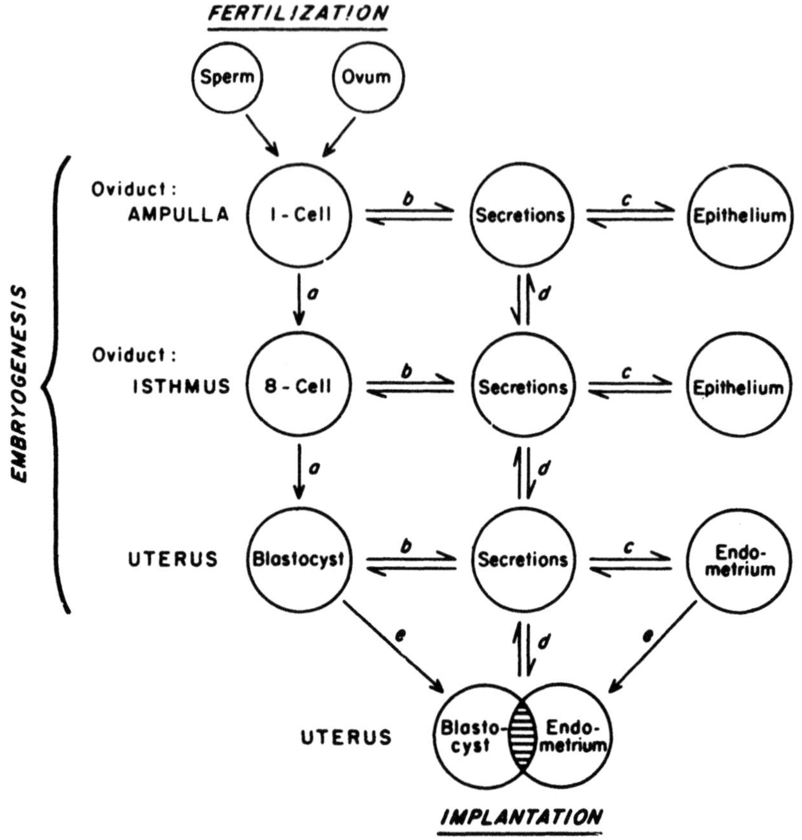

Figure 1. A theoretical model showing the changes between various compartments involved in the establishment of pregnancy.

the nature and timing of the cellular differentiations and interactions that occur, while Part III is concerned with the molecular processes that allow these cellular changes to be determined by a single genotype.

The cellular differentiation involved in the development of all multicellular organisms from a zygote—a single cell—may be described either in terms of cell lineages or in terms of the functions that are eventually subserved by particular parts of the organism. These approaches have a long history.

The study of cell lineage in early mammalian development began in 1875 with the independent work of Van Beneden and Rauber on the rabbit, and was continued by Kölliker [see Balfour (1880) for a review]. In the development of the blastocyst, following the lead from studies on nonmammalian forms, these workers recognized the formation of an outer layer of cells called the *epiblast* and an inner mass of cells that segregated into the inner *hypoblast* and the intermediate *mesoblast*. An argument soon developed over the existence of the hypoblastic mesoblast. The nature of the argument, which was limited by the techniques available to these early investigators, is illustrated by a comment of Balfour (1880):

> Kölliker does not believe in the existence of this stage, having never met with it himself. It appears to me, however, more probable that Kölliker has failed to obtain it, than that Van Beneden has been guilty of such an extraordinary blunder as to have described a stage which has no existence.

Hubrecht (1888) used the second approach, describing early mammalian development in functional terms. In his pioneer studies of the early development of the European hedgehog, he found it justifiable to distinguish that part of the embryo that plays only a nutritive function in embryonic development from the part that contributes cells to the individual that is eventually born. He proposed the name *trophoblast*. Specifically he wrote (1889):

> I propose to confer this name (*trophoblast*) to the epiblast of the blastocyst as far as it has a direct nutritive significance as indicated by proliferating processes, by immediate contact with maternal tissue, maternal blood, or secreted material.

Today we are still concerned with the analysis of the early differentiation of the embryo and with the transport of molecules, either in bulk to subserve anabolic and catabolic processes or as signals to facilitate communication between mother and embryo. We continue to study these phenomena because many questions are still unanswered. This work is justified because technical advances have provided the opportunity to design sophisticated experiments on early mammalian embryos that may elucidate these perennial, fundamental questions (for recent reviews, see Biggers and Borland, 1976; Rossant, 1977).

The technical advances that have made rapid progress possible in the study of early mammalian development are worth summarizing to show the importance of interdisciplinary contributions on which our present-day research rests. Some of the advances are: (1) the development of techniques for the culture of mouse preimplantation embryos *in vitro* (Hammond, 1949; Whitten, 1956) and the demonstration that normal young could be born after transfer of the cultured embryos into uterine foster mothers (McLaren and Biggers, 1958); (2) the development of techniques for superovulation in the mouse (Edwards and Gates, 1959); (3) the development of micromanipulative and microsurgical techniques for mouse embryos (Tarkowski, 1961; Mintz, 1962; Gardner, 1968); (4) the use of mutants in mice as genetic markers for the analysis of development (Mintz, 1962), a technique anticipated by Heape (1890) in the first mammalian embryo transfer experiments using rabbits, and (5) the application of rapid advances in ultramicrochemical techniques (see Biggers and Stern, 1973; Biggers and Borland, 1976, for reviews).

These techniques have resulted in three main areas of contemporary activity in the study of preimplantation mammalian embryos. First is the necessary continuation of the descriptive aspects of cleavage and blastocyst formation at the ultrastructural and molecular levels. Second is the experimental analysis of development at the cellular level. Third is the relatively new area that is concerned with the relation of the embryos with their environment; for example, the analysis of mechanisms involved in the transport of materials into and out of the embryo, and the properties of the cellular surfaces of the embryo.

There is perhaps a danger in the sophistication, and the inevitable specialization, of our present approaches to which our predecessors 100 years ago were not exposed. This danger arises from the predominant use of the mouse in recent experimental work, and the relative neglect of comparative studies in other species. In 1879, Walter Heape, who was destined to make major contributions to reproductive biology, became a pupil of Balfour and took up the investigation of the early development of the rabbit begun by Van Beneden and Kölliker. Also in 1879 Lieberkühn (quoted by Heape, 1881) described the early development of the dog and the European mole, thereby introducing a comparative approach. He concluded that the epiblast over the embryo is derived both from the original epiblast and from a large contribution of the inner mass of cells in the morula. This work stimulated Heape (1883) to make his own collection of mole embryos and to confirm Lieberkühn's interpretations. There is now a considerable body of comparative work on early development in mammals [see Boyd and Hamilton (1952) for a review]. Much of this work is confusing, and it is not clear, therefore, whether the recent work on mice has generality. We should thus examine critically the inferences made from investigations that are confined to a species that technical advances have enabled us to study in new ways. The new capabilities are not sufficient to claim that we now have a universal animal model, for such a representation can only be justified if we have enough evidence from comparative studies to know what we want to model (Biggers, 1979). To some investigators, the collection of these comparative data may seem unattractive work when contrasted with exciting, exploratory studies that generate new concepts. Nevertheless, this comparative work is essential to establish the generality of the exploratory studies. Fortunately, when the generality of studies on a technically convenient species is not substantiated, a formerly "pedestrian" study may find itself elevated to a pioneering project.

References

Balfour, F. M., 1880, *A Treatise on Comparative Embryology*, Vol. 2, p. 220, Macmillan, London.

Biggers, J. D., 1979, Fertilization and blastocyst formation, in: *Animal Models for Research on Contraception and Fertility* (N. J. Alexander, ed.), pp. 223–237, Harper & Row, Hagerstown, Md.

Biggers, J. D., and Borland, R. M., 1976, Physiological aspects of growth and development of the preimplantation mammalian embryo, *Annu. Rev. Physiol.* **38:**95–119.

Biggers, J. D., and Stern, S., 1973, Metabolism of the preimplantation embryo, *Adv. Reprod. Biol.* **6:**1–59.

Boyd, J. D., and Hamilton, W. J., 1952, Cleavage, early development and implantation of the egg, in *Marshall's Physiology of Reproduction*, Vol. II (A. S. Parkes, ed.), 3rd ed., pp. 1–126, Longmans, London.

Edwards, R. G., and Gates, A. H., 1959, Timing of the stages of maturation, divisions, ovulation, fertilization and the first cleavage of eggs of adult mice treated with gonadotropins, *J. Endocrinol.* **18:**292–304.

Gardner, R. L., 1968, Mouse chimaeras obtained by the injection of cells into the blastocyst, *Nature (London)* **220:**596–597.

Hammond, J., Jr., 1949, Recovery and culture of tubal mouse ova, *Nature (London)* **163:**28–29.

Heape, W., 1881, On the germinal layers and early development of the mole, *Proc. Roy. Soc.* **33:**190–198.

Heape, W., 1883, The development of the mole (*Talpa europea*). The formation of the germinal layers, and early development of the medullary groove and notochord. *Q. J. Microsc. Sci.* **23:**412–452.

Heape, W., 1890, Preliminary note on the transplantation and growth of mammalian ova within a uterine foster mother, *Proc. Roy. Soc.* **48:**457–458.

Hubrecht, A. A. W., 1888, Keimblatterbildung und Placentation des Igels, *Anat. Anz.* **3:**510–515.

Hubrecht, A. A. W., 1889, Studies in mammalian embryology. I. The placentation of *Erinaceus europoeus*, with remarks on the phylogeny of the placenta, *Q. J. Microsc. Sci.* **30:**283–404.

McLaren, A., and Biggers, J. D., 1958, Successful development and birth of mice cultivated *in vitro* as early embryos, *Nature (London)* **182:**877–878.

Mintz, B., 1962, Experimental recombination of cells in the developing mouse egg: Normal and lethal mutant genotypes, *Am. Zool.* **2:**145 (abstract).

Rossant, J., 1977, Cell commitment in early rodent development, in: *Development in Mammals*, Vol. 2 (M. H. Johnson, ed.), pp. 119–150, North-Holland, Amsterdam.

Tarkowski, A. K., 1961, Mouse chimaeras developed from fused eggs, *Nature (London)* **190:**857–860.

Whitten, W. K., 1956, Culture of tubal mouse ova, *Nature (London)* **177:**96.

3

The Origin of Trophoblast and Its Role in Implantation

J. Rossant and W. I. Frels

I. Introduction

The process of implantation in mammals involves the establishment of intimate contact between the embryo and the maternal tissues, leading to the formation of the mature placenta. Implantation marks the end of the free-living stages of embryonic growth and occurs at various stages of development in different species (Heap *et al.*, 1979). In the mouse, with which we are mostly concerned here, implantation occurs within the first few days of pregnancy and is marked by specific embryonic and uterine changes occurring over a brief period. Implantation in the mouse resembles quite closely that of the human embryo (Hamilton and Mossman, 1972) and differs from species such as the pig, where implantation is a late and poorly defined event (Heap *et al.*, 1979). In all species highly specialized tissues have developed to facilitate implantation; these are all tissues of the trophectoderm cell lineage. The purpose of this paper is to discuss what tissues arise from the trophectoderm cell lineage, how and when the lineage is established in development, and what its role is in implantation. The interaction of trophoblast (the post implantation derivatives of the trophectoderm) and uterus is critical for embryo survival, and this area has been intensively studied. Many studies have concentrated on *in vitro* model systems for implantation (Jenkinson, 1977), but we shall deal chiefly with the interaction of the embryo and uterus *in vivo*.

II. The Trophectoderm Cell Lineage

A. What Is Trophoblast?

The early development of the mouse has been extensively described elsewhere (Snell and Stevens, 1966; Rugh, 1968; Theiler, 1972), and we concentrate here on the trophec-

J. Rossant • Department of Biological Sciences, Brock University, St. Catharines, Ontario, Canada. *W. I. Frels* • Department of Molecular Biology, Roswell Park Memorial Institute, Buffalo, New York 14263. Present address of W.I.F.: The Jackson Laboratory, Bar Harbor, Maine 04609.

toderm and its derivatives. Trophectoderm is first distinguishable morphologically at 3½ days of development as the outer layer of the blastocyst, enclosing a compact group of cells, the inner cell mass (ICM) (Figure 1). Trophectoderm cells make up roughly three-fourths of the total cell number (Copp, 1978), which means that most cells of the early embryo are set aside for the task of establishing maternal–embryonic contact and do not give rise to the fetus itself. The cells of the trophectoderm appear much more specialized than those of the ICM, both by morphological and by functional criteria (Enders, 1971; Gardner, 1972).

At 4½ days of development, around the time of implantation, further changes occur within the blastocyst. A monolayer of primitive endoderm delaminates from the blastocoelic surface of the ICM (Snell and Stevens, 1966; Enders, 1971), and the remaining ICM cells constitute the primitive ectoderm. At the same time, changes occur within the trophectoderm cell population. Mural trophectoderm cells, away from the ICM, cease dividing and start to endoreduplicate their DNA to form the primary trophoblast giant cells (Duval, 1891; Dickson, 1963; Barlow & Sherman, 1972). Cells overlying the ICM continue to divide and form the polar trophoblast (Copp, 1978). Around the time of giant cell formation, attachment to the uterine epithelium occurs and the decidual cell response begins (Reinius, 1967; Potts, 1968; Enders and Schlafke, 1969). The exact mechanism of attachment of trophoblast and uterus is not known (Jenkinson, 1977).

Once implantation is achieved, further differentiation can occur within the blastocyst, resulting in the formation of the typical egg cylinder stage (Figure 2). The polar trophoblast continues to divide and forms the ectoplacental cone and extraembryonic ectoderm. The ICM derivatives are the embryonic ectoderm and the distal and proximal endoderm. Later in development, the ectoderm forms all three germ layers of the fetus, plus the amnion, allantois, and extraembryonic mesoderm, while the primitive endoderm gives rise only to

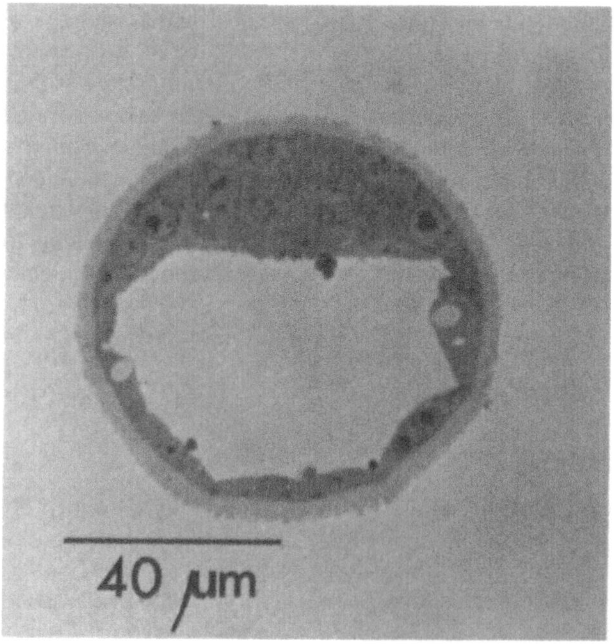

Figure 1. Section of 3½-day mouse blastocyst, showing outer trophectoderm layer enclosing the inner cell mass.

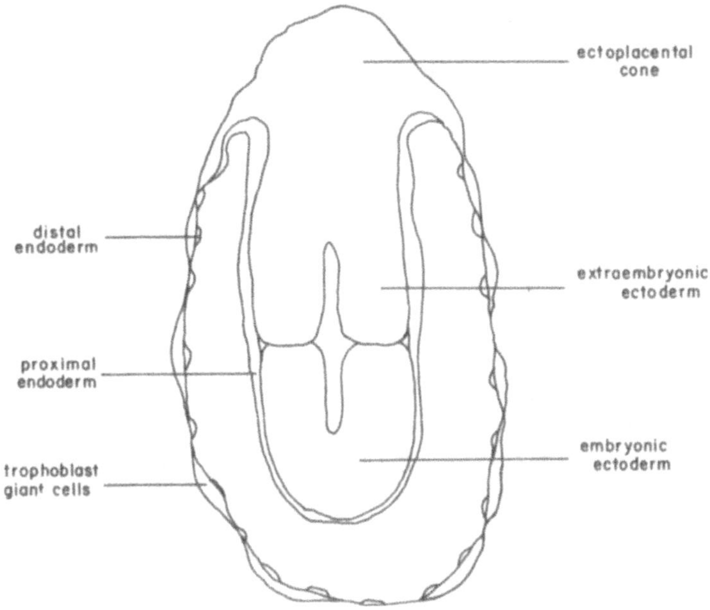

Figure 2. Diagram of a $5\frac{1}{2}$- or $6\frac{1}{2}$-day mouse egg cylinder.

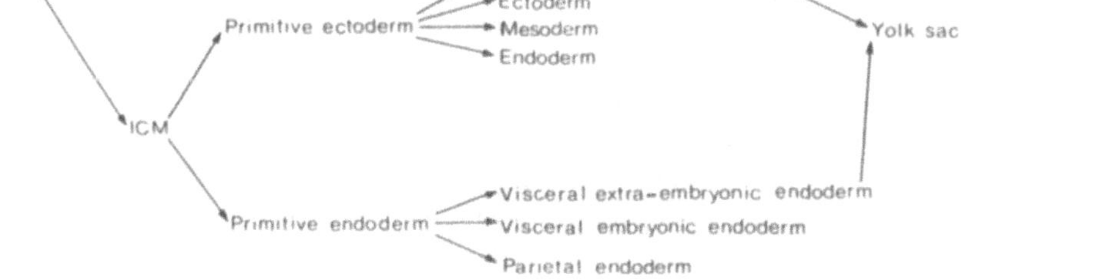

Figure 3. Tentative plan of cell lineage relationships in early mouse embryogenesis. From Rossant and Papaioannou (1977); reproduced by permission of MIT Press.

the endoderm of the visceral and parietal yolk sac (Figure 3) (reviewed by Gardner and Papaioannou, 1975; Rossant and Papaioannou, 1977). Secondary trophoblast giant cells arise at the periphery of the ectoplacental cone and migrate downwards to enclose the entire embryo and its membranes. These cells provide direct contact with the maternal environment until the mature placenta is established by fusion of maternal decidual tissue with ectoplacental cone, extraembryonic ectoderm, and allantois (Duval, 1891). Thus all cells of the trophectoderm lineage are solely involved with maternal–embryonic interactions and play no part in the development of the fetus itself.

B. How and When Is the Trophectoderm Cell Lineage Established?

ICM and trophectoderm cells can be distinguished morphologically and biochemically (Van Blerkom *et al.*, 1976) by 3½ days. There is also evidence to suggest that changes typifying ICM and trophectoderm may begin before overt signs of blastocyst formation. Tight junctions typical of trophectoderm first appear between the outside cells of the compacted 8- to 12-cell embryo (Ducibella *et al.*, 1975), and some ICM- and trophectoderm-specific proteins are apparently synthesized in inside and outside cells, respectively, before blastocyst formation (Johnson *et al.*, 1977; Handyside and Johnson, 1978). Such preliminary changes, however, do not necessarily mean that the two cell types are committed to their respective fates before blastulation; this can only be determined by examining the potential of the cells outside their normal embryonic environment (Weiss, 1939). A variety of such studies has been undertaken (reviewed by Rossant, 1977).

In vitro and *in vivo* studies on blastomeres isolated at early cleavage stages (Tarkowski and Wroblewska, 1967; Kelly, 1977) have suggested that cells are completely labile until at least the eight-cell stage. It is also clear that both ICM and trophectoderm are committed by the mature 64-cell blastocyst stage; no interconversion of one to the other has been detected after a variety of experimental manipulations (Gardner, 1972; Rossant, 1975a,b). Development from the eight-cell to the blastocyst stage appears to be a fairly ordered process in the intact embryo, with little cell mixing (Wilson *et al.*, 1972; Kelly *et al.*, 1978). Allocation of cells to ICM and trophectoderm lineages may be nonrandom (Graham and Deussen, 1978). The normal fate of cells enclosed in the morula as cleavage proceeds is thus almost certainly to become ICM, and the fate of the outside cells is to become trophectoderm.

Does this conclusion mean that the inside and outside cells are committed to these lineages well before blastulation? Several *in vivo* experiments showed that late morula cells could still aggregate and form blastocysts (Mintz, 1965; Stern and Wilson, 1972), suggesting that commitment was a late event, close to blastulation. However, such experiments relied solely on *in vitro* morphology and did not confirm that cells could be functionally interconverted from ICM to trophectoderm or vice versa; they also did not assess separately the potential of inside and outside cells, to see if commitment occurs at the same time in both tissues.

The development of the technique of immunosurgery (Solter and Knowles, 1975) led to a series of experiments designed to test the potential of ICMs or inside cells isolated at various stages of development. ICMs isolated from early blastocysts, just after blastulation, were still capable of producing apparent trophoblast giant cells *in vitro* (Handyside, 1978; Hogan and Tilly, 1978b; Spindle, 1978) and, more convincingly, functional trophectoderm derivatives *in vivo* (Rossant and Lis, 1979). Thus the ICM cell lineage is apparently not closed until after blastulation has begun.

Is the trophectoderm lineage closed at an earlier stage? Theoretical considerations regarding the need for recruitment of outside cells to the inside, to account for final ICM cell number (Gardner and Rossant, 1976; Rossant and Papaioannou, 1977), as well as the aggregation experiments mentioned previously, make this possibility unlikely. No direct analysis of the potential of outside cells has previously been undertaken, however, because of technical difficulties. Handyside (personal communication) isolated outside cells from late morulae using a fluorescent marker technique and showed that they can apparently reform a blastocyst *in vitro*. We used a microsurgical approach and subjected isolated outside cells to the same *in vivo* tests as we used for ICMs (Rossant and Vijh, 1980). When aggregated with cleavage-stage embryos, outside cells from late morulae (just before blastulation) formed chimeras at a fairly low rate, but among the 14 chimeras obtained, 12 showed contributions from the outside cell progency to the ICM derivatives. Only two showed unequivocal trophoblast contributions. This finding is somewhat surprising and suggests that the smaller aggregated outside cells preferentially move inside the embryo and contribute to the ICM. In the second assay, where morulae were "reconstituted" from only outside cells, structures resembling blastocysts were found after 24 hr in culture and 50% of the implants produced by transfer to pseudopregnant recipients contained complete egg cylinders. These experiments provide direct evidence for the lability of at least some outside cells up to the late morula stage. Technical difficulties have prevented the assay of the potential of the trophectoderm from the early blastocyst, but there is little reason now to suppose that ICM and trophectoderm are committed at different times.

It is not easy to be more precise about the time of cell commitment, because of the asynchrony of cell division in the embryo and the technical impossibility of ascertaining whether every cell in a given embryo is labile. Neverthelesss, these studies do give us insight into the likely mechanisms of establishing cell heterogeneity in the mouse embryo. Late lability of cells makes it unlikely that any simple segregation of cytoplasmic factors is involved (Dalcq, 1957), although a more complex process of segregating an "organizing center" (Denker, 1976) cannot be ruled out. The results are most easily accommodated by the "inside–outside" theory (Tarkowski and Wroblewska, 1967), which states that cells differentiate solely according to their position during cleavage. Blastocyst formation can then be seen as a complex event, involving progressive changes in cells brought about by their position in the embryo, until eventually they become committed to one lineage and lose the potential to form the other cell type (Rossant and Lis, 1979).

III. The Role of the Trophectoderm Cell Lineage in Implantation

A. Uterus–Trophectoderm–ICM Interactions

Direct evidence that trophectoderm alone is required for implantation comes from studies showing that isolated trophectoderm vesicles will initiate the decidual response when transferred to pseudopregnant mice (Gardner, 1972), although isolated mature ICMs will not (Gardner, 1972; Rossant, 1975b). Various artificial stimuli can also induce the decidual response (reviewed by Heap *et al.*, 1979), but the natural stimulus from the trophectoderm must be very potent, because embryos containing one-eighth the normal cell number can implant in the uterus (Rossant, 1976b). ICM tissue also is not required for maintaining embryos in implantational delay. Microsurgically isolated trophectoderm vesicles (Surani and Barton, 1977), or vesicles obtained after treatment of embryos with [^{3}H]thymidine (Snow *et al.*, 1976), can be held in experimental delay and will implant when given the correct hormonal stimulus. ICM tissue seems, therefore, to play no role in controlling the onset of implantation.

The ICM is required, however, for further development within the trophectoderm. Trophectoderm vesicles will implant in the uterus but no trophoblast proliferation occurs; only a few giant cells are formed (Gardner, 1972). If an ICM is placed in a trophectoderm vesicle, to "reconstitute" a blastocyst, normal development can ensue, with proliferation of trophoblast to form the ectoplacental cone (Gardner *et al.*, 1973). This dependence on the presence of ICM tissue for trophoblast proliferation seems to extend well into postimplantation development. Diploid extraembryonic ectoderm (Rossant and Ofer, 1977) and diploid ectoplacental cone cells (Rossant and Lis, unpublished) will cease division and

Table 1. Mitotic Index of 7½-Day Diploid Ectoplacental Cone under Different Culture Conditions

Type of culture conditions	Mitotic index after 1 day	Number of samples	Mitotic index after 2 days	Number of samples
Suspension	0.4 ± 0.2^a	12	0.08 ± 0.04	12
Collagen gel	0.2 ± 0.07	10	0.02 ± 0.02	7
Embryonic "pocket"	2.5 ± 0.45	13	2.5 ± 0.48	12

[a] S.E.

endoreduplicate when isolated from ICM derivatives as late as 8½ days of development. The proliferative stimulus from the ICM derivatives seems to be specific, although rat ICMs can also induce mouse trophoblast proliferation (Gardner and Johnson, 1975). The effect may depend on close cell contact in the intact embryo, as it is very hard to mimic *in vitro*. So far, the only way we have been able to maintain some trophoblast proliferation *in vitro* was by placing 7½-day diploid trophoblast inside the amniotic cavities of equivalent age embryonic portions (Rossant and Lis, unpublished). After 2 days in culture, the trophoblast could be rescued and its mitotic index was significantly higher than in similar tissues grown in suspension or in collagen gels (Table 1). The mitotic index, however, was still lower than in the cultured embryonic tissues and microdensitometry revealed that some trophoblast cells had begun endoreduplication even inside the embryonic sac.

ICM derivatives are required for further trophoblast development, but are trophectoderm derivatives required for further ICM development? *In vivo*, this may be so, for isolated ICMs do not apparently develop further than primitive endoderm formation in the reproductive tract (Rossant, 1975b). However, given the correct *in vitro* conditions, isolated ICMs have been shown to develop as far as mesoderm and blood island formation (Hogan and Tilly, 1978a,b; Wiley *et al.*, 1978). Thus, although trophectoderm–uterine interactions may mediate ICM activation *in vivo*, the effect is fairly easy to mimic *in vitro* and so may simply involve the provision of correct nutritional requirements rather than a specific signal. It is clear from this discussion that normal embryonic development beyond implantation depends on a complex series of interactions between trophectoderm, ICM, and uterus, with the trophectoderm–uterine interaction being the one primarily involved in establishing the embryo in the uterus.

B. Species Specificity of the Trophoblast–Uterine Interaction

Another area of research that has revealed the importance and specificity of the trophoblast–uterine interaction is the study of interpsecific embryo transfers and interspecific chimeras. The rat and the mouse have been the two species most extensively studied. Rat blastocysts transferred to the mouse reproductive tract can implant and induce the decidual cell response; they do not normally develop, however, beyond the early egg cylinder stage (Tarkowski, 1962; Rossant, 1976a), even when held in delay before implantation (Copp and Rossant, 1978). In these implants, ICM derivatives are highly disorganized and little trophoblast proliferation occurs. A recent EM study suggests that trophoblast–uterine interactions are abnormal at implantation and this may cause embryo death (Tachi and Tachi, 1979).

Direct evidence for the importance of the trophoblast is provided by studying interspecific chimeras. When rat ICMs are injected into mouse blastocysts, implantation and normal development can ensue (Gardner and Johnson, 1973). Rat cells were detected in the ICM derivatives until at least 9½ days of development, although embryos allowed to go to term contained little or no rat tissue (Gardner and Johnson, 1975). Rat/mouse chimeras made by aggregating morulae (Mulnard, 1973; Stern, 1973; Zeilmaker, 1973) can also implant in the mouse uterus but do not develop beyond implantation (Rossant, 1976a). The only difference between the two sorts of chimera is that the trophoblast is entirely mouse in the injection chimeras, whereas it may be both rat and mouse in the aggregation chimeras. Thus it seems clear that the survival of interspecific transfers or chimeras depends largely on the interaction of the trophoblast and uterus, although later selection against rat ICM derivatives may occur in the rat/mouse injection chimeras (Gardner and Johnson, 1975).

How specific is this trophoblast–uterine interaction? Recent studies we have undertaken using two more closely related rodent species, *Mus musculus* and *Mus caroli*, suggest that the interaction is highly species specific. *M. caroli* is a wild species from Southeast Asia (Marshall, 1972) that does not naturally interbreed with *M. musculus*. Attempts have been made to produce hybrids by artificial insemination as a means of introducing new genetic polymorphisms into the *M. musculus* gene pool (West *et al.*, 1977; West *et al.*, 1978). Successful hybrid production is very rare; many hybrid embryos apparently die after implantation (West *et al.*, 1977). There are many reasons why hybrid survival should be low. One possibility is that the expression of *M. caroli* genes on the hybrid trophoblast upsets normal maternal–fetal interactions. To investigate this possibility, we transferred *M. caroli* blastocysts to the *M. musculus* uterus (Frels *et al.*, 1980). In this interspecific transfer, the initial interactions of trophectoderm and uterus at implantation seemed to be unaffected, as normal egg cylinders were obtained. Resorption occurred later in pregnancy, however, and only 1 out of 69 embryos transferred survived to term. This survival rate is similar to that of the hybrid embryos, suggesting that failure to interact with the *M. musculus* uterus may be an important factor in hybrid death.

In this situation, as with the rat/mouse transfers, it is not clear whether the ICM also plays a role. Recent successful production of live *M. caroli*↔*M. musculus* chimeras by injection of *M. caroli* ICMs into *M. musculus* blastocysts (Rossant and Frels, 1980) strongly suggests that the trophoblast is of overriding importance in embryo survival. The injected blastocysts survived to term at a very high rate (48 of 52 transferred) and 38 out of the 48 mice born showed contributions of both *M. musculus* and *M. caroli* cells as judged by coat color and isozymal markers (Figure 4). No selection against *M. caroli* cells could be detected. Thus *M. caroli* ICMs can survive very well in the *M. musculus* uterus if pro-

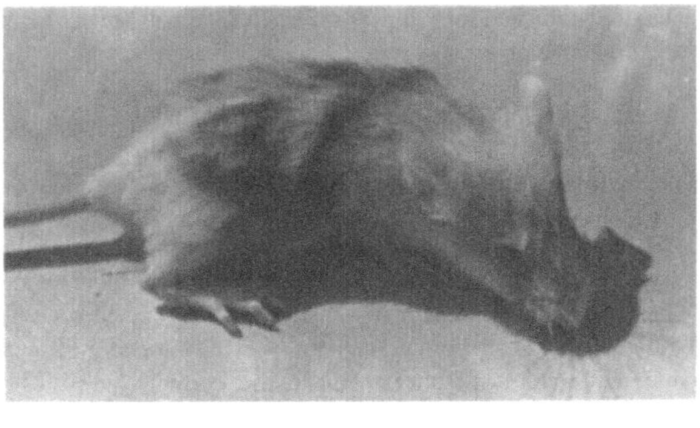

Figure 4. Interspecific chimera between *M. musculus* and *M. caroli*. Coat color pigment patches are derived from the *M. caroli* component; the *M. musculus* strain was albino.

tected by *M. musculus* trophoblast, but not if surrounded by *M. caroli* trophoblast. The failure of the interaction between *M. caroli* trophoblast and *M. musculus* uterus occurs not at implantation but later in development, perhaps due to a maternal immune response (Frels *et al.*, 1980). This finding emphasizes the importance of trophoblast in promoting correct embryo–uterine interactions not just at implantation but throughout pregnancy.

IV. Conclusions

The trophectoderm cell lineage contains the first tissues to show differentiated or specialized properties in mouse development and is committed to its restricted fate before implantation. It initiates the decidual response and brings about close alignment and attachment of embryo and uterus at implantation. Although very little trophoblast tissue is required to initiate these changes, the interaction is species specific. The decidual response can be induced by blastocysts of a variety of species, but correct uterine–trophoblast attachment leading to further development is rare. Although only trophoblast is required for implantation, no further trophoblast proliferation will occur without the presence of ICM derivatives. This interaction is not species specific: ICMs of other rodent species will stimulate mouse trophoblast development. Also, *in vivo* development of ICMs does not proceed very far without the presence of trophoblast. Thus a series of interactions between uterus, trophoblast, and ICM derivatives is required for embryo survival at implantation. Later in development the same is true, but again it is the trophoblast–uterine interaction that is critical for embryo survival and is highly species specific. Study of the trophoblast and its interactions with the maternal environment is very important, therefore, for a fuller understanding of implantation and embryonic development in mammals.

ACKNOWLEDGMENTS. We thank Dorothy Laughton for typing the manuscript and the Natural Sciences and Engineering Research Council of Canada for financial support.

Discussion

MANES (La Jolla): Short and Yoshinaga (*J. Reprod. Fertil.* **14**:287, 1967) reported that Walker carcinoma cells could implant in the rat uterus and give interactions with the uterine epithelium that looked like normal trophoblast interactions. Are there other cell types that can actually implant; these would be presumably derivatives from the inner cell mass?

ROSSANT: The decidual response can be mimicked by a variety of different agents, including cancer cells, but correct embryonic development is a highly specific process requiring trophoblast.

MANES: But the isolated ICMs do not induce this type of response?

ROSSANT: Mature ICMs do not implant, but earlier ICMs can regenerate trophoblast and then implant.

YOSHINAGA (NICHD): With Roger Short, I studied the hormonal environment which is suitable for implantation and also development of Walker carcinosarcoma in the uterus. The condition for the development of tumors was the same as for implantation. When the rats were castrated and treated with progesterone alone, tumor cells did not develop well, but when progesterone and estrogen were given, they developed quite well. I have a question for you. When you transferred inner cell mass it did not implant, is that correct?

ROSSANT: Yes, that is correct.

YOSHINAGA: In some experiments glass beads and killed sea urchin eggs have been shown to implant or cause a decidual reaction. What is the difference between the ICM and these nonspecific stimuli?

ROSSANT: It is strange that the inner cell mass does not cause a decidual response when so many other things do. I don't know why that is. Trophectoderm is a very potent initiator of the decidual response. A trophectoderm vesicle derived from a single blastomere at the eight-cell stage can implant. So one-eighth of the normal cell number will provoke a decidual response. Inner cell masses containing many more cells do not do so; this seems to be an important distinction between them.

SHERMAN (Roche Institute): Have you done the control experiment of putting an extra *Mus caroli* inner cell mass inside a *Mus caroli* blastocyst and then transplanting it into a *Mus musculus* uterus? Since you say that the initial events at implantation look normal, is there an abnormality in the inner cell mass segment, which could possibly be compensated for by a double-sized inner cell mass that might give the proper derivatives?

ROSSANT: You mean that a double-sized inner cell mass might allow a few more cells to survive? I haven't tried, but I suspect it would not work. There is a very variable response to transfer of *Mus caroli* blastocysts to *Mus musculus*. Some develop extensively and some stop quite early; the same sort of variability is seen with the hybrids.

SHERMAN: Do you see evidence of disruption of the trophoblast–uterine interaction?

ROSSANT: We have seen dead fetuses, but most of the recipients have been left to term in the hope of obtaining live young. Of course this procedure obscures what happens in between.

BEIER (Aachen): With rabbits, we take it as an indication of unsuccessful development if a blastocyst shows degeneration of the embryoblast and forms a trophoblast vesicle. If there is no implantation at all, we see these vesicles in the rabbit; there is no process like implantation if the embryoblast is lost. Regarding these experiments by Gardner (*J. Embryol. Exp. Morsphol.* **28**:279, 1972), retransferring so-called trophoblast vesicles into pseudo-pregnant recipients, do embryoblast or inner cell mass formations have to be established before something like implantation occurs, or is it possible in the mouse that something without any inner cell mass material can initiate implantation?

ROSSANT: The rabbit may well be different. Gardner worked originally with microsurgically isolated trophectoderm vesicles; he could find no evidence for any inner cell mass development. The vesicles attach to the uterus and provoke a decidual response but there is no further development; all you see are a few giant cells and eventually the tissue dies. Later in development there would just be a small mole in the uterus. Snow *et al.* (*J. Reprod. Fertil.* **48**:403, 1976) have done similar experiments with vesicles made using tritiated thymidine to kill the inner cell mass derivatives.

PEDERSEN (UCSF): I want to address the issue of the time at which trophectoderm cells become determined. At what cell stage were your donor cells obtained? It seems to me that the whole issue of the time of trophoblast determination is moot if the technical difficulty of getting the differentiated cell or the early pretrophoblast cell prevents putting it in an inside environment. What do you think about the time of trophectoderm determination in view of the fact that it does differentiate junctions which prevent you from dissociating blastocysts?

ROSSANT: The cells were from a late morula, so it would be about the 32-cell stage, well beyond compaction but at a stage where the cells can still be disaggregated. I used a very light disaggregation technique and suctioned off the outside cells. The early blastocyst, in many hands anyway, does not survive this manipulation, so I cannot exclude the possibility that the outer cells are committed at this stage.

JOHNSON (Cambridge): How confident are you that your cells are really from the outside? This issue is crucial in this sort of experiment. Alan Handyside in our laboratory has been using a different technique for taking off outside cells and has essentially confirmed the data you have given. At the 32-cell stage, outside cells do appear to be able to contribute to the inner cell mass.

ROSSANT: I never take more than about six cells from a single embryo, because if you keep taking outside cells from an embryo, after a while you don't know what's inside and what's outside. That means that the potential of all outside cells from a given embryo is not assessed.

GLASSER (Baylor): In the interspecies experiments, the finding of animals with 100% resorptions at term implies that the association of trophectoderm with the stromal cells has created at least a placental-like structure that is sufficient to produce the hormones necessary to carry these residua to term.

ROSSANT: Yes, whatever goes wrong is much later than implantation and indeed placental formation. These seems to be a peak of resorption at about 12–13 days of development.

BIGGERS (Harvard): This may not be just an endocrine problem, it may be a nutritional one. Systems may not develop properly.

C. TACHI (Tokyo): You mentioned that there was no species-specific segregation of the cells in the chimera between the *M. caroli* and the *M. musculus*. We have studied the distribution of blastomeres in chimeras between rats and mice and concluded that there was a segregation of the blastomeres in the blastocyst stage, because the plane of differentiation between the endoderm and the ectoderm changes a great deal.

ROSSANT: Even with an interspecific chimera there is not a great deal of cell mixing at the blastocyst stage. After aggregation of cleavage-stage mouse embryos, one can draw a more or less clearcut line down the middle of the blastocyst, separating the two components [Mintz, B., 1965, in: *Preimplantation Stages of Pregnancy* (G. E. W. Wolstenholme and M. O'Connor, eds.), pp. 194–207, Churchill, London]. However, there must be a stage in later development at which considerable cell mixing occurs, since adult chimeras show quite fine-grained mosaicism. Interspecific chimeras between *M. caroli* and *M. musculus* show the same evidence for cell mixing in the adult, unlike rat ↔ mouse chimeras where the rat cells appear to be selected out as development proceeds.

References

Barlow, P. W., and Sherman, M. I., 1972, The biochemistry of differentiation of mouse trophoblast: Studies on polyploidy, *J. Embryol. Exp. Morphol.* **27**:447–465.

Copp, A. J., 1978, Interaction between inner cell mass and trophectoderm of the mouse blastocyst. I. A study of cellular proliferation, *J. Embryol. Exp. Morphol.* **48**:109–125.

Copp, A. J., and Rossant, J., 1978, Effect of implantational delay on transfer of rat embryos to mice, *J. Reprod. Fertil.* **52**:119–121.

Dalcq, A., 1957, *Introduction to General Embryology*, Oxford Univ. Press, Oxford.

Denker, H.-H., 1976, Formation of the blastocyst, determination of trophoblast and embryonic knot, *Curr. Top. Pathol.* **62**:59–79.

Dickson, A. D., 1963, Trophoblastic giant cell transformation of mouse blastocysts, *J. Reprod. Fertil.* **6**:465–466.

Ducibella, T., Albertini, D. F., Anderson, E., and Biggers, J. D., 1975, The preimplantation mammalian embryo: Characterization of intercellular junctions and their appearance during development, *Dev. Biol.* **45**:231–250.

Duval, M., 1891, Le placenta des Rongeurs, *J. Anat. Physiol. Paris* **27**:279–476.

Enders, A. C., 1971, The fine structure of the blastocyst, in: *Biology of the Blastocyst* (R. J. Blandau, ed.), pp. 71–94, Univ. of Chicago Press, Chicago.

Enders, A. C., and Schlafke, S., 1969, Cytological aspects of trophoblast–uterine interaction in early implantation, *Am. J. Anat.* **125**:1–30.

Frels, W. I., Rossant, J., and Chapman, V. M., 1980, Intrinsic and extrinsic factors affecting the viability of *M. caroli* × *M. musculus* hybrid embryos, *J. Reprod. Fertil.*, **59**:387–392.

Gardner, R. L., 1972, An investigation of inner cell mass and trophoblast tissue following their isolation from the mouse blastocyst, *J. Embryol. Exp. Morphol.* **28**:279–312.

Gardner, R. L., and Johnson, M. H., 1973, Investigation of early mammalian development using interspecific chimaeras between rat and mouse, *Nature (London) New Biol.* **246**:86–89.

Gardner, R. L., and Johnson, M. H., 1975, Investigation of cellular interaction and deployment in the early mammalian embryo using interspecific chimaeras between the rat and mouse, in: *Cell Patterning* (Ciba Foundation Symposium), pp. 183–200, Associated Scientific Publishers, Amsterdam.

Gardner, R. L., and Papaioannou, V. E., 1975, Differentiation in the trophectoderm and inner cell mass, in: *The Early Development of Mammals* (M. Balls and A. E. Wild, eds.), pp. 107–132, Cambridge Univ. Press, London.

Gardner, R. L., and Rossant, J., 1976, Determination during embryogenesis, in: *Embryogenesis in Mammals* (Ciba Foundation Symposium), pp. 5–25, Associated Scientific Publishers, Amsterdam.

Gardner, R. L., Papaioannou, V. E., and Barton, S. C., 1973, Origin of the ectoplacental cone and secondary giant cells in mouse blastocysts reconstituted from isolated trophoblast and inner cell mass, *J. Embryol. Exp. Morphol.* **30:**561–572.

Graham, C. F., and Deussen, Z. A., 1978, Features of cell lineage in preimplantation mouse development, *J. Embryol. Exp. Morphol.* **48:**53–72.

Hamilton, W. J., and Mossman, H. W., 1972, *Hamilton, Boyd and Mossman's Human Embryology*, 4th ed., Heffer, Cambridge.

Handyside, A. H., 1978, Time of commitment of inside cells isolated from preimplantation mouse embryos, *J. Embryol. Exp. Morphol.* **45:**37–53.

Handyside, A. H., and Johnson, M. H., 1978, Temporal and spatial patterns of the synthesis of tissue-specific polypeptides in the preimplantation mouse embryo, *J. Embryol. Exp. Morphol.* **44:**191–199.

Heap, R. B., Flint, A. P., and Gadsby, J. E., 1979, Role of embryonic signals in the establishment of pregnancy, *Brit. Med. Bull.* **35:**129–136.

Hogan, B., and Tilly, R., 1978a, *In vitro* development of inner cell masses isolated immunosurgically from mouse blastocysts. I. Inner cell masses from 3.5 day p.c. blastocysts incubated for 24 hr before immunosurgery, *J. Embryol. Exp. Morphol.* **45:**93–105.

Hogan, B., and Tilly, R., 1978b, *In vitro* development of inner cell masses isolated immunosurgically from mouse blastocysts. II. Inner cell masses from 3.5 to 4.0 day p.c. blastocysts, *J. Embryol. Exp. Morphol.* **45:**107–121.

Jenkinson, E. J., 1977, The *in vitro* blastocyst outgrowth system as a model for the analysis of preimplantation development, in: *Development in Mammals*, Vol. 2 (M. H. Johnson, ed.), pp. 151–172, North-Holland, Amsterdam.

Johnson, M. H., Handyside, A. H., and Braude, P. R., 1977, Control mechanisms in early mammalian development, in: *Development in Mammals*, Vol. 2 (M. H. Johnson, ed.), pp. 67–98, North-Holland, Amsterdam.

Kelly, S. J., 1977, Studies of the developmental potential of 4- and 8-cell stage mouse blastomeres, *J. Exp. Zool.* **200:**365–376.

Kelly, S. J., Mulnard, J. G., and Graham, C. F., 1978, Cell division and cell allocation in early mouse development, *J. Embryol. Exp. Morphol.* **48:**37–51.

Marshall, J. T., 1972, Taxonomy of Thailand Sinda species of *Mus* (Rodentia, Muridae) *Mammal. Chrom. Newsl.* **13:**13–16.

Mintz, B., 1965, Experimental genetic mosaicism in the mouse, in: *Preimplantation Stages of Pregnancy* (G. E. W. Wolstenholme and M. O'Connor, eds.), pp. 194–207, Churchill, London.

Mulnard, J., 1973, Formation de blastocystes chimériques par fusion d'embryons de rat et de souris au stade VIII, *C. R. Acad. Sci.* **276:**379.

Potts, D. M., 1968, The ultrastructure of implantation in the mouse, *J. Anat.* **103:**77.

Reinius, S., 1967, Ultrastructure of blastocyst attachment in the mouse, *Z. Zellforsch. Mikrosk. Anat.* **77:**257–266.

Rossant, J., 1975a, Investigation of the determinative state of the mouse inner cell mass. I. Aggregation of isolated inner cell masses with morulae, *J. Embryol. Exp. Morphol.* **33:**979–990.

Rossant, J., 1975b, Investigation of the determinative state of the mouse inner cell mass. II. The fate of isolated inner cell masses transferred to the oviduct, *J. Embryol. Exp. Morphol.* **33:**991–1001.

Rossant, J., 1976a, Investigation of inner cell mass determination by aggregation of isolated rat inner cell masses with mouse morulae, *J. Embryol. Exp. Morphol.* **36:**163–174.

Rossant, J., 1976b, Postimplantation development of blastomeres isolated from 4- and 8-cell mouse eggs, *J. Embryol. Exp. Morphol.* **36:**283–290.

Rossant, J., 1977, Cell commitment in early rodent development, in: *Development in Mammals*, Vol. 2 (M. H. Johnson, ed.), pp. 119–150, North-Holland, Amsterdam.

Rossant, J., and Frels, W. I., 1980, Interspecific chimeras in mammals: Successful production of live chimeras between *Mus musculus* and *Mus caroli*, *Science* **208:**419–421.

Rossant, J., and Lis, W. T., 1979, Potential of isolated mouse inner cell masses to form trophectoderm derivatives *in vivo*, *Dev. Biol.* **70:**225–261.

Rossant, J., and Ofer, L., 1977, Properties of extra-embryonic ectoderm isolated from postimplantation mouse embryos, *J. Embryol. Exp. Morphol.* **39:**183–194.

Rossant, J., and Papaioannou, V. E., 1977, The biology of embryogenesis, in: *Concepts in Mammalian Embryogenesis* (M. I. Sherman, ed.), pp. 1–36, MIT Press, Cambridge, Mass.

Rossant, J., and Vijh, K. M., 1980, Ability of outside cells from preimplantation mouse embryos to form inner cell mass derivatives, *Dev. Biol.* **76:**475–482.

Rugh, R., 1968, *The Mouse: Its Reproduction and Development*, Burgess, Minneapolis, Minn.

Snell, G. D., and Stevens, L. C., 1966, Early embryology, in: *Biology of the Laboratory Mouse* (E. L. Green, ed.), 2nd ed., pp. 205–245, McGraw-Hill, New York.

Snow, M. H. L., Aitken, J., and Ansell, J. D., 1976, Role of the inner cell mass in controlling implantation in the mouse, *J. Reprod. Fertil.* **48**:403–404.

Solter, D., and Knowles, B. B., 1975, Immunosurgery of mouse blastocyst, *Proc. Natl. Acad. Sci. USA* **72**:5099–5102.

Spindle, A. I., 1978, Trophoblast regeneration by inner cell masses isolated from cultured mouse embryos, *J. Exp. Zool.* **203**:483–489.

Stern, M. S., 1973, Chimaeras obtained by aggregation of mouse eggs with rat eggs, *Nature (London)* **243**:472–473.

Stern, M. S., and Wilson, I. B., 1972, Experimental studies on the organisation of the preimplantation mouse embryo. I. Fusion of asynchronously cleaving eggs, *J. Embryol. Exp. Morphol.* **28**:247–254.

Surani, M. A. H., and Barton, S. C., 1977, Trophoblastic vesicles of preimplantation blastocysts can enter into quiescence in the absence of inner cell mass, *J. Embryol. Exp. Morphol.* **39**:273–277.

Tachi, S., and Tachi, C., 1979, Ultrastructural studies on maternal–embryonic cell interaction during experimentally induced implantation of rat blastocysts to the endometrium of the mouse, *Dev. Biol.* **68**:203–223.

Tarkowski, A. K., 1962, Interspecific transfers of eggs between rat and mouse, *J. Embryol. Exp. Morphol.* **10**:476–495.

Tarkowski, A. K., and Wroblewska, J., 1967, Development of blastomeres of mouse eggs isolated at the four- and eight-cell stage, *J. Embryol. Exp. Morphol.* **18**:155–180.

Theiler, K., 1972, *The House Mouse*, Springer-Verlag, Berlin.

Van Blerkom, J., Barton, S. C., and Johnson, M. H., 1976, Molecular differentiation in the preimplantation mouse embryo, *Nature (London)* **259**:319–321.

Weiss, P., 1939, *Principles of Development*, Holt, New York.

West, J. D., Frels, W. I., Papaioannou, V. E., Karr, J. P., and Chapman, V. M., 1977, Development of interspecific hybrids of *Mus*, *J. Embryol. Exp. Morphol.* **41**:233–243.

West, J. D., Frels, W. I., and Chapman, V. M., 1978, *Mus musculus* × *Mus caroli* hybrids: Mouse mules, *J. Hered.* **69**:321–326.

Wiley, L. M., Spindle, A. I., and Pedersen, R. A., 1978, Morphology of isolated mouse inner cell masses developing *in vitro*, *Dev. Biol.* **63**:1–10.

Wilson, I. B., Bolton, E., and Cuttler, R. H., 1972, Preimplantation differentiation in the mouse egg as revealed by microinjection of vital markers, *J. Embryol. Exp. Morphol.* **27**:467–479.

Zeilmaker, G. H., 1973, Fusion of rat and mouse morulae and formation of chimaeric blastocysts, *Nature (London)* **242**:115–116.

4

The Generation and Recognition of Positional Information in the Preimplantation Mouse Embryo

M. H. Johnson, H. P. M. Pratt, and A. H. Handyside

I. Introduction

Cell differentiation and commitment within the preimplantation mouse embryo depend on cell position (Mintz, 1965; Tarkowski and Wroblewska, 1967). Within the aggregate of cells that form the morula, the more centrally placed cluster of cells starts to differentiate as an inner cell mass (ICM), while the more peripherally placed layer of cells differentiates as trophectoderm. After these first overt signs of differentiation, the two diverging classes of cells become fixed on their courses of development and are therefore said to be committed. If cell position is changed before commitment by rearrangement of blastomeres, then both cell differentiation and fate are changed (see Rossant, 1977; Johnson, 1979; Rossant and Frels, this volume). The embryo can regulate itself.

Two questions therefore arise. How do cells arrive at different positions? How do cells recognize that they are at different positions? We approach these questions by analyzing preimplantation mouse development for evidence of positional differences as reflected in asymmetries of cell or tissue organization.

II. Asymmetries in the Oocyte

The unfertilized oocyte becomes polarized morphologically during the terminal phases of intrafollicular maturation (Nicosia *et al.*, 1978). At ovulation, the second metaphase spindle is placed eccentrically at one pole, displacing cell organelles, particularly the subcortically located cortical granules, from this region (Nicosia *et al.*, 1977). The overlying plasma membrane, unlike the remainder of the oocyte surface, is devoid of microvilli

M. H. Johnson, H. P. M. Pratt, and A. H. Handyside • Department of Anatomy, University of Cambridge, Cambridge, England.

(Eager *et al.*, 1976) and has a lower unit density of concanavalin A binding sites (Johnson *et al.*, 1975). After sperm attachment and fusion, which invariably occur at the microvillous surface of the oocyte, the constriction furrow separating the second polar body from the ovum forms along the interface between the smooth and the microvillous areas (Eager *et al.*, 1976). The membrane of the fertilized ovum thereby becomes entirely microvillous and structurally expanded and modified as a result of the fusion of cortical granules at fertilization, whereas the membrane of the second polar body is smooth and devoid of cortical granule membrane. There is no evidence to suggest that the asymmetry set up during oogenesis is conserved into early development. Thus, attempts to relate subsequent developmental events, such as the plane of first cleavage or the properties of one of the two-cell blastomeres, to the site of second polar body extrusion have not met with success (Graham and Deussen, 1978). Furthermore, the suppression of polar body extrusion in parthenogenetically activated oocytes does not prevent development to the blastocyst (Kaufman *et al.*, 1977). The morphological polarization of the secondary oocyte appears to be concerned primarily if not entirely with the internal economy of the ovum, conserving maximal cytoplasm with minimal residual loss to the polar body.

III. Asymmetries at Fertilization

There appears to be no preferred attachment or fusion site for spermatozoa within the microvillous portion of the oocyte surface. The act of sperm fusion at one restricted location on the oocyte, however, could serve as a reference center for subsequent developmental events. After fusion of the plasma membranes of mouse oocyte and spermatozoa, the cytoplasmic contents of the latter appear to disperse through the ooplasm. At the site of fusion, the surface membrane is devoid of microvilli and protrudes from the surface (Gwatkin, 1976). Recent evidence suggests that at least some of the components of the sperm membrane may be retained within this region. These molecules are evidently not free to diffuse laterally in the plane of the ovum membrane, but remain localized at the point of fertilization (Gabel *et al.*, 1979). A general restriction on free lateral diffusion of both lipid and glycoprotein components of the whole membrane of the newly fertilized egg has been measured quantitatively using fluorescent photobleaching (Johnson and Edidin, 1978). Could the residual stable focus of sperm membrane components serve as a molecular memory on which a subsequent polarity could be established?

General membrane fluidity as measured by fluorescent photobleaching relaxes about 12 to 15 hr postfertilization (Bearn and Johnson, unpublished), and this relaxation persists throughout cleavage (Table 1). The sperm membrane components, however, remain as a cluster during cleavage (Gabel *et al.*, 1979). These molecules must be held, therefore, in an aggregate either by mutual interaction within the plane of the membrane or by external or cytoplasmic cross-linking agents. Were cytoplasmic cross-linking to occur, for example via the cytoskeletal system, a basis for converting a surface focus into a cytoplasmic axis would exist and would be present in only one of the blastomeres of the two-cell stage. It would be of interest to determine the effects on localization of sperm membrane components of agents that disrupt internal cytoplasmic orientation.

However, it seems unlikely that the site of sperm fusion is an absolute requirement for normal generation of later embryonic structures, as parthenogenetically activated oocytes can develop to blastocysts and postimplantation stages (Van Blerkom and Runner, 1976; Kaufman *et al.*, 1977).

Table 1. Summary of Changes in Diffusion Constant (D) for a Lipid Probe, Di I–C18 and a Glycolipid/Glycoprotein Probe in Eggs [a]

	$D \times 10^{-8}$ (cm²/sec)			
Stage	Di I-C18	Glycolipid/glycoprotein[b]		
Unfertilized	1.9	0.01	0.12	1.41
Newly fertilized (3–15 hr postfertilization)	<0.1	<0.001	—	—
Late 1-cell/early 2-cell	0.4[c]	0.03	0.14	0.10
4-Cell/8-cell	0.2		ND[d]	

[a] Data summarized from Johnson and Edidin (1978) and Bearn and Johnson (unpublished).
[b] The Fab fraction of a polyspecific antiserum to mouse embryonic antigens. Three recovery phases probably reflect three general species of mobile antigens.
[c] Note that the high prefertilization *D* value for the lipid probe is not fully restored whereas those for glycolipid/glycoprotein molecules are.
[d] ND, not determined.

IV. Asymmetries in Cleavage

The first cleavage division in the mouse occurs between 20 and 22 hr after fertilization. During this extended one-cell stage, the two haploid sets of chromosomes are retained within separate pronuclear membranes, undergo DNA replication, and first come together briefly at the initiation of the first cleavage division. Thus, truly diploid embryonic nuclei do not form until the two-cell stage. DNA is reduplicated immediately, there being no G1 phase. It is probable that only at the mid two-cell stage (some 26 to 28 hr postfertilization) do the genes of the newly established zygote first become synthetically active, development hitherto being controlled posttranscriptionally (Braude *et al.*, 1979; Johnson, 1979; Schultz *et al.*, this volume; Van Blerkom, this volume). Cleavage then proceeds through two further divisions to the eight-cell stage at approximately 12-hr intervals. Throughout this period, a G1 phase is lacking or highly abbreviated, and net synthesis of macromolecules is low.

Consistent differences in blastomere size are not observed in cleaving embryos. Neither do blastomeres show a regular heterogeneity of morphology at either the light or the electron microscope level (Van Blerkom and Motta, 1979). Tests of developmental potential suggest equivalence of blastomeres (Kelly, 1977). Specialized cell junctions do not form between any adjacent blastomeres in any particular spatial array, as assessed morphologically, by electrical coupling and by dye injection experiments (Lo and Gilula, 1979). The two mitotic products of each blastomere at each of the first three divisions, however, do retain a lingering cytoplasmic continuity via the midbody for several hours (Lo and Gilula, 1979). In only one respect has any asymmetry been observed in cleavage stages, and this involves the timing of cleavage divisions and the pattern of cell contacts that results. Thus, one of the two-cell-stage blastomeres divides at a variable time in advance of the other (Graham and Deussen, 1978). Thereafter, throughout cleavage to the eight-cell stage and beyond, the progeny of the first cell maintain their temporal advantage over those of the second cell. Early division confers on some of the progeny a higher degree of physical contact with their neighbors. This greater degree of contact, it is argued, arises from the more advanced state of expansion of the membrane in a cell in late interphase and the greater number of microvilli that pull other cells around it (Graham and Lehtonen, 1979).

As a result, progeny of the first dividing two-cell blastomere eventually contribute disproportionately to inside cells of the postcleavage morula and thence to the ICM of the blastocyst (Kelly *et al.*, 1978).

These experiments appear to offer an explanation of how cells arrive at different positions. Two important features of the experiments require emphasis. First, it is not clear *why* one two-cell-stage blastomere divides ahead of the other; a relationship between early division and relative cell size, time of completion of DNA synthesis, possession of a unique organelle, site of polar body extrusion, or site of residual sperm membrane has not been sought or shown. Second, there is no *absolute* relationship between first division and contribution to the ICM. Thus, not all the progeny of the first two-cell-stage blastomere to divide end up in the ICM; neither are all ICM cells derived from division of the first two-cell-stage blastomere. Destruction of one two-cell-stage blastomere does not preclude or disturb ICM formation, and disaggregation and reaggregation of four- or eight-cell-stage blastomeres, which change the relative positions of progeny of early- and late-dividing two-cell-stage blastomeres, can reverse their relative contribution to the ICM. Thus, it is immediate cellular interactions within the morula that are considered important in determining contribution to the inside, and not some predetermined cellular memory of early division. These cellular interactions are, according to Graham and his colleagues, critically dependent upon cell surface properties and particularly on the relative area and microvillous state of the cell surface that permits cells to pull other cells over them. This conclusion, with its emphasis on cell surface properties, is important; it is essential, therefore, to remember that it is an inference rather than an observation, and is drawn from studies on fragments of embryo flattened into a monolayer under oil *in vitro* (Graham and Lehtonen, 1979). The possibility that cell surface properties may be highly abnormal under such conditions cannot be ignored.

V. Compaction

At the completion of cleavage in the mouse, the eight-cell embryo undergoes the process of compaction. Although this process has come to be considered a key event in the generation of positional information leading to blastocyst formation, there has been hitherto little direct evidence for this conclusion.

A. Description of Compaction

Compaction involves a number of major changes in embryo morphology. The eight hitherto spherical blastomeres undergo a change in cell shape that maximizes cell to cell contact. Cell membranes come to lie in close apposition over much of the cell's surface, and intercellular spaces are greatly reduced. It becomes impossible to see the gaps between the external (apical) faces of the cells at either the light or the scanning electron microscopic level, and indeed the boundaries themselves between cells can be discriminated only by use of the electron microscope. Specialized junctions develop between cells for the first time. Focal tight junctions form at the apical margins of adjacent cells, and by the 16- to 32-cell stage these have become zonular, "zipping" the cells together to form a permeability seal between the intercellular space at the core of the embryo and the external environment (Calarco and Epstein, 1973; Ducibella *et al.*, 1975; Ducibella, 1977; McLaren and Smith, 1977; Handyside, 1978a; Magnuson *et al.*, 1978).

Gap junctional connections also form between blastomeres at this stage. The most convincing demonstration of their presence comes from studies of electrical coupling and dye penetration, which show that intercellular continuity between all eight cells is first established as compaction is initiated (Lo and Gilula, 1979). Morphological evidence of gap junctional complexes is first obtained also at the eight-cell stage (Magnuson *et al.*, 1978; Ducibella *et al.*, 1975). Free communication between *all* cells of the embryo, presumably via gap junctions, persists until the late postattachment stages (Lo and Gilula, 1979).

In addition to these intercellular features of compaction, a marked polarization of individual cells also occurs. Thus, an early event in the process of compaction is the establishment of an axis within each cell, which changes it from a sphere to a columnarlike epithelial cell. Microvilli are restricted to the outer apical face of each cell and to a few sites at its base (Ducibella and Anderson, 1975). Tight junctional complexes are restricted to the circumference of the apical face. Microtubules come to lie parallel to the microvillous-free lateral walls of the cell (Ducibella and Anderson, 1975). A column of cytoplasmic organelles extends from the nucleus to the outer apical cytoplasmic face; this is particularly well seen in rabbit and rat blastomeres (Izquierdo, 1955; Austin and Bishop, 1959; Mazanec and Dvorak, 1963; Van Blerkom and Motta, 1979), and also occurs in the mouse (Reeve, 1981).

Individual cells also show polarization of surface membrane when stained with a variety of fluoresceinated ligands (Table 2). Prior to compaction, individual blastomeres of the two-, four-, or eight-cell stage stain evenly over their whole surface, and with time this even staining converts to a patchy staining pattern over the whole surface. Blastomeres stain in this way, whether incubated in the ligand before or after fixation with a variety of agents (Table 2). After compaction, the same treatments yield a different pattern. Individual blastomeres show a clear polarization of surface fluorescence (Figure 1b). The fluorescent pole is not induced by the ligands for (a) monovalent ligands also reveal polarization and (b) prefixation in 4% paraformaldehyde, which prevents patching, does not interfere with the polarization. Thus, the increased polar density of ligand receptors is a real feature of the compact cell. Moreover, the maintenance of this intracellular axis is not contingent upon continuing cell contact, as it persists for several hours after dissociation of embryos into individual blastomeres (Handyside, 1980). Following the next cleavage division, heterogeneity of 16-cell blastomeres, presumably due to division across the long axis of polarization, is clearly demonstrated by cells labeled with fluoresceinated concanavalin A (Figure 1d).

Thus it is possible to describe compaction in both cellular and intercellular terms. The full extent to which the various elements constituting the whole process of compaction can

Table 2. Surface Polarization of Embryonic Cells [a]

Cell stage	% ring stain	% polarization stain
2-Cell	100	0
4-Cell	100	0
Early 8-cell	100	0
Mid 8-cell	50	50
Late 8-cell	5	95

[a] Embryonic cells were disaggregated using 0.1% trypsin/EDTA, cytochalasin D, or Ca^{2+}-free medium, incubated in a fluoresceinated ligand (rabbit anti-mouse embryo homogenate, concanavalin A, peanut lectin, or rabbit anti-embryonal carcinoma), washed, and scored as fluorescent over the whole surface or only at one pole (see Figure 1b). The staining pattern was not affected by fixation, colcemid, cytochalasin, or the valency of the fluoresceinated ligand.

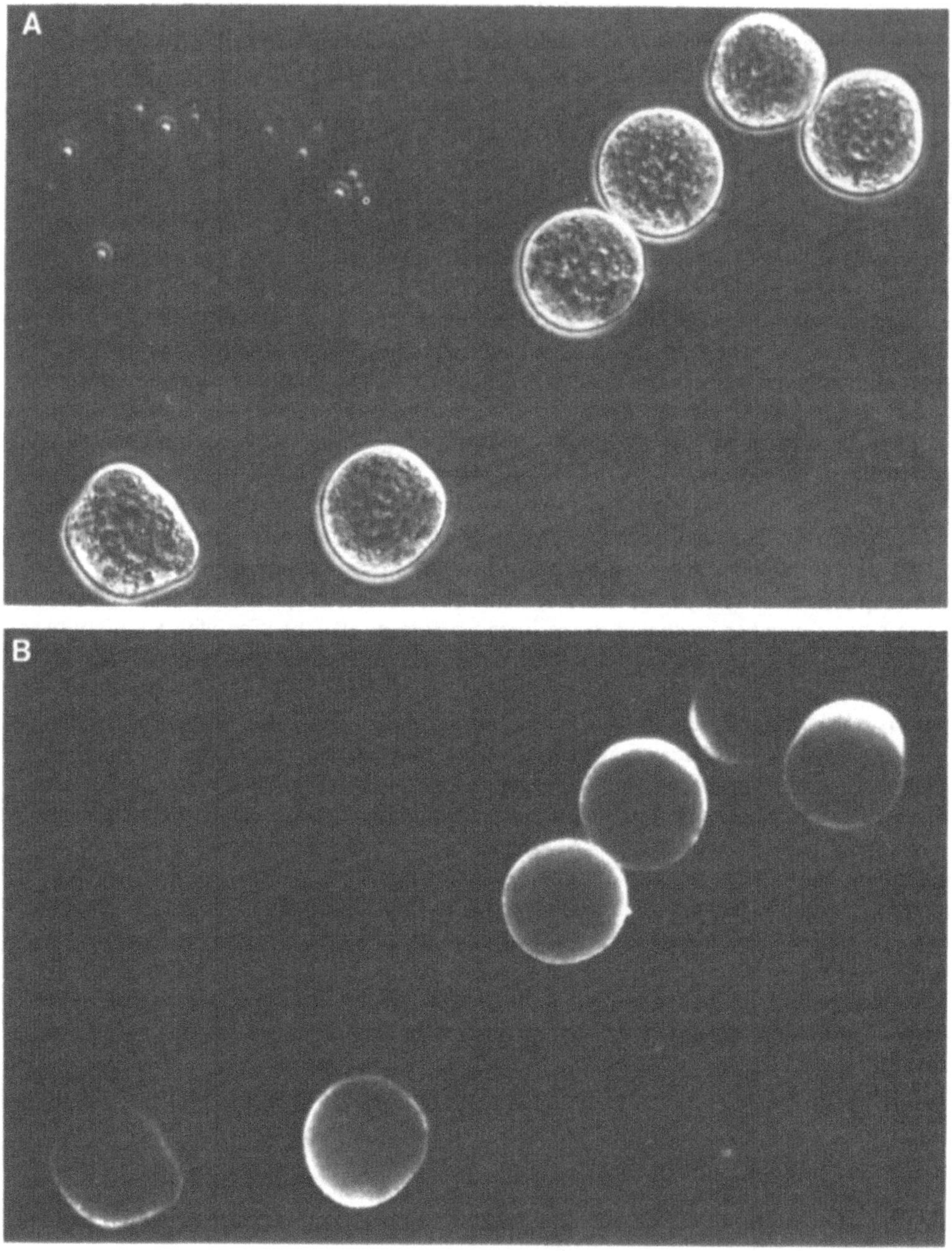

Figure 1. (a) Phase contrast and (b) fluorescence photomicrographs of paraformaldehyde-fixed blastomeres dissociated from compact 8-cell morulae and incubated in an antiserum made in a rabbit against mouse embryo homogenate followed by fluorescein-conjugated goat anti-rabbit IgG. Note polarized surface labeling. (c) Phase contrast and (d) fluorescence photomicrographs of nonfixed, partially dissociated blastomeres from a 16-cell compacted morula incubated in fluroescein-conjugated concanavalin A. Note the doublets of cells, one of which is surface labeled while the other is relatively unlabeled but contains a number of discrete internal vesicles.

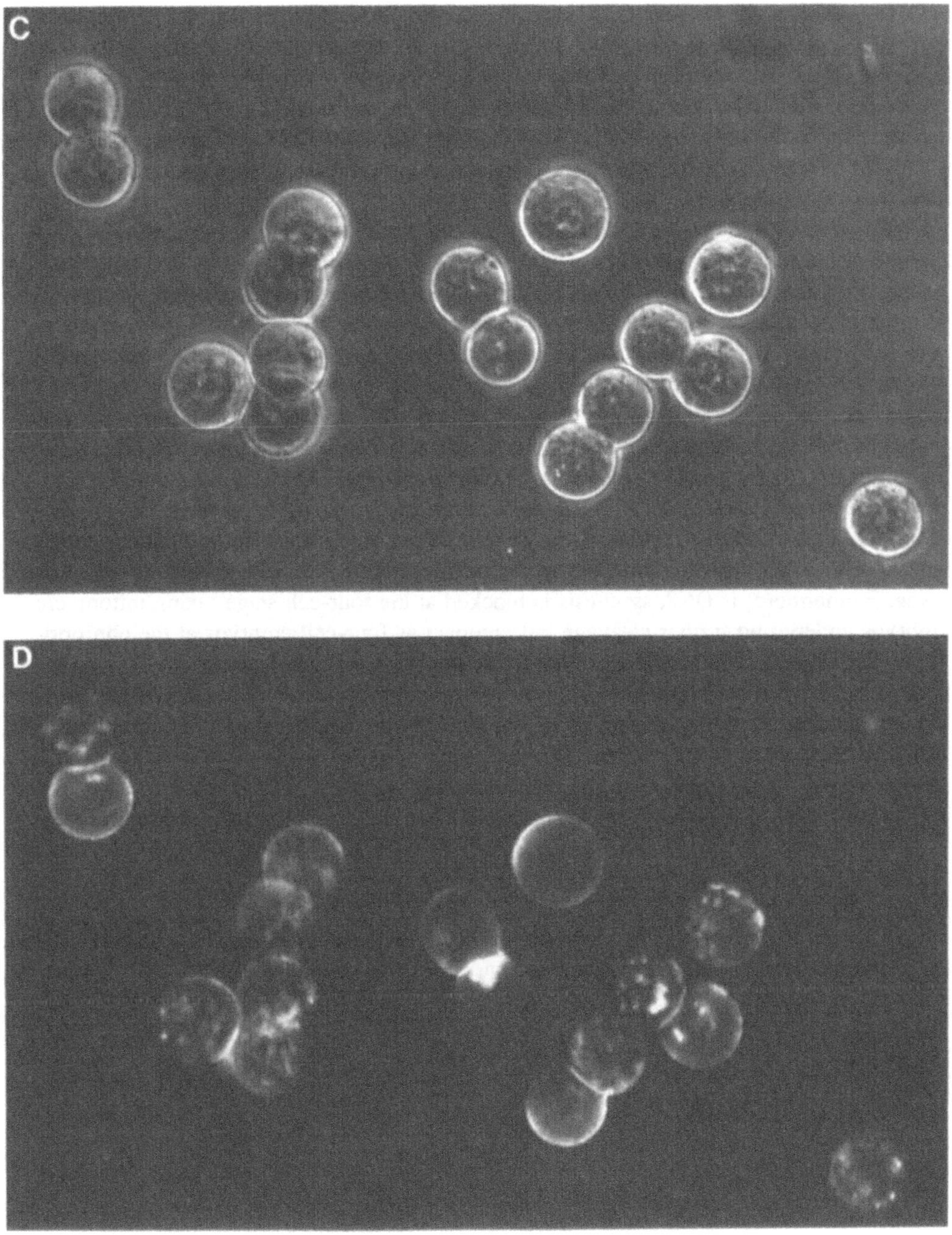

Figure 1. *(continued)*.

be dissociated and elicited separately is not yet clear. It is apparent, however, that some elements of compaction can develop or persist in the absence of others.

B. Control of Compaction

The cellular processes regulating compaction, and the molecular and supramolecular changes associated with it, have been studied using a variety of different experimental techniques. Two general conclusions may be drawn from the present evidence. (a) Compaction appears to occur at a fixed time after fertilization, as though it were a temporally programmed event. It has not been possible to advance compaction experimentally so that it occurs at an earlier time. Compaction appears to be independent of cytokinesis, nuclear: cytoplasmic ratios, cell number, and at least the later stages of DNA replication or expression. (b) The mechanics of the process of compaction are critically dependent upon the cell membrane and cytoskeleton.

Experiments in which normal four- or eight-cell embryos are disaggregated into pairs or quartete of blastomeres show that blastomeres will compact (as judged by cell flattening) only when they have undergone the equivalent of three cleavage divisions, irrespective of the number of cells in the aggregate at the time of compaction (Table 3). Cytokinesis per se and the partitioning of cytoplasm and surface membrane are also irrelevant to the process of compaction, as pulses of cytochalasin D sufficient to arrest cell division but not DNA replication of two- or four-cell embryos nevertheless permit compaction at the same time as normal controls. Thus compaction appears to occur at a defined time in development and seems to be under the control of some type of biological clock. One obvious source of timing in development is the cycle of DNA replication, but a variety of experiments suggest that this is unlikely to be the regulating factor. For example, embryos cultured under conditions that slow down cleavage compact at the normal time but as either two- or four-cell embryos. Furthermore, if DNA synthesis is blocked at the four-cell stage, using mitomycin C, embryos undergo no further cleavage but compact as four-cell embryos at the appropriate time. The absence of a requirement for DNA synthesis (at least from the four- to eight-cell stage) would also be compatible with the observation that four-cell embryos are capable of compacting in the presence of α-amanitin (Pratt, unpublished), an inhibitor of mRNA transcription.

Compaction clearly involves the cytoskeleton and cytocortex of the embryo, and a variety of agents that disrupt these structures have been used to probe the supramolecular

Table 3. Conditions Permitting Compaction at the Normal Time[a]

Experiment	Number of cells compacting	DNA replication prior to compaction	Cytokinesis prior to compaction
Control 8-cell embryo	8	–	–
2/4 blastomeres	4	+	+
2/8 blastomeres	2	–	–
Slowed cleavage due to modified culture conditions			
2-cell	2	–?	–
4-cell	4	–?	–
Parthenogenetic 8-cell embryos, diploid	8	–	–
15-hr pulse of cytochalasin D at			
2-cell	2 (polyploid)	+	–
4-cell	4 (polyploid)	+	–
Mitomycin C blocking DNA synthesis at			
4-cell	4	–?	–
8-cell	8	–?	–

[a] +, present; –, absent; –?, probably absent.

Table 4. Agents Inhibiting Compaction [a]

			Components of compaction remaining in treated embryos				
Agent	Action	Reversibility	Cell apposition	Junctions[b]	Polarization[c]	DNA synthesis	Cytokinesis
Cytochalasins B and D	Depolymerizes microfilaments	Reversible	−[d]	−	+	+	−
Colcemid	Inhibits polymerization of microtubules	Irreversible (after 6 hr)	−	−	+	−	−
Low Ca^{2+}	Destabilizes cytoskeleton Modifies (?) lipid fluidity	Reversible	−	±	+	+	+
7-Ketocholesterol	Inhibits sterol synthesis Modifies or reduces membrane sterol	Irreversible (after 6 hr)	−	+	+	+	+
Anti-embryonal carcinoma serum	Modifies membrane and cytoskeleton dynamics	Reversible	−	−	+	+	+
Tetracaine	Disrupts cytoskeleton Modifies (?) lipid fluidity	Reversible	−	−	ND	−	−

[a] Embryos exposed to agents for periods between 60 and 96 hr post-hCG.
[b] Tight junctions and other types of intercellular junctions.
[c] Indicated by a heterogeneous distribution of microvilli and cytoplasmic vesicles.
[d] +, present; −, absent; ±, trace indications; ND, not determined.

events of compaction in more detail (Table 4). Cytochalasin D, an agent that depolymerizes actin-containing microfilaments, reverses or inhibits the intercellular components of compaction in eight-cell embryos, indicating that changes in cellular shape, intercellular apposition, and junction assembly are all dependent upon intact microfilaments. However, some features of cell polarization appear to be resistant to the action of cytochalasin (Table 4). Colcemid, which prevents the polymerization of tubulin, another important cytoskeletal component, is not immediately effective in reversing or inhibiting compaction, but its longer term effects demonstrate that intact microtubules are necessary for maintaining intercellular apposition and junctions. Lowering the concentration of external Ca^{2+} or addition of the local anesthetic tetracaine are also rapid methods for decompacting embryos, presumably due to a destabilizing effect on the cytoskeleton, though effects on fluidity of membrane lipids may be also involved (Pratt, 1978). Disorganization of the cytoskeleton using low Ca^{2+}, unlike cytochalasin D or colcemid, may result in the persistence of some intercellular junctions. A further method of inhibiting compaction has been to incubate embryos in an antiserum to embryonal carcinoma cells. Cells continue to divide but do not appear to interact. The effects of the antiserum are presumably mediated by the surface antigens to which it binds (Kemler *et al.*, 1977; Johnson *et al.*, 1979).

An attempt to investigate the role of the lipid component of the cytocortex has been made by using 7-ketocholesterol, an agent that depletes or modifies the sterol composition of the membrane by inhibition of sterol synthesis (Pratt, 1978). Embryos grown in this agent continue through one or two cycles of cell division, but intercellular adhesion and junction formation are reduced compared with normal embryos, and fluid accumulation is inhibited. This effect is reversed by supplementing the medium with cholesterol when embryos develop into normal blastocysts (Pratt, 1978; Pratt *et al.*, 1980).

A survey of the effects of these agents has emphasized the role both of cytoskeleton and of membrane in compaction. Furthermore, it raises the possibility that agents can be found to interfere selectively with the different components of compaction (see earlier).

C. Consequences of Compaction for Blastocyst Formation

A clue to the role of compaction in blastocyst formation may be obtained by observing the consequences of its inhibition. Two approaches in this direction have been made in our laboratory. The first involves inhibition of both compaction and cytokinesis using cytochalasin D. The second involves the use of an antiserum to embryonal carcinoma cells that prevents compaction but permits normal cell division.

1. Cytochalasin D Treatment

Cytochalasin D has two obvious effects on the development of the preimplantation mouse embryo. It blocks cytokinesis and inhibits and rapidly reverses the intercellular features of compaction. Both actions presumably result from the disruption of polymerized actin-containing microfilaments (Wessells *et al.*, 1971). Treatment of embryos with cytochalasin for periods of less than 15 hr results in reversible delay of compaction and reversible inhibition of cytokinesis. Normal blastocysts develop with a normal ICM (although cell numbers will be reduced if one round of cytokinesis is lost). When embryos are incubated in cytochalasin D for periods exceeding 15 hr, cytokinesis is irreversibly inhibited, though the cycles of DNA replication continue with the same timing as in a normal embryo (Surani *et al.*, 1980). The nuclei replicate once and blastomeres of cytochalasin-D-treated embryos thus become binucleate and highly polyploid. If the drug is removed after 15 hr of treatment, but before 66 to 72 hr post-hCG (the time at which compaction occurs), the embryos do not resume cytokinesis but remain superficially unchanged until 66 to 72 hr post-hCG when they compact (for example as "two-cell" or "four-cell" embryos). These compacted embryos then accumulate fluid with the same timing as normal embryos and form "blastocystlike" vesicles that appear superficially identical to normal integrated blastocysts though consisting of only two or four cells and lacking an overt ICM.

If, however, cytochalasin D treatment of two-, four-, or eight-cell embryos is prolonged up to approximately 20 hr beyond the time of normal compaction and the drug is removed at 96 to 100 hr post-hCG, then the process of compaction and the formation of "blastocystlike" vesicles occur rapidly (Surani *et al.*, 1980). Embryos compact within 1 hr of removal of the drug, accumulate fluid within 4 to 6 hr, and form "blastocystlike" vesicles within 12 hr. These "blastocystlike" vesicles again look like normal blastocysts, though they also lack an obvious ICM and consist of only two, four, or eight cells that are binucleate and polyploid with the same DNA content as a normal embryo of equivalent age (Surani *et al.*, 1980). Eight-cell "blastocystlike" vesicles have been shown to implant *in vivo* though no embryonic derivatives were found (Surani *et al.*, 1980). Both four- and eight-cell "blastocystlike" vesicles can hatch and outgrow *in vitro*, while two-cell "blastocystlike" vesicles do not.

Despite the reduced cell numbers and the absence of an overt ICM, these "blastocystlike" vesicles are indistinguishable in molecular terms from their normal 32- to 64-

cell controls containing an ICM. For example, their polypeptide synthetic profile (as assessed from two-dimensional O'Farrell gels), cholesterol biosynthetic pattern, and development of various antigenic and enzymic activities are all similar to untreated controls. Thus, short-term inhibition of compaction for 20 to 25 hr beyond its normal time of occurrence is compatible with the formation of a blastocyst that is normal in molecular terms, but which lacks an obvious ICM.

When the cytochalasin D treatment is continued beyond approximately 96 to 100 hr post-hCG, fluid starts to accumulate in individual blastomeres of the 2-, 4-, or 8-cell cytochalasin-treated embryos. Removal of the drug at this stage fails to induce compaction as judged by intercellular interactions. Individual blastomeres remain independent within the aggregate and continue to accumulate fluid while the embryo as a whole appears to be unable to form a normal integrated "blastocystlike" vesicle. Nevertheless, the molecular maturation of these nonintegrated blastocyst aggregates is also identical to that of a normal 32- to 64-cell blastocyst containing an ICM.

2. Treatment with Antiserum to Embryonal Carcinoma Cells

An alternative approach to the inhibition of compaction is the use of antisera directed against certain cell surface specificities shared with embryonal carcinoma cells (Kemler *et al.*, 1977). A rabbit antiserum to LS5770 mouse nullipotent embryonal carcinoma cells does not affect cleavage, but does prevent compaction. Instead, an aggregate of blastomeres forms that lacks intercellular junctions (Johnson *et al.*, 1979).

If, after varying periods of culture in the antiserum, the embryos are rinsed and transferred to normal culture media, they compact after about 8 to 10 hr (presumably the time required for removal of inhibitory levels of antibody from the cell surface). The developmental status of the embryos treated in this way varies, as it does for cytochalasin treatment, with the time for which the embryos are incubated in the antiserum (Figure 2). Shorter incubations result in apparently normal blastocysts, but with more prolonged exposure to antiserum various abnormalities of blastocyst formation result (Table 5).

Two important features emerge from these data. First, a delay of compaction for up to 18 or 20 hr beyond the time of its normal occurrence appears to have little obvious effect

Table 5. Properties of "Blastocysts" [a,b]

	Control	True	False	Aggregate
Visible ICM	+	small	−	−
Mean cell No.	90	70	60	60
Mitotic index [c]	3	2	0.5	0.9
Inside:total [d]	0.23	0.13	0.006	0.001
Hatch	+	+	±	−
Attach	+	+	+(ZF) [e]	+(ZF)
Outgrow + ICM	+	±	−	−
Implant	+	±	−	−

[a] Summarized from Johnson *et al.* (1979).
[b] Superovulated embryos were recovered at the two-cell stage and cultured *in vitro* in varying regimes of control media and media containing antiserum to embryonal carcinoma. Embryos were classified as control, true, false, or aggregate (see Figure 2) and their properties assessed.
[c] Mitotic index = number of mitotic spreads divided by total number of nuclei in each embryo.
[d] Inside cell numbers determined by immunosurgery.
[e] ZF, Zona-free.

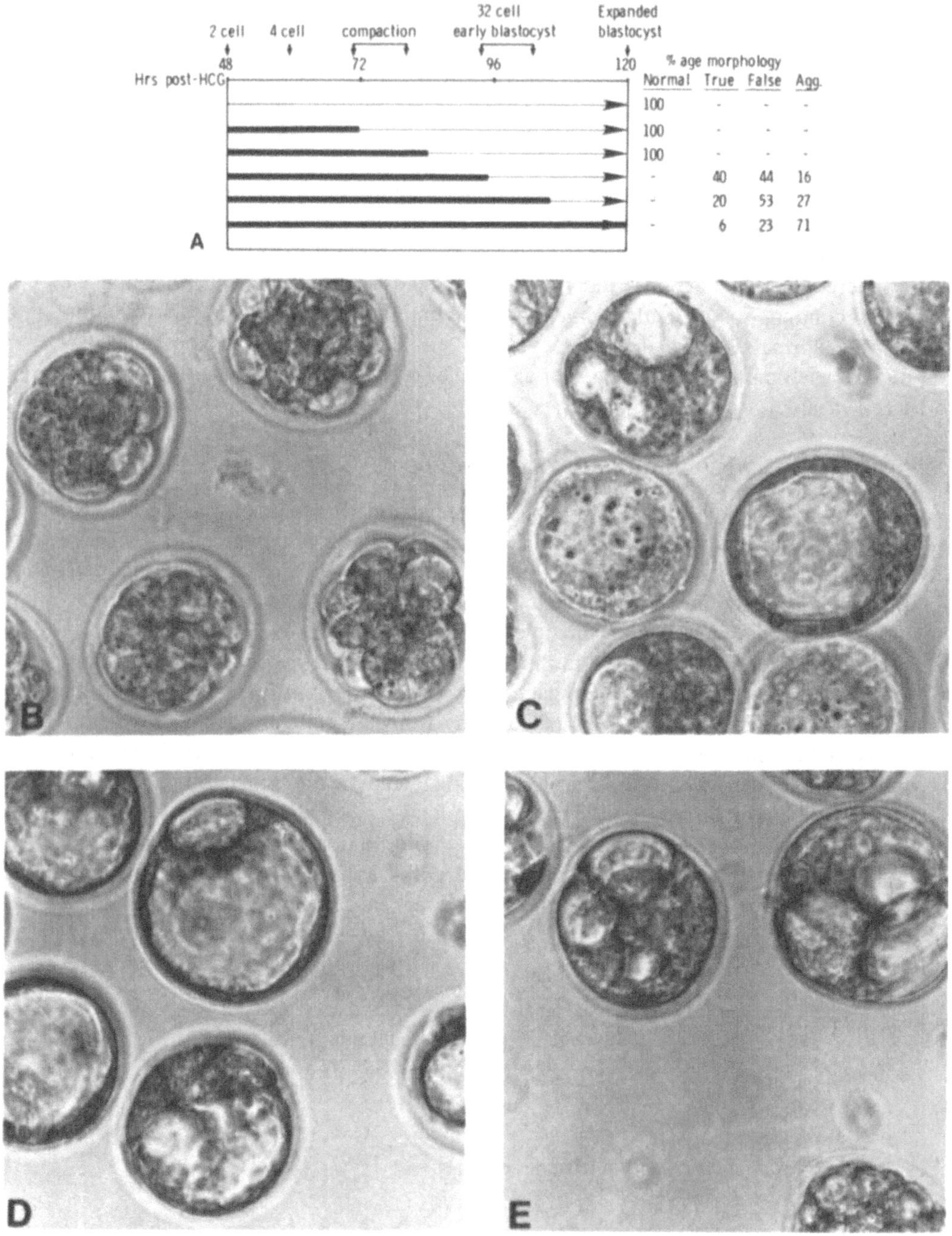

Figure 2. A: development of two-cell mouse embryos grown for varying periods in the presence (thick line) and absence (thin line) of antiserum to embryonal carcinoma. Time scale: hours post-hCG. Morphologies summarized on right. B: example of two-cell embryos cultured for 94 hr in antiserum. Note absence of blastocyst formation despite cell division. C: controls after 94 hr of culture in normal medium. D: example of false blastocyst showing no overt ICM cell cluster but either a vesicular or a very much reduced ICM. E: example of aggregate blastocyst showing absence of overt ICM and blastocoelic cavity. [True blastocyst (not shown) has an overt ICM that appears smaller than in control blastocysts.]

on blastocyst formation. Second, if the uncompacted state is prolonged beyond this time, the resulting blastocysts lack a clearly defined cluster of ICM cells (Johnson *et al.*, 1979). At first, a vesicle devoid of a clear ICM forms, which is somewhat similar to the "blastocystlike" vesicles obtained after cytochalasin treatment (but with a greater number of smaller cells). With more prolonged exposure an aggregate of fluid-accumulating cells develops, which again resembles the nonintegrated forms observed after long-term treatment with cytochalasin (but again with greater numbers of smaller cells).

These results with cytochalasin and with antiserum to embryonal carcinoma yield two conclusions. First, short-term decompaction is remarkably compatible with blastocyst formation. Second, the longer the period of decompaction, the more abnormal the morphology of the blastocyst. At first an overt functional ICM seems to be lost and later even the blastocoelic cavity fails to form properly. What is the explanation for these effects of prolonged decompaction on blastocyst formation?

A simple interpretation of these results is to assume that compaction is an important generator of positional cues, but that for up to 20 hr beyond the time that compaction would normally occur the inside cells are developmentally labile (Johnson *et al.*, 1979). Thus, if these cells do not "perceive" their position as being inside by this time, they lose the capacity to form an ICM. Such a conclusion is consistent with previous observations made on the period of developmental lability of groups of inside cells isolated from morulae and early blastocysts and studied *in vitro* (Johnson *et al.*, 1977; Handyside, 1978a,b). If such a conclusion is accepted, it argues in favor of a critical role for compaction in the generation of positional cues.

D. Mechanisms by Which Compaction Might Generate Positional Information during Blastocyst Formation

The formation of a blastocyst from a morula involves two types of processes. A process of cellular differentiation yields two distinctive tissues—the ICM and trophectoderm. A process of morphogenesis ensures that these tissues are appropriately arranged spatially and results in the formation of a blastocoele. A "microenvironmental hypothesis" has been proposed, relating compaction to blastocyst formation, and suggests that morphogenetic events play an important causal role in generating the conditions required for cytodifferentiation (Ducibella, 1977). This hypothesis proposes that apical zonular tight junctions prevent free paracellular passage of molecules, resulting in a distinctive internal microenvironment for the enclosed cells (Figure 3). These cells, it is argued, perceive and respond to microenvironmental changes by differentiating as ICM. In the absence of the appropriate microenvironment (as occurs with inhibition of the intercellular components of compaction), all cells should differentiate as trophectoderm.

Three experimental observations make this hypothesis less than totally convincing. First, the zonular tight junctions required for the generation of a distinctive microenvironment are only formed at the 32-cell stage, after the initial cytodifferentiation has commenced (Johnson, 1979). Second, it has not been shown that the vesicular and aggregate blastocysts that develop in the absence of compaction (see earlier) are purely trophectodermal, only that they lack an overt ICM. Third, the use of a number of variations in culture conditions for intact or fragmented embryos has never generated a microenvironment in which all the cells become ICM-like. Perhaps the most remarkable failure of an "internal microenvironment" to direct the differentiation of outer cells toward ICM is seen in the experiments of Pedersen and Spindle, (this volume). Injection of 8-cell cleavage-stage embryos or morulae into the blastocoelic cavity of giant blastocysts does not prevent their de-

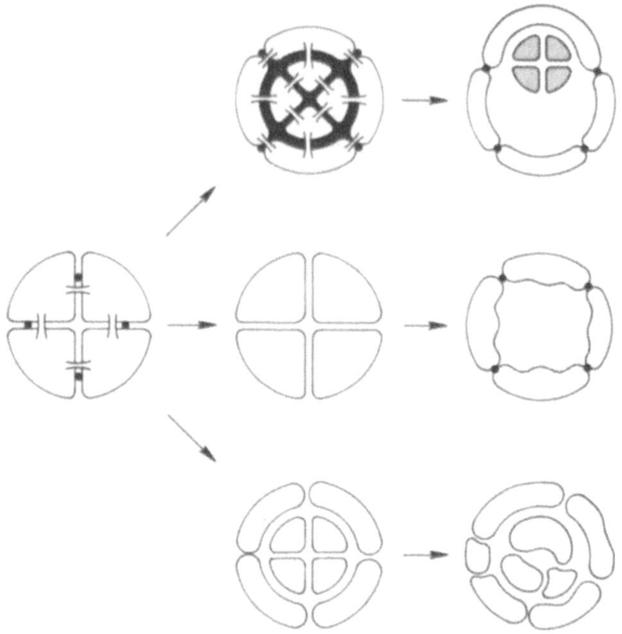

Figure 3. Summary of microenvironmental hypothesis. Upper: normal development; inner cells respond to microenvironment preserved by junctions and form ICM. Middle: cytochalasin treated; no inner cells or distinct microenvironment and thus only trophectoderm forms. Lower: antibody treated; inner cells but no distinctive microenvironment and thus, again, only trophectoderm forms.

velopment into complete blastocysts within the blastocoelic fluid. These reservations lead us to propose an alternative "polarization hypothesis" in which morphogenetic and cytodifferentiative processes occur in parallel (Figure 4).

There are five propositions central to this hypothesis. (1) With the establishment of intercellular communication at compaction, the cytoplasm of the 8-, 16-, and 32-cell morulae is reestablished as a single functional unit. (2) At compaction, a radial gradient of information is established within this cytoplasmic unit and is revealed in blastomeres at the 8-cell stage as an axial polarity. (3) This expression of polarity is preserved despite cytokinesis by virtue of maintained intercellular communication. (4) If intercellular communication is not maintained, each discrete cytoplasmic unit attempts to establish its own polarity. (5) Commitment (probably at the early 32-cell stage) coincides with the loss of ability of discrete cells to generate this polarity.

This hypothesis has features that superficially give it a resemblance to the earlier classical "segregation hypothesis" of Dalcq (1957). However, a strict "segregation hypothesis" cannot apply to the mouse embryo, as cells are developmentally labile until the 32-cell stage. Propositions 2, 4, and 5 distinguish the two hypotheses and suggest that polarization of the 8 cells at compaction results in nonequivalent cytoplasmic allocations, which are *not* determinative in the sense of fixing cell fates but *are* differentiative in the sense of guiding cell fates. The presumption here is that commitment *results from* rather than *leads to* an accretion of differentiative events (Johnson, 1979). The fundamental difference between this "polarization hypothesis" and the "microenvironmental hypothesis" is the assumption that the rounds of cell division following polarization at compaction generate two distinct populations of cells rather than generating identical cells that subsequently come to differ in response to external stimuli.

The hypothesis is not contradicted by present evidence and, as it makes predictions distinct from those of the "microenvironmental hypothesis," is open to experimental testing.

a. In normal intact morulae, it should be possible to show that features of the axial polarity of cells at the eight-cell compaction stage are conserved at division and distributed unequally to the progeny depending upon their relative position. Preliminary data are consistent with this prediction (see for example Figure 1c, d).

b. In disturbed morulae and the blastocysts that subsequently form, it should be possible to show that cells possess properties not just of trophectoderm alone but also of ICM. For example, if intercellular interactions are interrupted, as occurs by treatment with antiserum to embryonal carcinoma cells (Figure 4), each of the isolated cells in the 8-, 16-, and early 32-cell stages should attempt to recreate the polarity of an entire embryo. With prolonged treatment beyond the time of commitment, each cell will therefore become fixed as a unicellular unit of ICM and trophectoderm. Preliminary molecular analysis of false and aggregate blastocysts has revealed that, despite the absence of an overt ICM, the polypeptide synthetic profile of these embryos is not distinguishable from that of a normal blastocyst. Similarly, in cytochalasin-treated embryos, whether "blastocystlike vesicles" or nonintegrated forms, "intercellular interaction" is maximal as a result of failed cytokine-

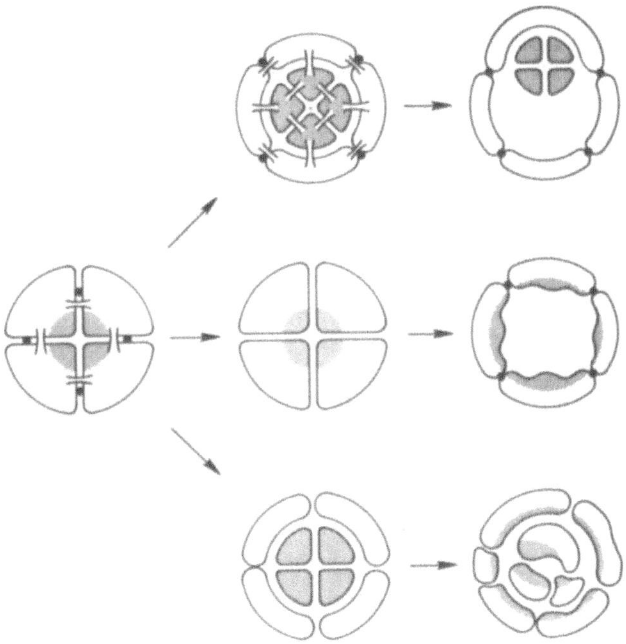

Figure 4. Summary of polarization hypothesis. Upper: normal development; cytoplasmic gradients and allocations preserved beyond time of commitment by cell interaction, ICM forms. Middle: cytochalasin treated; cytoplasmic gradients and allocations preserved beyond time of commitment by cytoplasmic continuity, causing cells of mixed constitution to form. Lower: antibody treated; distinct inner cells form but absence of cell interactions leads to reestablishment of gradients within individual cells and, consequently, cells are of mixed constitution.

sis, and so each embryo generates both ICM and trophectoderm components although not allocated to discrete units.

c. The signal underlying the formation of the initial axial cytoplasmic polarity may be susceptible to experimental analysis. For example, polarization of information could arise from the activity of cytocortical macromolecules that become nonrandomly distributed during membrane biogenesis at cleavage (Izquierdo, 1977). Alternatively, macromolecules distributed randomly throughout the cytocortex of precompaction blastomeres could be selectively activated or induced to cluster as a consequence of close apposition of some parts of the cell membranes at compaction. The consequences of such nonrandomizing of macromolecules in the cytocortex might then be expressed via, for example, polarized transport of molecular messengers, or generation of foci for microtubule assembly, which in turn would lead to polarization of cell activity.

VI. Summary and Conclusions

In this paper we have attempted to show:

a. That although developmental events prior to compaction may influence the position within the morula at which cytoplasmic and cytocortical zones subsequently arrive, they are not necessarily relevant to the mechanisms by which positional differences within the morula are generated.

b. That compaction is a complex event comprising several components each of which may be elicited or suppressed selectively and each of which may make contributions of varying importance to subsequent developmental events.

c. That blastocyst formation consists of morphogenetic and cytodifferentiative events each of which may involve distinct cellular and molecular processes.

d. That compaction has a role to play in blastocyst formation; its inhibition results first in the absence of an obvious ICM and, if prolonged further, leads to the absence of an overt blastocoelic cavity.

From these observations, we describe two hypotheses relating the events of compaction to the events of blastocyst formation, each hypothesis making distinct and experimentally testable predictions.

ACKNOWLEDGMENTS. We thank Gin Flach for excellent technical assistance. The work was supported by grants from the Medical Research Council, Cancer Research Campaign, and Ford Foundation to M.H.J., and from the Medical Research Council to H.P.M.P.

Discussion

BIGGERS (Harvard): Listening to your description of the experiments with cytochalasin, and the fact that the four-cell mouse embryo can be induced to make a trophoblastic vesicle, reminded me that there are some classes of mammals, such as the marsupials and the insectivores, that always make a trophoblastic vesicle.

JOHNSON: I was careful to describe all our vesicles as "blastocystlike," not trophoblasticlike. All our evidence suggests that these vesicles, although lacking an overt ICM, nonetheless have at least some ICM properties. To this extent of course they would resemble the embryos of those classes of mammals to which you refer, which, although lacking an overt ICM, clearly do form an embryonic egg cylinder.

SHERMAN (Roche Institute): Did the outgrowths from the cytochalasin-treated embryos contain more than four cells?

JOHNSON: No. If four-cell embryos are placed in cytochalasin D for longer than 15 hr, they remain as four cells throughout but they contain eight nuclei.

SHERMAN: In other words, there is no evidence of cell division after you remove the cytochalasin and after you get to what I call the trophoblastic vesicle stage?

JOHNSON: There is no evidence for that as long as the cytochalasin treatment is longer than about 15 hr. If the drug is removed before then, there is some division.

SURANI (Cambridge): With two-cell embryos cultured in cytochalasin D for 48 hr, the blastomeres become binucleated; if the drug is washed off, they form vesicles but there is no evidence of any cytokinesis after that (Surani *et al.*, 1980, *Exp. Cell Res.*, in press).

A. KAYE (Weizmann Institute): The polarity aspect of the compaction, as demonstrated by toluidine blue staining, is a fascinating phenomenon. Is there any information on other types of staining which will demonstrate compaction? Might this be a suggestion of polarity in the synthesis and processing of either a protein or mucopolysaccharide, which has to get to the outer surface and which is essential, therefore, for the remainder of the process of implantation?

JOHNSON: Polarity of staining is not limited to toluidine blue as John Reeve in our laboratory has shown (Reeve, 1981). However, it is difficult to relate light-microscope-staining patterns to specific organelles. There is some work in the literature suggesting that multivesicular bodies are involved and there are morphological correlates at the ultrastructural level in the right position.

DENKER (Aachen): Your findings on blastomere polarity are fascinating because they should enable us to reconcile some of the older data from the histochemical work of the Dalcq school (Dalcq, 1954, *Rev. Gen. Sci. Pures Appl.* **61**:19). Some of your pictures reminded me very much of those in recent electron microscopical investigations of Dvorak and co-workers (Dvorak 1978, *Adv. Anat. Embryol. Cell Biol.* **55,** Part 2). I would like to point out that there might be some species differences. In the rabbit and the cat, for example, there is some evidence that one of the first signs of differentiation of some blastomeres into trophoblast is the display of cell polarity. Polarity is seen, however, only in part of the population of outside blastomeres, which appears to contradict the inside–outside theory (Denker, 1970, *Zool. Jahrb. Physiol.* **75**:246; Denker, 1976, *Curr. Top. Pathol.* **62**:59; Denker *et al.*, 1978, *Anat. Anz.* **144**:457). Thus, the first differentiating trophoblast cells are found grouped together on one pole of the embryo, the future abembryonic pole. Such a polar organization of the whole embryo (different from the cytoplasmic polarity mentioned above) is found consistently only in some species; in the mouse and the rat, it is not as well expressed or may be lost at later cleavage stages which may display radial symmetry (Denker, 1972, *Anat. Anz. Suppl.* **130**:267).

JOHNSON: I wouldn't want to make a general statement about embryos of all species. Neither would I claim that our polarization hypothesis is in any way generally novel, although I think that this particular form is probably novel. The hypothesis, as we point out, is quite distinct from that of Dalcq. I agree, one has to use more than cytochemical markers in order to establish polarity. Our work on surface markers suggests that subsequent divisions do conserve cytoplasmic differences in a clearly defined way. Histochemical markers may not allow one to follow labeled cells in this way. So I agree with your cautions. Our main aim is to reduce the emphasis on the microenvironment that has been adopted somewhat uncritically and to encourage the testing of alternative explanations.

PEDERSEN (UCSF): I am bothered by the conclusion that polarity may be the most promising hypothesis since the embryo that you treated with the inhibitor does not go on to divide. Your two types of inhibitor seem to have some specificity, but still there is a problem of interpreting their effect because of the lack of cell division. Perhaps the cells are able to make the required messenger RNAs and proteins for both inner cell mass and trophoblast, but because they are not dividing there are not enough cells to form an inner cell mass.

JOHNSON: I think you are agreeing with us. We are arguing that embryos are on a developmental program during which specific developmental events occur at certain times. The extent to which you parcel up the cytoplasm by progressive cytokinesis will determine how you distribute the consequences of those events. The events themselves appear to go on regardless and so when we talk about four-cell blastocystlike vesicles, we mean just that. We must now develop more markers for ICM and trophectoderm so that we can say unequivocally whether we are dealing with pure trophectoderm or pure ICM or cells which have a bit of both. Let's be very cautious in our interpretation of these abnormal morphologies.

PEDERSEN: Another way of looking at the gels and the inhibitors might be that the markers for inner cell mass do not truly reflect inner cell mass development, since they can be found in these vesicles.

JOHNSON: All we can say about the data at present is that in molecular and biochemical terms it is not possible to distinguish between aggregates, vesicles, and blastocysts. Since some of the molecular and biochemical features are normally limited either to ICM or trophectoderm, it is reasonable to assume that in aggregates and vesicles there are regions both of ICM- and trophectodermlike cytoplasm. We should also not be blinded by the idea that a cell *must* be one thing or another. A cell could be confused; its confusion may help us to elucidate normality.

ROBERTS (Florida): How many polypeptides do distinguish the trophectoderm from the inner cell mass?

JOHNSON: Not very many; perhaps ten or so out of several hundred of the polypeptides are reasonable markers. These tissues are remarkably similar and the differences are often quantitative; one can very rarely be certain of the complete absence of a given peptide in the other tissue.

RICHARDSON (Boston): You alluded, I think, to cholesterol. Could you expand a little on that allusion?

JOHNSON: Hester Pratt used 7-ketocholesterol and other oxygenated sterols to try to inhibit the synthesis of cholesterol (Pratt, 1978, in: *Development in Mammals* (M. H. Johnson, ed.), Vol. 3, pp. 83–129 North-Holland, Amsterdam). 7-Ketocholesterol prevents compaction. Supplemental treatment with exogenous cholesterol overcomes the block to compaction. A tentative interpretation is that a cholesterol domain is required for the intercellular features of compaction, although not for cytokinesis. These are difficult experiments to interpret because one cannot preclude a direct effect of 7-ketocholesterol itself. The results could be taken as evidence that membrane fluidity is an important component of compaction, and that if the membrane is made less fluid it is less able to take part in the compaction process.

References

Austin, C. R., and Bishop, M. W. H., 1959, Differential fluorescence in living rat eggs treated with acridine orange, *Exp. Cell Res.* **17**:35–43.

Braude, P. R., Pelham, M. R. B., Flach, G., and Lobatto, R., 1979, Posttranscriptional control in the early mouse embryo, *Nature (London)* **282**:102–105.

Calarco, P. G., and Epstein, C. J., 1973, Cell surface changes during preimplantation development in the mouse, *Dev. Biol.* **32**:208–213.

Dalcq, A. M., 1957, *Introduction to General Embryology,* Oxford Univ. Press (Clarendon), London/New York.

Ducibella, T., 1977, Surface changes in the developing trophoblast cell, in: *Development in Mammals,* Vol. 1 (M. H. Johnson, ed.), Elsevier/North-Holland, Amsterdam.

Ducibella, T., and Anderson, E., 1975, Cell shape and membrane changes in the eight cell mouse embryo: Prerequisites for morphogenesis of the blastocyst, *Dev. Biol.* **47**:45–58.

Ducibella, T., Albertini, D. F., Anderson, E., and Biggers, J. D., 1975, The preimplantation mammalian embryo: Characterisation of intercellular junctions and their appearance during development, *Dev. Biol.* **45**: 231–250.

Eager, D. D., Johnson, M. H., and Thurley, K. W., 1976, Ultrastructural studies on the surface membrane of the mouse egg, *J. Cell Sci.* **22**:345–353.

Gabel, C. A., Eddy, E. M., and Shapiro, B. M., 1979, After fertilization sperm surface components remain as a patch in sea urchin and mouse embryos, *Cell* **18**:207–216.

Graham, C. F., and Deussen, Z. A., 1978, Features of cell lineage in preimplantation mouse development. *J. Embryol. Exp. Morphol.* **48**:53–72.

Graham, C. F., and Lehtonen, E., 1979, Formation and consequences of cell patterns in preimplantation mouse development, *J. Embryol. Exp. Morphol.* **49**:277–294.

Gwatkin, R. B. L., 1976, Fertilisation, in: *The Cell Surface in Animal Embryogenesis and Development* (G. Poste and G. L. Nicolson, eds.), Elsevier/North-Holland, Amsterdam.

Handyside, A. H., 1978a, Time of commitment of inside cells isolated from preimplantation mouse embryos, *J. Embryol. Exp. Morphol.* **45**:37–53.

Handyside, A. H., 1978b, *The Origin of Cell Lineages in the Preimplantation Mouse Embryo,* Ph.D. thesis, Cambridge University.

Handyside, A. H., 1980, Distribution of antibody and tectin binding sites on dissociated blastomeres from mouse morulae: evidence for polarization at compaction, *J. Embryol. Exp. Morph.* (in press).

Izquierdo, L., 1955, Fixation des oeufs de Rat colorés vitalement par le bleu de Toluidine. Technique et observations cytologiques, *Arch. Biol. (Liege)* **66**:403–438.

Izquierdo, L., 1977, Cleavage and differentiation, in: *Development in Mammals*, Vol. 2 (M. H. Johnson, ed.), pp. 99–118, (M. H. Johnson, ed.), pp. 99–118, Elsevier/North-Holland, Amsterdam.

Johnson, M. H., 1979, Intrinsic and extrinsic factors in preimplantation development, *J. Reprod. Fertil.* **55**:255–265.

Johnson, M. H., and Edidin, M., 1978, Lateral diffusion in plasma membrane of mouse egg is restricted after fertilisation, *Nature (London)* **272**:448–450.

Johnson, M. H., Eager, D. D., Muggleton-Harris, A. L., and Grave, M. H., 1975, Mosaicism in the organisation of concanavalin A receptors on surface membrane of mouse egg, *Nature (London)* **257**:321–322.

Johnson, M. H., Handyside, A. H., and Braude, P. R., 1977, Control mechanisms in early mammalian development, in: *Development in Mammals*, Vol. 2 (M. H. Johnson, ed.), pp. 67–97, Elsevier/North-Holland, Amsterdam.

Johnson, M. H., Chakraborty, J., Handyside, A. H., Willison, K., and Stern, P., 1979, The effect of prolonged decompaction on the development of the preimplantation mouse embryo, *J. Embryol. Exp. Morphol.* **54**:263–275.

Kaufman, M. H., Barton, S. C., and Surani, M. A. H., 1977, Normal postimplantation development of mouse parthenogenetic embryos to the forelimb bud stage, *Nature (London)* **265**:53–55.

Kemler, R., Babinet, C., Eisen, M., and Jacob, F., 1977, Surface antigen in early differentiation, *Proc. Natl. Acad. Sci. USA* **74**:4449–4452.

Kelly, S. J., 1977, Studies of the developmental potential of 4- and 8-cell stage mouse blastomeres, *J. Exp. Zool.* **200**:365–376.

Kelly, S. J., Mulnard, J. G., and Graham, C. F., 1978, Cell division and cell allocation in early mouse development, *J. Embryol. Exp. Morphol.* **48**:37–51.

Lo, C. W., and Gilula, N. B., 1979, Gap junctional communication in the preimplantation mouse embryo, *Cell* **18**:399–409.

McLaren, A., and Smith, R., 1977, Functional test of tight junctions in the mouse blastocyst, *Nature (London)* **267**:351–352.

Magnuson, T., Jacobson, J. B., and Stackpole, C. W., 1978, Relationship between intercellular permeability and junction organisation in the preimplantation mouse embryo, *Dev. Biol.* **67**:214–224.

Mazanec, K., and Dvorak, M., 1963, On the submicroscopical changes of the segmenting ovum in the albino rat, *Cesk. Morfol.* **11**:103–108.

Mintz, B., Experimental genetic mosaicism in the mouse, in: *Preimplantation Stages of Pregnancy* (G. W. Wolstenholme & M. O'Connor, eds.), pp. 194–207, Churchill, London.

Nicosia, S. V., Wolf, D. P., and Inoue, M., 1977, Cortical granule distribution and cell surface characteristics in mouse eggs, *Dev. Biol.* **57**:56–74.

Nicosia, S. V., Wolf, D. P., and Mastroianni, L., Jr., 1978, Surface topography of mouse eggs before and after insemination, *Gamete Res.* **1**:145–156.

Pratt, H. P. M., 1978, Lipids and transitions in embryos, in: *Development in Mammals*, Vol. 3 (M. H. Johnson, ed.), pp. 83–129, Elsevier/North-Holland, Amsterdam.

Pratt, H. P. M., Keith, J., and Chakraborty, J., 1980, Membrane sterols and the development of the preimplantation mouse embryo, *J. Embryol. Exp. Morphol.* (in press).

Reeve, W. J. R., 1981, Cytoplasmic polarity develops at compaction in rat and mouse embryos, *J. Embryol. Exp. Morph.* (in press).

Rossant, J., 1977, Cell commitment in early rodent development, in: *Development in Mammals*, Vol. 2 (M. H. Johnson, ed.), pp. 119–150, Elsevier/North-Holland, Amsterdam.

Surani, M. A. H., Barton, S. C., and Burling, A., 1980, Differentiation of 2-cell and 8-cell mouse embryos arrested by cytoskeletal inhibitors, *Exp. Cell Res.* **125**:275–286.

Tarkowski, A. K., and Wroblewska, J., 1967, Development of blastomeres of mouse eggs isolated at the 4- and 8-cell stage, *J. Embryol. Exp. Morphol.* **18**:155–180.

Van Blerkom, J., and Motta, P., 1979, *The Cellular Basis of Mammalian Reproduction*, Urban & Schwarzenberg, Munich.

Van Blerkom, J., and Runner, M. N., 1976, The fine structural development of preimplantation mouse parthenotes, *J. Exp. Zool.* **196**:13–123.

Wessells, N. K., Spooner, B. S., Ash, J. F., Bradley, M. O., Ludena, M. A., Taylor, E. L., Wrenn, J. J., and Yamada, K. M., 1971, Microfilaments in cellular and developmental processes, *Science* **171**:135–143.

Additional References

Johnson, M. H., and Zoimek, C. A., 1981, The foundation of two distinct cell lineages within the mouse morulae, *Cell* (in press).

Pratt, H. P. M., Chakraborty, J., and Surani, M. A. H., 1981, Molecular and morphological differentiation of the mouse blastocyst after manipulations of compaction using cytochalasin D, *Cell* (in press).
Reeve, W. J. D., and Ziomek, C. A., 1981, Distribution of microvilli on dissociated blastomeres from mouse embryos: evidence for surface at compaction, *J. Embryol. Exp. Morph.* (in press).
Ziomek, C. A., and Johnson, M. H., 1981, Cell surface interaction induces polarization of mouse 8-cell blastomeres at compaction, *Cell* (in press).

5

Relationship between the Programs for Implantation and Trophoblast Differentiation

Michael I. Sherman, Martin H. Sellens, Sui Bi Atienza-Samols, Anna C. Pai, and Joel Schindler

I. Introduction

Trophoblast cells are specialized cells that appear to be involved at one time or another during pregnancy in several functions essential to the well-being of the fetus. These functions include the attachment of the conceptus to the uterine wall, the acquisition of nutrients from the mother, and the protection of the fetus from immunologic rejection. As trophoblast cells are capable of metabolizing steroids (albeit to varying degrees and beginning at different times, depending upon the species), they might also play a role in fetal endocrinology. In studies on murine trophoblast, a number of biochemical properties have been observed that are exhibited neither by the early embryo per se nor by most other extraembryonic cell types (Table 1). We therefore consider these properties to be indicators of trophoblast differentiation, and we assume that some or all of these biochemical markers are involved in the specialized functions of these cells.

We have endeavored to determine the earliest time of appearance of markers of differentiation in trophoblast cells. This is difficult to do with mouse conceptuses developing continuously *in utero* as limited numbers of embryos are available and they are particularly difficult to obtain in the peri-implantation period because of their small size. Furthermore, trophoblast cells are tightly adherent to decidual cells and are bathed in maternal blood, both of which can contain enzymes similar or identical to some trophoblast markers (see Table 1). Indeed, trophoblast cells are capable of taking up enzymes in active form from the surrounding milieu (Sherman and Chew, 1972; Sherman, 1975). Because of these complications, we have turned to culture studies in our efforts to learn more about trophoblast differentiation. By inducing superovulation, we are able to obtain relatively large

Michael I. Sherman, Martin H. Sellens, Sui Bi Atienza-Samols, Anna C. Pai, and Joel Schindler • Roche Institute of Molecular Biology, Nutley, New Jersey 07110.

Table 1. Biochemical Markers of Differentiation of Mouse Trophoblast Cells

Trophoblast differentiation marker	Presence in other embryonic or extra-embryonic cells	Earliest time of observation *in utero*	Earliest time of observation *in vitro*[a]	References
Polyploidization	None	5th day	EGD 5	Barlow *et al.* (1972); Barlow and Sherman (1972); Sellens and Sherman (1980)
Plasminogen activator	Parietal endoderm[b]	Not determined[c]	EGD 5–6	Strickland *et al.* (1976); Jetten *et al.* (1979); Sherman (1981)
Δ^5,3β-Hydroxysteroid dehydrogenase	None	9th day	EGD 6	Deane *et al.* (1962); Chew and Sherman (1975); Sherman and Atienza (1977)
Esterase A[d]	Visceral endoderm	8th day[e]	EGD 7	Sherman (1972a, 1975); Sherman and Atienza-Samols (1979)
Alkaline phosphatase[f]	Yolk sac	8th day[e]	EGD 8	Sherman (1972b)

[a] In general, blastocysts removed from uteri on the fourth day of gestation were used as the starting material in culture studies. The earliest time of appearance of the markers *in vitro* is given in terms of the equivalent gestation day (EGD), i.e., the age that the embryo would have reached had it been left *in utero*. Thus, a marker detected on EGD 5 has appeared on the day after blastocyst culture was initiated.

[b] Other investigators (Bode and Dziadek, 1979) have claimed recently that embryonic and extraembryonic cells other than trophoblast and parietal endoderm possess plasminogen activator activity. We have been unable to confirm this report. It is, perhaps, possible that the activity observed in the tissue clumps used by Bode and Dziadek had been trapped from the fluids bathing the tissues (see, e.g., footnote *c*).

[c] Because serum possesses both plasminogen activator activity and inhibitors of this enzyme, we have felt that trophoblast cells, which are bathed in serum, cannot be assayed meaningfully for the production and secretion of this enzyme, upon being taken directly from the uterus.

[d] Esterase A is a nonspecific esterase isozyme characterized in the cited papers by its electrophoretic mobility.

[e] Assays for these enzymes were not carried out at earlier stages because adequate amounts of tissue could not be obtained and separated from decidual cells, which also possess these activities.

[f] Trophoblast, decidua, and yolk sac, but not embryo proper, possess an alkaline phosphatase activity that is extracted with butyric acid and can be characterized electrophoretically.

numbers of preimplantation mouse blastocysts as the starting material and, if need be, we can eliminate the concern of contaminating serum factors by the use of a serum-free medium (Rizzino and Sherman, 1979).

As Table 1 indicates, all of the trophoblast differentiation markers observed in studies on cells developing *in utero* can also be detected in trophoblast cells developing from blastocysts *in vitro*. The time of appearance of these markers *in vitro* is at least as early as in the uterus (for technical reasons, including the availability of assay material, we have been able to detect the production of several markers at earlier stages in culture than *in utero*). These studies, with markers that we have identified and for which we have sensitive assays, indicate that trophoblast differentiation begins no later than the fifth day of pregnancy. From the electrophoretic analyses of Johnson and his colleagues (Van Blerkom *et al.*, 1976; Johnson *et al.*, 1977; Handyside and Johnson, 1978), there is reason to suspect that trophoblast differentiation begins at earlier stages: Prior to implantation, trophectoderm cells and even earlier precursors of trophoblast synthesize polypeptides that are not produced by the other cells in the embryo. However, because the identity and function of these polypeptides are unknown, it is premature to consider them unequivocally as markers of trophoblast differentiation. Nevertheless, these studies have been important in establishing the concept of trophoblast development as a continuous and progressive program beginning as early as the time of inside–outside cell formation (Rossant and Papaioannou, 1977).

Implantation is a complex phenomenon during which both mother and conceptus must undergo changes. In preparation for this event, the mouse blastocyst sheds its zona pellucida. The surface trophectoderm cells then become adherent to the uterine wall as they convert to trophoblast cells. A period of invasiveness follows, during which trophoblast cells displace underlying maternal endometrial cells and establish a firm attachment site in the uterus. Little is known about the events at the molecular level in the blastocyst that are involved in implantation per se. However, because the blastocyst undergoes events *in vitro* that are analogous to implantation *in utero* (Sherman and Wudl, 1976; Shalgi and Sherman, 1979), we have attempted to learn more about the molecular aspects of implantation by studies on cultured embryos. In the following sections, we consider how information gained from *in vitro* investigations bears upon changes that must occur in the blastocyst as implantation proceeds.

II. What Is the Nature of Blastocyst Factors Involved in Implantation?

Most of the events analogous to implantation *in utero* occur *in vitro* only under certain conditions. Blastocysts can hatch from the zona pellucida at high frequency in nutritionally simple preimplantation media lacking amino acids and containing albumin as the only source of protein (Whitten, 1971; Sellens and Sherman, 1980; Table 2). The rate of hatching is delayed, however, by 10 hr or more in these media compared to more complex nutritive media containing serum, in our case NCTC-109 supplemented with 10% heat-inactivated fetal calf serum (cNCTC; Sellens and Sherman, 1980).

In order for blastocysts to attach to the substratum and give rise to trophoblast outgrowths, both the exogenous source of amino acids and an appropriate macromolecular fraction are required (Cole and Paul, 1965; Gwatkin, 1966a,b; Jenkinson and Wilson, 1973;

Table 2. Effect of Culture Conditions upon Postblastocyst Development [a]

Culture medium[b]	Hatching	Attachment	Outgrowth	Polyploidization	Plasminogen activator	Δ^5,3β-Hydroxysteroid dehydrogenase
cNCTC	Yes	Yes	Yes	Yes	Yes	Yes
cNCTC/agarose	Yes	No	No	Yes	Yes	Yes
PCM + AVF	Yes	Delayed	Delayed	Delayed	Yes	Very low; delayed
PCM + AV	Delayed	No	No	Delayed	Yes	Very low; delayed
PCM + F	Delayed	Delayed	No	Delayed	Yes	No
PCM	Delayed	Delayed; transient	No	No	Yes	No

[a] These data and the methods involved are presented in detail by Sellens and Sherman (1980). Embryos were placed into culture at the blastocyst stage, on the fourth day of gestation. Generally, enzyme activities and the degree of polyploidization in the suboptimal media were lower than values observed in cNCTC, the optimal culture medium in these experiments (see below); values indicated as "very low" were just above detection limits. If the average time for hatching, attachment, or outgrowth of embryos in any of the suboptimal media was delayed by 10 hr or more compared to that of embryos in cNCTC medium, this is so noted in the table. Similarly, the expression of biochemical markers by blastocysts in suboptimal media was considered to be delayed if these markers were first detected one or more days after initial observation in blastocysts in cNCTC medium.

[b] cNCTC medium is NCTC-109 medium supplemented with antibiotics and 10% heat-inactivated fetal calf serum. Embryos were also cultured in this medium on pads of 2% agarose, which provides a substratum precluding adhesion and outgrowth (Sherman, 1978). PCM is the preimplantation culture medium described by Goldstein *et al.* (1975). This medium was supplemented where indicated with amino acids (A; Spindle and Pedersen, 1973), vitamins (V; Spindle and Pedersen, 1973), and/or fetuin (F; Rizzino and Sherman, 1979), an α-globulin extracted from serum.

Spindle and Pedersen, 1973; Sherman and Salomon, 1975; Rizzino and Sherman, 1979; Sellens and Sherman, 1980). The macromolecular requirement can be satisfied by whole serum (fetal calf serum is usually used) or by crude preparations of collagen (Jenkinson and Wilson, 1973; Spindle and Pedersen, 1973) or fetuin (Gwatkin, 1966a; Rizzino and Sherman, 1979), and serves primarily for conditioning of the substratum, as opposed to acting intracellularly (Jenkinson and Wilson, 1973; Rizzino and Sherman, 1979). In the absence of appropriate macromolecular fractions, transient blastocyst attachment to the substratum might occur, but outgrowth will not take place (Table 2; Sellens and Sherman, 1980).

There is a hiatus *in vitro* between shedding of the zona and blastocyst attachment to the substratum (Sherman and Salomon, 1975; Rizzino and Sherman, 1979; Sellens and Sherman, 1980). Sherman and Atienza-Samols (1978) reported that premature removal of the zona does not notably accelerate blastocyst attachment. These observations led to the conclusion that at some time after hatching there is a programmed change in the blastocyst surface from nonadhesive to adhesive (Sherman and Wudl, 1976). This change occurs independently of the presence of a suitable substratum (Sherman, 1978; Sherman and Atienza-Samols, 1978). As cycloheximide treatment delays the acquisition of surface adhesiveness (K. Matthaei, unpublished observations), one might propose that this phenomenon requires the synthesis of new proteins and their insertion into the surface membrane. However, the existence and nature of these hypothetical adhesive proteins on the blastocyst surface have yet to be established.

A report that extended periods of exposure to trypsin could inhibit blastocyst attachment to glass in a reversible manner also suggests that certain proteins involved in adhesiveness are present on the blastocyst surface (Glass *et al.*, 1979). These effects were not observed, however, in similar investigations carried out in this laboratory (Sherman and Atienza-Samols, 1978). On the other hand, in the latter studies, we observed that collagenase treatment of blastocysts could transiently block attachment (Sherman and Atienza-Samols, 1978). The further observation that certain types of collagen were associated with blastocysts (Sherman *et al.*, 1979) led us to propose the participation of these proteins in blastocyst adhesiveness. Although more extensive experiments have revealed the association of collagens long before embryos become adhesive (Sherman *et al.*, 1980), it is possible that collagens help to promote adhesiveness by undergoing changes in conformation, organization, or localization. Alternatively, these collagens could effect adhesiveness by interacting with other macromolecules, e.g., fibronectin or fibronectinlike proteins, that are produced just prior to or during the adhesive period.

It appears as though the surface glycoprotein composition of mouse embryos changes through pre- and peri-implantation stages (Sherman, 1979). Because surface glycoprotein alterations have been correlated with profound behavioral changes by various cell types (Sherman and Atienza-Samols, 1978), we and others considered that such changes are responsible for the acquisition of blastocyst adhesiveness. In initial studies aimed at altering the oligosaccharide composition of blastocyst surface glycoproteins, we failed to detect any involvement of the carbohydrate moiety of glycoproteins in blastocyst attachment (Sherman and Atienza-Samols, 1978). Some of these experiments were repeated by Glass *et al.* (1979) with similar results. On the other hand, Glass *et al.* (1979) reported that diazo-oxo-norleucine treatments had adverse effects upon blastocyst attachment and trophoblast outgrowth. Because this antimetabolite interferes with the glycosylation of proteins, Glass *et al.* (1979) raised the possibility that glycoproteins were involved in implantation-related events. However, diazo-oxo-norleucine can disrupt cellular metabolism at several other levels (Greene and Pratt, 1977).

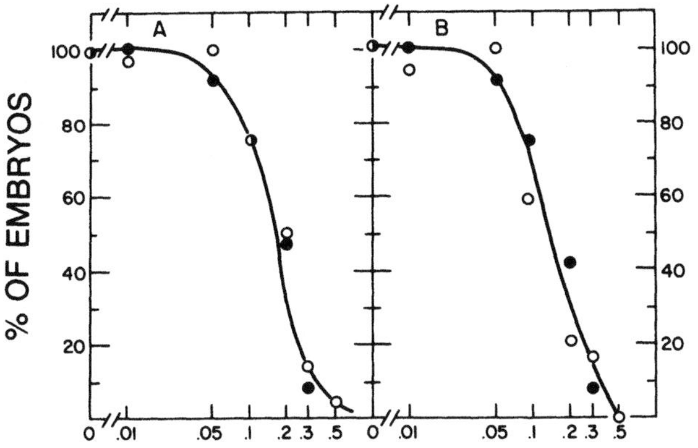

Figure 1. Effect of tunicamycin upon development of eight-cell embryos. Embryos were removed from mice on the third day of gestation and in some cases were treated with Pronase to remove the zona pellucida (Mintz, 1962). Embryos with (O) or without (●) zonas were placed into preimplantation culture medium (Goldstein *et al.*, 1975) containing the indicated concentrations of tunicamycin. After 48 hr, the embryos were scored to determine the percentage that reached the blastocyst stage (A). The embryos were then transferred to cNCTC medium containing the same concentrations of tunicamycin and scored for hatching, attachment, and outgrowth (B). All embryos that were capable of hatching also attached and gave rise to trophoblast outgrowths. Thus, the curve in B describes all three events. The curves in A and B are normalized to 100% for untreated embryos. The actual percentage of untreated embryos reaching blastocyst and outgrowth stages was 94 and 90%, respectively.

We have used another antimetabolite, tunicamycin (Takatsuki and Tamura, 1971), to interfere with protein glycosylation in pre- and peri-implantation mouse embryos. This antimetabolite blocks the condensation of core oligosaccharides onto asparaginyl residues of proteins by preventing the synthesis of the *N*-acetylglucosaminyl pyrophosphoryl polyisoprenol intermediate (Takatsuki *et al.*, 1975; Tkacz and Lampen, 1975; Struck and Lennarz, 1977). In some cells the effect of tunicamycin on protein glycosylation is relatively specific, whereas in others the presence of the antimetabolite leads also to inhibition of protein synthesis and of sulfation of glycosaminoglycans (Duksin and Bornstein, 1977; Damsky *et al.*, 1979; Olden *et al.*, 1979; Pratt *et al.*, 1979.

Tunicamycin is toxic to preimplantation embryos at concentrations as low as those observed for the most sensitive cell types (Olden *et al.*, 1979; Atienza-Samols *et al.*, 1980). In order to determine whether there is a specific effect of this antimetabolite on implantation-related events, we plotted dose–response curves of embryos placed into culture on the second, third, and fourth days of pregnancy.* We observed that when two-cell (second-day) embryos were treated continuously with tunicamycin, a concentration of 0.025 μg/ml was adequate to reduce by 50% the proportion of embryos reaching the hatched blastocyst stage. This 50%-effective dose (ED_{50}) was found also to apply to attachment and outgrowth, i.e., if the embryo developed to the hatched blastocyst stage in the presence of the drug, it would invariably also become adhesive and give rise to trophoblast outgrowths (Atienza-Samols *et al.*, 1980). Third-day (four- to eight-cell) embryos could also be prevented from developing normally by continuous exposure to tunicamycin (Figure 1), but the ED_{50} for blastocyst formation, hatching, attachment, and trophoblast outgrowth (approximately 0.15 μg/ml in every case; cf. Figure 1A,B) was substantially higher than that

* The day of observation of the sperm plug is considered the first day of pregnancy. The term equivalent gestation day (EGD) refers to the age of cultured embryos had they been left *in utero*.

for two-cell embryos. Premature removal of the zona pellucida did not influence the ED_{50} values (Figure 1). Tunicamycin at concentrations up to 0.5 μg/ml had no effect upon hatching, attachment, or outgrowth when treatment was begun on the fourth day (late morula to midblastocyst). However, tunicamycin at concentrations only one-tenth as high had profound effects upon trophoblast development subsequent to the initiation of outgrowth (Atienza-Samols *et al.*, 1980).

Despite efforts undertaken so far to identify the surface components responsible for blastocyst adhesiveness and trophoblast outgrowth, it is premature to draw any firm conclusions concerning the nature of the molecules or macromolecules involved. Although a requirement for exogenous amino acids implies a role for new proteins, their participation might be secondary, e.g., as enzymes required to synthesize phospholipids or glycosaminoglycans. As mentioned above, a specific role of collagens is brought into question by their association with embryos long before the onset of adhesiveness. Finally, none of our experiments directly supports a role for glycoproteins. Our investigations with tunicamycin suggest that protein glycosylation is necessary for blastocyst formation. However, we have been unable to find a dosage of the antimetabolite that will specifically block blastocyst adhesion and trophoblast outgrowth, even if we obviate the need for hatching by removing the zona manually (Figure 1). In any event, results with antimetabolites such as tunicamycin and diazo-oxo-norleucine must be interpreted with caution because of the side effects noted above.

The possible role of factors other than surface proteins in implantation should not be ignored. Surface membranes contain components other than proteins, such as glycosaminoglycans and lipids. We have observed changes in the pattern of glycosaminoglycan synthesis through peri-implantation stages (Cantor *et al.*, 1976), but we do not know what relationship these changes bear to implantation per se. Lipids appear to play an important role in behavior and interaction of other cell types (Prives and Shinitzky, 1977), and nothing is presently known about the lipid composition of surface membranes of the early mammalian embryo. The possibility also exists that events unrelated to changes in the surface membrane must occur in the blastocyst in preparation for implantation. One indication that this might be the case is the observation that blastocysts cultured in preimplantation medium supplemented with an adequate supply of exogenous amino acids and macromolecular factors suitable to promote adhesion are, nevertheless, substantially delayed in attaching to the substratum and giving rise to trophoblast outgrowths (Table 2; Sellens and Sherman, 1980).

III. What Is the Relationship between Implantation and Other Differentiative Events?

If several events are to occur in the elaboration of a particular differentiative program, it is conceivable that the occurrence of any of these events must await the expression of all of the previous ones in that program (Figure 2a). Alternatively, the program might consist of two or more independent pathways that together lead to the final differentiative phenotype (e.g., Figure 2b). A third possibility is that the differentiative program involves multiple pathways with varying degrees of interaction (e.g., Figure 2c).

We have, over the years, carried out a series of experiments designed to assess the interrelationship between implantation and other events in the differentiative program for trophoblast. In the earliest of these studies, we observed that blastocysts that fail to hatch from the zona but remain viable are capable of giving rise to differentiated trophoblast cells

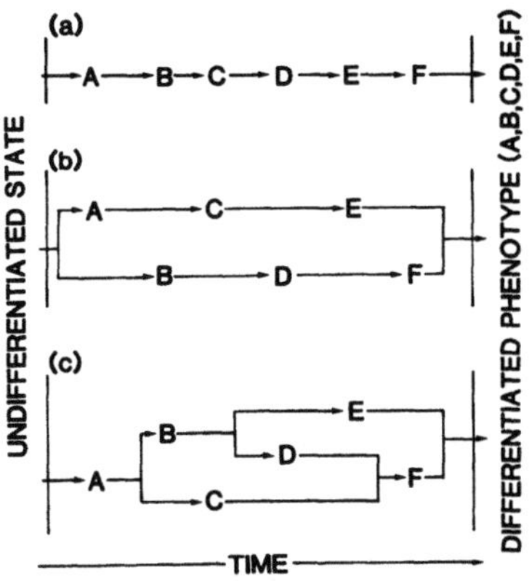

Figure 2. Alternative schemes describing the differentiative program of mouse trophoblast cells. In these schemes, we have hypothesized that specific gene products, identified as A to F, must be expressed in the conversion of undifferentiated progenitor cells (presumably at cleavage stages) to cells with a mature trophoblast phenotype. In (a), the expression of each gene depends upon the prior expression of the previous genes in the pathway. In (b), there are independent pathways for various processes leading to the final phenotype. In (c), there is some, but not complete, interdependence among the gene products. Thus, gene product F requires for its expression the prior appearance of gene products A, B, C, and D, but not E. Similarly, gene product D requires the prior expression of gene products A and B but not C.

as determined by production of trophoblast-specific alkaline phosphatase and esterase and by polyploidization (Barlow and Sherman, 1972; Sherman, 1972a,b). More recently, we have demonstrated that blastocysts prevented from adhering to the substratum by the use of agarose-coated dishes also produce plasminogen activator and Δ^5,3β-hydroxysteroid dehydrogenase activities (Table 2; Sellens and Sherman, 1980).

By culturing blastocysts in suboptimal media, we have been able to interfere with the expression of various differentiative markers (Table 2). Under these circumstances, we again observe no obvious interdependence between implantation and the appearance of the other markers (e.g., the biochemical markers are expressed equally well in supplemented preimplantation medium with and without fetuin, despite the fact that attachment and outgrowth occur in the former medium but not in the latter). Indeed, there is divergence among the biochemical markers themselves: Whereas plasminogen activator is produced even in the least nutritive medium, polyploidization and Δ^5,3β-hydroxysteroid dehydrogenase activity appear to be more susceptible to suboptimal metabolic conditions.

There is one marker whose appearance has been correlated with implantation *in utero* or attachment to a substratum *in vitro,* and that is the production of lactate dehydrogenase A subunits (Auerbach and Brinster, 1967, 1968; Monk and Petzholdt, 1977). However, Spielmann *et al.* (1978) found that blastocysts failing to hatch did produce A subunits. On the basis of these reports and our own work (Sellens and Sherman, 1980), we believe that lactate dehydrogenase A subunits are produced in response to a cellular interaction, but one that is not necessarily related to implantation; in fact, it appears as though even the contact of ICM cells with the inner trophoblast surface, as occurs during blastocoel collapse in free-floating embryos, is adequate to trigger A subunit production (Monk and Petzholdt, 1977; Sellens and Sherman, 1980).

In all the above studies, we have examined the effects of interfering with implantation-related events upon the expression of biochemical markers of trophoblast differentiation. The reciprocal experiments have proven to be far more difficult to carry out and interpret because they usually involve the use of antimetabolites that affect the overall well-being of the embryo. However, from preliminary experiments with DNA polymerase inhibitors, we suspect that we can interrupt polyploidization of trophoblast cells without interfering with blastocyst attachment and trophoblast outgrowth (P. Pine and M. Sherman, unpublished observations).

On the basis of the experiments described in this section, we can eliminate pathway (a) in Figure 2 as the differentiative program for trophoblast cells because prevention of implantation-related events does not preclude the expression of later markers. The scheme in Figure 2b seems, at present, to define the program best (i.e., the program for implantation-related events, A→C→E, is carried out independently of the production of other differentiation markers, B→D→F); however, it is possible that when we have achieved a better understanding of the molecular events governing implantation, the interactive scenario in Figure 2c may prove to be a closer approximation.

IV. When Are Implantation-Related Gene Products Produced?

We have observed in several independent studies (Sherman and Salomon, 1975; Rizzino and Sherman, 1979; Sellens and Sherman, 1980) that blastocysts placed in culture on the fourth day of gestation become adherent to the substratum and give rise to trophoblast outgrowths on equivalent gestation day (EGD) 6. This fact alone does not provide any information concerning the time at which the gene products required for these events are produced. We do have results, however, from several different kinds of experiments that bear upon this issue. The first, which has already been discussed, is that blastocysts deprived of an exogenous source of amino acids from the fourth day are delayed from acquiring adhesive properties until EGD 7 and fail to give rise to trophoblast outgrowths (Table 2; Sellens and Sherman, 1980). Presumably, then, the synthesis of some protein(s) beyond the fourth day is required for the occurrence of these implantation-related events. On the other hand, as we have noted above, fourth-day blastocysts exposed to relatively large doses of tunicamycin are not impeded from attaching to the substratum or initiating trophoblast outgrowth at the expected time. Therefore, if normally glycosylated glycoproteins are required for attachment and outgrowth, these glycoproteins must have been synthesized prior to the blastocyst stage.

Another series of experiments, the results of which we believe relate to the time of programming of implantation-related genes, involves t^6-mutant embryos. Embryos that are homozygous for this recessive mutation fail to develop beyond the early egg cylinder stage *in utero*. *In vitro*, blastocysts hatch, attach, and outgrow; however, the ICM degenerates by EGD 6 and trophoblast cells fail to polyploidize normally thereafter (Erickson and Pedersen, 1975; Wudl *et al.*, 1977; Wudl and Sherman, 1978; for a general review of *t* mutations, see Sherman and Wudl, 1977). When a population of blastocysts containing a mixture of normal ($+/+$ or $+/+^6$) and mutant (t^6/t^6) embryos is cultured from the fourth day, the embryos give rise to trophoblast outgrowths as a single population, i.e., there is no observable effect of t^6 homozygosity upon the timing of this implantation-related event (Table 3, Experiment 1). The same can be said of fourth-day blastocysts delayed from outgrowing by culture in fetuin-supplemented preimplantation medium (PCM + F) for several days prior to being placed in cNCTC medium, which supports outgrowth (Table 3, Experiment 2). On the other hand, if third-day embryos are removed from the genital tract and placed into PCM + F for several days, there is a clear distinction between normal and mutant embryos with regard to the time of outgrowth after transfer to cNCTC medium (Table 3, Experiment 3). The interpretation of these results is complicated by the pleiotropic nature of *t* mutations (Sherman and Wudl, 1977). We believe that the data are most easily explained by the proposal that the program for trophoblast outgrowth includes the expression by mouse embryos of one or more gene products between the third and the fourth days of gestation followed by one or more gene products between EGD 4 and EGD 6. We shall suppose for the sake of simplicity that single gene products are involved in each case, X being synthesized between the third and the fourth days and Y being synthesized between EGD 4 and EGD 6.

The results could then be explained as follows: In Experiment 1 (Table 3), both nor-

Table 3. Time of Outgrowth of Trophoblast Cells from Normal and t^6/t^6-Mutant Embryos [a]

Experiment	Initial age of embryos (days)	Days in delay medium	T_{50} for outgrowth (hr)		ΔT_{50}
			+/+ and $+/t^6$ embryos	t^6/t^6 embryos	
1	4	0	38.0	37.0	−1.0
2	4	4	31.0	34.0	+3.0
3	3	4	31.5	46.0	+14.5

[a] In these experiments embryos were obtained from $(+/t^6)$ mice mated *inter se*. They were removed from the genital tract either on the third or on the fourth day of gestation, as indicated in the second column, and placed either into cNCTC medium (which permits outgrowth) in Experiment 1 or preimplantation culture medium (Goldstein *et al.*, 1975) supplemented with fetuin (this medium does not permit outgrowth as indicated in Table 2) in Experiments 2 and 3. After 4 days, embryos in Experiments 2 and 3 were transferred to cNCTC medium. Outgrowth was considered to have begun when at least two cells and their nuclei could be seen flattened against the substratum. Blastocysts could be phenotyped either as normal (+/+ or $+/t^6$) or mutant (t^6/t^6) within 4 days of outgrowth by the state of the ICM and the size of trophoblast nuclei (Sherman and Wudl, 1977; Wudl and Sherman, 1978). T_{50} refers to the time at which 50% of the embryos had begun outgrowth *from the time of placement into cNCTC medium.* ΔT_{50} is calculated as (T_{50} for mutant embryos) − (T_{50} for normal embryos). The total number of embryos scored in each experiment is: 1, 64; 2, 79; and 3, 181.

mal and mutant embryos collected on the fourth day have already synthesized X and upon culture in cNCTC, both types of embryos produce Y, leading to the initiation of outgrowth. In Experiment 2, both mutant and normal embryos will have synthesized X at the time of collection, but will be unable to synthesize Y in PCM + F. Accordingly, production of Y will occur only (simultaneously in normal and mutant embryos) upon transfer to cNCTC medium. In Experiment 3, embryos will not have produced X at the time they are collected and placed in PCM + F. Because of the nature of the t^6 mutation, t^6/t^6 embryos will not be able to synthesize X in this medium, whereas this gene product will be produced by +/+ and $+/t^6$ embryos. Thus, upon transfer to cNCTC, normal embryos need only produce Y in order for outgrowth to occur, and the time required for outgrowth will then be the same as that for the embryos in Experiment 2 (which had already produced X prior to being placed into culture). On the other hand, t^6/t^6 embryos would have to produce both X and Y upon transfer into cNCTC, and this would account for their delay in initiating outgrowth.

If the scheme presented above is correct, then product Y cannot be produced until product X has been synthesized. It is possible that X and Y are discrete gene products in the outgrowth program. However, two other alternatives are feasible: X might be a precursor of Y, or X might be a mRNA and Y its translational product. In order to investigate the latter alternative, we cultured embryos continuously from the fourth day in the presence of α-amanitin, an antimetabolite that inhibits RNA polymerase II activity, and thus prevents transcription of mRNA (Stirpe and Fiume, 1967; Jacob *et al.*, 1970; Lindell *et al.*, 1970). We observed that even in the presence of α-amanitin at a concentration of 1 μg/ml, blastocysts hatch, attach, and grow out along the dish at the same time as untreated controls (Figure 3). Continuous treatment with α-amanitin at this concentration should be adequate to inhibit virtually all polymerase II activity (Levey and Brinster, 1978): in fact, it causes a severe depression in the rate of synthesis of total RNA over the 72-hr period under study (Schindler and Sherman, submitted for publication). Thus, these observations are consistent with the view that mRNAs critical for hatching, attachment, and outgrowth are synthesized prior to the fourth day of gestation.

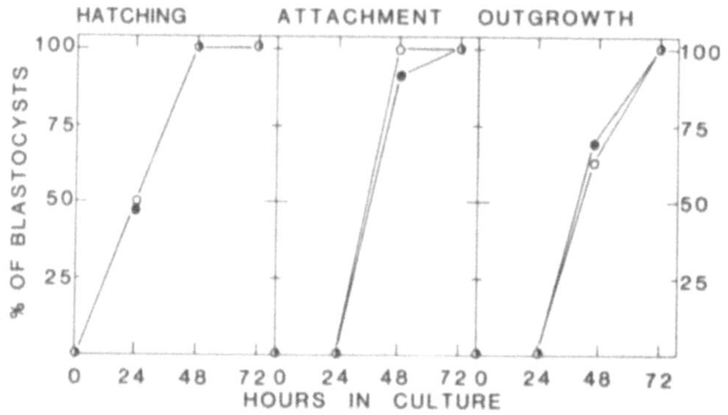

Figure 3. Behavior of blastocysts treated with α-amanitin. Fourth-day blastocysts were placed in cNCTC medium with (○) and without (●) 1 μg/ml α-amanitin. Blastocysts were scored daily for hatching, attachment to the substratum, and outgrowth of trophoblast cells.

V. Conclusions and Speculations

Implantation is a cooperative event that involves mother and conceptus and requires careful timing and a proper degree of interaction. It is likely that numerous molecular changes occur in the mother and in the conceptus to prepare both for implantation. Even when the peri-implantation blastocyst is studied in isolation from maternal factors, it is difficult to define, at the molecular level, events that are directly related to implantation, in part because the implantation process is temporally superimposed upon other differentiative events that are taking place in developing trophoblast cells. Whereas it is likely that the appearance of certain proteins, perhaps some of them glycoproteins, on the trophoblast surface is responsible for the acquisition of adhesiveness and contributes to the property of invasiveness, we do not as yet have direct evidence for the existence of such proteins. Indeed, only recently have we begun to learn in a general way about the molecular mechanisms involved in adhesiveness, migration, and invasiveness of cells.

Although we envisage implantation as an integral part of the overall differentiative program of trophoblast cells, it is not immediately evident how the internal controls over implantation-related processes relate to those governing the appearance of the other trophoblast markers that we have studied. The program for implantation might be relatively independent of that for other trophoblast characteristics, either because it is initiated at an earlier time or because it is subject to control by different factors. It is tempting to speculate from the experiments we have described that the program for implantation begins long before the event itself and that the necessary mRNAs are transcribed before the embryo reaches the blastocyst stage. Studies by Johnson *et al.* (1977) with α-amanitin have indicated that there is in fact a population of trophoblast-specific proteins whose mRNAs appear to be transcribed prior to blastocyst formation.

ACKNOWLEDGMENTS. Tunicamycin, originally prepared by Dr. G. Tamura, was obtained from the Drug Evaluation Branch of the National Cancer Institute via Dr. Robert Pratt, National Institute of Dental Research. We wish to thank Drs. J. Monahan and H. Ennis for comments on the manuscript.

Discussion

BIGGERS (Harvard): You showed dose–response curves of inhibition by tunicamycin of blastocyst formation and hatching from the zona pellucida . . .

SHERMAN: . . . hatching, attachment, and outgrowth. If hatching occurs, attachment and outgrowth will inevitably follow under these conditions with these treatments.

BIGGERS: Does the inhibitor prevent cavity formation?

SHERMAN: Yes, the concentration of 0.15 μg/ml from the eight-cell stage will block formation of blastocysts in half the embryos.

BIGGERS: Perhaps the inhibitor interferes with compaction.

SHERMAN: Tunicamycin can block compaction but only at high doses.

BIGGERS: Compaction is connected with a constellation of events concerned with establishment of the blastocoele cavity. The hatching phenomenon, on the other hand, has to do with expansion and escape from the zona pellucida, which seems to be a later event that is contingent on the earlier one. I have a bit of trouble relating one set of results to the other.

SHERMAN: The experiment shows that the concentration of tunicamycin needed to block cavitation is the same as that needed to block hatching. In other words, we can block blastocyst formation at a lower concentration of tunicamycin than needed to block compaction, but we can't do that for hatching as opposed to blastocyst formation.

BIGGERS: If the inhibitor blocks the compaction stage, how did you get embryos to test for hatching?

SHERMAN: We used the ones that were not blocked. At concentrations of tunicamycin below 0.3 μg/ml, some embryos in the population will undergo compaction and cavitation. These embryos also hatch, attach, and outgrow.

BIGGERS: To test the inhibitor on hatching then, you used a sample of embryos that is insensitive to the inhibitor.

SHERMAN: That's right, and none of those that were insensitive becomes sensitive at a later stage.

BIGGERS: You thus have a highly selected set of blastocysts for determining the dose–response for hatching.

SHERMAN: If you wish.

ROBERTS (Florida): Although tunicamycin is a very potent inhibitor of glycosylation, it also inhibits protein synthesis to some extent. Invariably to reduce glycosylation by something like 90% at least a 30% inhibition of protein synthesis occurs. Second, tunicamycin often causes intracellular accumulation and lack of movement of molecules to the surface. The effect you observe could be a lack of polypeptides at the surface rather than carbohydrate groups per se.

SHERMAN: We have measured these parameters. The trophoblast cells in particular are extremely sensitive to tunicamycin in regard to their ability to incorporate amino acids into protein; they behave like transformed cells. In fact, polypeptide synthesis is inhibited not by 30% but more like 60 to 70%. The inner cell mass cells are relatively resistant to the tunicamycin; protein synthesis in these cells is only inhibited by 20 to 30%. Secretion of protein by blastocysts is suppressed by tunicamycin, primarily because of its selective effect on trophoblast cells, which normally secrete more of their protein into the medium than do inner cell mass cells.

ROBERTS: In that case, is it valid to draw any conclusion about glycosylation?

SHERMAN: Yes, I think it is valid to draw conclusions if they are negative conclusions. If tunicamycin had selectively blocked hatching, attachment, or outgrowth, we would not know whether this was a primary effect on glycosylation or a secondary inhibition of protein synthesis. The fact that we do not see a block of these developmental stages leads me to feel that one can make a valid interpretation of these results.

ROBERTS: When the conceptus attaches to the monolayer of epithelium, is that effect specific for the mass of uterine epithelial cells? Is it valid to draw conclusions about a cell monolayer of a dividing line which is presumably not at an equivalent stage of differentiation to the intact uterus?

SHERMAN: I can answer both questions at once by saying that you can use any cell types you wish and the trophoblast will attach to, and ultimately displace, the cells.

SURANI (Cambridge): We have quite extensive data on the effect of tunicamycin (Surani, 1979, *Cell* **18**:217) and I agree with you that different stages of embryos respond differently to the drug. I must point out that the amount of drug you are using does not cause complete inhibition of glycosylation.

SHERMAN: That is correct, 85% approximately.

SURANI: I think it is much less than that.

SHERMAN: That depends on the preparation of tunicamycin.

SURANI: With your concentration of tunicamycin, we find incorporation of mannose is inhibited by only about 30%.

SHERMAN: It is 85% with the tunicamycin preparation we used.

SURANI: In our studies on four-day blastocysts with tunicamycin at 1 μg/ml, the incorporation of leucine is inhibited by about 17% and that of mannose by about 80%. Under these conditions the effect of the drug is entirely reversible so that, after washing and placing in fresh medium, the embryos will show trophoblast adhesion and giant cell outgrowth.

SHERMAN: How long is your exposure?

SURANI: The exposure is about 24 hr.

SHERMAN: What was your source of tunicamycin?

SURANI: Eli Lilly.

SHERMAN: Our tunicamycin was prepared by Tamura, using his original method (Takatsuki *et al.*, 1971, *J. Antibiot.* **24**:215). An exposure to 0.5 μg/ml of this preparation for 24 hr is somewhat reversible at the blastocyst stage.

SURANI: Concerning the translocation of unglycosylated polypeptide to the cell surface, we have done cell surface iodination experiments and found that unglycosylated glycoproteins are translocated normally to the cell surface; this would add weight to the argument that the oligosaccharide moieties are probably important in adhesion and outgrowth of trophectoderm.

SHERMAN: How does this add weight to the importance of the oligosaccharide portion?

SURANI: Because the polypeptides which were translocated to the cell surface were presumably free of carbohydrate moieties.

SHERMAN: Did attachment occur?

SURANI: Under the conditions where the antibiotic is present, attachment does not occur.

SHERMAN: I see. That is very different from our results.

BEIER (Aachen): In the control experiments on protein synthesis, you also saw the secretion of proteins from the trophoblast and from the inner cell mass cells. There has not been too much described in the literature to my knowledge about secretion by inner cell mass.

SHERMAN: Both trophoblast cells and inner cell mass cells secrete proteins into the medium. The latter cells, however, secrete a much smaller proportion of their newly synthesized protein than do the former.

DICKMANN (Kansas): You used hatching as an end point and presumably you extrapolate to the situation *in vivo*. I feel that your conclusions are invalid, because there is a basic difference between hatching *in vitro* and hatching *in vivo*.

SHERMAN: I did not mean to give the impression that I was using hatching as an end point. One cannot get attachment and outgrowth without getting hatching. Hatching is not an end point but a starting point for implantation and later events. The tunicamycin data I presented refer to hatching, attachment, and outgrowth because all these events are blocked by the same concentration of the antimetabolite. I prefer not to take issue with you about whether hatching is the same or different *in vivo* or *in vitro*.

BIGGERS: I would like to rise to the defense of the study of hatching. It might not be relevant to what happens *in vivo* but, as in *in vitro* assy, it is a very good system for measuring fluid formation as far as blastocoele fluid is concerned. The argument about whether it is relevant *in vivo* or not is not important as an end point for fluid movement.

C. WARNER (Iowa State): You showed that treatment of blastocysts with α-amanitin allowed hatching, attachment, and outgrowth, then they died. The interpretation that messenger RNA synthesis is important requires caution. We showed a couple of years ago that α-amanitin inhibits DNA synthesis as well as RNA synthesis, possibly by inhibiting an RNA primer of the DNA synthesis (Warner and Hearn, 1977, *Differentiation* **7**:89). Have you taken this possibility into account?

SHERMAN: Again, my answer is that if α-amanitin had blocked hatching, attachment, or outgrowth, I would be very worried about the interpretation. We are aware of your work and that of others showing that there are other effects of α-amanitin. Because we do not see a block to hatching, attachment, and outgrowth, regardless of the side effects of α-amanitin, we conclude there is no evidence of participation of newly synthesized messenger RNA. I did not say that our results *indicated* a role for preformed messages in the event of implantation, I said they were consistent with that possibility.

WARNER: But the embryos may die because of inhibition of DNA synthesis.

SHERMAN: At this point, we are not so much concerned by the amanitin-induced embryo death as we are about the inability of amanitin to block hatching, attachment, and outgrowth; that is the important part of the experiment for us.

VAN BLERKOM (Colorado): Have you tried to transfer the blastocysts, that were exposed to α-amanitin, into a uterus, to see if they undergo any of the changes associated with implantation *in vivo*?

SHERMAN: No, we have not done that yet.

References

Atienza-Samols, S. B., Pine, P., and Sherman, M. I., 1980, Effects of tunicamycin upon glycoprotein synthesis and development of early mouse embryos, *Dev. Biol.* **79**:19.

Auerbach, S., and Brinster, R. L., 1967, Lactate dehydrogenase isozymes in the early mouse embryo, *Exp. Cell Res.* **46**:89.

Auerbach, S., and Brinster, R. L., 1968, Lactate dehydrogenase isozymes in mouse blastocyst cultures, *Exp. Cell Res.* **53**:313.

Barlow, P. W., and Sherman, M. I., 1972, The biochemistry of differentiation of mouse trophoblast cells: Studies on polyploidy, *J. Embryol. Exp. Morphol.* **27**:447.

Barlow, P. W., Owen, D. A. J., and Graham, C., 1972, DNA synthesis in the preimplantation mouse embryo, *J. Embryol. Exp. Morphol.* **27**:431.

Bode, V. C., and Dziadek, M. A., 1979, Plasminogen activator secretion during mammalian embryogenesis, *Dev. Biol.* **73**:272.

Cantor, J., Shapiro, S. S., and Sherman, M. I., 1976, Chondroitin sulfate synthesis by mouse embryonic, extraembryonic and teratoma cells *in vitro*, *Dev. Biol.* **50**:367.

Chew, N. J., and Sherman, M. I., 1975, Biochemistry of differentiation of mouse trophoblast: Δ^5,3β-hydroxysteroid dehydrogenase, *Biol. Reprod.* **12**:351.

Cole, R. J., and Paul, J., 1965, Properties of cultured preimplantation mouse and rabbit embryos, and cell strains derived from them, in: *Preimplantation Stages in Pregnancy* (G. E. W. Wolstenholme and M. O'Connor, eds.), pp. 82–112, Academic Press, New York.

Damsky, C. H., Levy-Benshimol, A., Buck, C. A., and Warren, L., 1979, Effect of tunicamycin on the synthesis, intracellular transport and shedding of membrane glycoproteins in BHK cells, *Exp. Cell Res.* **119:**1.

Deane, H. W., Rubin, B. L., Driks, E. C., Lobel, B. L., and Leipsner, G., 1962, Trophoblastic giant cells in placentas of rats and mice and their probable role in steroid-hormone production, *Endocrinology* **70:**407.

Duksin, D., and Bornstein, P., 1977, Changes in surface properties of normal and transformed cells caused by tunicamycin, an inhibitor of protein glycosylation, *Proc. Natl. Acad. Sci. USA* **74:**3433.

Erickson, R. P., and Pedersen, R. A., 1975, *In vitro* development of t^6/t^6 embryos, *J. Exp. Zool.* **193:**377.

Glass, R. H., Spindle, A. I., and Pedersen, R. A., 1979, Mouse embryo attachment to substratum and interaction of trophoblast with cultured cells, *J. Exp. Zool.* **208:**327.

Goldstein, L. S., Spindle, A. I., and Pedersen, R. A., 1975, X-Ray sensitivity of the preimplantation mouse embryo *in vitro*, *Radiat. Res.* **62:**276.

Greene, R. M., and Pratt, R. M., 1977, Inhibition by diazo-oxo-norleucine (don) of rat palatal glycoprotein synthesis and epithelial cell adhesion *in vitro*, *Exp. Cell Res.* **105:**27.

Gwatkin, R. B. L., 1966a, Defined media and development of mammalian eggs *in vitro*, *Ann. N.Y. Acad. Sci.* **139:**79.

Gwatkin, R. B. L., 1966b, Amino acid requirements for attachment and outgrowth of the mouse blastocyst *in vitro*, *J. Cell. Physiol.* **68:**335.

Handyside, A. H., and Johnson, M. H., 1978, Temporal and spatial patterns of the synthesis of tissue-specific polypeptides in the preimplantation mouse embryo, *J. Embryol. Exp. Morphol.* **44:**191.

Jacob, S., Sadjel, E., Muecke, W., and Munro, H., 1970, Soluble RNA polymerases of rat liver nuclei: Properties, template specificity and amanitin responses *in vitro* and *in vivo*, *Cold Spring Harbor Symp. Quant. Biol.* **35:**681.

Jenkinson, E. J., and Wilson, I. B., 1973, *In vitro* studies on the control of trophoblast outgrowth in the mouse, *J. Embryol. Exp. Morphol.* **30:**21.

Jetten, A. M., Jetten, M. E. R., and Sherman, M. I., 1979, Analyses of cell surface and secreted proteins of primary cultures of mouse extraembryonic tissues, *Dev. Biol.* **70:**89.

Johnson, M. H., Handyside, A. H., and Braude, P. R., 1977, Control mechanisms in early mammalian development, in: *Development in Mammals,* Vol. 2 (M. H. Johnson, ed.), pp. 67–97, Elsevier/North-Holland, Amsterdam.

Levey, I., and Brinster, R. L., 1978, Effects of α-amanitin on RNA synthesis by mouse embryos in culture, *J. Exp. Zool.* **203:**351.

Lindell, T. F., Weinberg, F., Morris, P., Roeder, R., and Rutter, W., 1970, Specific inhibition of nuclear RNA polymerase II by α-amanitin, *Science* **170:**447.

Mintz, B., 1962, Experimental study of the developing mammalian egg: Removal of the zona pellucida, *Science* **138:**594.

Monk, M., and Petzholdt, U., 1977, Control of inner cell mass development in cultured mouse blastocysts, *Nature (London)* **265:**338.

Olden, K ., Pratt, R. M., and Yamada, K. M., 1979, Selective cytotoxicity of tunicamycin for transformed cells, *Int. J. Cancer* **24:**60.

Pratt, R. M., Yamada, K. M., Olden, K., Ohanian, S. H., and Hascall, V. C., 1979, Tunicamycin-induced alterations in the synthesis of sulfated proteoglycans and cell surface morphology in the chick embryo fibroblast, *Exp. Cell Res.* **118:**245.

Prives, J., and Shinitzky, M., 1977, Increased membrane fluidity precedes fusion of muscle cells, *Nature (London)* **268:**761.

Rizzino, A., and Sherman, M. I., 1979, Development and differentiation of mouse blastocysts in serum-free medium, *Exp. Cell Res.* **121:**221.

Rossant, J., and Papaioannou, V. E., 1977, The biology of embryogenesis, in: *Concepts in Mammalian Embryogenesis* (M. I. Sherman, ed.), pp. 1–36, MIT Press, Cambridge, Mass.

Sellens, M. H., and Sherman, M. I., 1980, Effects of culture conditions on the developmental programme of mouse blastocysts, *J. Embryol. Exp. Morphol.* **56:**1.

Shalgi, R., and Sherman, M. I., 1979, Scanning electron microscopy of the surface of normal and implantation-delayed mouse blastocysts during development *in vitro*, *J. Exp. Zool.* **210:**69.

Sherman, M. I., 1972a, The biochemistry of differentiation of mouse trophoblast cells: Esterase, *Exp. Cell Res.* **75:**449.

Sherman, M. I., 1972b, The biochemistry of differentiation of mouse trophoblast cells: Alkaline phosphatase, *Dev. Biol.* **27:**337.

Sherman, M.I., 1975, Esterase isozymes during mouse embryonic development *in vivo* and *in vitro*, in: *Isozymes* (C. L. Markert, ed.), Vol. III, pp. 83–98, Academic Press, New York.

Sherman, M.I., 1978, Implantation of mouse blastocysts *in vitro*, in: *Methods in Mammalian Reproduction* (J. C. Daniel, ed.), pp. 247–257, Academic Press, New York.

Sherman, M. I., 1979, Developmental biochemistry of preimplantation mammalian embryos, *Annu. Rev. Biochem.* **48:**443.

Sherman, M. I., 1981, Embryo-associated plasminogen activator prior to and during implantation in the mouse, in: *Reproductive Endocrinology, Proteins and Steroids in Early Mammalian Development* (H. M. Beier and P. Karlson, eds.), Springer-Verlag, Berlin (in press).

Sherman, M. I., and Atienza, S. B., 1977, Production and metabolism of progesterone and androstenedione by cultured mouse blastocysts, *Biol. Reprod.* **16:**190.

Sherman, M. I., and Atienza-Samols, S. B., 1978, *In vitro* studies on the surface adhesiveness of mouse blastocysts, in *Human Fertilization* (H. Ludwig and P. F. Trauber, eds.), pp. 179–183, Thieme, Stuttgart.

Sherman, M. I., and Atienza-Samols, S. B., 1979, Enzyme analyses of mouse extraembryonic tissues, *J. Embryol. Exp. Morphol.* **52:**127.

Sherman, M. I., and Chew, N. J., 1972, Detection of maternal esterase in mouse embryonic tissues, *Proc. Natl. Acad. Sci. USA* **69:**2551.

Sherman, M. I., and Salomon, D. S., 1975, The relationships between the early mouse embryo and its environment, in *The Developmental Biology of Reproduction* (C. L. Markert and J. Papaconstantinou, eds.), pp. 277–309, Academic Press, New York.

Sherman, M. I., and Wudl, L. W., 1976, The implanting mouse blastocyst, in: *The Cell Surface in Animal Embryogenesis and Development* (G. Poste and G. L. Nicolson, eds.), pp. 81–125, Elsevier/North-Holland, Amsterdam.

Sherman, M. I., and Wudl, L. R., 1977, T-Complex mutations and their effects, in: *Concepts in Mammalian Embryogenesis* (M.I. Sherman, ed.), pp. 136–234, MIT Press, Cambridge, Mass.

Sherman, M. I., Shalgi, R., Rizzino, A., Sellens, M. H., Gay S., and Gay, R., 1979, Changes in the surface of the mouse blastocyst at implantation, *Ciba Found. Symp.* **64:**33.

Sherman, M. I., Gay, R., Gay, S., and Miller, E. J., 1980, Association of collagen with preimplantation and peri-implantation mouse embryos, *Dev. Biol.* **74:**470.

Spielmann, H., Eibs, H.-G., Jacob-Müller, U., and Bischoff, R., 1978, Expression of lactate dehydrogenase isozyme 5 (LDH-5) in cultured mouse blastocysts in the absence of implantation and outgrowth, *Biochem. Genet.* **16:**191.

Spindle, A. I., and Pedersen, R. A., 1973, Hatching, attachment and outgrowth of mouse blastocysts *in vitro:* Fixed nitrogen requirements, *J. Exp. Zool.* **186:**305.

Stirpe, F., and Fiume, L., 1967, Studies on the pathogenesis of liver necrosis by α-amanitin, *Biochem. J.* **105:**779.

Strickland, S., Reich, E., and Sherman, M. I., 1976, Plasminogen activator in early embryogenesis: Enzyme production by trophoblast and parietal endoderm, *Cell* **9:**231.

Struck, D. K., and Lennarz, W. J., 1977, Evidence for the participation of saccharide-lipids in the synthesis of the oligosaccharide chain of ovalbumin, *J. Biol. Chem. 252:*1007.

Takatsuki, A., and Tamura, G., 1971, Tunicamycin, a new antibiotic. III. Reversal of the antiviral activity of tunicamycin by aminosugars and other derivatives, *J. Antibiot.* **24:**232.

Takatsuki, A., Kohno, K., and Tamura, G., 1975, Inhibition of biosynthesis of polyisoprenol sugars in chick embryo microsomes by tunicamycin, *Agr. Biol. Chem.* **39:**2089.

Tkacz, J. S., and Lampen, J. O., 1975, Tunicamycin inhibition of polyisoprenol *N*-acetyglucosaminyl pyrophosphate formation of calf liver microsomes, *Biochem. Biophys. Res. Commun.* **65:**248.

Van Blerkom, J., Barton, S. C., and Johnson, M. H., 1976, Molecular differentiation in the preimplantation mouse embryo, *Nature (London)* **259:**319.

Whitten, W. K., 1971, Nutrient requirements for the culture of preimplantation embryos *in vitro, Adv. Biosci.* **6:**129.

Wudl, L. R., and Sherman, M. I., 1978, *In vitro* studies of mouse embryos bearing mutations in the T complex: t^6, *J. Embryol. Exp. Morphol.* **48:**127.

Wudl, L. R., Sherman, M. I., and Hillman, N., 1977, Nature of lethality of *t* mutations in embryos, *Nature* (London) **270:**137.

6

Cellular and Genetic Analysis of Mouse Blastocyst Development

Roger A. Pedersen and Akiko I. Spindle

I. Introduction

The major morphogenetic events that occur during preimplantation development of the mammalian embryo are formation of a compact mass of cells (the morula) and formation of the blastocyst. Development of the blastocyst involves differentiation of a unique cell type, trophectoderm, that is necessary for implantation and nutrition of the embryo in the maternal environment, and the inner cell mass (ICM), which differentiates into primary endoderm and ectoderm before implantation (Snell and Stevens, 1966; Enders, 1971; Enders *et al.*, 1978). In the mouse, trophectoderm can form even when there are few or no ICM cells, and hollow vesicles of trophectoderm can implant *in vivo* or grow out *in vitro* (Gardner, 1972; Ansell and Snow, 1975; Sherman, 1975; Sherman and Atienza-Samols, 1979). However, subsequent proliferation of trophectoderm cells to form the ectoplacental cone and extraembryonic ectoderm depends on the presence of the ICM. The ICM itself gives rise to the fetus and to the extraembryonic membranes but cannot implant or survive in the uterus in the absence of trophectoderm (Gardner, 1972; Gardner *et al.*, 1973). Thus, the early postimplantation development of mouse embryos, and perhaps other mammalian embryos, depends on the successful differentiation and interaction of the two cell populations that form in the preimplantation period.

We have studied the development of mouse embryos by both cellular and genetic approaches. In the cellular analysis we have studied the determination of cell fate in blastocysts and in cell populations derived from them in an attempt to estimate the time that these cells become committed to their fate. In the genetic analysis, we have taken advantage of existing mutations that are lethal to mouse embryos to discern essential features of early development. In this review we consider briefly the timing of cell determination in the ICM and the primary ectoderm, and we consider in detail the manifestation of defects in mouse embryos that are homozygous for the A^y allele of the agouti locus.

Roger A. Pedersen • Laboratory of Radiobiology and Department of Anatomy, University of California, San Francisco, California 94143. *Akiko I. Spindle* • Laboratory of Radiobiology, University of California, San Francisco, California 94143.

II. Cell Determination during Early Mouse Embryogenesis

During the late morula (16- to 32-cell) stage the mouse embryo undergoes its first overt differentiation: blastomeres of the embryo differentiate into the trophectoderm and the ICM. On the basis of her studies with chimeras, Mintz (1965) suggested that the microenvironment of early embryonic cells plays an important role in trophectoderm–ICM differentiation. This idea was developed further by Tarkowski and Wróblewska (1967) to account for their observations of disaggregated 4- and 8-cell embryos. Their hypothesis was that blastomeres occupying outside positions at the morula stage form the trophectoderm and those in inner positions form the ICM.

The major prediction of this "epigenetic" hypothesis is that blastomeres of early embryos should be developmentally labile, i.e., capable of forming either trophectoderm or ICM, before the stage at which some come to occupy internal positions. This prediction has been amply confirmed experimentally for two-, four-, and eight-cell blastomeres of mouse embryos (Tarkowski, 1959; Hillman *et al.*, 1972; Wilson *et al.*, 1972; Kelly, 1975, 1977; reviewed by Rossant, 1977; Rossant and Papaioannou, 1977).

Cell determination, or restriction in the developmental potential of cells, has been studied in early mouse embryos with microsurgical and immunosurgical techniques in an attempt to define the time that inner or outer cells become committed to their fate. By studying microsurgically isolated ICMs or trophectoderm tissues, Gardner and co-workers found that both tissues were already committed to mutually exclusive developmental fates by 3.5 days of gestation (dg) (Gardner, 1972; Gardner, 1975; Gardner and Papaioannou, 1975; Rossant, 1975a,b; Gardner and Rossant, 1976). With the advent of the immunosurgical technique (Solter and Knowles, 1975) it was possible to isolate inner cells from larger numbers of embryos and at earlier stages than with the technically more difficult microsurgical procedures. It is now clear from such studies that ICM determination occurs shortly before 3.5 dg. By contrast, cell determination in the trophectoderm has not been studied extensively, probably because of the technical difficulty of disaggregating the cells and placing them in inside positions. However, it has been shown that polar trophoblast moves into mural trophectoderm positions as blastocysts develop (Gardner and Papaioannou, 1975; Copp, 1978, 1979) and that mural trophoblast can be converted into polar trophoblast in early blastocysts by placing a donor ICM in contact with host mural trophectoderm (Gardner *et al.*, 1973). Cell determination of primary ectoderm has been studied by both microsurgical and immunosurgical techniques, but with apparently contradictory results, which we will discuss.

A. Cell Determination in the Inner Cell Mass

To assess the state of cell determination or commitment in a tissue, such as the ICM, it is necessary to move the cells away from their normal spatial or temporal environment. Rossant carried out two studies to assess the developmental potential of inner cells exposed to an outside environment. In the first study (Rossant, 1975a), ICMs isolated microsurgically from 3.5-dg blastocysts were aggregated with morulae, and the aggregates were transferred to the uteri of foster mothers for further development. Analysis of glucose phosphate isomerase showed a weak ICM contribution to the trophoblastic fraction in 3 of 13 conceptuses. Rossant attributed this minor contribution to contamination from chorionic cells, which are derived from the ICM, and concluded that the ICM cells are committed to their fate by 3.5 dg. In the second experiment (Rossant, 1975b), two microsurgically isolated ICMs were aggregated and placed in an empty zona pellucida, then transferred to the oviduct of a foster mother for further development. ICMs recovered 1 to 3 days later had endodermlike outer layers but trophectoderm was absent, confirming that ICM is committed by 3.5 dg.

In other studies ICMs isolated by immunosurgery from 3.5-dg blastocysts that had been cultured an additional 24 hr (3.5 dg + 24 hr) formed two-layered structures that resembled those seen by Rossant (1975a,b): an inner core of primary ectoderm surrounded by a layer of primary endoderm (Hogan and Tilly, 1978a; Wiley *et al.*, 1978). However, during our study of immunosurgically isolated ICMs (unpublished observations) we observed that ICMs isolated from blastocysts (3.5 dg + 24 hr) on rare occasions formed blastocystlike fluid-filled structures after overnight culture. This suggested that at least a few ICM cells could still be developmentally labile at this stage.

In further experiments we studied the stage dependence of trophectoderm formation by isolated inner cells. When ICMs were isolated from early blastocysts developing *in vitro* (69 hr of culture from the two-cell stage), one-third of the ICMs were capable of forming blastocystlike structures within 24 hr of isolation (Spindle, 1978). Upon further cultivation these structures formed small but typical trophoblast outgrowths consisting of giant cells. This formation of trophectoderm by isolated ICMs did not appear to depend on the number of cells they contained; ICMs composed of two to three cells could form fluid-filled structures (A. I. Spindle, unpublished observations). The subsequent survival of ICMs, on the other hand, was strongly dependent on their size, as suggested by Snow (1976). The poor survival of inner cells isolated from early blastocysts could also be due to the toxicity of antiserum or guinea pig serum used in immunosurgery (Hogan and Tilly, 1978b; A. I. Spindle, unpublished observations).

We solved the ICM survival problem by the use of aggregation chimeras. All ICMs from five-embryo chimeras survived and most formed trophectoderm and trophoblast outgrowths when inner cells were isolated from late morulae or early blastocysts that were beginning to cavitate (Spindle, 1978) (Table 1). By contrast, this ability was reduced to low levels when ICMs were isolated at 69 hr of culture from blastocysts that were partially expanded. ICMs isolated from fully expanded late blastocysts at 93 hr of culture appear to have lost their capacity for forming trophectoderm (Spindle, 1978). A comparison of blastocysts that developed *in vitro* and *in vivo* revealed that the determinative state of ICMs of cultured embryos at 60 hr of culture from the late two-cell stage (113 hr after injection of

Table 1. Stage Dependence of Trophectoderm Formation by Isolated Inner Cells [a]

		Developmental fate of inner cells	
Stage of embryo	*n*	Trophectodermal vesicle formation	Trophoblast outgrowth
Late morula[b]	82	59	65
Early blastocyst[b]	116	73	73
Midblastocyst[b]	51	1	5
Late blastocyst[c]	27	0	0

[a] Inner cells were isolated by immunosurgery from chimeras of five embryos at the stage indicated and scored for vesicle formation (24 hr later) or trophoblast outgrowth and giant cell formation (120 hr later). Data from Spindle (1978).

[b] Inner cells were isolated at 69 hr of culture from the two-cell stage. The transition between early blastocyst and midblastocyst was arbitrarily taken as blastocoele expansion equal to or greater than one-half the volume of the blastocyst, as estimated in the dissecting microscope.

[c] Inner cells were isolated at 93 hr of culture from the two-cell stage.

human chorionic gonadotropin) is similar to that of embryos obtained from the uterus at 10 A.M. on the fourth day of gestation (93 hr after human chorionic gonadotropin). (This observation suggests that development of embryos is delayed 1 day during culture from the two-cell to the blastocyst stage, as noted by Spindle, 1978.)

Trophoblast formation by immunosurgically isolated ICMs was also independently observed by Johnson *et al.* (1977), Handyside (1978), Hogan and Tilly (1978b), and Rossant and Lis (1979). Rossant and Lis (1979) also found that regenerated trophoblast cells were capable of inducing a decidual reaction in the uterus. The possibility that trophectoderm is regenerated by rapid proliferation of a few contaminating trophectoderm cells appears to be remote (Handyside, 1978; Spindle, 1978). Thus, it is now clear that at least some cells of the mouse ICM remain developmentally labile after trophectoderm differentiation.

Cell determination in the ICM appears to coincide with expansion of the blastocoele from a small to a large cavity. One major cellular event that occurs during this period is the sixth cleavage division, which doubles the total cell number of the embryo and increases the number of cells in the ICM (Handyside, 1978; Spindle, 1978). The relationship between cell division and commitment during this period of development deserves further attention, particularly in view of the indication (Braude, 1979) that gene expression necessary for blastocyst formation occurs just before the fifth cleavage division.

Because ICM cells of early blastocysts do not differentiate into trophectoderm when left inside the embryo, there appears to be a mechanism for them to receive information regarding their position in the intact embryo. One possible vehicle for the transfer of positional cues between outer and inner cells is the blastocoele fluid. This fluid is contained by the trophectoderm (Ducibella *et al.*, 1975; Magnuson *et al.*, 1977) and has an ionic composition distinct from that of the culture medium (Borland *et al.*, 1977). We examined the role of the blastocoele microenvironment on trophoblast–ICM differentiation (Pedersen and Spindle, 1980) and found that eight-cell embryos or morulae injected into the blastocoele of giant chimeras could differentiate into morphologically normal blastocysts, provided that cell–cell contact between the donor embryo and the recipient chimeras was prevented by leaving the donor zona pellucida intact. However, when donor–recipient cell contact was allowed by removing zonae from the donor embryos, the donor embryos attached to the host trophectoderm and formed compact structures instead of blastocysts. Labeling donor cells with [^{3}H]thymidine excluded the possibility that trophoblast cells of donor embryos were incorporated into the recipient trophectoderm. From our results it is clear that blastocoele fluid alone does not prevent trophoblast differentiation in donor embryos. Whether cell–cell contact alone or both blastocoele fluid and cell contact are involved in inhibiting donor blastocoele formation remains to be determined.

The nature of the compact structures formed by zona-free donor embryos is not yet clear. As judged from their ability to form trophoblast outgrowths in culture when they were removed from the recipients, the compact structures may be merely delayed or physically prevented from accumulating blastocoele fluid, or they may be ICMs that were not yet committed at the time of removal from the host. Further analysis is under way using two-dimensional gel electrophoresis to identify cell types in these structures. This system may also be useful for studying the time of trophectoderm determination because it permits outer cells to be placed in an internal environment without disaggregating the donor embryo.

B. Cell Determination in the Primary Ectoderm

Determination of primary ectoderm was studied initially by microsurgical techniques. Gardner and Rossant (1976, 1979) investigated the stage of determination in primary ectoderm and primary endoderm by dissecting them from 4.5-dg blastocysts and injecting them into early blastocysts. Their results indicate that primary endoderm cells from 4.5-dg blastocysts contribute exclusively to the endodermal layer of the visceral yolk sac, whereas primary ectoderm cells contribute to the mesodermal layer of the visceral yolk sac and to the fetus, but not to yolk sac endoderm. These results imply either that both cell types are determined by 4.5 dg, that donor cells migrate to their appropriate position in the host ICM, or that donor cells can proliferate only in the correct position (Gardner and Rossant, 1979). It is impossible to distinguish between these alternatives without further experiments because microsurgically isolated ectoderm and endoderm are placed back in approximately the same positions that they would occupy in normal blastocysts.

To test the developmental potential of primary ectoderm that is moved out of its normal environment, ectoderm was isolated from late blastocysts by double immunosurgery and its developmental capacity was studied *in vitro* (Strickland *et al.*, 1976; Hogan and Tilly, 1977). In our experiments (Pedersen *et al.*, 1977) immunosurgery was first performed on late blastocysts at 93 hr of culture from the two-cell stage. At the time of isolation most of the ICMs were partially covered by primary endoderm, and after overnight culture in suspension they were completely enclosed by endoderm. The second immunosurgery (to remove endoderm) was performed at 24, 48, or 72 hr after the first. Ectoderm isolated at 24 and 48 hr (but not at 72 hr) regenerated a layer of endoderm during an additional 24 to 48 hr culture in suspension (Table 2). The survival of ectoderm in culture depended on the mass of tissue; ectoderm from single embryos or chimeras of two embryos died, whereas ectoderm from four- eight-, and ten-embryo chimeras survived and developed a second layer of endoderm. Hogan and Tilly (1977) reported that ectoderm that was isolated from single blastocysts and cultured on fibroblast feeder cells attached to the substratum and differentiated into various tissues, including skin, nerve, beating muscle, cartilage, and fibroblasts, but did not regenerate primary endoderm. When ectoderm was cultured in suspension it did not survive. From these observations it appears that attachment and the presence of feeder cells are necessary conditions for survival and extensive differentiation. A major difference between the experimental conditions used by Hogan and Tilly (1977) and ours (Pedersen *et al.*, 1977) is the size of embryos from which ectoderms were obtained; they isolated ectoderms from normal-size blastocysts, whereas we obtained ours from chimeras made by aggregating eight to ten embryos. The increased size of ectoderm obtained from chimeras appears to ensure its survival in the absence of feeder cells; furthermore, suspension culture seems to be a necessary condition for regeneration of endoderm. Atienza-Samols and Sherman (1979) found that ectoderm cores isolated from single ICMs 48 hr after the initial immunosurgery could regenerate a layer of endoderm. They found that culturing the ectoderm cores in conditioned medium improved their sur-

Table 2. Time Dependence of Endoderm Formation by Ectoderm Isolated from ICMs [a]

Time of ICM culture (hr)	*n*	Ectoderm with endoderm
24	28	22
48	14	14
72	31[b]	1

[a] ICMs were isolated from late blastocyst chimeras of ten embryos using immunosurgery and cultured for 24 to 72 hr; ectoderm was then isolated by immunosurgery, cultured for 48 to 72 hr, and scored for endoderm formation. Data from Pedersen *et al.* (1977).
[b] Most ectoderms degenerated within 48 hr without forming endoderm.

vival and doubled the incidence of endoderm regeneration from 10 to 20%. Their results likewise indicate a cell feeding requirement for ectoderm cell survival. Recently, Dziadek (1979) reported that endoderm could be regenerated by ectoderm isolated from chimeras of three embryos 24 to 40 hr after the initial immunosurgery. The regenerated endoderm synthesizes α-fetoprotein, which is a characteristic product of visceral endoderm cells. As noted by Dziadek (1979) and by Gardner and Rossant (1979), the definitive identification of these endoderm cells requires tissue-specific markers. However, the morphological similarity of regenerated endoderm and visceral endoderm (Pedersen *et al.*, 1977) strongly implies that they are identical. Furthermore, it seems unlikely that the regenerated layer is the definitive endoderm of the fetus, as suggested by Gardner and Rossant (1979), because the definitive endoderm would not be expected to form until 7 to 8 dg (Snell and Stevens, 1966; Diwan and Stevens, 1976), and the ectoderm isolated in our studies (Pedersen *et al.*, 1977), by Atienza-Samols and Sherman (1979), and by Dziadek (1979) is probably equivalent to an earlier developmental stage.

The development of tissues isolated from 6.5- to 8-dg mouse and rat embryos and transferred to ectopic sites suggests that embryonic ectoderm gives rise to the definitive endoderm of the mammalian fetus; the definitive endoderm does not seem to arise from the primary (or visceral) endoderm (Grobstein, 1952; Levak-Svajger and Svajger, 1974; Diwan and Stevens, 1976). The implication of these and other (biochemical) observations (Johnson *et al.*, 1977) is that cell differentiation precedes cell commitment in early mouse embryos (see Johnson *et al.*, and Rossant and Frels, this volume).

III. Mutations at the a Locus Affecting Implantation

The ease of observing coat color has facilitated study of the *a* locus in mice and has been partially responsible for the extensive use of this locus for genetic, embryological, and physiological studies. The control of coat color by this and other loci has been recently reviewed (Silver, 1979). The early identification of the yellow phenotype by mouse fanciers was followed by the discovery that yellow mice had both yellow and black progeny (Cuénot, 1905). Early workers (Cuénot, 1905; Castle and Little, 1910) deduced that yellow mice are heterozygous for the dominant yellow allele and that the homozygote is lethal. This was the first mutant allele to be identified in mice and led to further analysis of this locus and the genetic basis for coat color. Although our interest in mutations at the *a* locus is in what the alleles affecting early development can tell us about the early embryo, we also consider the genetics of the *a* locus because this subject has not been reviewed.

A. Genetics

The *a* locus is in linkage group V, which has been identified with chromosome 2 (Miller *et al.*, 1971). The G bands on this chromosome are colinear with the linkage order and the *a* locus appears to be located in G-band 2HI (Searle *et al.*, 1979). There are 14 *a* locus alleles affecting coat color, as reviewed by Green (1975). Only two of these are known to affect early embryonic development: A^y (yellow) and a^x, which is dark brown with a slightly paler belly. The yellow allele is dominant to agouti and the other alleles with respect to coat color but is recessive for lethality. The a^x allele, which was derived from irradiated spermatogonia, is recessive to A^y, A^w, A, and a^t, is dominant to a for its coat color effects, and is recessive for lethality (Russell *et al.*, 1963). Closely linked genes that have been used in studies of the *a* locus include Wellhaarig (*we*), waved hair, 12 centimorgans (cM) proximal; undulated (*un*), wavy tail and abnormal vertebral column, 5 cM proximal; brachypodism (*bp*), short, abnormal feet, 0.3 cM distal; and kreisler (*kr*), circling and head-shaking behavior, 3 cM distal. The minor histocompatibility locus, H-13, which is 2 cM proximal, is also a potentially useful marker (Snell *et al.*, 1967).

In a study of a balanced lethal stock (A^y/a^x), Russell *et al.* (1963) reported a low level of recombination between the A^y and the a^x alleles, showing that a^x is pseudoallelic to A^y and a. These authors concluded that the gene order was kr–a^x–A^y–un and that a^x and A^y are separated by 0.5 cM; they also suggested that A^y may be pseudoallelic to a because an agouti offspring was recovered from matings between A^y/a parents. Similarly, Phillips (1966) found 0.6% recombination between A^s and A^w or a^t. The proximity of *bp*, which is only 0.3 to 0.4 cM from a (Runner, 1959; Phillips, 1966), raises the possibility that *bp* lies between a and a^x or A^s; however, Phillips (1966) discounted this possibility because a crossover was obtained with *bp* that did not separate A^s and A^w. Because of the crossovers observed between a and A^s or a^x, Wallace (1965) and Phillips (1966) suggest that the a locus is a complex, multigenic locus.

B. Embryology

1. Histological Observations

The earliest embryological studies of yellow mice showed that a portion of them died early in gestation, just after implantation. Kirkham (1917, 1919) serially sectioned oviducts and uteri of yellow mice and compared their embryos to those of white mice. Neither strain of mouse was characterized. The two-cell embryos of yellow mice were normal, but 2 of 11 morulae in two litters had indistinct cell boundaries at 3 dg. In six matings studied at 4 to 6 dg, 11 of 43 embryos were "degenerate." Kirkham (1919) identified the abnormal blastocysts by their shrunken appearance, their small, crowded cells, and the small blastocyst cavity. In 21 matings of yellow mice studied at 3 to 20 dg, Kirkham (1919) classified 39 of 94 embryos (29%) as degenerate, but found only 2 of 189 degenerate embryos (1%) in white mice. Despite their rapid degeneration, abnormal embryos induced a normal uterine decidual response. Kirkham (1919) concluded that the blastocysts that are lost during implantation in yellow mice are the homozygous embryos, and that they are defective in implantation. The use of white, rather than sibling nonyellow, mice as controls in Kirkham's studies decreases the certainty of his identification, because strains vary in their rates of early embryo mortality, some showing higher spontaneous rates of preimplantation death than others (Leonard *et al.*, 1971). In a similar study carried out on 13- to 19-dg embryos, Ibsen and Steigleder (1917) used the appropriate control matings ($A^y/a \times a/a$, $a/a \times A^y/a$, and $a/a \times a/a$) of an unspecified strain to show a 15.6% excess of early resorptions in 33 $A^y/a \times A^y/a$ matings. There were no differences in the frequency of late resorptions (death after 13 dg) in experimental and control matings. Ibsen and Steigleder (1917), as well as Little (1919), thus verified Kirkham's findings that abnormal embryos could induce a decidual reaction, but the late times chosen for histological analysis precluded a precise estimate of the time of A^y/A^y embryo death or the nature of their abnormalities.

Robertson (1942a,b) carried out the first study of yellow mouse embryos using an inbred strain, 101-A^y. His histological observations were concentrated on 2- to 6.5-dg embryos. Only 3 of 56 cleavage and early blastocyst stage embryos (5.4%) were abnormal. One of these was at the eight-cell stage, one was at the morula stage, and one was at the early blastocyst stage. Because of their rarity it seems unlikely that the two early abnormalities were related to A^{y}/A^{y} homozygosity, but Robertson (1942b) considered the abnormal blastocyst to be in transition between normality and the degeneration seen at later stages. This blastocyst had crowded cells and was small, with a collapsed blastocoele filled with trophoblastic processes. By comparison, 8 of 29 implanting blastocysts (27.6%) in six litters studied at 5 to 5.5 dg were small, lacked a blastocoele, had no obvious trophectoderm, endoderm, or ectoderm, and had pycnotic nuclei. Robertson (1942b) concluded that A^{y}/A^{y} embryos developed normally up to the late blastocyst stage but that their trophectoderm degenerated on contact with the uterine epithelium, leaving the ICM exposed. This view implies that the primary defect of A^{y}/A^{y} embryos is in the trophectoderm and that ICM is affected only indirectly, through loss of trophoblast functions that are essential for implantation or that protect or nourish the inner cells within the uterine environment. However appropriate this view may be, Robertson's results should be interpreted cautiously because only a small number of embryos were studied and there were no control observations to support the conclusion that the abnormalities observed were specifically related to A^{y} homozygosity.

In another study using three inbred strains, Eaton and Green (1962) sectioned embryos from eight experimental matings ($A^{y}/a \times A^{y}/a$) and nine control matings ($a/a \times A^{y}/a$ or $A^{y}/a \times a/a$) at 5.5 to 7.5 dg. In four experimental matings of C57BL/6J mice, 15 of 36 embryos (41.6%) were dead and in the other two strains, 13 of 34 embryos (38.2%) were dead at 5.5 to 6.5 dg. Uterine crypts containing viable embryos were deeper than those with degenerate ones. In control matings of C57BL/6J mice, 6 of 55 embryos (10.9%) were dead at 5.5 to 6.5 dg. Because the abnormal embryos evoked a normal decidual reaction and persisted in the uterus until after implantation (which occurred at 4.5 to 5.25 dg.), Eaton and Green (1962) concluded that death of A^{y}/A^{y} embryos was not due to faulty implantation. Their conclusion is based in part on the advanced development of the abnormal embryos they observed at 6.5 dg. Correcting for the control deaths, there were 30.7% excess deaths in the embryos from C57BL/6J-$A^{y}/a \times A^{y}/a$ matings, which is slightly above the 25% Mendelian expectation. However, the criteria for distinguishing between abnormalities characteristic of control matings and those specific to experimental matings were not given. The authors did carry out matings, however, to show that heterozygous yellow embryos survived to birth as well as did wild-type embryos, so it is unlikely that any excess deaths specifically represent A^{y}/a embryos.

In their companion study, presumably carried out with the same strains of mice, Eaton and Green (1963) examined 262 embryos of 20 $A^{y}/a \times A^{y}/a$ matings at 4.33 to 5.5 dg., finding 71 abnormal embryos (27%). These fell into three classes: (1) expanded blastocysts with an expanded ICM and undifferentiated trophoblast giant cells, lying free in the uterine crypt (2%); (2) collapsed blastocysts with pycnotic nuclei in the ICM, partially attached at the abembryonic pole (13%); (3) blastocysts with partial blastocoele expansion, partially attached by a variable number of equatorial trophoblast giant cells (12%). Eaton and Green interpreted their results as indicating a correlation between the extent of equatorial trophoblast giant cell attachment and the survival of class 3 embryos to early egg cylinder stages. The more advanced embryos (with more than 50% of their trophoblast attached to the endometrium) developed ectoplacental cone, embryonic and extraembryonic endoderm, and ectoderm; however, they did not increase in length during implantation, had virtually no mitotic figures, and began degenerating at 6 dg. Eaton and Green concluded that trophoblast giant cell differentiation is the limiting factor in development of these abnormal embryos. This conclusion hardly seems justified on the basis of their histological observations because the hardier embryos may simply progress further in trophoblast development and die for other than the reasons suggested. Furthermore, they state that the three classes of abnormal embryos were not found in control matings, but they presented no numerical data to support this conclusion, even though they found nearly 11% degenerating embryos in control matings in their earlier study (Eaton and Green, 1962). Therefore, caution is warranted in accepting their interpretations for the cause of A^{y}/A^{y} embryo lethality.

In a recent study of C57BL/6J-A^{y}/a mice, Cizadlo and Granholm (1978a) compared cell numbers and embryo sizes of experimental and control litters. They found that 3 of 24 embryos from $A^{y}/a \times A^{y}/a$ matings were abnormal at 4.5 dg and that 1 of 22 from the $A^{y}/a \times a/a$ control matings was abnormal. The abnormal embryos had cell numbers below the mean values, which were 123 ± 7 (S.E.) and 141 ± 11 for experimental and control matings, respectively. Although mean embryo cell numbers and trophoblast cell numbers were not significantly different between experimental and control matings, ICM cell (nuclei) numbers were lower in experimental matings (30 ± 2 versus 42 ± 5). Other quantitative features, such as number of corpora lutea, implantation sites per female, and number of sections per embryo, were similar in experimental and control matings. Also, a lower proportion of the embryos in experimental matings underwent abembryonic trophoblast proliferation, as judged by the appearance of cells in this region. Even assuming that the observed differences are due to the A^{y} allele, which cannot be stated with certainty, the authors' conclusion that A^{y} homozygosity preferentially interferes with ICM cells does not appear to be justified, because of the effects they observed on trophoblast proliferation and effects on hatching of blastocysts.

2. *Ultrastructural Observations*

The histological observations of presumed homozygous yellow embryos are complemented by three electron microscopic studies of embryos from $A^{y}/a \times A^{y}/a$ matings. Calarco and Pedersen (1976) examined 17 morphologically abnormal embryos obtained at 3 or 4 dg from several C57BL/6J-$A^{y}/a \times A^{y}/a$ matings with particular attention to those showing excluded blastomeres and thought to be A^{y}/A^{y} because of their development in culture. They found that the excluded blastomeres resembled earlier cleavage stage embryos in having crystalloid inclusions; intracisternal A particles; small, vacuolated mitochondria; dense, rounded nucleoli; few polysomes; and little rough endoplasmic reticulum. The remaining cells of presumed homozygotes were ultrastructurally normal, and inner cells were found in all abnormal blastocysts. No observations were made of normal or abnormal embryos from control matings, and the identification of A^{y}/A^{y} embryos by blastomere exclusion has been questioned by other workers. Nowel and Chapman (1976) reported their observations of two grossly normal and one retarded, abnormally small blastocyst from matings of C57BL/6J-A^{y}/a mice *inter se*. Normal cells at 4.5 dg had abundant free ribonucleoprotein particles, contorted nuclei with prominent nucleoli, junctional complexes between adjacent trophectoderm cells, and other ultrastructural characteristics of normal embryos at the late blastocyst stage (Enders, 1971; Enders *et al.*, 1978). Cells of the abnormal embryo and of one presumed normal embryo had enlarged nuclei, empty spaces in the cytoplasm, and contorted and vacuolated mitochondria. Because of their limited observations one cannot safely conclude that any of the features noted by Nowel and Chapman (1976) actually characterize the homozygous yellow embryo at the late blastocyst stage. Using the same C57BL/6J-A^{y}/a strain, Cizadlo and Granholm (1978b) studied 24 embryos obtained from four $A^{y}/a \times A^{y}/a$ matings (20 of 24 obtained at 3.5 dg and 4 of 24 at 4.25 dg) and five embryos from five $a/a \times A^{y}/a$ matings. The six abnormal embryos from $A^{y}/a \times A^{y}/a$ matings included one degenerate cleavage stage embryo, two morulae with isolated blastomeres that had condensed cytoplasmic material, two morulae with increased numbers of intracisternal A particles and relatively unreticulated nuclei, and one blastocyst with a degenerating trophoblast cell and other vacuolated blastomeres. These embryos were ultrastructurally normal in other respects, like the embryos from control matings. It is puzzling that all the advanced embryos in their experimental sample (morulae or blastocysts) showed ultrastructural abnormalities, because only 25% would be predicted to be A^{y}/A^{y} according to Mendelian expectations, and no abnormalities were observed in the small control sample. Because of these limitations and the previously stated reservations regarding embryo genotype, the ultrastructural features of homozygous yellow embryos remain poorly understood; therefore, the existing observations, which are all on preimplantation stage embryos, must be interpreted with caution. The uncertainty in identifying A^{y}/A^{y} embryos by their ultrastructural features before their death reflects the inherent limitations in static observations of developing systems. These problems have been overcome to some extent by experimental approaches.

3. *Experimental Observations*

A variety of experimental approaches have been used in attempts to identify A^{y}/A^{y} embryos and to determine the mechanism of lethality, including ovarian transplantation, hormone treatment, embryo culture, and chimera formation.

The earliest experimental analysis was by Robertson (1942b), who transferred ovaries of A^{y}/a mice to ovariectomized agouti littermates and histologically studied the embryos obtained by mating these agouti females to A^{y}/a males. Among 57 embryos in 18 matings, there were no abnormalities in cleavage stage or preimplantation blastocysts, but 5 of 17 implanting blastocysts (29%) and 9 of 28 egg cylinders (32%) were abnormal. The abnormalities appeared at the same time (5 dg) as in embryos developing in A^{y}/a females, but the affected embryos survived longer, developed approximately twice the number of cells, formed Reichert's membrane and a rudimentary ectoplacental cone, and completely eroded the uterine epithelium. These features were not seen with the abnormal embryos in A^{y}/a mothers. Thus, by comparison with the usual $A^{y}/a \times A^{y}/a$ matings, the nonyellow uterine environment appeared to be slightly more favorable for the development of A^{y}/A^{y} embryos; nevertheless these died shortly after implantation and were disorganized masses of cells undergoing resorption by 6.5 dg (Robertson, 1942b). As noted, however, positive identification of A^{y}/A^{y} abnormalities was not possible in Robertson's studies owing to lack of control observations.

In an extension of earlier work, Eaton (1968) reported that administering progesterone (5 mg/mouse) at 4.5 dg to A^{y}/a females that were mated to A^{y}/a males led to an increase in the number of "escaper" embryos at 5.5 dg. The "escapers" constituted only 2% of the embryos in untreated mothers, but the incidence increased to 7, 13, and 23% in those treated with 5, 7.5, and 15 mg progesterone, respectively. Although Eaton suggests that the progesterone treatment was rescuing embryos that would otherwise fail to attach and would die before 5.5 dg, the small number of embryos in each treatment group does not permit such a conclusion. It seems equally possible that the additional "escapers" were drawn from the normal class of embryos, particularly in view of his observations that the higher amounts of progesterone (15 to 20 mg) given at 4 dg + 20 hr led to whole litters of "escapers," but 5 to 16 mg progesterone administered over several days produced no increase in "escapers." This issue could have been resolved by observations on control matings, but unfortunately these were not reported.

Another category of experimental studies on mutant embryos involves their culture and analysis *in vitro* (reviewed by Bennett, 1975; Wudl *et al.*, 1977; Gluecksohn-Waelsch,

1979). Pedersen (1974) first used this approach with yellow mouse embryos by obtaining eight-cell, morula, or blastocyst stage embryos from C57BL/6J or C57BL × SEC F1 hybrid mice and culturing them for 24 to 120 hr. No abnormalities were consistently seen in eight-cell embryos; however, after 24 hr of culture, there were 11% more abnormal embryos in $A^{y}/a \times A^{y}/a$ matings than in $A^{y}/a \times a/a$ matings. Also, Pedersen (1974) found 15% more abnormal morulae and blastocysts obtained without culture from experimental matings than from control matings. These abnormal embryos, like those found after 24-hr culture of eight-cell embryos, had delayed or arrested cleavage, partial arrest and exclusion of blastomeres, and extensive fragmentation or disorganization. Approximately two-thirds of the abnormal embryos had exclusion of some blastomeres, and it was suggested that this defect characterizes the A^{y}/A^{y} embryo in C57BL mice. The only abnormalities seen in embryos from hybrid mice were the small, collapsed blastocysts found in experimental matings. Pedersen's (1974) evidence for a defect expressed in A^{y}/A^{y} embryos at preimplantation stages is weakened by the small numbers of matings studied in both experimental (10 to 13) and control (5 to 10) matings, and by the relatively low frequency of excess abnormalities in experimental matings after the incidence of abnormalities in appropriate controls was subtracted. A more solidly established feature of embryos from $A^{y}/a \times A^{y}/a$ matings was the failure of approximately one-fourth of cultured blastocysts to escape from the zona pellucida. Pedersen (1974) reported that removing the zona pellucida with pronase enabled blastocysts from experimental matings to attach and grow out, but there was an 18% excess of abnormal outgrowths, as compared to controls. The principal abnormality observed was the small size and number of nuclei in the outgrown trophoblast monolayer and the absence of ICM cells in these outgrowths, in contrast to normal embryos.

In another attempt to detect early defects in A^{y}/a embryos, Pedersen and Spindle (1976) compared the cell numbers in morphologically normal and abnormal embryos of $A^{y}/a \times A^{y}/a$ matings. The morphologically abnormal embryos had half as many cells ($n = 33$) as normal ($n = 60$), but their labeling index with [^{3}H]thymidine was the same. The smaller number of cells in abnormal embryos may be related to the delayed cleavage of embryos that developed morphological abnormalities. The time-lapse analysis cited by Pedersen and Spindle (1976) suggested that abnormal embryos divided from the two- to four-cell stages 2 to 4 hr later than normal embryos. For confirmation of these preliminary observations it would be necessary to carry out further time-lapse analysis of both control and experimental matings. It was apparent from our results (data not shown) that except for the delays noted in early cleavage, both normal and abnormal embryos cleaved with 11- to 17-hr cell cycle times, and that there was variation between embryos in the asynchrony of blastomere divisions. These results are consistent with those obtained for normal embryos by Kelly *et al.* (1978) with time-lapse or periodic examination in several inbred strains of mice.

Studying the *in vitro* development of embryos from C57BL-A^{y}/a mice, Granholm and Johnson (1978) also found a correlation between early cleavage delay and the incidence of hatching failure in embryos from experimental and control matings. The excess of unhatched embryos in experimental matings was greatest (26%) among those that lagged in development at the four- to eight-cell stage, perhaps owing to cleavage retardation in A^{y}/A^{y} embryos. However, there was also a 15% excess of unhatched embryos in the advanced group, which suggests that under their experimental conditions, neither hatching failure nor lagging could be used to identify A^{y}/A^{y} embryos exclusively. In other work, Johnson and Granholm (1978) reported that among control matings of C57BL-A^{y}/a mice, there were as many embryos with excluded blastomeres (17.5%) as in experimental matings (15.9%). More morphologically abnormal embryos were found at the morula stage in 61 control matings than in 16 experimental matings, although there was a 15.6% excess of abnormal embryos at the blastocyst stage in the experimental matings. These observations led Johnson and Granholm (1978) to conclude that morphological abnormalities are not adequate for identifying individual A^{y}/A^{y} embryos.

Papaioannou and Gardner (1976, 1979) came to a similar conclusion from studies of the A^{y} allele in AG/Cam mice. They transferred morphologically normal and abnormal embryos to foster mothers and found no correlation between morphological abnormalities and early postimplantation death of embryos from experimental matings. However, the high incidence of excluded blastomeres in both their control (32%) and their experimental (37%) matings suggests that they may have been recognizing persistent polar bodies of morphologically normal embryos, or that abnormalities were induced by the superovulation protocol. Also, the manifestation of early abnormalities in A^{y}/A^{y} embryo development may be highly strain dependent, as suggested by the comparison of embryos from C57BL and C57BL × SEC F1 hybrid mice (Pedersen, 1974).

Despite their inability to recognize A^{y}/A^{y} mice before implantation, Papaioannou and Gardner (1979) made chimeras between embryos of experimental ($A^{y}/a \times A^{y}/a$) and control matings. Their results, obtained by transferring the ICM from embryos of experimental matings into a control blastocyst (or vice versa), show that death of homozygous yellow mouse embryos is not due to cell lethality or to effects exclusively in the ICM. On the contrary, their results suggest that the trophectoderm of presumed A^{y}/A^{y} embryos may be specifically defective, and that ICM derivatives may survive for longer periods than previously realized when they are in chimeras with normal ICM and trophectoderm cells. Papaioannou and Gardner (1979) concluded from the observed rates of survival and chimerism that presumed A^{y}/A^{y} ICM derivatives are able to develop at least until 10.5 dg.

IV. Conclusions

From the studies described here on cell commitment in the mouse ICM and primary ectoderm it is apparent that inner cells become committed to their fate sometime after the beginning of outer cell differentiation. ICM becomes committed during a period of several hours between the beginning and the end of blastocoele formation. After this time the isolated inner cells lose the ability to form trophectoderm and instead form primary endoderm when exposed to outside conditions. Primary ectoderm, which is enclosed by polar trophectoderm and primary endoderm in the intact late blastocyst, retains the ability to form primary endoderm for 40–48 hr in ICMs isolated from chimeric embryos. After this time, the fate of isolated ectoderm may still be labile, but inner cells seem to require cell feeding from endoderm or other sources for survival. The apparent differentiation of outer cells before the commitment of inner cells may imply that outer cells play a role in the commitment of inner cells as originally suggested by Tarkowski and Wróblewska (1967). The ICM is known to affect the fate of trophectoderm differentiation (Gardner, 1972; Gardner *et al.*, 1973); however, the nature of positional cues in the early embryo is poorly understood. It appears that blastocoele fluid by itself does not mediate the commitment of pluripotent cells to an inner fate. In further analysis of cell commitment in the early mouse embryo, the role of cell–cell contact and even cell–extracellular matrix interaction should not be overlooked (Hay and Meier, 1976; Wolpert, 1978). It may be necessary to define the substance(s) that indicates position in early mouse embryos in order to study the biochemical basis for cell commitment and for the transition from one decision-making state to another during subsequent stages of development.

Although the analysis of the *a* locus has been carried out from a very different perspective than the analysis of cell commitment, several relevant points can be made from the study of the mutant embryos. First, although no single definitive study has been carried out to describe the time and mode of death of A^y/A^y embryos *in utero*, a comparison of all the studies done with yellow mouse embryos leads to the conclusion that the homozygotes die at approximately 5.5 to 6.0 dg. This is after the commitment of ICM and may approximately coincide with the time of commitment of primary ectoderm cells. However, the timing of A^y/A^y embryo death probably has little to do with commitment of ICM or ectoderm, in view of the interpretation from Gardner and Papaioannou's (1975) chimeric analysis that ICMs of A^y/A^y embryos can survive as late as 10.5 dg. Further work is clearly necessary to determine whether A^y/A^y cells can survive throughout gestation.

The intriguing possibility that trophectoderm of A^y/A^y embryos is specifically defective also deserves further work. Such a defect could account for the preponderance of hatching failures in matings containing A^y/A^y embryos. A carefully controlled morphological and biochemical analysis of yellow mouse embryos at 4.5 to 5.5 dg could further elucidate the effect of homozygosity on trophectoderm and ICM tissues before resorption occurs. Such a study would require identifying the homozygotes before extensive deterioration occurs. The cleavage delay in A^y/A^y embryos may not be large enough to serve this purpose, given the variation in the timing of early developmental events in normal embryos. Also, hatching failure is a retrospective means of identifying presumed homozygotes, and embryos that fail to hatch may have already begun to deteriorate. A means for identifying preimplantation stage A^y/A^y embryos would clearly facilitate these studies. Similar analyses of embryos homozygous for the a^x allele could also add perspective to the pattern of defects found in A^y/A^y embryos.

The relevance to implantation of the studies reviewed here derives mainly from the fact that successful postimplantation development *in utero* requires both trophectoderm and ICM survival and differentiation. The understanding of cell commitment in early mammalian embryos should be greatly facilitated by the application of genetic analysis, as in other developing systems. Because it is difficult to select for recessive mutations that affect mammalian embryos (Pedersen and Goldstein, 1979), existing mutations at the *a* locus and other loci should continue to be valuable for understanding preimplantation and early postimplantation development.

ACKNOWLEDGMENTS. This work was supported by the U.S. Department of Energy. We thank Dr. Hanna Balakier and Dr. Gerald Kidder for their comments, and Ms. Mary McKenney and Ms. Naomi Sinai for their assistance with the manuscript.

Discussion

ORSINI (Wisconsin): When you obtained development of the trophoblast in your donor embryos injected into chimeric blastocysts, did you find hatching or shedding of the zona pellucida and was there any further development?

PEDERSEN: We found hatching at about 48 hr after injection of morulae, so they are apparently hatching on schedule in the blastocoele. It is technically difficult to study further development *in vitro*, because the recipient is at a somewhat more advanced stage than the donor and must be kept in a somewhat deprived condition (without serum or adequate amino acids for normal postimplantation development) in order to prevent it from attaching and flattening out, and thereby obscuring the blastocoele.

BULLOCK (Baylor): Can you recover these donor blastocysts from their giant host and put them into recipient uteri?

PEDERSEN: I have not yet done any transfers to foster mothers. One can simply recover the blastocysts by tearing open the host or, in the case of the compact structure, by doing immunosurgery. When I recovered either the blastocysts or the compact structures, they were able to outgrow normally. It would be interesting to transfer the chimeras containing the compact donor structures to recipient uteri to see whether the compact structure would differentiate into an inner cell mass with fetal derivatives.

JOHNSON (Cambridge): I take issue with your conclusion that outside cells become committed substantially earlier than inside cells. While it is not possible to say that outer cells stay labile for as long as inner cells, the difference is only a matter of hours, not much more, and this very short gap needs to be tested experimentally (see Rossant and Frels, this volume). Secondly, to say that outer cells differentiate when inner cells do not, implies a qualitative judgement which I think we cannot make at the moment. There is evidence that inner and outer cells each have distinctive properties from the moment they first appear.

The available evidence suggests that outside and inside cells both start their differentiation at about the same time, and then both appear to become committed at about the same time. Marilyn Monk's notion of a totipotent, stem cell line from which differentiated committed cells branch off is a very attractive one, but it is based on one property only, that of X-chromosome inactivation. Most other data, however, are more consistent with a series of symmetrical decisions occurring in early development, in which two populations of cells undergo divergent differentiation and commitment. I think your conclusion is rather strong.

Your injection experiments, particularly of embryos with the zona intact, provide the strongest evidence yet against the microenvironmental hypothesis. In the injected zona-free embryos that attached to the inner surface of the trophectoderm, are there any specialized junctions with the trophectoderm and do those cells divide during the period of days over which they are cultured?

PEDERSEN: With respect to the last question, although we haven't done any extensive cell counts, the sections do appear to have substantially more cells than at the time of injection. With respect to the junctions, we have not done any ultrastructural analysis so I do not know what type of junctions form. It is interesting that the donor embryos attach to the inside of the host trophectoderm, whereas the outside of the trophectoderm no longer agglutinates with other embryos after the blastocoele forms. Concerning the definition of determination, I am simply talking about the time of commitment of the inner cells in relation to differentiation of the outer cells, because there are no good data on the commitment of the outer cells at blastocyst formation.

ROSSANT (Brock): When you put in a whole embryo and it attached to the trophoblast, did it form an inner cell mass with endoderm?

PEDERSEN: The point of our two-dimensional gels is to answer that question. We have not done enough histology to know whether there is endoderm forming; if so, it is certainly not fully differentiated, visceral endoderm. The embryos certainly retain the ability to form trophoblast.

ROSSANT: I did some unpublished experiments in which I injected a single blastomere from an eight-cell embryo into a blastocyst and found no evidence of chimeras; the injected cell often formed a clump of cells at the abembryonic pole in early postimplantation embryos. These structures appeared similar to yours and I also see no obvious endoderm; it is a rather strange phenomenon.

PEDERSEN: We have not excluded a possible role of the internal environment; to do so would require removing the blastocoele fluid, which means inverting the embryo and that is technically very difficult.

ROSSANT: Gardner (personal communication) has done that. Inverted embryos blastulate again, with the inner cell mass on the outside. Those embryos will implant after transfer; somehow they manage to turn themselves the right way round, although I am not sure how.

PEDERSEN: The few chimeric blastocysts that I have inverted seem also to turn themselves around to the right side again, at least judging from the ultimate position of microspheres placed in contact with the inner surface of the blastocyst.

VAN BLERKOM (Colorado): Considering that the mouse blastocyst expands and collapses, would you expect the environment to remain stable? If there was a short-lived substance involved in commitment of the inner cell mass, would you expect it to persist throughout the blastocyst stage?

PEDERSEN: The rhythmic collapse and reexpansion of the blastocyst is very poorly understood; it is not even known whether it is a passive or an active contraction. There is certainly adequate actin and myosin in the early blastocyst to enable it to undergo an active process of contraction, as shown by Dr. Sobel's work.

VAN BLERKOM: What happens to the environment inside when the blastocyst collapses?

PEDERSEN: As I look at the time-lapse movies, the blastocoele fluid is never totally lost, but there could be electrical continuity between the inside and the outside at that moment. That fact would not alter my conclusion, because I am not arguing for a decisive role of the fluid, quite the opposite.

BIGGERS (Harvard): Cole and Paul [*in: Preimplantation Stages of Pregnancy* (G. E. W. Wolstenholme and M. O'Connor, eds.), pp. 82–122, Churchill, London, 1965] showed that the amount of fluid lost from the blastocoele is about one-third of the total volume. The fluid is lost relatively suddenly and then is slowly reformed. If there is resynthesis of fluid by the trophectoderm, I do not see why the composition should not change. As far as I know, nobody has looked at this question experimentally. The rhythmic contractions observed in culture, seen in the rat and rabbit as well, are very strange. Whitten (personal communication) found, under very carefully controlled culture conditions, that the pulsations sometimes did not occur at all. Leonov (personal communication) finds very irregular pulsations, so they may be an artifact.

PSYCHOYOS (Bicêtre): I do not agree that the contraction and dilation process which is observed *in vitro* can be an artifact, because microcinematography shows the rhythmic way in which the normal blastocyst performs this movement. A blastocyst in the delayed condition, however, followed *in vitro* by time-lapse cinematography, shows no movement, for the first 12 hr; the blastocyst remains in the dilation phase. Normal blastocysts apparently require a specific protein to manifest this phenomenon, which could be important for implantation. Perhaps the synthesis of actin in the blastocyst may be the key for this pumping movement of the blastocyst. In your homozygotes, where there is a modification in hatching, is there a failure of the blastocyst to synthesize a specific protein that may be involved in this phenomenon of motility?

PEDERSEN: That is possible. It would be fascinating to look at actin and other proteins in mutant embryos, but I would be reluctant to do so until we have a sure means of identifying the homozygotes.

References

Ansell, J. D., and Snow, M. H. L., 1975, The development of trophoblast *in vitro* from blastocysts containing varying amounts of inner cell mass, *J. Embryol. Exp. Morphol.* **33:**177–185.

Atienza-Samols, S. B., and Sherman, M. I., 1979, *In vitro* development of core cells of the inner cell mass of the mouse blastocyst: Effects of conditioned medium, *J. Exp. Zool.* **208:**67–72.

Bennet, D., 1975, The T-locus of the mouse, *Cell* **6:**441–454.

Borland, R. M., Hazra, S., Biggers, J. D., and Lechene, C. P., 1977, The elemental composition of the environments of the gametes and preimplantation embryo during the initiation of pregnancy, *Biol. Reprod.* **16:**147–157.

Braude, P. R., 1979, Time-dependent effects of α-amanitin on blastocyst formation in the mouse, *J. Embryol. Exp. Morphol.* **52:**193–202.

Calarco, P. G., and Pedersen, R. A., 1976, Ultrastructural observations of lethal yellow (A^y/A^y) mouse embryos, *J. Embryol. Exp. Morphol.* **35:**73–80.

Castle, W. E., and Little, C. C., 1910, On a modified Mendelian ratio among yellow mice, *Science* **32:**868–870.

Cizadlo, G. R., and Granholm, N.H., 1978a, *In vivo* development of the lethal yellow (A^y/A^y) mouse embryo at 105 hours post coitum, *Genetica* **48:**89–93.

Cizadlo, G. R., and Granholm, N. H., 1978b, Ultrastructual analysis of preimplantation lethal yellow (A^y/A^y) mouse embryos, *J. Embryol. Exp. Morphol.* **45:**13–24.

Copp, A. J., 1978, Interaction between inner cell mass and trophectoderm of the mouse blastocyst. I. A study of cellular proliferation, *J. Embryol. Exp. Morphol.* **48:**109–125.

Copp, A. J., 1979, Interaction between inner cell mass and trophectoderm of the mouse blastocyst. II. The fate of the polar trophectoderm, *J. Embryol. Exp. Morphol.* **51:**109–120.

Cuénot, L., 1905, Les races pures et leurs combinaisons chez les souris (4me note), *Arch. Zool. Exp. Gen. 4e Sér.* **3:**123–132.

Diwan, S. B., and Stevens, L. C., 1976, Development of teratomas from the ectoderm of mouse egg cylinders, *J. Natl. Cancer Inst.* **57:**937–942.

Ducibella, T., Albertini, D. F., Anderson, E., and Biggers, J. D., 1975, The preimplantation mammalian

embryo: Characterization of intercellular junctions and their appearance during development, *Dev. Biol.* **45**:231–250.
Dziadek, M., 1979, Cell differentiation in isolated inner cell masses of mouse blastocysts *in vitro:* Onset of specific gene expression, *J. Embryol. Exp. Morphol.* **53**:367–379.
Eaton, G. J., 1968, Stimulation of trophoblastic giant cell differentiation in the homozygous *yellow* mouse embryo, *Genetica* **39**:371–378.
Eaton, G. J., and Green, M. M., 1962, Implantation and lethality of the *yellow* mouse, *Genetica* **33**:106–112.
Eaton, G. J., and Green, M. M., 1963, Giant cell differentiation and lethality of homozygous *yellow* mouse embryos, *Genetica* **34**:155–161.
Enders, A. C., 1971, The fine structure of the blastocyst, in: *Biology of the Blastocyst* (R. J. Blandau, ed.), pp. 71–94, Univ. of Chicago Press, Chicago.
Enders, A. C., Given, R. L., and Schlafke, S., 1978, Differentiation and migration of endoderm in the rat and mouse at implantation, *Anat. Rec.* **190**:65–78.
Gardner, R. L., 1972, An investigation of inner cell mass and trophoblast tissues following their isolation from the mouse blastocyst, *J. Embryol. Exp. Morphol.* **28**:279–312.
Gardner, R. L., 1975, Analysis of determination and differentiation in the early mammalian embryo using intra- and interspecific chimeras, in: *The Developmental Biology of Reproduction* (C. L. Markert and J. Papaconstantinou, eds.), pp. 207–236, Academic Press, New York.
Gardner, R. L., and Papaioannou, V. E., 1975, Differentiation in the trophectoderm and inner cell mass, in: *Early Development of Mammals* (M. Balls and A. E. Wild, eds.), pp. 107–132, Cambridge Univ. Press, Cambridge.
Gardner, R. L., and Rossant, J., 1976, Determination during embryogenesis, in: *Embryogenesis in Mammals,* Ciba Foundation Symposium 40 (new series), pp. 5–25, Elsevier/Excerpta Medica/North-Holland, Amsterdam.
Gardner, R. L., and Rossant, J., 1979, Investigation of the fate of 4.5 day *post-coitum* mouse inner cell mass cells by blastocyst injection, *J. Embryol. Exp. Morphol.* **52**:141–152.
Gardner, R. L., Papaioannou, V. E., and Barton, S. C., 1973, Origin of the ectoplacental cone and secondary giant cells in mouse blastocysts reconstituted from isolated trophoblast and inner cell mass, *J. Embryol. Exp. Morphol.* **30**:561–572.
Gluecksohn-Waelsch, S., 1979, Genetic control of morphogenetic and biochemical differentiation: Lethal albino deletions in the mouse, *Cell* **16**:225–237.
Granholm, N. H., and Johnson, P. M., 1978, Enhanced identification of lethal yellow (A^y/A^y) mouse embryos by means of delayed development of four-cell stages, *J. Exp. Zool.* **205**:327–333.
Green, M. C., 1975, The laboratory mouse, *Mus musculus,* in: *Handbook of Genetics,* Vol. 4 (R. C. King, ed.), pp. 203–241, Plenum Press, New York.
Grobstein, C., 1952, Intra-ocular growth and differentiation of clusters of mouse embryonic shields cultured with and without primitive endoderm and in the presence of possible inductors, *J. Exp. Zool.* **119**:355–380.
Handyside, A. H., 1978, Time of commitment of inside cells isolated from preimplantation mouse embryos, *J. Embryol. Exp. Morphol.* **45**:37–53.
Hay, E. D., and Meier, S., 1976, Stimulation of corneal differentiation by interaction between cell surface and extracellular matrix, *Dev. Biol.* **52**:141–157.
Hillman, N., Sherman, M. I., and Graham, C., 1972, The effect of spatial arrangement on cell determination during mouse development, *J. Embryol. Exp. Morphol.* **28**:263–278.
Hogan, B., and Tilly, R., 1977, *In vitro* culture and differentiation of normal mouse blastocysts, *Nature (London)* **265**:626–629.
Hogan, B., and Tilly, R., 1978a, *In vitro* development of inner cell masses isolated immunosurgically from mouse blastocysts. I. Inner cell masses from 3.5-day *p.c.* blastocysts incubated for 24 h before immunosurgery, *J. Embryol. Exp. Morphol.* **45**:93–105.
Hogan, B., and Tilly, R., 1978b, *In vitro* development of inner cell masses isolated immunosurgically from mouse blastocysts. II. Inner cell masses from 3.5- to 4.0-day *p.c.* blastocysts, *J. Embryol. Exp. Morphol.* **45**:107–121.
Ibsen, H. L., and Steigleder, E., 1917, Evidence for the death *in utero* of the homozygous yellow mouse, *Am. Natur.* **51**:740–752.
Johnson, L. L., and Granholm, N. H., 1978, *In vitro* analysis of pre- and early postimplantation development of lethal yellow (A^y/A^y) mouse embryos, *J. Exp. Zool.* **204**:381–389.
Johnson, M. H., Handyside, A. H., and Braude, P. R., 1977, Control mechanisms in early mammalian development, in: *Development in Mammals,* Vol. 2 (M. H. Johnson, ed.), pp. 67–97, North-Holland, Amsterdam.
Kelly, S. J., 1975, Studies of the potency of the early cleavage blastomeres of the mouse, in: *The Early Development of Mammals* (M. Balls and A. E. Wild, eds.), pp. 97–105, Cambridge Univ. Press, Cambridge.

Kelly, S. J., 1977, Studies of the developmental potential of 4- and 8-cell stage mouse blastomeres, *J. Exp. Zool.* **200**:365–376.
Kelly, S. J., Mulnard, J. G., and Graham, C. F., 1978, Cell division and cell allocation in early mouse development, *J. Embryol. Exp. Morphol.* **48**:37–51.
Kirkham, W. B., 1917, Embryology of the yellow mouse, *Anat. Rec.* **11**:480–481.
Kirkham, W. B., 1919, The fate of homozygous yellow mice, *J. Exp. Zool.* **28**:125–135.
Leonard, A., Deknudt, G., and Linden, G., 1971, Ovulation and prenatal losses in different strains of mice, *Exp. Anim.* **4**:1–6.
Levak-Svajger, B., and Svajger, A., 1974, Investigation on the origin of definitive endoderm in the rat embryo, *J. Embryol. Exp. Morphol.* **32**:445–459.
Little, C. C., 1919, A note on the fate of individuals homozygous for certain color factors in mice, *Am. Natur.* **53**:185–187.
Magnuson, T., Demsey, A., and Stackpole, C. W., 1977, Characterization of intercellular junctions in the preimplantation mouse embryo by freeze-fracture and thin-section electron microscopy, *Dev. Biol.* **61**:252–261.
Miller, D. A., Kouri, R. E., Dev, V. G., Grewal, M. S., Hutton, J. J., and Miller, O. J., 1971, Assignment of four linkage groups to chromosomes in *Mus musculus* and a cytogenetic method for locating their centromeric ends, *Proc. Natl. Acad. Sci. USA* **68**:2699–2702.
Mintz, B., 1965, Experimental genetic mosaicism in the mouse, in: *Preimplantation Stages of Pregnancy*, Ciba Foundation Symposium (G. E. W. Wolstenholme and M. O'Connor, eds.), pp. 194–216, Churchill, London.
Nowel, M. S., and Chapman, G. B., 1976, The ultrastructure of implanted trophoblast cells of the yellow agouti mouse, *J. Anat.* **122**:177–188.
Papaioannou, V. E, and Gardner, R. L., 1976, Discussion (Table 1), in: *Embryogenesis in Mammals*, Ciba Foundation Symposium 40 (new series), pp. 232–233, Elsevier/Excerpta Medica/North-Holland, Amsterdam.
Papaioannou, V., and Gardner, R. L., 1979, Investigation of the lethal yellow A^y/A^y embryo using mouse chimeras, *J. Embryol. Exp. Morphol.* **52**:153–163.
Pedersen, R. A., 1974, Development of lethal yellow (A^y/A^y) mouse embryos *in vitro, J. Exp. Zool.* **188**:307–320.
Pedersen, R. A., and Goldstein, L. S., 1979, Detecting mutations expressed during early development of cultured mammalian embryos, *Genetics* **92**:(Suppl.) 141–151.
Pedersen, R. A., and Spindle, A. I., 1976, Genetic effects on mammalian development during and after implantation, in: *Embryogenesis in Mammals*, Ciba Foundation Symposium 40 (new series), pp. 133–149, Elsevier/Excerpta Medica/North-Holland, Amsterdam.
Pedersen, R. A., and Spindle, A. I., 1980, Role of the blastocoele microenvironment in early mouse embryo differentiation, *Nature (London)* **284**:550–552.
Pedersen, R. A., Spindle, A. I., and Wiley, L. M., 1977, Regeneration of endoderm by ectoderm isolated from mouse blastocysts, *Nature (London)* **270**:435–437.
Phillips, R. J. S., 1966, A *cis–trans* position effect at the *A* locus of the house mouse, *Genetics* **54**:485–495.
Robertson, G. G., 1942a, Increased viability of homozygous yellow mouse embryos in new uterine environments (abstract), *Genetics* **27**:166–167.
Robertson, G. G., 1942b, An analysis of the development of homozygous yellow mouse embryos, *J. Exp. Zool.* **89**:197–231.
Rossant, J., 1975a, Investigation of the determinative state of the mouse inner cell mass. I. Aggregation of isolated inner cell masses with morulae, *J. Embryol. Exp. Morphol.* **33**:979–990.
Rossant, J., 1975b, Investigation of the determinative state of the mouse inner cell mass. II. The fate of isolated inner cell masses transferred to the oviduct, *J. Embryol. Exp. Morphol.* **33**:991–1001.
Rossant, J., 1977, Cell commitment in early rodent development, in: *Development in Mammals*, Vol. 2 (M. H. Johnson, ed.), pp. 119–150, North-Holland, Amsterdam.
Rossant, J., and Lis, W. T., 1979, Potential of isolated mouse inner cell masses to form trophectoderm derivatives *in vivo, Dev. Biol.* **70**:255–261.
Rossant, J., and Papaioannou, V. E., 1977, The biology of embryogenesis, in: *Concepts in Mammalian Embryogenesis* (M. I. Sherman, ed.), pp. 1–36, MIT Press, Cambridge, Mass.
Runner, M. N., 1959, Linkage of brachypodism. A new member of linkage group V of the house mouse, *J. Hered.* **50**:81–84.
Russell, L. B., McDaniel, M. N. C., and Woodiel, F. N., 1963, Crossing-over within the *a* "locus" of the mouse (abstract), *Genetics* **48**:907.
Searle, A. G., Beechey, C. V., Eicher, E. M., Nesbitt, M. N., and Washburn, L. L., 1979, Colinearity in the mouse genome: A study of chromosome 2, *Cytogenet. Cell Genet.* **23**:255–263.

Sherman, M. I., 1975, The role of cell–cell interaction during early mouse embryogenesis, in: *The Early Development of Mammals* (M. Balls and A. E. Wild, eds.), pp. 145–165, Cambridge Univ. Press, Cambridge.

Sherman, M. I., and Atienza-Samols, S. B., 1979, Differentiation of mouse trophoblast does not require cell–cell interaction, *Exp. Cell Res.* **123:**73–77.

Silver, W. K., 1979, *The Coat Colors of Mice: A Model for Mammalian Gene Action and Interaction*, Springer-Verlag, New York.

Snell, G. D., and Stevens, L. C., 1966, Early embryology, in: *Biology of the Laboratory Mouse* (E. L. Green, ed.), 2nd ed., pp. 205–245, McGraw–Hill, New York.

Snell, G. D., Cudkowicz, G., and Bunker, H. P., 1967, Histocompatibility genes of mice. VII. H-13, a new histocompatibility locus in the fifth linkage group, *Transplantation* **5:**492–503.

Snow, M. H. L., 1976, The immediate postimplantation development of tetraploid mouse blastocysts, *J. Embryol. Exp. Morphol.* **35:**81–86.

Solter, D., and Knowles, B. B., 1975, Immunosurgery of mouse blastocyst, *Proc. Natl. Acad. Sci. USA* **72:**5099–5102.

Spindle, A. I., 1978, Trophoblast regeneration by inner cell masses isolated from cultured mouse embryos, *J. Exp. Zool.* **203:**483–489.

Strickland, S., Reich, E., and Sherman, M. I., 1976, Plasminogen activator in early embryogenesis: Enzyme production by trophoblast and parietal endoderm, *Cell* **9:**231–240.

Tarkowski, A. K., 1959, Experiments on the development of isolated blastomeres of mouse eggs, *Nature (London)* **184:**1286–1289.

Tarkowski, A. K., and Wróblewska, J., 1967, Development of blastomeres of mouse eggs isolated at the 4- and 8-cell stage, *J. Embryol. Exp. Morphol.* **18:**155–180.

Wallace, M. E., 1965, Pseudoallelism at the agouti locus in the mouse, *J. Hered.* **56:**267–271.

Wiley, L. M., Spindle, A. I., and Pedersen, R. A., 1978, Morphology of isolated mouse inner cell masses developing *in vitro, Dev. Biol.* **63:**1–10.

Wilson, I. B., Bolton, E., and Cuttler, R. H., 1972, Preimplantation differentiation in the mouse egg as revealed by microinjection of vital markers, *J. Embryol. Exp. Morphol.* **27:**467–479.

Wolpert, L., 1978, Gap junctions: Channels for communication in development, in: *Intercellular Functions and Synapses* (J. Feldman, N. B. Gilula, and J. D. Pitts, eds.), pp. 83–96, Chapman & Hall, London.

Wudl, L. R., Sherman, M. I., and Hillman, N., 1977, Nature of lethality of *t* mutations in embryos, *Nature (London)* **270:**137–140.

III

Macromolecular Synthesis in the Developing Egg

Introduction
Macromolecular Synthesis in the Developing Egg

Charles J. Epstein

Scarcely fifteen years have passed since the first quantitative study of an enzyme (lactate dehydrogenase) in a mammalian embryo was published by Brinster (1965), and just about the same time has elapsed since analysis of embryonic protein and nucleic acid synthesis was first performed qualitatively by Mintz (1964), and quantitatively by Monesi and Salfi (1967). In the intervening years there have been rapid advances in our ability to apply biochemical and molecular techniques to the investigation of oogenesis and the very early stages of embryonic development. Although application of these techniques to mammalian eggs and embryos has tended to lag behind their application to similar stages from invertebrates and lower vertebrates, this discrepancy seems to be rapidly narrowing. As the papers in this section illustrate, it has now become possible to directly assess the genetic control of early mammalian development rather than having to extrapolate from the results obtained with other classes and phyla. While much still remains to be done, it is clear that there are both important similarities and differences between mammalian and nonmammalian oocytes and embryos and among the different types of mammalian eggs and embryos.

In the chapters that follow, several of the recent advances in the study of the synthesis and function of two major classes of macromolecules, proteins and nucleic acis, in mammalian oocytes and preimplantation embryos are considered. These include the assay of nucleic acid polymerases, the detailed analysis of the synthesis of a large variety of polypeptides, the determination of the number of DNA sequences that are represented in embryonic messenger RNA, and the *in vitro* translation of oocyte and embryonic messenger RNAs. While directed primarily at developmental stages before the time of implantation, these studies provide a necessary foundation on which a description of the events that initiate and govern the process of implantation must be built.

Charles J. Epstein • Departments of Pediatrics and of Biochemistry and Biophysics, University of California, San Francisco, California 94143.

References

Brinster, R. L., 1965, Lactate dehydrogenase activity in the preimplanted mouse embryo, *Biochim. Biophys. Acta* **110**:439.

Mintz, B., 1964, Synthetic processes and early development in the mammalian egg, *J. Exp. Zool.* **157**:85.

Monesi, V., and Salfi, V., 1967, Macromolecular synthesis during early development in the mouse embryo, *Exp. Cell Res.* **46**:632.

7

Mobilization of Genetic Information in the Early Rabbit Trophoblast

Cole Manes, Michael J. Byers, and Andrew S. Carver

I. Introduction

Any broad understanding of the cellular and molecular aspects of embryonic implantation in the mammalian uterus would be expected to benefit from experimental evidence obtained from as great a variety of mammals as possible. Not only have two sharply divergent strategies been evolved for regulating the implantation event—allowing a division between those mammals capable of inducing a reversible growth arrest in the blastocyst and those incapable of doing so—but it is apparent that further refinements in implantation mechanisms characterize individual families or even genera of mammals—such as the single implantation site available to the several hundred embryos of the elephant shrew (Van der Horst and Gillman, 1942). Furthermore, as the implantation event requires the interaction of two competent tissues, the achievement of this competence by each tissue must be understood before a satisfactory comprehension of the event can be achieved.

Although studies of implantation in rodents have contributed, and no doubt will continue to contribute, a large share of the information in this area, it is advisable from the standpoint of generalizing these studies to consider the implantation in a mammal employing a quite different strategy, such as the rabbit. Rodents and rabbits, while they are phylogenetically closely related, differ in their ability to induce embryonic diapause. At least with respect to its inability to undergo a reversible growth arrest, the rabbit blastocyst is more akin to primates, which are presumably of principal relevance to the understanding of human reproduction.

This chapter is concerned with the rabbit blastocyst as it approaches implantation. The studies in our laboratory described briefly here were motivated by a desire to understand the interaction of genetic and environmental factors that result in the onset of cellular differentiation during early embryogenesis. These studies sought to identify those molecules that play pivotal roles during this critical developmental period, as contrasted with all other

Cole Manes, Michael J. Byers, and Andrew S. Carver • La Jolla Cancer Research Foundation, La Jolla, California 92037.

developmental periods. The generation of a competent trophoblast cell layer is an essential and unique requirement that must be met by the mammalian embryo during the first few days of development; the rabbit blastocyst just before implantation possesses almost 10^5 cells, greater than 90% of which are trophoblast (Daniel, 1964). Although our studies are of the whole blastocyst, the contribution of embryoblast cells to the resulting data is relatively minor.

Our report will consider the mobilization of genetic information in this early rabbit trophoblast, from the level of gene transcription to the level of polysome assembly. We shall also present some recent findings with regard to nucleic acid metabolism whose significance, in terms of genetic expression, is not yet clear. Much of the information presented must be considered preliminary, for these studies are still in progress; they may serve to provide a state-of-the-art summary in one implantation system, and may point the way toward potentially important areas for future investigation.

II. Results and Discussion

A. Unique-Sequence DNA Transcription

The theory of differential gene expression proposes that different cellular phenotypes are generated from identical genotypes through selective utilization of a common endowment of genetic information. According to this model, it would be logical to expect that a phenotype such as the trophoblast cell could be characterized by virtue of the specific genetic information expressed, either qualitatively or, at the very least, quantitatively. The vast majority of proteins synthesized by any given cell type are known to be genetically specified by that portion of the genome consisting of virtually unique DNA sequences. Up to two-thirds of the genome may be made up of unique-sequence DNA; the role of the remaining "repeat-sequence" DNA (Britten and Kohne, 1968) is not yet established.

We have denatured rabbit DNA and allowed it to reanneal to an extent sufficient for all of the repetitive DNA to re-form double-stranded hybrids. These hybrids can then be removed from the total DNA preparation by passage over hydroxyapatite (Britten and Kohne, 1968) and the eluted "single-copy DNA" used to ask questions regarding how much of this DNA is represented by RNA transcripts in a given cell, and to compare transcript populations from different cell types.

Schultz *et al.* (1973) reported that total RNA from the late rabbit blastocyst is homologous to 1.8% of the unique-sequence DNA in the rabbit genome. As a preliminary to further studies on the reprsesentation of the unique-sequence DNA in various early stages of rabbit embryogenesis, we have repeated that assay, with the refinement of assaying nuclear and cytoplasmic transcripts separately. It is also interesting to compare the overall complexity of blastocyst transcription with that of midgestation embryos and several adult tissues. Table 1 shows the results of these preliminary studies. While the values for the blastocyst are somewhat lower than in the previous study, all other values are quite comparable, and all show a relatively low level of complexity. The interesting exception is the brain.

These percentages can be converted into more meaningful terms when the total DNA content of rabbit single-copy DNA (3.15×10^9 base pairs; Schultz *et al.*, 1973) and the size of the average messenger RNA (2000 nucleotides; Lewin, 1975) are taken into account. The 0.41% representation in the cytoplasm of the rabbit trophoblast could potentially specify the synthesis of some 6400 diverse proteins. Certain qualifications must be applied to this

Table 1. RNA Homologous to Single-Copy Rabbit DNA [a]

RNA source	Percentage single-copy DNA hybridized to RNA at saturation R_0t	Schultz *et al.* (1973)
6-Day blastocyst		1.81
Nucleus	0.86	
Cytoplasm	0.41	
12-Day embryo	2.29	2.53
12-Day placenta	2.60	
Adult liver	2.85	2.90
Adult brain		6.60
Adult kidney	0.91	
Adult spleen	1.54	
Adult ovary	1.74	

[a] The [^{3}H]-DNA used in these determinations was prepared from rabbit blastocysts at a specific activity of approximately 400,000 cpm/μg and sonicated in 0.3 N NaOH to an average fragment size of 300 bases. It was self-reacted to a C_0t of 370, at which time 33.7% of the DNA was in double-stranded form and could be removed by passage over hydroxyapatite. The remaining unreacted DNA was considered to be the single-copy fraction; it reacted to 85% completion with unlabeled rabbit liver DNA, and the RNA/DNA reaction values reported here were corrected accordingly.

RNA for these reactions was prepared by homogenizing tissues in lysis buffer (100 mM NaCl, 10 mM Tris, pH 8.0, 1 mM EDTA, 0.1% 2-mercaptoethanol, 5 mM dithiothreitol, 1 mM aurintricarboxylic acid, 0.5% sodium dodecyl sulfate (SDS), and 0.5% sarkosyl). Homogenates were extracted with phenol:chloroform (1:1), then with chloroform:isoamyl alcohol (24:1), and precipitated under 2 vol of 95% ethanol at −20°C. The precipitates were treated with DNAse I (Worthington) twice, with phenol:chloroform extraction between DNAse digestions.

In a typical RNA/DNA reaction, RNA was present at 1000-fold excess and was reacted to a R_0t of 13,000–15,000 to achieve saturation. Twenty micrograms of RNA was mixed with 0.02 μg [^{3}H]-DNA; the mixture was boiled in 10 μl of 0.48 M phosphate buffer, pH 6.8, in a capped microcentrifuge tube, overlayered with paraffin oil, and cooled to 68°C. Completion of the reaction required approximately 4 days. Five- to ten-microliter samples were diluted into 2.2 ml of 0.12 M phosphate buffer. Four 500-μl aliquots were taken from this solution, two of which were made 0.02% SDS, and passed over hydroxyapatite at 60°C. Fifty micrograms of *E. coli* tRNA was added to each of the other two samples, which were then treated with RNAse A for 2 hr at 50°C. The samples were then made 0.02% SDS and passed over hydroxyapatite. The latter duplicate samples contain only DNA–DNA hybrids, and the radioactivity values from these hybrids were subtracted from the untreated hybrid values to yield the values for RNA–DNA hybrid formation.

type of calculation, however. It is evident that, at best, no more than 1000 to 1200 labeled peptide "spots" can be resolved in the blastocyst by two-dimensional polyacrylamide gel electrophoresis (Van Blerkom and Manes, 1977). The peptides so visualized are almost certainly synthesized from RNA templates that are members of the abundant and midabundant classes, which are not the templates forming the majority of the hybrids reported here. It is not clear yet just how biologically important these infrequent templates, less than ten copies being present per cell, may be in determining cellular phenotypes, although in sea urchin development, stage-specific changes appear to occur in this class (Galau *et al.*, 1976). A more profitable way to search for cell-specific messenger RNAs may be to prepare cDNA copies to total cytoplasmic message, which then allows analysis of the abundant and midabundant classes; this effort is currently under way in our laboratory.

We consider that we are measuring the complexity of the rare class of messenger RNAs only in the trophoblast cell; the embryoblast cells comprise 5% or less of the blastocyst and would presumably contribute a proportionally small share of the bulk of the RNA analyzed. Thus, we are analyzing the template RNA population of essentially a single cell type, and its complexity is low in comparison to organs and tissues composed of several cell types; these in turn are low in comparison to brain. It is not clear whether the higher complexity of brain is due to its possession of more distinct cell types than the other tissues, or whether brain somehow requires more complex genetic information per cell to carry out its functions.

B. Polysome Assembly and Ternary Complex Formation

When polysomes are prepared from rabbit blastocysts radiolabeled for 1 to 4 hr and analyzed by centrifugation through sucrose gradients, it becomes apparent that as much as 40–45% of the newly synthesized poly A-containing RNA and 60–70% of the newly synthesized ribosomes are *not* immediately taken up in the task of synthesizing proteins (Manes, unpublished data). Admittedly, these findings reflect the situation obtained after a period of *in vitro* maintenance and exposure to radioisotopes, during which unknown factors may have altered the *in vivo* situation. However, electron micrographs of rabbit trophoblast (Van Blerkom *et al.*, 1973) display an abundance of free ribosomes in the cytoplasm and render the above results more believable.

Thus, it appears that the ongoing rate of protein synthesis in these cells is not limited by the supply of ribosomes or messenger RNA, but rather by factors that influence the efficiency of polysome assembly. The point at which such regulation occurs in most eukaryotic cells is clearly at the *initiation* of the polypeptide chain, which requires the binding of messenger RNA to the smaller ribosomal subunit in the presence of initator Met-tRNA$_i$ (Lodish, 1976). Before this stable binding can occur, a "ternary complex" consisting of Met-tRNA$_i$, GTP, and a eukaryotic initiation factor (eIF-2) must be assembled (Weissbach and Ochoa, 1976). The ability of eIF-2 to participate in ternary complex formation is dependent upon its degree of phosphorylation, the phosphyorylated form being relatively inactive (Ernst *et al.*, 1978). In sum, the rate of protein synthesis in eukaryotic cells is largely dependent upon the state of phosphorylation of eIF-2.

The activity of eIF-2 in dormant embryos of the brine shrimp *Artemia salina* is very low, but markedly increases as development resumes (Filipowicz *et al.*, 1975). By anal-

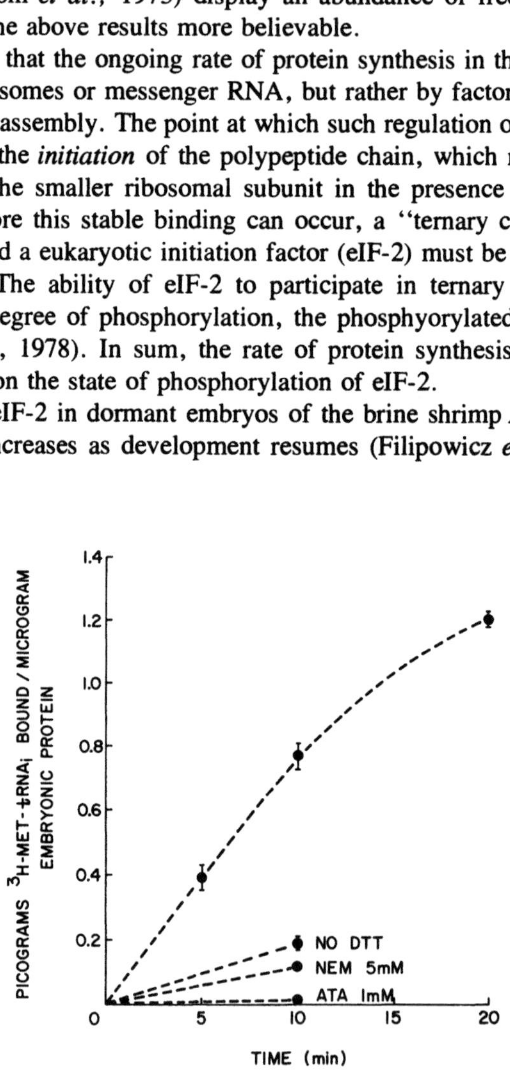

Figure 1. Ternary complex formation catalyzed by rabbit blastocyst cytoplasm. Blastocysts were washed twice in cold isotonic saline immediately after recovery from the uterus, and removed from the zona pellucida. They were then homogenized in 200–400 μl of 20 mM HEPES, pH 7.3, 100 mM KCl, 3 mM Mg acetate, 1 mM dithiothreitol (DTT) in ice. The homogenate was centrifuged at 12,000*g* for 10 min at 0–4°C. Fifty-microliter aliquots of the supernatant were then assayed for eIF-2 activity by the method of Filipowicz *et al.* (1975). L-[*methionyl-methyl-*^{3}H]tRNA$_i$ was obtained from New England Nuclear Corporation. NEM, *N*-Ethylmaleimide; ATA, aurintricarboxylic acid.

ogy, it would be of great interest to compare the activity of eIF-2 in growing and diapausing mammalian blastocysts. The rabbit blastocyst offers the quantitative advantage of providing more biological material, but the obvious disadvantage is that it does not undergo diapause. We can report here that it is possible to measure the specific activity of eIF-2 in the rabbit blastocyst: 1 μg cytoplasmic protein catalyzes the binding of 1.2 pg L-[*methionyl-methyl*-^{3}H] Met-tRNA$_i$ in a 20-min reaction. The reaction is linear for 10 min, is dependent upon added dithiothreitol (DTT) and GTP, and is strongly inhibited by *N*-ethylmaleimide and aurintricarboxylic acid (ATA) (Figure 1). Pretreatment of blastocysts with either ATA or diamide, a compound known to convert reduced glutathione quantitatively to the oxidized form (Zehavi-Willner *et al.*, 1971), lowers the specific activity of the eIF-2 by about one-half.

Calculation reveals that some 500 mouse blastocysts would be required to furnish sufficient embryonic protein for a single determination. Those numbers are not unthinkable, but such an undertaking would require the conviction that this regulatory aspect of blastocyst development is worth exploring.

C. Cytoplasmic DNA

When rabbit blastocysts are maintained for more than 4 hr in the presence of low concentrations of [^{3}H]thymidine and with ethidium bromide at 1 μg/ml, labeled fragments of DNA begin to accumulate in the cytoplasm. The critical reader will at once want assurance that this DNA is indeed "cytoplasmic" and not the result of nuclear leakage, cell death, or the fraction of cells undergoing mitosis. The method of cell fractionation into "nuclear" and "cytoplasmic" fractions is obviously important at this point, and must be dealt with in some detail. Our procedure is to wash the radiolabeled blastocysts in cold isotonic saline and to remove zona pellucida manually. The washed embryonic cells are transferred by capillary pipet to 2 ml sterile RSB buffer (10 mM Tris, pH 8.5, 10 mM NaCl, 1.5 mM $MgCl_2$, 0.3% NP40) in ice and allowed to stand 10 min. The sample is then vortexed at high speed for 30 sec and centrifuged at 3000*g* for 10 min. The supernatant is carefully withdrawn with a pasteur pipet and recentrifuged. The final supernatant is considered to be the "cytoplasmic" fraction. Microscopic examination of the initial pellet shows that it consists of unbroken nuclei with occasional cytoplasmic tags. Electrophoretic analysis of these fractions obtained from embryos radiolabeled for 30 min with [^{3}H]uridine reveals only the expected low-molecular-weight RNA species in the cytoplasm, while the nuclear fraction contains much high-molecular-weight material (Figure 2). Furthermore, during the first 2 hr of radiolabeling with [^{3}H]thymidine, the cytoplasmic fraction yields radioactive DNA of high molecular weight that can be shown to be entirely of mitochondrial origin, in spite of the presence of ethidium bromide. These results strengthen the argument that the cytoplasmic fraction is negligibly contaminated by nuclear contents.

Figure 3 shows the electrophoretic behavior of overnight-labled cytoplasmic DNA in a denaturing polyacrylamide gel. Size standards were provided by fragments of known length produced from polyoma DNA by Hpa II digestion. The major peak of cytoplasmic DNA averages 150 to 160 base pairs in length; it is totally resistant to S1 nuclease, indicating that it is double stranded. In cesium chloride gradients containing propidium iodide, this DNA bands entirely in the region of linear or relaxed circular DNA (Hudson *et al.*, 1969).

The 150- to 160-base-pair peak has not yet been analyzed in isolation. The total cytoplasmic DNA sample self-reacts after denaturation with virtually the same kinetics as nuclear DNA; thus it contains both repetitive and unique-sequence DNA, and is not viral or mycoplasmal. The DNA reacts with cytoplasmic and nuclear RNA to the same extent as

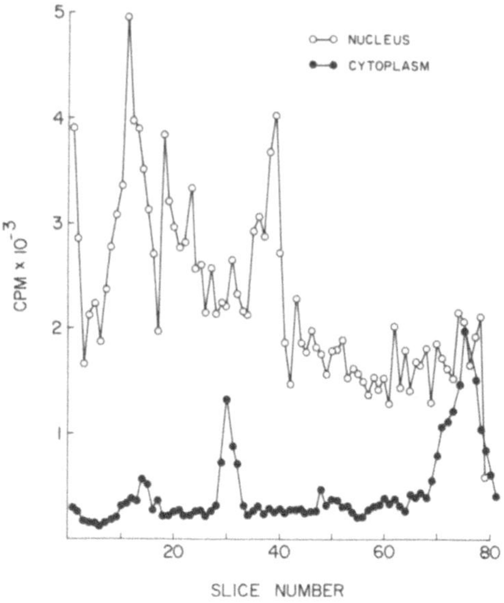

Figure 2. Electrophoretic separation of embryonic RNA radiolabeled for 35 min. Four 6-day rabbit blastocysts were exposed to [5-^{3}H]uridine at 40 μCi/ml, then separated into nuclear and cytoplasmic fractions as described in the text. The crude nuclear pellet was resuspended in 2 ml of RSB buffer; both fractions were made 0.2 M NaCl and precipitated overnight at −20°C under 2.5 vol of 95% ethanol. Precipitates were collected by centrifugation at 12,000*g* for 60 min and dried under vacuum. RNA was extracted and electrophoresis was performed as described previously (Manes, 1977). Gels were sliced, dissolved in 30% hydrogen peroxide, and counted in 10 ml Beta-Phase (West Chem Products).

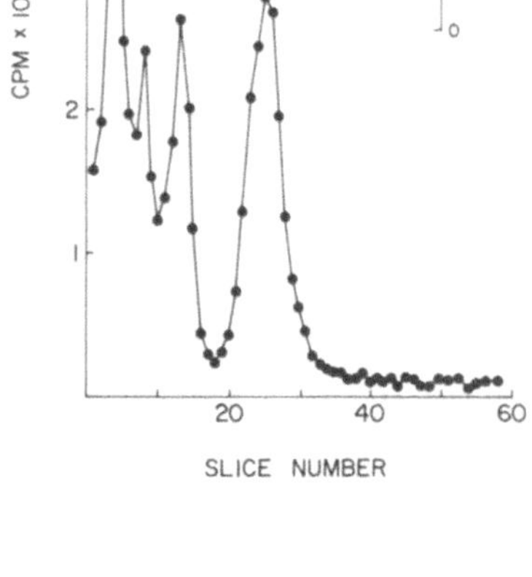

Figure 3. Electrophoretic separation of radiolabeled DNA obtained from the embryonic cytoplasm after 20-hr exposure to [*methyl*-^{3}H]thymidine. Blastocysts were maintained in modified F-12 medium (Manes, 1977) containing 5% rabbit serum, ethidium bromide at 1 μg/ml, and [*methyl*-^{3}H]thymidine at 5 μCi/ml (0.1 μM). Nuclear and cytoplasmic fractions were prepared as described in the text. The cytoplasmic fraction was made 100 mM in NaCl, 10 mM in EDTA, and 0.6% in SDS and extracted with an equal volume of phenol:chloroform (1:1), followed by a second extraction with an equal volume of chloroform:isoamyl alcohol (24:1). Nucleic acids were precipitated from the final aqueous phase with 2 vol of 95% ethanol at −20°C overnight. Precipitates were collected by centrifugation at 12,000*g* for 60 min and dried under vacuum. Glyoxal denaturation was carried out as described by McMaster and Carmichael (1977), and the denatured samples were then electrophoresed in the buffer system described previously (Manes, 1977). Size standards were Hpa II fragments of polyoma [^{3}H]-DNA electrophoresed in a parallel gel; this DNA was generously supplied by Dr. E. Linney. The gels were sliced, dissolved in 30% hydrogen peroxide, and counted in 10 ml Beta-Phase (West Chem Products). b.p., Base pairs.

does total nuclear DNA, indicating that it is not being generated preferentially from transcription sites.

The function of this DNA is not known. If the labeling medium from the overnight culture is extracted, the 150- to 160-base-pair peak is recovered predominantly, with a small peak seen in the 350-base-pair region. The fact that it appears as homogeneous peaks, and that it is presumably at least transiently associated with the cell surface, is provocative. Cytoplasmic DNA of discrete size classes has been reported in several other cell types (Bell, 1971; Koch, 1973; Meinke *et al.*, 1973). Phytohemagglutinin-stimulated human lymphocytes are also reported to "excrete" DNA (Rogers, 1976). Moreover, in leukemic AKR mice, a population of T cells can be identified as suppressors of the immune reaction, and these cells require cell-surface DNA to carry out this suppressor function (Russell and Golub, 1978). It is possible, therefore, that this phenomenon in the blastocyst approaching implantation is involved in the as yet poorly understood mechanism whereby the embryo and fetus resist immunological rejection during gestation. Further investigation is obviously required to reveal the biological significance, if any, of this cytoplasmic DNA.

D. RNA-Directed DNA Polymerase

Studies in our laboratory on the buoyant characteristics of pulse-labeled cytoplasmic RNA in the rabbit blastocyst have suggested that there may be at least a transient associa-

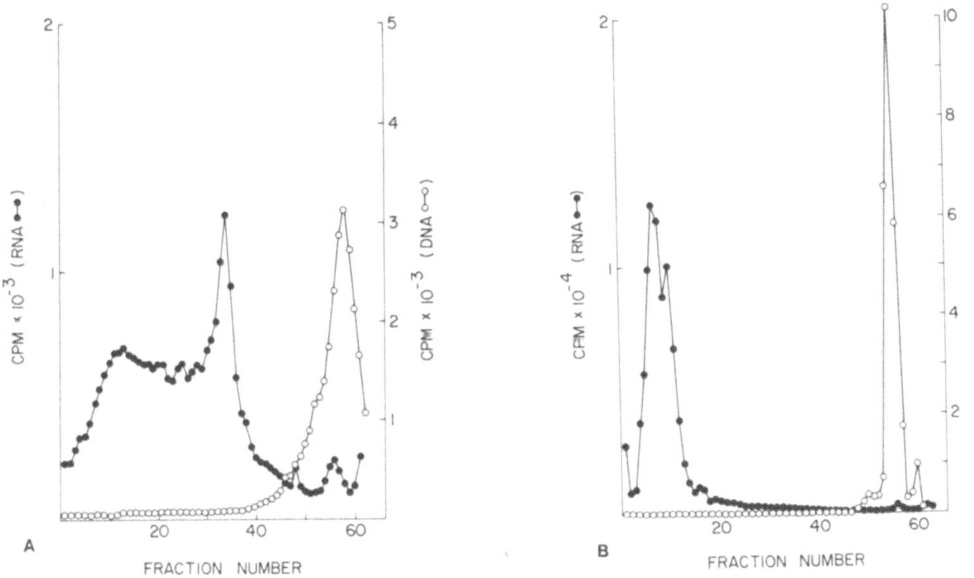

Figure 4. A: Isopycnic analysis of pulse-labeled cytoplasmic RNA in the rabbit blastocyst. Embryos were exposed to [5-^{3}H]uridine at 200 μCi/ml for 15 min, then separated into cytoplasmic and nuclear fractions as described in the text. Nucleic acids were extracted from each fraction as described elsewhere (Manes, 1977), precipitated, and dried under vacuum. They were then added to 4.5-ml cesium chloride–guanidinium chloride solutions in polyallomer tubes prepared according to Enea and Zinder (1975). Centrifugation was carried out in a Beckman SW 50.1 rotor at 35,000 rpm for 66 hr at 15°C. Eight-drop fractions were collected by puncturing the bottom of the centrifuge tube. Each fraction was diluted with 0.5 ml water and counted in 10 ml Beta-Phase (West Chem Products). (B) Isopycnic analysis of pulse-labeled nuclear RNA in the rabbit blastocyst. Procedures were as described under (A). The radioactive peaks near the top of the gradient (fractions 52–60) in both figures were from [^{3}H]-DNA centrifuged separately.

tion of this material with DNA. Earlier studies of the kinetics of cytoplasmic RNA labeling had shown that the first appearance of labeled 18 S ribosomal RNA occurred at 30 to 35 min after exposure to radiolabel; it was followed within about 5 min by the appearance of 28 S ribosomal RNA. Embryos pulse-labeled for 15 min yield cytoplasmic RNA that is rather heterogeneous, with a broad peak at about 18 S, as well as 4 S RNA. When this 15-min pulse-labeled RNA is centrifuged to equilibrium in cesium chloride containing guanidinium chloride (Enea and Zinder, 1975), a variable portion of it displays densities characteristic of RNA/DNA hybrids (Figure 4A). By way of contrast, the nuclear RNA obtained from the same embryos consistently behaves as "pure" RNA (Figure 4B).

This buoyant behavior of the pulse-labeled cytoplasmic RNA can only be accounted for by the association of single-stranded RNA with single-stranded DNA. This finding, along with kinetic considerations to be presented elsewhere, prompted a search for RNA-directed DNA polymerase(s) in the embryonic cytoplasm. One or several such enzymes stimulated by oligo $(dT)_{10}\cdot$poly rA are easily demonstrable. Using total cytoplasmic extracts prepared in 0.3% NP40, no endogenous reaction has yet been detected. The enzyme(s) will polymerize [^{3}H]thymidine triphosphate equally well on an oligo $(dT)_{10}\cdot$poly dA template, and is strongly inhibited by 25 μM ethidium bromide. It therefore has many of the characteristics of DNA polymerase γ, which is responsible for, among other things, mithochondrial DNA replication (Bolden *et al.*, 1976), and which is capable of generating long regions of single-stranded DNA (Bjursell *et al.*, 1979). The embryonic enzyme does not fractionate exclusively with the mitochondrial pellet, however. Nor have efforts been successful to date to associate this activity with poly A-containing particles sedimenting at viral densities. Nonetheless, it would be premature to rule out a viral origin for at least a portion of this activity. Van Blerkom has demonstrated what appear to be both extracisternal A-type particles as well as budding C-type particles in the rabbit blastocyst by electron microscopy.* The polymerase activity is not detectable in blastocoelic fluid, however, and appears to be an intrinsic component of embryonic cells. Again, an understanding of its biological role, if any, will have to await further investigation.

III. Summary

We have presented here several aspects of nucleic acid metabolism in the rabbit blastocyst during the immediate preimplantation period. These findings may be viewed as relatively discrete pieces of a puzzle that, when assembled with other appropriate pieces, will disclose the essential activities required of the embryo to generate a cell type capable of implantation. We do not yet know how early in development these properties appear in the embryo, nor how many of them may be shared with other cell types. It is thought-provoking that malignant tumor cells appear to be capable of something like "implantation" in the hormonally prepared uterus (Short and Yoshinaga, 1967), and that viral-like RNA-directed DNA polymerase activity is a component of the normal placenta (Nelson *et al.*, 1978). It can be expected that new insights regarding the mobilization of genetic information from several sources during the evolution of viviparity will result from future investigation.

ACKNOWLEDGMENT. The work reported here was supported by Grant HD 11654 from the U.S. Public Health Service.

* See his photographs reproduced in the review article by Manes (1974).

Discussion

SHERMAN (Roche Institute): What is the proportion of labeled DNA in the cytoplasm to the labeled DNA in the nucleus?

MANES: It is consistently about 8 to 10% of the total cellular DNA.

A. KAYE (Weizmann Institute): I have a comment and a question. Retrospectively we have taken your suggestion of looking at eIF-2 in developmental stages. Specifically, while following up the report of Liang *et al.* (*J. Biol. Chem.* **252:**5692, 1977) of hormonal influences on eIF-2 after castration and readministration of testosterone, Schimke and I thought that the chick oviduct system, which shows an exaggerated hormonal response in so many other ways, would be ideal for studying a change in eIF-2 concentration. Unfortunately, using the assay of the formation of the ternary complex in the absence of RNA that you mentioned, we could find no difference in the apparent concentration of eIF-2, in terms of cytosol protein concentration, in immature chick oviducts stimulated by estradiol benzoate, in oviducts which had been withdrawn from stimulation or in oviducts of laying hens. I hope this is not the discouraging bridge alluded to between insects and mammals. Control of initiation at the level of the apparent concentration of eIF-2 may not be the panacea. As you say, it may be something much more subtle concerning phosphorylation of the initiation factor. My question concerns the characterization of the DNA, which is a fascinating problem. Have you considered looking at methylation patterns of DNA, which can now be done easily and which becomes a very subtle way of distinguishing very similar DNAs? Have you considered whether there is any possibility of gene amplification and/or possible loss of genetic elements into the cytoplasm?

MANES: Well, I could talk the rest of the afternoon on your comments and questions. I will briefly say that the initiation factor in the reticulocyte lysate system is known to be affected by oxidized glutathione and oxidized sulfhydryl groups; this is also true, in fact, of the mammalian embryo. At least it is with the rabbit. I am not sure if people have studied it in the mouse. Protein synthesis in the rabbit is very sensitive to the presence of free sulfhydryls and this was another parallel that seemed to be intriguing. It turns out that it is also true that Diamide, which is a compound that reduces glutathione to oxidized glutathione, will depress the activity of this initiation factor in the intact blastocyst. There are so many factors involved that eIF-2 is not likely to be the only one.

Methylation is certainly of current interest in many laboratories. We are in fact digesting this material and the nuclear DNA with the endonucleases Hpa II and Msp 1, to compare the patterns. I do not yet have results, but it would be very interesting if there are different methylation patterns in the two categories of DNA. We have done one preliminary experiment in which we asked how much of the cytoplasmic RNA is homologous to this cytoplasmic DNA compared to the total nuclear DNA. Unfortunately, it turns out to be homologous to about the same extent, so it does not look as though there is overrepresentation of transcription sites.

EPSTEIN (UCSF): Have you tried hybridizing back to the nuclear DNA?

MANES: We have not done that experiment.

EPSTEIN: It would give some idea of whether it is a specific product or just nonspecific breakdown.

SCHULTZ (Calgary): In the RNA–DNA hybrids that are visualized after a pulse-label with uridine from the cytoplasm, in terms of characterization of that hybrid, is it true that the RNA is heterogenous?

MANES: The pulse-labeled RNA is a heterogenous population with a broad peak around 18 S, but there is also a fair amount of 4 S RNA that is labeled in the 15-min pulse period. I have not yet sized the two categories, namely the free RNA on one side and the hybrids in the middle of the gradient, so I do not know whether we can account for all the free RNA as the 4 S material; that is one possibility.

SCHULTZ: I see. The DNA would appear to be the discrete hundred-base-pair unit that you visualized initially?

MANES: Yes. There is a major peak at about 150–160 base pairs and another peak at about 350, a bit too large to be a dimer but that is possible.

SCHULTZ: In terms of visualizing transcription with the EM, are there any experiments relating to whether transcription occurs in a one-cell newly fertilized embryo?

MANES: There is transcription in the early cleavage stages; it is all very large transcript, as a matter of fact, much larger than in later stages. There are fewer initiations and the gradients show more scatter, but the transcripts are there. Our inability to detect them by radiolabeling probably means that they are very labile nuclear RNA species that are not stabilized and are not detected in, say, a 4-hr accumulated labeling time, but there is transcription very definitely going on.

JOHNSON (Cambridge): Can you rule out the possibility of cell death in your system, and of the cytoplasmic DNA being degradation products? When we pulse-label mouse uteri with [^{3}H]thymidine when cell death is occurring, we see incorporation of label into some of the breakdown products in the cells.

MANES: That is obviously a relevant question. If this is degradation, which it might turn out to be, it is occurring at very specific sites. These things are being clipped in very specific ways, which could mean that there are site-specific nucleases in the nucleus, which would in itself be an interesting finding.

KIDDER (UCSF): I am not sure if I understood how you set up these experiments where you detect what you think are DNA–RNA hybrids. Was that a cesium gradient?

MANES: It was a cesium gradient containing guanidinium chloride, which complexes enough with the RNA so that its density is sufficient to float off the bottom of the tube.

KIDDER: I have not worked with the guanidinium, but I have done plenty of cesium sulfate and chloride gradients, and cesium salts alone are not sufficient to completely deproteinize RNP particles. Did you consider that possibility?

MANES: Yes, that is one of the beauties of the guanidinium system—that you can actually put in RNP particles and get out pure nucleic acids from this system.

KIDDER: So the middle peak is not incompletely deproteinized RNA?

MANES: Not to the best of our knowledge. There was a report from Atardi's lab (*Proc. Natl. Acad. Sci. USA* **74:**1348, 1977) on initiation complex in mitochondria, that he thought was DNA complexed to membranes, which could not be removed by any of the detergents or proteases he used and which did float near the middle of that gradient. So it is possible that these are nucleic acids complexed with membrane, but we used lots of detergents and proteases in our preparation.

KIDDER: I suggest you run the experiment again and use an amino acid label in conjunction with uridine label to see if there is any protein there.

MANES: That is a good point.

EPSTEIN: Having mentioned mitochondria, where do the mitochondrial DNAs show up in your isolation system?

MANES: Those are of high molecular weight and they would stay at the top of the gels. During the first hour or two of labeling, in spite of the fact that we have ethidium bromide in the medium, we still get some labeling of mitochondrial DNA which self-hybridizes at very low complexity so we can see that it is not nuclear DNA and is an early product of cytoplasmic DNA synthesis.

KIESSLING (Oregon): You showed us DNA profiles that are very reminiscent of Okasaki fragments. Do you know it is not RNA-primed DNA? Tseng and Goulian, who isolated Okasaki fragments from human lymphocytes (*J. Mol. Biol.* **99:**339, 1975), found it difficult to distinguish covalent RNA–DNA association from true hybrids using cesium gradients with high concentrations of nucleic acid. How have you ruled out the possibility that the RNA–DNA "hybrid" is not two equal-length pieces of covalently bound RNA and DNA?

MANES: It will come apart by heating.

KIESSLING: Does it come completely apart?

MANES: Yes, it comes completely apart. The DNA that is involved with this material for some reason is not sufficiently labeled for it to be analyzed as part of the hybrid. We can get little peaks of labeled material in the middle of the gradient when we label with high concentrations of thymidine, but it labels much more slowly than the RNA. Perhaps the uridine pool is more easily saturated than the thymidine pool, or nuclear DNA is competing with cytoplasmic DNA for the label, but I would like very much to get enough label into this material to allow analysis. Its density is all the information I have so far.

CHILTON (Pennsylvania): Have you looked for RNA-dependent DNA polymerase and would you care to speculate on the role of your RNA–DNA hybrids?

MANES: Oh, I would like to speculate. The RNA-directed DNA polymerase is present at fairly high specific activity in a total cytoplasmic extract; it does not pellet with mitochondria, which was a possibility since it is inhibited almost totally by ethidium bromide, as is true of the DNA polymerase gamma. This material, however, does depend upon divalent cations and is sulfhydryl group-dependent. It will copy, or at least one enzyme in the cytoplasm will copy, the poly rA oligo dT template; there is another enzyme that will copy the poly dA oligo dT template just about as efficiently. We have tried a number of times to pellet material that should be of viral densities and have not succeeded. Van Blerkom and Motta (*The Cellular Basis of Mammalian Reproduction,* p. 201, Urban & Schwarzenberg, Baltimore) showed viral particles in the rabbit trophoblast cytoplasm several years ago by electron microscopy and also what looked like C particles budding from the surface of the rabbit trophoblast. As yet we have not been able to identify these biochemically. This enzyme activity is not present at any detectable level in the blastocoelic fluid, which I would think it should be if it were budding off.

CHILTON: Are you sure that you separated trophoblast from endometrium, where this polymerase is active?

MANES: The embryo in the rabbit is still inside the zona pellucida, so we can recover the embryo free of endometrial components, wash it, and remove the zona pellucida. I do not think we are getting any endometrial contamination. We have analyzed endometrium and also find reverse transcriptase there.

SHERMAN (Roche Institute): Have you shown that the labeled cytoplasmic DNA is DNAse sensitive and RNAse resistant?

MANES: No, I haven't in the sense that you are asking. It is S1 nuclease resistant, but that just means it is double stranded.

SHERMAN: Have you looked for transcriptase activity with oligo dG poly rC?

MANES: Yes, we have and there is no effect. It does not seem to copy that template, although there was a slight bit of copying of the methylated cytosine poly rC.

SHERMAN: That would be cause for caution in stating that a reverse transcriptase really exists in those embryos.

MANES: I agree. However, the fact that it can copy at least the poly rA template from an oligo dT primer is helpful.

SHERMAN: But DNA-directed DNA polymerases can do that under the right salt conditions, and they can do it quite substantially. I believe Weisbach and his colleagues have shown this (Knopf *et al.*, 1976, *Biochemistry* **15:**4540).

References

Bell, E., 1971, I-DNA: Its packaging into I-somes and its relation to protein synthesis during differentiation, *Nature (London)* **224:**326.

Bjursell, G., Gussander, E., and Lindahl, T., 1979, Long regions of single-stranded DNA in human cells, *Nature (London)* **280:**420.

Bolden, A., Noy, G. P., and Weissbach, A., 1976, DNA polymerase of mitochondria is a γ-polymerase, *J. Biol. Chem.* **252:**3351.

Britten, R. J., and Kohne, D. E., 1968, Repeated sequences in DNA, *Science* **161:**529.

Daniel, J. C., Jr., 1964, Early growth of rabbit trophoblast, *Am. Natur.* **98:**85.

Enea, V., and Zinder, N. D., 1975, Guanidinium–CsCl density gradients for isopycnic analysis of nucleic acids, Science **190:**584.

Ernst, V., Levin, D. H., and London, I. M., 1978, Inhibition of protein synthesis initiation by oxidized glutathione: Activation of a protein kinase that phosphorylates the α subunit of eukaryotic initiation factor 2, *Proc. Natl. Acad. Sci. USA* **75:**4110.

Filipowicz, W., Sierra, J. M., and Ochoa, S., 1975, Polypeptide chain initiation in eukaryotes: Initiation factor MP in *Artemia salina* embryos, *Proc. Natl. Acad. Sci. USA* **72:**3947.

Galau, G. A., Klein, W. H., Davis, M. M., Wold, B. J., Britten, R. J., and Davidson, E. H., 1976, Structural gene sets active in embryos and adult tissues of the sea urchin, *Cell* **7:**487.

Hudson, B., Upholt, W. B., Devinny, T., and Vinograd, J., 1969, The use of an ethidium analogue in the dye-buoyant density procedure for the isolation of closed circular DNA: The variation of the superhelix density of mitochondrial DNA, *Proc. Natl. Acad. Sci. USA* **62**:813.

Koch, J., 1973, Cytoplasmic DNA's consisting of unique nuclear sequences in hamster cells, *FEBS Lett.* **32**:22.

Lewin, B., 1975, Units of transcription and translation: Sequence components of heterogeneous nuclear RNA and mRNA, *Cell* **4**:77.

Lodish, H. F., 1976, Translational control of protein synthesis, *Annu. Rev. Biochem.* **45**:39.

McMaster, G. K., and Carmichael, G. G., 1977, Analysis of single- and double-stranded nucleic acids on polyacrylamide and agarose gels by using glyoxal and acridine orange, *Proc. Natl. Acad. Sci. USA* **74**:4835.

Manes, C., 1974, Phasing of gene products during development, *Cancer Res.* **34**:2044.

Manes, C., 1977, Nucleic acid synthesis in preimplantation rabbit embryos. III. A "dark period" immediately following fertilization, and the early predominance of low molecular weight RNA synthesis, *J. Exp. Zool.* **201**:247.

Meinke, W., Hall, M., Goldstein, D. A., Kohne, D. E., and Lerner, R. A., 1973, Physical properties of cytoplasmic membrane-associated DNA, *J. Mol. Biol.* **78**:43.

Nelson, J., Leong, J., and Levy, J. A., 1978, Normal human placentas contain RNA-directed DNA polymerase activity like that in viruses, *Proc. Natl. Acad. Sci. USA* **75**:6263.

Rogers, J. C., 1976, Characterization of DNA excreted from phytohemagglutinin-stimulated lymphocytes, *J. Exp. Med.* **143**:1249.

Russell, J. L., and Golub, E. S., 1978, Leukemia in AKR mice: A defined suppressor cell population expressing membrane-associated DNA, *Proc. Natl. Acad. Sci. USA* **75**:6211.

Schultz, G. A., Manes, C., and Hahn, W. E., 1973, Estimation of the diversity of transcription in early rabbit embryos, *Biochem. Genet.* **9**:247.

Short, R. V., and Yoshinaga, K., 1967, Hormonal influences on tumour growth in the uterus of the rat, *J. Reprod. Fertil.* **14**:287.

Van Blerkom, J., and Manes, C., 1977, The molecular biology of the preimplantation embryo, in: *Concepts in Mammalian Embryogenesis* (M. I. Sherman, ed.), pp. 37–94, MIT Press, Cambridge, Mass.

Van Blerkom, J., Manes. C., and Daniel, J. C., Jr., 1973, Development of preimplantation rabbit embryos *in vivo* and *in vitro*. I. An ultrastructural comparison, *Dev. Biol.* **35**:262.

Van der Horst, C. J., and Gillman, J., 1942, Pre-implantation phenomena in the uterus of *Elephantulus*, *S. Afr. J. Med. Sci.* **7**:47.

Weissbach, H., and Ochoa, S., 1976, Soluble factors required for eukaryotic protein synthesis, Annu. Rev. Biochem. **45**:191.

Zehavi-Willner, T., Kosower, E. M., Hunt, T., and Kosower, N. S., 1971, Glutathione. V. The effects of the thiol-oxidizing agent diamide on initiation and translation in rabbit reticulocytes, *Biochim. Biophys. Acta* **228**:245.

8

Activity of RNA and DNA Polymerases in Delayed-Implanting Mouse Embryos

H. M. Weitlauf and A. A. Kiessling

I. Introduction

Mouse blastocysts may enter either of two distinctly different developmental pathways before implantation. In normal pregnancies their development is characterized by high rates of metabolic activity and cell division as they prepare to implant. Alternatively, in pregnancies complicated by concurrent lactation or castration, the embryos have decreased levels of metabolic activity; cell division slows and finally stops altogether and development stops at the blastocyst stage. Furthermore, these embryos do not implant but remain quiescent in the uterine lumen for days or even weeks; this is referred to as facultative delayed implantation. The embryos can be reactivated by removal of the suckling young or by injection of estradiol-17β along with progesterone. A similar metabolic activation appears to occur when the dormant embryos are removed from the uterus and placed *in vitro* (McLaren, 1973; Weitlauf, 1974; Van Blerkom and Manes, 1977).

In attempting to determine how embryonic metabolism is regulated in delayed implantation, several investigators have examined the synthesis of RNA by delayed-implanting embryos and those undergoing the process of activation. In those experiments blastocysts were usually incubated with [^{3}H]uridine and its incorporation into RNA was determined. This approach showed that less labeled uridine is incorporated by delayed-implanting embryos than is incorporated by activated or implanting embryos (Weitlauf, 1976; Chavez and Van Blerkom, 1979). However, the actual rate of RNA synthesis under these conditions could not be determined because the specific activity of the intracellular pool of [^{3}H]-UTP was not known. To overcome this difficulty in the present experiments, the specific activities of the [^{3}H]-UTP pools as well as the rate of [^{3}H]uridine incorporation into RNA were determined. In this way the level of overall RNA polymerase activity was measured *in situ* and valid comparisons could be made between rates of RNA synthesis in delayed-implanting and activated embryos.

H. M. Weitlauf and A. A. Kiessling • Department of Anatomy, School of Medicine, University of Oregon Health Sciences Center, Portland, Oregon 97201.

In addition to the general reduction in metabolism during the dormant phase associated with delayed implantation, the embryonic cells do not synthesize DNA or undergo division until the blastocysts are activated (McLaren, 1968; Sherman and Barlow, 1972). In our experiments total DNA polymerase activity in the embryos was measured to determine whether the level of this enzyme could be limiting and thus regulate cell division during delayed implantation.

II. Materials and Methods

A. Preparation of Embryo Donors

Virgin white Swiss mice were induced to ovulate with gonadotropins (PMSG and hCG) and placed with fertile males overnight (Fowler and Edwards, 1957). The finding of a vaginal plug on the following morning confirmed mating and that day was designated as Day 1 of pregnancy. Animals to be used as donors of normal blastocysts were left intact; those to be used as donors of either dormant or activated delayed-implanting embryos were ovariectomized bilaterally between 11 and 12 A.M. on Day 4 of pregnancy and allowed to recover for 2 days. The females used to provide the dormant blastocysts were injected with progesterone (2.0 mg/day s.c.) from Day 7 until they were killed; those used to provide activated embryos were injected with progesterone as above and, in addition, they received estradiol-17β (25 ng s.c.) on Days 9–10 or 14–15 (Weitlauf, 1969). The specific schedules for collecting embryos are described below in the various experiments.

B. Radiolabeling of Embryos for the Determination of RNA Polymerase Activity

Twenty to thirty dormant or active embryos were placed in 100–200 μl basal Eagle's medium (BME; Eagle, 1955) containing 5 mg/ml BSA and 25 μCi/ml [5,6-^{3}H]uridine (41.3 Ci/mmol, New England Nuclear Corp.). The embryos were incubated for 2–24 hr at 37°C in a moist atmosphere of 5% CO_2 in air. At the end of the incubation, the embryos were removed and washed in several changes of Tris-buffered saline (TBS; 145 mM NaCl; 10 mM Tris–HCl, pH 7.6) containing 5 mg/ml BSA. Five to ten embryos from each group were placed directly on pieces of glass-fiber filter paper (GF/C, Whatman) and washed extensively with cold trichloroacetic acid (10%). The filters were dried and radioactivity in the acid-insoluble material (RNA) was estimated with a scintillation counter. The amount of [^{3}H]uridine incorporated into RNA with respect to time *in vitro* was reduced to a polynomial by regression analysis (Draper and Smith, 1966); the rates of [^{3}H]uridine incorporation were calculated from the derivatives of those expressions.

C. Specific Activity of the [^{3}H]-UTP Pools

The specific activities of the [^{3}H]-UTP pools were determined by means of a double-isotope technique using *E. coli* RNA polymerase to synthesize an equal molar copolymer on poly dA·dT template with [^{3}H]-UTP from the embryos and exogenous [^{14}C]-ATP (modified from Clegg and Piko, 1977; see Weitlauf and Kiessling, 1980, for details). Briefly, labeled embryos were transferred in 1–2 μl TBS to 0.5 M perchloric acid (PCA) in a small centrifuge tube and frozen and thawed five times in a slurry of dry ice and ethanol. The PCA-precipitable material was removed by centrifugation. The supernatant fluid was removed and neutralized to pH 8.0 by adding 0.83 N KOH. Potassium perchlorate was allowed to precipitate at 4°C and was removed by centrifugation. The assay was carried out with the following ingredients in a final volume of 50 μl: neutralized PCA extract containing ^{3}H-labeled nucleotides from the embryos; poly dA·dT (P-L Biochemicals); [^{14}C]-ATP (460–501 mCi/mmol, New England Nuclear Corp.); RNA polymerase (fraction IV *E. coli* 4000 IU/mg, Gibco); and buffer (40 mM Tris–HCl, pH 8.0; 10 mM $MgCl_2$; 1 mM DTT). Incubations were carried out in individual wells of a microtiter plate at 37°C for 60 min. The reaction mixture was absorbed onto glass-fiber filter paper (GF/C, Whatman) previously coated with carrier DNA. The filters were washed extensively with cold TCA (10%) followed by cold HCl (0.1 N). The filters were dried and the radioactivity in acid-insoluble RNA was estimated with a scintillation counter. Counting efficiency was 30% for ^{3}H and 60% for ^{14}C; spillover from the ^{14}C to the ^{3}H channel was 40% and was corrected for in all calculations. The specific activities of the [^{3}H]-UTP pools were calculated from the equation

$$\frac{{}^{3}\text{H cpm}}{{}^{14}\text{C cpm}} \cdot K = \frac{\text{specific activity } {}^{3}\text{H}}{\text{specific activity } {}^{14}\text{C}}$$

where K is the ratio of counting efficiencies for ^{14}C and ^{3}H (i.e., 60% and 30%, respectively). With the rates of incorporation of [^{3}H]uridine into RNA and the specific activity of the [^{3}H]-UTP pools known, the overall rates of incorporation of uridine were calculated. One unit of RNA polymerase activity was defined as 1 fmol of uridine incorporated into RNA per hour.

D. DNA Polymerase Assay

Embryos were assayed for total DNA polymerase activity as previously described (Kiessling and Weitlauf, 1979). Briefly, 4–16 embryos in 1–2 μl fresh buffer A (0.01 M Tris–HCl, pH 7.4; 0.25 M sucrose; 0.1 M KCl; 1 mg/ml BSA) were transferred into individual wells of a microtiter plate. The embryo cells were lysed by the addition of an equal volume of a solution of 0.5% Nonidet-P-40, 0.01 M DTT. Enzyme activity was assayed di-

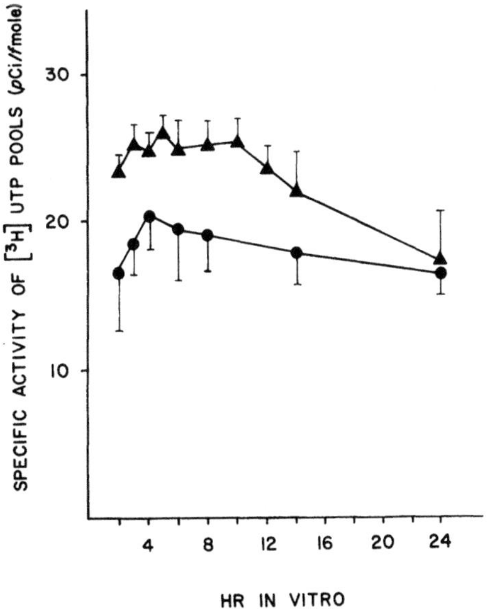

Figure 1. Specific activity of [^{3}H]-UTP in active and dormant delayed-implanting mouse blastocysts incubated with [^{3}H]uridine *in vitro*. Specific activity of [^{3}H]-UTP was determined in active (●) and dormant (▲) embryos by the double-isotope method described by Clegg and Piko (1977). The assay involves the use of RNA polymerase (*E. coli*) to transcribe poly dA·dT into poly rU·rA with [^{3}H]-UTP from the embryos and exogenous [^{14}C]-ATP. Values shown are mean ± S.E.M. of five to ten determinations.

rectly in the well in a mixture (total volume 15 μl) that contained 30 mM Tris–HCl, pH 8.3; 5 mM $MgCl_2$; 1 mg/ml BSA; 1 μg poly dC·oligo dG (P-L Biochemicals); 0.2 mM dCTP; 4 μCi [^{3}H]-dGTP (34.5 Ci/mmol, New England Nuclear Corp.) plus sufficient nonlabeled dGTP to bring the final concentration to 12 μM (8 cpm/fmol). The reaction proceeded for 30 min at 37°C. The reaction mixture was absorbed onto glass-fiber filter paper (GF/C, Whatman) that was pretreated with carrier DNA. The DNA product was precipitated with 10% TCA (4°C) and the filters were washed extensively with cold 0.1 N HCl. Radioactivity in the acid-insoluble DNA product was estimated with a scintillation counter. One enzyme unit was defined as 1 fmol of dGTP incorporated into acid-precipitable DNA product in 30 min.

E. Analysis of Data

The specific activities of the UTP pools were treated by analysis of variance followed by Duncan's multiple range test where appropriate (Steel and Torrie, 1960; $p<0.05$ was considered statistically significant). The variance in estimates for RNA polymerase activity (i.e., $1/\bar{x}$) was determined using Taylor's series approximation (Wilks, 1962).

III. Results

A. Specific Activities of the [^{3}H]-UTP Pools

Specific activities of the embryonic pools of [^{3}H]-UTP in active and dormant embryos are summarized in Figure 1. In both types of embryos the specific activity was found to

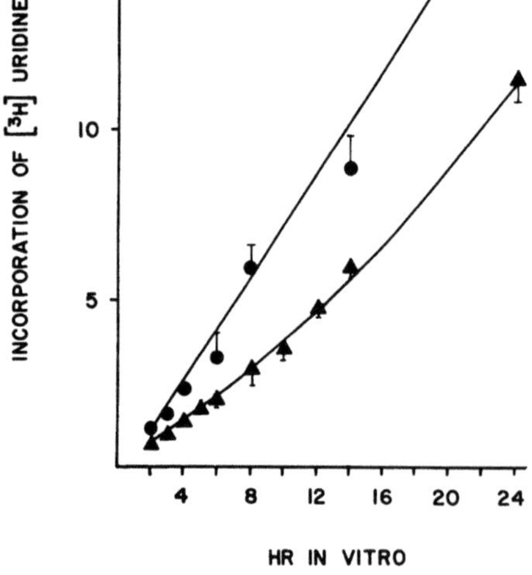

Figure 2. Incorporation of [^{3}H]uridine into RNA by active and dormant delayed-implanting mouse embryos incubated *in vitro* with [^{3}H]uridine. Incorporation of [^{3}H]uridine into acid-insoluble RNA by active (●) embryos increased steadily with time (770 cpm/100 cells/hr). Incorporation of [^{3}H]uridine by dormant embryos (▲) increased as a function of time [$y=(14x+309)$ (cpm/100 cells/hr), where x = hours *in vitro* and y = rate of [^{3}H]uridine incorporation].

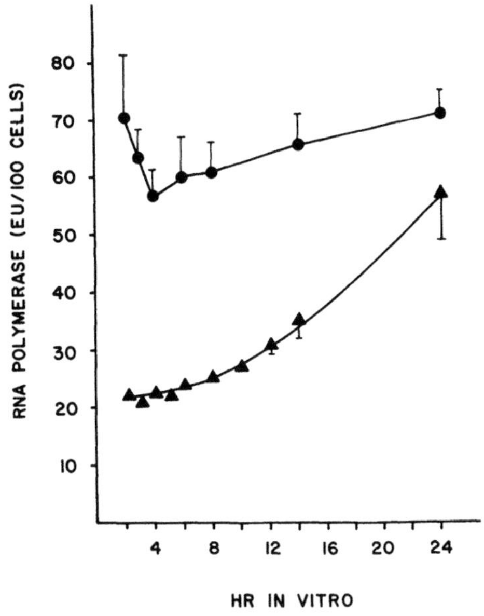

Figure 3. RNA polymerase activity in active and dormant delayed-implanting mouse embryos incubated *in vitro*. Active (●) and dormant (▲) embryos were incubated for 2 to 24 hr *in vitro*. One enzyme unit is 1 fmol of uridine incorporated per hour. Values shown are mean ± S.E.M. of five to ten determinations.

increase for the first few hours *in vitro* and then to gradually decrease; the differences with time were not statistically significant. Specific activity of the [^{3}H]-UTP pool in delayed-implanting embryos was slightly greater than that in implanting embryos through 12–14 hr, but again this difference was not statistically significant.

B. Rate of Incorporation of [^{3}H]Uridine into RNA

Cumulative incorporation of [^{3}H]uridine into RNA by implanting and delayed-implanting embryos *in vitro* is summarized in Figure 2. Incorporation by the implanting embryos was linear for 24 hr; the rate of incorporation was constant at 770 cpm/100 cells per hr. In contrast, incorporation of [^{3}H]uridine into RNA by delayed-implanting embryos was initially low and increased with time; the rate of incorporation was a function of time *in vitro* ($dy/dx = 14x + 309$).

C. RNA Polymerase Activity

The amount of RNA polymerase activity in the embryos is summarized in Figure 3; in implanting embryos it was found to be relatively constant throughout the incubation period (i.e., between 57 and 71 EU/100 cells, $p > 0.05$). In contrast, the amount of activity in delayed-implanting embryos increased from 20–22 EU/100 cells when they were first removed from the uterus to 57 EU/100 cells after 24 hr *in vitro* ($p < 0.05$).

D. DNA Polymerase Activity

Total DNA polymerase activity increased in normal embryos between noon of Day 4 and the time of implantation (i.e., midnight of Day 5, Figure 4). The total amount of enzyme activity in the delayed-implanting embryos also increased, at the normal rate, for the first 12 hr after the ovariectomy but decreased (by 30%) in the next 36 hr. This decrease in enzyme did not seem to affect DNA synthesis for a few hours, and cell division in the

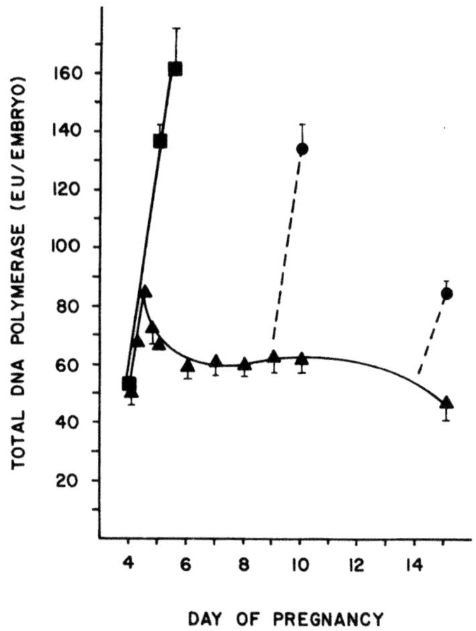

Figure 4. The total amount of DNA polymerase activity in normal and delayed-implanting mouse embryos. Normal embryos (■) were recovered from intact females at the times indicated. Delayed-implanting (▲) and activated embryos (●) were recovered from ovariectomized females treated with hormones as outlined in the text.

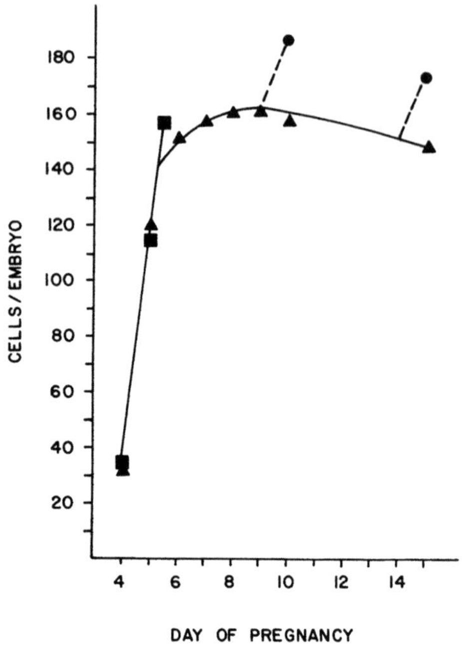

Figure 5. The number of cells in normal and delayed-implanting embryos. Normal embryos (■) recovered from intact females at the times indicated. Delayed-implanting (▲) and activated embryos (●) were recovered from ovariectomized females treated with hormones as outlined in the text. Cell counts were done on 12 to 60 embryos for each point by the method of Tarkowski (1966).

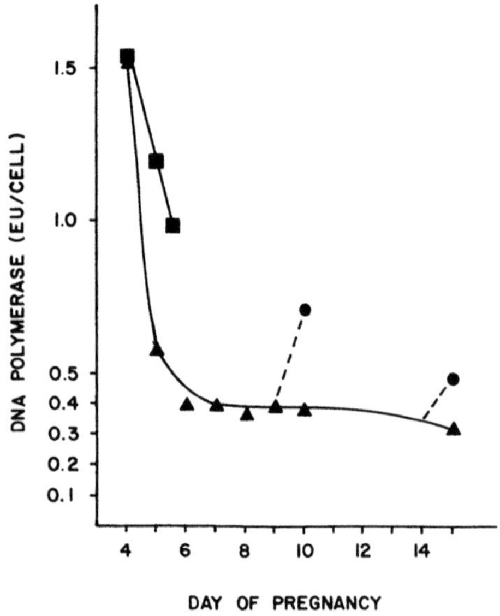

Figure 6. The amount of DNA polymerase per cell in normal and delayed-implanting mouse embryos. (■) Normal embryos; (▲) delayed-implanting embryos; (●) activated embryos. Ratios of total DNA polymerase activity per embryo vs. number of cells per embryo were calculated from the data contained in Figures 4 and 5. In addition, DNA polymerase activity in spleen was determined (0.04 EU/cell) for purposes of comparison with activity in the embryos; the spleen was shown (by autoradiography following injection of [^{3}H]thymidine) to be actively synthesizing DNA.

delayed-implanting embryos continued at the normal rate in the 24 hr following ovariectomy; it then slowed but did not stop altogether for another 48 hr (Figure 5). The amount of DNA polymerase remained relatively constant on a per cell basis in the dormant embryos during the quiescent phase (i.e., 0.32–0.39 EU/cell, Figure 6), but following activation (on either Days 9–10 or Days 14–15), it increased as cell division resumed. Thus, the amount of DNA polymerase per cell was greatest in embryos where cell division was occurring and lowest in the dormant embryos where cell division did not occur. The amount of DNA polymerase activity per cell in the active embryos was between 12 and 25 times greater than that in adult spleen (i.e., a mature organ with a relatively high labeling index 0.04 EU/cell; Figure 6); the amount in dormant embryos was between 8 and 9 times greater than that in adult spleen.

IV. Summary

These results demonstrate that there is less RNA polymerase activity in dormant embryos than in those that are activated and in the process of implanting. Thus the pattern of change in RNA polymerase activity is similar to that already shown for protein synthesis and metabolism in general as delayed-implanting mouse embryos became activated (McLaren, 1973; Weitlauf, 1974; Chavez and Van Blerkom, 1979). The present experiments do not determine if changes in RNA polymerase are responsible for activation or are a consequence of that process.

The assay used here measures activity of RNA polymerase *in situ*. Therefore, any of several conditions within the cells could be responsible for the different levels of activity observed: (1) a limited size of pools of rNTPs in the dormant embryos; (2) actual differences in the amount of enzyme in the embryos; (3) the presence of factors that inhibit, or the absence of factors that stimulate one or another of the RNA polymerases; and (4) the availability of DNA template. The sizes of the total pools of UTP in implanting and delayed-implanting embryos have been estimated under the same conditions described here (Weitlauf and Kiessling, 1979). After 2 hr *in vitro* activated embryos were found to have a total UTP pool of 140 fmol/100 cells and dormant embryos, 138 fmol/100 cells. Therefore, although total UTP pools in the embryos were similar, the level of RNA polymerase activity was three times greater in the active embryos. It seems unlikely that the size of other rNTP pools would be limiting, although this has not been determined. Siracusa (1973), Siracusa and Vivarelli (1975), and Warner (1977) have estimated the actual amount of RNA polymerase activity in normal embryos with direct measurements *in vitro,* but no such observations have been reported for delayed-implanting or activated embryos; therefore, the possibility of different amounts of enzyme being present has not been determined. Although factors that either stimulate or inhibit RNA polymerase activity have been reported in other mammalian cells (Ueno *et al.*, 1979; Stewart and Krueger, 1976; Sasaki *et al.*, 1974; Antonoglou *et al.*, 1977), there have been no reports of such factors in preimplantation mouse embryos. Therefore, the question of how the level of RNA polymerase activity is regulated in the blastocysts cannot yet be answered.

The level of DNA polymerase activity in normal mouse blastocysts increased up to the time of implantation. In contrast, in the delayed-implanting embryos it increased for only a few hours after the mother had been castrated and then decreased. The level of activity remained low throughout the dormant phase associated with delayed implantation and then increased as the embryos were activated following an injection of estradiol-17β. Thus, activity of this enzyme followed the general pattern already established for metabolic and nitotic activity in delayed-implanting embryos (McLaren, 1973).

Although these findings are compatible with the possibility that the amount of DNA polymerase is important in regulating DNA synthesis and thus cell division in the embryos, it does not provide proof of a causal relationship. Indeed, if the ratio of enzyme to DNA is significant in this regard, then a critical value for cell division must lie between 0.39 and 0.48 EU/cell, as these are the maximum dormant-embryo and the minimum active-embryo values, respectively. Both of these ratios are several times greater than that in adult spleen. Therefore, even though there is a statistically significant difference in the ratio of DNA polymerase to DNA in quiescent and active embryos, it seems unlikely that it is responsible for regulating cell division in the blastocysts unless there is a markedly different critical ratio in embryonic and adult cells.

If it is true that the ratio of DNA polymerase to DNA is not responsible for regulating mitosis, then there is presumably an excess of enzyme in both dormant and quiescent embryos, and the question of why DNA is not synthesized during the quiescent period is left unanswered. It is possible that *in vivo* the embryonic pools of dNTPs, the availability of template or RNA necessary for initiation are limiting; the enzyme may be located in the cytoplasm rather than the nucleus and thus be unavailable for DNA synthesis; or the form of the enzyme responsible for DNA replication (usually thought to be polymerase α) may be absent. These possibilities cannot be distinguished by the present method because: (1) although the *in vitro* assay used here is optimal for DNA-polymerase-α-like activity in other cells, it does not identify the various forms of polymerase activity (i.e., α, β, or γ); (2) whole-cell lysates were used and the polymerase could be located in either the nucleus or the cytoplasm; and (3) the assay requires the addition of exogenous dNTPs and template primer and therefore does not determine whether or not these are limiting *in vivo*.

Aside from the problem of whether the amount of DNA polymerase regulates the rate of DNA synthesis and therefore cell division in delayed-implanting embryos, these observations demonstrate that the level of enzyme activity in the embryos does change with the hormonal status of the mother and raise the question of how this occurs. The change in activity as the embryos become quiescent could be due to decay of enzyme along with reduced synthesis as metabolism decreases. It is equally possible that changes in other factors, such as those that stimulate or inhibit DNA polymerase activity in other cells, are responsible (Lee and Lucas, 1977, 1978; Sarngadharan *et al.*, 1978). However, there is no information on any of these points and further experiments will be necessary to resolve this question.

ACKNOWLEDGMENTS. This work was carried out with the assistance of Ms. L. Buschman, Ms. K. Lantz, and Ms. C. Williams; it was supported in part by grants from the NICHD (HD08496 and HD0020), the NSF (880), and the Medical Research Foundation of Oregon (11.10).

Discussion

EPSTEIN (UCSF): Why do you equate RNA synthesis with polymerase activity? You mentioned that the activity of the polymerase is just one of many factors that can influence the rate of synthesis.

WEITLAUF: The rate of RNA synthesis defines the level of polymerase activity *in situ;* it does not provide information about the control of RNA synthesis, however, because other factors may be involved.

C. WARNER (Iowa State): Your talk was based on measuring different enzyme activities in normal vs. delayed-implanting embryos and you have done a lot of work on measuring metabolic activities. I was wondering if you have looked at surface antigens on normal vs. delayed-implanting embryos?

WEITLAUF: No.

WARNER: I think that you might go on measuring metabolic activities forever and never see any meaningful differences if it is a membrane effect, which is my personal prejudice about normal vs. delayed-implanting embryos. There may be a specific protein on the surface of delayed embryos which prevents their implantation.

CHĀVEZ (UC–Davis): You compared normal embryos with delayed-implantation embryos that had been cultivated for 24 hr *in vitro*. Did you mention how you were able to distinguish between embryos that might possibly have been reactivated by *in vitro* culture and those that were still dormant after 24 hr in culture?

WEITLAUF: In the RNA polymerase experiments, all the embryos were removed from the uterus, placed in culture, and they were then removed from culture after various periods. All the embryos were activated by 24 hr, as I showed. Normal embryos do not change their apparent rate of metabolic activity under these conditions.

DICKMANN (Kansas): Do you have data on delayed implantation in which you do not give progesterone?

WEITLAUF: No, not in this study. We have looked at the effect of uterine fluid from animals that did not receive progesterone, and we have looked at overall incorporation of [^{3}H]uridine in animals that did not receive progesterone, but we have not examined the UTP pool size or the specific activity in animals treated in that way.

GLASSER (Baylor): Have you looked at the quality of the RNA produced by the polymerase activity?

WEITLAUF: We have not yet been able to do that.

GLASSER: The conditions under which you ran the assay, although they give maximum enzyme activity, also maximize degradation of the RNA that is produced. Spelsberg and I (*Biochem. J.* **130:**947, 1972) ran our polymerase assays at 15 to 20°C, rather than 37°C, and so prevented degradation of the RNA.

EPSTEIN: Excuse me. I don't understand that point because again I thought the RNA polymerase was measured by *in vitro* incorporation.

GLASSER: That's right, but if you run it at 37 . . .

EPSTEIN: Which is usually the temperature . . .

GLASSER: . . . which is usually the temperature for polymerase assays because a maximum amount of RNA is transcribed. But, under those conditions, the RNA is degraded as quickly as it is transcribed. Have you looked at the differential pattern of polymerase I and II activity over time?

WEITLAUF: That's the obvious next step and we have not yet done it.

SHERMAN (Roche Institute): In order to discriminate between the different RNA polymerase activities, have you used antibodies specified against the specific polymerases?

WEITLAUF: We have undertaken some of these experiments in normal embryos but have not yet done them in the delayed embryos.

VILLEE (Harvard): You made the interesting point that embryos are turned on by changes in maternal endocrine milieu. Is that really endocrine or is the estrogen getting right through to the embryos? Can you get the same effect by, let's say, injecting estradiol into the uterine lumen, or taking these delayed embryos and adding estradiol *in vitro?*

WEITLAUF: There have been several attempts to see whether or not steroids directly influence the uptake or incorporation of amino acids and uridine by the embryos *in vitro*. The only report I am aware of which shows some increase in any of these indices with embryos incubated directly with estrogen is a paper by Smith and Smith (*Biol. Reprod.* **4:**66, 1971), in which it was shown that estrogen caused a slight increase in the amount of amino acid incorporation. We have not been able to demonstrate that with amino acid or with uridine; Warner and Tollefson (*Biol. Reprod.* **19:**334, 1978) also found that estrogens do not affect uridine uptake or incorporation directly. That does not really say that that does not happen in the uterine lumen.

EPSTEIN: Is there any way to activate *in vitro?*

WEITLAUF: Embryos do become active *in vitro,* in terms of RNA synthesis, amino acid incorporation, and carbon dioxide production, whether or not BSA is present in the medium.

MARTIN (ICRF, London): I was intrigued to see that the blastocysts continue to grow after ovariectomy. In fact, the cell numbers only decline at a time when one would normally expect to see a well-developed decidual reaction in process. I wonder whether the blastocysts are simply switching off because of a uterine deficiency, notably a vascular one? In the ovariectomy experiments, do you see any signs of an abortive pontamine blue reaction? When you restimulate, is there a pontamine blue reaction *before* the changes in polymerase and cell numbers?

WEITLAUF: When the animsls are ovariectomized as we have done, there is no pontamine blue reaction on Days 5, 6, or 7 with or without progesterone. If we add estrogens, we can show the beginning of the pontamine blue reaction within 8 hr; it is maximal about 15 or 16 hr after estrogen. We measured the DNA polymerase activity at only one point after estrogen, a full 24 hr later, so I do not know intermediate values.

BULLOCK (Baylor): You have shown that in the dormant embryos RNA synthesis and DNA polymerase activity are depressed, just as are most of the other metabolic parameters that have been looked at. As you point out, this leaves open the question of what it is that is regulating the metabolic dormancy of these embryos. Have you considered approaching the question from the opposite direction and ask what can one do to an active embryo to make it dormant in terms of these sorts of parameters?

WEITLAUF: Incorporation of [^{3}H]uridine can be depressed in normal embryos placed in certain fractions of uterine fluid recovered from delayed-implanting animals. A similar reduction in uridine incorporation, however, occurs when the fluid is recovered from normal animals, where one would not expect to see an inhibition of RNA synthesis. The inhibitory effect, however, is greater with fluid from delayed-implanting animals than from normal animals (Weitlauf, 1978, *J. Reprod. Fertil.* **58**:321). The problem is to decide what the concentration of a putative uterine inhibitor should be in a tissue culture situation. Only recently has Roger Hoversland in our laboratory been able to determine the volume of uterine fluid *in situ* and therefore give us some idea of what kind of concentration we might have to use.

Using incorporation of [^{3}H]uridine in that kind of experiment is fraught with other difficulties. One does not know, for example, whether decreased incorporation of uridine is caused by an effect at the membrane or somewhere within the cell. We should be able to answer these questions by measuring the specific activities of the pools. The answer to the question of whether or not anything is necessary in the milieu to stimulate embryos seems to be no. We find exactly the same amount of RNA synthesis when bovine serum albumin is replaced by 10% serum. With 10% serum, the embryos will outgrow but with BSA they will not. I think this probably means that if outgrowth *in vitro* is a valid model for implantation, the metabolic aspect and the outgrowth aspect are under slightly different control. Proponents of inhibitors and proponents of stimulators are probably both going to have a say in the answer.

SURANI (Cambridge): I will reinforce what Dr. Warner said, that the membrane properties during delay are quite important, including changes in the membrane transport systems. I would like to ask if there is any increase in cell numbers after the embryos are released from delay?

WEITLAUF: There is no increase in cell numbers in the first 24 hr when the embryos are incubated *in vitro* under the conditions used here or when serum is present in the medium, even though they will start to outgrow by 24 hr.

ARMSTRONG (W. Ontario): If embryos are activiated in culture, can they be transferred back into delayed recipients and will they implant?

WEITLAUF: After transfer into delayed recipients, they will turn dormant again; after transfer into normal recipients, they will implant. That whole process takes between 24 and 36 hr. I do not think transfer measures much of what has happened to the embryo at the time of removal from culture. The embryo has a recuperative power and it is very difficult to dissect that out.

BARKER (Nebraska): How similar is the specific activity of the added tritiated uridine to the specific activity of the UTP measured in the active and inactive situations?

WEITLAUF: The specific activity of the tritiated uridine was 36 Ci/mmol and the specific activity in the embryos was between 17 and 26 Ci/mmol.

BARKER: Did this vary with the delayed vs. the active embryos?

WEITLAUF: The size of the total UTP pool, that is, endogenous plus exogenous UTP, was very similar in the two cases after 2 hr and stayed the same throughout. Endogenous UTP pools in the delayed embryo started out slightly smaller than those in implanting embryos and increased in size more slowly. Taken together, these differences led to the results I showed in Figure 2.

BARKER: Measuring the UTP pools, as you have done, is an elegant first step in quantifying polymerase activity. It is possible, however, that there is a differential dilution of the specific radioactivity of the UTP at the site of polymerase activity, namely within the nucleus, compared to the cytoplasm. I would expect the extranuclear UTP to be a significant portion of the total cellular UTP. If exchange between the nucleus and cytosol, where phosphorylation of uridine is likely to occur, is slow, you may be overestimating the specific activity of the UPT pool which is used for RNA synthesis.

References

Antonoglou, O., Salakido, H., Haralmbidou, E., and Trakatellis, A., 1977, A macromolecular inhibitor of rat liver DNA-dependent RNA polymerases, *Biochim. Biophys. Acta* **474:**467.

Chavez, D. J., and Van Blerkom, J., 1979, Persistence of embryonic RNA synthesis during facultative delayed implantation in the mouse, *Dev. Biol.* **70:**39.

Clegg, K. B., and Piko, L., 1977, Size and specific activity of the UTP pool and overall rates of RNA synthesis in early mouse embryos, *Dev. Biol.* **58:**76.

Draper, N. R., and Smith, H., 1966, *Applied Regression Analysis,* Wiley, New York.

Eagle, H., 1955, Nutrition needs of mammalian cells in tissue culture, *Science* **122:**501.

Fowler, R. E., and Edwards, R. G., 1957, Induction of superovulation and pregnancy in mature mice by gonadotropins, *J. Endocrinol.* **15:**374.

Kiessling, A. A., and Weitlauf, H. M., 1979, DNA polymerase activity in preimplantation mouse embryos, *J. Exp. Zool.* **208:**347.

Lee, S. C., and Lucas, L. J., 1977, Regulatory factors produced by lymphocytes. II. Inhibition of cellular DNA synthesis associated with a factor inhibiting DNA polymerase activity, *J. Immunol.* **118:**88.

Lee, S. C., and Lucas, L. J., 1978, Properties of an inhibitor of DNA synthesis in supernatants of activated lymphocytes, *Cell. Immunol.* **39:**250.

McLaren, A., 1968, A study of blastocysts during delay and subsequent implantation in lactating mice, *J. Endocrinol.* **42:**453.

McLaren, A., 1973, Blastocyst activation, in: *The Regulation of Mammalian Reproduction* (S. Segal, R. Crozier, P. Corfman, and P. Gondliffe, eds.), pp. 321–328, Thomas, Springfield, Ill.

Sarngadharan, M. G., Robert-Guroff, M., and Gallo, R. C., 1978, DNA polymerases of normal and neoplastic mammalian cells, *Biochim. Biophys. Acta* **516:**419.

Sasaki, R., Goto, H., Arima, K., and Sasaki, Y., 1974, Effect of polyribonucleotides on eukaryotic DNA-dependent RNA polymerases, *Biochim. Biophys. Acta* **366:**435.

Sherman, M. I., and Barlow, P. W., 1972, Deoxyribonucleic acid content in delayed mouse embryos, *J. Reprod. Fertil.* **29:**123.

Siracusa, G., 1973, RNA polymerase during early development in mouse embryo, *Exp. Cell Res.* **78:**460.

Siracusa, G., and Vivarelli, E., 1975, Low-salt and high-salt RNA polymerase activity during preimplantation development in the mouse, *J. Reprod. Fertil.* **43:**567.

Steel, R. G. D., and Torrie, J. H., 1960, *Principles and Procedures of Statistics,* McGraw–Hill, New York.

Stewart, L. E., and Krueger, R. C., 1976, Nuclear ribonucleoproteins as inhibitors of mammalian RNA polymerase, *Biochim. Biophys. Acta* **425:**322.

Tarkowski, A. K., 1966, An air-drying method for chromosome preparations from mouse eggs, *Cytogenetics* **5:**394.

Ueno, K., Sekimizu, K., Mizuno, D., and Natori, S., 1979, Antibody against a stimulatory factor of RNA polymerase II inhibits nuclear RNA synthesis, *Nature (London)* **277:**145.

Van Blerkom, J., and Manes, C., 1977, The molecular biology of the preimplantation embryo, in: *Concepts in Mammalian Embryogenesis* (M. J. Sherman, ed.), pp. 37–94, MIT Press, Cambridge, Mass.

Warner, C. M., 1977, RNA polymerase activity in preimplantation mammalian embryos, in: *Development in Mammals,* Vol. 1 (M. H. Johnson, ed.), pp. 99–136, North-Holland, New York.

Weitlauf, H. M., 1969, Temporal changes in protein synthesis by mouse blastocysts transferred to ovariectomized recipients, *J. Exp. Zool.* **171:**481.

Weitlauf, H. M., 1974, Metabolic changes in the blastocysts of mice and rats during delayed implantation, *J. Reprod. Fertil.* **39:**213.

Weitlauf, H. M., 1976, Effect of uterine flushings on RNA synthesis by 'implanting' and 'delayed implanting' mouse blastocysts *in vitro, Biol. Reprod.* **14:**566.
Weitlauf, H. M., and Kiessling, A. A., 1980, Comparison of overall rates of RNA synthesis in implanting and delayed implanting mouse blastocysts *in vitro, Dev. Biol.* **77:**116–129.
Wilks, S., 1962, *Mathematical Statistics*, Wiley, New York.

9

A Reexamination of Messenger RNA Populations in the Preimplantation Mouse Embryo

Gilbert A. Schultz, Jeremy R. Clough, Peter R. Braude, Hugh R. B. Pelham, and M. H. Johnson

I. Introduction

Regulation of gene expression and the flow of genetic information from the nucleus to the cytoplasm in eukaryotic cells are controlled at a number of levels, beginning with the transcription of DNA to produce a primary RNA transcript. These transcripts are subjected to a number of posttranscriptional modifications that can include processing through splicing and ligation of different regions of the RNA molecule, capping and methylation of the 5′ end, and polyadenylation of the 3′ end of the RNA. Additional posttranscriptional controls related to association of mRNAs with ribonucleoproteins (mRNP) and translational controls at the level of binding of mRNA to ribosomes and subsequent steps in the synthesis of polypeptides also are involved. Similar types of controls can be expected to operate in differential gene expression during preimplantation development. The problem resides in assessing the extent to which the various controls are involved in regulating the genetic activity of different cells at various stages of the early developmental process.

A recent review of a considerable body of literature encompassing both genetic and biochemical studies is consistent with the interpretation that the first and/or first few cleavages of the fertilized mouse egg are largely independent of nuclear gene activity and are characterized by posttranscriptional control, relying, at least in part, on maternally derived mRNA in the oocyte (Johnson, 1979). In contrast, the transition of the compacted morula to the blastocyst seems to be more tightly coupled to transcriptional events. Treatment of embryos in the early part of this transitional period with α-amanitin to inhibit RNA poly-

Gilbert A. Schultz • Division of Medical Biochemistry, University of Calgary, Calgary, Alberta, Canada. *Jeremy R. Clough, Peter R. Braude, and M. H. Johnson* • Department of Anatomy, University of Cambridge, Cambridge, England. *Hugh R. B. Pelham* • Department of Biochemistry, University of Cambridge, Cambridge, England.

merase II activity leads to a block in cavitation and blastocyst expansion, prevents the quantitative increase in net protein synthesis characteristic of this period of development, and also interferes with the qualitative changes in protein synthetic pattern that accompany this phase of development (Braude, 1979a,b).

In this paper, we reexamine some of the parameters related to the function of stable mRNA molecules in preimplantation mouse embryo development. The recent development of the micromethod for extraction and *in vitro* translation of mouse embryo mRNA (Braude and Pelham, 1979) has allowed comparison of polypeptides directed by template mRNA molecules free of cellular control versus those synthesized by template mRNAs in intact embryos. Evidence is presented that suggests that some protein synthesis at the early two-cell stage is not dependent on transcription, but rather is controlled at some posttranscriptional level in mRNAs synthesized prior to fertilization.

Attention has also been directed toward the nature of the 5′ end of mouse embryo mRNA molecules as one possible level of posttranscriptional control. Most eukaryotic mRNAs contain at their 5′ end a cap structure of the form $m^7G^{5'}$ $ppp^{5'}X^mpY$ . . . that is added posttranscriptionally to the primary transcript (Shatkin, 1976; Revel and Groner, 1978). In addition to protecting the mRNA at its 5′ terminus from enzymatic degradation, this modification appears to be functionally important in mRNA translation, for analogs such as m^7GMP and m^7GTP competitively inhibit translation of capped message (Canaani *et al.*, 1976; Weber *et al.*, 1976). Chemical and enzymatic excision of cap structures from mRNA molecules (Both *et al.*, 1975; Zan-Kowalczewska *et al.*, 1977; Gedamu and Dixon, 1978) reduces the translation of mRNA *in vitro* and *in vivo* (Lockard and Lane, 1978). Because newly fertilized mouse eggs incubated in media containing [^{3}H]guanosine appear to incorporate a small amount of this precursor into cap structures (Young, 1977), and because modifications of the 5′ ends could provide one way in which translational control might be exercised, the degree of capping in mRNA molecules in mouse eggs and embryos was analyzed through (1) reduction in the degree of translation *in vitro* in the presence of the cap analog m^7GTP and (2) enzymatic cap removal and end-labeling with [γ-^{32}P]-ATP and polynucleotide kinase.

II. Materials and Methods

All procedures have been described previously and are summarized as follows:

1. Superovulation procedures, methods of recovery of eggs and embryos, and *in vitro* culture methods in the presence of radioactive precursors were according to Braude (1979a,b) and Braude and Pelham (1979).

2. Total RNA was extracted from eggs or embryos exactly as described by Braude and Pelham (1979) except that the EDTA in the extraction buffer was reduced from 0.01 to 0.001 M.

3. Translation of egg or embryo RNA in message-dependent reticulocyte lysate was performed according to Braude and Pelham (1979). Inhibition of translation with the cap analog m^7GTP was conducted as described by Schultz *et al.* (1980).

4. Methods for resolution of radio-labeled polypeptides by one- or two-dimensional electrophoresis in polyacrylamide gels were as described by Handyside and Johnson (1978) and Braude and Pelham (1979).

5. Removal of 5′-end cap structures, dephosphorylation, and subsequent end-group labeling of RNA with [γ-^{32}P]-ATP and polynucleotide kinase were performed using slight

modifications of the procedure of Efstratiadis *et al.* (1977) as described by Schultz *et al.* (1980). Purification of end-labeled poly A-containing mRNA (poly A RNA) by affinity chromatography on oligo dT cellulose was accomplished as described by Schultz *et al.* (1980).

6. End-labeled RNA products were resolved on 2.5% polyacrylamide gels according to Loening (1967) and on gels of higher acrylamide concentration according to Sanger and Coulson (1975). Where desired, gels were stained with methylene blue as described by Peacock and Dingman (1967).

III. Results and Discussion

A. mRNA in the Morula–Blastocyst Transition Period

When early compacted morulae at 72 hr post-hCG are cultured *in vitro* for a period of 24 hr, most of them reach various stages of blastocyst formation. If these embryos are cultured through this period in the presence of 11 μg/ml α-amanitin to inhibit transcription by RNA polymerase II, virtually none of them cavitate (Braude, 1979a). In addition, rates of incorporation of [^{35}S]methionine into protein by control embryos increase two- to fivefold during the 24-hr culture period, whereas incorporation by embryos in α-amanitin remains more or less constant. The increased rate of incorporation of [^{3}H]uridine into RNA that normally occurs during this developmental period is also blocked by treatment with α-amanitin (Braude, 1979a). In a more detailed study of the necessity for continuous transcription for the transition of morulae to blastocysts, Braude (1979b) demonstrated that if embryos are treated with α-amanitin prior to 80 hr post-hCG, blastocyst formation and expansion are blocked. Treatment of embryos at various intervals after the critical period of about 80 hr post-hCG leads to increasing yields of mature and expanded blastocysts. These results suggest that a critical transcriptional event concerned with blastocyst formation occurs around 80 hr post-hCG and may be associated with the fifth cleavage division. As overt signs of blastocyst formation first occur around 83 to 85 hr post-hCG, these findings argue for a tight coupling of developmental and transcriptional events.

While the synthesis of all major classes of RNA is active during this developmental period, it is difficult to make direct qualitative comparisons of mRNA populations themselves from one stage to the next because of their heterogeneous nature. A more sensitive method for analysis of qualitative differences in the kinds of mRNAs being utilized has been the application of high-resolution two-dimensional electrophoretic procedures to resolve radiolabeled polypeptides coded for by the mRNAs in these embryos.

Figure 1 shows the qualitative patterns of some of the proteins synthesized by 72-hr post-hCG morulae, 72-hr morulae cultured for 24 hr *in vitro* to the blastocyst stage, and 72 hr morulae cultured for 24 hr in the presence of 11 μg/ml α-amanitin. We wish to emphasize two major features of these patterns. First, the majority of polypeptides (all unmarked spots, Figure 1) show little change between the morula and the blastocyst stages and are largely unaffected if morulae are cultured for 24 hr in the presence of α-amanitin. Second, there are two classes of polypeptides that are affected by α-amanitin. One set (◢ arrows) is represented by polypeptides faint or absent in the morula, present in the blastocyst, and absent or very faint in amanitin-treated morulae. The other set (➚ arrows) is represented by polypeptides that are strongly present in the morula, faint or decreased in the blastocyst, but strong in amanitin-treated morulae. The numbering system of these

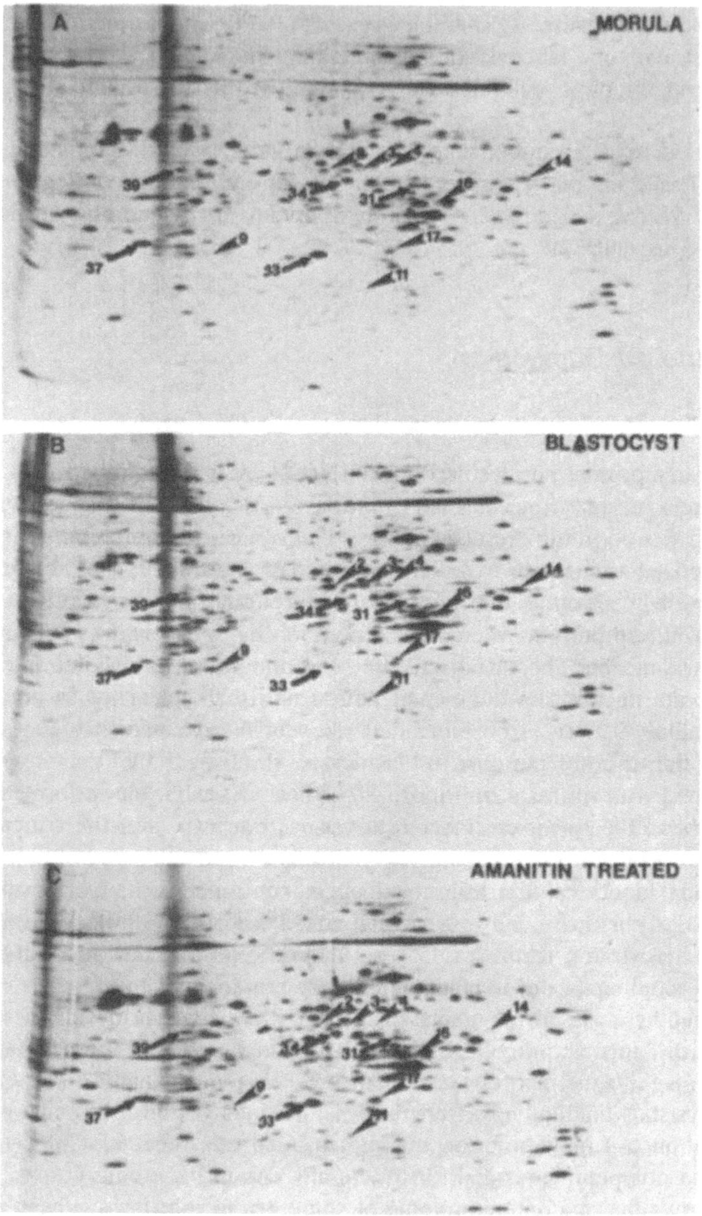

Figure 1. Effect of α-amanitin on qualitative patterns of polypeptides synthesized during the morula to blastocyst transition period in mouse embryos. (A) Fluorograph of radioactively labeled polypeptides synthesized by 72-hr post-hCG morulae incubated for 3 hr in medium containing [^{35}S]methionine. (B, C) Fluorographs of 72-hr morulae cultured for 24 hr in the absence (B) or presence (C) of 11 μg/ml α-amanitin prior to radioactive labeling as in (A). Approximately 40 embryos were utilized in each case. Separation was on a 7 to 15% exponential gradient polyacrylamide gel over a pH range of 4.5 to 7.0. Spots marked by ➤ are faint or absent in the morula, have appeared more strongly in the 24-hr period leading to blastulation, but fail to increase in intensity in the presence of inhibitor (C). Spots marked by ➶ are those clearly detectable at the morula stage, only relatively weakly detectable in the blastocyst pattern, but continue to be synthesized 24 hr later in the presence of transcriptional inhibition. The numbering system is that of Braude (1979b).

amanitin-sensitive polypeptides, which normally either increase or decrease in intensity with blastocyst formation, follows that of Handyside and Johnson (1978) and Braude (1979b).

Possible secondary effects of α-amanitin on the embryo along with inherent difficulties in obtaining fluorographs of comparable intensity come into play when making interpretations from such figures. Precautions required to minimize such problems have been discussed previously (Handyside and Johnson, 1978; Braude, 1979a). Taking these precautions, it is clear that the majority of polypeptides produced in morulae are translated from messages of relatively high stability, for they continue to be synthesized 24 hr later despite the presence of a transcriptional inhibitor shown previously to suppress synthesis of poly A-RNA within 30 min of exposure (Levey *et al.*, 1978). On the other hand, the relatively few polypeptides (▲ , Figure 1) that fail to increase in intensity in the presence of α-amanitin appear to be translated from mRNAs that are dependent on transcription during this 24-hr development period. Finally, the class of polypeptides (➚ , Figure 1) in which synthesis normally ceases during the period leading to blastocyst formation but apparently continues in the presence of α-amanitin may indicate transcription of some regulatory RNA that itself, or via its protein product, speeds the degradation of existing message or selectively switches off their translation.

It seems unlikely that the quantitative increase in net protein synthesis that normally occurs during the morula to blastocyst transition could be due to the few amanitin-sensitive polypeptides described above. More probably, this increase is due to the transcription and/or translation of further quantities of predominantly the same species of mRNA (those coding for unmarked polypeptides) that had been synthesized up to the morula stage. Taken together, these results suggest that transcriptional activity is important in blastocyst formation. However, it is not possible to preclude posttranscriptional controls in the quantitative increase in protein synthesis. It is also possible that the mRNA species that code for the novel polypeptides synthesized during this transition are synthesized prior to 80 hr post-hCG, but are only activated selectively later. This latter possibility is being tested experimentally by *in vitro* translation of mRNA species extracted prior to and after 80 hr post-hCG. This approach has already yielded evidence of posttranscriptional control of polypeptides synthesized at earlier stages.

B. Stable mRNA and Posttranscriptional Control in Early Cleavage

Previous comparison of protein synthetic patterns between the one- and two-cell-stage mouse embryo has revealed that qualitative changes do occur (Van Blerkom and Brockway, 1975), although the overall rate of synthesis remains low and relatively constant throughout this period (Brinster, 1971). On two-dimensional gels there are few notable differences between unfertilized eggs, newly fertilized one-cell eggs, and early two-cell embryos (40 hr post-hCG). However, at the late one-cell and early two-cell stages, there is one set of polypeptides with molecular weight of about 35,000 that is produced in greatly increased quantity compared to the unfertilized egg (Figures 2A, D).

When eggs are fertilized *in vitro* and cultured over a 24-hr period in the presence of 11 μg/ml α-amanitin, the majority of polypeptides continue to be synthesized much as in the case of the majority of polypeptides of Figure 1 (Braude *et al.*, 1979). These polypeptides appear to be translated from stable mRNAs already present and utilized in the egg. The appearance of the characteristic early two-cell polypeptides also occurs in the presence of α-amanitin (Braude *et al.*, 1979), suggesting that control of the synthesis of these polypeptides is due to some form of selective activation of mRNAs at a posttranscriptional level.

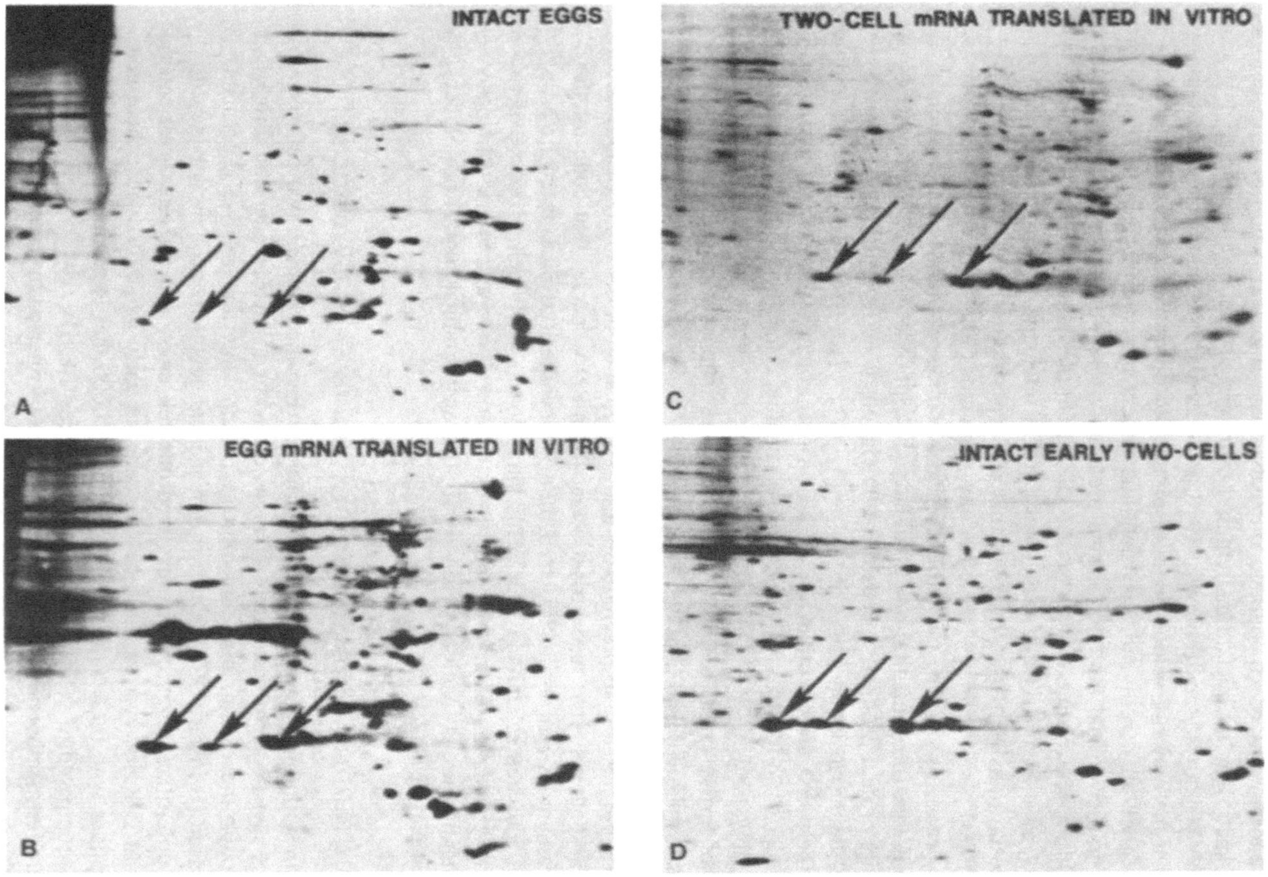

Figure 2. Two-dimensional fluorographs of [^{35}S]methionine-labeled polypeptides from one- and two-cell mouse embryos. (A) Polypeptides from 50 intact unfertilized eggs (15 hr post-hCG). (B) Polypeptides from *in vitro* translation of mRNA extracted from 400 unfertilized eggs. (C) Polypeptides from *in vitro* translation of mRNA extracted from 112 two-cell embryos (40 hr post-hCG). (D) Polypeptides from 70 intact two-cell embryos. The arrows show polypeptides that are not strongly labeled until the late one-cell or early two-cell stage but are translated *in vitro* from egg mRNA.

This conclusion is emphasized by the observation that the same polypeptides may also be transiently expressed during intrafollicular maturation of oocytes (Van Blerkom, this volume).

To assess more conclusively whether a posttranscriptional level of control was operating, total RNA was extracted from about 400 unfertilized eggs and was translated *in vitro* in message-dependent reticulocyte lysate. The two-cell marker proteins (arrows, Figure 2D) are prominent among the cell-free products (Figure 2B) even though they are virtually undetectable in the intact unfertilized egg pattern (Figure 2A). Thus, the mRNAs coding for these 35,000-dalton polypeptides are indeed present in unfertilized eggs but are not translated *in vivo*. The increased intensity of these spots in the fluorograph of the translational products from unfertilized eggs is probably not due to selective translation of these mRNAs by the lysate because they are not more prominent among the translation products of RNA from two-cell embryos (Figure 2C). Consequently, it is reasonable to conclude that increased appearance of these synthetic products *in situ* is due to some form of posttranscriptional regulation that is distinct from other mechanisms operating on the majority of stable mRNAs within the egg.

We do not know the nature or function of these polypeptides. Only a few are involved and their similar molecular weight suggests they may be structurally related and under coordinate control. One can speculate that the developing embryo requires large amounts of these proteins at a time when rapid accumulation of newly synthesized mRNA from the one-cell zygote genome may not be possible. Any number of mechanisms may be involved in regulating the appearance of the products of these mRNAs. As mRNA structure at the 5′ terminus itself is important for translation, we turned our attention to analyzing whether mRNAs in preimplantation mouse embryos contained cap structures.

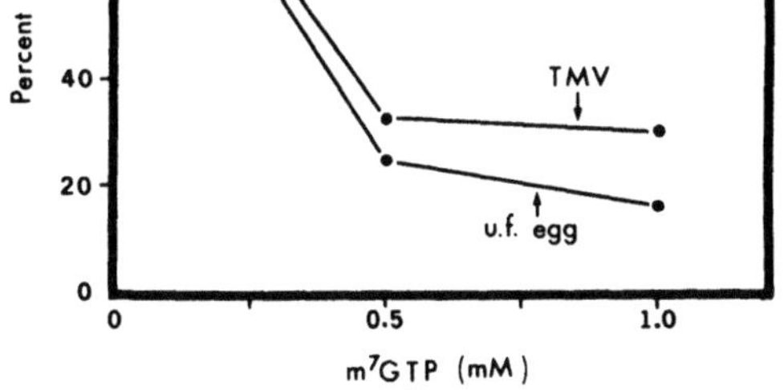

Figure 3. Effect of m^7GTP on translation of TMV, CPMV, and unfertilized mouse egg (u.f. egg) mRNA. TMV and CPMV mRNA were present at a concentration of 0.4 μg/20 μl lysate and were translated in the presence of 0, 0.5, and 1.0 mM m^7GTP, respectively. Protein synthesis is expressed as a percentage of control value (approximately 1.2×10^5 cpm/2 μl lysate for both TMV and CPMV mRNA) and the tRNA background (approximately 8×10^3 cpm/2 μl lysate) was subtracted. For the u.f. egg sample, RNA was prepared from an equivalent of 825 eggs, and, after addition of [^{35}S]methionine and lysate, the RNA preparation was divided into three equal fractions and translated as above. The control value was 25,605 cpm/2-μl sample and the tRNA background was 11,634 cpm/2-μl sample of lysate.

C. Inhibition of Translation with Cap Analogs

Considerable evidence exists to indicate that cap structures on mRNA molecules in eukaryotic cells appear to augment efficiency of initiation in translation and that cap analogs like m^{7}GTP can reduce the translation of capped mRNAs *in vitro* (Shatkin, 1976; Kozak, 1978; Revel and Groner, 1978). However, the degree of discrimination between translation of capped and uncapped mRNAs and the specificity of action of cap analogs for the inhibition of translation of capped messages depends on a number of parameters including the choice of *in vitro* system (Lodish and Rose, 1977), the temperature and concentration of monovalent ions during translation (Kemper and Stolarsky, 1977; Weber *et al.*, 1978; Chu and Rhoads, 1978), and sources of initiation factors (Bergmann *et al.*, 1979).

We have established the nature of the specificity of inhibition of translation by the cap analog m^{7}GTP in message-dependent reticulocyte lysate by comparing its effect on the translation of mRNA from tobacco mosaic virus (TMV), which has capped mRNA (Keith and Fraenkel-Conrat, 1975), and cowpea mosaic virus (CPMV), which has uncapped mRNA (Klootwijk *et al.*, 1977). Both contain mRNA components of similar molecular weight (1.3 to 2.2×10^6) and incorporate similar amounts of [^{35}S]methionine into protein per microgram mRNA under standard conditions in the message-dependent reticulocyte system (Table 1). Translation of capped TMV RNA was reduced to about 30% of control values over the range of cap analog concentrations utilized, whereas the uncapped CPMV mRNA was largely unaffected (Figure 3 and Table 1). The translation of mRNA extracted from unfertilized eggs was also reduced to about 20% of control values in the presence of 0.5 to 1.0 mM m^{7}GTP (Figure 3). The same is true for mRNAs extracted from two-cell embryos and blastocysts, results of which are summarized in Table 1. In addition, the importance of the methyl group in the m^{7}GTP cap analog was reconfirmed as GTP itself had a slight stimulatory rather than inhibitory effect on translation (Table 1).

Table 1. Inhibition of mRNA Translation by Cap Analog

	[^{35}S]Methionine incorporation			
RNA	tRNA background (cpm)	Control (cpm)	Experimental[a] (cpm)	Percentage of control[b]
TMV[c]	8,572	123,018	32,720	30.7
CPMV[c]	8,385	113,925	106,795	93.2
Egg (460)[d]	10,454	27,305	10,665	1.3
Egg (546)	11,234	25,355	13,834	18.4
Egg (490)	12,383	20,268	22,400*	127.0
Egg (412)	13,052	22,144	23,020*	109.6
2-Cell (313)	10,841	27,482	14,213	20.3
2-Cell (224)	11,538	23,447	14,110	21.5
Blastocyst (204)	13,740	18,989	14,235	9.4
Blastocyst (222)	14,213	18,829	14,998	17.0

[a] The cap analog m^{7}GTP was added to a final concentration of 1 mM except in the two cases designated by *, where GTP was substituted and also added to a final concentration of 1 mM.
[b] Calculated after extraction of tRNA background.
[c] TMV and CPMV mRNA were present at a concentration of 0.4 μg/20 μl lysate.
[d] Number in parentheses designates number of mouse eggs extracted.

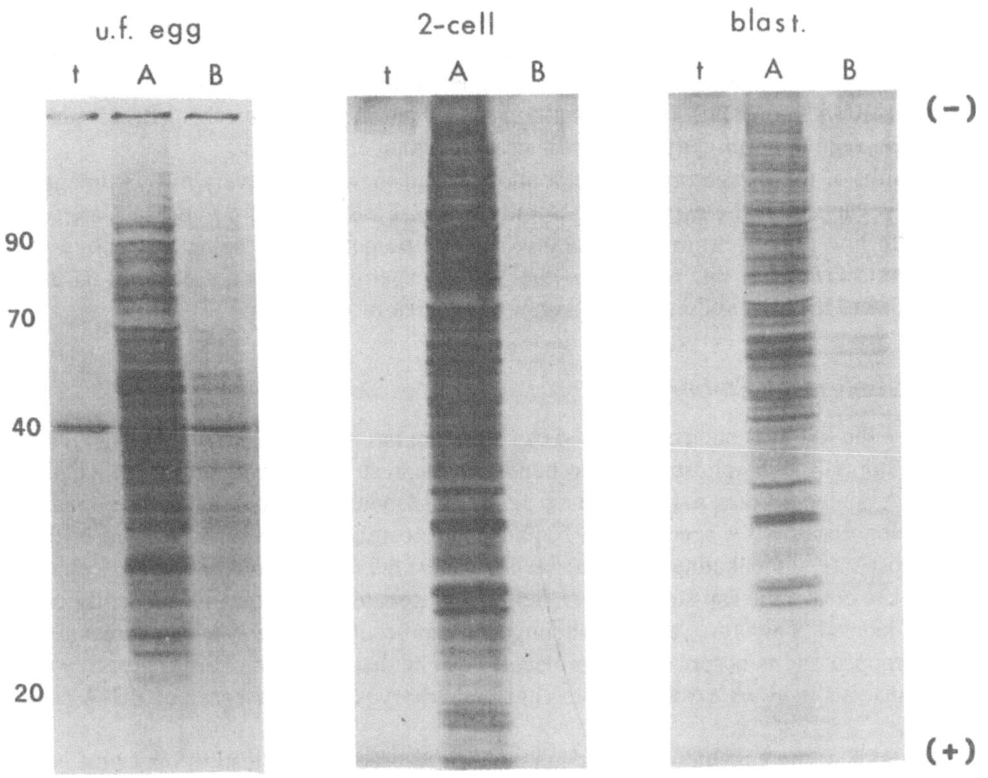

Figure 4. Fluorographs of one-dimensional SDS polyacrylamide gels of the [^{35}S]methionine-labeled translation products from unfertilized eggs (u.f. egg), two-cell embryos, and blastocysts. RNA was prepared from 200 to 400 eggs or embryos and after addition of [^{35}S]methionine and lysate, the RNA prepartion was divided into two equal fractions. One fraction (A) was translated as a control, the second (B) in the presence of 1 mM m^{7}GTP; t is the background activity of the lysate with added tRNA but without egg or embryo RNA in each case. Approximate molecular weight ($\times 10^{-3}$) is given on the left-hand scale. Four microliters of translate was applied to the gel in each case. Exposure times were 6 to 8 days.

Qualitative comparisons of the translation products directed by egg and embryo mRNA in the presence or absence of 1 mM m^{7}GTP were obtained from fluorographs of one-dimensional polyacryladmide gels (Figure 4). Clearly a reduction in the synthesis of all types of polypeptides coded for by mRNA from all stages of development analyzed occurs in the presence of the cap analog. The radioactive band present even in the tRNA background samples (most notable in the unfertilized egg preparation at about 40,000 molecular weight) is the consequence of addition of [^{35}S]methionine to an endogenous reticulocyte polypeptide by a ribosome-independent reaction (Pelham and Jackson, 1976) and consequently is insensitive to inhibition by m^{7}GTP. The intensity of this band varies from one batch of lysate to another as well as with the type of manipulations utilized in preparing translation products for analysis on gels.

The fact that cap analogs do not reduce translation completely to background values is a common observation in these types of experiments and partially reflects the fact that initiation may occur at reduced efficiency by mechanisms not involving the cap structure (Kozak, 1978). Because all the polypeptide translation products resolved on gels were reduced in intensity by the cap analog, it appears that virtually all the translatable mRNA species in the embryos are capped. Included within this set would be the "masked" mRNAs coding for the characteristic 35,000-molecular-weight two-cell polypeptides outlined above. For these messages, our results argue against addition of cap structures to preexisting mRNA molecules of maternal origin as an important mechanism of posttranscriptional regulation in early postfertilization developmental events.

The results of these experiments do not allow us to conclude that every mRNA in eggs or embryos is capped. If the egg contains some uncapped mRNAs that do not translate (or translate with low efficiency) *in vitro*, they would not be detected as being insensitive to inhibition by m^7GTP. For this reason, another set of experiments that did not rely on the translation of mRNA was undertaken to assess the degree of capping.

D. End-Labeling of Embryo RNA

Because the terminal nucleotide of the cap structure has an inverted 5′–5′ linkage relative to the normal 3′–5′ phosphodiester bonds in the rest of the polynucleotide chain, capped RNA molecules are not capable of being end-labeled with polynucleotide kinase and ATP. Noncapped RNA species with 5′-phosphates can be labeled after removal of the terminal phosphate with alkaline phosphatase. However, the enzyme tobacco acid phosphatase (TAP) can cleave the cap structures to yield 5′-phosphate termini that subsequently can also be end-labeled. Therefore, by end-labeling RNA molecules with and without treatment with TAP, capped and noncapped RNA molecules can be distinguished. This approach was applied to analysis of RNA extracted from eggs and embryos in the presence of tRNA carrier.

To assess that the enzymes and reaction conditions used did not lead to major mRNA degradation, end-labeling was performed after TAP removal of cap structures of tobacco rattle virus (TRV) mRNA, and the electrophoretic mobility of this labeled material was compared to untreated TRV mRNA (A1 and A2 of Figure 5). Under the gel conditions utilized, two high-molecular-weight mRNA components were resolved that remained intact after end-labeling procedures were employed. If significant degradation had occurred, it would have been detected as a number of smaller-molecular-weight components in the autoradiograph.

The same procedures were applied to egg and embryo RNA extracted in the presence of tRNA carrier, and the poly A RNA component (putative message) was purified by affinity chromatography on oligo dT cellulose. An example of such an experiment conducted

Table 2. Purification of End-Labeled Poly A RNA from Unfertilized Mouse Eggs on Oligo dT Cellulose [a]

RNA	cpm [^{32}P]-RNA applied ($\times 10^{-6}$)	cpm [^{32}P]-RNA bound ($\times 10^{-3}$)	% bound	Δ % bound	Ratio −TAP/+TAP	% capped
Egg + TAP	29.36	29.35	0.099	0.085	0.223	77.7
tRNA + TAP	23.71	3.32	0.014			
Egg − TAP	28.06	8.36	0.030	0.019		
tRNA − TAP	25.36	2.79	0.011			

[a] RNA was prepared from 584 unfertilized eggs along with 5-μg tRNA carrier and divided into two parts; one part was end-labeled after enzymatic decapping by tobacco acid phosphatase (+TAP) and the other was end-labeled without decapping (−TAP). Background was determined using a control of 5-μg carrier tRNA. All samples were subjected to affinity chromatography on oligo dT cellulose as described in Materials and Methods.

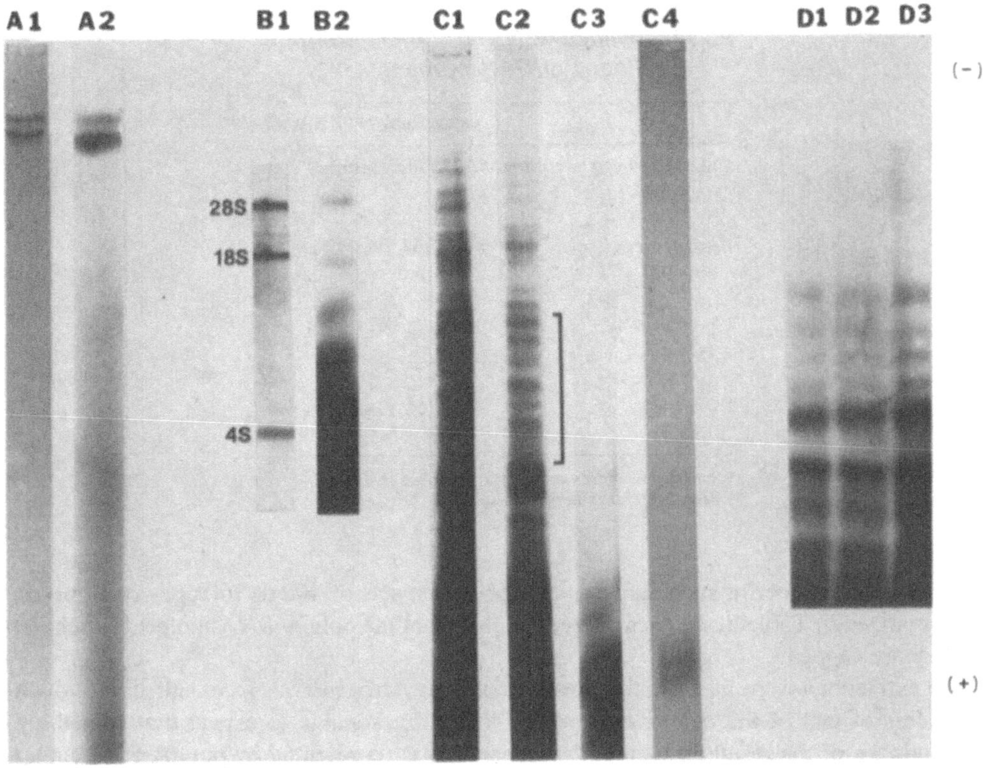

Figure 5. Analysis of end-labeled RNA on polyacrylamide gels. A1, photograph of stained pattern of 18 μg TRV RNA not subjected to end-labeling procedures and resolved on a 3.5% polyacrylamide gel. A2, autoradiograph of 3 μg TRV RNA end-labeled after TAP treatment and run on an adjacent slot of the gel to material in A1. B1, photograph of stained pattern of 50 μg total cytoplasmic RNA from mouse liver resolved on a 2.5% polyacrylamide gel. B2, autoradiograph of TAP-treated and end-labeled RNA extracted from unfertilized eggs in the presence of tRNA carrier and resolved under the same gel conditions utilized for B1. C1, autoradiograph of TAP-treated and end-labeled RNA extracted from unfertilized eggs in the presence of tRNA carrier and resolved on a 4.0% polyacrylamide gel. C2, autoradiograph of TAP-treated and end-labeled RNA extracted from fertilized eggs in the presence of tRNA and resolved as in C1. C3, end-labeled tRNA carrier resolved as in C1 and C2. C4, stained pattern of 25 μg tRNA carrier resolved as in C1, C2, and C3. D1, D2, and D3, autoradiographs of a region of a 5% polyacrylamide gel containing TAP-treated, end-labeled, poly A RNA derived from either unfertilized eggs (D1), two-cell embryos (D2), or blastocysts (D3). The poly A RNAs resolved correspond roughly to the molecular weight region designated by the bracket in C2.

on RNA extracted from 584 unfertilized eggs is summarized in Table 2. Prior to purification on oligo dT columns, between 20×10^6 and 30×10^6 cpm of [$\gamma-^{32}$P]-ATP was transferred to 5′ ends of RNA molecules. End-labeled tRNA alone (whether TAP treated or not) bound nonspecifically to the oligo dT cellulose at levels of 0.014 and 0.011%, respectively. Egg RNA extracted in the presence of tRNA carrier bound at a level of 0.085% above the background value when end-labeled after "cappase" treatment with TAP, while in an equivalent aliquot of RNA labeled in the absence of "cappase" treatment, only 0.019% (after background subtraction) of the ^{32}P label copurified with poly A RNA. The egg + TAP sample should represent end-labeled poly A RNA molecules that are both capped and uncapped while the egg−TAP sample should represent only end-labeled uncapped molecules. The ratio between the two values leads to the calculation that 77.7% of the poly A RNA molecules of the unfertilized-egg RNA preparation are capped (Table 2).

Table 3. Extent of Capping of Polyadenylated RNA from Different Stages of Development

Stage	% capped poly A RNA
Unfertilized egg ($n = 6$)[a]	82.27 (74.51–90.92)[b]
Fertilized egg ($n = 4$)	81.97 (68.29–94.51)
Late two-cell ($n = 4$)	78.45 (73.94–85.89)
Blastocyst ($n = 4$)	77.12 (64.83–90.14)

[a] Number of samples analyzed as described in Table 2.
[b] Mean, with range in parentheses.

The results of a number of such analyses carried out on sets of 400 to 700 eggs or embryos are summarized in Table 3. At each stage, about 80% of the poly A RNA molecules behave as if they are capped.

As extractions were made in the presence of 5 μg carrier tRNA from calf liver, which is in excess of that of the egg or embryo RNA, it is reasonable to expect that something less than 0.5% of the resulting total RNA preparation is represented by poly A RNA molecules. The specific activity of the [γ-^{32}P]-ATP used in end-labeling was 6.5×10^3 cpm/fmol. In the example of Table 2, the egg + TAP sample represented RNA from 292 eggs and contained 26×10^3 cpm of ^{32}P label in poly A RNA. On the basis of previous calculations, 300 eggs can be expected to contain 3 to 6 ng (or on a molar basis, 6 to 12 fmol) of poly A RNA with an average length of 1500 nucleotides (Schultz *et al.*, 1980). If every 5′ end was labeled, a yield of approximately 39×10^3 to 78×10^3 cpm was theoretically possible. The reduced level observed is not unexpected, as recovery of RNA during extraction must be less than 100% and enzymatic decapping and end-labeling procedures cannot be expected to go to completion even when enzymes and ATP substrate are provided in excess. If extensive fragmentation of RNA molecules occurred during the reactions, many additional 5′ termini available for end-labeling would be generated, a result that is not consistent with the amount of ^{32}P incorporation we observed. That a small amount of RNA cleavage has occurred during the procedures cannot be ruled out. The consequence of such events would be the generation of 5′ termini that would behave as uncapped molecules. Therefore, the observation that about 80% of the poly A RNA molecules are capped must be regarded as a minimum estimate; conceivably all of the molecules could be capped. Taken together with results on translational inhibition with m^{7}GTP, it appears that most of the mRNA in mouse eggs and embryos is capped.

E. Qualitative Analysis of End-Labeled RNA

Egg and embryo RNA preparations subjected to decapping and end-labeling procedures were analyzed electrophoretically to determine whether any qualitative differences in such *in vitro*-labeled steady-state RNA populations were detectable from one stage to another (Figure 5). Under conditions where the major RNA classes can be resolved on the same gel (B1, Figure 5), end-labeled RNA extracted from unfertilized eggs in the presence of tRNA carrier contains a small amount of radioactivity in the 28 and 18 S rRNA species, but the vast majority of label, as expected, is in the tRNA carrier (B2, Figure 5). When similar preparations are resolved on longer gels of higher acrylamide concentration, a number of labeled bands of greater molecular weight than the carrier tRNA are detectable (C1–C4, Figure 5). The high-molecular-weight species are clearly of egg origin. However, it was not possible to detect any reproducible differences between unfertilized and fertilized egg RNA (C1 and C2) or between other stages (not presented) in these types of gels. When decapped, end-labeled, purified poly ARNA from eggs and embryos was resolved electrophoretically on 5% gels, again several radioactive RNA species were resolved (D1–D3, Figure 5). However, even under these conditions, no differences between egg, two-cell, and blastocyst putative mRNAs were observed. These results are perhaps not surprising because the number of RNA bands resolved electrophoretically is limited. The gels are run under denaturing conditions and resolution is based largely on molecular weight differences. As mRNA species coding for different proteins can have similar molecular weights, potential differences could be obscured as each band can be expected to contain a number of different mRNA molecules of similar length.

An additional consideration is that concentrations of mRNAs within mammalian cells vary from abundant through intermediate to rare. On average, 65 to 90% of the mRNA populations are made up of a few hundred intermediate and abundant mRNA species, each of which is present in several hundred to more than a thousand copies per cell (Tobin, 1979). The rare class makes up only 10 to 35% of the total mRNA population but represents some 8000 to 10,000 different species present in only a few copies per cell (Tobin, 1979). Because end-labeling occurs on a molar basis, the most abundant mRNA classes will contain most of the label. These mRNAs are most likely those responsible for providing the structural proteins and enzymes necessary for maintenance of all cell types and would not be expected to vary significantly from one to another. Many of the polypeptides synthesized by the stable mRNAs of Figures 1 and 2, which are common to all stages of preimplantation development, would be expected to fall within this abundant mRNA class. The technique is not sufficiently sensitive to allow visualization of rare mRNA species that could arise through either a decreased rate of transcription or a rapid turnover rate. While it is possible that such mRNAs could play regulatory roles in cell function and differentiation, we simply do not have sufficient data to evaluate their role in preimplantation development at present.

IV. Concluding Remarks

The first cleavage of the mouse zygote can occur in the apparent absence of nuclear gene activity and appears to be directed by mRNAs of maternal origin. Evidence has been presented that the messages for a set of polypeptides of 35,000 molecular weight, synthesis of which is not normally evident until the early two-cell stage, are already present in abundance in the unfertilized egg. We do not know how the expression of these mRNAs is selectively regulated at the posttranscriptional level. The possibility that they are packaged in mRNP particles and are inactive in translation, as in sea urchin eggs (Jenkins *et al.*, 1978), might be entertained. Conceivably, destabilization of such particles by ionic changes, protein modification, or proteolytic cleavage as a consequence of fertilization events could lead to activation. Any of several other mechanisms could also be involved. The activation is unlikely to be due to modification of the 5′ terminus of the mRNA through addition of a cap structure, for virtually all translatable mRNAs in the unfertilized egg and other stages of development already possess cap structures. This property is not unexpected, as cap structures provide protection from 5′ exonuclease degradation (thereby increasing stability) and appear to augment translational efficiency in eukaryotic cells. The mRNA of unfertilized sea urchin eggs has also been shown to be predominantly capped (Hickey *et al.*, 1976). Preliminary experiments indicate that the "masked mRNA" is also polyadenylated (Johnson, Schultz, and Braude, unpublished results). Thus there appears to be no unique structural feature of these mRNAs at either the 5′ or the 3′ terminus; future investigations should be directed at more physiological aspects of their utilization.

The morula–blastocyst transition, in contrast, is susceptible to arrest by α-amanitin and relies on continuous transcription for development to proceed normally. Appearance of blastocyst-specific polypeptides is also transcriptionally dependent, although we do not as yet know if a posttranscriptional regulatory event is also involved. Whether the transcriptionally dependent blastocyst marker polypeptides themselves are regulatory in blastocyst formation or arise as a consequence of other phenomena is not yet resolved. Efforts to analyze mRNA populations after *in vitro* end-labeling with [γ-^{32}P]ATP and polynucleotide kinase failed to provide a method with sufficient sensitivity to detect qualitative differences in steady-state mRNA populations from one stage to another during preimplantation development. Nonetheless, continued application and development of microbiochemical methods will undoubtedly bring us closer to a more detailed understanding of mechanisms of control of gene expression in early mammalian embryogenesis in the near future.

ACKNOWLEDGMENTS. This work was supported by grants from the Ford Foundation and the Medical Research Council (England) to M.H.J., and by the Medical Research Council (Canada) to G.A.S. The technical assistance by Ms. G. Flach in many phases of the work is greatly appreciated. Tobacco acid phosphatase was a gift from A. Efstratiadis, Biological Laboratories, Harvard University. Polynucleotide kinase was a gift from H. van de Sande, Division of Medical Biochemistry, University of Calgary.

Discussion

EPSTEIN (USCF): You have shown that capping is essentially the same quantitatively in the egg as it is in the embryo, yet you were able to translate messages from the egg that do not normally appear in the embryos until later. Is there something else involved?

SCHULTZ: All messages in eggs and embryos appear to be capped, based on qualitative analysis of translation products after cap analog inhibition. Capping is probably not the mechanism, therefore, by which the message is selectively activated. Messages may be sequestered in an inactive state as ribonucleoprotein particles, or altered protein synthesis at different stages of development may be caused by changes in factors associated with the translation apparatus.

EPSTEIN: A few years ago we took unfertilized eggs after 1 or 2 days in the oviduct and labeled them. We found a prominent protein in the two-cell embryo, so the mechanism that shifts from the egg pattern to the first cleavage pattern is independent of fertilization. Not all the peptides shifted, but only that one; maybe it is related to ovulation, or to time.

MANES (La Jolla): I agree that there is evidence that capping may have an effect on the efficiency with which a message is translated. There is also evidence, at least in some systems, that capping does affect the stability of a message. Since one class of mammals is able reversibly to arrest their blastocysts in growth and another class cannot, rabbits naturally cannot, I wonder if there is any difference in stability of messages between one class of mammals and the other? Have you looked at rabbit protein synthesis with your cap analog to see if the capping picture is the same?

SCHULTZ: No, I have not done anything on rabbits. I hope to be able to repeat at least some of these experiments in the rabbit system. I would guess that all translatable mRNAs from rabbit embryos will also be subject to inhibition by a cap analog.

CHANG (Worcester Foundation): This conference is on implantation, but I am interested in the fertilizing life of the egg. Is there any difference between the unfertilized egg recovered early and recovered late?

SCHULTZ: In terms of whether the mRNA is capped or not, I have looked mostly at eggs ovulated between 14 and 16 hr post-hCG, so I cannot answer that directly. The fact that there are qualitative differences in the types of proteins being made during oocyte maturation and accompanying very early postfertilization events suggests that the exact time of inspection could well be important, even in the unfertilized egg. It is possible that the major proteins we see at the early or late one-cell or two-cell stage may vary in their abundance, even in the unfertilized egg.

BULLOCK (Baylor): I almost hate to ask this question, particularly in view of some of the people in the audience, but I worry about the quantitative aspects of two-dimensional gels and the measurements that were made on the basis of spot density. I know people claim that this can be done fairly reliably and quantitatively, yet I wonder whether you are really confident of the quantitative aspects of these two-dimensional gels, particularly when the data are used to make arguments of the order of posttranscriptional regulation?

SCHULTZ: I think anyone who works with two-dimensional gels has a problem of how to assess two gels and how to expose them to exactly equal intensities, so that there are fair comparisons of the relative synthesis from one experiment to another. Unless one has a standard spot that one can always quantitate densitometrically, which every cell and embryo makes in an equivalent amount, it is very hard. In this particular case, those polypeptides that we are looking at are very faint or absent in the unfertilized egg, and in the translation products they are predominant amongst the products of synthesis. I think there is unquestionably a major difference there. I do not think this is due simply to the reticulocyte cell-free system selectively translating these RNAs more efficiently than the egg itself. If that was the case, they should be extremely abundant in the translation products because the two-cell egg makes lots of these proteins. They are not more abundant in the two-cell translation products than in the unfertilized egg translation products.

EPSTEIN: One can do a reasonably good job of quantitating two-dimensional patterns within the linear range of response of the gels, by calibrating the gels to establish the linear range, and by using different peptides as standards. In our experience, correlating several standard peptides one with another and then with the unknowns detects differences of 25 to 50%.

KIDDER (University of California, San Francisco) What concentration of α-amanitin do you use, how long do you treat, and how did you arrive at those conditions?

SCHULTZ: The concentration of α-aminitin was 11 μg/ml. We used Whittingham's medium M16 along with BSA, for 24 hr. According to Levey and Brinster (*J. Exp. Zool.* **203:**351, 1978) this inhibits all the poly A-RNA synthesis; we found it also affects total RNA synthesis.

KIDDER: The Levey and Brinster paper concerns 3½-day blastocysts. Have you checked if those conditions are appropriate for earlier stages, for the morula stage for example?

SCHULTZ: Braude (*Dev. Biol.* **68:**440, 1979) has shown that this concentration of α-amanitin also reduces RNA synthesis in morulae. I suppose the most effective answer to that is that the morula does not develop any further. Braude has done another set of experiments (*J. Embryol. Exp. Morphol.* **52:**193, 1979) in which he recovered morulae and put them into α-amanitin at increasing times of development. There is a point at about 80 hr when these embryos become resistant to the amanitin blockage with respect to cavitation and blastocyst formation and begin to show the increase in protein synthetic pattern characteristic of blastocyst formation. This is consistent with what Michael Sherman mentioned earlier, that there might be two synthetic phases, one between 3 and 4 days of development and one after. At least in terms of formation of blastocyst, there is a period prior to 80 hr that is very sensitive to α-amanitin and a period later that is not.

KIDDER: If I can add just one more point, it bothers me that we conclude, from experiments such as this, that we are seeing the effect of interrupting the transcription of some critical structural genes. Unless a careful characterization of the effects of that concentration and that treatment time have been done on that stage, one cannot rule out the result of a partial block of polymerase I.

SCHULTZ: Yes, I think it is very hard to remove the secondary effects of the inhibitor. A better approach nowadays is to use *in vitro* translation systems to try to answer the same sort of questions and to get away from using the inhibitors. Regarding posttranscriptional control in early cleavage, however, both approaches lead to the same conclusion.

SHERMAN (Roche Institute): As we embark with some trepidation upon the *in vitro* translation systems that you have developed, we are concerned about the problem of selective translation and preferential translation of different messages. What information can you offer on that score?

SCHULTZ: Every cell-free system has its own biases as to which messages it will translate most efficiently. Globin message will translate somewhat differently in the wheat germ system than in the reticulocyte system, and so forth. We have done no experiments to compare translation of mouse egg message in a reticulocyte system with a different system to see if there are differences in efficiency of translation.

MANES (La Jolla): A word of caution about amanitin might be useful. Commercial amanitin is not very pure; there are traces of pholloidin, which is also made by the same mushroom, which could interact with actin filaments, for example. As Mike Sherman said, when something happens in spite of an inhibitor you feel relatively comfortable; when something does not happen, to try to interpret that is another problem altogether.

EPSTEIN: I think anybody who works with amanitin and looks at the cells knows that there are a lot of effects in the nuclei even though supposedly the type I polymerase is not affected. Can I come back to the point of what is responsible for the increased rate of synthesis and blastulation even though the protein pattern does not change very much? Do you have any further thoughts on whether it is increased efficiency of translation of preexistence messages?

SCHULTZ: We do not know whether there is a stockpile of the same messages already existing in the cells and not being utilized. We do know there are transcription-dependent events which may or may not be related to increased synthesis of the same kinds of mRNAs. There may be a number of other translation control factors as well.

PEDERSEN (UCSF): In the translation product of the unfertilized egg, did you see any of the proteins that characterize the trophoblast or the inner cell mass?

SCHULTZ: No.

EPSTEIN: In the *in vitro* translation system, how many of the peptides are actually from the egg or embryo messages compared to background?

SCHULTZ: We see a small number of background bands, varying from lysate to lysate. Often there is essentially nothing and the main one that does occur is due to some ribosome-independent reaction, in which a methionine is added to the N-terminal end of a reticulocyte protein, but not on the translation apparatus. On two-dimensional gels there are hardly any background spots; there is just one big streak representing the reticulocyte protein that is quite easy to distinguish from all the other polypeptides. Examples are presented in Braude and Pelham's original paper (*J. Reprod. Fertil.* **56:**153, 1979).

EPSTEIN: How many eggs or embryos are involved in these kinds of experiments?

SCHULTZ: Most of what I showed was from approximately 200 eggs or embryos, but it can be scaled down. In Peter Braude and Hugh Pelham's paper they report dilutions down to the equivalent of 12½ eggs, which still reveals translation products.

References

Bergmann, J. E., Trachsel, H., Sonenburg, N., Shatkin, A., and Lodish, H. F., 1979, Characterization of rabbit reticulocyte factor(s) that simulates the translation of mRNAs lacking 5′-terminal 7-methylguanosine, *J. Biol. Chem.* **254:**1440.

Both, G. W., Banerjee, A. K., and Shatkin, A. J., 1975, Methylation-dependent translation of viral messenger RNAs *in vitro*, *Proc. Natl. Acad. Sci. USA* **72:**1189.

Braude, P. R., 1979a, Control of protein synthesis during blastocyst formation in the mouse, *Dev. Biol.* **68:**440.

Braude, P. R., 1979b, Time-dependent effects of α-amanitin on blastocyst formation in the mouse, *J. Embryol. Exp. Morphol.* **52:**193.

Braude, P. R., and Pelham, H. R. B., 1979, A microsystem for the extraction and *in vitro* translation of mouse embryo mRNA, *J. Reprod. Fertil.* **56:**153.

Braude, P. R., Pelham, H. R. B., Flach, G., and Lobatto, R., 1979, Post-transcriptional control in the early mouse embryo, *Nature (London)* **282:**102.

Brinster, R. L., 1971, Uptake and incorporation of amino acids by the preimplantation mouse embryo, *J. Reprod. Fertil.* **27:**329.

Canaani, D., Revel, M., and Groner, Y., 1976, Translational discrimination of "capped" and "non-capped" mRNAs: Inhibition by a series of chemical analogues of ^{7}mGpppX, *FEBS Lett.* **64:**326.

Chu, L. Y., and Rhoads, F. E., 1978, Translational recognition of the 5′-terminal 7-methylguanosine of globin messenger RNA as a function of ionic strength, *Biochemistry* **17:**2450.

Efstratiadis, A., Vournakis, J., Doris-Keller, H., Chaconas, G., Dougall, D., and Kafatos, F. C., 1977, End-labelling of enzymatically decapped mRNA, *Nucleic Acid Res.* **4:**4165.

Gedamu, L., and Dixon, G. H., 1978, Effect of enzymatic "decapping" on protamine translation in wheat germ S-30, *Biochem. Biophys. Res. Commun.* **85:**114.

Handyside, A. H., and Johnson, M. H., 1978, Temporal and spatial patterns of synthesis of tissue-specific polypeptides in the pre-implantation mouse embryo, *J. Embryol. Exp. Morphol.* **44:**191.

Hickey, E. D., Weber, L. A., and Baglioni, C., 1976, Translation of RNA from unfertilized sea urchin eggs does not require methylation and is inhibited by 7-methylguanosine-5′-phosphate, *Nature (London)* **261:**71.

Jenkins, N. A., Kaufmeyer, J. F., Young, E. M., and Raff, R. A., 1978, A test for masked message: The template activity of messenger ribonucleoprotein particles from sea urchin eggs, *Dev. Biol.* **63:**279.

Johnson, M. H., 1979, Intrinsic and extrinsic factors in preimplantation development, *J. Reprod. Fertil.* **55:**255.

Keith, J., and Fraenkel-Conrat, H., 1975, Tobacco mosaic virus RNA carries 5′-terminal triphosphorylated guanosine blocked by 5′-linked 7-methylguanosine, *FEBS Lett.* **57:**31.

Kemper, B., and Stolarsky, L., 1977, Dependence on potassium concentration of the translation of messenger ribonucleic acid by 7-methylguanosine-5′-phosphate, *Biochemistry* **16:**5676.

Klootwijk, J., Klein, I., Zabel, P., and Van Kammen, A., 1977, Cowpea mosaic virus RNAs have neither m^{7}GpppN . . . nor mono-, di-, or triphosphates at their 5′-ends, *Cell* **11:**73.

Kozak, M., 1978, How do eucaryotic ribosomes select initiation regions in messenger RNA?, *Cell* **18:**1109.

Levey, I. L., Stull, G. B., and Brinster, R. L., 1978, Poly(A) and synthesis of polyadenylated RNA in the preimplantation mouse embryo, *Dev. Biol.* **64:**140.

Lockard, R. E., and Lane, C., 1978, Requirement for 7-methyl-guanosine in translation of globin mRNA *in vivo*, *Nucleic Acid Res.* **5:**3237.

Lodish, H. F., and Rose, J. K., 1977, Relative importance of 7-methylguanosine in ribosome binding and translation of vesticular stomatitis virus mRNA in wheat-germ and reticulocyte cell-free systems, *J. Biol. Chem.* **252:**1181.

Loening, U. E., 1967, The fractionation of high-molecular-weight ribonucleic acid by polyacrylamide-gel electrophoresis, *Biochem. J.* **102:**251.

Peacock, A. C., and Dingman, C. W., 1967, Resolution of multiple ribonucleic acid species by polyacrylamide gel electrophoresis, *Biochemistry* **6:**1818.

Pelham, H. R. B., and Jackson, R. J., 1976, An efficient mRNA-dependent translation system from reticulocyte lysates, *Eur. J. Biochem.* **67:**247.

Revel, M., and Groner, Y., 1978, Post-transcriptional and translational controls of gene expression in eukaryotes, *Annu. Rev. Biochem.* **47:**1079.

Sanger, F., and Coulson, A. R., 1975, A rapid method for determining sequence in DNA by primed synthesis with DNA polymerase, *J. Mol. Biol.* **94:**441.

Schultz, G. A., Clough, J. R., and Johnson, M. H., 1980, Presence of cap structures in the messenger RNA of mouse eggs, *J. Embryol. Exp. Morphol.* **56:**139.

Shatkin, A. J., 1976, Capping of eukaryotic mRNAs, *Cell* **9:**645.

Tobin, A. J., 1979, Evaluating the contribution of post-transcriptional processing to differential gene expression, *Dev. Biol.* **68:**47.

Van Blerkom, J., and Brockway, G. O., 1975, Qualitative patterns of protein synthesis in the preimplantation mouse embryo, *Dev. Biol.* **44:**148.

Weber, L. A., Feman, E. R., Hickey, E. D., Williams, M. C., and Baglioni, C., 1976, Inhibition of HeLa cell messenger RNA translation by 7-methylguanosine-5′-monophosphate, *J. Biol. Chem.* **251:**5657.

Weber, L. A., Hickey, E. D., and Baglioni, C., 1978, Influence of potassium salt concentration and temperature on inhibition of mRNA translation by 7-methylguanosine-5′-monophosphate, *J. Biol. Chem.* **253:**178.
Young, R. J., 1977, Appearance of 7-methylguanosine-5′-monophosphate in the RNA of mouse 1-cell embryos three hours after fertilization, *Biochem. Biophys. Res. Commun.* **76:**32.
Zan-Kowalczewska, M., Bretner, M., Sierakowska, H., Szczesna, E., Filipowicz, W., and Shatkin, A. J., 1977, Removal of 5′-terminal m^7G from eukaryotic mRNAs by potato nucleotide pyrophosphatase and its effect on translation, *Nucleic Acid Res.* **4:**3065.

10

Intrinsic and Extrinsic Patterns of Molecular Differentiation during Oogenesis, Embryogenesis, and Organogenesis in Mammals

Jonathan Van Blerkom

I. Introduction

Two central goals for investigations of modern developmental biology are (1) an understanding of the mechanisms by which genomic information is selectively retrieved, processed, and expressed as the structural proteins and enzymes that ultimately define cell phenotype and function and (2) a characterization of the events controlling normal and abnormal differentiation. Studies of translational patterns during early embryogenesis have been particularly useful in this respect because they have provided not only "stage-related" molecular markers associated with cellular development and differentiation but also have suggested insights into mechanisms by which genomic information may be retrieved and selectively expressed during embryogenesis.

It seems clear from available evidence that "stage-related" protein synthesis accompanies development in all invertebrate and vertebrate species subjected to analysis, from slime molds (Francis, 1976; Alton and Lodish, 1977a; Giri and Ennis, 1978), insects (Arking, 1978; Seybold and Sullivan, 1978), and echinoderms (Davidson, 1977) through amphibians (Ballantine *et al.*, 1979) and mammals (Levinson *et al.*, 1978; Van Blerkom and McGaughey, 1978b). From these analyses, two fundamental generalizations relevant to stage-related protein syntheses may be attempted: (1) as the term implies, such translations are usually short-lived when compared to the bulk of proteins synthesized by a developing embryo and (2) work from nonmammalian species has demonstrated that certain components essential for early embryogenesis, including template RNA, are synthesized and stored in the cytoplasm of the developing oocyte (Davidson, 1977; and Kuo and Garen, 1978).

Jonathan Van Blerkom • Department of Molecular, Cellular, and Developmental Biology, University of Colorado, Boulder, Colorado 80309.

One of the major experimental questions to have emerged from more detailed analyses of protein synthesis during preimplantation embryogenesis concerns the extent to which changes in translational profiles reflect "intrinsic" and/or "extrinsic" programs of development (Johnson, 1979; Van Blerkom, 1979; Van Blerkom et al., 1979). Simply defined, an intrinsic or endogenous program would encompass those molecular and cellular events that are apparently independent of the "state" of the blastomere or embryo. An example of this type is the persistence of normal ontogenic shifts in protein synthesis associated with preimplantation embryogenesis but which occur also in unfertilized, postovulatory oocytes or in embryos in which mitosis has ceased. Such a program of development would be entrained within the embryo and would be expected to be supported by oocyte components including RNA templates and the machinery for their translation. An activating stimulus, such as the breakdown of the oocyte nucleus during the resumption of arrested meiosis, may be one of a series of methods by which an intrinsic program could be initiated (Schultz and Wassarman, 1977).

Extrinsic influences on preimplantation development, although somewhat speculative at present, could include microenvironments within the embryo (Borland, 1977), social interactions (intercellular contact and communication) among blastomeres (Ducibella *et al.*, 1975; Alton and Lodish, 1977b), geophysical position of blastomeres (Hillman *et al.*, 1972; Graham and Lehtonen, 1979), as well as biochemical and morphophysiological changes in the environment provided for the embryo by the reproductive tract (Van Blerkom and Motta, 1979; Van Blerkom *et al.*, 1979). The existence of extrinsic signals at particular stages of preimplantation embryogenesis would be necessarily correlated with the acquisition by the embryo of the capacity to perceive and respond to such signals; perhaps in acquiring such a capacity, significant input from the genome is required for the first time during development.

A simplistic model of preimplantation development evolved from the above discussion suggests that the initial postfertilization period may be supported and maintained primarily by intrinsic processes that have been previously "set in motion" by a specific event(s), such as the breakdown of the oocyte nucleus. Embryonic gene expression would be expected to be minimal at the outset of development. Extrinsic factors would then become increasingly important in the selection and flow of genomic information, especially as development progressed through the later stages of cleavage and the initial stages of blastocyst formation (Johnson *et al.*, 1977; Van Blerkom and McGaughey, 1978a,b; Johnson, 1979; Van Blerkom, 1979). A correlate of such a model is that extrinsic factors may be proximate causes in restricting developmental potentialities (totipotency) of individual blastomeres and consequently may be involved in the ill-defined process by which embryonic cells become increasingly directed toward specific differentiative fates. The research pre-

→

Figure 1. Patterns of newly synthesized, L-[^{35}S]methionine-labeled polypeptides at three stages of resumed meiosis in the mouse oocyte (*in vivo* maturation). (A) Early post-germinal-vesicle breakdown, (B) early metaphase I, (C) arrested at metaphase II. Polypeptides were separated by high-resolution, two-dimensional polyacrylamide gel electrophoresis with isoelectric focusing (IEF) between approximately pI 7.0 and 4.5 in the first dimension and electrophoresis in slab gels containing sodium dodecyl sulfate (SDS) in the second dimension. Approximate molecular weights ($\times 10^4$) are presented on the far right. Polypeptides indicated by unmarked arrows are related to the particular stage. Polypeptides a and t represent actin and tubulin, respectively, while some of the ribosomal proteins synthesized by maturing mouse oocytes are enclosed in the region indicated by R. Arrow 4 indicates a polypeptide that is prominent in the patterns of maturing oocytes and early postfertilization embryos (Figure 2). Two spots noted by asterisks in (B) are detected in the autoradiographic patterns for approximately 2 hr during early metaphase I. In (C), polypeptides 8 and 12 are detected for the first time during oocyte maturation and continue to be translated in the newly fertilized egg.

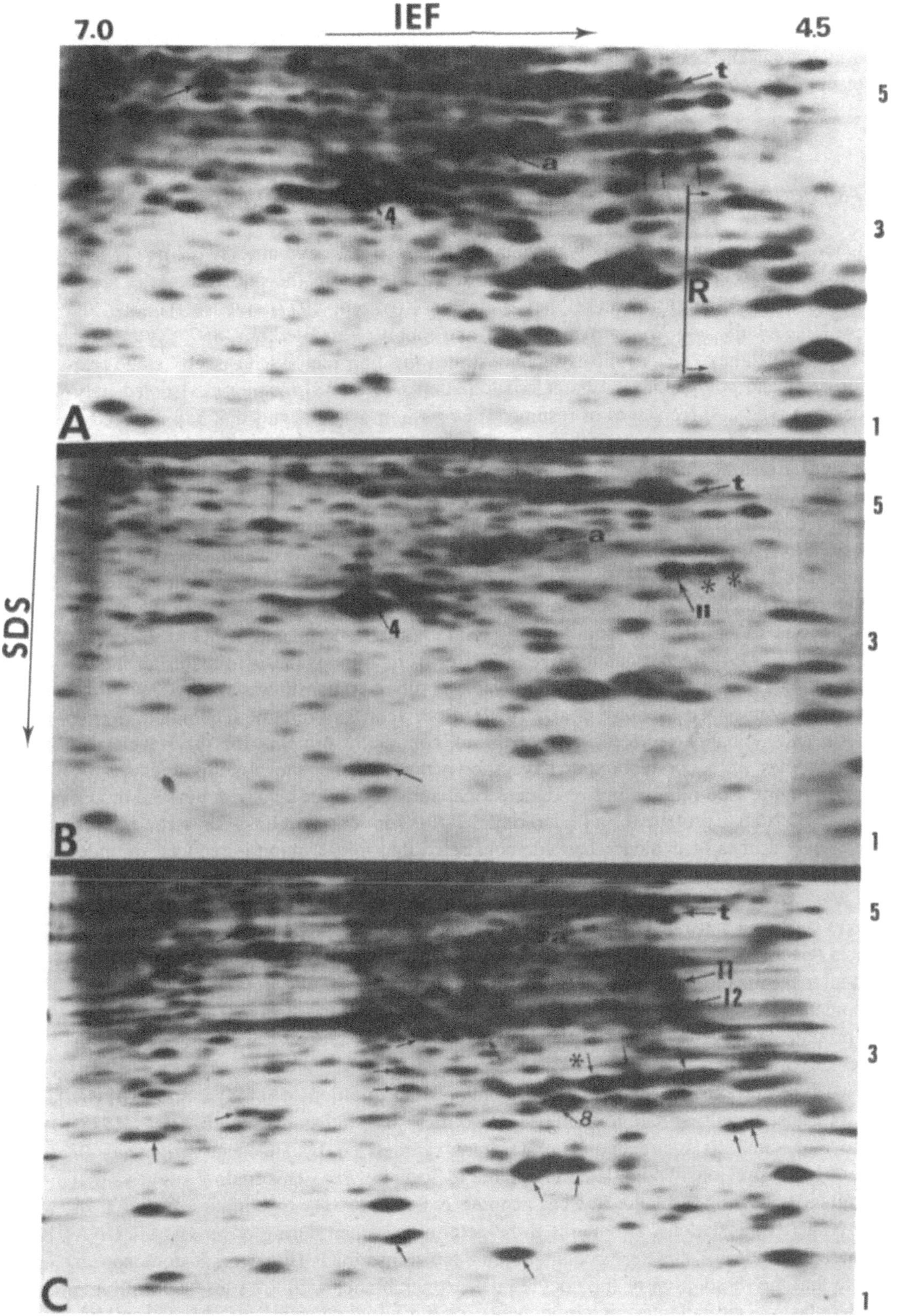

7.0
IEF
4.5
t
5
a
4
3
R
A
1
t
5
a
* *
11
4
3
SDS
B
1
t
5
11
12
3
*
8
C
1

sented in this chapter focuses on the question of extrinsic and intrinsic factors, as they relate to the above model, at four stages of mammalian development: (1) resumption of arrested meiosis in the oocyte, (2) cleavage of the fertilized egg to the morula, (3) reactivation of the blastocyst from facultative delay of implantation, and (4) during organogenesis of the limb.

II. Translational Patterns of Oocytes Undergoing Resumed Meiosis

One- and two-dimensional polyacrylamide gel electrophoretic analysis of newly synthesized proteins during the reinitiation of arrested meiosis in the oocytes of the sheep (Warnes *et al.*, 1977), pig (McGaughey and Van Blerkom, 1977), mouse (Golbus and Stein, 1976; Schultz and Wassarman, 1977), and rabbit (Van Blerkom, 1977; Van Blerkom and McGaughey, 1978a) clearly demonstrates the transient synthesis of "stage-related" proteins and polypeptides. Characteristic patterns of L-[^{35}S] methionine-labeled polypeptides observed at three stages of resumed meiosis in mouse oocytes that had matured *in vivo* are shown in Figure 1. Only the most apparent changes in polypeptide synthesis observed during resumed meiosis in the mouse are indicated in Figure 1 (unmarked arrows). The noted changes are highly reproducible among numerous samples collected and subjected to electrophoretic examination at very different times. Of those polypeptides whose synthesis is constitutive during oogenesis and embryogenesis, two classes deserve special mention. The contractile proteins actin (arrow a) and tubulin (arrow t) are synthesized in relatively large quantities during the entire period of oocyte maturation and preimplantation embryogenesis (Van Blerkom and Manes, 1974; Van Blerkom and Brockway, 1975a; Schultz *et al.*, 1979). Recent quantitative estimates of tubulin synthesis in maturing mouse oocytes indicate that this particular protein (α and β subunits) constitutes about 1.3% of the total protein synthesized (Schultz et al., 1979). Both actin and tubulin accumulate during oogenesis and probably serve as a reservoir of contractile proteins for the developing embryo. Studies from our laboratory have shown that some of the proteins enclosed by arrow R in Figure 1 co-migrate in two-dimensional polyacrylamide gels with purified mouse blastocyst ribosomal proteins (Van Blerkom, 1980). This observation is of some interest because several lines of evidence have demonstrated that transcription is reduced to undetectable levels in post-germinal-vesicle-breakdown oocytes (Rodman and Bachvarova, 1976; Wassarman and Letourneau, 1976; Motlik *et al.*, 1978). In addition, mouse oocytes appear to contain a sizeable population of ribosomes that are apparently inactive in protein synthesis (Bachvarova and DeLeon, 1977); presumably "stored" oocyte ribosomes are activated after fertilization and support embryonic protein synthesis during the early stages of development. Therefore, it seems that the coordination between ribosomal RNA and ribosomal protein synthesis is not as strictly regulated in the mouse oocyte as it appears to be in other eukaryotic cells.

Two types of experimental approaches address the question of intrinsic and extrinsic modes of translational regulation during resumed meiosis. Studies by Schultz *et al.* (1978) indicate that some "stage-related" proteins are synthesized by anucleate fragments of mouse oocytes, previously stimulated to resume meiosis *in vitro*. This finding suggests that concomitant transcription of the nuclear genome is not necessary to initiate a program of "stage-related" translations. It remains to be determined when during oogenesis the RNA templates that code for these particular proteins are transcribed. However, it does appear probable that the dissolution of the oocyte nuclear membrane, with the subsequent mixing of nucleoplasmic and cytoplasmic components, or the initial "signal" to resume meiosis, is sufficient to trigger a series of "stage-related" protein syntheses that is, apparently, more dependent on time following nuclear breakdown than the stage of maturation attained (i.e., metaphase I, abstriction of first polar body, metaphase II).

Comparisons of polypeptide synthesis in rabbit oocytes matured *in vivo* and *in vitro* have demonstrated essentially identical patterns of stage-related translation (Van Blerkom and McGaughey, 1978). Furthermore, *in vitro*-matured rabbit (Van Blerkom and McGaughey, 1978b) and mouse oocytes (Cross and Brinster, 1970) are capable of fertilization and development to the expanded blastocyst and advanced fetal stages, respectively. This observation indicates strongly that the follicular environment in which the germinal vesicle stage oocyte is located is not instructive with respect to molecular and cellular changes that occur subsequent to the reinitiation of meiosis. Therefore, once meiosis is resumed, it appears that a program of development entrained within the oocyte unfolds progressively and leads to the attainment of the fertilizable state of the female gamete. The above conclusion may not be valid, however, for all mammals because significantly different patterns of protein synthesis have been noted for sheep oocytes, depending upon whether the resumption of meiosis occurred within the follicle or in an extrafollicular environment (Warnes *et al.*, 1977).

The detection of reproducible molecular markers of oocyte maturation serves a critical function in experiments designed to relate cell and molecular processes occurring during this stage of oogenesis. For some species, such functions include the progressive acquisition of (1) the ooplasm to decondense spermatozoal chromatin (Thibault, 1977; Yanagimachi, 1977) and (2) the plasma membrane of the oocyte to recognize and bind with cell surface components of spermatozoa (Overstreet and Bedford, 1974). Little is known about the detection and actual identification of cell surface proteins during mammalian oogenesis, especially during the period of resumed meiosis. Experiments in progress in our laboratory are designed to examine changes in cell surface proteins during oogenesis and, in particular, to assess the effects of altering the structure of cell surface glycoproteins on the ability of the oocyte to complete requirements necessary for fertilization and, possibly, postfertilization development. The availability of a population of molecular markers related to the later phases of oogenesis provides the basis for more precise investigation into the process(es) by which an oocyte gains competence to resume meiosis and ultimately to be fertilized, and also the extent to which the molecular component of such competence is intrinsic to the oocyte.

III. Translational Patterns of Preimplantation Embryos

To a greater degree than during resumed meiosis, the various stages of preimplantation embryogenesis are also accompanied by a progressive series of well-defined, reproducible, stage-related protein and polypeptide syntheses (Epstein and Smith, 1974; Van Blerkom and Manes, 1974, 1977; Van Blerkom and Brockway, 1975a; Levinson *et al.*, 1978; Van Blerkom and McGaughey, 1978b). By means of electrophoresis in polyacrylamide slab gels containing sodium dodecyl sulfate, followed by autoradiography, our laboratory was the first to demonstrate that complex and rapid changes in protein synthesis occurred during cleavage, rather than during the blastocyst stages (as had been previously thought) of rabbit (Van Blerkom and Manes, 1974) and mouse preimplantation development (Van Blerkom and Brockway, 1975a). Further study revealed, however, that when mouse blastocysts were microdissected into inner cell mass and trophectodermal components, some differential polypeptide synthesis was taking place in these two populations of

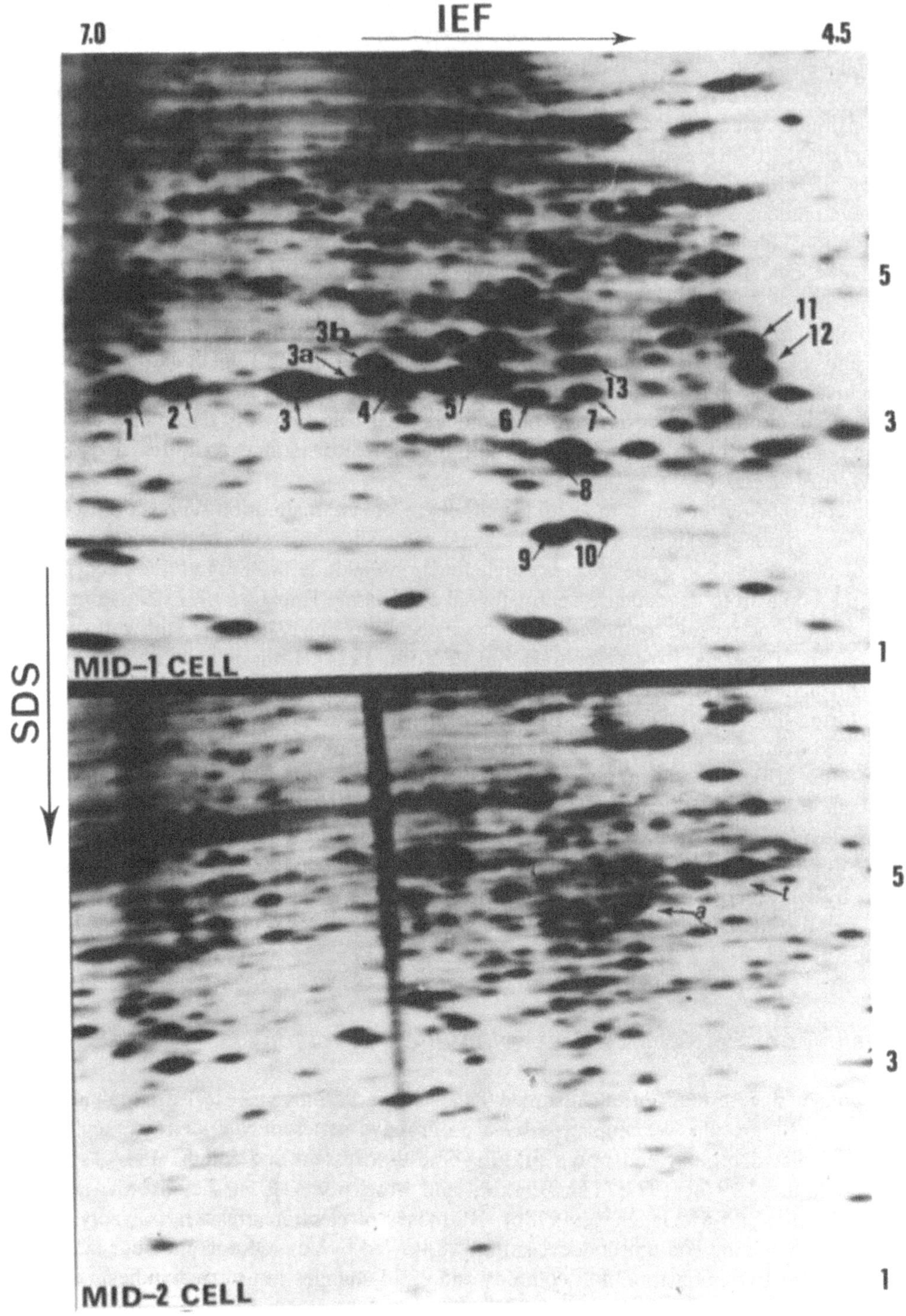

Figure 2. Demonstration of reproducible changes in the pattern of polypeptide synthesis between the mid-one-cell and the mid-two-cell stages of preimplantation development in the mouse. Polypeptides representing "carry-over" translation from the oocyte, as well as those only detected in fertilized eggs, are described in the text. Embryos were exposed to medium containing L-[^{35}S]methionine for 1 hr.

cells (Van Blerkom *et al.*, 1976). Since that time, other investigators have detected additional polypeptide markers of these two cell types (Handyside and Barton, 1977; Dewey *et al.*, 1978). The availability of a population of polypeptides that is apparently related to either inner cell mass or trophectoderm has proven to be quite useful in tracing cell lineages during pre- (Handyside, 1978; Handyside and Johnson, 1978) and postimplantation development (Martin *et al.*, 1978), as well as in studies comparing molecular differentiation in embryonal carcinoma and normal pre- and postimplantation cells (Dewey *et al.*, 1978).

In general, the bulk of the proteins detected in early rabbit (Van Blerkom and McGaughey, 1978b) and mouse embryos (compare Figures 1C and 2) (Van Blerkom and Brockway, 1975a; Levinson *et al.*, 1978) are the same as those observed in the mature metaphase II oocyte. Some of the more prominent polypeptides that fall into this category are numbered 4, 5, 8, 9, 10, 11, and 12 in Figures 1 and 2. However, as exemplified by the early mouse embryo, the transient synthesis of specific polypeptides does occur. In this class, the more prominent polypeptides are numbers 1, 2(?), 3, 3a, 3b, 6, 7, and 13 in Figures 2 and 4.

Recently, we have shown that six autoradiographically prominent polypeptides, previously thought to be "specific" to the 12- to 16-cell rabbit embryo (Van Blerkom and McGaughey, 1978b), are also detected for the first time in unfertilized oocytes of the same chronological age as the embryos (i.e., 36 hr postcoitum or 48 hr postovulation) (Van Blerkom, 1979). As noted in Figure 3, the translation of these particular polypeptides takes place in oocytes that had undergone maturation *in vivo* or *in vitro,* and were then allowed to reside for an additional 36 hr in culture or in the reproductive tract. A similar situation prevails in the mouse when postovulatory oocytes and late 1-cell embryos are compared. For example, polypeptides 1, 2, 3a, 3b, and 13 (Figures 2 and 4) are first detected in unfertilized oocytes at approximately 18 hr after ovulation. In both studies, oocytes were carefully screened in order to exclude from analysis those cells exhibiting signs of parthenogenetic activation (second polar body or presence of female pronucleus, for example). While it may be argued that concomitant transcription of the oocyte genome is involved in the expression of these particular polypeptides, several studies have failed to detect RNA synthesis in postovulatory oocytes (Young *et al.*, 1978) and indeed, cytogenetic analysis of chromosomes in aged rabbit oocytes indicated that they were highly condensed, some pycnotic, but that most were still associated with the second metaphase spindle (Van Blerkom, 1979). However, it cannot be excluded at present that uptake mechanisms in aged oocytes are defective and, therefore, that some transcription, utilizing endogenous pools of uridine, is occurring. In order to examine this possibility, *in situ* assays, which measure RNA polymerase activity directly, and which circumvent problems of percursor uptake and incorporation (Moore and Ringertz, 1973; Chavez and Van Blerkom, 1979), are in progress in our laboratory. In any event, the above findings do lead to the conclusion that the transient synthesis of some polypeptides in early mouse and rabbit embryos is probably more closely related to time following the breakdown of the oocyte nucleus than to ovulation, fertilization, or the number of cell or nuclear divisions. These results are further evidence of the existence of an endogenous program in early mammalian development.

A. Support of Early Embryogenesis by Oocyte Components

Evidence from several lines of investigation demonstrates that significant transcription of embryonic mouse (Young *et al.*, 1978) and rabbit genomes (Manes, 1977; Cotton *et al.*, 1978) is not detectable until approximately the 2- and 16-cell stages, respectively. That both mouse and rabbit 1-cell embryos are capable of one or two cleavage divisions in the

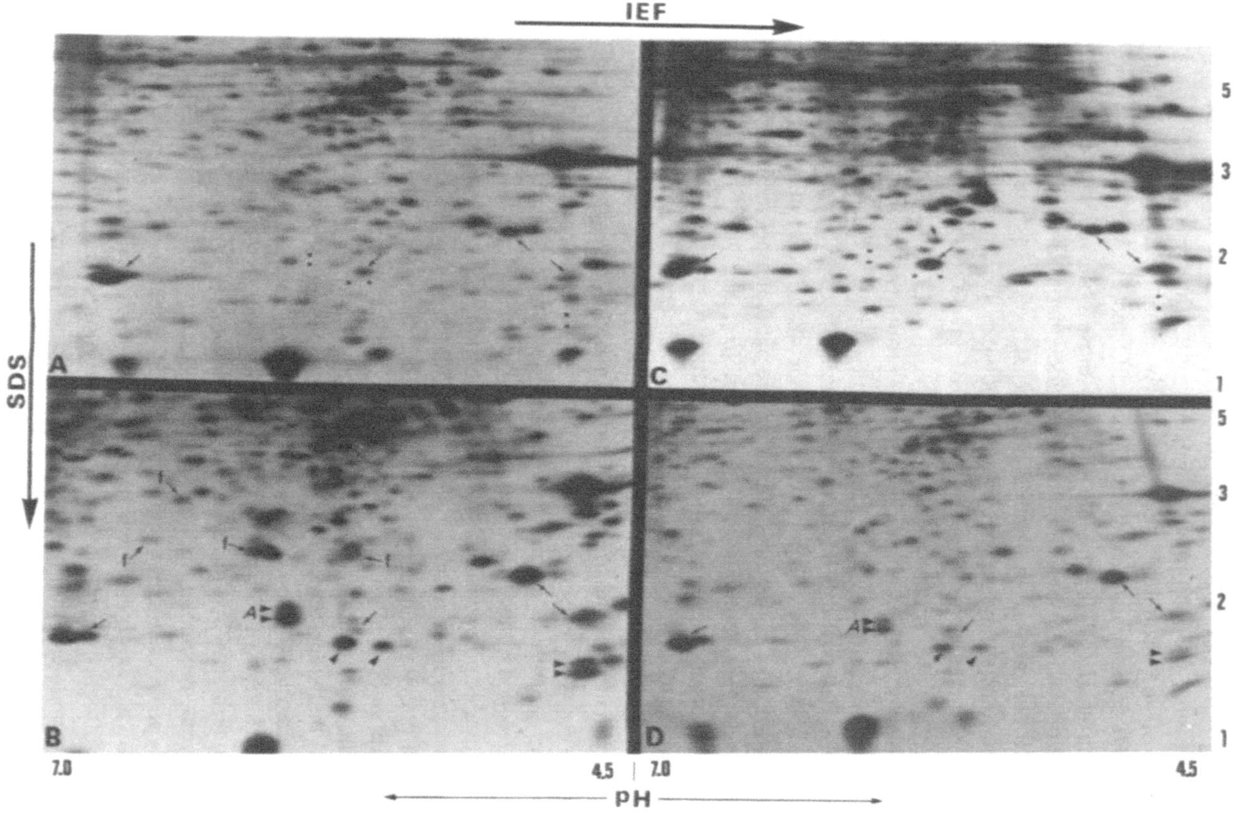

Figure 3. Two-dimensional autoradiographic patterns of L-[^{35}S]methionine-labeled polypeptides of (A) newly fertilized rabbit eggs, (B) 12- to 16-cell rabbit embryos cultured *in vitro* for 36 hr from the 1-cell stage, (C) newly ovulated oocytes, and (D) oocytes that had undergone meiotic maturation to metaphase II *in vitro*, followed by an additional 36 hr of culture. Polypeptides indicated by unmarked arrows serve as positional markers for comparison of the patterns. Polypeptides denoted by an f were detected only in the patterns of fertilized eggs. Spots noted by triangles were detected in cleavage stage embryos and ovulated (or *in vitro* matured and "aged") but unfertilized oocytes of the same chronological age as the embryos. The positions in the autoradiographs of newly ovulated oocytes or newly fertilized eggs where these polypeptides will appear with time are indicated by small squares (Van Blerkom, 1979).

presence of α-amanitin (Golbus *et al.*, 1973; Manes, 1973), at doses that abolish transcription, suggests that protein synthesis in the early embryo is supported by oocyte components, including mRNA (Young *et al.*, 1978; Bachvarova and DeLeon, 1980) and ribosomes (Bachvarova and DeLeon, 1977). From these observations has come the assumption that some stage-related translations associated with the immediate postfertilization period are directed by RNA templates present, but not translated, in detectable quantities by mature oocytes (Braude *et al.*, 1979; Johnson, 1979; Van Blerkom, 1979).

A more direct test for the presence of RNA templates in mature oocytes that are not translated in detectable amounts until after fertilization (putative "stored" or "masked" mRNA) is available by means of cell-free translation. The results of such a study, in which mouse oocyte RNA was used to direct translation in rabbit reticulocyte lysates (Pelham and Jackson, 1976), are shown in Figure 4. A comparison of polypeptide synthetic patterns of (a) oocytes between the stage of germinal vesicle breakdown and metaphase I, (b) cell-free translates directed by total RNA from these oocytes, and (c) late one-cell embryos indicates two basic findings: RNA templates for at least two polypeptides only detected in mature oocytes (arrested at metaphase II) are present in earlier stage oocytes (polypeptides 8 and 12); at least two polypeptides whose synthesis is first detected in aged oocytes or newly fertilized eggs (3a and 3b) are observed in the cell-free translates of post-germinal-vesicle-breakdown oocytes. In some of the reticulocyte lysate translations, polypeptides 1 and 2 (Figure 2) are detected as relatively prominent spots. However, it is possible that these two polypeptides are present in low abundance in oocytes as early as the germinal vesicle stage. The work is still in progress; additional and detailed study therefore is required to establish definitively the identity of specific polypeptides which may be derived from preexisting mRNA.

Recently, Bachvarova and DeLeon (1980) demonstrated that poly A-containing RNA in mouse oocytes was approximately 8% of the total RNA and that 40% of the total RNA in the mature oocyte was lost during the first day of embryonic development. The proportion of polyadenylated RNA remained somewhat constant during this period. These investigators suggested that the early turnover of oocyte RNA may underlie the major and rapid change in polypeptide synthesis observed during early embryogenesis in the mouse. Activation of oocyte RNA templates was also indicated by the studies of Young (1977), who noted that a 7-methylguanosine residue was added to the 5′ terminus of RNA shortly after fertilization in the mouse. However, a recent study by Schultz *et al.* (1980) has demonstrated that the majority of mRNA in mouse oocytes contains 5′-terminal 7-methylguanosine cap structures and, more significantly, that there is no detectable difference in the degree of "capping" in unfertilized and fertilized mouse egg mRNA. Therefore, it appears that if stored mRNA is translated after fertilization, capping of these species may not be involved.

From the above observations, it is tempting to conclude that, to an as yet undetermined extent, stage- or time-dependent translations in newly fertilized mouse eggs are derived from preexisting mRNAs, a situation that is common in invertebrates and lower vertebrates (Davidson, 1977). However, until more detailed analyses are performed, this conclusion may be premature. For example, it may be argued that a low efficiency of translation of a set of oocyte RNA templates, coupled with a limited pool of translationally active ribosomes and polysomes in the mature oocyte (Bachvarova and DeLeon, 1977), may preclude the synthesis of certain polypeptides from electrophoretic detection. Changes in translational efficiencies after fertilization (Brandis and Raff, 1978), the continued synthesis of ribosomal proteins, and the mobilization of previously inactive ribosomes could all combine to make certain species of mRNA more readily translated and thus their prod-

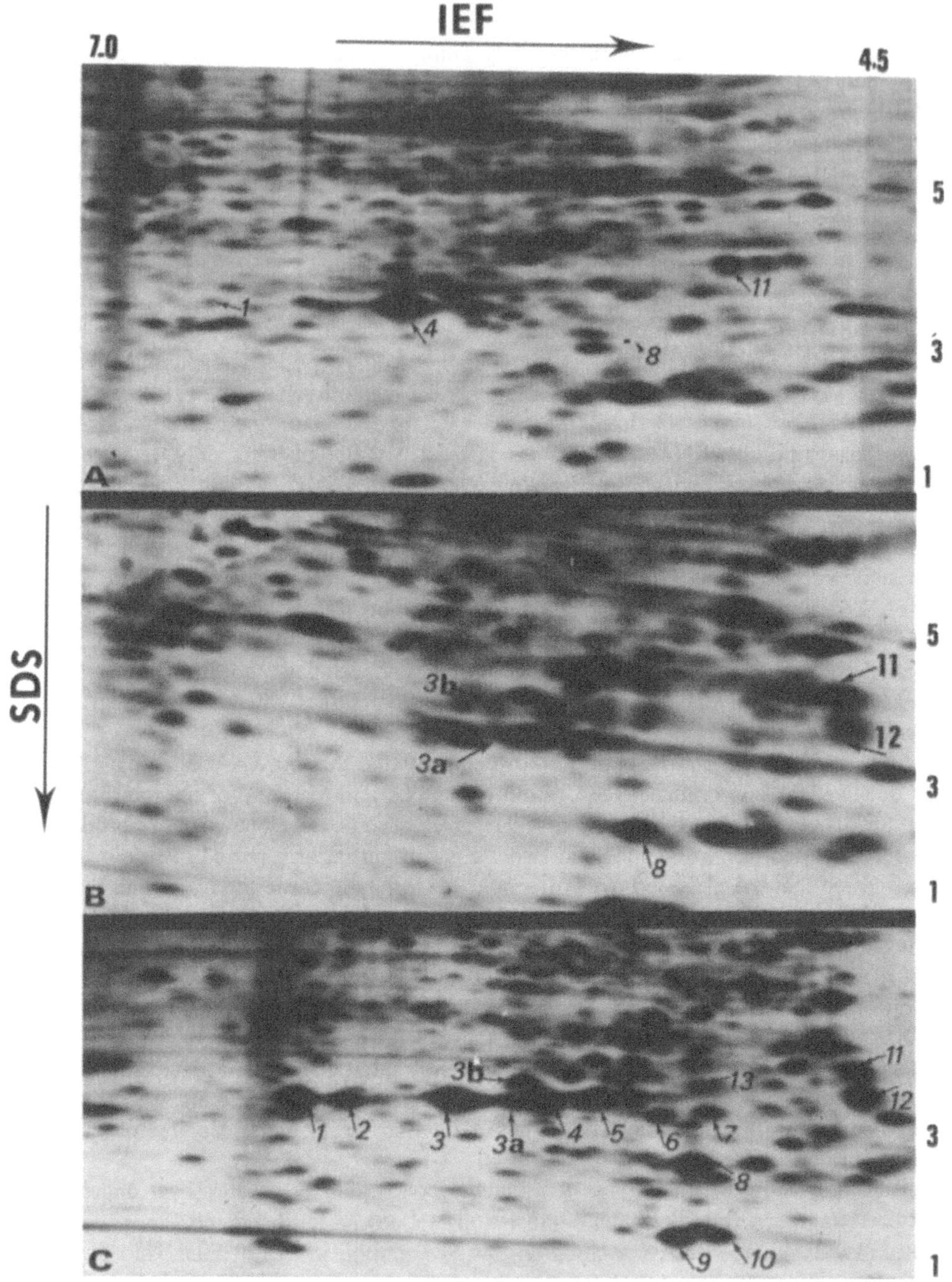

Figure 4. Comparison of polypeptides synthesized during early metaphase I in the mouse oocyte (A), with cell-free translation (reticulocyte lysates) of total, early metaphase I RNA (B), and mid-one-cell embryos (C). Polypeptides 3a, 3b, 8, and 12 are normally first detected in oocytes arrested at metaphase II (see Figure 1C) but, apparently, the RNA templates for these polypeptides are present in the oocyte at an earlier stage of maturation. Two polypeptides (3a, 3b) are detected in the patterns of mid/late one-cell embryos or in unfertilized oocytes of the same postovulatory age. In some of the cell-free translates, polypeptides 1 and 2 are prominent spots but they may be synthesized by maturing oocytes in relatively low abundance and therefore not readily detectable in the patterns.

ucts more readily visualized by electrophoresis and fluorography or autoradiography. Furthermore, those mRNAs coding for polypeptides specific to resumed meiosis may no longer be competing for available ribosomes (Van Blerkom and Manes, 1977). The presence of stage- and time-specific molecular markers does provide a basis for more definitive experimentation into the origin(s) of the template RNAs that code for these proteins, as well as a basis for the study of the regulation of their expression during early embryogenesis.

B. Differential Protein Synthesis at the Blastomere Level

It has been proposed that the geophysical location of blastomeres within a developing mammalian embryo, as well as the concurrent establishment of channels of intercellular communication, may be involved in the differential activation of RNA templates, or differential retrieval of genomic information (Handyside, 1978; Handyside and Johnson, 1978; Johnson, 1979). By extension, this proposition implies that factors extrinsic to the cells of an embryo could be important in directing blastomeres toward certain pathways of differentiation. In part, such a hypothesis has served as the basis for analyses of translational patterns in inner and outer cells of the mouse morula (Handyside, 1978; Handyside and Johnson, 1978) and also in inner cell mass and trophectodermal cells of the mouse blastocyst (Van Blerkom *et al.*, 1976; Handyside and Barton, 1977). From these studies, it is apparent that nonuniformity of translation (quantitative and/or qualitative) is taking place within the component parts of the developing mouse embryo.

In Figure 5, patterns of polypeptide synthesis derived from two individual blastomeres of a 16-cell rabbit embryo are compared. The autoradiographs were obtained from ultrathin, two-dimensional polyacrylamide gel slabs, prepared as previously described for samples containing low levels of incorporated radioactivity (Van Blerkom, 1978). Examination of the patterns reveals that many of the polypeptides synthesized by the two blastomeres are identical. However, prominent qualitative and quantitative differences do exist. It should be noted that the duration of exposure of the slab gels to X-ray film lasted 8 months. As a consequence of such protracted exposure times, additional spots representing polypeptides possibly present in relatively low abundance are detected. These polypeptides are generally not observed in the patterns of whole embryos owing to the higher levels of incorporated radioactivity and the resultant reduction in exposure time required to obtain satisfactory autoradiographic or fluorographic patterns.

It is of particular interest to note that four of the polypeptides whose synthesis appears to be time- rather than fertilization-dependent (Figure 3) are not translated in detectable amounts by one of the blastomeres shown in Figure 5. Other differences between synthetic patterns are evident upon close examination of Figure 5. In these experiments, 16-cell embryos were radiolabeled with a mixture of ^{14}C-labeled amino acids and then microdissected into component blastomeres. Comparisons of electrophoretic profiles of newly synthesized polypeptides in other blastomeres of 16-cell rabbit embryos indicated both qualitative and quantitative variation, but not to the extent exhibited by the cells shown in Figure 5. As this work is still in the preliminary stage, further study is required in order to establish (1) whether the actual isolation of blastomeres perturbs the pattern of polypeptide synthesis, (2) the reproducibility of differential blastomere polypeptide synthesis, and (3) the precise geographical location of the blastomeres within the embryo prior to preparation for electrophoretic analysis.

As the embryonic genome of the rabbit is transcriptionally active during the mid-morula stage (Manes, 1977; Cotton *et al.*, 1978), differential polypeptide synthesis among

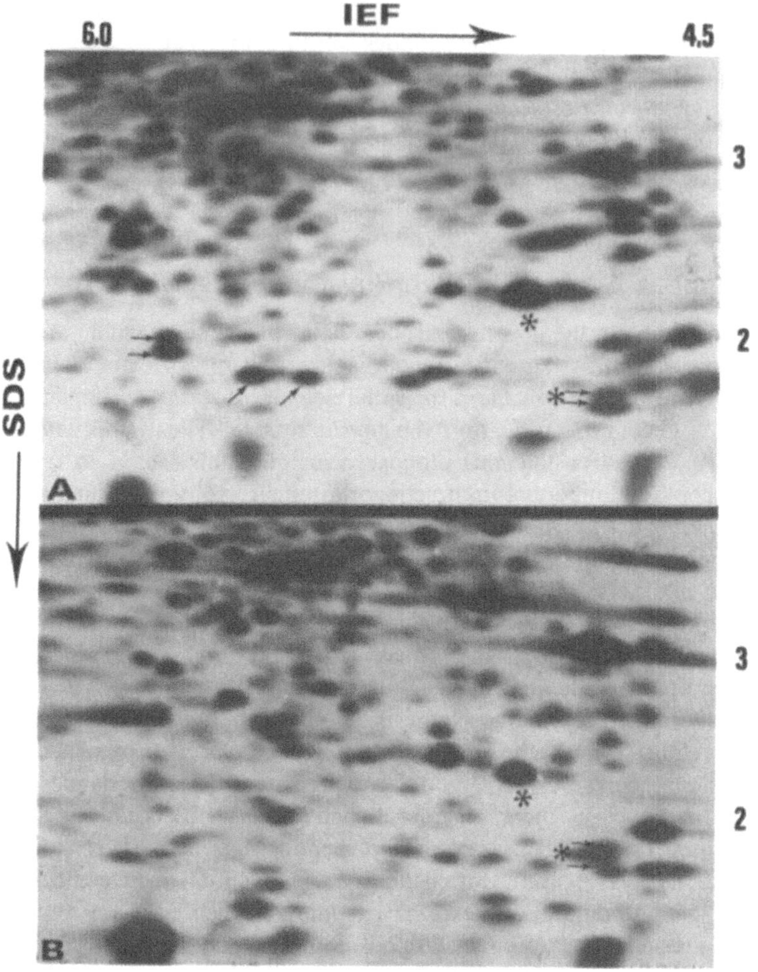

Figure 5. Comparison of two-dimensional, polyacrylamide gel autoradiographs of two individual blastomeres derived from 16-cell embryos labeled with a mixture of ^{14}C amino acids. Exposure to X-ray film lasted 8 months and some of the comparatively less abundant species of polypeptides are thus revealed (compare with Figure 3). Prominent quantitative and apparent qualitative differences exist between the two blastomeres. However, four spots that were time- rather than fertilization-dependent (arrows; see also Figure 3) are not translated in detectable amounts by one of the blastomeres (B). On the other hand, two polypeptides that were also time- rather than fertilization-dependent are translated by both of the blastomeres (asterisks with arrows). The single polypeptide indicated by an asterisk serves as a reference marker for comparison of the two patterns. Approximately 800 TCA-insoluble cpm was loaded onto the first-dimension gel. The second-dimension slab gel was 0.1 mm thick. Approximate molecular weights ($\times 10^4$) are given on the far right.

blastomeres could be explained by the flow of similar amounts of new mRNA into differently sized and differently composed pools of preexisting mRNAs; consequently, such a process could be an early and somewhat indirect determinant of some physiological differences upon which cell differentiation may depend (Senger and Gross, 1978). It could also be argued that preformed RNA templates are compartmentalized or sequestered during oogenesis or early embryogenesis (Jeffry and Capco, 1978) and therefore are not uniformly distributed or translated during subsequent cell divisions. A simplier explanation suggests that stage-related polypeptide synthesis terminates earlier in some blastomeres than in others. Although the above alternatives are still speculative, they do form a variety of experimental approaches to the study of the regulation of gene expression during early mammalian embryogenesis.

A final set of preliminary observations that may ultimately be relevant to differential protein synthesis within an early mammalian embryo is derived from mouse embryos arrested at the two-cell stage. Arrest of cell division at the two-cell stage occurs in many oocytes after parthenogenetic activation (Van Blerkom and Runner, 1976) or in one-cell embryos cultured *in vitro* or exposed to cytochalasin B at the late one-cell stage. Initial two-dimensional analysis of polypeptide synthesis in two-cell embryos arrested for between 24 and 30 hr indicated that some of the polypeptides whose synthesis was "related" to the four- and eight-cell embryo were also translated at approximately the same time as in the normally developing embryo. Tentatively, it may be concluded that the regulation of the expression of these proteins follows an intrinsic time-dependent program similar to that described earlier for the occyte–embryo transition (Van Blerkom, 1979). On the other hand, some four- and eight-cell-related polypeptides are not detected in the patterns of mouse embryos arrested at the two-cell stage; perhaps extrinsic factors, such as cell position and/or the nature and extent of intercellular contact and communication, may be important at this time. Alternatively, while embryos arrested at the two-cell stage are still capable of transcription, such RNA synthesis may differ significantly in quality and/or quantity from normal four- and eight-cell embryos.

The availability of molecular markers in (1) whole embryos at different stages of development, (2) inner and outer cells of the morula, (3) inner cell mass and trophectodermal cells of the blastocyst, and (4) among the component blastomeres of an embryo provides the conceptual framework upon which experiments addressed to such questions as intrinsic vs. extrinsic regulation of embryonic gene expression may be designed. For example, continued analyses should be able to distinguish, in a complex and changing pattern of polypeptide synthesis, those translations that may be dependent upon time following the reinitiation of arrested meiosis, or following fertilization, from those that may be a function of egg activation, number of cell or nuclear divisions, location of blastomeres within the embryo, or the establishment and maintenance of intercellular contact and communication.

IV. Translational Patterns of Blastocysts during Activation from Facultative Delayed Implantation

One of the most prominent examples of the external control of mammalian development is observed during delayed implantation. When intrauterine conditions inconsistent with continued embryonic development and implantation occur, the blastocyst stage embryos of numerous mammals enter into a period of developmental arrest of varying duration. In general, delayed-implanting blastocysts are characterized by reduced rates of metabolism and macromolecular synthesis and by a cessation of mitosis (see reviews by McLaren, 1973; Van Blerkom *et al.*, 1979). Facultative delay of implantation in the mouse may be brought about experimentally if ovariectomy is performed prior to the fourth day of embryogenesis (with the first day being indicated by detection of a vaginal plug) and may be terminated by either administration of estrogen, in which case implantation begins after approximately 24 hr, or removal of blastocysts from uteri experiencing delayed implantation, followed by culture *in vitro*. *In vitro* culture of delayed-implanting blastocysts is a very useful experimental system for studying molecular and cellular processes occurring during the peri-implantation period (Jenkinson, 1977) because trophectodermal outgrowth, giant cell transformation, and delamination of endoderm will take place in a fashion that approximates the *in vivo* situation (Glass *et al.*, 1979; Van Blerkom *et al.*, 1979).

As noted by Johnson (1979) and Van Blerkom *et al.* (1979), embryos in delay appear to be under a uterine control that is primarily, if not solely, due to the deprivation of key metabolic substrates or cofactors that regulate the *rate* at which existing intrinsic programs advance, rather than fundamentally diverting embryos along a distinctive path. The uterine milieu is envisaged as being permissive or nonpermissive to continued growth and implantation, but not instructive in terms of molecular differentiation of the embryo. Analyses of translational patterns before, during, and after delayed implantation in the mouse is only one approach to testing the validity of this hypothesis. While one-dimensional SDS–polyacrylamide slab gel electrophoretic analysis of protein synthesis during reactivation of delayed mouse blastocysts failed to reveal the synthesis of new species of protein (Van Blerkom and Brockway, 1975b), recent two-dimensional analyses have indicated that subtle changes in translation do occur (Van Blerkom *et al.*, 1979). Figure 6 illustrates such changes at three intervals following the estrogen-mediated termination of delay. While most of the new species of polypeptides translated by reactivated mouse embryos become constitutive after they are initially detected, some appear in the patterns for only a relatively brief period.

At the cellular level, it is of interest to note that delayed implantation in the mouse is accompanied by the disassembly of polysomes into free ribosomes and by a significant decline in the relative rate of protein synthesis (Van Blerkom *et al.*, 1979). By contrast, relative rates of uptake and incorporation of radiolabeled uridine decline by approximately 60 and 36%, respectively, from those observed in a normal preimplantation blastocyst (Chavez and Van Blerkom, 1979). While reactivation of embryos is not associated with a change in the relative rate of RNA synthesis, it is accompanied by an increased relative rate of protein synthesis and by a progressive return to polysomal configurations (Van Blerkom *et al.*, 1979). The persistence of RNA synthesis during delayed implantation in the mouse and wallaby (which undergoes obligate delay of implantation) (Moore, 1978) indicates that the maintenance of basic cellular functions may be required in order for the embryo to resume a program of development when conditions consistent with implantation appear in its milieu.

The detection of reactivation-associated translations offers an experimental approach to the question of whether or not development at the molecular level is truly in a state of arrest. For example, we might ask whether the template RNAs that code for these species of polypeptides are present in but not translated by predelay blastocysts, or are transcribed but not translated during delay, and whether their transcription is concurrent with reactivation. It is of additional interest to ask whether or not reactivation-associated polypeptide synthesis is occurring differentially in inner cell mass and trophectoderm and, if so, whether the synthesis of these polypeptides can be traced through the derivative tissues of postimplantation embryos. Ultimately, these types of questions are designed to examine how the interplay between intrinsic and extrinsic factors regulates embryonic development during the periimplantation period.

Figure 6. Autoradiographs of two-dimensional polyacrylamide gels showing the polypeptide synthetic patterns of newly synthesized ^{35}S-labeled polypeptides in mouse blastocysts at three periods after the estrogen-mediated termination of facultative delayed implantation (4–5, 8–9, 16–17 hr). Spots indicated by arrow 1 are not detected in predelay blastocysts but do appear and then disappear from the patterns between 0.5 and 6 hr after reactivation. Arrow 2 denotes one of the proteins that appears and then disappears from the autoradiographic patterns between 7 and 12 hr after reactivation. Arrow 3 indicates some of the proteins detected only between 14 and 18 hr. Arrow 4 notes those polypeptides that appear in the patterns at a particular time following reactivation and then are synthesized constitutively for the first 30 hr following reactivation (Van Blerkom *et al.*, 1979).

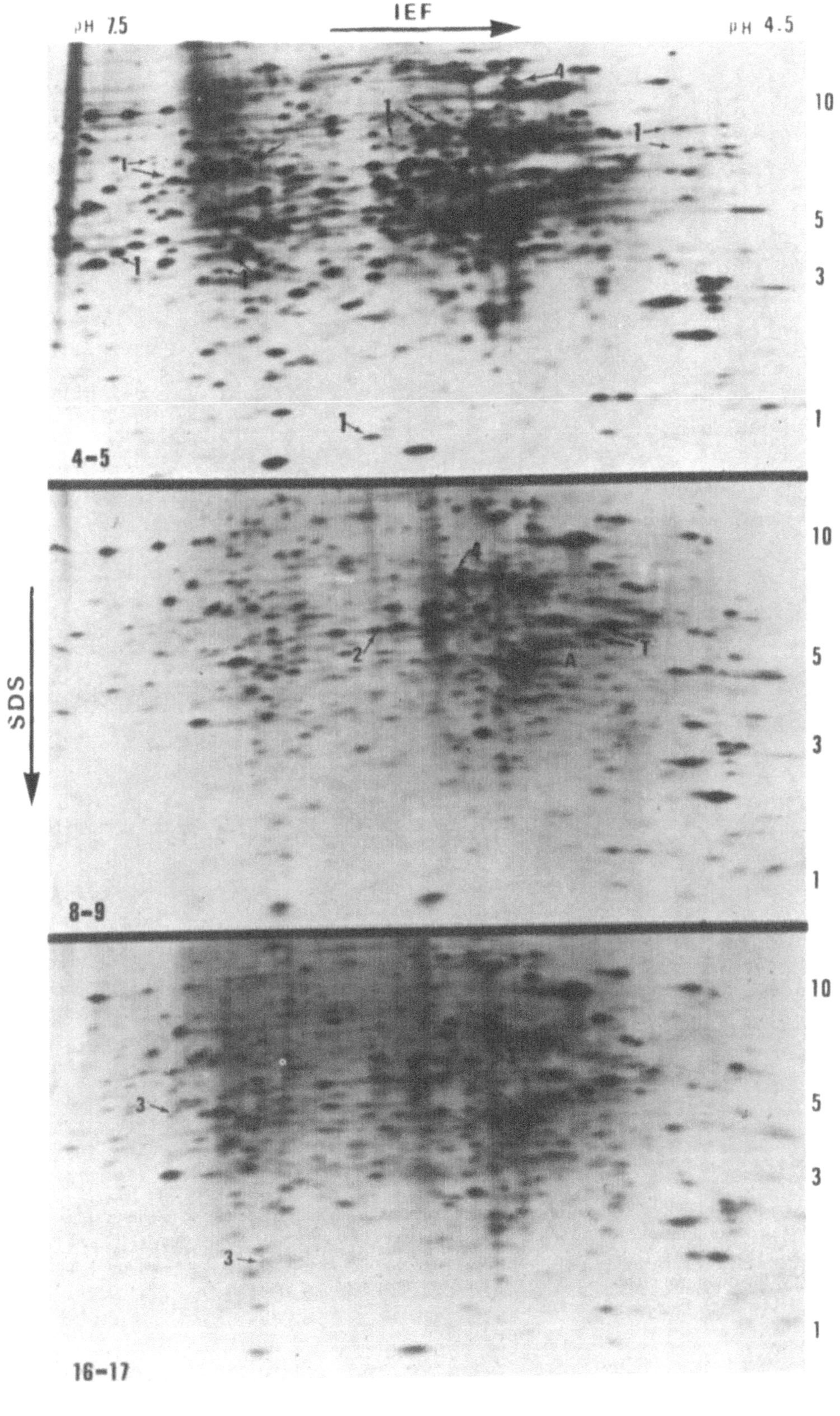

pH 7.5
IEF
pH 4.5
SDS
10
5
3
1
4-5
10
5
3
1
8-9
10
5
3
1
16-17

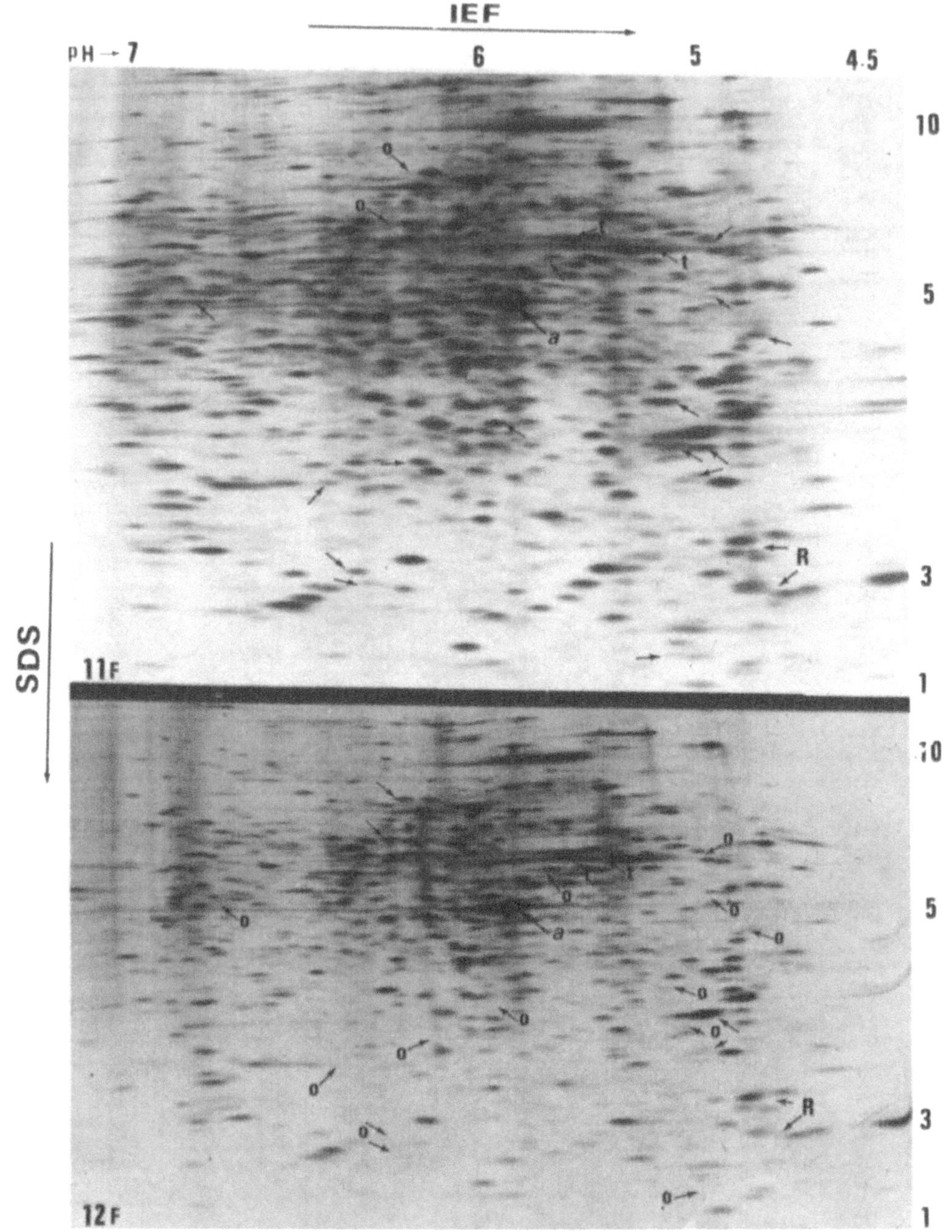

Figure 7. Comparison of autoradiographs of polypeptide synthesis in whole, excised mouse forelimbs cultured *in vitro* in the presence of a mixture of ^{14}C amino acids. Differences in synthesis between 11-day (11F) and 12-day forelimbs (12F) are indicated both by unmarked arrows and by arrows associated with 0. a = actin, t = tubulin, R = some of the ribosomal proteins detectable in this two-dimensional gel system.

V. Translational Patterns during Limb Organogenesis

Owing to the fact that embryonic morphogenesis is characterized by complex, mutual interactions among cells and tissues in three dimensions, the mouse limb bud is exceptionally well-suited to analyses of intrinsic and extrinsic patterns of gene expression during

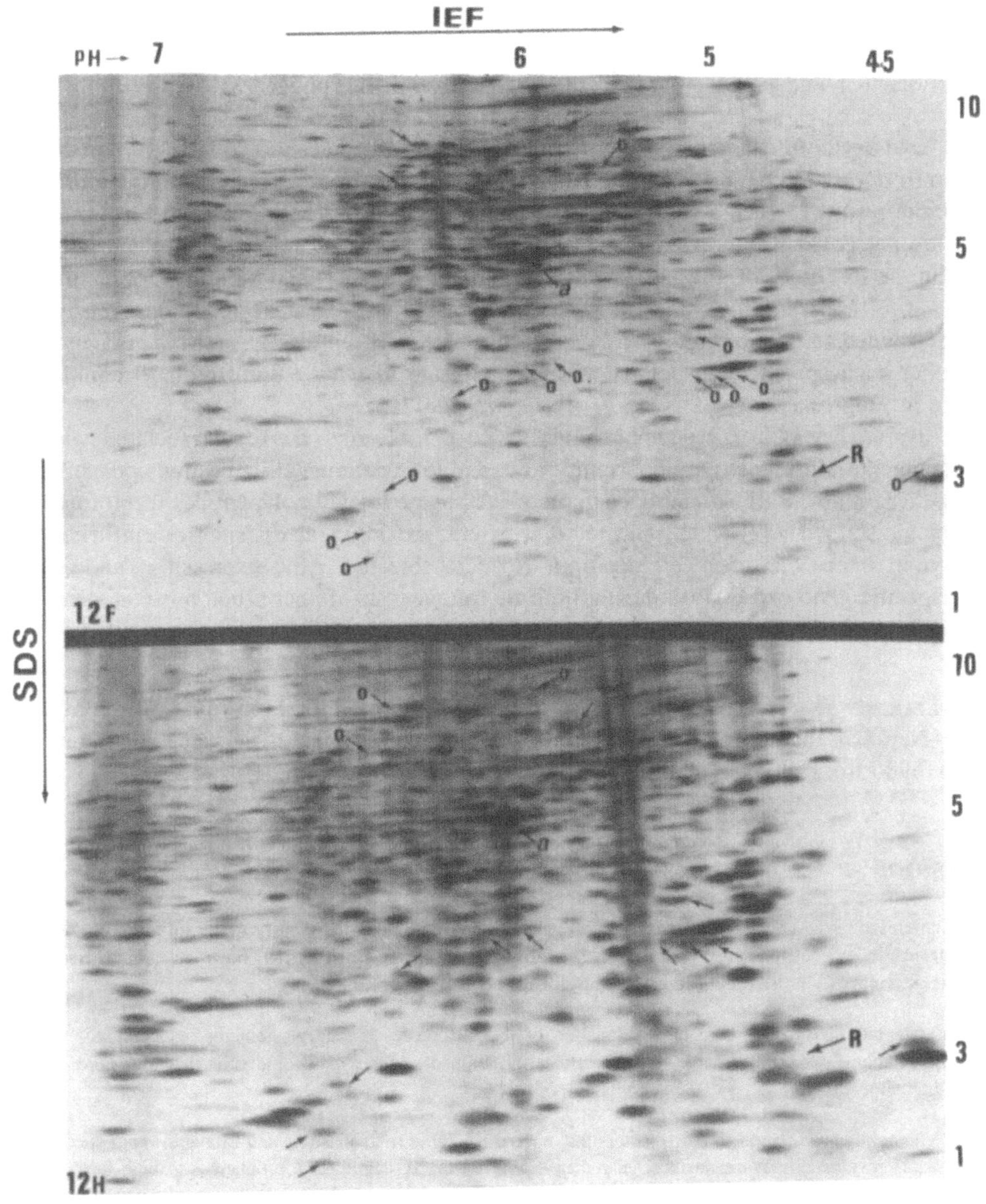

Figure 8. Comparison of autoradiographs of polypeptide synthesis in whole, excised mouse limbs cultured *in vitro* in the presence of a mixture of ^{14}C amino acids. Differences in synthesis between 12-day (12F) forelimbs and 12-day hind limbs (12H) are indicated both by unmarked arrows and by arrows associated with a 0. a = actin, R = some of the ribosomal proteins detectable in this two-dimensional polyacrylamide gel system of protein separation.

organogenesis. That a wide range of exposures of the mother to teratogenic agents alters limb morphogenesis in predictable ways indicates the responsiveness of this developing structure to environmental signals (Kochhar and Aydelotte, 1974; Kochhar and Agnish, 1977). Some of the extrinsic factors known to be involved in limb morphogenesis include positional information and mesenchymal cell aggregation (Gruneberg and Lee, 1973; Elmer and Selleck, 1975), and oxygen and carbon dioxide in chondrocyte differentiation (Caplan and Stollmiller, 1973). An intrinsic program of development would be demonstrated in part by the ability of the three components of a developing limb, epithelium, prechondrogenic core, and stem cell periphery, to express in isolation (i.e., during *in vitro* culture) molecular and cellular parameters associated with morphogenesis of the limb *in vivo*.

We have begun to characterize patterns of translation by intact and component parts of mouse limbs during the period when it changes from a rudiment to an organ with contoured digits. Initial results of such a study, utilizing intact, excised limbs radiolabeled for 2 hr in culture, are presented in Figures 7 and 8. From a comparison of newly synthesized polypeptides in the forelimbs of 11- and 12-day embryos, and in fore- and hindlimbs of 12-day embryos, it is apparent that stage/limb-related translations exist. More comprehensive analyses designed to determine the distribution of these polypeptides in limb components, and whether such specific translations will occur during *in vitro* culture of limb components, are in progress.

The availability of a reproducible population of molecular markers associated with limb organogenesis presents the opportunity to examine experimentally questions concerning (1) the origin of the RNA templates that code for stage-related polypeptides (preformed vs. newly transcribed), (2) when, and to what extent, extrinsic factors such as epithelial–mesenchymal interactions, or cell position, regulate selective gene expression, and (3) whether specific gene expressions during limb morphogenesis are perturbed by teratogenic agents.

Acknowledgments. This research was supported by Grants HD 11654 and HD 13500 from the National Institutes of Health, United States Public Health Service, and Grant PCM 78-15986 from the National Science Foundation.

Discussion

Surani (Cambridge): I was very interested in your single-cell analysis; this is obviously going to be very useful in analyzing inside and outside cells. There is considerable asynchrony in cell division, however, and I wonder whether the differences you have noticed are because of differences in the cell cycle?

Van Blerkom: I think so. One possible explanation of the differences among blastomeres, espcially for the spots related to the 12- to 16-cell stage, is the early termination of synthesis in some cells that are more advanced in the cell cycle.

Soupart (Vanderbilt): In relation to those factors that appear briefly at metaphase II, and I have in mind specifically the factors involved in sperm nucleus decondensation, and the so-called male pronucleus growth factor, I wonder if your methods could show, in the polyspermic situation for example, the disappearance of some of those factors, and therefore attach to them a functional significance?

Van Blerkom: I believe the analytical methods are sensitive enough to examine the disappearance of a particular protein after fetilization. Presumably a factor such as the male pronucleus growth factor is no longer required by the egg after fertilization; if it could be identified in the maturing oocyte, I would expect its synthesis to terminate in the egg. In this manner, I believe that functional significance could be associated with particular stage-related polypeptides.

A. KAYE (Weizmann Institute): To get back to the question of quantitation of these two-dimensional gels, operationally you have shown quite reproducible and quite consistent results, which enable you to define stage-specific proteins on the basis of appearance or nonappearance. Now, taking the extreme point of view of some molecular biologists that all the genes are expressed at all times in mammalian cells, only at differing rates, can you make any sort of estimate, on the basis of these extremely long exposures of a year that you have had the patience to endure, of the difference between the spot which just appears and a spot which would not appear? Would this be a difference of an order of magnitude, half an order of magnitude? What sort of a guess could you make at this point?

VAN BLERKOM: That is always a difficult question to answer. Using standards of known specific radioactivity, with ultrathin gels and fluorographic amplification, the lightest spot probably represents one molecule in 10^7. Greater resolution is usually obtainable by immunoprecipitation (one molecule in 10^9). As far as the question of whether or not all genes are turned on, and the only difference in the detection of their products is the level of sensitivity of the technique, I probably would not rely exclusively on gel electrophoresis for a definitive answer. Fluorescent antibody techniques are also quite sensitive and I suppose if you had an antibody to a known protein, then its detection by this procedure may provide definitive information about whether or not a particular gene is expressed.

EPSTEIN (UCSF): You have ranged over several different species, and I wonder what kinds of inferences one can draw from studies on any one species? Each of us tends to think in terms of the system we work with and then rapidly extrapolates from it. Ultimately I think the interest is neither in the mouse nor the rabbit nor the pig, but in human reproduction. The unknown question is: Is any of this extrapolatable or how is it extrapolatable?

VAN BLERKOM: I would say that the important focus of these studies is the similarities in synthesis among the different species. If you observe stage-related syntheses that fall into families of polypeptides (i.e., some approximate molecular weight and isoelectric points), and if you can relate the time course of synthesis to known functions occurring in the oocyte or embryo, then experiments to establish specific synthesis–function relationships become realistic. For example, two known functions in several mammalian oocytes during the resumption of meiosis are the ability of the cytoplasm to decondense spermatozoal chromatin and of the plasma membrane to bind with a spermatozoon. These are two functions to which protein synthetic changes among various species, that appear to be similar in position in the gels and in the time of appearance during maturation, could be related. I would suspect that the apparent stage-independent syntheses will be found in other mammals aside from the rabbit and mouse. I would also suspect that this class of protein syntheses would also occur in later stages of embryogenesis and morphogenesis. Clearly, there are differences in the patterns that are related to the fact that different species are being examined. Yet I believe that similarities in synthesis that are not related to constitutive proteins exist among the few species we have examined, and that these similarities may prove to be quite important in relation to concomitant cell function.

EPSTEIN: That is the question. How real are the differences? A few years ago a lot was said about the differences in the time of turn on of the embryonic genome in the rabbit vs. the mouse. With the mouse we have evidence from mutants (Epstein, 1975, *Biol. Reprod.* **12**:82) that the genome turns on quite early, and in the rabbit it was thought to be later. Is that thinking going to be revised in terms of these kinds of studies?

VAN BLERKOM: If we return to the oocyte again, resumed meiosis from the germinal vesicle to metaphase II stage requires about 40 hr in the pig compared to only 12 hr in the mouse. In both species, the protein patterns show the appearance of similar stage-related spots. These spots appear at very different times because maturation requires different times in different species. In the same fashion, a comparison of the fine structure of preimplantation embryos of the mouse and rabbit demonstrates that a 2-cell mouse embryo is comparable to an 8- to 12-cell rabbit embryo in terms of nucleolar and mitochondrial development. Alterations in fine structure for the rabbit require, apparently, a longer time than for the mouse. If one normalized all species in terms of specific developmental events related to time, they would probably be very comparable.

EPSTEIN: So you think it is time scales rather than real differences?

VAN BLERKOM: For specific ultrastructural changes, such as nucleolar maturation, I suspect time scales are important. But the fine structure of rabbit and mouse embryos is quite different in other respects, and these differences are not time-scale dependent but species dependent. The same would hold true, I believe, for the protein patterns among the species. There are some stage-related proteins that may be quite similar; there are some, such as actin and tubulin, that would be expected to be identical, and there should be many others that are definitely species specific. It is the sorting out of these categories that is the real task. If and when the sorting out occurs, your question of real differences vs. apparent or time-scale-related differences may be answered more clearly.

SURANI: You mentioned briefly detection of glycoproteins in oocytes. Do you have any more information on that?

VAN BLERKOM: The problem with using most sugar labels for glycoproteins is that, in order to get enough incorporation, you have to do long-term labeling. It is really not clear after long-term labeling that the label will be incorporated into glycoproteins. With fairly short-term exposures, say 30-min periods, most of the proteins labeled with glucosamine or mannose are probably glycoproteins. In these experiments, oocytes were labeled for 30 min with glycosamine or mannose, and the spots that appeared radioactive after fairly long exposure, and after using quite a few oocytes, were the ones that exhibited the multiple spotting and seemed to streak across the gels as a series of spots; those are the ones I indicated also appeared first in early metaphase II.

References

Alton, T. H., and Lodish, H., 1977a, Developmental changes in messenger RNAs and protein synthesis in *Dictyostelium discoideum, Dev. Biol.* **60:**180–206.

Alton, T. H., and Lodish, H., 1977b, Synthesis of developmentally regulated proteins in *Dictyostelium discoideum* which are dependent on continued cell–cell interaction, *Dev. Biol.* **60:**207–216.

Arking, R., 1978, Tissue-, age-, and stage-specific patterns of protein synthesis during the development of *Drosophila melanogaster, Dev. Biol.* **63:**118–127.

Bachvarova, R., and DeLeon, V., 1977, Stored and polysomal ribosomes of mouse ova, *Dev. Biol.* **58:**248–254.

Bachvarova, R., and DeLeon, V., 1980, Polyadenylated RNA of mouse ova and loss of maternal RNA in early development, *Dev. Biol.* **74:**1–8.

Ballantine, J. E. M., Woodland, H. R., and Sturgess, E. A., 1979, Changes in protein synthesis during the development of *Xenopus laevis, J. Embryol. Exp. Morphol.* **51:**137–153.

Borland, R. M., 1977, Transport processes in the mammalian blastocysts, in: *Development in Mammals,* Vol. 1 (M. H. Johnson, ed.), pp. 31–68, North-Holland, Amsterdam.

Brandis, J. W., and Raff, R. A., 1978, Translation of oogenetic mRNA in sea urchin eggs and early embryos. Demonstration of a change in translational efficiency following fertilization, *Dev. Biol.* **67:**99–113.

Braude, P., Pelham, H., Flach, G., and Lobatto, R., 1979, Post-transcriptional control in the early mouse embryo, *Nature (London)* **283:**102–105.

Caplan, A. I., and Stollmiller, A. C., 1973, Control of chondrogenic expression in mesodermal cells of embryonic chick limb, *Proc. Natl. Acad. Sci. USA* **70:**1713–1717.

Chavez, D. J., and Van Blerkom, J., 1979, Persistence of embryonic RNA synthesis during facultative delayed implantation in the mouse, *Dev. Biol.* **70:**39–49.

Cotton, R. W., Manes, C., and Hamkalo, B. A., 1978, Electron microscopic analysis of transcription during mammalian embryogenesis, *J. Cell Biol.* **79**(part 2):abstract, F906.

Cross, P. C., and Brinster, R. L., 1970, *In vitro* development of mouse oocytes, *Biol. Reprod.* **3:**298–307.

Davidson, E., 1977, *Gene Activity in Early Development,* 2nd ed., Academic Press, New York.

Dewey, M. J., Fuller, R., and Mintz, B., 1978, Protein patterns of developmentally totipotent mouse teratocarcinoma cells and normal early mouse cells, *Dev. Biol.* **65:**171–182.

Ducibella, T., Albertini, D. F., Anderson, A., and Biggers, J., 1975, The preimplantation mammalian embryo: Characterization of intercellular junctions and their appearance during development, *Dev. Biol.* **45:**231–250.

Elmer, W. A., and Selleck, D. K., 1975, *In vitro* chondrogenesis of limb mesoderm from normal and brachypod mouse embryos, *J. Embryol. Exp. Morphol.* **33:**371–386.

Epstein, C. J., and Smith, S. A., 1974, Electrophoretic analysis of proteins synthesized by preimplantation mouse embryos, *Dev. Biol.* **40:**233–244.

Francis, D., 1976, Changes in protein synthesis during alternative pathways of differentiation in the cellular slime mold *Polysphondylium pallidum, Dev. Biol.* **53:**62–72.

Giri, J. G., and Ennis, H. L., 1978, Developmental changes in RNA and protein synthesis during germination of *Dictyostelium discoideum* spores, *Dev. Biol.* **67:**189–201.

Glass, R. H., Spindle, A. I., and Pedersen, R. A., 1979, Mouse embryo attachment to substratum and interaction of trophoblast with cultured cells, *J. Exp. Zool.* **208:**327–336.

Golbus, M. S., and Stein, M. P., 1976, Qualitative patterns of protein synthesis in the mouse oocyte, *J. Exp. Zool.* **198:**337–342.

Golbus, M. S., Calarco, P. G., and Epstein, C. J., 1973, The effects of inhibitors of RNA synthesis (alpha-amanitin and actinomycin D) on preimplantation mouse embryogenesis, *J. Exp. Zool.* **186:**207–216.

Graham, C. F., and Lehtonen, E., 1979, Formation and consequences of cell patterns in preimplantation mouse development, *J. Embryol. Exp. Morphol.* **49:**277–294.

Gruneberg, H., and Lee, A. J., 1973, The anatomy and development of brachypodism in the mouse, *J. Embryol. Exp. Morphol.* **30:**119–141.

Handyside, A. H., 1978, Time of commitment of inside cells isolated from preimplantation mouse embryos, *J. Embryol. Exp. Morphol.* **45**:37–53.

Handyside, A. H., and Barton, S. C., 1977, Evaluation of the technique of immunosurgery for the isolation of inner cell masses from mouse blastocysts, *J. Embryol. Exp. Morphol.* **37**:217–226.

Handyside, A. H., and Johnson, M. H., 1978, Temporal and spatial patterns of synthesis of tissue specific polypeptides in the preimplantation mouse embryo, *J. Embryol. Exp. Morphol.* **44**:191–199.

Hillman, N., Sherman, M. I., and Graham, C. F., 1972, The effect of spatial arrangement on cell determination during mouse development, *J. Embryol. Exp. Morphol.* **28**:263–278.

Jeffery, W. R., and Capco, D. G., 1978, Differential accumulation and localization of maternal poly(A)-containing RNA during early development in the ascidian, Styela, *Dev. Biol.* **67**:152–166.

Jenkinson, E. J., 1977, The *in vitro* blastocyst outgrowth system as a model for the analysis of peri-implantation development, in: *Development in Mammals,* Vol. 2 (M. H. Johnson, ed.), pp. 151–172, North-Holland, Amsterdam.

Johnson, M. H., 1979, Intrinsic and extrinsic factors in preimplantation development, *J. Reprod. Fertil.* **55**:255–265.

Johnson, M. H., Handyside, A. H., and Braude, P. R., 1977, Control mechanisms in early mammalian development, in: *Development in Mammals,* Vol. 2 (M. H. Johnson, ed.), pp. 67–98, North-Holland, Amsterdam.

Kochhar, D., and Agnish, N. D., 1977, "Chemical surgery" as an approach to study of morphogenetic events in embryonic mouse limb, *Dev. Biol.* **61**:388–394.

Kochhar, D., and Aydelotte, M. B., 1974, Susceptible changes and abnormal morphogenesis on the developing mouse limb, analyzed in organ culture after transplacental exposure to vitamin A (retinoic acid), *J. Embryol. Exp. Morphol.* **31**:721–734.

Kuo, C. H., and Garen, A., 1978, Analysis of the coding activity and stability of messenger RNA in *Drosophila* oocytes, *Dev. Biol.* **67**:237–242.

Levinson, J., Goodfellow, P., Vadeboncoeur, M., and McDevitt, H., 1978, Identification of stage specific polypeptide synthesis during murine preimplantation development, *Proc. Natl. Acad. Sci. USA* **75**:3332–3336.

McGaughey, R. W., and Van Blerkom, J., 1977, Patterns of polypeptide synthesis of porcine oocytes during maturation *in vitro, Dev. Biol.* **56**:241–254.

McLaren, A., 1973, Blastocyst activation, in: *The Regulation of Mammalian Reproduction* (S. J. Segal, R. Crozier, and P. A. Corfman, eds.), pp. 321–334, Thomas, Springfield, Ill.

Manes, C., 1973, The participation of the embryonic genome during cleavage in the rabbit, *Dev. Biol.* **32**:453–459.

Manes, C., 1977, Nucleic acid synthesis in preimplantation rabbit embryos. III. A "dark period" immediately following fertilization, and the early predominance of low molecular weight RNA synthesis, *J. Exp. Zool.* **201**:247–258.

Martin, G. R., Smith, S., and Epstein, C. J., 1978, Protein synthetic patterns in teratocarcinoma stem cells and mouse embryos in early stages of development, *Dev. Biol.* **66**:8–16.

Moore, G. P. M., 1978, Embryonic diapause in the marsupial *Macropus eugenii.* Stimulation of nuclear RNA polymerase activity during resumption of development, *J. Cell. Physiol.* **95**:31–36.

Moore, G. P. M., and Ringertz, N. R., 1973, Localization of DNA-dependent RNA polymerase activities in fixed human fibroblasts by autoradiography, *Exp. Cell Res.* **76**:223–228.

Motlik, J., Kopency, V., and Pivko, J., 1978, The fate and role of macromolecules synthesized during mammalian oocyte meiotic maturation. I. Autoradiographic topography of newly synthesized RNA and proteins in the germinal vesicle of pig and rabbit, *Ann. Biol. Anim. Biochim. Biophys.* **18**:735–746.

Overstreet, J. W., and Bedford, J. M., 1974, Comparison of the permeability of the egg vestments in follicular oocytes, unfertilized and fertilized ova of the rabbit, *Dev. Biol.* **41**:185–192.

Pelham, H. R. B., and Jackson, R. J., 1976, An efficient mRNA-dependent translation system from reticulocyte lysates, *Eur. J. Biochem.* **67**:247–256.

Rodman, T. C., and Bachvarova, R., 1976, RNA synthesis in preovulatory mouse oocytes, *J. Cell Biol.* **70**:251–257.

Schultz, G. A., Clough, J. R., and Johnson, M. H., 1980, Presence of cap structures in the messenger RNA of mouse eggs, *J. Embryol. Exp. Morphol.* **56**:139–156.

Schultz, R. M., and Wassarman, P. M., 1977, Specific changes in the pattern of protein synthesis during meiotic maturation of mammalian oocytes *in vitro, Proc. Natl. Acad. Sci. USA* **74**:538–541.

Schultz, R. M., Letourneau, G. E., and Wassarman, P. M., 1978, Meiotic maturation of mouse oocytes *in vitro.* Protein synthesis in nucleate and anucleate oocyte fragments, *J. Cell Sci.* **30**:251–264.

Schultz, R. M., Letourneau, G. E., and Wassarman, P. M., 1979, Program of early development in the mammal: Changes in patterns and absolute rates of tubulin and total protein synthesis during oogenesis and early embryogenesis in the mouse, *Dev. Biol.* **68**:341–359.

Senger, D. R., and Gross, P. R., 1978, Macromolecule synthesis and determination in sea urchin blastomeres at the sixteen-cell stage, *Dev. Biol.* **65:**404–415.

Seybold, W. D., and Sullivan, D. T., 1978, Protein synthetic patterns during differentiation of imaginal discs *in vitro*. *Dev. Biol.* **65:**69–80.

Thibault, C., 1977, Are follicular maturation and oocyte maturation independent processes?, *J. Reprod. Fertil.* **51:**1–15.

Van Blerkom, J., 1977, Molecular approaches to the study of oocyte maturation and embryonic development, in: *Immunobiology of the Gametes* (M. Edidin and M. H. Johnson, eds.), pp. 187–206, Cambridge Univ. Press, London/New York.

Van Blerkom, J., 1978, Methods for the high resolution analysis of protein synthesis. Applications to studies of early mammalian development, in: *Methods in Mammalian Reproduction* (J. C. Daniel, ed.), pp. 67–109, Academic Press, New York.

Van Blerkom, J., 1979, Molecular differentiation of the rabbit ovum. III. Fertilization-autonomous polypeptide synthesis, *Dev. Biol.* **72:**188–194.

Van Blerkom, J., 1980, The cellular and molecular biology of resumed meiosis in the mammalian oocyte, in: *The Ovary* (P. Motta, and E. S. E Hafez, eds.), North-Holland, Amsterdam.

Van Blerkom, J., and Brockway, G., 1975a, Qualitative patterns of protein synthesis in preimplantation mouse embryos. I. Normal pregnancy, *Dev. Biol.* **44:**148–157.

Van Blerkom, J., and Brockway, G., 1975b, Qualitative patterns of protein synthesis in preimplantation mouse embryos. II. During release from facultative delayed implantation, *Dev. Biol.* **46:**446–451.

Van Blerkom, J., and McGaughey, R. W., 1978a, Molecular differentiation of the rabbit ovum. I. During oocyte maturation *in vivo* and *in vitro, Dev. Biol.* **63:**139–150.

Van Blerkom, J., and McGaughey, R. W., 1978b, Molecular differentiation of the rabbit ovum. II. During the preimplantation development of *in vivo* and *in vitro* matured oocytes, *Dev. Biol.* **63:**151–164.

Van Blerkom, J., and Manes, C., 1974, Development of preimplantation rabbit embryos *in vivo* and *in vitro*. II. A comparison of qualitative aspects of protein synthesis, *Dev. Biol.* **40:**40–51.

Van Blerkom, J., and Manes, C., 1977, The molecular biology of preimplantation embryogenesis, in: *Concepts in Mammalian Embryogenesis* (M. I. Sherman, ed.), pp. 38–93, MIT Press, Cambridge, Mass.

Van Blerkom, J., and Motta, P., 1979, *The Cellular Basis of Mammalian Reproduction,* Urban & Schwarzenberg, Baltimore.

Van Blerkom, J., and Runner, M. N., 1976, The fine structure of parthenogenetically activated preimplantation mouse embryos, *J. Exp. Zool.* **196:**113–124.

Van Blerkom, J., Barton, S., and Johnson, M. H., 1976, Molecular differentiation of the preimplantation mouse embryo, *Nature (London)* **259:**319–321.

Van Blerkom, J., Chavez, D. J., and Bell, M., 1979, Molecular and cellular aspects of facultative delayed implantation in the mouse, in: *Maternal Recognition of Pregnancy,* Ciba Symposium Series 64 (J. Whelan, ed.), pp. 141–172, Excerpta Medica, Amsterdam.

Warnes, G. M., Moor, R. M., and Johnson, M. H., 1977, Changes in protein synthesis during maturation of sheep oocytes *in vivo* and *in vitro, J. Reprod. Fertil.* **49:**331–335.

Wassarman, P. M., and Letourneau, G. E., 1976, RNA synthesis in fully grown mouse oocytes, *Nature (London)* **361:**73–74.

Yanagimachi, R., 1977, Specificity of sperm–egg interactions, in: *Immunobiology of the Gametes* (M. Edidin, and M. H. Johnson, eds.), pp. 255–295, Cambridge Univ. Press, London/New York.

Young, R. J., 1977, Appearance of 7-methylguanosine-5′-phosphate in the RNA of mouse 1-cell embryos three hours after fertilization, *Biochem. Biophys. Res. Commun.* **76:**32–39.

Young, R. J., Sweeney, K., and Bedford, J. M., 1978, Uridine and guanosine incorporation by the mouse one-cell embryo, *J. Embryol. Exp. Morphol.* **44:**133–148.

IV

Uterine Preparation for Implantation

Introduction

Uterine Preparation for Implantation

M. A. H. Surani

The heterogeneous populations of cells in different uterine tissues are target cells for the ovarian steroids, progesterone and estrogen. These cells tend to respond differently to the changing endocrine stage of the female. The differentiation of stroma in preparation for decidualization and the induction of the receptive state of the uterine epithelium for blastocyst implantation must be coordinated under the influence of the ovarian steroids. Despite the obvious recognition of the stromal–epithelial interactions for blastocyst implantation, little is yet known about how such interactions are mediated. The bulk of the early work on the uterus was centered around determining the biochemical action of steroid hormones, with significant contributions from Gorski, Jensen, O'Malley, and Baulieu among many others. The question of the physiological implications of this action of hormones on the uterus for blastocyst implantation was seldom asked directly. Psychoyos provided some classical studies on the hormonal requirements for implantation and decidualization. Decidualization of stroma was extensively studied by Shelesnyak and colleagues and some outstanding contributions were made by De Feo. The functional aspects of uterine differentiation were emphasized further by the studies of Finn, Martin, and Leroy, who illustrated selective recruitment of uterine cells during early pregnancy and cell proliferation under different endocrine conditions. It is still not understood precisely how selective recruitment of cells occurs in the uterus. Glasser and colleagues and Heald and O'Grady have endeavored to change the emphasis from the biochemical action of hormones to establishing how such action is translated into uterine receptivity for blastocyst implantation.

Some of these points are examined in this part. Finn and Publicover extend their observations on cell proliferation and cell death in the endometrium by utilizing [^{3}H]thymidine incorporation into the uterine cells as a parameter. The changes in pool size of the precursor in the uterus under different endocrine conditions require attention in these studies, which otherwise provide a satisfactory approach to the problem. Glasser and McCormack provide an important contribution as they describe separation of endometrial cell types and their morphological and biochemical properties. This approach is long overdue

M. A. H. Surani • A.R.C. Institute of Animal Physiology, Cambridge, England.

and should provide new concepts of uterine preparation for implantation. Some account will have to be taken of the fact that the separated cell types may behave slightly differently compared to their responsiveness in the intact uterus. However, many of the fundamental and outstanding problems can be tackled by this approach. Padykula describes the influence of macrophages in remodeling the uterine matrix. Her contention is that progesterone in high concentrations may be immunosuppressive and has an anti-inflammatory action. When progesterone levels and binding to uterine receptors decline nearer term, uterine collagenase activity is enhanced and the breakdown products may provide an antigenic stimulus for attracting monocytes to the uterine stroma. Villee and colleagues discuss changes in steroid receptors, especially the decline in estrogen and progesterone receptors in decidual tissues. A relationship between these changes in receptor content and eventual decidual regression is suggested. An important point during the discussions concerns the possibility of steroid hormone action that is not mediated via the receptors but acts directly on cells and their membranes. The circumstances and significance of this type of steroid hormone action also deserve further attention.

Many problems still remain and perhaps will only be solved by a radical change in the approach both experimentally and in the types of questions that are asked. Uterine preparation for blastocyst implantation therefore remains a challenging system for further investigation, not only because it is inherently a fascinating biological problem but also because of its practical importance in devising new methods of chemical contraception and alleviating infertility.

11

Cell Proliferation and Cell Death in the Endometrium

C. A. Finn and Mary Publicover

I. Introduction

Implantation of ova into the wall of the uterus involves complex cellular changes in the endometrium resulting in the formation of the placenta. In preparation for these changes the cells of each of the three main tissues of the endometrium undergo a cycle of proliferation and differentiation under the control of the ovarian hormones that are secreted in a characteristic pattern during the estrous cycle and early pregnancy.

The factors controlling the buildup of the endometrium, at least in the mouse and rat, have been fairly well elucidated (Martin and Finn, 1968). Proliferation can be studied simply by arresting cells in metaphase of mitosis, for example, by injecting colchicine and counting the number of cells going into division over a set period, or by labeling proliferating cells during the S phase of the cycle with tritiated thymidine and counting the number of labeled nuclei in histological sections after autoradiography.

Differentiation of the cells following mitosis varies according to the type of tissue. The lumenal epithelial differentiation, for attachment of the blastocyst, can be seen with the electron microscope as a characteristic change of the cell surface, producing the so-called closure reaction. This change occurs in two stages (Pollard and Finn, 1972), in the first of which the opposing surfaces come together with interdigitation of microvilli, followed in the second stage by an apparent change in the structure of the surface membranes, bringing those from opposite sides of the lumen into very close contact. When a blastocyst is present, a similar change brings the surfaces of the trophoblast and endometrium into very close contact (Nilsson, 1967), the so-called attachment reaction. The glandular epithelial cells differentiate to synthesize and secrete a mucopolysaccharide substance that can be seen with the light microscope especially after PAS staining (Finn and Martin, 1976). What determines whether an endometrial epithelial cell will differentiate for attachment or secretion is not known; presumably it is something associated with the position of the cell. In

C. A. Finn and Mary Publicover • Departments of Veterinary Physiology and Pharmacology, University of Liverpool, Liverpool, England.

most species the stromal cells do not differentiate very far unless blastocyst attachment takes place; indeed in some species very little differentiation of stromal cells is apparent even after implantation. The exception is the human in whose uterus some differentiation of the stromal cells takes place during the menstrual cycle with the formation of predecidual cells.

Obviously there will be many occasions when cell proliferation and differentiation will not be followed by implantation, and unless the cell numbers are to increase dramatically with every cell cycle, there must be considerable death and removal of cells. This phenomenon is most obvious in women, when the menstrual discharge signals the breakdown of the endometrium. Furthermore, there are many cases during development of embryos when cell death plays an important part in organogenesis (Saunders, 1966). It therefore seemed appropriate to study the pattern of cell death in the uterus of the mouse especially with regard to whether the ovarian hormones play any part in its control.

II. Assessment of Cell Death

Dying cells can be identified in histological sections by the shape and staining properties of their nuclei. A count of these altered cells gives some indication of the extent of cell death (Martin *et al.*, 1973). It is not known, however, for how long these dead cells remain in the tissues; they are presumably removed, mainly by macrophages, and will therefore only be available for counting for a short period. Furthermore, their rate of removal may not necessarily be constant, thus giving another variable to be taken into consideration.

An indirect method for assessing cell death would be to measure changes in total cell numbers. A fall in number would represent an excess of cell death over cell proliferation; if the latter were known to be zero, the changes in total cell number would represent cell death. An estimate of cell numbers can be achieved either histologically or biochemically. Counting cells in histological sections is very tedious and time consuming, however, and, even if several sections were counted along the length of the uterus, only a very small part of the organ would be assessed. Biochemical estimates of total DNA, although giving a more complete estimate of changes in the number of nuclei along the entire length of the uterus, do not distinguish changes in cells from different tissues. It is not possible physically to separate accurately all the various tissues of the endometrium, or indeed to separate the endometrium from the myometrium with absolute certainty. Furthermore, in our laboratory, attempts at measuring total DNA of uteri, under conditions in which cells are known to be dividing, failed to show significant changes in DNA content.

We therefore attempted to develop a method for assessing the extent of cell death under various hormonal and other conditions that would be reasonably simple to carry out and that would separate changes in the various tissues of the uterus. As mentioned earlier, in response to the ovarian hormones during early pregnancy, there is a characteristic pattern of cell proliferation in the uterus such that a peak of mitosis occurs in the three main tissues of the uterus at different times. If one therefore injects tritiated thymidine into animals during the S phase, before one of these peaks, one should label a predominantly lumenal, glandular, or stromal cell population. Having a particular population of labeled cells, one can follow the decline in cell numbers under various conditions. If none of the cells died, we would expect the amount of radioactivity to remain constant, even if some or all of the cells undergo further division. If the amount of radioactivity falls, it is a reasonable assumption that cells have died and been removed from the uterus. We originally followed the decline in radioactivity by carrying out autoradiography and counting the labeled cells.

This procedure demonstrated that it was possible to get the separate labeled populations we had assumed (although the separation of lumenal and glandular cells was not as good as the separation of epithelial and stromal cells) and that changes in the number of labeled cells could be assessed. However, the method was rather tedious and gave an indication of what was happening at only one point along the length of the uterus.

An alternative method for measuring the radioactivity is to carry out scintillation counting of the whole uterus. This technique has the disadvantage that any radioactivity not incorporated into the nuclei will also be measured, which will limit the accuracy of the method and small but real changes may be missed. The method has the advantage, however, of being simple to perform and of measuring changes in the whole uterine horn. Most of the results presented in this paper are based on scintillation counting, although we have a lot of data from experiments using autoradiography, without which we would be less confident of our observations using scintillation counting.

To carry out the scintillation counting the uteri were removed, separated into the two horns, weighed, and placed in counting vials. The tissue was dissolved in Protosol, Omniflour dissolved in toluene was added, and the radioactivity was counted using a liquid scintillation counter. Quench corrections were made using a sample channels ratio method (Neame and Homewood, 1974). As would be expected, we found a close inverse relationship between weight of tissue in the sample and counting efficiency, so that the results from large uteri, for example, after decidualization, are relative underestimates of the radioactivity present compared to smaller uteri.

III. Results

The pattern of cell division during pregnancy in the mouse is shown in Figure 1 (Finn and Martin, 1967). The lumenal epithelium proliferates during diestrus and proestrus, the glandular epithelium on Day 3 of pregnancy, and the stroma on Days 4 and 5. Subsequent work (Finn and Martin, 1969, 1970) showed that a similar pattern of cell division in ovariectomized mice was obtained by injecting estradiol and progesterone on the schedule shown in Table 1.

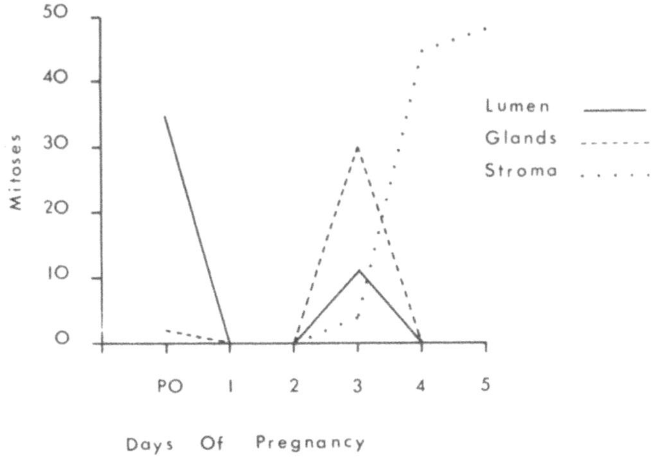

Figure 1. Number of cells undergoing mitosis in a cross section of mouse uterus during early pregnancy (data from Finn and Martin, 1967).

Table 1. Schedule of Hormone Injections in Ovariectomized Mice[a]

Day −2	Day −1	Day 0	Day 1[b]	Day 2	Day 3	Day 4	Day 5
E	E	—	—	—	P	Pe	Pe

[a] E = 100 ng estradiol; e = 20 ng estradiol (nidatory dose); P = 500 μg progesterone.
[b] Day 1 is approximately equivalent to the first day of pregnancy.

Table 2. Radioactivity in Uterine Horns of Mice after Indicated Treatments, Killed on Day −1 or Day 0[a]

Day −2	Day −1	Day 0	Mean cpm/horn (±S.E.M.)
E	TK		49,826 ± 3113
E	TE	TK	115,930 ± 6052
E	E	TK	71,471 ± 5332

[a] E = 100 ng estradiol; T = 20 μCi [^{3}H]thymidine; K, day on which killed (1 hr after last or only thymidine injection).

Table 3. Radioactivity in Uterine Horns of Mice after Indicated Treatments, Killed on Days Specified[a]

Day −2	Day −1	Day 0	Days 1 & 2	Day 3	Day 4	Day 5	Day 6	Day 7	Mean cpm/horn (±S.E.M.)
E	TE	TK							117,044 ± 8735
E	TE	T	—	K					31,509 ± 4130
E	TE	T	—	Pr	K				31,797 ± 2397
E	TE	T	—	Pr	e	K			24,973 ± 2758
E	TE	T	—	Pr	—	K			22,656 ± 1347
E	TE	T	—	Pr	e	e*	—	K	(L) 21,601 ± 1154
									(R) 15,907 ± 631
E	TE	T	—	Pr	—	—*	—	K	(L) 16,609 ± 2531
									(R) 16,549 ± 1684

[a] E = 100 ng estradiol; e = 20 ng estradiol; T = 20 μCi [^{3}H]thymidine; Pr = 1 mg medroxyprogesterone acetate; * = 0.02 ml oil injected into left horn, K, animals killed; L, left horn; R, right horn.

Injections of 100 ng estradiol on the first and second days of the schedule produce a peak of cell division in the lumenal epithelium on the second and third days with a peak of glandular epithelial division on the fifth and sixth days. The 3 days free from hormone treatment are necessary for this glandular division to take place. If progesterone treatment is started immediately after the estrogen treatment, the glands do not divide. Following the injection of progesterone on the sixth and seventh days, there is pronounced mitotic activity in the stroma on the seventh day; if a very small dose of estradiol (10 ng) is then given on the seventh and eighth days, more cells undergo division and the endometrium becomes sensitive to implantation or a dedidual stimulus, such as intralumenal injection of oil. Progesterone works in combination with estrogen to bring about stromal division; alone it produces no stromal division.

Vaginal smears from mice injected with hormones on this schedule are cornified on the second day after the two estrogen injections. This day is therefore equivalent to the day on which mating occurs, that is, Day 1 of pregnancy. So as to be able to relate results from experiments using this schedule to normal pregnancy, we date our treatment from this day forwards or backwards. Using this protocol the peak of glandular mitosis occurs on Day 3 and stromal mitosis on Days 4 and 5, as is found in pregnancy. In some cases the daily injections of progesterone were replaced by a single injection of the long-acting progestin medroxyprogesterone acetate (Provera, Upjohn). This compound acts for at least 6 days in the mouse (Finn and Martin, 1979) and therefore obviates the necessity for daily injections and presumably gives a more constant level of progestin.

Earlier work (Martin *et al.*, 1973) has shown that uptake of tritiated thymidine (measured autoradiographically) is at a maximum about 8 hr before the peak of mitosis. It should be possible, therefore, to label the cells of the various tissues by injecting thymidine at that time.

In the first experiment, we measured the uptake of tritiated thymidine into the uterus after 1 and 2 days of adminstration of 100 ng estradiol (Table 2). Injections of 20 μCi [^{3}H]thymidine were given 16 hr after the first and/or second estrogen injections, and the animals were killed 1 hr after the last or only thymidine injection. The counts show that more label is taken up when thymidine is administered after the second estrogen injection than after the first. When both thymidine injections are given, there is approximately three times as much label than after only the first injection. This result suggests that many of the cells responding to the first injection divide again in response to the second.

In the following experiments we labeled the cells 16 hr both after the first and after the second estradiol injection and killed the mice on various days of the standard schedule as shown in Table 3. Animals in some groups received a nidatory dose of estrogen and in

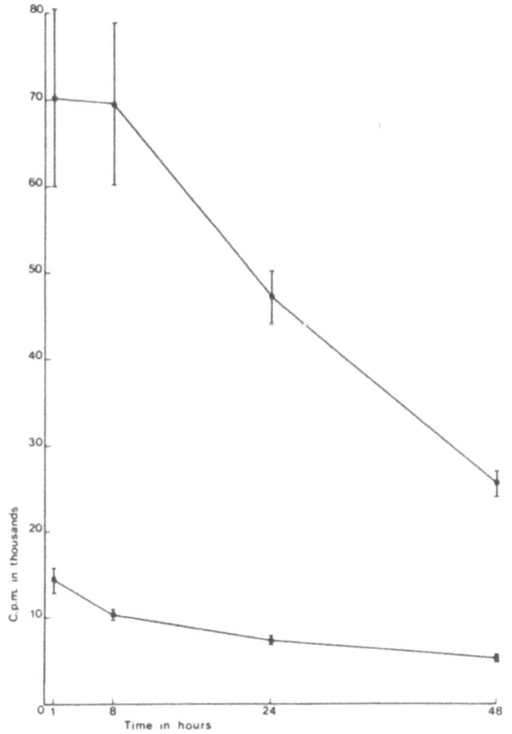

Figure 2. Uptake of radioactivity into uterine horns at times indicated after an injection of [^{3}H]thymidine 16 hr after the second estrogen injection (E) of the standard schedule (Table 1) (top graph) or after no hormone treatment (bottom graph). Vertical lines show ± S.E.M.

others progestin only, as indicated. Mice in groups 6 and 7 had oil introduced into the lumen of the left horn on Day 5, as indicated by the asterisk. This treatment initiated decidual cell formation in most of the injected horns when estradiol had been given (group 7) but not when progesterone was given alone (group 6).

The most surprising finding was the very large drop in labeling between Day 0 and Day 3. After this time there was a much steadier decline, which is mainly independent of the treatment, although the fall in counts may be slightly halted by the induction of decidualization.

In the next experiment we attempted to assess the background radioactivity by measuring the uptake of [^{3}H]thymidine into the uterus of ovariectomized untreated mice killed at the same time as the treated animals. On this occasion only one injection of thymidine was given, 16 hr after the second injection of estradiol, and the animals were killed 1, 8, 24, and 48 hr after the thymidine injection.

The results are shown graphically in Figure 2. Clearly there is a small uptake of label into the uterus of the ovariectomized untreated mice, the value falling slowly for 48 hr following the thymidine injection; this is in contrast to the larger uptake in the treated animals, which falls abruptly 8 hr after injection. The uteri of the treated animals are of course considerably heavier than those of the untreated animals, and uterine weight falls off somewhat by 48 hr. Correction for weight-related quenching would therefore increase the difference shown.

The very large drop in thymidine uptake after cessation of estrogen injection seemed worthy of further investigation. We first set out to determine if an equivalent drop occurred after the estrus preceding an anticipated pregnancy. This experiment is somewhat difficult to organize as the S phase of the proestrous peak of mitosis would be expected to occur at the end of the preceding diestrus and one has therefore to label the cells during the diestrus before mating. In our colony the mice cycle very irregularly, which makes it very difficult to predict the last day of diestrus. To circumvent this difficulty we adopted a technique we use regularly to obtain females that have mated on a specified day. We have found that if mice are given two daily injections of 500 μg progesterone and are then placed with males, they will usually come into estrus and mate 3 or 4 days later. Thus by adminstering [^{3}H]thymidine on the second day after the progesterone injection, we hoped we would catch many of the lumenal epithelial cells in the S phase of the proestrous division. In order to increase this probability, we gave two injections of [^{3}H]thymidine 8 hr apart. Ten mice were killed 1 hr after the second thymidine injection and the remainder were left with males; most of these mated 2 days later and were killed on Day 3 of pregnancy. The results of Table 4 show that there was a large uptake of label, confirming that the cells were probably in the S phase of the division when the thymidine was injected, and the large drop on Day 3 is consistent with the earlier experiments (Tables 2 and 3).

Table 4. Radioactivity in Uterine Horns of Mice to which Two Injections of [^{3}H]Thymidine were Administered during Diestrus

Time of sacrifice	Mean cpm/horn (±S.E.M.)
Diestrus	59,146 ± 10,618
Day 3	25,607 ± 3,803

Table 5. Radioactivity in Uterine Horns of Mice after Indicated Treatments, Killed on Days Specified [a]

Day −2	−1	0	1	2	3	4	5	6	7	8	9	Mean cpm/horn (±S.E.M.)
E	TE	TK										78,700 ± 6617
E	TE	T	—	K								30,766 ± 2812
E	TE	TE	E	K								83,100 ± 9144
E	TE	T	—	—	Pr	e	e*	—	K			(L) 31,105 ± 1936 (R) 19,195 ± 1522
E	TE	T	—	—	Pr	e	e*	—	—	—	K	(L) 17,499 ± 2219 (R) 13,742 ± 387

[a] Abbreviations as in Table 3.

Table 6. Radioactivity in Uterine Horns of Mice after Indicated Treatments, Killed on Day 4 or Day 10 [a]

Day 3	Day 4	Day 5	—	Day 10	n	Mean cpm/horn (±S.E.M.)
P	TK				8	33,175 ± 5421
P	T	—	—	K	9	4,973 ± 481
P	TPr	—	—	K	9	9,590 ± 1316
P	TPr	e	—	K	9	8,459 ± 986

[a] P = 500 μg progesterone; T = 30 μCi[^{3}H]thymidine; Pr = 1 mg Provera®; e = 20 ng estradiol.

This presumed cell death in the lumenal epithelium is occurring at a time when the lumen is reorganizing from the swollen, fluid-filled cavity characteristic of the time of insemination to the slitlike structure found in the uterus ready for implantation. This change of shape in the uterus occurs only if the influence of estrogen is removed. In the next experiment (Table 5) we therefore included a group in which the estrogen injections were continued and compared it with a group treated as before (Table 3). Both groups were killed on Day 2. To check the previous result, we also included a group in which decidualization had been induced. These mice were killed on either Day 7 or Day 9.

In the absence of estrogen the drop in counts was considerable (Table 5). However, when the estrogen was continued no drop in counts occurred, indicating that over this period cell death does not occur unless estrogen treatment ceases. In the decidualization experiment the animals killed on Day 7 again showed a larger drop in radioactivity in the unstimulated horn, although on Day 9 the difference was very small. In view of the problems associated with the method, we would not at present wish to attach much importance to this observation.

These results show clearly that there is a remarkable rapid turnover of cells in the endometrial epithelium both during the estrous cycle and after hormone treatment. This turnover is presumably associated with the rapid changes of volume and shape of the lumen as it is reorganized from a fluid-filled cavity for sperm transport to a receptacle for blastocyst attachment and implantation. Thus it seems reasonable that cell death plays a part in the differentiation of the lumen.

Table 7. Radioactivity in Uterine Horns of Mice after Indicated Treatments, Killed on Day 5 or Day 10[a]

Day 3	Day 4	Day 5	—	Day 10	n	Mean cpm/horn (±S.E.M.)
Pr	T	K			9	19,876 ± 1825
Pr	T	—	—	K	9	9,056 ± 815
Pr	T	e	—	K	9	7,713 ± 446

[a] Abbreviations as in Table 6.

Table 8. Radioactivity in Uterine Horns of Mice after Indicated Treatments, Killed on Day 8[a]

Day 3	Day 4	Day 5	—	Day 8	n	Mean cpm/horn (±S.E.M.)
Pr	T	—	—	K	9	8,794 ± 828
Pr	Te	e	—	K	12	7,918 ± 458
Pr	Te	e*	—	K	9	(L) 28,890 ± 2569 (R) 10,031 ± 965

[a] Abbreviations as before except e = 10 ng estradiol.

We next turned our attention to the endometrial stromal cells. Finn and Martin (1970) showed that the stromal cells divide in response to a single injection of progesterone given 3 days after the estrous estrogen injections. By injecting [^{3}H]thymidine on Day 4 of our schedule, we were thus able to label a population of stromal cells. Autoradiographs confirmed that the label was taken up almost entirely by the stromal cells.

We particularly wished to determine the effect on stromal cell numbers of continuing or discontinuing progestin treatment after administration of the thymidine label, with or without injections of nidatory estrogen, and of stimulating the stromal cells to decidualize. Five separate experiments were carried out.

The results of the first experiment are shown in Table 6. All animals were given 500 μg progesterone on Day 3 and all were injected with 30 μCi [^{3}H]thymidine on Day 4. Animals in group 1 were killed on Day 4, 1 hr after the thymidine injection, and those in group 2 on Day 10 with no further treatment. There was clearly a very large fall in radioactivity. Groups 3 and 4 received an injection of medroxyprogesterone on Day 4 to maintain progestin level and those in group 4 also received an injection of 20 ng estradiol on Day 5. The drop in these groups was still considerable, although less than when no progestin was given on Day 4. The effect of the single dose of nidatory estrogen was to reduce the count slightly but not significantly (group 4 vs. group 3). We have also looked at this effect of nidatory estrogen autoradiographically and again found a consistent but not statistically significant fall.

This effect of estrogen was investigated further in the next experiment (Table 7), where again a small but nonsignificant difference was shown.

In the next experiment (Table 8) we tried dividing the nidatory estrogen into two doses of 10 ng on Days 4 and 5: this produces optimal conditions for decidual receptivity in these mice. In a third group we induced decidualization by injecting oil into the left horn of the uterus about 6 hr after the second estradiol injection. The effect of the estrogen was in the same direction, but again was not clearcut; the effect of decidualization, however, was pronounced. The horns containing the decidual tissue had many more counts (28,890) than the unstimulated horns (10,031). As the decidualized horns were of course much larger than the unstimulated horns, the efficiency of counting was less, so the difference is presumed to be even greater than that shown. We have also looked at similar decidualized uteri autoradiographically and confirmed that there are very large numbers of stromal cells labeled. In a group of five animals killed on Day 8, the mean number of labeled cells was 634 in the decidualized horn compared to 154 in the nondecidualized horn.

In view of the very high [^{3}H]thymidine counts in the decidualized stroma, we set out in the next experiment (Table 9) to follow more closely the changes in radioactivity occurring between sensitization and full decidual development. Whereas the counts in the noninjected horns dropped to about a third, those in the decidualized horns dropped very little. This result was confirmed in the final experiment (Table 10), which also confirmed the effect of omitting progestin treatment entirely (group 4). Finally, in calculating the mean weight for the decidualized horns, we have included all responding horns. However, there was usually a considerable variation in the degree of decidual response. Nevertheless, if one plots counts vs. uterine weight for individual decidualized uteri, one obtains a fairly good regression curve (Figure 3).

These results show that the population of cells that divides on Day 4 of pregnancy is probably maintained until the endometrium is fully parepared for decidualization (Day 6) but that after this time the numbers drop. Although the drop in cell number is more precipitous when no progesterone is given, it is considerable even when the normal schedule of hormones is administered. There is a slight indication in the fully sensitized uterus, i.e., those receiving nidatory estrogen in addition to progesterone, that the drop is greater, but the more interesting finding is that, in conditions akin to delayed implantation (proges-

Table 9. Radioactivity in Uterine Horns of Mice after Indicated Treatments, Killed on Days Specified [a]

Day 3	Day 4	Day 5	Day 6	Day 7	Day 8	*n*	Mean cpm/horn (±S.E.M.)
Pr	Te	K				9	43,157 ± 3338
Pr	Te	e*	K			10	(L) 36,703 ± 2622 (R) 30,240 ± 2717
Pr	Te	e*	—	—	K	8	(L) 36,223 ± 3087 (R) 12,775 ± 996

[a] Abbreviations as before except e = 10 ng estradiol.

Table 10. Radioactivity in Uterine Horns of Mice after Indicated Treatments, Killed on Days Specified [a]

Day 3	Day 4	Day 5	—	Day 8	*n*	Mean cpm/horn (±S.E.M.)
P	TK				10	31,095 ± 3046
P	TPre	K			10	28,073 ± 1862
P	TPre	e*	—	K	10	(L) 34,100 ± 5004 (R) 11,481 ± 699
P	T	—	—	K	10	6,279 ± 606

[a] Abbreviations as before except e = 10 ng estradiol.

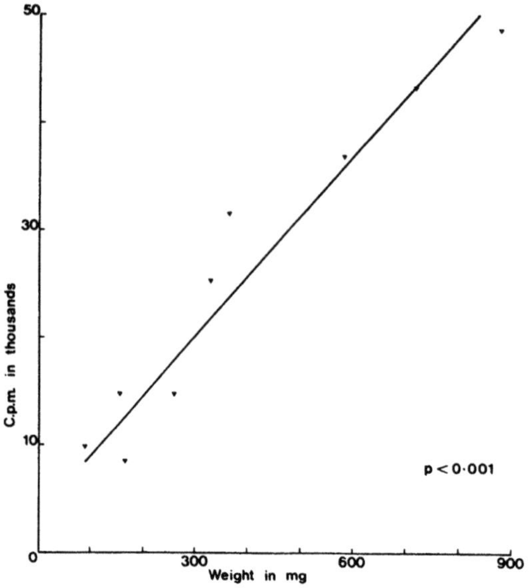

Figure 3. Graph showing the relationship between weight of decidualized horns and radioactivity. The equation of the line estimated by regression analysis is $y = 56x + 3399$. The regression is significant at the 5% level of probability.

terone only) in which full sensitivity can be induced, there is still a very considerable degree of cell death. From the fact that the cell numbers are maintained after decidualization, we conclude that many of the cells on Days 5 and 6 are in an unstable condition in which they can either decidualize or die, but maintenance of a large population of undecidualized stromal cells is not favored.

IV. Conclusions

In summary, we make the following suggestions regarding cell turnover in the endometrium. The lumenal epithelial cells proliferate rapidly in preparation for estrus and then a large proportion of them die during the first 2 or 3 days of pregnancy. Proliferation of the stromal cells occurs on Days 4 and 5 of pregnancy. These cells are maintained until the time of implantation provided progesterone secretion is maintained, but after this time, regardless of hormone secretion, the cell numbers drop. If decidualization is induced, however, the cell death is prevented and the cells either divide again or differentiate.

The rapid changes in cell numbers in the various tissues of the endometrium are, we think, an important factor to be taken into consideration when evaluating experiments measuring biochemical changes following treatment with ovarian hormones.

Discussion

Leroy (Brussels): The three cell populations in the endometrium behave quite differently from the point of view of cell kinetics. We have been looking at them as separate entities, but I am concerned with the relationship between the surface epithelium and the glands. While you have increased cell death in the lumenal epithelium in the experimental setup, as well as during the estrous cycle, you would have glands proliferating, wouldn't you? So I wonder how far glands contribute to the regeneration of this dying epithelium at that time. In the small peak of uptake in the epithelium at the same time that you have very active gland proliferation, are the labeled cells located uniformly around the uterine lumen or are they concentrated near the gland collars?

FINN: Yes, I think it is a very interesting fact that the glands appear to be proliferating after the main peak of lumenal epithelial cell death or maybe about the same time. Regarding your point of whether labeling is even or at gland mouths, all I can say is I have not looked specifically but I've never noticed any difference. That does not mean to say that I didn't look carefully, because I find it is very difficult to see where the gland is coming from. Sometimes you do cut across the mouth, but in many of the glands you do not see many gland mouths in a particular section.

LEORY: Did you look at what happens on Day 6 in animals stimulated with oil?

FINN: Day 6? Yes, I think one group.

LEROY: The problem I'm thinking about is the alleged G-2 block of stromal cells before decidualization, which I don't believe is true because we have cytophotometric data which disprove this. Have you any opinion on this problem in relation to your data?

FINN: No, I can't see how my data can contribute to that problem.

LEORY: I mean, if there is a sharp increase in uptake of thymidine on Day 6, that would go against the G-2 block, wouldn't it?

FINN: Yes, but all the thymidine was injected on Day 4, well before decidualization started; we have not injected on Day 6.

GLASSER (Baylor): Is there a difference between the effect of these treatments on cell death, in terms of the fact that more cells are present and therfore more cells eventually die, or is there a direct effect on the life span of an individual cell type? I would like Azim Surani to think whether or not he can provide a correlation in terms of the secretory capbility of stimulated cells that eventually die. What might this mean for the sensitivity of the uterus to decidualization and implantation and whether these are functionally related in terms of implantation per se?

FINN: That is a very interesting point. Do the cells die simply because their number exceeds a certain level, so they die to get back to a more stable number, or is it controlled in some way? I must say I am rather surprised at the numbers dying. I think the answer is we don't yet know.

SURANI (Cambridge): As far as the secretory function of the uterus is concerned, the main problem is that we have not yet been able to identify the cell types from which secretion occurs. There are two cell types which may lead to secretion, one is the glandular cells and the other is the epithelial cells; until we have some information on how these two cell types are involved in secretion, it will be difficult to answer that question.

A. KAYE (Weizmann Institute): Perhaps you didn't go into the techniques for lack for time, but it wasn't clear to me how you detected uptake of thymidine into DNA in these experiments, particularly in view of the rapid and large-scale changes in uptake after estrogen, for example.

FINN: Yes, that is a problem. We assume that when we pulse label, the majority will go into the nuclei. Autoradiography shows that, by killing after an hour, most of it by then is in the nucleus. But I think we must improve the method and either separate the nuclei or separate the DNA and measure the radioactivity. This was a first attempt, and I think now we have shown that the cell numbers do change so rapidly, presumably from cell death, we must go further into the methodology and make sure that there is incorporation into the nuclei.

SURANI: I would like to follow up that question. After you inject the thymidine intraperitoneally, do you know if the pool size of thymidine that reaches the uterus is the same in every case? This point is important because of increased vascularization during estrus, where more thymidine might reach the uterus.

FINN: That factor would not be relevant for most of the experiments, because all groups received the same treatment up to the time thymidine was given.

HEAP (Cambridge): Could you explain what you mean by the term cell death? You said something about macrophages removing cells. Could you also say something about whether cells are released into the uterine lumen and whether that contributes to some extent to the cell death you describe?

FINN: Dead cells are visible in light microscope or electron microscope sections from the shape of the nuclei, which become karyorrhectic, and also there are droplets which can be seen very easily. Regarding the epithelial cells, I think many of them do go into the lumen, although we never see many at any one time, but we do see odd cells there and I imagine they get there very quickly. We also see plenty of macrophages in the stroma, but to say that the macrophages take up the eipthelial cells is at present an assumption.

HEAP: Can you put figures on that? Do you know what proportion of cells are actually released into the lumen, or is this a situation similar to the cell turnover one sees in the intestine, for example?

FINN: I don't know.

HEAP: What happens with actinomycin?

FINN: You mean trying to determine whether there is programmed cell death? That is one thing we were thinking of doing, but I don't have the answer at present.

C. TACHI (Tokyo): Why are the subepithelial stromal cells which are replenished after progesterone and estrogen, different from other stromal cells, in being so disposed towards cell death?

FINN: Our results do not give that information; they only provide data on the cells which are dividing on Day 4. In the mouse, there is no indication that the subepithelial stromal cells are replenished more than the others; they may be in the rat.

MARSTON (Birmingham): Implantation is normally preceded by mating and the mouse is unusual in that ejaculation occurs directly into the uterus. The role of the female genital tract before estrus is to provide seminal plasma as a pool in the uterus into which ejaculation occurs. Surely this endometrial surface epithelium change is more related to the needs of sperm as challenge to the genital tract, than anything to do with implantation? What is the effect of mating on cell death, I think in most of your experiments the animals had not been mated?

FINN: There was one experiment in which they had been mated.

MARSTON: But do you agree with that idea?

FINN: Yes. I did mention that the uterus goes from a balloon structure with a large number of cells back to a thin structure. I think the surprising thing is that, unlike the bladder which has to balloon up and then go down again, the uterus seems to produce a tremendous number of cells and then they die off. This seems a strange way of going from a bladder-shaped uterus to a thin one ready for implantation.

KENNEDY (W. Ontario): Keith Brown-Grant (*Horm. Behav.* **8**:62, 1977) has shown that rats kept in constant estrus do not have ballooned uteri, yet they will mate and become pregnant. So a fluid within the uterine lumen is not essential for sperm transport and fertilization in rodents.

FINN: To show that sperm can pass up the uterus under those experimental conditions does not disprove that the situation found in normal estrus is the most favorable for sperm transport.

BEIER (Aachen): We are also interested in the function of the secretory activities of the lumenal epithelium and did a number of studies on hormonal interference with the normal timing of the inherent epithelial program before implantation. If we treat rabbits with an overdose of estrogen (Beier and Mootz, 1979, in: *Maternal Recognition of Pregnancy* (J. Whelan, ed.), Ciba Foundation Series 64 (new series), pp. 111–140, Excerpta Medica, Amsterdam) or with a synthetic estrogenic compound, 1-β-acetoxy-norgestrel (Elgar and Petzoldt, 1978, *Acta Endocrinol.* **87**(Suppl. 215):40; Beier *et al.*, 1979, 5th Congr. Anat. Eur, Prague, Argum, Commun. Univ. Carolinal Proga, p. 28), we can selectively destroy the epithelium. Figure 1 shows, in the right upper corner, fairly normal epithelium. Figure 2 shows a scanning EM with normal surfaces at the upper right corner and, in the middle of the picture, vacuolar degeneration with release of material into the lumen. After another 4 to 8 days, Figure 3, we see phenomenon. formed epithelial cells that may be derived from the islets that remained and an accumulation of plasma cells in the stroma underneath this new epithelium. These rabbits underwent a normal pregnancy after this astonishing phenoenon. This cell population has an amazing potency to give rise to a new normal preimplantation phase, and implantation and normal pregnancy.

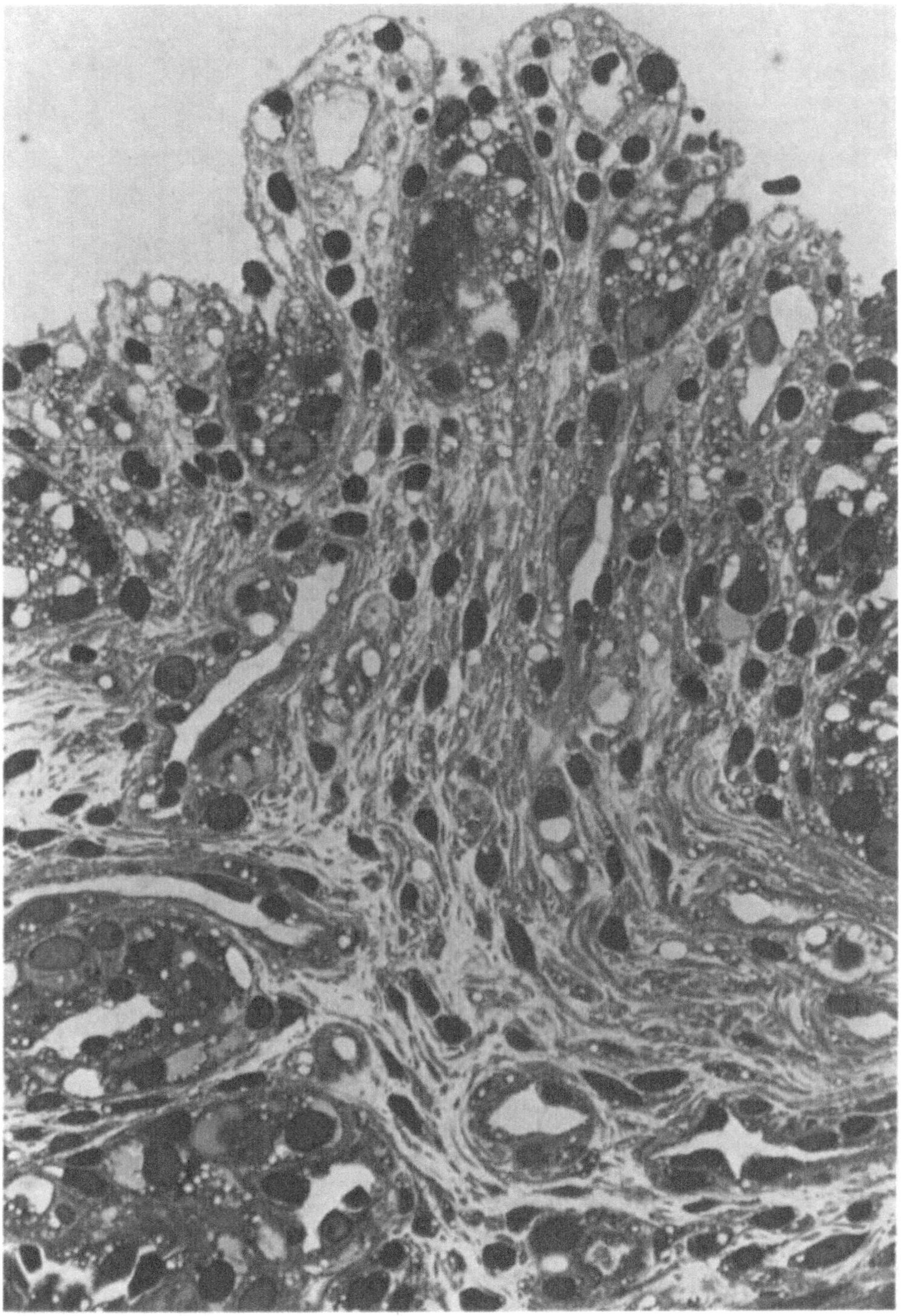

Figure D1. Semithin section of a rabbit endometrium showing epithelial degeneration. The upper parts of the lumenal epithelium are pinched off, and only the epithelial cells of the deeper folds are still present. The vacuolar degeneration is not restricted to lumenal endometrial cells but also occurs in endothelial cells of the capillaries. Specimen taken from an ovariectomized rabbit treated with progesterone (Day 0, 10.0 mg; Day 1, 1.5 mg; Day 2, 1.5 mg; Days 3 and 4, each 3.0 mg) and with estradiol-17 β-benzoate (Day 0, 0.1 mg; Day 1, 0.15 mg) and sacrificed on Day 4 of the experiment. Toluidine blue staining, $\times 840$. From Beier and Mootz (1979).

Figure D2. Scanning electron microscope view of rabbit endometrium showing epithelial damage, as in Figure 1. This specimen was obtained after the same treatment as described in Figure 1. ×1950. From Beier and Mootz (1979).

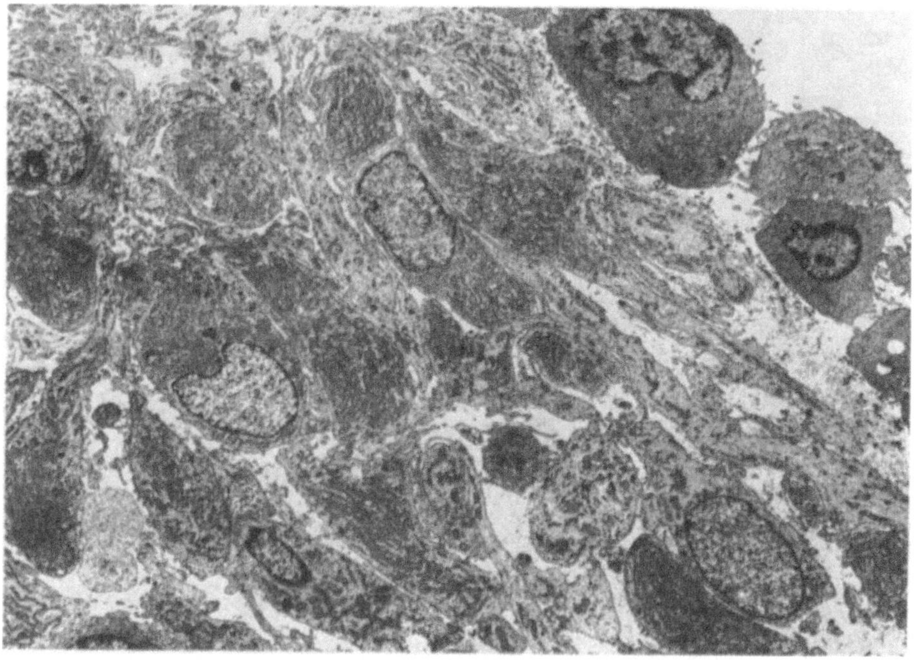

Figure D3. Transmission electron micrograph of endometrial tissue of a rabbit that was treated with 1-β-acetoxy-D-norgestrel for 8 days (1.0 mg/day per rabbit subcutaneously) and kept without treatment for the next 4 days. ×3250. From Beier *et al.* (1979).

References

Finn, C. A., and Martin, L., 1967, Patterns of cell division in the mouse uterus during early pregnancy, *J. Endocrinol.* **39:**593–597.

Finn, C. A., and Martin, L., 1969, Hormone secretion during early pregnancy in the mouse, *J. Endocrinol.* **45:**57–65.

Finn, C. A., and Martin L., 1970, The role of the oestrogen secreted before oestrus in the preparation of the uterus for implantation in the mouse, *J. Endocrinol.* **47:**431–438.

Finn, C. A., and Martin, L., 1976, Hormonal control of the secretion of the endometrial glands in the mouse, *J. Endocrinol.* **71:**273–274.

Finn, C. A., and Martin, L., 1979, Effect of a long-acting progestin on reproductive function in female mice. *J. Endocrinol.* **79:**235–238.

Martin, L., and Finn, C. A., 1968, Hormonal regulation of cell division in epithelial and connective tissues of the mouse uterus, *J. Endocrinol.* **41:**363–371.

Martin, L., Finn, C. A., and Trinder, G., 1973, Hypertrophy and hyperplasia in the mouse uterus following oestrogen treatment: An autoradiographic study, *J. Endocrinol.* **56:**133–144.

Neame, K. D., and Homewood, C. A., 1974, *Introduction to Liquid Scintillation Counting,* Butterworths, London.

Nilsson, O., 1967, Attachment of rat and mouse blastocysts onto uterine epithelium, *J. Fertil.* **12:**5–13.

Pollard, R. M., and Finn, C. A., 1972, Ultrasturcture of the uterine epithelium during the hormonal induction of sensitivity and insensitivity to a decidual stimulus in the mouse. *J. Endocrinol.* **55:**293–298.

Saunders, J. W., 1966, Death in embryonic systems, *Science,* **154:**604–612.

12

Shifts in Uterine Stromal Cell Populations during Pregnancy and Regression

Helen A. Padykula

I. Introduction

The blastocyst of primates and rodents interacts directly with the uterine stroma during implantation and placentation to establish critical functional relationships. This confrontation between tissues of two different genotypes is tolerated during normal pregnancy. This phenomenon imparts distinction on the uterine stroma and sets it apart from most stromal compartments in the maternal system, which respond forcefully to allogeneic stimuli. Yet the uterine stroma, except for decidual cells, has been relatively little studied in terms of other cell populations and the nature of the extracellular matrix (ECM).

This deficiency in knowledge may be explained in part by the difficulty of precise identification in routine light microscopic preparations of the 11 or more stromal cell types that may occur in the endometrium during cyclic and gestational states (Table 1). These cell types are plainly visible when the stroma is observed with the electron microscope (Figure 1). At the light microscopic level, histochemical–cytochemical procedures are useful for viewing the size and distribution of certain stromal cell populations. For example, the lysosome-rich macrophages are easily visualized by cytochemical procedures that demonstrate acid phosphatase or other lysosomal enzymes (Figure 2). Uterine heterophils (neutrophils) become conspicuous in a periodic acid–Schiff preparation of the rat uterus because only they among these stromal cells carry rich stores of glycogen (Padykula and Campbell, 1976). Such selective light microscopic histochemical identifications are necessary to recognize the overall distribution of each cell type.

Shifts in stromal cell populations occur during the menstrual or estrous cycle and pregnancy and are accompanied by steady remodeling of the endometrial and myometrial extracellular matrix; these two events are most likely interrelated. Interpretation of the composition of the ECM is made difficult not only by progressive cyclic and gestational change but also by puzzling phylogenetic differences in the relative amounts of collagen and

Helen A. Padykula • Department of Anatomy, University of Massachusetts Medical School, Worcester, Massachusetts 01605.

Table 1. Uterine Stromal Cell Populations [a]

Resident cells	Transient cells
Fibroblasts	Decidual cells
Mast cells	Monocyte macrophages
Lymphocytes	Lymphocytes (T and B)
Stem cells	Eosinophils
	Heterophils (neutrophils)
	Plasma cells
	Endometrial granular cells

[a] These 11 cell types occur in marsupial (opossum), rodent (rat), and primate (human) uteri during the nonpregnant cycle, pregnancy, and/or regression. The monocyte and macrophage are here listed as modulations of a single cell type but may also be viewed as separate types. Similarly B lymphocytes and plasma cells may be considered modulations of one cell type.

ground substance among species. Overall, the ground substance has eluded morphologists because it is difficult to preserve this material of high water content; however, its origin and degradation in terms of cellular involvement in proteoglycan synthesis are approachable.

The uterine stroma during the estrous or menstrual cycle is a panorama that varies on a day-by-day and even hourly basis in terms of kinds and differentiative states of cell populations, as well as in the migratory activity of some cell types. This cellular activity occurs within an ECM that appears also to be progressively modified. During the anabolic phase, these changes unfold subtly under the cyclic shifts in relative stimulation by estrogen and progesterone; however, the catabolic changes at the close of a cycle or gestation are more rapid and conspicuous and thus are more easily recognized by morphologic parameters.

Emphasis will be placed here on the fibroblast and the monocyte–macrophage as a contrasting pair of cells in terms of their structure and functional roles in the "management" of the ECM. The invasion of blood leukocytes into the peripartum stroma of the rat (Day 16 gestation to Day 4 postpartum) will be described and interpretations of this transient phenomenon will be proposed. Finally, cellular evidence for the presence of a local immune response in the early postpartum rat uterus and early human gestational uterus (5–12 weeks) will be presented and discussed.

II. Involvement of the Uterine Fibroblast and Monocyte–Macrophage in Modeling of the Extracellular Matrix

Fibroblasts are established as the synthesizers of the precursors of the protein macromolecular framework (collagen, elastin, and proteoglycans) of the ECM (Figures 2 and 3). As resident uterine cells, the cyclic and gestational activity of these protein producers is modulated by estrogen and progesterone. A single injection of estradiol into an immature or ovariectomized rat causes enlargement of the cytoplasmic protein-synthesizing and transporting membrane system (rough endoplasmic reticulum and Golgi complex), an increased amount of collagen fibrils and ground substance in the ECM, and increased permeability of endometrial microvasculature. The uterine stroma of the rat is rich in collagen, particularly during late pregnancy (Harkness and Harkness, 1956; Harkness and Moralee, 1956).

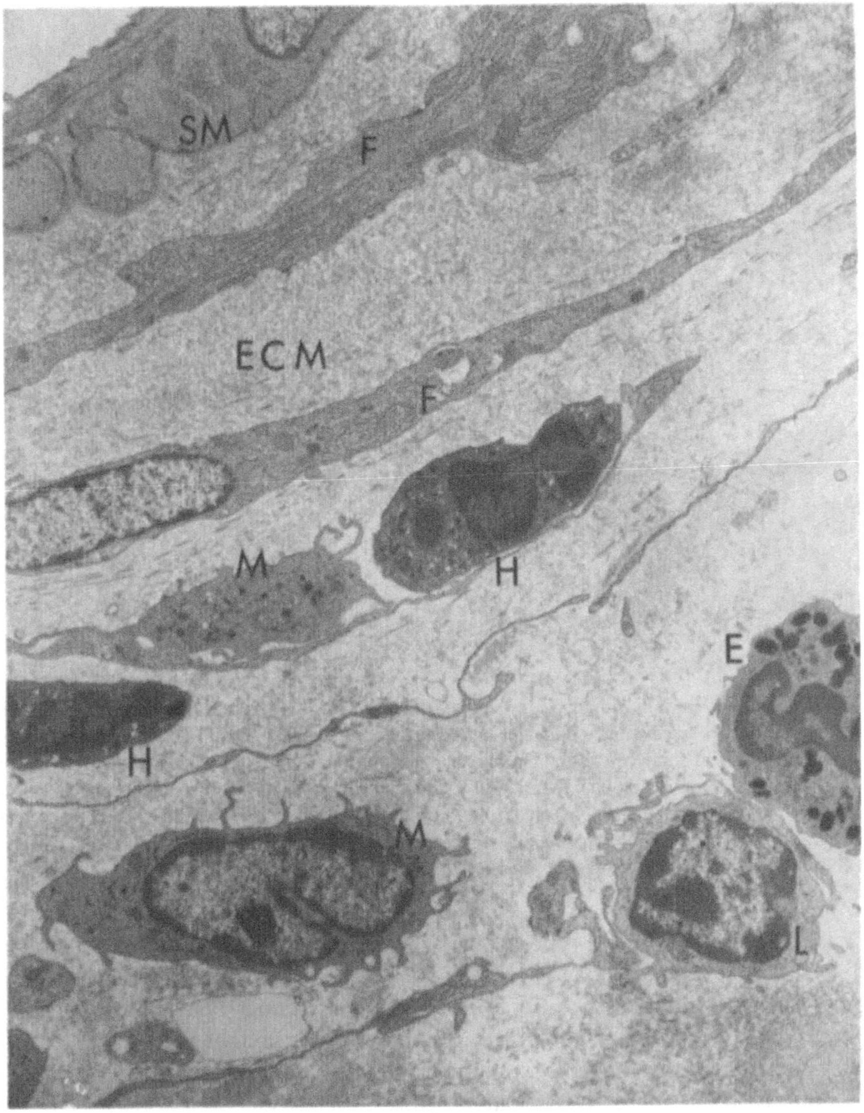

Figure 1. Leukocytic populations of the normal rat endometrial stroma, Day 21 of pregnancy. During the last week of gestation tissue leukocytic populations are composed of: heterophils (H), eosinophils (E), monocytes (M), and lymphocytes (L). These transient stomal cells occur among fibroblasts (F) within an abundant extracellular matrix (ECM). A small artery (SM, smooth muscle) occurs at the upper left corner. ×4100. From Padykula and Tansey (1979).

Along with this important anabolic activity, fibroblasts may also play a catabolic function; in monolayer culture they secrete neutral collagenase and proteoglycan-splitting enzymes into the medium (Werb and Burleigh, 1974; Werb *et al.*, 1978). Collagenase cleaves collagen molecules and thus dismantles intact collagen fibrils (see Figures 13A and B). Uterine procollagenase is activated during regression and causes an unusually rapid degradation of collagen fibrils for a normal tissue (see review by Gross, 1974). However, the cellular origin or mode of activation of uterine collagenase is uncertain because the monocyte–macrophage in cell culture is also a producer of neutral procollagenase (Werb

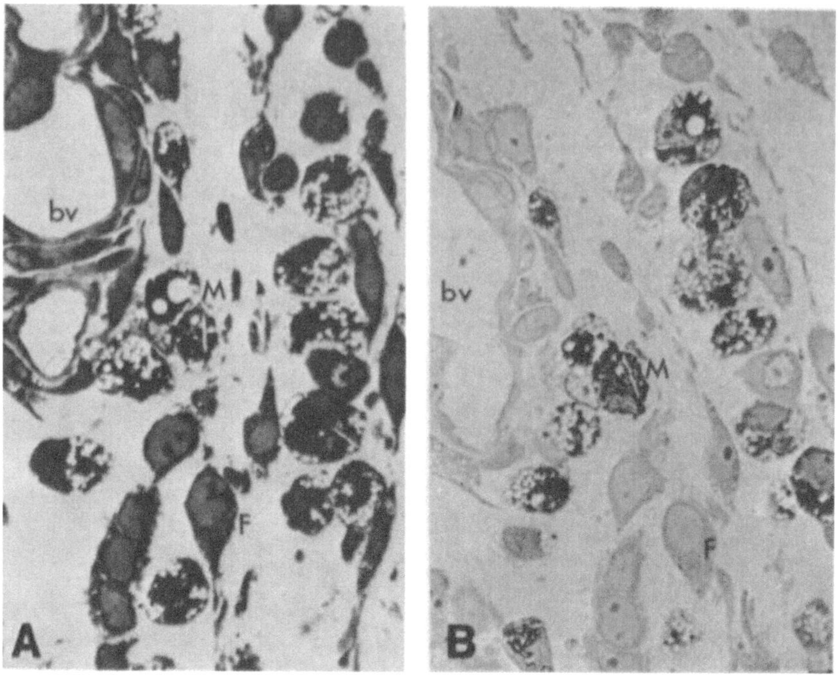

Figure 2. Rat endometrial strom at 3 days 21 hr postpartum, two adjacent 1-μm plastic sections. (A) Stained with toluidine blue. Numerous transient macrophages (M) occur among resident fibroblasts (F). The macrophages contain many lipid droplets and some crystalline inclusions that have been removed, and this creates the conspicuous cytoplasmic vacuolization. bv, blood vessel. (B) Cytochemical localization of acid phosphatase activity (Gomori procedure). The high acid phosphatase activity of the phagolysosomal system of the macrophages isolates them visually as a population of specialized phagocytes. The low activity of these fibroblasts is barely evident here. Both ×900. From Padykula and Campbell (1976).

and Gordon, 1975a), and this cell type appears in increasing numbers during regression. Thus the fibroblast and/or macrophage are potential sources of uterine procollagenase. The macrophage is distinguished from the fibroblast by its conspicuous lysosomal–phagosomal system (Figures 2 and 3); it has cytoplasmic membrane systems for protein synthesis that are less voluminous than those of the stimulated fibroblast, and are most likely involved, at least in part, in the production of lysosomal enzymes.

Because macrophages are phagocytic cells it has been generally assumed that they are involved in the cyclic renewal of the extracellular matrix, for they appear transiently in large numbers during regression (Lobel and Deane, 1962; Parakkal, 1972). Also, collagen fibrils have been reported within smooth-membrane-limited vacuoles (phagosomes?) of rat uterine macrophages during regression (Parakkal, 1969). However, the presence of collagen fibrils within uterine macrophages has not been substantiated (Schwarz and Güldner, 1967), although uterine fibroblasts, in certain differentiative states, contain intact collagen fibrils as well as fibrils in various stages of degradation (Figures 4 and 5). During normal postpartum regression (rat) only fibroblasts located close to the endometrial–myometrial junction and in the myometrium contain intracellular collagen-containing vacuoles (Figure 6) (Schwarz and Güldner, 1967). However, when postpartum regression and collagen degradation are slowed by administration of exogenous progesterone during the first two postpartum days (Jeffrey and Koob, 1973), fibroblasts throughout the endometrium contain numerous collagen-containing vacuoles (Figures 4 and 5). In an ultrastructural investigation of the rat uterus at metestrus, Dyer and Peppler (1977) reported a widespread distribution of stromal cells (not identified as to type) that contain intracellular collagen fibrils; most occurred immediately beneath the uterine epithelium but were also widely distributed throughout the endometrium. This stromal differentiation was not influenced by the presence of an intrauterine device.

Ultrastructural evidence for the presence of collagen-containing vacuoles in normal and experimentally modified uterine fibroblasts is harmonious with that obtained in other tissue systems undergoing normal remodeling of the ECM, as in the periodontal ligament

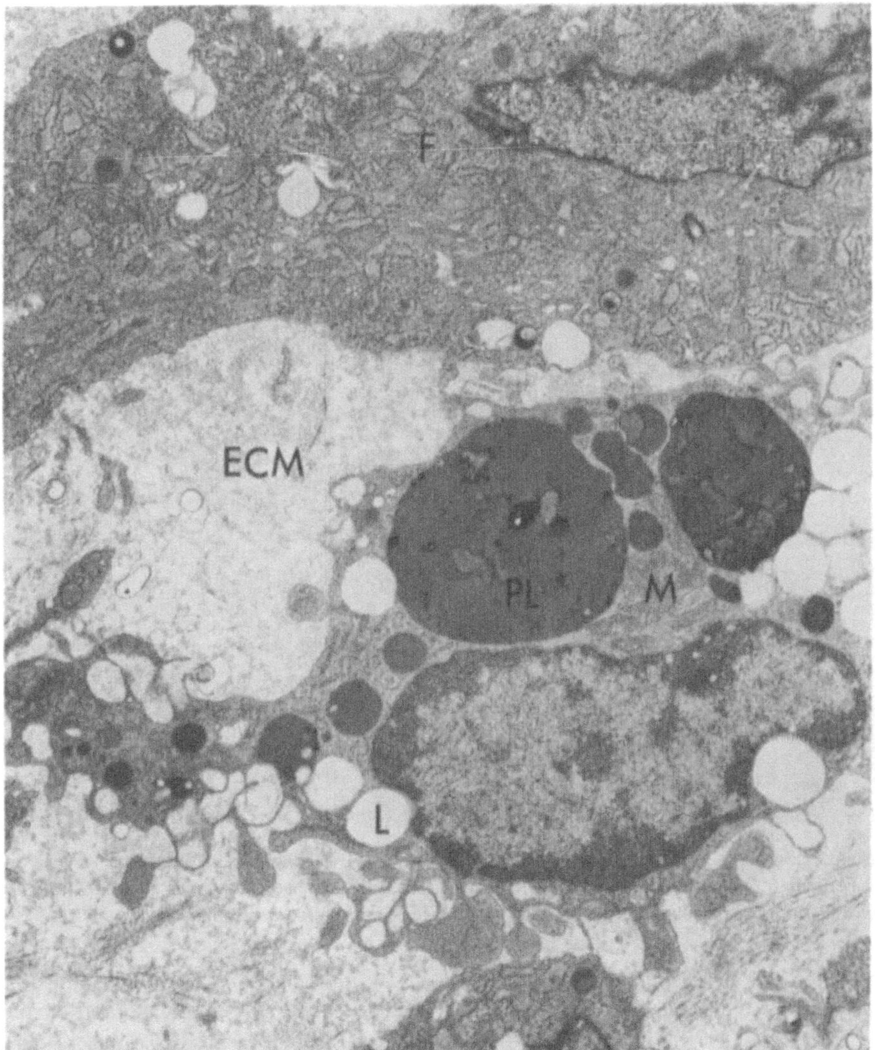

Figure 3. Electron micrograph illustrating a fibroblast (F) and a macrophage (M) in the early postpartum rat endometrial stroma. The fibroblast is rich in cytoplasmic membrane systems (rough endoplasmic reticulum and Golgi complex); the macrophage has numerous phagolysosomal derivatives (PL), lipid droplets (L), and a highly motile cell surface, ECM, extracellular matrix. ×8000.

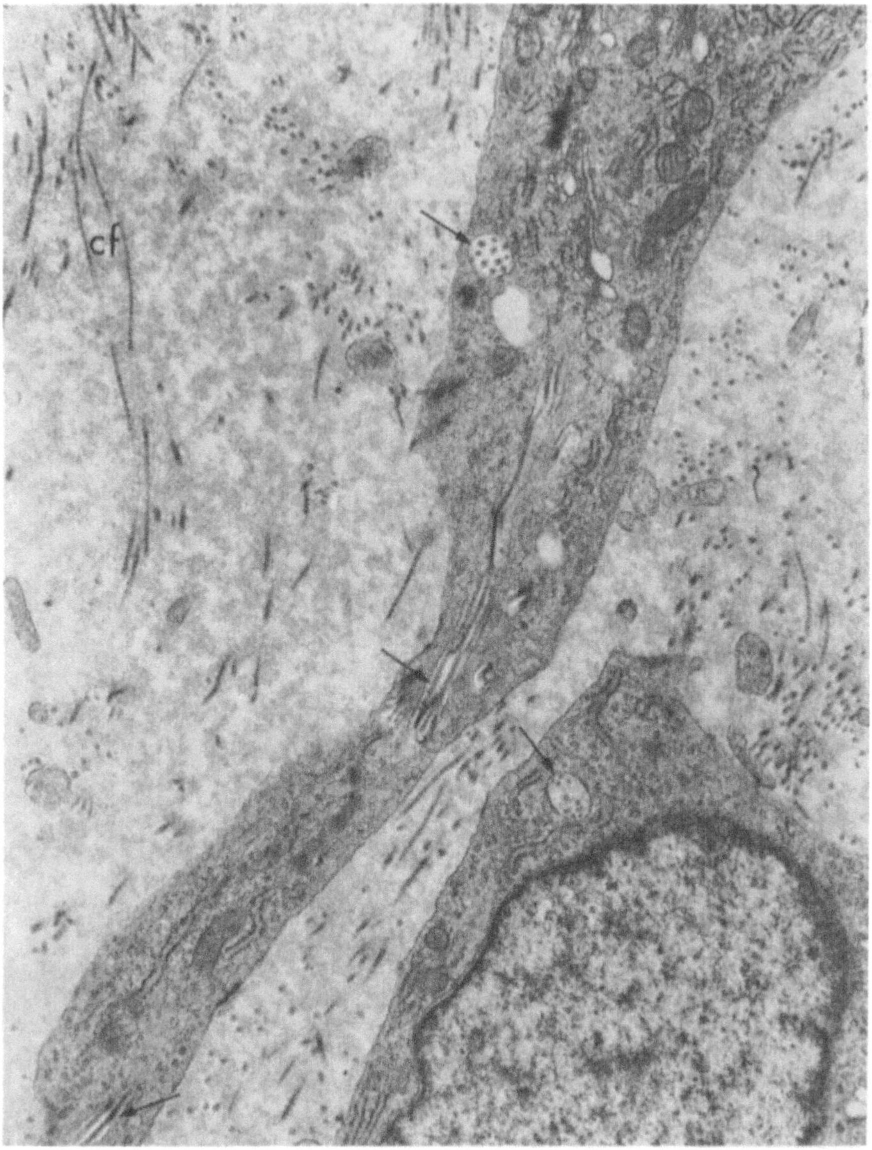

Figure 4. Uterine fibroblasts in the deep endometrial stroma of a rat given 40 mg progesterone per day starting at parturition. Portions of two fibroblasts are shown, each of which has collagen-containing vacuoles (arrows). Note collagen fibrils in longitudinal and cross section. cf, collagen fibrils in the extracellular matrix. ×14,300.

during physiologic movement of teeth (Deporter and Ten Cate, 1973; Ten Cate and Syrbu, 1974; Ten Cate *et al.*, 1976), in response to skin wounds (Ten Cate and Freeman, 1974), and in disease states (Perez-Tamayo, 1970; Renteria and Ferrans, 1976; Levine *et al.*, 1978). Existing evidence suggests that various cells of mesenchymal origin (fibroblasts, smooth muscle cells, osteoblasts) involved in the production and maintenance of the ECM may contain intracellular collagen during certain normal (Brandes and Anton, 1969) and abnormal (Levine *et al.*, 1978) states. This cellular mechanism, which is expressed in the

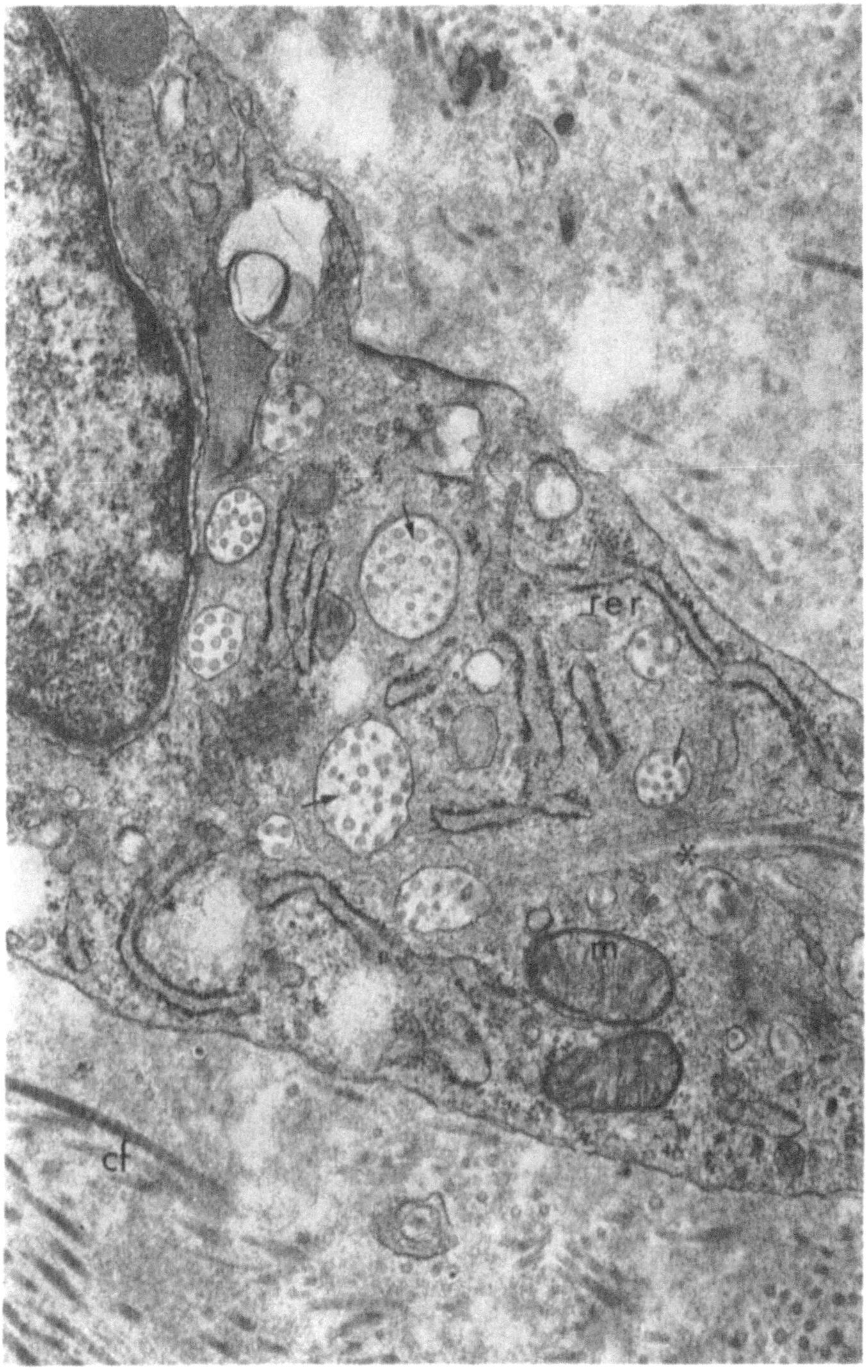

Figure 5. Uterine fibroblasts in the deep endometrial stroma of a rat given 40 mg progesterone per day starting at parturition. Most of the cytoplasmic vacuoles have intracellular collagen fibrils cut in cross section (arrows). At the asterisk, intracellular collagen fibrils occur in longitudinal section. m, mitochondrion; rer, rough endoplasmic reticulum.

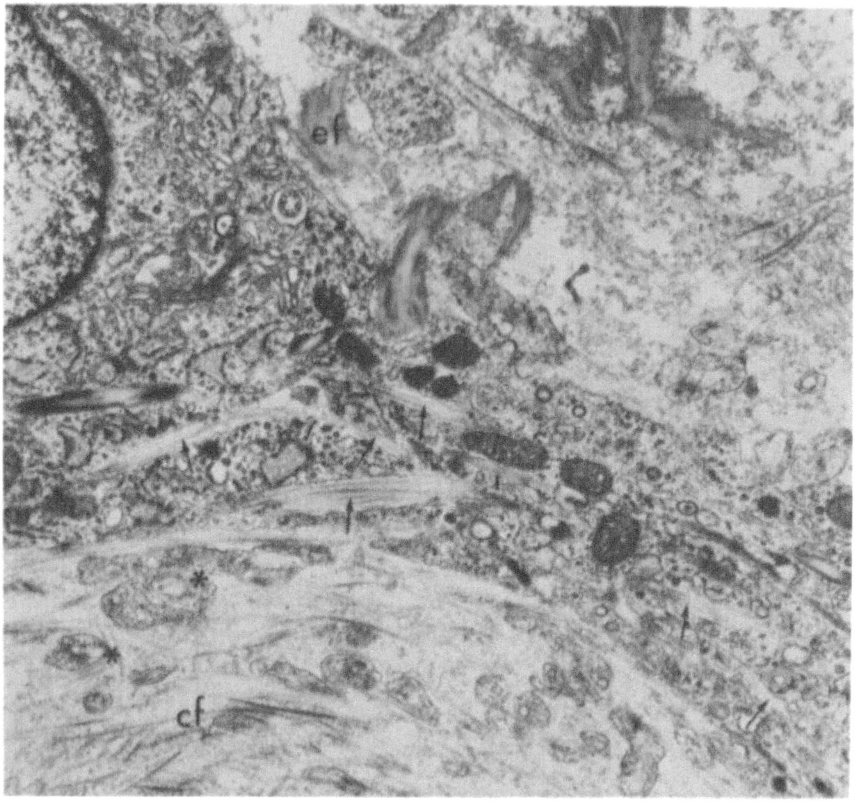

Figure 6. Uterine fibroblast, 3 days postpartum normal rat. In the normal postpartum uterus, collagen-containing vacuoles (arrows) occur in fibroblasts located near the endometrial–myometrial junction. Note also the close association of elastic fibers (ef) with the fibroblastic surface. cf, collagen fibrils in the extracellular matrix. ×12,300.

uterus during normal remodeling of the ECM at the close of a cycle or pregnancy, is thus operative also during pathologic states.

The involvement of the fibroblast in intracellular degradation of collagen fibrils is largely a morphologic concept that has only partially surfaced. It gains some support from a recent study that demonstrated *in vitro* "phagocytosis" of collagen fibrils by periodontal fibroblasts (Svoboda *et al.*, 1979). Their ultrastructural images presented in serial section indicate that extracellular protease activity may not always be an essential prelude to intracellular collagen degradation. Recent biochemical findings provide a new angle for viewing the role of the fibroblast in the regulation of the ECM. A significant amount (30–40%) of newly synthesized collagen is degraded intracellularly by fibroblasts in culture (Bienkowski *et al.*, 1978). That is, fibroblasts may possess a posttranslational mechanism for regulating collagen release from the cell, as is true for other types of protein-secreting cells (Smith and Farquhar, 1966). This intracellular system for collagen degradation would most likely be localized in the fibroblastic lysosomes, which may, as in the hepatocyte, contain proteases and peptidases that hydrolyze collagen at acid pH (Coffey *et al.*, 1976). Such an intracellular enzymatic lysosomal system in uterine fibroblasts along with the widespread distribution of extracellular inactive (latent) collagenase along collagen fibrils (Woessner, 1976) and in basement membranes (Montfort and Perez-Tamayo, 1975) may provide a dual enzymatic system for cyclic remodeling of the uterine ECM as well as for hormonal regulation of this fundamental process.

Evidence for the function of the uterine macrophage in the regulation of the ECM is largely circumstantial at this time. Its transient appearance in large numbers during the catabolic phases of the uterus suggests involvement in phagocytic reduction of cellular and extracellular components. It is likely that uterine macrophages originate largely from blood monocytes that enter the rat uterine extracellular compartment before regressive events such as collagen degradation commence (Padykula and Tansey, 1979). It is reasonable to assume that monocyte–macrophages would function in the dismantling of the macromolecular collagen and proteoglycan scaffolding of the ECM. Thus far, there has been no convincing demonstration of the presence of intracellular collagen in uterine macrophages. As has already been mentioned, macrophages in cell culture are known to secrete collagenase (Werb and Gordon, 1975a; Birkedal-Hansen *et al.*, 1976) and elastase (Werb and Gordon, 1975b). Besides secretion of these proteases, macrophages may be involved in the conversion of inactive collagenase into an active form (Werb *et al.*, 1977). Macrophages in cell culture secrete plasminogen activator (Unkeless *et al.*, 1974; Vassalli and Reich, 1977), a known activator of collagenase activity (Werb *et al.*, 1977). Thus, besides their well-known phagocytic activity, such secretory function may be stimulated during uterine regression.

An amazing new role for the uterine macrophage has recently been reported (Sandow *et al.*, 1979). During the period of relatively low circulating titers of estrogen and progesterone in the estrous cycle of the hamster, intraepithelial macrophages phagocytose lumenal epithelial cells that are undergoing autolysis. Extrapolating to our observation that monocytes enter the rat lumenal epithelium before obvious regression, it is likely that monocytes possess collagenase activity necessary to penetrate the collagen-rich basal lamina. Thus, monocyte–macrophages may be involved in both epithelial and stromal regression. Their disappearance from the uterus at the close of the catabolic phase may be effected through transepithelial emigration into the uterine cavity and/or into lymphatic vessels located at the endometrial–myometrial junction (Padykula, 1976; Padykula and Campbell, 1976; Padykula and Taylor, 1976).

III. Transient Invasion of Blood Leukocytes into the Rat Uterine Stroma in the Peripartum Period (Day 16 Gestation through Day 3 Postpartum)

During the last week of gestation, the level of plasma progesterone declines steadily (Wiest, 1970) while plasma estrogen rises on Days 21–22 (Yoshinaga, 1976) (Figure 7). Maximal concentrations of uterine nuclear progesterone receptors exist between Days 9 and 15 of pregnancy and thereafter decline to a very low or undetectable level at parturition (Figure 8) (Vu Hai *et al.*, 1978). This hormonal profile of low uterine binding of progesterone and rising plasma estrogen creates a shifting endocrine milieu in which monocytes, heterophils (neutrophils), and eosinophils appear in the prepartum uterine stroma (Padykula and Tansey, 1979) by emigration from the blood. In addition, the monocytes and heterophils assume intraepithelial positions, particularly in the lumenal epithelium. Although the extravascular presence of heterophils and monocytes is generally interpreted as a sign of inflammation, there is evidence that the gestational appearance of these blood leukocytes may represent a programmed intrinsic differentiation that is probably under hormonal control. The presence of these leukocytes may represent prepartum preparation for tissue

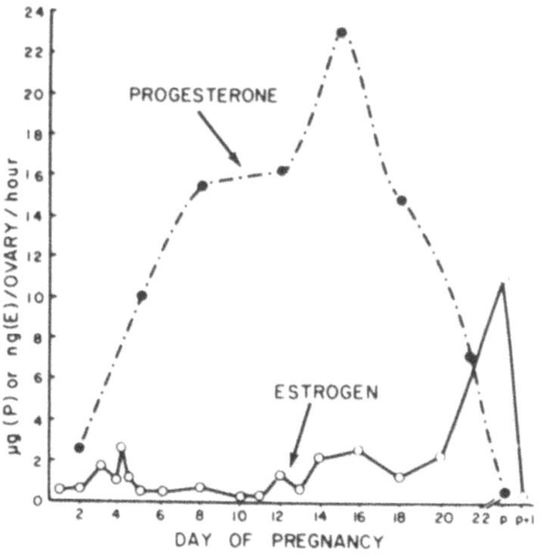

Figure 7. Ovarian hormonal secretion during pregnancy in the rat. Progesterone secretion declines steadily during the last week of pregnancy to reach a nadir on the day of parturition; concomitantly estrogen secretion rises steadily to a peak of parturition. From Yoshinaga (1976); reprinted by kind permission of Harvard University Press.

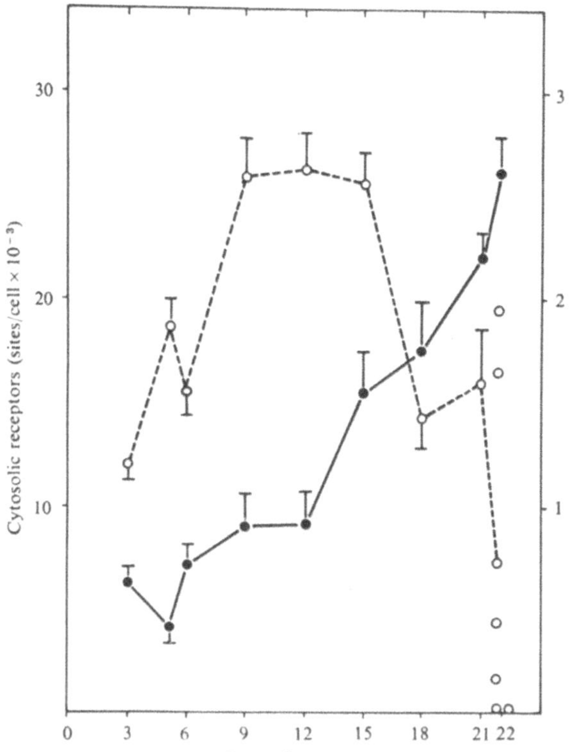

Figure 8. Variations in the concentration of cytoplasmic and nuclear progesterone receptors during pregnancy in the rat. Note particularly the drop in concentration of nuclear receptors (dashed line) during the last week of pregnancy that occurs concomitantly with a rise in the concentration of cytoplasmic receptors. From Vu Hai *et al.* (1978); reprinted by kind permission of the *Journal of Endocrinology*.

regression after birth and for the postpartum estrus. The prepartum appearance of eosinophils most likely reflects rising estrogen levels, as demonstrated experimentally in ovariectomized rats (Ross and Klebanoff, 1966). In this late gestational period, uterine macrophages and lymphocytes are sparse.

At birth, breakdown of the rich collagen framework of the gestational uterus commences as collagenase activity rises (Figure 9) (Eisen *et al.*, 1970). This extracellular event is accompanied by the continued presence of tissue heterophils, monocytes, and eosinophils and also by monocytic–macrophagic conversion (Figure 10) (Padykula and Campbell, 1976). At 48 hr postpartum, histologic disorganization of the collagen bundles is evident, lymphocytes and macrophages are now numerous, and some plasma cells are present. Also monocytic–macrophagic conversion is evident in the stromas as well as in intraepithelial loci (Figure 11A). By 72 hr, macrophages and lymphocytes are abundant and widely distributed, plasma cells are conspicuous (Figure 12), but the heterophils and eosinophils are sparse. Thus the principal cell types associated with inflammation and/or humoral immune response are present transiently during a peripartum period spanning at least 10 days.

Interpretation of this shifting stromal differentiation is difficult because of its cellular complexity and the rapidity of attendant hormonal changes. Moreover, the ECM is undergoing rapid remodeling as uterine collagenase activity rises quickly after birth to a peak at 24 hr and declines rapidly thereafter (Woessner, 1976). The unusually rapid degradation of the triple helix of collagen molecules into two polypeptide fragments that may subsequently denature (Figures 13A and B) would create a powerful transient local antigenic stimulus to the maternal organism. Isolated α-chains as well as smaller peptides derived from the collagen molecule are known to be chemotactic for human monocytes *in vitro* (Postlethwaite and Kang, 1976). Such endogenous antigenic stimulation by fragments of collagen molecules would be local, strong, and temporary, and might lead to the cellular manifestations of transient inflammation and/or humoral immune response (Padykula, 1976). This interpretation receives further support from recent evidence indicating a widespread occurrence of immunoglobulin-secreting cells that are producing autoantibodies in

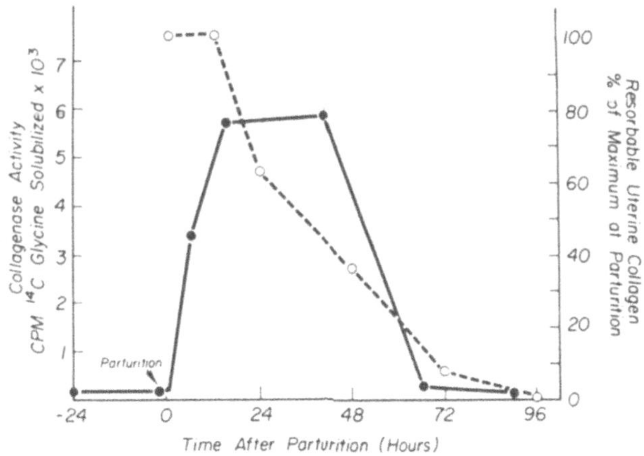

Figure 9. Collagenase activity (solid line) and collagen degradation (dashed line) in the rat uterus during the first 96 hr postpartum. Activation of collagenase occurs at parturition and gives rise to rapid collagenolysis during the first 72 hr postpartum. From Eisen *et al.* (1970); reprinted by kind permission of the Williams & Wilkins Company.

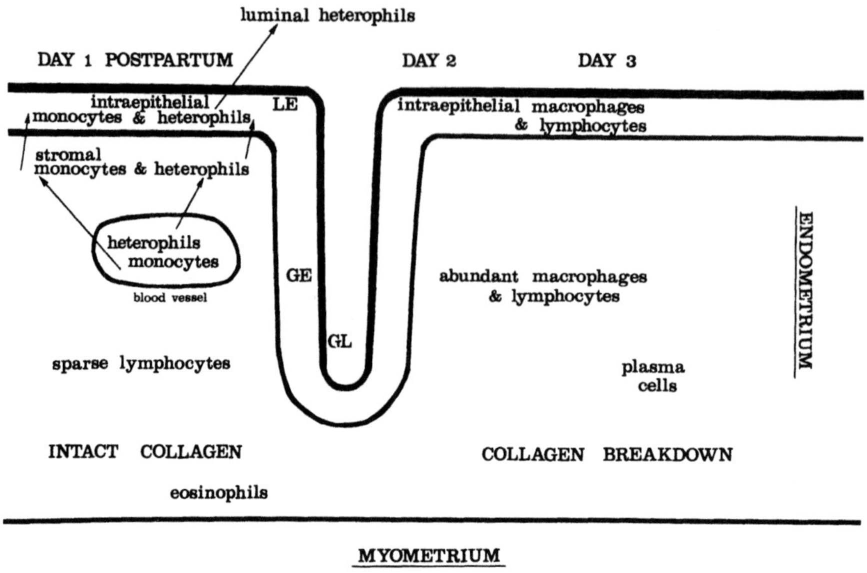

Figure 10. Diagrammatic interpretation of the program of leukocytic migrations and differentiations in the normal early postpartum rat uterus. Day 1 postpartum is marked primarily by the presence of tissue granulocytes and monocytes in the uterine stroma. This emigration from blood to tissue commenced during the last week of pregnancy (see Figure 1). The eosinophils tend to localize more in the deep endometrial stroma while the heterophils prevail in the superficial endometrium. Monocytes and heterophils occur in intraepithelial positions as well as in the stroma, whereas eosinophils are confined to the stroma. By Days 2 and 3 postpartum, monocytic–macrophagic conversion is mostly complete, as macrophages become abundant in the stroma and occupy intraepithelial sites as well. The population of lymphocytes increase; plasma cells appear. During this 72-hr period, the rich collagenous framework of the uterus is being destroyed by the events that accompany the sudden activation of collagenase at birth (see Figure 9).

normal animals (Steele and Cunningham, 1978). According to this hypothesis, the cellular and tissue manifestations of inflammation and local humoral response might represent normal regulatory mechanisms involved in the cyclic renewal of the extracellular compartment of the stroma. Moreover, these regulatory mechanisms are most likely influenced by plasma and tissue levels of estrogen and progesterone.

IV. Experimental Analysis of Cellular Aspects of Peripartum Uterine Differentiation

To test the hypothesis that the inflammatory and local humoral immunologic responses of the postpartum uterus might be related to rapid collagen degradation, experiments were performed (Tansey and Padykula, 1978) that were based on biochemical evidence indicating that progesterone strongly inhibits uterine collagenase activity *in vitro* and *in vivo* (Jeffrey *et al.*, 1971; Jeffrey and Koob, 1973; Koob and Jeffrey, 1974; Halme and Woessner, 1975). In the first experiment, the condition of prolonged gestation was produced (progesterone, 10 mg/day starting on Day 19 gestation) to prevent the prenatal drop in maternal plasma progesterone. This treatment preserved the prepartum state of cellular and tissue differentiation, prevented conversion of the tissue monocytes into macrophages, and blocked immigration of eosinophils and lymphocytes into the uterine stroma through Day 26 of prolonged gestation (parturition occurs normally on Day 22).

Progesterone administered *in vivo* after birth results in only partial inhibition of collagenolytic activity (Jeffrey and Koob, 1973; Halme and Woessner, 1975). Such treatment delayed but did not block stromal differentiation during the first 48 hr postpartum. For example, on Day 2 postpartum, macrophages are widely distributed in the normal endometrium (Figure 11A), whereas in the Day 2 progesterone-treated uterus, macrophages occur primarily in the superficial stroma and are sparse in the deep stroma (Figure 11B). In addition, much collagen retention is evident in the deep stroma after progesterone treatment and the fibroblasts are filled with numerous collagen-containing vacuoles (Figures 4 and 5). Eosinophils and heterophils are present in diminished numbers. Plasma cells were not observed in the progesterone-treated uteri in either experiment, and the population of lymphocytes did not change noticeably from that of control uteri.

Overall, these experiments demonstrate that administration of exogenous progesterone during the peripartum period delays the temporal pattern of leukocytic emigration into the uterus concomitant with the delay in collagen resorption. The low progesterone levels at birth appear to be essential for normal regression.

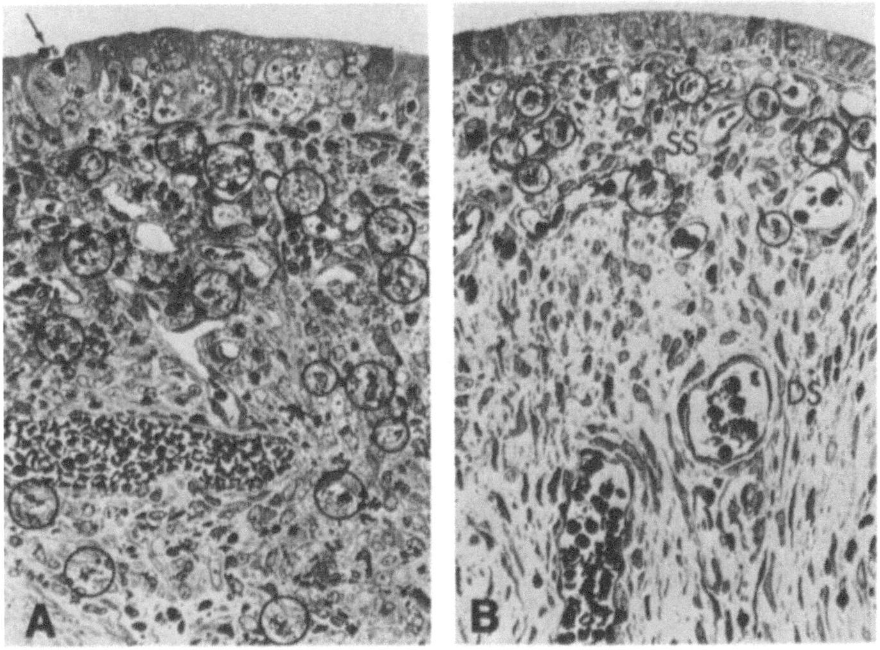

Figure 11. Postpartum rat endometrium, toluidine blue, 1-μm plastic sections. (A) Two days postpartum, normal control rat. The lumenal epithelium (E) is infiltrated with heterophils and macrophages. Arrow indicates a heterophil entering the uterine lumen. Stromal macrophages (encircled) are prevalent and tend to be more numerous in the superficial stroma than in the deep endometrium. Note also the high cellularity of the stroma and the relatively small size of the extracellular compartment compared to that in (B). (B) Two days postpartum, rat administered 40 mg progesterone per day starting at parturition. The lumenal epithelium (E) is shorter and less infiltrated than that of the control (A). Also, macrophages (encircled) are less numerous and occur primarily in the superficial stroma (SS). The deep stroma (DS) consists primarily of fibroblasts, and infiltrating cells are much less frequent than in the control. The larger extracellular compartment of the deep stroma reflects collagen retention. The ultrastructure of these deep stromal fibroblasts is illustrated in Figures 4 and 5. ×300. From Tansey and Padykula (1978).

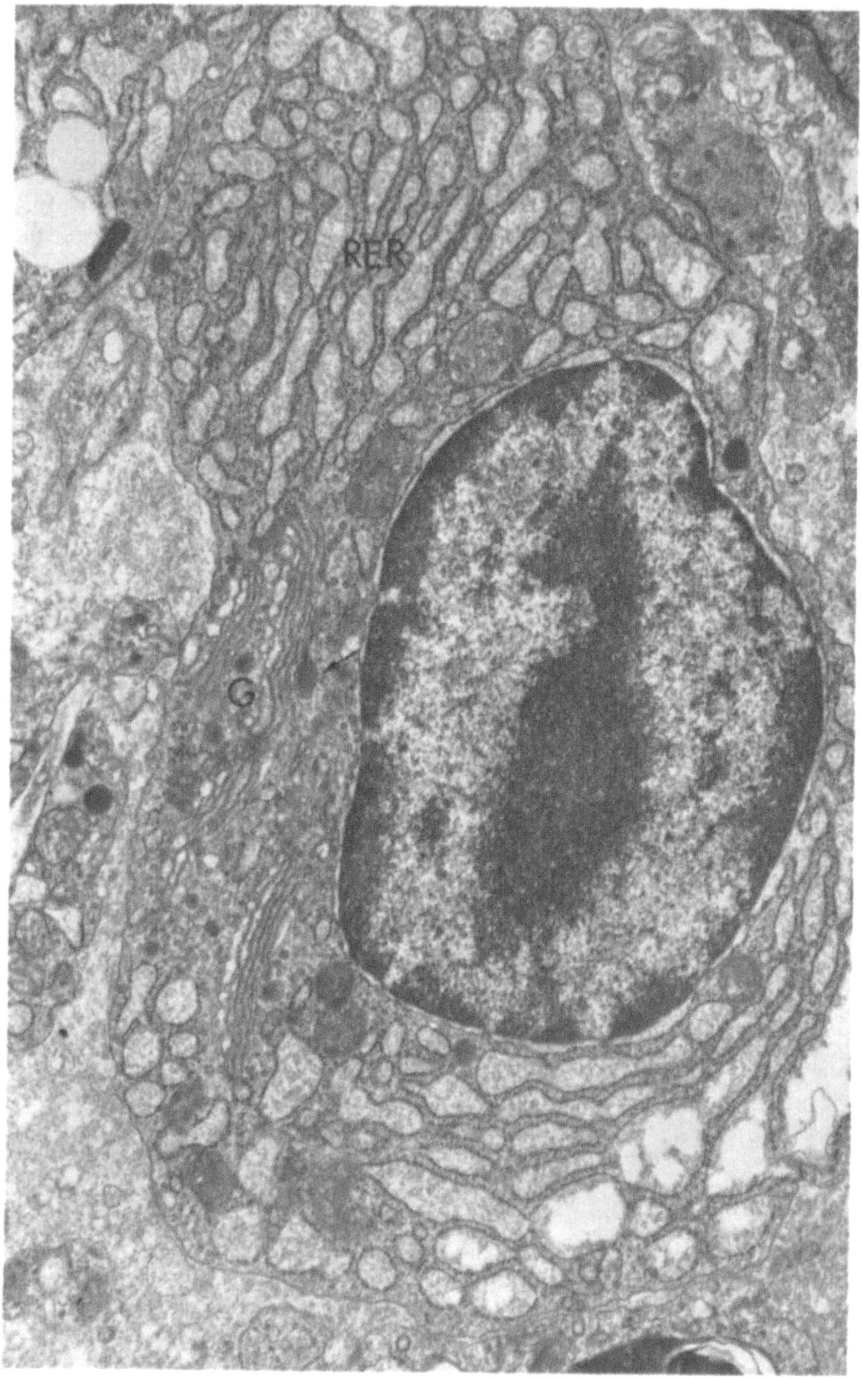

Figure 12. Plasma cell in the rat postpartum uterus (3 days 8 hr). The ultrastructure of this uterine plasma cell indicates active involvement in protein synthesis for export with its elaborately expanded cisternal rough endoplasmic reticulum (RER) and abundant Golgi membranes (G). Arrow indicates site of packaging of secretion at an expanded tip of a Golgi saccule. ×20,600.

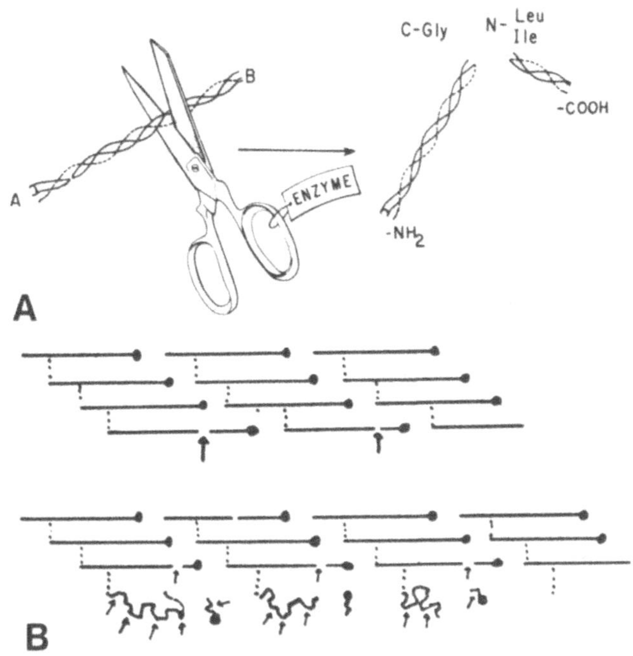

Figure 13. Mechanism of degradation on the intact collagen fibril. (A) Diagrammatic representation of the cleavage of the triple helix of the collagen molecule by collagenase into two unequal fragments. (B) Hypothetical scheme to explain the dismantling of an intact collagen fibril, which is a polymer of rod like asymmetric collagen molecules linearly arranged with a one-quarter stagger. The site of cleavage by collagenase is indicated by arrows. The two products of enzymatic action denature progressively and lose intermolecular associations. From Gross (1974); reprinted by kind permission of Academic Press.

V. Cellular Evidence Related to the Postulated Immunosuppressant Action of Progesterone

During the last nine days of pregnancy in the rat, progesterone dominance is diminished steadily and brought to a close at term. The steady decrease in the concentration of nuclear progesterone–receptor complex in uterine cells (Figure 8) (Vu Hai *et al.*, 1978) correlates temporarily with the progressive invasion of leukocytes into the uterine stroma. These observations on normal and progesterone-treated peripartum uteri conform to the hypothesis put forth by Siiteri *et al.* (1977) that progesterone in high local concentrations exerts an anti-inflammatory or immunosuppressive effect. However, the stimulus for the local uterine inflammatory response remains unidentified. In the previous section, one possible explanation was offered that relates to cyclic endogenous antigenic stimulation arising from rapid degradation of extracellular macromolecules. Another possibility arises from the recent demonstration that the uterus is capable of secreting immunoglobulins and that this process is regulated by estrogen and progesterone (Wira and Sandoe, 1977). Their evidence indicates that estrogen stimulates whereas progesterone suppresses the secretion of immunoglobulins. Experimental evidence for the migration of B immunoblasts into the mouse uterus has recently been obtained by McDermott and Bienenstock (1979).

Recent immunocytochemical analyses of mouse implantation sites indicate widespread distribution of maternal immunoglobulins within the ECM of the uterus, along the surface of the trophoblast, and within the embryo (Bernard *et al.*, 1977). This phenomenon was tentatively interpreted as reflecting the local synthesis of "enhancing" antibodies, i.e., immunoglobulins with immunosuppressive properties. Alternatively, it was proposed that some of the maternal immunoglobulins may represent nutritive protein to the embryo.

A recent investigation from our laboratory has demonstrated a surprising association of lymphoid cells with decidualization in the normal early human gestational uterus (6–11 weeks) (Padykula *et al.*, 1978). Decidualization (parietalis region) was accompanied by a lymphoid infiltration composed of lymphocytes, monocytes, macrophages, and plasma cells. This period is characterized by a transient drop in serum progesterone (Yoshimi *et al.*, 1969; Mishell *et al.*, 1973) during the switch over ovarian to placental production of progesterone. Finally, it has long been recognized that premenstrual changes in the human and rhesus monkey uterus include leukocytic infiltration. It seems likely that menstrual sloughing may represent a highly "evolved" system of regression that includes inflammation and an immunologic component. The cumulative evidence at this time suggests that cellular mechanisms long known to participate in pathologic conditions may also participate in the normal cyclic remodeling of the uterus.

ACKNOWLEDGMENT. This work was supported by Research Grant HD09209 from the U.S. Public Health Service.

Discussion

MOULTON (Cincinnati): Did you see any evidence of autophagic activity involving the lysosomal system in the various tissue regressions which you studied?

PADYKULA: I saw autophagy within the lumenal epithelial cells, but macrophages were also present in an intraepithelial position.

LEROY (Brussels): How much bacterial invasion can one see in the postpartum period in the intrauterine cavity?

PADYKULA: I have not seen evidence of bacterial invasion. The invasion of these leukocytic cells into the postpartum uterus is hormonally controlled. I think leukocytic invasion is also going on in the estrous cycle on a smaller scale; that has not been mapped out as thoroughly as the postpartum situation. This event may be related to the arrival of sperm during estrus. The uterus has a mucosal local immune response, as do the respiratory and digestive systems.

LEROY: Part of the response, such as lymphocyte and macrophage mobilization, could perhaps be related to incipient infection because, certainly in the human, swabs from a uterine cavity would give positive cultures even at early postpartum.

PADYKULA: One possible answer to that point is that we first observed it in the opossum, where there is no destruction of the endometrial lining, and it was even more dramatic in the opossum.

LEROY: But there is vast opening of the reproductive tract which favors the ascension of bacteria.

PADYKULA: So you are proposing that it's entirely a local exogenous stimulus?

LEROY: No, I just wonder how much of the reaction you see could be due to this aspect. I am by no means implying that the whole picture would relate to that.

PADYKULA: That might be one biological function for the whole differentiation process, to prepare for the possibility of infection. The entry of monocytes and heterophils into epithelial and stromal compartments before birth, however, is a strong factor indicating that this leukocytic invasion is evidently preparatory to a postpartum event.

MARTIN (ICRF, London): To return to the question Dr. Moulton asked about autophagosomes, we do not see invasion by macrophages in the mouse uterus during hyperplasia and regression after treatment with estrogen alone. The relatively small amount of cell death seems to be, as Colin Finn said, by karyorrhexis and then takeup by adjoining epithelial cells. After an estrous cycle, we see much more cell death, with cells sloughing off into the lumen and macrophages in the epithelium. Do you think that there might be different modes of cell death, depending on the amount of cell death?

PADYKULA: Do you mean there are different modes of getting rid of them?

MARTIN: Yes. Are macrophages killing the cells or are the cells dying and them simply being phagocytosed by macrophages?

PADYKULA: The complication in the postpartum uterus that limits interpretation is the preparation for estrus which occurs at the same time as catabolic changes. There is an increase in mitotic rate in the lumenal epithelium. So the whole scene is a mixture of an anabolic event related to the postpartum estrus and the catabolic removal of the cells of pregnancy. Vacuolation of the lumenal epithelium occurs, as you have seen also, which reflects autolytic and/or phagocytic activity. I think the macrophages then remove those "spent" cells while the anabolism for the next cycle is starting up.

MARTIN: As regards the effects of hormones on some of these wandering cell populations, there is work by Tchernitchin *et al.* [*Nature* (*London*) **248:**142, 1974] on eosinophils in the rat uterus, showing rapid invasion by eosinophils after estrogen treatment.

PADYKULA: Ross and Klebanoff (*J. Exp. Med.* **124:**653, 1966) before that, in a beautiful paper, demonstrated that estradiol "drives" blood eosinophils into the rat uterine stroma.

MARTIN: One of the early events in the vaginal cornification response to estrogen is the disappearance of polymorphs. In the uteri of mice bearing IUDs, there are many leukocytes. When we treat those animals with estrogen or with progesterone, leukocytes disappear. These are further examples of hormones affecting the wandering cells.

PADYKULA: The evidence that eosinophils come into the uterine stroma with rising estrogen is strong. Yanagimachi and Chang (*J. Reprod. Fertil.* **5:**389, 1963) studied the hamster uterus and showed that leukocytes migrated into the lumen at estrus, whether or not the animal was mated. When mated, leukocytic emigration was at a higher level, but these authors never identified the cell types; they're probably the heterophils. It seemed to me that these granulocytes emigrate from the blood to the uterine stroma during a regular cycle and then, if mating occurs, the extent of migration is greater. Dr. Chang is agreeing with my interpretation. May I say though that the monocytes may emigrate because of chemotaxis, which might be related to the breakdown of extracellular matrix; they may be attracted into the tissue in a more traditional way than by hormonal stimulation.

ORSINI (Wisconsin): That was a beautiful paper but, may I ask, did you look at the mesometrium as well?

PADYKULA: No, but I've looked into the myometrium. This infiltration occurs through the connective tissue of the whole uterus; there are macrophages and monocytes all through the connective tissue. Harkness and Harkness (*J. Physiol.* **132:**492, 1956) indicated that a large percentage of the collagen is in the myometrium, so its degradation is going on there as well as in the endometrium. These leukocytes occur throughout the serosal connective tissue as well as the connective tissue in the myometrium and in the endometrium.

ORSINI: In the postpartum rats did you notice any infusion of fluid within the mesometrium? In the hamster at the time of parturition, the mesometrium is thick, edematous, gelatinous, and turgid, with three vascular channels, lymphatics, arterial, and venous (Orsini, 1957, *J. Morphol.* **100:**565); it certainly supports your view that there is thick involvement there.

HEAP (Cambridge): Perhaps this might be an opportunity to turn the discussion towards ideas about progesterone as one of Nature's immunosuppressive agents. There has been a number of reservations about this hypothesis, particularly because of the high concentrations of progesterone required to demonstrate an effect (Siiteri *et al.*, 1977, *Ann. N.Y. Acad. Sci.* **286:**384). The doubt that many people have had is whether this can be considered in any way a general mechanism of action of progesterone, in view of those species in which placental progesterone production is very low (e.g., goat, pig). However, a recent paper by Neifeldt and Tormey (*Transplantation* **27:**309, 1979) shows that progesterone suppression of PHA–lymphocyte stimulation is concentration dependent. The concentration of progesterone needed for a 50% reduction in stimulation is almost exactly the same as the concentration of progesterone found, for example, in early trophoblast at the time of attachment in the pig. Obviously this gives greater credence to the hypothesis. Could you enlarge on the hypothesis that products from collagen may be functioning as an antigenic stimulus?

PADYKULA: The collagen molecule is cleaved into two parts by a neutral collagenase. There would be quick release of new antigenic sites by this proteolytic cleavage as well as by the opening of the triple helix. Then there is a denaturation of the two collagen fragments, which could release more antigenic sites. Such rapid destruction of collagen does not happen very often in a normal body, except in the uterus; it happens frequently in disease. I thought that these changes in the collagen molecule then could be an endogenous local antigenic stimulus that was strong and transient. The apparent uterine inflammation and/or hormonal immune response might be, in part, a response to this massive, rapid extracellular proteolysis. I made this interpretation before Siiteri advanced his idea of progesterone being an immunosuppressant, and his hypothesis fits well with this cellular evidence. There is recent evidence indicating that normal sera in mice have a high percentage of autoantibodies. So a mechanism which is usually associated with pathologic states, i.e., inflammation/hormonal immune response, might be used in this normal situation for cyclic removal of extracellular matrix and epithelial cells.

SURANI (Cambridge): I think that the immunosuppressive role of progesterone is very interesting. There is a need, however, to work out a mechanism, because the experiments which Brian Heap quoted on PHA and lymphocytes are on a known target cell and we are dealing with a target organ.

References

Bernard, O., Ripoche, M. A., and Bennett, D., 1977, Distribution of maternal immunoglobulins in the mouse uterus and embryo in the days after implantation, *J. Exp. Med.* **145:**58.

Bienkowski, R. S., Baum, B. J., and Crystal, R. G., 1978, Fibroblasts degrade newly synthesised collagen within the cell before secretion, *Nature (London)* **276:**413.

Birkedal-Hansen, H., Cobb, C. M., Taylor, R. E., and Fullmer, H. M., 1976, Synthesis and release of procollagenase by culture fibroblasts, *J. Biol. Chem.* **251:**3162.

Brandes, D., and Anton, E., 1969, Lysosomes in uterine involution: Intracytoplasmic degradation of myofilaments and collagen, *J. Gerontol.* **24:**55.

Coffey, J. W., Fiedler-Nagy, C., Georgiadis, A. G., and Salvador, R. A., 1976, Digestion of native collagen, denatured collagen, and collagen fragments by extracts of rat liver lysosomes, *J. Biol. Chem.* **251:**5280.

Deporter, D. A., and Ten Cate, A. R., 1973, Fine structural localization of acid and alkaline phosphatase in collagen-containing vesicles of fibroblasts, *J. Anat.* **114:**457.

Dyer, R. F., and Peppler, R. D., 1977, Intracellular collagen in the nonpregnant and IUD-containing rat uterus, *Anat. Rec.* **187:**241.

Eisen, A. Z., Bauer, E. A., and Jeffrey, J. J., 1970, Animal and human collagenases, *J. Invest. Dermatol.* **55:**359.

Gross, J., 1974, Collagen biology: Structure, degradation, and disease, in: *The Harvey Lectures, 1972–1973,* pp. 351–432, Academic Press, New York.

Halme, J., and Woessner, J. F., Jr., 1975, Effect of progesterone on collagen breakdown and tissue collagenolytic activity in the involuting rat uterus, *J. Endocrinol.* **66:**357.

Harkness, M. L. R., and Harkness, R. D., 1956, The distribution of the growth of collagen in the uterus of the pregnant rat, *J. Physiol.* **132:**492.

Harkness, R. D., and Moralee, B. E., 1956, The time-course and route of loss of collagen from the rat's uterus during postpartum involution, *J. Physiol.* **132:**502.

Jeffrey, J. J., and Koob, T. J., 1973, Hormonal regulation of collagen catabolism in the uterus: Endocrinology, *Excerpta Med. Int. Congr. Ser.* **273:**1115.

Jeffrey, J. J., Coffey, R. J., and Eisen, A. Z., 1971, Studies on uterine collagenase in tissue culture. II. Effect of steroid hormones on enzyme production, *Biochim. Biophys. Acta* **252:**143.

Koob, T. J., and Jeffrey, J. J., 1974, Hormonal regulation of collagen degradation in the uterus: Inhibition of collagenase expression by progesterone and cyclic AMP, *Biochim. Biophys. Acta* **354:**61.

Levine, A. M., Reddick, R., and Triche, T., 1978, Intracellular collagen fibrils in human sarcomas, *Lab. Invest.* **39:**531.

Lobel, B. L., and Deane, H. W., 1962, Enzymic activity associated with postpartum involution of the uterus and with its regression after hormone withdrawal in the rat, *Endocrinology* **70:**567.

McDermott, M. R., and Bienenstock, J., 1979, Evidence for a common mucosal immunologic system. I. Migration of B immunoblasts into intestinal, respiratory, and genital tissues, *J. Immunol.* **122**(5)**:**1892.

Mishell, D. R., Jr., Thorneycroft, I. H., Nagata, Y., Murata, T., and Nakamura, R., 1973, Serum gonadotropin and steroid patterns in early human gestation, *Am. J. Obstet.* **117**(5):631.

Montfort, I., and Perez-Tamayo, R., 1975, The distribution of collagenase in the rat uterus during postpartum involution. An immunohistochemical study, *Connect. Tissue Res.* **3**:245.

Padykula, H. A., 1976, Cellular mechanisms involved in cyclic stromal renewal of the uterus. III. Cells of the immune response, *Anat. Rec.* **184**:49.

Padykula, H. A., and Campbell, A., 1976, Cellular mechanisms involved in cyclic stromal renewal of the uterus. II. The albino rat, *Anat. Rec.* **184**:27.

Padykula, H. A., and Tansey, T. R., 1979, The occurrence of uterine stromal and intraepithelial monocytes and heterophils during normal late pregnancy in the rat, *Anat. Rec.* **193**:329.

Padykula, H. A., and Taylor, J. M., 1976, Cellular mechanisms involved in cyclic stromal renewal of the uterus. I, The opossum, *Didelphis virginiana, Anat. Rec.* **184**:5.

Padykula, H. A., Driscoll, S. G., and Cardasis, C. A., 1978, Decidual cell differentiation in the normal early gestational human uterus includes lymphoid infiltration, Anat. Rec. **190**:500.

Parakkal, P. F., 1969, Involvement of macrophages in collagen resorption, *J. Cell Biol.* **41**:345.

Parakkal, P. F., 1972, Macrophages: The time course and sequence of their distribution in the postpartum uterus, *J. Ultrastruct. Res.* **40**:284.

Perez-Tamayo, R., 1970, Collagen resorption in carrageenin granulomas. II. Ultrastructure of collagen resorption, *Lab. Invest.* **22**:142.

Postlethwaite, A. E., and Kang, A. H., 1976, Collagen and collagen peptide-induced chemotaxis of human blood monocytes, *J. Exp. Med.* **143**:1299.

Renteria, V. G., and Ferrans, V. J., 1976, Intracellular collagen fibrils in cardiac valves of patients with the Hurler syndrome, *Lab. Invest.* **34**:263.

Ross, R., and Klebanoff, S. J., 1966, The eosinophilic leukocyte. Fine structure studies of changes in the uterus during the estrous cycle, *J. Exp. Med.* **124**:653.

Sandow, B. A., West, N. B., Norman, R. L., and Brenner, R. M., 1979, Hormonal control of apoptosis in hamster uterine luminal epithelium, *Am. J. Anat.* **156**:15.

Schwarz, W., and Güldner, R. H., 1967, Elektronenmikroskopische Untersuchungen der Kollagenabbaus im Uterus der Ratte nach der Schwangerschaft, *Z. Zellforsch. Mikrosk. Anat.* **83**:416.

Siiteri, P. K., Febres, F., Clemens, L. E., Chang, R. J., Gondos, B., and Stites, D., 1977, Progesterone and maintenance of pregnancy: Is progesterone nature's immunosuppressant?, *Ann. N.Y. Acad. Sci.* **286**:384.

Smith, R. E., and Farquhar, M. G., 1966, Lysosome function in the regulation of the secretory process in cells of the anterior pituitary gland, *J. Cell Biol.* **31**:319.

Steele, E. J., and Cunningham, A. J., 1978, High proportion of Ig-producing cells making autoantibody in normal mice, *Nature (London)* **274**:483.

Svoboda, E. L. A., Brunette, D. M., and Melcher, A. H., 1979, *In vitro* phagocytosis of exogenous collagen by fibroblasts from the periodontal ligament: An electron microscopic study, *J. Anat.* **128**(2):301.

Tansey, T. R., and Padykula, H. A., 1978, Cellular responses to experimental inhibition of collagen degradation in the postpartum rat uterus, *Anat. Rec.* **191**:287.

Ten Cate, A. R., and Freeman, E., 1974, Collagen remodelling by fibroblasts in wound repair. Preliminary observations, *Anat. Rec.* **179**:543.

Ten Cate, A. R., and Syrbu, S., 1974, A relationship between alkaline phosphatase activity and the phagocytosis and degradation of collagen by the fibroblast, *J. Anat.* **117**:351.

Ten Cate, A. R., Departer, D. A., and Freeman, E., 1976, The role of fibroblasts in the remodeling of periodontal ligament during physiologic tooth movement, *Am. J. Orthod.* p. 155.

Unkeless, J. C., Gordon, S., and Reich, E., 1974, Secretion of plasminogen activator by stimulated macrophages, *J. Exp. Med.* **139**:834.

Vassalli, J. D., and Reich, E., 1977, Macrophage plasminogen activator: Induction of products of activated lymphoid cells, *J. Exp. Med.* **145**:429.

Vu Hai, M. T., Logeat, F., and Milgrom, E., 1978, Progesterone receptors in the rat uterus: Variations in cytosol and nuclei during the oestrous cycle and pregnancy, *J. Endocrinol.* **76**:43.

Werb, Z., and Burleigh, M. C., 1974, A specific collagenase from rabbit fibroblasts in monolayer culture, *Biochem. J.* **137**:373.

Werb, Z., and Gordon, S., 1975a, Secretion of a specific collagenase by stimulated macrophages, *J. Exp. Med.* **142**:346.

Werb, Z., and Gordon, S., 1975b, Elastase secretion by stimulated macrophages, *J. Exp. Med.* **142**:361.

Werb, Z., Mainardi, C. L., Vater, C. A., and Harris, E. D., Jr., 1977, Endogenous activation of latent collagenase by rheumatoid synovial cells. Evidence for a role of plasminogen activator, *N. Engl. J. Med.* **296**:1017.

Werb, Z., Dingle, J. T., Reynolds, J. J., and Barrett, A. J., 1978, Proteoglycan-degrading enzymes of rabbit fibroblasts and granulocytes, *Biochem. J.* **173**:949.

Wiest, W. B., 1970, Progesterone and 20α-hydroxpyregen-4-en-3-one in plasma, ovaries and uteri during pregnancy in the rat, *Endocrinology* **87**:43.

Wira, C. P., and Sandoe, C. P., 1977, Sex steroid hormone regulation of IgA and IgG in rat uterine secretions, *Nature (London)* **268**:534.

Woessner, J. F., Jr., 1976, Proteolytic control of collagen breakdown, in: *Proteolysis nand Physiological Regulation,* Miami Winter Symposium (D. W. Ribbon and K. Brew, eds.), pp. 357–369, Academic Press, New York.

Yoshimi, T., Strott, C. A., Marshall, J. R., and Lipsett, M. B., 1969, Corpus luteum function in early pregnancy, *J. Clin. Endocrinol. Metab.* **29**:225.

Yoshinaga, K., 1976, Ovarian hormone secretion and ovum implantation, in: *Implantation of the Ovum* (K. Yoshinaga, R. K. Meyer, and R. O. Greep, eds.), pp. 3–17, Harvard Univ. Press, Cambridge, Mass.

13

Separated Cell Types as Analytical Tools in the Study of Decidualization and Implantation

Stanley R. Glasser and Shirley A. McCormack

I. Historical Perspective

Various strategies used by different species to assure successful blastocyst–endometrial interaction have evolved. The advantages and limitations of the decidual cell reaction (DCR) as a model for studying these events have been recognized (Psychoyos, 1973; Glasser and Clark, 1975; Glasser and McCormack, 1980a,c). Concise descriptions of the sequential relationships of the steroid hormones (Psychoyos, 1973; Glasser and Clark, 1975) and the different hormonal responses of epithelial and stromal cells have been provided (Martin and Finn, 1968; Tachi *et al.*, 1972; Martin *et al.*, 1973a,b). These researches have validated the determinant role of progesterone in the development of uterine sensitivity. Certain species require estrogen to complete the maturation of the sensitive uterus to the final stages of uterine receptivity for ovum implantation. Detailed cytological, ultrastructural (Nilsson, 1970; Enders and Schlafke, 1971; Schlafke and Enders, 1975; Sherman and Wudl, 1976), and physiological correlates (Meyers, 1970; Psychoyos, 1973; Glasser and Clark, 1975; Glasser and McCormack, 1979) of these events are also part of the literature.

In the intact cycling rat, implantation is initiated only after the progesterone-induced sensitive uterus becomes receptive to the blastocyst. This step depends on estrogen and is linked closely to the blastocyst-stimulated transformation of uterine stromal cells to decidual cells. The information required to direct these differentiative processes is short-lived and is believed to be a product of estrogen-induced gene expression (Glasser and Clark, 1975). In a recent study of the biochemical mechanisms by which this hormone can initiate implantation and decidualization (Glasser and McCormack, 1979), data were presented to suggest that progesterone acts at the level of transcription to sensitize the uterus. Data showing *restriction* of gene expression produced by the intervention of estrogen (Fig-

Stanley R. Glasser and Shirley A. McCormack • Department of Cell Biology, Baylor College of Medicine, Houston, Texas 77030.

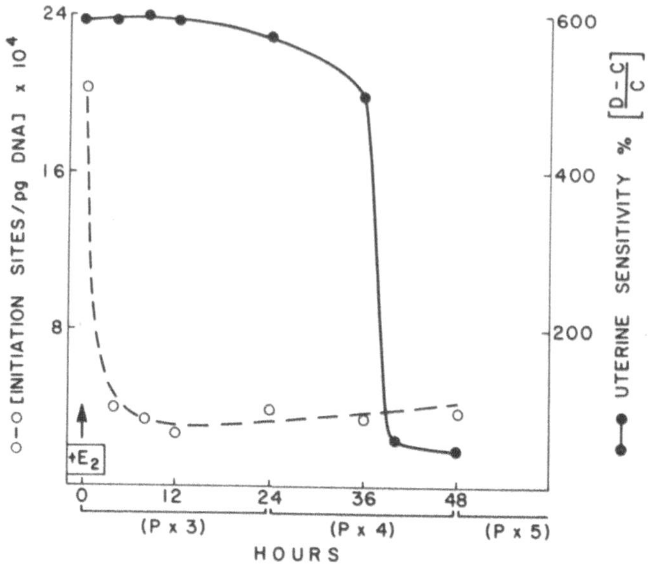

Figure 1. Changes in the index of transcriptive activity of uterine chromatin prepared from progesterone (P)-treated ovariectomized rats compared to the time course describing the estradiol-17β (E_2)-induced loss of uterine sensitivity to deciduogenic stimuli. A minimum of three daily injections (P × 3) of P (2 mg) were required to sensitize the uterus to decidualization and to increase the number of RNA polymerase binding sites available for the initiation of RNA synthesis from the castrate level of 4000/pg uterine chromatin DNA to 20,000/pg DNA. Intervention by a single dose of E_2 (0.2 μg) resulted in a rapid decrease in the number of RNA initiation sites. Significant loss of uterine sensitivity was not noted until 36 hr after E_2. Continued injection of P (P × 4, P × 5) did not reverse these responses. Uterine sensitivity (%) is obtained from the weight (mg) of the decidualized horn (D) less the weight of the nonstimulated horn (c) divided by (C) × 100 measured 96 hr after uterine stimulation (Glasser and McCormack, 1979).

ure 1) were a unique finding of this study. The estrogen that initiates implantation was thus thought to restrict gene expression both qualitatively and quantitatively, in order to regulate the final stages of uterine differentiation for ovum implantation.

Because it is based on analyses of the whole uterus, the usefulness of this hypothesis is defining the events of the immediate preimplantation period is limited. The experiments did not identify the specific target cell of estrogen modulation. A temporal shift in target organ specificity from epithelium to stroma, related to the interactions between progesterone and estrogen, is peculiar to the uterus (Tachi *et al.*, 1972); Martin *et al.*, 1973a,b). There is a consensus that the epithelial cell plays an obligatory role in the initial steps of implantation and that the nature of subsequent information, e.g., the program for stromal cell differentiation, depends on specific influences from the epithelium (Finn, 1974). The uterine surface of the mouse and rat is still intact at the time implantation is being initiated (Sherman and Wudl, 1976). Thus the stimulatory influence of the blastocyst, particularly with respect to stroma, should be exerted via the epithelial cells.

It remains to be proven if the role of the epithelial cell in triggering these events is that of a transducer (Leroy and Lejeune, this volume) or, in terms of the hypothesis presented, the synthetic site of a compulsory intermediary (Psychoyos, 1974). Methods to separate the individual cell types of the uterus would avoid the constraints placed on interpretation due to complexity of the *in vivo* condition, which does not distinguish between the cell types and their individual hormonal sensitivities.

II. Cell Separation Techniques

Attempts to identify and characterize the different responses of individual uterine cell types have included both *in situ* and physical methods of separating the component tissues. *In situ* methods have been histochemical (Peel and Bulmer, 1975; Leroy *et al.*, 1976; West *et al.*, 1978) or autoradiographic (Galand *et al.*, 1971; Martin *et al.*, 1973a,b). These procedures have yielded useful information about the order in which the individual cell types respond to estrogenic stimulation, especially with regard to cell division. However, these methods are severely limited because they can only provide terminal, "fixed-frame" data and they restrict the opportunities for biochemical investigation or control of a variety of *in vivo* influences.

Physical methods of separation employ mechanical or enzymatic means. Mechanical methods have included squeezing (Heald *et al.*, 1975), scraping (Alberga and Baulieu, 1968; Martel and Psychoyos, 1978), shearing (Smith *et al.*, 1970; Pollard and Martin, 1975), dissection (Jackson and Chalkley, 1974; Makler and Eisenfeld, 1974; Koligan and Stormshak, 1977), and the use of surgical curetting (Tseng, 1978). Enzymatic methods have used trypsin (Vladimirsky *et al.*, 1977; Sananes *et al.*, 1978), collagenases (Pietras and Szego, 1975a,b; Kirk *et al.*, 1978), pronase (Williams and Gorski, 1973), and hyaluronidase (Gerschenson *et al.*, 1974; Gerschenson and Berliner, 1976; Liszczak *et al.*, 1977) singly or in a variety of combinations.

The utility of these methods is limited by the completeness of separation (purity, quantity) they produce and the condition of the separated cells. Only recently have methods been developed that separate individual epithelial, stromal, and myometrial cell populations. These methods should replace procedures that simply dissociate endometrium from myometrium. With the exception of two methods (Smith *et al.*, 1970; Heald *et al.*, 1975), most mechanical separations do not separate epithelium from stroma but remove them together as endometrium. Enzymatic methods have been modified so the separated cells are attainable in reasonably good yield and purity (Kirk *et al.*, 1978; Glasser and McCormack, 1980; McCormack and Glasser, 1980a). These techniques must be combined with convenient and reliable methods of evaluating the structural and functional integrity of the separated cells.

The viability of the separated cells has been assessed by their ability to incorporate labeled precursors into proteins (Smith *et al.*, 1970; Williams and Gorski, 1973; Pollard and Martin, 1975) and to exclude nigrosin dye (Williams and Gorski, 1973; Pietras and Szego, 1975a,b). The intracellular concentrations of Na^+, K^+, Cl^-, and water have also been monitored (Pietras and Szego, 1975a). Vladimirsky *et al.* (1977) and Sananes *et al.* (1978) found that their preparations decidualized during short-term culture, providing evidence of the integrity of the cells after separation with trypsin. Similar proof is found in the survival of primary cultures of normal human endometrial epithelial cells (10–15 days) and the longer but still finite life of stromal cells (Kirk *et al.*, 1978). Investigators have frequently failed to test the condition of their separated cells; freshness of the specimen was their only guarantee of its physiological integrity.

A. Method for Separating Individual Cell Types from the Rat Uterus

The procedure we are currently using is our modification of a mechanical method (L. Martin, personal communication) and an enzymatic method (M. I. Sherman, personal communication) generously shared with us by their developers. Our modification satisfies our criteria for purity, integrity, viability, and yield necessary for studies of steroid hormone receptor complexes and *in vitro* models of implantation.

The method consists of four main steps during which epithelial cells (Step A), stromal cells (Step B), and myometrial cells (Step C) are consecutively separated from whole uteri (Figure 2). Step D clarifies the individual fractions and outlines procedures for cell counting and testing for viability (Figures 3, 5, 7, 9).

1. Epithelial Cells

The epithelial cell fractions derived from Step A (Figure 3) appear as plaques rather than individual cells (Figure 4a). The preparations contain few, if any, cells of any other kind. Further evidence for the efficiency of separation is provided by: (1) light microscopic examination of slit whole uteri from which epithelium has been removed (Figure 4b); (2) the lack of fibroblastlike colonies in epithelial cell monolayers (Figure 4c); (3) significantly different uptake of [^{3}H]cytidine compared to stromal and myometrial cell preparations (McCormack and Glasser, 1980a); and (4) differences in the concentration and dynamics of the estrogen cytoplasmic (RcE) and nuclear (RnE) receptor complexes (Glasser and McCormack, 1980b; McCormack and Glasser, 1980a).

2. Stromal Cells

Stromal cell fractions (Step B; Figure 5) were estimated to contain approximately 10% epithelial cells. This estimate was based on the phase microscopic appearance of stromal

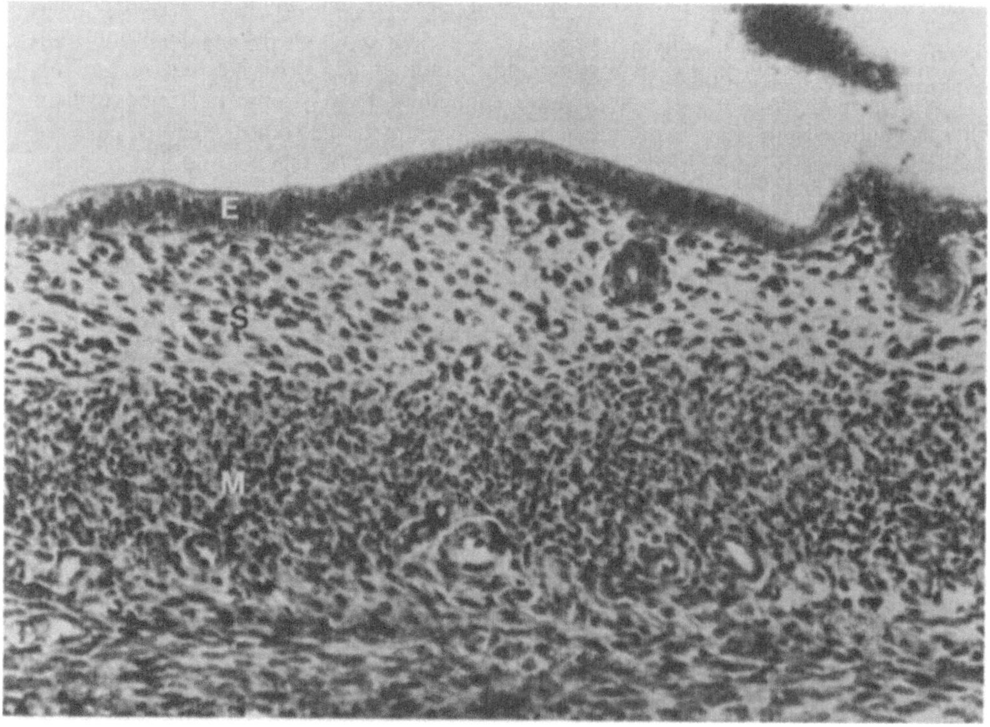

Figure 2. Routine histological cross section through the whole immature rat uterus 24 hr after a single s.c. injection of 2.5 μg E_2. The lumen is lined by a single layer of cuboidal epithelial cells (E). Gland cross sections may be present in the stroma (S). Myometrial (M) circular and longitudinal layers are present. (H and E) $\times$ 63.)

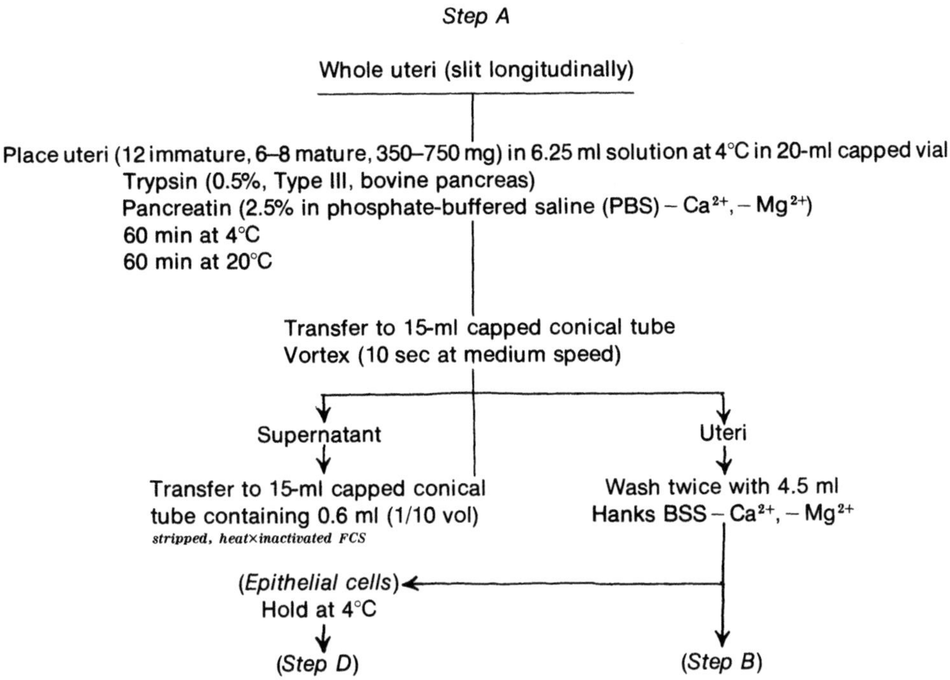

Figure 3. Procedures for the preparation of rat uteri and the isolation of epithelial cells.

cell preparations, which contain single cells (Figure 6a) and occasional tubular structures (probably glandular epithelium) but rarely any epithelial plaques, and on histological examination of the remaining whole uterine tissue (Figure 6b). When cultured, these suspensions of single cells produced fibroblastic colonies (Figure 6c). The occasional epithelial colony that appeared was probably derived from the presumed glandular epithelial structures. The efficiency of stromal cell separation was also validated by the concentration and dynamics of RcE and RnE (see Figure 10; Glasser and McCormack, 1980b; McCormack and Glasser, 1980a) and [^{3}H]cytidine uptake and incorporation (McCormack and Glasser, 1980a), which were different from those of epithelial cells.

3. *Myometrial Cells*

Myometrial cell fractions (Step C; Figure 7) did not contain epithelial cell plaques but are assumed to be contaminated with stromal cells (Figure 8A). Assessment of stromal cell contamination was difficult because the deep stroma is enmeshed in myometrium. Myometrial cells appear single and rounded like stromal cells; they also produce fibroblastic colonies in culture (Figure 8B). Histological examination of uterine tissue remaining after removal of epithelial and stromal cells did not show any stroma remaining (Figure 6b) nor did glandular epithelium appear in outgrowths of myometrial cell suspensions (Figure 8B). Furthermore, the uptake and incorporation of [^{3}H]cytidine by myometrial cell suspensions were different from stromal and epithelial cell fractions (McCormack and Glasser, 1980a).

4. *Viability and Yield*

Estimation of the viability of cells (Step D; Figure 9) as determined by vital dye exclusion proves to depend on the individual investigator. In the best hands we routinely

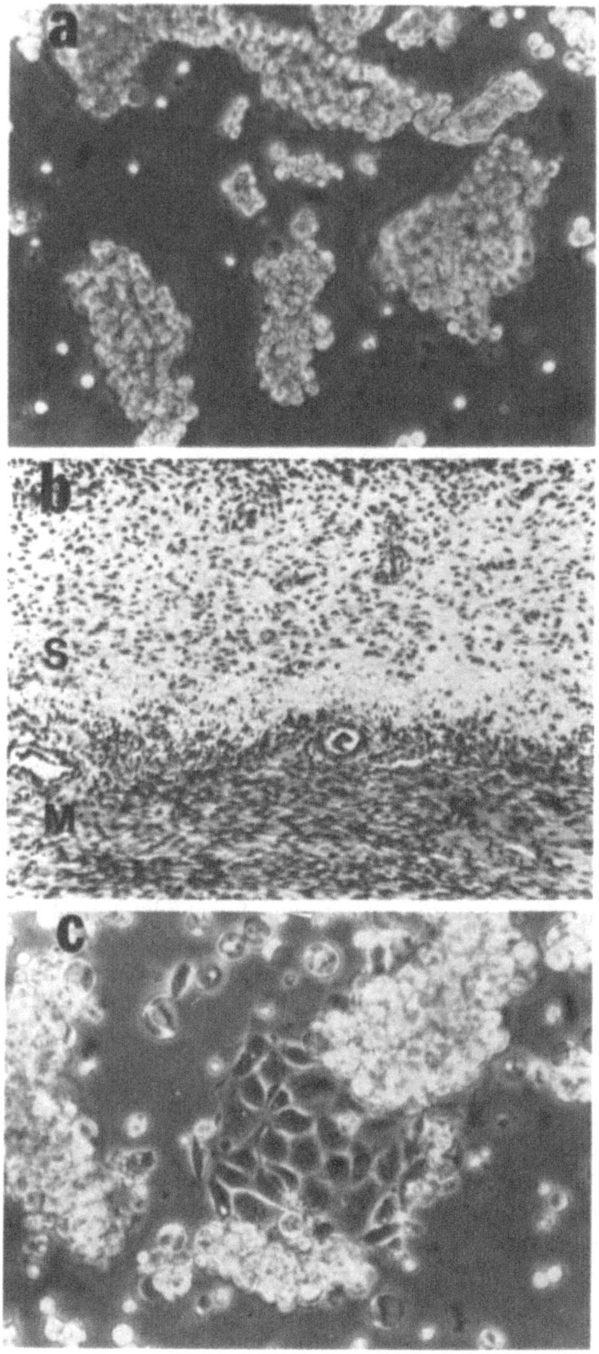

Figure 4. (a) Typical plaques of epithelial cells still joined to each other after separation from the uterus (phase contrast, ×200); (b) cross section of the uterus after removal of the epithelial cells; stroma (S) and myometrium (M) still intact (H and E, ×63); (c) monolayer outgrowth of epithelial cells after 16 hr in culture (phase contrast, ×200).

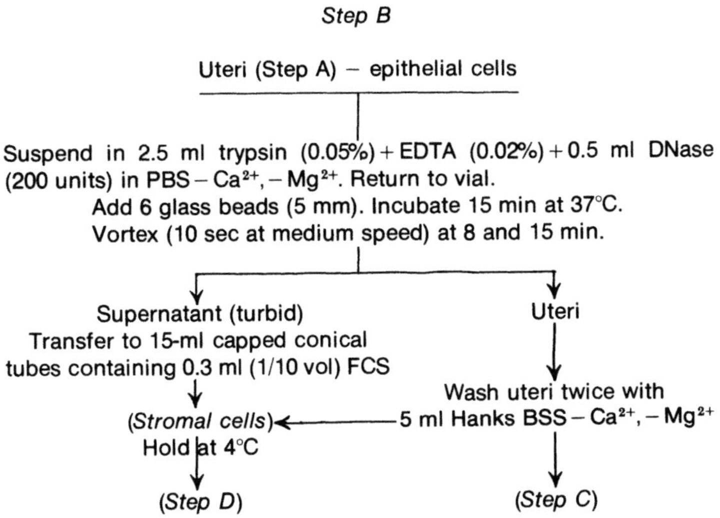

Figure 5. Procedures for the isolation of stromal cells from uteri stripped of epithelial cells.

obtained viability indices of over 90%, whereas certain personnel, otherwise skilled, never produced counts from the same samples in excess of 65% live cells.

Routine histological examination of fresh cell suspensions and comparable sections of uterus representing either different intervals following estradiol-17β (E_2) injections to immature female rats (McCormack and Glasser, 1980a) or different times during the preimplantation period of the pregnant rat (Glasser and McCormack, 1980b) demonstrated that the separate cells were intact and possessed the various characteristics of their specific type at a specific time. These data were further verified by transmission (TEM) and scanning electron microscopy (SEM) of whole sections of uterus and pellets prepared from suspensions of individual cell types.

SEM showed epithelial cells from pregnant uteri to be intact and still joined together in groups or plaques. On Day 0 microvilli were prominent on the lumenal aspects of cells but became less numerous and were condensed by Day 4. No other cell types were observed. TEM also showed the microvilli. Separated epithelial cells were joined by junctional complexes with desmosomes. The cells contained lipid droplets; dense heterochromatin was found within intact nuclear membranes.

Stromal cells appeared to be separate and intact. SEM showed stromal cells covered with a profusion of short cellular projections and blebs. TEM provided evidence that the separated stromal cells were engaged in the production of collagen fibers; cell and nuclear membranes were not altered by these separation methods.

Separated myometrial cells were elongate and SEM showed microfibrils, less numerous than in whole uterus, protruding from the cell surface. TEM showed membrane thickenings indicative of subsarcolemmal filament-attachment areas, internal stress fibers, and glycogen granules.

The structural integrity of individual cell types separated by methods described here is substantiated by the comparative appearance of whole uterus or separated cell preparations seen by SEM and TEM. Separated cells appeared to be intact and to possess the characteristics of their respective types. TEM and SEM micrographs will be published elsewhere

Figure 6. (a) A suspension of individual stromal cells after separation from the myometrium; an occasional epithelial gland (gl) may be present (phase contrast, ×200); (b) the remaining myometrium after epithelial and stromal cells have been removed from the uterus (H and E, ×63); (c) monolayer outgrowth of stromal cells after 40 hr in culture (phase contrast, ×200).

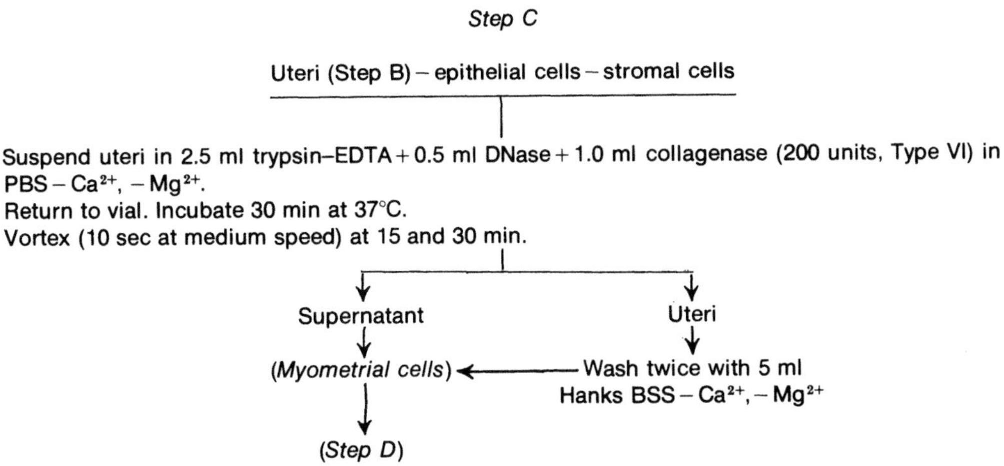

Figure 7. Procedures for the isolation of myometrial cells from uteri stripped of epithelial and stromal cells.

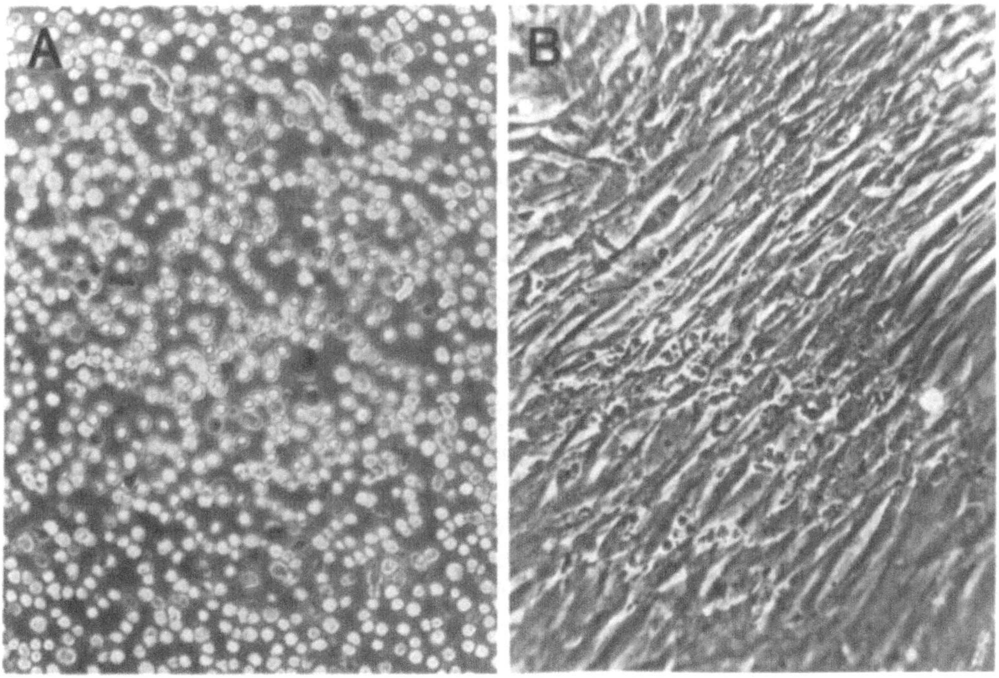

Figure 8. (A) Individual myometrial cells enzymatically dissociated from myometrium. Note similarity to individual stromal cells (Figure 6a) (phase contrast, ×200); (B) monolayer outgrowth of myometrial cells after 48 hr in culture (phase contrast, ×160).

(E_2-injected immature rats, McCormack and Glasser, 1980a; preimplantation uterus, Glasser and McCormack, in preparation).

Total yield of uterine cells, calculated from the sum of DNA in whole suspensions of the three cell types compared to the amount of DNA in whole homogenates of intact uteri, ranged from 73 to 101%. In E_2-injected immature rats, epithelial cells accounted for 1.0%,

Step D

↓

Epithelial (Step A), stromal (Step B), and myometrial (Step C) cells

↓

Centrifuge individual cell fractions
(500*g*, 5 min)

→ Discard supernatant

↓

Wash with 15 ml Hanks BSS $-Ca^{2+}, -Mg^{2+}$

→ Remove fibrous fragments

↓

Resuspend individual fractions in 5–6 drops medium (DMEM + 10% FCS [h.i.] + penicillin/streptomycin [100 units/10 μg]). Finally, take up in 10 ml medium.

↓

Put through glass bead column

↓

Rinse column with 5 ml medium

↓

Resuspend *gently* to homogeneity

→ Remove 100 μl (Trypan blue viability test)

→ Remove 100 μl to culture (if cells are to be used fresh)

→ Remove drop to stain (permanent slide)

↓

Cells are ready to pellet and suspend in desired buffer or dilute with additional medium for plating and culture, biochemistry, receptor assay, etc.

Figure 9. Purification of isolated rat uterine epithelial, stromal, and myometrial cells; cell counting methods and viability tests.

stromal cells for 3.2%, and myometrial cells for 69% of the DNA measured in whole uteri at zero time. At 24 hr after E_2, the respective percentages were 2.7, 4.1, and 86.4 (McCormack and Glasser, 1980a). The discrepancy between these sums and 100% at any time reflects the fact that recovery of cells from the uterus was not 100%. We presume that most of our loss occurred in the myometrial fraction, for we have yet to carry the enzymatic digestion to complete dissolution of the uteri as we wish to avoid possible degradation of RcE and RnE (which occur in myometrium at exceedingly low concentrations). Other cell loss occurred through cell breakage.

III. Estrogen Receptor Assays

Estrogen receptor (RE) assays were used to provide functional evaluations of the cell separation technique. We chose the E_2-stimulated immature female rat to compare the estradiol receptor dynamics between individual cell types and the whole uterus. The immature rat uterus was selected initially because it promised to be an excellent model and is the subject of an extensive literature (O'Malley and Means, 1974; Clark *et al.*, 1977). The specific methods for the measurement of RE in both cytosol and nuclei have been described extensively in previous publications (McCormack and Glasser, 1976, 1978).

A. Estrogen Receptor in Individual Cell Types from Immature Rat Uterus

Sampling times of 0.5, 1, and 2 hr after estrogen injection were selected to bridge the early response patterns (increased amino acid and nucleotide uptake, stimulated nuclear RNA polymerase activity, increased water imbibition and uterine weight) characteristic of the immature rat uterus to a single injection of 2.5 μg E_2. The later responses of the uterus to E_2 (sustained RNA polymerase activity, increased DNA, RNA, and protein synthesis, cell hyperplasia and hypertrophy, increased uterine dry weight) were covered by observations at 8, 12, 24, and 48 hr postinjection.

Figure 10 shows levels of RE in epithelial, stromal, and myometrial cells. At zero time epithelial cell RnE was 5.5 fmol/μg DNA and RcE was 0.5 fmol/μg DNA. Epithelial cell RnE was significantly reduced at 2 and 24 hr after E_2. Following a transient decrease at 2 hr, the concentration of RcE rose significantly up to 48 hr. Stromal RnE was not significantly different from zero-time concentrations. RcE, however, increased significantly in stromal cells ($p < 0.05$) from zero to 24 hr. RE in myometrial cells was determined in undissociated myometrium. Prior to E_2 injection myometrial cells contained significantly less

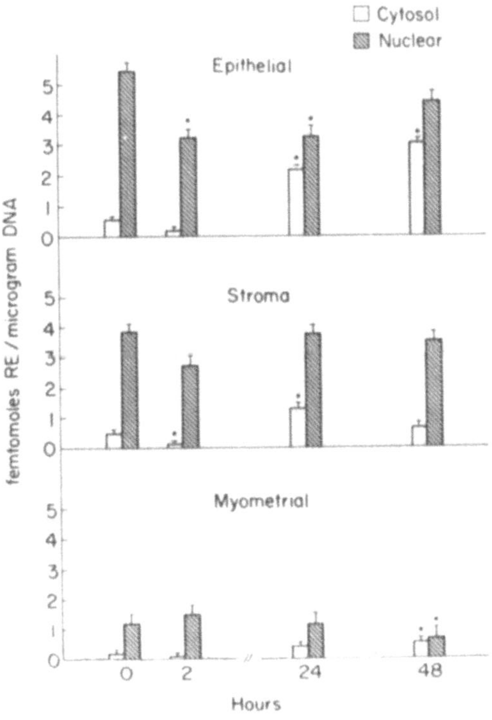

Figure 10. Estrogen receptor concentration in the high-speed cytosolic and nuclear fractions of cell types separated from the immature rat uterus after injection of 2.5 μg E_2. Myometrial nuclear estrogen receptor concentrations were significantly lower ($p<0.01$) than either epithelial or stromal nuclear estrogen receptors at all times. * Significantly different ($p<.05$) from corresponding zero-time value. From McCormack and Glasser (1980a).

($p < 0.01$) RnE and RcE than either epithelial or stromal cells. At 48 hr following E_2, the concentration of RnE fell significantly ($p < 0.05$) while RcE showed a significant ($p < 0.01$) rise.

Table 1 shows these data expressed as the percentage of total receptor in the cytosolic and nuclear compartments of the three cell types. RcE in the epithelium rose from less than 10% at early times to 40% at 24 and 48 hr postinjection, while RnE showed a corresponding decline in percentage. In the stroma the relative receptor distribution shifted from 4 to 25% RcE at 24 hr. By 48 hr, both the distribution and the amount of stromal RE (cf. Figure 10) had returned to zero-time levels. The fall in myometrial RnE and the rise in RcE by 48 hr created receptor compartments of essentially the same size in this tissue.

The percentage of total uterine RE recovered in each cell type (Table 2) showed that 48 hr after E_2 RE had increased 4% in myometrium, had decreased 6% in stroma, and had increased 3% in epithelium. Expressed on this basis, only the stromal change was statistically significant ($p < 0.01$) compared to pretreatment values. The biological significance of these shifts between cell types and within their compartments is not obvious. Each cell type seems to respond to E_2 individually. These shifts in RE distribution could be considered as integral to the late expressed responses to E_2.

B. Comparison with Data from Whole Immature Rat Uterus

The total RE concentrations found in separated cell types are similar in magnitude to those previously reported for the whole uterus of the immature rat (Gorski *et al.*, 1971; Clark and Peck, 1976). The data differ with respect to the low concentrations of cytosol RE found in separated cells. For purposes of comparison, we have converted the data of Clark *et al.* (1979) from picomoles per uterus to femtomoles per microgram DNA (McCormack and Glasser, 1980a). When compared to our data (Figure 10), these recalculations demonstrate that the total RE concentration of the immature rat uterus is similar whether measured in separated cells or in intact whole uterus. The calculations reinforce our observation that the partition of RE between cytosolic and nuclear compartments in individual cells is different from that measured in whole uterus.

Data of this type are not unique to our studies. Low cytosol concentrations of RE in mechanically separated uterine cell fractions have also been recorded under somewhat different experimental conditions (Martel and Psychoyos, 1978). The concentration of RcE in endometrial and myometrial fractions of untreated castrate uterus was 1.0 and 0.25 fmol/μg DNA, respectively. Twenty-four hours after the last of three E_2 injections (0.25 μg), the myometrial RcE concentration rose to 1.0 fmol/μg DNA. RnE was not measured.

The reason for the differential partitioning of RE to the epithelial cell nuclear compartment is not clear. Our representative zero-time data for whole uterus (RcE, 4.02 fmol/μg DNA; RnE, 0.40 fmol/μg DNA) are similar to the zero-time data of Clark *et al.* (1979) (4.80 and 0.80 fmol/μg DNA in RcE and RnE, respectively). Thus, we reason that our method of RE assay was not at fault. The possibility that the period of elevated temperature used in Step A (Figure 3) to increase the yield of epithelial cells might artificially translocate RE to the epithelial cell nucleus, independent of E_2 administration, prompted us to analyze RE in epithelial cells kept at 4°C throughout the entire procedure. An identically low RcE (0.6 fmol/μg DNA) was measured in separated epithelial cells under both sets of temperature conditions. While RnE was lower at 4°C (2.0 fmol/μg DNA), the concentration was still three times greater than RcE in the uteri of uninjected immature rats.

We have also eliminated possible E_2 contamination of pancreatin, possible damage to cell and/or nuclear membranes, and an increase in cell death as factors responsible for the

Table 1. Partition of the Estrogen Receptor (RE) between Cytosol and Nucleus in Uterine Cell Types Following Subcutaneous Injection of 2.5 μg E_2 in Immature Rats [a]

Time (hr)	Percentage of total cellular RE					
	Epithelial cells		Stromal cells		Myometrial cells	
	Cytosol	Nucleus	Cytosol	Nucleus	Cytosol	Nucleus
0	8.5	91.5±3.0[b]	11.3	88.7±2.1	13.5	86.5±3.0
2	4.3	95.7±3.4	*4.0	96.0±2.4	*1.0	99.0±3.4
24	†40.0	60.0±3.4	†25.3	74.7±2.4	*26.0	74.0±3.4
48	†41.0	59.0±3.4	15.3	84.7±2.4	†44.0	56.0±3.4

[a]From McCormack and Glasser (1980a).
[b]S.E.M.
*$p<0.05$, †$p<0.01$: significantly different from corresponding zero-time value.

Table 2. Percentage of the Total Uterine Estrogen Receptor (RE) in Immature Rat Uterine Cell Types Following Subcutaneous Injection of 2.5 μg E_2 [a]

Time (hr)	Epithelial cells	Stromal cells	Myometrial cells
0	5.4±0.8[b]	10.2±0.8	84.2±1.9
2	3.9±1.0	*7.4±0.9	88.7±2.2
24	5.8±1.0	7.6±0.9	84.0±2.2
48	7.8±1.0	†4.0±0.9	87.7±2.2

[a]From McCormack and Glasser (1980a).
[b]S.E.M.
*$p<0.05$, †$p<0.01$: significantly different from corresponding zero-time value.

partitioning effect. We continue to study this phenomenon. It is important to note that the concentration of RcE can be increased experimentally in epithelial cells 24 and 48 hr after the injection of E_2 (Figure 10).

The apparent K_d for the cytosolic and nuclear RE in the different cell types ranged from 1 to 10×10^{-10} M. There were no significant differences in K_d between cell type, cytosolic and nuclear RE, or whole uteri in either E_2-treated immature rats (McCormack and Glasser, 1980a) or pregnant rats (Glasser and McCormack, 1980b). The single exception was that the K_d for epithelial RcE ($5.6\pm2.9\times10^{-10}$ M) was significantly greater ($p<0.005$) than for RnE ($2.7\pm0.9\times10^{-10}$ M) by Student's paired-*t* test. Martel and Psychoyos (1978) have also reported a significant difference ($p<0.005$) between the K_d of endometrium (epithelium + stroma) (3×10^{-10} M) and myometrium (6×10^{-10} M) in ovariectomized adult rats.

C. Estrogen Receptors in the Preimplantation Uterus

Previous attempts to measure the concentration of uterine RE during the period from conception to implantation have not been in agreement (Glasser and Clark, 1975; Martel and Psychoyos, 1976; Glasser and McCormack, 1979). Our most recent study of the cytosolic and nuclear compartments in the uterus of pregnant rats of established cyclicity and proven fertility was done under rigorous kinetic conditions using a 12-point Scatchard anal-

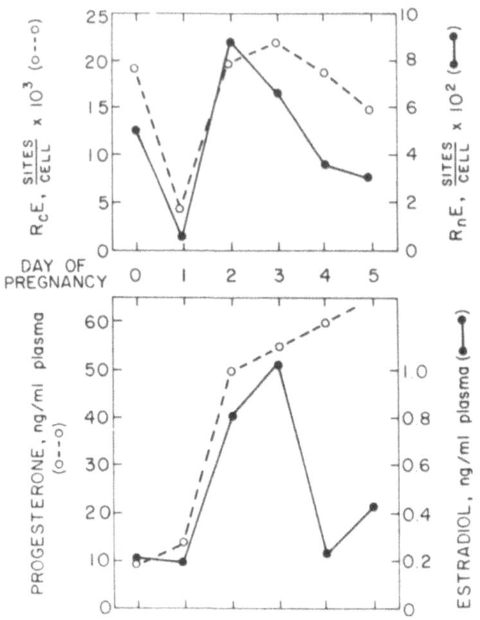

Figure 11. (Top) Comparison of cytosolic (RcE) and nuclear (RnE) estrogen receptor complexes in the uterus of pregnant rats. Day 0 of pregnancy designated by presence of vaginal sperm. Data are expressed as sites per cell. (Bottom) Comparative plasma levels of progesterone and estradiol from the same rats. Steroid levels were determined by specific radioimmunossay. From Glasser and McCormack (1980c).

ysis (McCormack and Glasser, 1978) under exchange conditions (Anderson *et al.*, 1972). The presence of a sperm-positive vaginal smear designated Day 0 of pregnancy.

As shown in Figure 11, the concentrations of RcE and of RnE in whole pregnant uteri on Day 0 were interpreted as reflecting the previously elevated proestrous plasma estrogen (E) levels. The lower concentrations on Day 1 were considered to be the consequence of low diestrous plasma E. Plasma progesterone (P) was low on both days. Between Days 1 and 2, both plasma E and P rose markedly (Figure 11). This change was accompanied by a 4-fold increase in RcE and a greater than 30-fold increase in RnE. We assigned these changes in RE to the influence of E on the synthesis, translocation, and replenishment of its own receptor. We made the further assumption, based on the work of Hsueh *et al.* (1976), that a primary action of increasing P titers was on the depression of RE replenishment and thus on the measurable RE.

From its peak on Day 2, the concentration of RnE fell to a plateau on Days 4 to 5. The fall in RnE preceded the fall in RcE, which began on Day 3 after a continued rise from Day 2. These changes were based on the interaction between rising plasma P (causing diminished replenishment) and falling plasma E (causing depressed synthesis). We could not reasonably account for the fall of RnE while RcE was still rising but considered that a number of factors could be involved, including the influence of rising plasma P on the dynamics and concentration of both RE and RP and the functional responses of the uterine epithelium as distinct from the responses of the stromal cells (Tachi *et al.*, 1972). Failing to confirm or refute the data of Martel and Psychoyos (1976) utilizing their own protocols, we concluded that the mechanism and site of steroid hormone action and the role of their receptors in the maturation of the preimplantation uterus remain to be defined.

D. Estrogen Receptors in Individual Cell Types from the Preimplantation Uterus

Analysis of receptors in individual cell types of the preimplantation uterus revealed that the elevated concentrations of RcE and RnE described for the whole uterus (Figure 11), and attributed to proestrous plasma E, actually represented only the RE concentration of epithelial cells (Figure 10). On Day 0 the stromal cell concentrations of RcE and RnE were less than 30% of those in the epithelial cells (Figure 12) and stromal RnE was higher than RcE.

Epithelial RcE continued to fall up to Day 4 of pregnancy (Figure 12) even after plasma E reached its diestrous nadir on Day 1 and rebounded (Figure 11). Epithelial RnE fell between Day 0 and Day 1, recovered transiently on Day 2 as plasma E increased, but fell again although the elevated titers of E were sutained through Day 3. There were no statistically significant changes in the concentration of stromal cell RcE during Days 0 to 2 (Figure 12), but stromal RnE increased from Day 0 to Day 3 despite the complex shifts in plasma P and E (Figure 11).

Epithelial and stromal cells of the endometrium thus respond to a given hormonal milieu in a dissimilar if not independent manner. While epithelial cells, purportedly stimulated by proestrous E, proceed through the mitotic cycle (Martin and Finn, 1968; Tachi *et al.*, 1972; Martin *et al*, 1973a,b,c), the concentrations of RcE and RnE begin to fall in these same cells and continue to do so regardless of rising plasma E levels. Thus synthesis and replenishment of epithelial RE do not appear to be later responses to the preovulatory E that stimulates epithelial cell division. The epithelial cell is not without sensitivity to E_2 during the preimplantation period. There is presumptive evidence of translocation, as an increasing proportion of the total RE remaining in epithelial cells can be measured in the nucleus. The specific effects of the continuously rising P are not easily explained at present.

The response of the stromal cell during the preimplantation period is even more difficult to understand. Total RE remains low during Days 0 to 3, apparently unaffected by the high proestrous titers of E. The low plasma E of the early preimplantation period (dies-

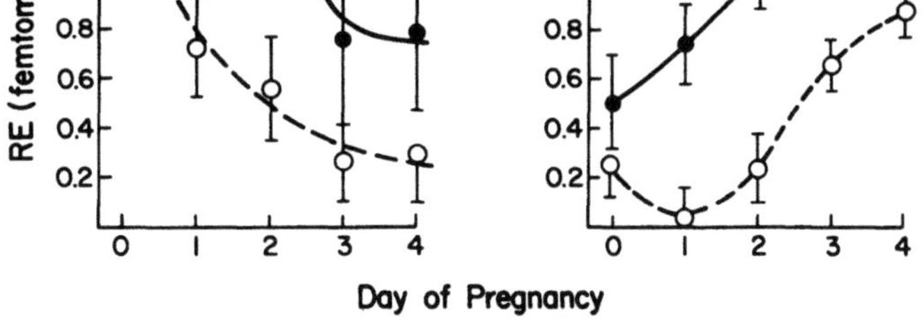

Figure 12. Estrogen receptor concentrations (fmol/μg DNA$\pm$SEM) in high-speed cytosolic (HSC) and nuclear fractions of epithelial and stromal cells separated from pregnant rat uteri on various days of gestation. These data should be compared with Figure 10, top panel.

trus) appears to have only a modest regulatory influence on the stromal cell. Only after 72 hr of rising E (peak Day 3) is there a late rise in stromal RE (which is actually measured during a transient fall in E on Day 4). Beginning while plasma E is low, the proportion of total RE that is in the nucleus rises progressively from 67 to 84%. It is not yet possible to determine if this accumulation of RE by the stromal cell nucleus is due primarily to translocation or to a decrease in the degradation of the nuclear hormone receptor complex. Only after Day 2 does RE begin to increase in the stromal cytosolic compartment, in the face of a rapidly increasing plasma P level. Stromal RcE continues to increase while epithelial RcE and RnE fall and the stromal cell nucleus no longer accumulates RE (Figure 12).

We are without the appropriate data to analyze the causes of epithelial cell total RE depletion. Before we can explain this change, we need to consider the influence of each factor that contributes to the net change in the size of an E-stimulated epithelial cell population (Martin *et al.*, 1976). Minimally we shall have to know how the process of cell replication influences the RcE and RnE of daughter cells and something of the dynamics of RE in cells programmed for death (Smith and Martin, 1974).

The shifts in the concentration of RE in the individual cells of the preimplantation uterus (Figure 12) were not obvious in our previous analysis of whole uterus (Figure 11). The rules we presently follow to analyze receptor dynamics, viz. E enhances the replenishment of its own receptor, P inhibits that replenishment process, are really too rudimentary to be effective. Before we can understand the consequences of the data reported here, we shall have to identify the interval between a change in the rate of steroid hormone secretion and the specific intermediary response to that change, e.g., replenishment, degradation, or translocation. We will also require knowledge of how each hormone (P, E) influences the ontogeny of its own receptor, the receptors of other hormones, and the interaction of these hormones on the different receptor populations in a single species of target cell. Until we can obtain that information, our interpretation of cause and effect regarding steroid hormone action and the response of the preimplantation uterus will remain rudimentary.

There is no discernible difference in the physiochemical nature of RE whether it is found in epithelium or stroma, cytoplasm or nucleus (McCormack and Glasser, 1980a). Thus, the role of cell–cell communication must be defined if we are to attempt to explain why the same hormonal environment, which will charge the stromal cell nucleus and subsequently its cytoplasmic compartment with increasing concentrations of RE, will not sustain the RE concentration of the epithelial cells. Evidence of a shift in the target of steroid hormone action is supported by cytological data (Martin and Finn, 1968; Tachi *et al.*, 1972; Martin *et al.*, 1973a,b). We are still without a fair clue as to how this shift in the site of steroid hormone action is accomplished. With respect to the implanting blastocyst, what is the strategic importance of this shift and of the consistent partitioning of RE, in both preimplantation epithelium and stroma, which favors the nuclear compartment?

There has yet to be a comprehensive and systematic study in any of the laboratory rodents to integrate the relationships of cell division with the molecular biology of implantation. Although the regulation of uterine target cell division is only one manifestation of the regulatory action of P and E, studies of the cell cycle should no longer be considered separately from the general principles that describe the mechanism of hormone action. The complexity of this *in vivo* situation suggests that the use of *in vitro* as well as *in vivo* models would serve to recognize the various factors that could be involved in integrating these processes.

However, a single model will not provide the information required. The patterns of endometrial cell division that now serve as our references for this process have been derived from models based on the estrous cycle. The data extracted from these models best represent the influence of proestrous E on the epithelial cells during the first 24 to 30 hr of gestation (Finn and Martin, 1973; Martin *et al.*, 1973a,b; Kirkland *et al.*, 1979). These models minimize the determinant action of P on uterine sensitization for decidualization and implantation (Psychoyos, 1973; Glasser and Clark, 1975) and are only of limited use in the interpretation of the periimplantation period.

Estrous-type models have provided useful basic information. The pharmacodynamics of various estrogens and their temporal influences on the relationship between cell death and the kinetics of cell division were the subjects of a significant paper (Martin *et al.*, 1976). Earlier papers recognized three specific E-directed responses: (1) increased rate of DNA synthesis, (2) formation of large prominent nucleoli, and (3) increased [^{3}H]uridine incorporation (Tachi *et al.*, 1972). Less emphasis was placed on findings that indicated that prior exposure to P (10 to 12 hr before E) altered or abolished the E-induced responses. If P was given at least 30 hr prior to E, the responses previously noted in the epithelium were not diminished but were redirected to the stromal cells (Clark, 1973; Martin *et al.*, 1973c). Models based on the pregnant or pseudopregnant animal (E modulation of P rather than vice versa) may be more appropriate to develop these important data.

Models based on pregnant and pseudopregnant animals, however, including the decidual cell response, are not without their limitations. Studies concerned with cell division in the uteri of pregnant rats show that the mitotic index of epithelial cell types is twice that of the pseudopregnant rat. If these differences prove real, they suggest a role for an embryonic signal or maternal recognition factor; they also restrict the use of the pseudopregnant rat or models derived from it.

IV. Culture of Monolayers of Individual Uterine Cell Types

The ability to establish individual uterine cell types in monolayer cultures provides *in virto* models which circumvent the problems inherent in organ cultures and *in vivo* systems. *In vitro* tissue culture affords the opportunity of analyzing the individual components of the implantation process either separately or in various experimental reassociations. These methods allow the investigator to select vigorous experimental conditions and regulatory options that cannot be applied to other models but have utility in well-defined culture systems.

Compared with mouse and rat ova and blastocysts, little attention has been devoted to *in vitro* culture of the endometrium or its component cells. Rat and rabbit endometrium have been successfully cultured *in vitro* (Gerschenson *et al.*, 1974; Sonnenschein *et al.*, 1974). E sustained and enhanced the growth of cultured cells in both species but P inhibited epithelial-like components of rabbit endometrium (Gerschenson and Berliner, 1976). In the rat E_2 was implicated as a neoplastic transforming factor for a derivative tumor line that contained E_2-binding proteins with receptor characteristics. E_2 sensitivity could not be shown in epithelial cells from human endometrium. Cultured human glandular epithelium nevertheless maintained the entire spectrum of morphological characteristics of fresh epithelium for periods of 50 days (Liszczak *et al.*, 1977). Established, growing cultures of human endometrial epithelium, insensitive to E, P, and insulin, were also reported by Kirk *et al.* (1978). However, the limited life span of these cells was significantly extended by establishing dense primary cultures.

The term stroma is used in a generic sense. There are at least 11 cell species that comprise this endometrial component. At present we cannot distinguish these cells with our analytical tools. Thus we are not able to assign specific roles to individual types of stromal cells in the implantation process. Human endometrial stromal cells in culture are sensitive to the presence of P (Kirk *et al.*, 1978), as are uterine stromal cells isolated from rats sensitized by P *in vivo* (Vladimirsky *et al.*, 1977; Sananes *et al.*, 1978). P proved essential to the maintenance of *in vitro* decidualization.

These preliminary studies highlight methodological problems still to be solved regarding the culture of isolated endometrial cells, e.g., selection and preparation of cells, species, culture conditions, differential sensitivity to hormones, and other regulatory factors. Nevertheless, the *in vitro* culture of epithelial and stromal cell monolayers is unique for the purpose of defining the cell biology of these cells in the implantation process and should be encouraged. Significant progress in the study of implantation depends on the use of these models to obtain knowledge concerning: (1) the cellular microenvironment and its role in establishing the homeostasis of epithelial and stromal cells (Lamerton, 1973); (2) the cooperative influence of different cell populations and densities on growth and differentiation of these cellular interactions (Lamerton, 1973), and (3) mesenchymal (stromal)–epithelial cell-to-cell interactions (Cunha, 1976) during the differentiative maturation of the hormonally regulated endometrium.

V. Coculture of Blastocysts with Monolayers of Specific Uterine Cell Types

One of the purposes of developing cell separation methods was to provide specific cell monolayers with which to coculture rat blastocysts. *In vitro* coculture models of implantation are very useful in permitting the display of the functional capabilities of the endometrial cell types, the trophectoderm and embryonic cells. The model is not a perfect analog for *in vivo* implantation (Enders *et al.*, this volume). Results from coculture models of implantation should be used with the utmost caution in interpreting events relating to invasion of the uterus. Spatial rearrangements also limit the use of this system as an adhesion model. Nevertheless, if used judiciously this coculture system can yield productive data.

We have monitored the interaction between monolayers of uterine cell types and rat blastocysts by measuring the rate and success of hatching and attachment as well as the daily steroid production by the blastocysts. In these experiments (Figure 13) the monolayers of epithelial or stromal cells were prepared from immature rat uteri. Blastocysts were recovered from the uteri of pregnant rats on Day 4 of gestation and cocultured with the established monolayers. The conditions for successful *in vitro* culture of peri- and postimplantation rat blastocysts are described elsewhere (McCormack and Glasser, 1980b). The data from the coculture experiments were compared with the steroid secretory patterns (progesterone, testosterone, estradiol) of blastocysts outgrowing in culture on plastic dishes.

The data in the coculture experiments depicted in Figure 13 were derived by comparing P production by blastocysts at similar times (equivalent to their gestation days, EGD). Media were changed daily. Control values were obtained by assay of coincubated cell-free media. This value was subtracted from the experimental value and the result was expressed as micrograms of steroid per blastocyst per day.

Coculture of developing blastocysts with established stromal cell monolayers suppressed the secretion of P by the blastocysts to nondetectable levels. Growth on epithelial cell monolayers reduced P production from EGD 4 + 0 to 4 + 7 to 52% of that produced by contemporary blastocysts cultured in plastic dishes for the same time.

We intend to extend the usefulness of this model by the study of species with other types of implantation (noninvasive, syncytial). More detailed, timed studies of the func-

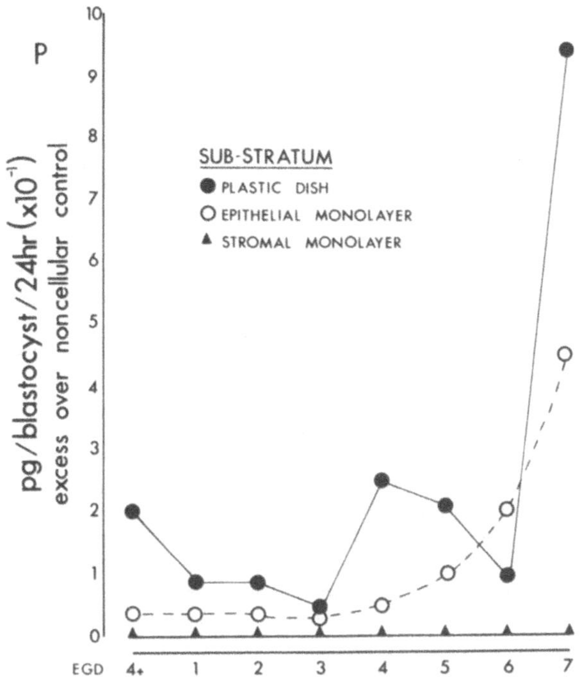

Figure 13. Daily production of progesterone by rat blastocysts cocultured with monolayers of uterine epithelial and stromal cells or on plastic dishes. Monolayers were derived from cells separated from immature rat uteri. Equivalent gestation days (EGD) are the days in culture added to Day 4 of pregnancy when the blastocysts were collected. Steroid measurements are corrected for control values obtained by analysis of incubated cell-free media (Glasser and McCormack, 1980b).

tional cytology of implantation in a variety of species, both *in vivo* and *in vitro*, are also required in order to identify those regulatory events unique to the implantation process.

The data presented in this paper validate the use of viable, essentially homogeneous populations of individual cell types as effective probes in the definition of target cell responses to hormones and other regulatory agents. The use of these cell populations makes it possible to differentiate target cell responses while avoiding the indirect or retrospective relationships of functions and cytology that necessarily arise from analysis of whole organs. We continue to improve these methods so that their usefulness as probes will be recognized by morphologists and biochemists in order that they may be applied to a broader range of investigations.

ACKNOWLEDGMENTS. The technical assistance of Ms. JoAnne Julian and Mr. Robert A. May was invaluable to this study. We are indebted to them for their gracious support. These investigations were supported by NIH Grants HD-12964 and HD-07495 and by National Institute of Cancer Grant CA-20853.

Discussion

FINN (Liverpool): I was most interested in the fact that the level of receptor in the epithelium seemed to go down and you therefore suggested that estrogen acted on the stromal cells. I wonder whether this is entirely true? I'm sure you know that epithelium, though it doesn't divide during this time, does undergo a change in the cell surface. The interesting thing is whether that change is due to estrogen acting directly on the epithelium or whether it is acting via the stroma.

GLASSER: Referring to Gerald Cunha's work on mesenchymal induction of epithelium (*Int. Rev. Cytol.* **47:**137, 1976), I think if we attribute to the stroma much more functional importance than has been attributed before, and analyze the systems as Cunha analyzed his system, we would find stromal induction of epithelium may be a productive hypothesis.

MARTIN (ICRF, London): I agree entirely that the stroma may be the determining factor. I know of no concrete evidence that any changes in epithelial function are mediated by direct effects of the hormones on the epithelium. There is controversy as to whether estrogen binds to the epithelium. Certainly in the untreated castrate rat there is evidence that estrogens bind to the nuclei of the epithelial cells. In the progesterone-treated animal, it's equivocal. Tachi *et al.* (*J. Reprod. Fertil.* **31:**59, 1977) indicate that there is no estrogen binding in the epithelium. I think this has also been shown by Stumpf and Sor (*Meth. Enzymol.* **36A:**135, 1975). We isolated uterine epithelial cells as homogenates from castrate and progesterone-treated mice. We found the same amount of estrogen binding, though there are problems of contamination from other tissues. Apropos separating cell types, we have a simple method that works very easily. It does not produce viable cells, but if you want simply to measure biochemical parameters in epithelial cells, you can dissect out the mouse uterus, open it longitudinally, put it in a tube of buffer with five glass balls, 5 mm in diameter, and, if I may use the neologism (which I'm sure *will* eventually get into the Oxford English Dictionary), vortex for 2 min. We get a homogenate of "epithelium" in which about 80% of the nuclei are derived from the epithelium. We submitted a paper on the method, under the title of "Two Horns, Five Balls, Two Minutes" (there were actually two females and three male authors, making for some interesting counts!) and we subtitled it "Depitheliation in a Flash." The paper was published (Fagg *et al.*, 1979, *J. Reprod. Fertil.* **57:**335), but not under that title, I'm afraid.

BEIER (Aachen): You interpreted the arena for hormonal action to be in the stroma rather than in the epithelium. I would like to ask if the interpretation can be excluded that there is a triggering of an inherent epithelial cell program, with no necessity for any further replenishment or translocation of the receptor?

GLASSER: Our data do not exclude the possibility, because at present it is impossible for us to test that possibility. My operational prejudice is that there may very well be an epithelial trigger, but there is a stromal finger.

BULLOCK (Baylor): While the overall impression from your data is a marked decrease in cytoplasmic receptor levels, I think the picture looks promising because there is translocation going on, as you pointed out, first in the stroma and then in the epithelium. This phenomenon provides some biochemical correlation with the views that are being expressed now for the determining influence of the stroma on the events in the epithelium. In terms of how the receptors are working, when there don't seem to be very many cytoplasmic receptors around, I think the relatively low but stable level of nuclear receptor is sufficient to account for the changes that subsequently take place. Whether this low persistent level of nuclear receptor is a static or a dynamic state, we don't know. It's probably a dynamic one, but it does make us rethink our views about how cytoplasmic and nuclear receptor levels relate to physiological events.

GLASSER: I sometimes wake in a cold sweat that all this factoring of cytoplasmic numbers is extraneous, because we cannot answer the question of how few receptors are necessary for response. I think that is a very real question, what is the actual number of receptors required for a response?

SURANI (Cambridge): I'd like to bring up another point from your talk and that is the restriction of gene expression after you injected estrogen. Of course, there are species, like hamsters and guinea pigs, where under certain experimental conditions you can induce implantation by giving progesterone alone. Would you like to comment on that?

GLASSER: I think the answer lies in some work by Dr. McCormack having to do with minute levels of secretion of progesterone and estrogen from trophoblast or, as Gabori suggested, sometimes decidual cells (Gabori *et al.*, 1977, *Endocrinology* **100:**1483). We must be concerned with the hormonal titers at the interface between trophoblast and stromal cells. At a time when peripheral progesterone titers are 80 ng/ml, the trophoblast suddenly begins to secrete progesterone (approximately 2–3 pg/trophoblast/24 hr). At that time trophoblast and decidual E receptors begin to decline rapidly. These data describe an entirely different level of regulatory organization. When we discuss animals that do not require E, we may have to define what the cells are doing at the point of interaction.

RAHMAN (Cleveland): Have you looked at progesterone receptors in your preparation?

GLASSER: No. With the exception of the work done in Kraicer's laboratory in Tel Aviv, I can think of no progesterone receptor assay for the rat in which I have any confidence. There is work in the hamster that I can accept, but my personal bias is that I don't come to rest with ease with any of the progesterone receptor assays that have been advanced for the rat. This is not to say that there are not any good ones, I just don't feel comfortable with them.

RAHMAN: Have you tested any other biological parameters in your preparation, for example glucose uptake or oxygen uptake, in terms of biological integrity of these cells?

GLASSER: We have preliminary evidence on incorporation of cytidine that is far from complete but leans in the right direction.

ARMSTRONG (W. Ontario): I wasn't able to see whether there are any qualitative differences in the receptors of the epithelial cells and the stroma. I ask the question because it may be that there are differences. For instance, there is evidence that androgens will displace estrogen from receptors, but androgens have qualitatively different effects on the stroma and epithelial cells compared to the effects of estrogens. Androgens will stimulate growth only of the stroma, whereas estrogens stimulate both stromal and mucosal growth. It would be interesting to know if this androgen action is mediated via estrogen receptors or androgen receptors. Have you done specificity studies on the receptors of the stromal and epithelial cells to see whether there is any qualitative difference in the way they respond to different steroids?

MCCORMACK: We did specificity studies of the receptor in trophoblast cells but have not repeated these studies in the separated endometrial cell types.

References

Alberga, A., and Baulieu, E.-E., 1968, Binding of estradiol in castrated rat endometrium *in vivo* and *in vitro*, *Mol. Pharmacol.* **4**:311.

Anderson, J. M., Clark, J. H., and Peck, E. J., 1972, The relationship between nuclear receptor–oestrogen binding and uterotrophic responses, *Biochem. Biophys. Res. Commun.* **48**:1460.

Clark, B. F., 1973, The effect of oestrogen and progesterone on uterine cell division and epithelial morphology in spayed–hypophysectomized rats, *J. Endocrinol.* **56**:341.

Clark, J. H., and Peck, E. J., 1976, Nuclear retention of receptor–estrogen complex and nuclear acceptor sites, *Nature (London)* **260**:635.

Clark, J. H., Peck, E. J., and Glasser, S. R., 1977, Mechanism of action of sex steroids in the female, in: *Reproduction in Domestic Animals* (H. H. Cole and P. T. Cupps, eds.), 3rd ed., p. 143, Academic Press, New York.

Clark, J. H., McCormack, S. A., Padykula, H., Markaverich, B., and Hardin, J. W., 1979, Biochemical and morphological changes stimulated by the nuclear binding of the estrogen receptor, in: *Effects of Drugs on the Cell Nucleus* (H. Busch, S. T. Crooke, and Y. Daskal, eds.), p. 381, Academic Press, New York.

Cunha, G. R., 1976, Epithelial–stromal interactions in development of the urogenital tract, *Int. Rev. Cytol.* **47**:137.

Enders, A. C., and Schlafke, S., 1971, Penetration of the uterine epithelium during implantation in the rabbit, *Am. J. Anat.* **132**:219.

Finn, C. A., 1974, the induction of implantation in mice by actinomycin D, *J. Endocrinol.* **60**:199.

Finn, C. A., and Martin, L., 1973, Endocrine control of gland proliferation in the mouse uterus, *Biol. Reprod.* **8**:585.

Galand, P., Leroy, F., and Cretien, J., 1971, Effect of oestradiol on cell proliferation and histological changes in the uterus and vagina of mice, *J. Endocrinol.* **49**:243.

Gerschenson, L. E., and Berliner, J. A., 1976, Further studies on the regulation of cultured rabbit endometrial cells by diethylstilbestrol and progesterone, *J. Steroid Biochem.* **7**:159.

Gerschenson, L. E., Berliner,, J., and Yang, J., 1974, Diethystilbestrol and progesterone regulation of cultured rabbit endometrial cell growth, *Cancer Res.* **34**:2873.

Glasser, S. R., and Clark, J. H., 1975, A determinant role for progesterone in the development of uterine sensitivity to decidualization and ovoimplantation, in: *The Developmental Biology of Reproduction* (C. Markert and J. Papaconstantinou, eds.), p. 311, Academic Press, New York.

Glasser, S. R., and McCormack, S. A., 1979, Estrogen-modulated uterine gene transcription in relation to decidualization, *Endocrinology* **104:**1112.

Glasser, S. R., and McCormack, S. A., 1980a, Functional development of rat trophoblast and decidual cells during establishment of the hemochorial placenta, in: *The Development of Responsiveness to Steroid Hormones* (A.M. Kaye and M. Kaye, eds.), p. 165, Pergamon, London.

Glasser, S. R., and McCormack, S. A., 1980b, Analyses of hormonal responses of the rat endometrium by the use of separated uterine cell types, in: *The Endometrium: Eighth Brook Lodge Conference on Problems of Reproductive Physiology* (F. A. Kimball, ed.), Spectrum Publications, p. 173.

Glasser, S. R., and McCormack, S. A., 1980c, Cellular and molecular aspects of decidualization and implantation, in: *Proteins and Steroids in Early Mammalian Development* (H. Beier and P. Karlson, eds.), Springer-Verlag Berlin (in press).

Gorski, J., DeAngelo, A. B., and Barnea, A., 1971, Estrogen actions: The role of specific RNA and protein synthesis, in: *The Sex Steroids* (K. W. McKerns, ed.), p. 181, Appleton–Century–Crofts, New York.

Heald, P. J., Govan, A. D. T., and O'Grady, J. E., 1975, A simple method for the preparation of luminal epithelial and stromal cells from rat uterus, *J. Reprod. Fertil.* **42:**593.

Hsueh, A. J. W., Peck, E. J., and Clark, J. H., 1976, Control of uterine estrogen receptor levels by progesterone, *Endocrinology* **98:**438.

Jackson, V., and Chalkley, R., 1974, The binding of estradiol-17β to the bovine endometrial nuclear membrane, *J. Biol. Chem.* **249:**1615.

Kirk, D., King, R. J. B., Heyes, J., Peachey, L., Hirsch, P. J., and Taylor, W. T., 1978, Normal human endometrium in cell culture. I. Separation and characterization of epithelial stromal components, *In Vitro* **14:**651.

Kirkland, J. L., LaPointe, L., Justin, E., and Stancel, G. M., 1979, Effects of estrogen on mitosis in individual cell types of the immature rat uterus, *Biol. Reprod.* **21:**269.

Koligan, K. B., and Stormshak, F., 1977, Nuclear and cytoplasmic estrogen receptors in ovine endometrium during the estrous cycle, *Endocrinology* **101:**524.

Lamerton, L. F., 1973, The mitotic cycle and cell population control, in: *The Cell and Cancer* (A. R. Currie, ed.), *J. Clin. Pathol.* Suppl. 7, p. 19.

Leroy, F. C., Bogaert, C., and Van Hoek, J., 1976, Stimulation of cell division in the endometrial epithelium of the rat by uterine distention, *J. Endocrinol.* **70:**517.

Liszczak, T. M., Richardson, G. S., MacLaughlin, D. T., and Kornblith, P. L., 1977, Ultrastructure of human endometrial epithelium in monolayer culture with and without steroid hormones, *In Vitro* **13:** 344.

McCormack, S. A., and Glasser, S. R., 1976, A high affinity estrogen binding protein in rat placental trophoblast, *Endocrinology* **99:**701.

McCormack, S. A., and Glasser, S. R., 1978, Ontogeny and regulation of a rat placental estrogen receptor, *Endocrinology* **102:**273.

McCormack, S. A., and Glasser, S. R., 1980a, Differential response of individual uterine cell types from immature rats treated with estradiol, *Endocrinology* **106:**1634.

McCormack, S. A., and Glasser, S. R., 1980b, Hormone production by rat blastocysts and mid-gestation trophoblast *in vitro*, in: *The Endometrium: Eighth Brook Lodge Conference on Problems of Reproductive Physiology* (F. A. Kimball, ed.), Spectrum Publications, p. 145.

Makler, A., and Eisenfeld, A. J., 1974, *In vitro* binding of ^{3}H-estradiol to macromolecules from the human endometrium, *J. Clin. Endocrinol. Metab.* **38:**1974.

Martel, D., and Psychoyos, A., 1976, Endometrial content of nuclear estrogen receptor and receptivity for ovoimplantation in the rat, *Endocrinology* **99:**470.

Martel, D., and Psychoyos, A., 1978, Progesterone-induced oestrogen receptors in the rat uterus, *J. Endocrinol.* **76:**145.

Martin, L., and Finn, C. A., 1968, Hormonal regulation of cell divisron in epithelial and connective tissues of the mouse uterus, *J. Endocrinol.* **41:**363.

Martin, L., Finn, C. A., and Trinder, G., 1973a, Hypertrophy and hyperplasia in the mouse uterus after oestrogen treatment: An auto-radiographic study, *J. Endocrinol.* **56:**133.

Martin, L., Finn, C. A., and Trinder, G., 1973b, DNA synthesis in the endometrium of progesterone-treated mice, *J. Endocrinol.* **56:**303.

Martin, L., Das, R. M., and Finn, C. A., 1973c, The inhibition by progesterone of uterine proliferation in the mouse, *J. Endocrinol.* **57:**549.

Martin, L., Pollard, J. W., and Fagg, B., 1976, Oestriol, oestradiol-17β and the proliferation and death of uterine cells, *J. Endocrinol.* **69:**103.

Meyers, K., 1970, Hormonal requirements for the maintenance of oestradiol-induced inhibition of uterus sensitivity in the ovariectomized rat, *J. Endocrinol.* **46:**341.

Nilsson, O., 1970, Some ultrastructural aspects of ovo-implantation, in: *Ovo-implantation, Human Gonadotropins and Prolactin* P. O. Hubinont, F. Leroy, C. Robyn, and P. Leleux, eds.), p. 52, Karger, Basel.

O'Malley, B. W., and Means, A. R., 1974, Female steroid hormones and target cell nuclei, *Science* **183:**610.

Peel, S., and Bulmer, D., 1975, A study of proliferative activity of the uterine epithelium of the pregnant rat in relationship to morphogenesis of the new lumen, *J. Reprod. Fertil.* **42:**189.

Pietras, R. J., and Szego, C. M., 1975a, Steroid hormone responsive isolated endometrial cells, *Endocrinology* **96:**946.

Pietras, R. J., and Szego, C. M., 1975b, Surface modifications evoked by estradiol and diethylstilbestrol in isolated endometrial cells: Evidence from lectin probes and extracellular release of lysomal protease, *Endocrinology* **97**1447.

Pollard, J. W., and Martin, L., 1975, Cystoplasmic and nuclear non-histone proteins and mouse uterine cell proliferation, *Mol. Cell. Endocrinol.* **2:**183.

Psychoyos, A., 1973, Endocrine control of egg implantation, in: *Handbook of Physiology* (Sect. 7, Endocrinology, Vol. II, Part 2) (R. O. Greep and E. B. Astwood, eds.), p. 187, Am. Physiol. Soc., Washington, D.C.

Psychoyos, A., 1974, Hormonal control of ovo-implantation, *Vitam. Horm.* **32:**201.

Sananes, N., Weiller, S., Baulieu, E./E., and LeGaoscagne, C., 1978, *In vitro* decidualization of rat endometrial cells, *Endocrinology* **103:**86.

Schlafke, S., and Enders, A. C., 1975, Cellular basis of interaction between trophoblast and uterus at implantation, *Biol. Reprod.* **12:**41.

Sherman, M. I., and Wudl, L. R., 1976, The implanting mouse blastocyst, in: *The Cell Surface in Animal Embryogenesis and Development* (G. Poste and G. L. Nicholson, eds.), p. 81, North-Holland, Amsterdam.

Smith, J. A., and Martin, L., 1974, Regulation of cell proliferation, in: *Cell Cycle Controls* (G. M. Padilla, I. L. Cameron, and A. Zimmerman, eds.), p. 43, Academic Press, New York.

Smith, J. A., Martin, L., King, R. J. B., and Vertes, M., 1970, Effects of oestradiol-17β and progesterone on total and nuclear-protein synthesis in epithelial and stromal tissues of the mouse uterus, and of progesterone on the ability of these tissues to bind oestradiol-17β, *Biochem. J.* **119:**773.

Sonnenschein, C., Weiller, S., Farookhi, R., and Soto, A. M., 1974, Characterization of an estrogen-sensitive cell line established from normal rat endometrium, *Cancer Res.* **34:**3147.

Tachi, C., Tachi, S., and Lindner, H. R., 1972, Modification by progesterone of oestradinol-induced cell proliferation, RNA synthesis and oestradiol distribution in the rat uterus, *J. Reprod. Fertil.* **31:**59.

Tseng, L., 1978, Steroid specificity in the stimulation of human endometrial estradiol dehydragenase, *Endocrinology* **102:**1398.

Vladimirsky, F., Chen, L., Amsterdam, A., Zor, U., and Lindner, H. R., 1977, Differentiation of decidual cells in cultures of rat endometrium, *J. Reprod. Fertil.* **49:**61.

West, N. B., Norman, R. L., Sandow, B. A., and Brenner, R. M., 1978, Hormonal control of nuclear estradiol receptor content and the luminal epithelium in the uterus of the golden hamster, *Endocrinology* **103:**1732.

Williams, D., and Gorski, J., 1973, Preparation and characterization of free cell suspensions from the immature rat uterus, *Biochemistry* **12:**297.

14

Decidual Cell Function
Role of Steroid Hormones and Their Receptors

Claude A. Villee, E. Glenn Armstrong, Jr., Deanna J. Talley, and Hiroshi Hoshiai

I. Introduction

The decidual cells in the pseudopregnant rat uterus undergo many of the structural and functional changes exhibited by endometrial cells in normal pregnancy. Thus the decidual reaction in pseudopregnancy has been studied as a model of the implantation process. The proliferation and differentiation of stromal fibroblasts into deciduomal cells require both estrogen and progestin in addition to some kind of trauma to the endometrium to initiate the decidual reaction. The trauma may be a needle scratch, a thread tied through the endometrium, or the transcervical intralumenal instillation of olive oil. The deciduomal tissue of pseudopregnant rats contains high-affinity, low-capacity receptor proteins specific for estrogens (Talley *et al.*, 1977) and other receptor proteins specific for progesterone (Armstrong *et al.*, 1977). The binding of the steroid to the receptor proteins is believed to be an obligatory first step in the action of steroid hormones. Estrogen receptors have been demonstrated in the cytosol (Mester *et al.*, 1974) and in the nucleus (Martel and Psychoyos, 1976) of the endometrium during pregnancy.

II. Induction of Pseudopregnancy and Decidualization

In our experiments pseudopregnancy was induced in virgin Sprague–Dawley rats weighing 180 to 300 g by vaginocervical vibration for 1 min on the morning of estrus (De Feo, 1966; Tobert, 1976). Uterine decidualization was induced unilaterally by the transcervical intralumenal installation of 50 μl olive oil on the afternoon of Day 4 of

Claude A. Villee, E. Glenn Armstrong, Jr., Deanna J. Talley, and Hiroshi Hoshiai • Department of Biological Chemistry and Laboratory of Human Reproduction and Reproductive Biology, Harvard Medical School, Boston, Massachusetts 02115.

pseudopregnancy (Finn and Keen, 1963; Tobert, 1976). Rats were killed on successive days after the instillation of oil and the decidualized horn and the untreated contralateral horn were removed and weighed. The untreated horns weighed approximately 0.15 g; the decidualized horns ranged from 0.6 g on Day 2 to 4.5 g on Day 8 (Figure 1). Decidual horns were slit open and the deciduomal tissue was removed from the myometrium by gentle scraping with a glass slide. Tissues were minced and homogenized at 0°C in 20 vol TED–BSA buffer (0.01 M Tris–HCl, pH 7.5; 1.5 mM EDTA–Na_2; 1 mM dithiothreitol; 0.1% bovine serum albumin). The homogenates were centrifuged for 1 hr at 105,000*g* and the cytosol was treated for 15 min with dextrancoated charcoal. The cytosol preparations were diluted to appropriate protein concentrations and incubated with $10–30 \times 10^{-11}$ M [3,4,6,7-^{3}H]estradiol-17β. To correct for nonspecific binding other samples of the cytosol were incubated with labeled estradiol-17β and a 100-fold excess of radioinert estradiol. After incubation to equilibrium (24 hr) the unbound steroid was removed by treatment with dextran-coated charcoal for 15 min. The amount of labeled estradiol bound to the cytosol receptor was measured and the data were expressed according to Scatchard (1949).

Deciduomal tissue from uterine horns 4 days after decidualization was homogenized in 6 vol of TED buffer containing 0.15 M NaCl and centrifuged at 105,000*g* for 1 hr. The cytosol preparation was incubated 2 hr with 1 nM [3,4,6,7-^{3}H]estradiol-17β or with labeled hormone plus 100 nM radioinert estradiol-17β. The cytosols were treated for 15 min with dextran-coated charcoal, layered on linear gradients of 5–20% sucrose prepared in TED–NaCl buffer, and centrifuged at 39,000 rpm for 17 hr using an SW50.1 rotor. Bovine serum albumin and ^{14}C-labeled gammaglobulin were used as markers to determine the

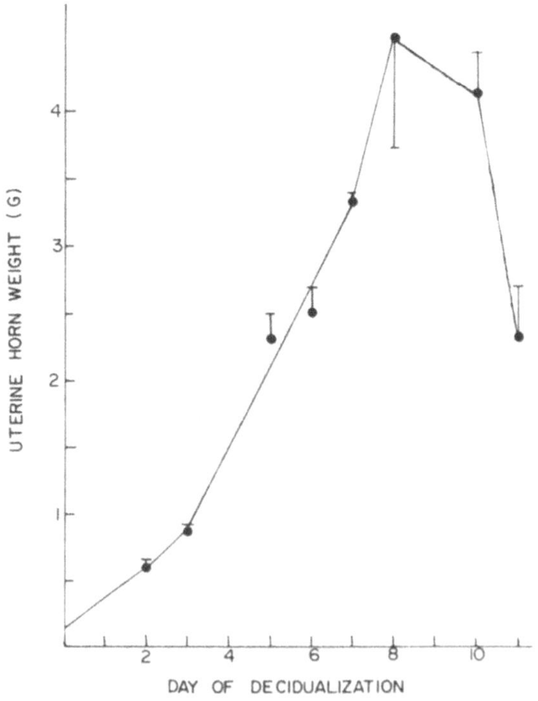

Figure 1. Changes in uterine horn weight on successive days after induction of decidualization by injection of oil into the uterine lumen. The values are mean ± standard deviation for three to eight measurements. From Talley *et al.* (1977).

sedimentation coefficient (Martin and Ames, 1961). The DNA content of pellets from homogenates was determined by the method of Giles and Myers (1965), using calf thymus DNA as standard. The DNA:cytosol protein ratio was 1:25 for these tissues. Assuming that 10 μg DNA represents approximately 10^6 cells, 1 mg of cytosol protein represented about 4×10^7 cells. This relationship has been used to calculate the number of binding sites per cell.

When stimulated by the injection of oil (Day 4 of pseudopregnancy), the rat's uterine horn increases in weight to approximately 30 times that of the untreated horn by the seventh to eighth day of decidualization (Figure 1). The increased weight is due to the induction, proliferation, and growth of the deciduomata from transformed fibroblasts of the endometrium, as well as to hypertrophy of the myometrial cells. By the eighth day the decidual mass is reduced and the onset of necrosis is apparent.

III. Estrogen Receptors in Decidual Tissue

Deciduomal tissue removed from uterine horns on Day 4 after the instillation of oil contained a specific estogen-binding entity with a sedimentation coefficient of 6 S in buffer containing 0.15 M NaCl. The presence of a 100-fold excess of radioinert estrogen during the incubation completely abolished the binding of the radioactive steroid. On Day 2 of decidualization the concentrations of high-affinity estrogen receptor in the untreated and decidualized uterine horns were similar. By Day 3 the deciduomal tissue comprised approximately one-half of the mass of the horn. The concentration of high-affinity estrogen receptor in the cytosols of deciduomal tissue on Day 3 was $1.75 \pm 0.20 \times 10^{-10}$ M, approximately the same as in the cytosols from the untreated uterine horn and from the uteri of rats ovariectomized at estrus. By Day 7 of decidualization the concentration of high-affinity receptor in the deciduomata had decreased to $0.53 \pm 0.10 \times 10^{-10}$M. Thus, during the period of deciduomal differentiation (Days 2 and 3 of decidualization), the concentration of estrogen receptor corresponded to the concentration in nongravid uteri of rats ovariectomized at estrus. By Day 5 of decidualization the level of deciduomal estrogen receptor was reduced despite the fact that the weight of the decidualized uterine horns continued to increase until Day 8 (Figure 2). The number of receptor sites per decidual cell on Day 3 of decidualization was approximately 30,000, similar to that reported in the nongravid uterus (Williams and Gorski, 1974; Chan and O'Malley, 1976). In contrast, the number of receptor sites per cell by Day 7 of decidualization had decreased to about 6000.

These experiments demonstrate that deciduomata contain estrogen receptors with an affinity constant of 10^{10} M^{-1}, identical to that of the estrogen receptor in the nongravid uterus of cycling rats (Toft and Gorski, 1966). The concentration of cytosol binding sites measured during decidualization represents more than 90% of the total receptor concentration. At estrus no more than 10% of the high-affinity estradiol receptors are saturated with circulating endogenous hormones (Williams and Gorski, 1974). As the amount of circulating estradiol during the decidualization process does not exceed the concentration at estrus, it is likely that most of the high-affinity cytoplasmic receptors are being measured in this study. The finding that the concentrations of binding sites in the cytosols of untreated uterine horns on Days 2, 3, and 7 of decidualization are similar to those of uteri from estrous rats after ovariectomy is further support for this assumption. The number of estrogen receptors in the deciduomata has decreased and continues to drop despite continued growth of the tissue until Day 7. The decidual tissue begins to regress on Day 7 and necrotic changes are extensive by Day 10.

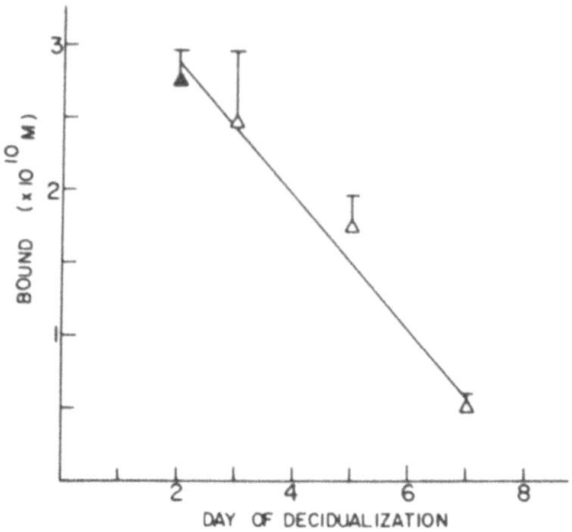

Figure 2. Change in concentration of estradiol binding sites (normalized to 1 mg protein/ml cytosol) as a function of time after decidualization. Decidualized horn (▲), isolated deciduomata (△). From Talley *et al.* (1977).

IV. Progesterone Receptors in Decidual Tissue

Comparable studies were undertaken to measure progesterone receptors in decidual tissue. In all of these experiments cortisol was added to a final concentration of 1 μM to prevent the binding of labeled progesterone to cortisol-binding globulin (Toft and O'Malley, 1972). Two peaks of bound radioactivity were observed following sucrose density gradient centrifugation of decidual cytosol labeled with 2 nM [^{3}H] progesterone in the presence of 1 μM cortisol (Figure 3). A large peak of radioactivity sedimented in the 4–5 S region of the gradient and a somewhat smaller peak in the 6–7 S region. A 100-fold concentration of radioinert progesterone eliminated the binding of [^{3}H] progesterone to 6–7 S macromolecules whereas binding in the 4–5 S region was slightly increased. This increase may represent the greater availability of [^{3}H]progesterone for binding to unsaturable 4–5 S macromolecules in the presence of excess unlabeled progesterone. An equivalent concen-

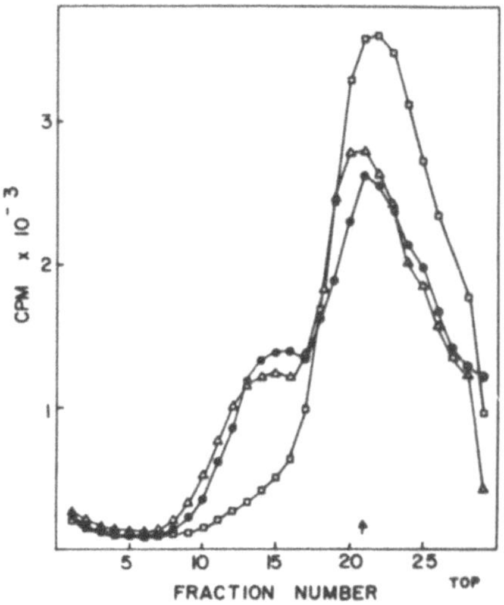

Figure 3. Sucrose gradient sedimentation profiles of the radioactivity in 6-day deciduomal cytosol labeled with 2 nM [^{3}H]progesterone in the absence (△) or presence of 3×10^{-7} M unlabeled progesterone (□) or estradiol-17β (●). The cytosol protein concentration was 26.2 mg/ml. The arrow indicates the sedimentation of bovine serum albumin marker (4.6 S). From Armstrong *et al.* (1977).

tration of estradiol-17β had no effect on the binding of [^{3}H]progesterone to either the 4–5 or the 6–7 S macromolecules. Thus rat deciduomal cytosol contains 6–7 S macromolecules that specifically bind progesterone. Incubation of the deciduomal cytosol at 37°C for 30 min or incubation with N-ethylmaleimide at 0°C inhibited the binding of [^{3}H] progesterone in the 6–7 S region of sucrose density gradients but not in the 4–5 S region. Treating the cytosol with pronase completely eliminated the binding of [^{3}H]progesterone to both the 6–7 S and the 4–5 S macromolecules. These findings support the inference that the 6–7 S macromolecule that binds progesterone in rat decidual cytosol is a heat-labile protein that requires free sulfhydryl groups for binding activity. Progesterone binding was linear with respect to protein concentration to a receptor site concentration of 4.5×10^{-10} M in the cytosol of the proestrous rat uterus, and to a concentration of 1.7×10^{-10} M in deciduomal cytosol.

The concentration of specific progesterone binding sites and their affinity constants were determined by analysis of data according to the method of Scatchard (1949). The affinity constants for progesterone binding were 10^9 M^{-1} in the uteri of estrous and proestrous rats, in the untreated horns on Days 3 and 6 of decidualization, and in the decidualized horns. The average concentrations of progesterone receptor sites in the cytosols from Day 3, 5, and 6 deciduomata were 3.37×10^{-10}, 3.57×10^{-10}, and 2.49×10^{-10} M, respectively. Assuming that 10 μg DNA corresponds to 10^6 cells and that deciduomal polyploidy introduces only a minor error in estimating cell number, there are approximately 50,000 progesterone receptor sites per decidual cell on Days 3 and 5 of decidualization. However, after the fifth day of decidualization, the concentration of progesterone receptor sites decreases linearly and at Day 7 the concentration of receptor sites is 1.63×10^{-10} M, a level approximately one-half that of Day 3 or 5 (Figure 4). This change in receptor concentration in deciduomata from Days 5 to 7 is statistically significant ($P<0.01$).

The development of deciduomata is dependent upon the concerted action of estrogen

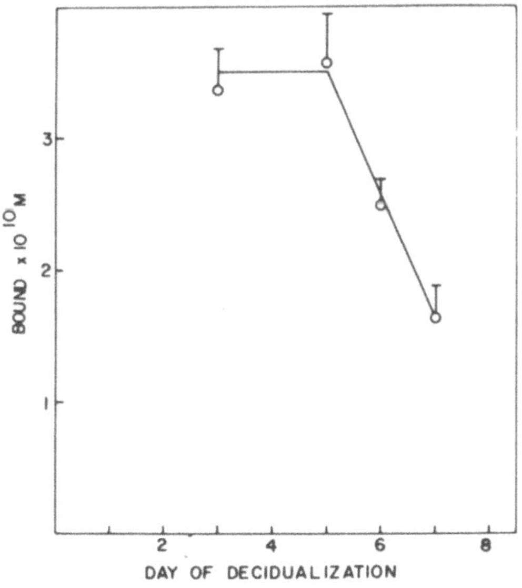

Figure 4. Change in the concentrations of progesterone binding sites (expressed for 1 mg protein/ml cytosol) as a function of days after decidualization. The vertical bars represent the standard error of measurements. The circles represent the means of six measurements on Days 3, 5, and 6 and three on Day 7. From Armstrong *et al.* (1977).

and progestin. A priming dose of estrogen or the estrogen secreted at proestrus, followed by an increase in the concentration of circulating progesterone, is required for decidualization to occur (Nelson and Pfiffner, 1930; Yochim and De Feo, 1962; Finn and Martin, 1972). Progesterone plays a major role in regulating the sensitivity of the uterus to decidualization, but estrogen modulates the changes that occur. Maximal growth of the deciduomata in the rat requires the presence of a continued low level of estrogen in addition to an elevated level of progesterone (Nelson and Pfiffner, 1930; Yochim and De Feo, 1962). When progesterone levels remain elevated for a prolonged period, the deciduomata achieve their maximal growth potential and then degenerate in spite of the continued presence of the hormone (Finn and Porter, 1975). These findings suggest that factors other than the circulating levels of steroid hormones have a regulatory role in deciduomal regression.

The concentration of receptors for estradiol-17β in the cytosol from isolated deciduomata undergoes a linear decrease from Day 3 to Day 7 of decidualization (Finn and Porter, 1975; Talley *et al.*, 1977). In the cytosols of whole decidualized uterine horns from pseudopregnant rats, McCormack and Glasser (1976a) observed a similar but more gradual decrease in specific estrogen binding. The sedimentation coefficient (6–7 S), the affinity constant, and other properties of the progesterone receptor of the decidual tissue are similar to those described for progesterone receptors in the chick oviduct (Toft and O'Malley, 1972) and in the mammalian uterus (Leavitt *et al.*, 1974). The concentrations of progesterone receptors in the cytosols of deciduomata from Days 3 and 5 are similar to those in uterine cytosols of estrous rats and in the contralateral, untreated horn at Days 3 and 6 of decidualization. The concentration of receptor sites in deciduomal cytosols undergoes a linear decrease between Days 5 and 7 of decidualization. This decrease is preceded by a linear decrease in estrogen receptor concentration from Day 2 until Day 7 (Talley *et al.*, 1977). These decreases appear to be independent of changes in the circulating levels of estrogen and progesterone, which remain relatively constant during this period (Yoshinaga *et al.*, 1969; Pepe and Rothchild, 1974). The concentration of progesterone receptor sites in the deciduomata remains constant from Days 3 to 5. As there is a concomitant rapid tissue growth, this finding indicates that the rate of synthesis for the progesterone receptor parallels that for total soluble protein through Day 5 of decidualization. This maintains a constant cellular receptor concentration of about 50,000 sites per cell. After Day 5 of decidualization there is a linear decrease in the progesterone receptor site concentration, even though tissue growth continues through Day 7. The cessation of synthesis of progesterone receptors and the subsequent dilution of the receptor by continued synthesis of total soluble protein could account for the decrease.

The factors causing the decrease in progesterone receptor concentration are unknown, but it has been shown in both birds (Toft and O'Malley, 1972) and mammals (Leavitt *et al.*, 1974) that the level of progesterone receptor is under estrogenic control. the decreased estrogen receptor concentration at Day 5 of decidualization (less than 15,000 sites per cell) may be ineffective in maintaining the concentration of progesterone receptors. The subsequent regression of the deciduomata may result from the decrease in concentration of progesterone receptors. These observations of a decreased progesterone receptor concentration might explain the failure of others to prolong decidualization by administration of progesterone (Finn and Porter, 1975).

V. Nuclear Receptors

Pseudopregnant rats that had been given a decidualizing stimulus 3 days earlier were castrated and injected subcutaneously with 20 mg Depo-provera. Twenty-four hours later

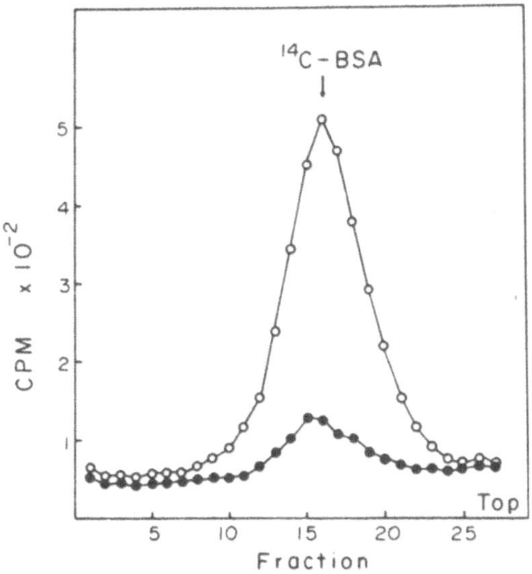

Figure 5. Sucrose gradient profiles of 0.4 M KCl extracts of deciduomal nuclei from 4-day decidualized rats injected with 0.2 μg radioinert estradiol-17β (●). The nuclear extracts contained 0.8 and 1.2 mg protein/ml, respectively. From Armstrong and Villee (1978).

the animals were injected intraperitoneally with 0.2 μg [^{3}H]estradiol-17β with or without 20 μg radioinert estradiol-17β in 1 ml 0.15 M NaCl containing 4% ethanol. One hour later the animals were sacrificed, the deciduomata were removed, and the nuclei were isolated using the method of McCormack and Glasser (1976b) as modified by Armstrong and Villee (1978).

Sucrose density gradient centrifugation of a 0.4 M KCl extract of nuclei isolated from the deciduoma of 4-day pseudopregnant rats injected with [^{3}H]estradiol-17β showed that the estrogen receptor complex sediments at 4.6 S (Figure 5). The labeled hormone appears to be complexed to a macromolecule of low capacity, for the simultaneous injection of a 100-fold excess of unlabeled estradiol-17β markedly reduced the amount of bound radioactivity. Extracting the purified nuclei with 0.4 M KCl solubilized 85% of the specifically bound labeled estrogen. The nuclear exchange assay described by McCormack and Glasser (1976b) was used to quantitate the number of estradiol-17β receptor sites in purified deciduomal nuclei. The binding of estradiol-17β to nuclear receptors was maximal after 2 hr at 20°C and remained stable up to at least 4 hr of incubation.

The data obtained from saturation analysis of the binding of estradiol-17β to purified nuclei were expressed in Scatchard plots. A single class of high-affinity sites with limited capacity for estradiol-17β is present in purified deciduomal nuclei during the third to seventh days of decidualization. The mean values for the concentration of nuclear estrogen receptor sites decreased from Day 3 to Day 7 of decidualization (Figure 6). On Day 3 the concentration of specific nuclear estrogen receptor sites was 0.21 pmol/mg DNA, or approximately 1300 sites per deciduomal cell, based on a DNA content of 10.3 pg per nucleus (McCormack and Glasser, 1976b). The association constants for the binding of estradiol-17β by nuclei from Day 3, 5, and 7 deciduomata did not differ significantly.

The characteristics of the high-affinity, low-capacity estrogen receptor from deciduomal nuclei are quite similar to those reported for the nuclear estrogen receptor of other estrogen target tissues. The estrogen receptor complex from decidual nuclei sediments at 4.6 S on high-salt sucrose density gradients. This value is in reasonable agreement with the 5 S

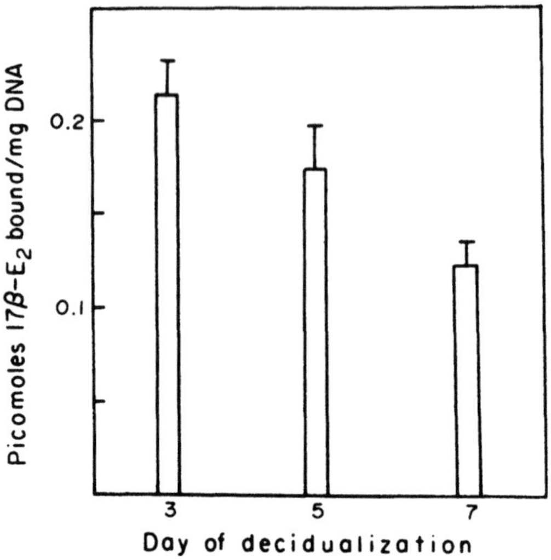

Figure 6. The concentration of specific estradiol-17β receptor sites in nuclei from decidua taken 3, 5, and 7 days after decidualization. The mean values ± standard error of three determinations for each day were: Day 3, $n = 2.1 \pm 0.2 \times 10^{-13}$ mol/mg DNA; Day 5, $n = 1.7 \pm 0.2 \times 10^{-13}$ mol/mg DNA; Day 7, $n = 1.2 \pm 0.1 \times 10^{-13}$ mol/mg DNA.

value obtained by other workers for the nuclear estrogen receptor of rat uterus (Jensen and DeSombre, 1973; Juliano and Stancel, 1976). The apparent association constant, K_a, for the binding of estradiol-17β by purified decidual nuclei was $4 \pm 0.2 \times 10^9$ M^{-1}. McCormack and Glasser (1976b) reported a K_a of 5.4×10^9 M^{-1} for the binding of estradiol-17β by nuclei isolated from the basal-zone trophoblast of pregnant rats. These K_a values are approximately an order of magnitude higher than those reported for the binding of estradiol-17β by the rat uterine nuclear receptor (Clark and Peck, 1976; Juliano and Stancel, 1976). Whether this dissimilarity is due to differences in the actual affinities of the receptors for estradiol-17β or in the techniques employed is not known. The concentration of estrogen receptor sites in decidual nuclei from 3-day decidualized rats, approximately 1300 sites per cell, is quite similar to the level measured by Martel and Psychoyos (1976) in the endometrium of 7-day-pregnant rats, about 1400 sites per cell. The close agreement in the number of nuclear estrogen receptor sites in endometrial cells and in deciduomal cells on Day 7 of pregnancy and pseudopregnancy, respectively, seems particularly relevant, as decidual cells are derived from the stromal fibroblasts of the endometrium.

The concentration of nuclear estrogen receptor in deciduomata on Day 3 is comparable to that in uteri at metestrus or estrus, but considerably less than that at diestrus or proestrus (Clark *et al.*, 1972). The fluctuation in nuclear estrogen receptor concentration in rat uteri during the estrous cycle closely parallels the rate of ovarian estrogen secretion (Yoshinaga *et al.*, 1969). Until Day 10 the concentration of estradiol in the plasma of the pseudopregnant rat is similar to the concentration at metestrus (Welschen *et al.*, 1975). On Day 12 of pseudopregnancy the concentration of estradiol in the plasma is about twice that on Day 10. Thus the low levels of nuclear estrogen receptors present during decidual development are associated with low concentrations of circulating estrogen.

VI. Hormonal Control of Decidualization

The number of nuclear estrogen receptor sites per cell decreases from approximately 1300 on Day 3 to 800 on Day 7 of decidualization. This decrease occurs during the period of rapid formation of deciduomal cells by proliferation and transformation of stromal fibroblasts of the endometrium. The deciduoma of pseudopregnant rats achieves its maximal growth potential on Day 7 of decidualization, after which the tissue regresses (Talley *et al.*, 1977). Clark and Peck (1976) have concluded that the induction of uterotropic responses requires the long-term retention of a low number, about 1400 per nucleus, of estrogen receptor sites in rat uterine nuclei. A similar mechanism may be operative in deciduomata so that the decreased levels of nuclear estrogen receptor sites on Day 7 of decidualization are unable to maintain decidual development and the tissue regresses.

Although there are decreases in the concentration of cytoplasmic estrogen and progesterone receptor sites in decidua following Days 3 and 5 of decidualization, respectively, there is no significant change in the concentrations of estrogen and progesterone receptors in the contralateral untreated uterine horns (Armstrong *et al.*, 1977; Talley *et al.*, 1977). Thus the decreases in receptor levels in the decidua were not the result of a general alteration in the endocrine status of the animal. The decreases in cytoplasmic and nuclear estrogen receptors occur during the same period of decidualization, but the number of cytoplasmic sites decreases more rapidly than that in the nucleus. The number of cytoplasmic estrogen receptor sites is approximately 30,000 per cell on Day 3 and decreases to about 6000 per cell by Day 7.

We could speculate that the decrease in nuclear sites is due to a decreased availability of cytoplasmic receptors for binding of estrogen and subsequent translocation of the receptor–ligand complex into the nucleus. However, a decreased level of nuclear receptor sites might also lead to a decrease in the concentration of cytoplasmic receptors, for estrogen induces the synthesis of its own receptors (Sariff and Gorski, 1971). Why the concentration of estrogen receptors decreases during decidual development is not clear, but it is apparently not due to changes in the concentrations of circulating estradiol-17β or progesterone, as these remain relatively constant during this period (Pepe and Rothchild, 1974; Welschen *et al.*, 1975). The mechanism by which estrogen controls the synthesis of specific proteins appears to involve the regulation of nuclear RNA synthesis by the estrogen receptor complex (Chan and O'Malley, 1976). The decreased concentration of estrogen receptor complexes within decidual nuclei after Day 3 of decidualization may be insufficient to maintain the synthesis of progesterone receptor and the depressed level of progesterone receptor might, in turn, lead to regression of the deciduomata.

Discussion

FINN (Liverpool): Did I understand, when working out the number of receptor sites per cell, that you ignored the polyploidy?

VILLEE: We had to assume that polyploidy was minimal, so we based the number on the amount of DNA in the preparation.

FINN: Since there was a lot of polypoidy, the number of receptors per cell might not have gone down. Does a 60N cell require 60 times the number of receptor sites as a 2N cell, for instance?

VILLEE: I'm sure I can't answer that question and I'm not sure that anyone can, but it would be an interesting thing to think about.

A. KAYE (Weizmann Institute): I'd like to return to a point that Stan Glasser made in reference to the conference at the Weizmann Institute on *Development of Responsiveness to Steroid Hormones* (A. Kaye and M. Kaye, eds., Pergamon Press, Oxford, 1980), in which we considered at length the problem of the correlation of the number of receptor sites to the effective dose of steroids and the response which one obtains. In addition to the familiar example of Clark and Peck (*Nature* (*London*) **260**:635, 1976), in which they showed that one-tenth of the maximal number of receptors which were able to enter the uterus of the immature rat was capable of eliciting the maximal biological response, therefore giving something like a 90% spare receptor capacity, we discussed several other examples in which the correlation between the number of apparently filled active nuclear receptor sites and biological activity of the steroid was minimal. Pending the time at which someone is able to devise an assay which will not only measure the binding of the steroid to the chromatin but will also be able to show some effective activity which directly results from this binding, I think the question that Dr. Finn has just asked and the reservations that Stan Glasser made must be kept at the forefront of our minds. We should not be worried or suspicious if the simple correlation between the number of receptor sites that are filled and the activity of the steroid is not necessarily fulfilled in every case. Looking at the other side of the coin, I don't think we should be particularly pleased if we find that there is a clear correlation between the number of receptor sites filled and biological activity, pending further clarification of just how many of these receptor sites are necessary for the biological activity.

VILLEE: Your point is very well taken and we are aware of these problems. I wouldn't say that we were pleased with ourselves when we found this correlation, but we found it a rather interesting output of our experiments and it did seem to provide at least a reasonable explanation as to why the addition of exogenous progesterone to decidualized animals does not prolong the decidua beyond a certain day. I wouldn't want to go any further than that.

GLASSER (Baylor): Have you ever done these studies in animals maintained on progesterone alone?

VILLEE: These animals are not castrated, they are intact animals. We have done other studies in which we castrate and maintain on Provera but that was a different sort of study and we were not looking at receptors.

GLASSER: I recall a paper by George Marcus (*Biol. Reprod.* **10**:447, 1974) about populations of subepithelial stromal cells which responded to progesterone alone with only a limited amount of decidual growth. With estrogen added to the system, an additional number of cells, called a metastable population of stromal cells, was recruited. These cells decidualized and accounted for the spurt in growth at Days 8 and 9 that really makes the large decidual reaction. I wonder if the drop in relative receptor population is due to the fact that the recruitment of what Marcus called metastable stromal cells and their decidualization produces a subpopulation of cells that do not synthesize any receptor whatsoever?

SHELESNYAK (Washington): I believe the facts as Stan Glasser gave them are correct. May I take this opportunity to make a few comments not specifically related to Claude Villee's excellent paper? I believe it is reputed that some 30 years ago I started a series of studies that can be considered as having stimulated a great deal of research in the field of nidation. I implore investigators for their own sake, for their contemporaries', and for their students' sakes to go back seriously and examine a good deal of the work which has been done in the last two or three decades. There may be some ideas and some data that will be supportive and helpful; there may be some that may be challenging.

The second point I'd like to make is that we are dealing with a biological problem. Modern technology has given us some fantastic tools to isolate and examine specific areas, but if we forget that we are dealing within a total organism and a total system, we do not clarify a great deal. We should do experiments that mimic biological phenomena, but to do so we have to understand the basic biological phenomena. Anyone who says "we treated an animal with hormones to give a normal sequence of events" is talking out of his hat. When we have situations where we can use the normal animal, as Claude Villee has done, then let's not try to play God by giving estrogen and progesterone to mimic something, when there is a great deal more going on in the uterus and in the animal and in the environment than the estrogens and the progesterone which are responsible for the phenomena we are studying. Finally, let us reflect on the fact that we are really seeking to understand mechanisms of action, which means we have to take all the bits and pieces, and try to synthesize them, and place the meaning of our results within the framework of the total picture.

SURANI (Cambridge): I agree with some of what Dr. Shelesnyak has said and I think there is a need to combine different disciplines to understand what is actually happening in the *in vivo* situation. However, I think we should perhaps have a few questions on the paper by Dr. Villee.

DICKMANN (Kansas): I would like Dr. Villee or anyone else in the audience to comment on the mode of action of estrogen via a vehicle other than receptors. I know that some people have opinions that not everything goes through receptors. There are examples in which cells respond to estrogen and receptors, with all the effort that was put into the experiment, have not been found. I think it's also possible to have a cell which has receptors, and the estrogen may work via the receptors, but it may in addition work through another mechanism. Perhaps it would be helpful if there were some comments on some alternative routes that the estrogen may take in order to facilitate its action.

VILLEE: It is remarkable that exactly this question arose in the meeting on Estrogens in the Environment at Raleigh and in the Biochemical Endocrinology meeting up in Maine. There seem to be examples of estrogen action not involving receptors, or happening too fast for receptors to be involved. To go back to ancient history, there is the estrogen-dependent pyridine nucleotide transhydrogenase (Villee and Hagerman 1958, *J. Biol. Chem.* **233**:42) that responds to estradiol in an *in vitro* system which contains no particles, mitochondria, ribosomes, or nuclei. There is no way for receptors to be involved in that reaction, so there is one specific example of an estrogen response that does not involve receptors. I think the hyperemia of a rat uterus is another case that probably does not involve estrogen receptor.

BULLOCK (Baylor): Not everything one observes in a tissue that responds to estrogen need be mediated by binding to the receptors. We should not forget, however, that many tissues, perhaps every tissue in the body, are exposed to the same estrogen that causes the physiological response in the so-called target tissues. If we abandon the idea that one of the prime features of the mechanism of steroid hormone action is the cytoplasmic receptor, then we have to come up with an alternative explanation of why tissues are able to discriminate, so that some of them respond to circulating estrogen and some of them do not. The idea that this is due to the presence in these cells of specific receptors is a very powerful one. The other comment I'd make is that the involvement of receptor steroid binding in the so-called mechanism of hormone action is really only considered in a very proscribed area of the hormone's action, that is in the stimulation of protein synthesis. The hormone may well be doing many other things that do not involve protein synthesis.

RICHARDSON (Harvard): Clark and Gorski (*Science* **169**:76, 1970) documented the ontogeny of the estrogen receptor during early uterine development in the rat, noting that its appearance precedes estrogen secretion from the ovary and is probably not an estrogen-dependent process, but an autonomous property of the cells. Since in older animals the estrogen receptor is an estrogen-dependent protein, one must suppose that both hormone-independent and hormone-dependent mechanisms for receptor production must exist at the level of the genome. Similarly the proliferation of endometrium in humans is clearly estrogen dependent, and yet its earliest phase, which is already visible while menstruation is still occurring, is a rapid regrowth at a time when estrogen is at its nadir (Ferenczy, 1976, *Am. J. Obstet. Gynecol.* **124**:65). In addition, the sensitivity of human endometrium to an estrogen stimulus may not be constant under all conditions. King *et al.* (*Cancer Res.* **39**:1094, 1979) observed that the level of the estrogen-dependent cytosol receptor for progesterone in human endometria is higher for a given level of estradiol receptor in the nucleus in samples of hyperplastic tissue, and especially atypical hyperplasia, than it is in the normal tissue—suggesting that hyperplastic tissue is more estrogen sensitive.

References

Armstrong, E. G., Jr., and Villee, C. A., 1978, Estrogen receptors in purified nuclei from deciduomata of the pseudopregnant rat, *J. Steroid Biochem.* **9**:1149.

Armstrong, E. G., Jr., Tobert, J. A., Talley, D. J., and Villee, C. A., 1977, Changes in progesterone receptor levels during deciduomata development in the pseudopregnant rat, *Endocrinology* **101**:1545.

Chan, L., and O'Malley, B. W., 1976, Mechanism of action of the sex steroid hormones, *N. Engl. J. Med.* **294**:1322.

Clark, J. H., and Peck, E. J., Jr., 1976, Nuclear retention of receptor–oestrogen complex and nuclear acceptor sites, *Nature (London)* **260**:635.

Clark, J. H., Anderson, J., and Peck, E. J., Jr., 1972, Receptor–estrogen complex in the nuclear fraction of rat uterine cells during the estrous cycle, *Science* **176**:528.

De Feo, V. J., 1966, Vaginal-cervical vibration: A simple and effective method for the induction of pseudopregnancy in the rat, *Endocrinology* **79**:440.

Finn, C. A., and Keen, P. M., 1963, The induction of deciduomata in the rat, *J. Embryol. Exp. Morphol.* **11**:673.

Finn, C. A., and Martin, L., 1972, Endocrine control of the timing of endometrial sensitivity to a decidual stimulus, *Biol. Reprod.* **7**:82.

Finn, C. A., and Porter, D. G., 1975, The decidual cell reaction, in: *The Uterus,* p. 81, Flek, London.

Giles, K. W., and Myers, A., 1965, An improved diphenylamine method for the estimation of deoxyribonucleic acid, *Nature (London)* **206**:93.

Jensen, E. V., and DeSombre, E. R., 1973, Estrogen–receptor interaction, *Science* **182**:126.

Juliano, J. V., and Stancel, G. M., 1976, Estrogen receptors in the rat uterus: Retention of hormone–receptor complexes, *Biochemistry* **15**:916.

Leavitt, W. W., Toft, D. O., Strott, C. A., and O'Malley, B. W., 1974, A specific progesterone receptor in the hamster uterus: Physiologic properties and regulation during the estrous cycle, *Endocrinology* **94**:1041.

McCormack, S. A., and Glasser, S. R., 1976a, The developmental biology of a trophoblastic estrogen binding protein, The 58th Meeting of the Endocrine Society, Abstract 43.

McCormack, S. A., and Glasser, S. R., 1976b, A high affinity estrogen-binding protein in rat placental trophoblast, *Endocrinology* **99:**701.

Martel, D., and Psychoyos, A., 1976, Endometrial content of nuclear estrogen receptor and receptivity for ovoimplantation in the rat, *Endocrinology* **99:**470.

Martin, R., and Ames, B., 1961, A method for determining the sedimentation behavior of enzymes: Application to protein mixtures, *J. Biol. Chem.* **236:**1372.

Mester, I., Martel, D., Psychoyos, A., and Baulieu, E.-E., 1974, Hormonal control of oestrogen receptor in uterus and receptivity for ovimplantation in the rat, *Nature (London)* **250:**776.

Nelson, W. O., and Pfiffner, J. J., 1930, Experimental production of deciduomata in rat by extract of corpus luteum, *Proc. Soc. Exp. Biol. Med.* **27:**863.

Pepe, G. J., and Rothchild, I., 1974, A comparative study of serum progesterone levels in pregnancy and in various types of pseudopregnancy in the rat, *Endocrinology* **95:**275.

Sariff, M., and Gorski, J., 1971, Control of estrogen-binding protein concentration under basal conditions and after estrogen administration, *Biochemistry* **10:**2557.

Scatchard, G., 1949, The attraction of proteins for small molecules and ions, *Ann. N.Y. Acad. Scit.* **51:**660.

Talley, D. J., Tobert, J. A., Armstrong, E. G., Jr., and Villee, C. A., 1977, Changes in estrogen receptor levels during deciduomata development in the pseudopregnant rat, *Endocrinology* **101:**1538.

Tobert, J. A., 1976, A study of the possible role of prostaglandins in decidualization using a nonsurgical method for the instillation of fluids into the rat uterine lumen, *J. Reprod. Fertil.* **47:**391.

Toft, D., and Gorski, J., 1966, A receptor molecule for estrogens: Isolation from the rat uterus and preliminary characterization, *Proc. Natl. Acad. Sci. USA* **55:**1574.

Toft, D. O., and O'Malley, B. W., 1972, Target tissue receptors for progesterone: The influence of estrogen treatment, *Endocrinology* **90:**1041.

Welschen, R., Osman, P., Dullart, J., DeGreef, W. J., Uilenbroek, T. J., and DeJong, F. H., 1975, Levels of follicle-stimulating hormone, luteinizing hormone, oestradiol-17β and progesterone and follicular growth in the pseudopregnant rat, *J. Endocrinol.* **64:**37.

Williams, D., and Gorski, J., 1974, Equilibrium binding of estradiol by uterine cell suspensions and whole uteri *in vitro, Biochemistry* **13:**5537.

Yochim, J. M., and De Feo, V. J., 1962, Control of decidual growth in the rat by steroid hormones of the ovary, *Endocrinology* **71:**134.

Yoshinaga, K., Hawkins, R. A., and Stocker, J. F., 1969, Estrogen secretion by the rat ovary during the estrous cycle and pregnancy, *Endocrinology* **85:**103.

V

Gene Expression in the Uterus

Introduction
Gene Expression in the Uterus

Cole Manes

The uterine lumen is for the most part an inhospitable place for cells to grow, even cells of highly malignant tumors (Short and Yoshinaga, 1967). Only for a relatively short period following ovulation does the uterus tolerate the proliferation of nonuterine cells, and when these cells are those of the developing embryo, there may be even more subtle restrictions imposed that are stage specific, at least in some species (Chang, 1950). An understanding of the hormonally regulated alteration between "hostility" and "receptivity" in the uterus is a crucial question and a topic of primary importance to the field of mammalian implantation.

The papers in the preceding section have dealt with cellular changes that can be correlated with uterine receptivity. This section approaches the specific molecular changes accompanying the acquisition of uterine receptivity that provide evidence for alterations in uterine gene expression in the postovulatory phase of the reproductive cycle. The relationship between these alterations and the onset of uterine receptivity is correlative, as it is not yet obvious what exact biological role these molecules play in the reproductive phenomena under examination. One question that continues to overshadow all these investigations is whether the uterus must in some sense "promote" embryonic development, or whether the provision of at least a minimally "permissive" environment is all that the embryo requires.

Davidson and Britten (1979) proposed that structural gene expression displays a basic dichotomy. The majority of structural genes appear to be transcribed continuously and at relatively constant, low rates in all cells. The appearance of their transcripts as functioning messenger RNAs in the cytoplasm is regulated by a variety of post-transcriptional processing mechanisms about which we are just beginning to learn. The expression of some structural genes, however, appears to be regulated primarily at the level of transcription itself. Examples of this latter class include genes that are demonstrably under hormonal control. We may therefore expect that the structural genes whose expression changes in the postovulatory uterus will show characteristics of this class, although, as these papers illustrate, post-

Cole Manes • La Jolla Cancer Research Foundation, La Jolla, California 92037.

transcriptional controls are also operative. It is likely that important new aspects of gene regulation will be uncovered by future investigations in this field.

References

Chang, M. C., 1950, Development and fate of transferred rabbit ova or blastocyst in relation to the ovulation time of recipients, *J. Exp. Zool.* **114:**197.

Davidson, E. H., and Britten, R. J., 1979, Regulation of gene expression: Possible role of repetitive sequences, *Science* **204:**1052.

Short, R. V., and Yoshinaga, K., 1967, Hormonal influences on tumour growth in the uterus of the rat, *J. Reprod. Fertil.* **14:**287.

15

Mechanisms of Induction of Uterine Protein Synthesis

Hormonal Regulation of Uteroglobin

D. W. Bullock, L. W. L. Kao, and C. E. Young

I. Introduction

While "gene expression" is used optimistically in the title of this section, the work in this paper does not deal directly with gene expression but rather with the products of gene expression in the uterus. It would be possible to treat this subject without once mentioning implantation, for at first sight the work seems to be far removed from this process. The connection lies in the need to bridge the gap that exists between the elegant morphological and endocrinological studies that other contributors to this volume have presented, and the biochemical and molecular changes that accompany these events. We know, in other words, a good deal about what hormones do to bring about implantation; the question we now address is how do they do it? Thus we arrive at questions of the mechanism of hormone action on the uterus. The work reported in this and other papers in this section is a first approach to understanding the complexities of the biochemical events related to implantation that are induced by the steroid hormones estradiol and progesterone.

Progress in this area has been hampered by the lack of suitable marker proteins that could be used as biochemical indices of hormone action in the uterus. We have adopted the rabbit uterine protein uteroglobin as a marker for the action of progesterone. Uteroglobin is secreted during the preimplantation period of pregnancy in the rabbit, when it is the dominant protein in uterine secretions (Beier, 1976; Daniel, 1976). The protein exists as a dimer of about 15,000 molecular weight, composed of two identical subunits, 70 amino acids in length, joined by two disulfide bridges (Buehner and Beato, 1978; Mornon *et al.*, 1978, 1979; Ponstingl *et al.*, 1978).

The normal pattern of increase in uteroglobin secretion can be reproduced in nonpregnant rabbits by injections of progesterone (Arthur and Daniel, 1972). Superimposing es-

D. W. Bullock, L. W. L. Kao, and C. E. Young • Departments of Cell Biology and of Obstetrics and Gynecology, Baylor College of Medicine, Houston, Texas 77030.

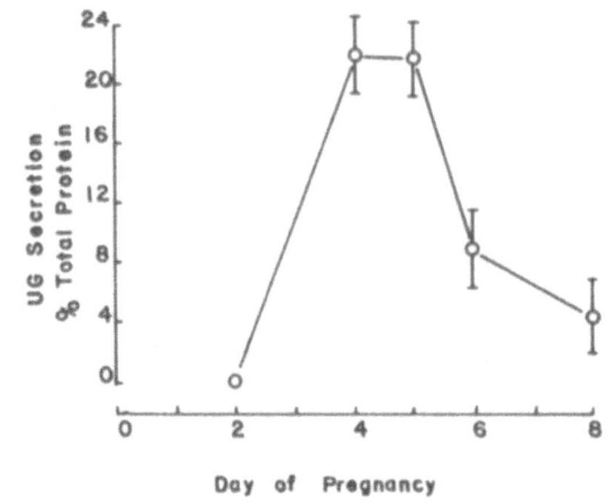

Figure 1. Pattern of uteroglobin (UG) secretion as a percentage of total protein in uterine flushings collected on different days of early pregnancy. Vertical lines show ± S.E.M. (*df* 9).

tradiol on the progesterone treatment reverses the induction of uteroglobin in a manner resembling the decline in its secretion during pregnancy (Bullock and Willen, 1974). The protein binds progesterone (Beato, 1976; Fridlansky and Milgrom, 1976; Beato *et al.*, 1977), but the physiological usefulness of this feature remains speculative.

Despite intensive efforts, there is no reliable evidence for a function of uteroglobin in implantation. Attempts to prevent implantation by passive immunization with anti-uteroglobin antibodies (Krishnan, 1971; Johnson, 1974) have been shown to yield nonspecific results (Bhatt and Bullock, 1978). Nevertheless, uteroglobin is an important component of uterine secretions, which are known to be crucial for blastocyst development and implantation. If the normal pattern of uterine secretion, which can be monitored by uteroglobin determination, is disrupted, implantation is prevented and the blastocysts die. Disruption can be achieved either by delaying secretion through treatment with estradiol (Beier, 1976) or by advancing secretion through treatment with progesterone (McCarthy *et al.*, 1977).

In the absence of a defined function, the usefulness of uteroglobin, therefore, lies in its suitability as a marker protein for the action of progesterone. In this paper, we present data from studies in which the secretion of uteroglobin, the rate of synthesis, the activity of uteroglobin messenger RNA ($mRNA_{UG}$), and the levels of nuclear progesterone receptor have been examined in endometrial tissue collected on different days of early pregnancy.

II. Secretion of Uteroglobin

The pattern of secretion of uteroglobin has been reported from studies on uterine flushings in which uteroglobin has been identified and measured by physical methods. These studies show that the protein begins to accumulate in the secretions on Day 3, reaches a peak on Days 4 to 6, and thereafter declines to virtually nondetectable levels on Day 9 (Beier, 1976). The mass of uteroglobin secreted rises steadily until Day 6 (Daniel, 1976), an observation confirmed by radioimmunoassay (Mayol and Longenecker, 1974). No information is available about the turnover time of the protein in the secretions, but it appears to be secreted rapidly after synthesis (Murray and Daniel, 1973; Joshi and Ebert, 1976; Beato, 1977).

Determination of uteroglobin in uterine flushings collected on different days of early pregnancy, using a specific radioimmunoassay developed in this laboratory (Kao and Bullock, unpublished), gave the results shown in Figure 1, where uteroglobin is expressed as a percentage of the total recovered protein, which was measured by the Folin-phenol method. On Day 2 of pregnancy, uteroglobin accounts for a negligible proportion of the protein in the flushings; it rises to a plateau of about 22% on Days 4 and 5 (Day 0 is the day of mating). After Day 5, uteroglobin declines and has returned to nonpregnant levels by Day 9. This pattern is similar to that reported by others, but the percentage of uteroglobin measured by radioimmunoassay is lower than that found by others using physical methods of separation. The discrepancy may be due in part to calculating the percentage from immunoreactive uteroglobin and total protein determined chemically. Other methods of measuring uteroglobin, however, may not be specific. There is thus an approximately 20- to 25-fold increase in the cumulative amount of uteroglobin found in uterine flushings at its maximum level.

III. Synthesis of Uteroglobin

To determine the rate of synthesis of uteroglobin, we were obliged to turn to *in vitro* methods because the large endogenous amount of this protein made immunoprecipitation impossible and greatly diluted the specific radioactivity after *in vivo* incorporation of radioactive amino acids. Rates of synthesis for uteroglobin after incorporation of radioactive amino acids *in vivo* have been reported (Daniel, 1976), but details of how the measurements were made were not given.

We collected endometrium under sterile conditions from rabbits during early pregnancy. The tissue was cut into small pieces (about 1-mm cubes) and placed on lens paper supported by stainless-steel grids in 35-mm plastic petri dishes. Organ culture was carried out in 2 ml Eagle's Minimum Essential Medium supplemented with 5 μg/ml insulin, 4 mg/ml glucose, 100 IU/ml penicillin, and 100 μg/ml streptomycin. Protein synthesis was

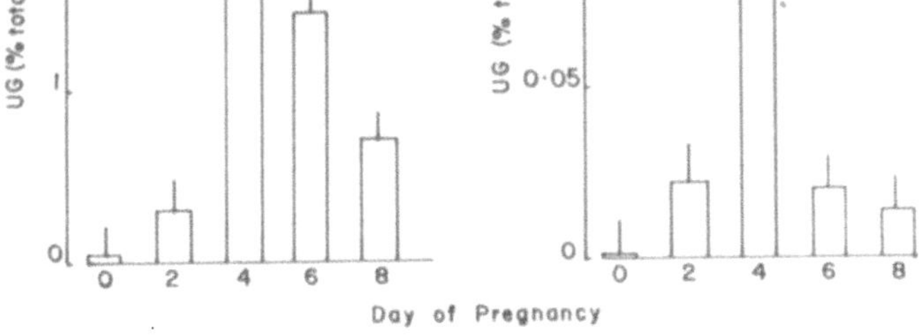

Figure 2. Uteroglobin as a percentage of total protein in medium and tissue after 24 hr of organ culture of pieces of endometrium collected on different days of pregnancy. The lines at the top of the bars show 1 S.E.M. (*df* 16).

assessed by the incorporation of [^{3}H]leucine (5 μCi/ml, specific activity 53 Ci/mmol, Amersham/Searle), with nonradioactive leucine added to a final concentration of 200 μM, for 24 hr; this period was shown in preliminary work to be in the linear range of incorporation. Total protein synthesis was assessed by TCA precipitation of medium and tissue proteins and uteroglobin synthesis was assessed by immunoprecipitation. The mass of uteroglobin in tissue and medium was determined by radioimmunnoassay and compared to the total protein measured by the Folin-phenol method.

Figure 2 shows uteroglobin as a proportion of total protein in tissue and medium after 24 hr of culture. Accumulation of uteroglobin in the medium was discernible on Day 2 and rose to a peak of slightly more than 2% of total protein on Day 4. The pattern is similar to that of uteroglobin secretion *in vivo* (Figure 1). Similar changes were seen in the tissue, but the percentage of uteroglobin was an order of magnitude lower than in the medium. The data confirm the results of others (Beato, 1977) that uteroglobin is secreted rapidly after synthesis. Large amounts of uteroglobin are never found in tissue, either *in vivo* or *in vitro*.

Total protein synthesis in tissue did not change significantly over the days of pregnancy examined (Figure 3). Newly synthesized protein appearing in the medium rose steadily to reach a peak on Day 4 and declined thereafter. In the medium, newly synthesized uteroglobin began to increase on Day 2 and accounted for 60% of the secreted newly synthesized total protein on Day 4 (Figure 4). Incorporation of radioactivity into tissue uteroglobin was below the background of the immunoprecipitation reaction. Thus most of the increase in synthesis of secreted proteins between Day 0 and Day 4 is accounted for by an increased rate of synthesis of uteroglobin. The fact that the mass of uteroglobin accounts for a much smaller percentage of total protein in the medium may be due to leakage of tissue proteins, a common feature of organ culture experiments. By the same token, the ap-

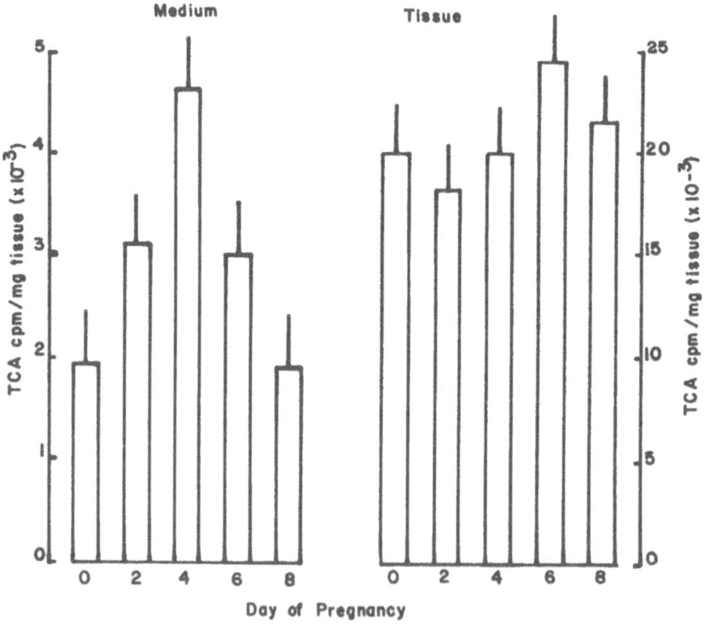

Figure 3. Incorporation of [^{3}H]leucine into total protein of medium and tissue after 24 hr of organ culture of pieces of endometrium collected on different days of pregnancy. The lines at the top of the bars show 1 S.E.M. (*df* 16).

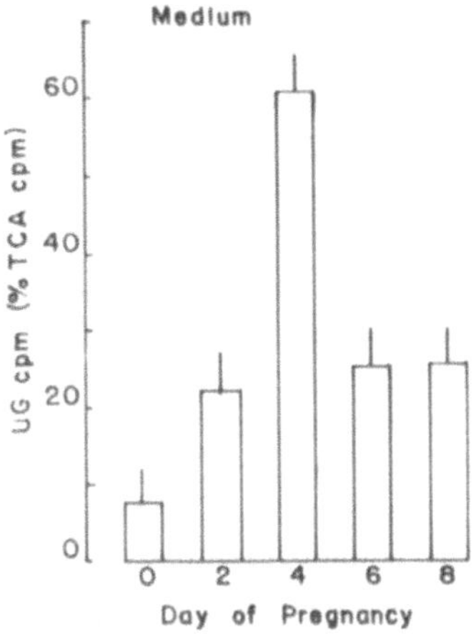

Figure 4. Incorporation of [^{3}H]leucine into uteroglobin as a percentage of incorporation into total protein secreted into the medium during 24 hr of organ culture of pieces of endometrium from different days of pregnancy. The lines at the top of the bars show 1 S.E.M. (*df* 16).

pearance of radioactive uteroglobin in the medium may not give an accurate estimate of the rate of secretion.

Because the amount in the tissue appeared negligible, radioactive uteroglobin in the medium can be taken as a measure of uteroglobin synthesis. There is thus an apparent 12-fold increase in the rate of synthesis of uteroglobin as a percentage of total secreted protein synthesis between Day 0 and Day 4 of pregnancy. The increase in the rate of uteroglobin synthesis precedes the increase in the rate of accumulation of uteroglobin in uterine flushings, which shows about a 20-fold rise over the same period. It is important to note that these changes in rates are expressed relative to total protein.

IV. Activity of Uteroglobin mRNA and Levels of Nuclear Progesterone Receptor

The induction of uteroglobin is correlated with a preferential increase in $mRNA_{UG}$ activity during early pregnancy (Bullock *et al.*, 1976) or after treatment with progesterone (see review by Beato, 1977). Current concepts of steroid hormone action suggest that an increase in specific mRNA activity results, at least in part, from a hormonally induced increase in the rate of transcription of the corresponding gene (O'Malley and Means, 1974). The increased transcription is believed to follow the binding of the steroid–receptor complex to specific sites on nuclear chromatin (O'Malley *et al.*, 1977).

Receptors for progesterone have been identified in the rabbit uterus (Faber *et al.*, 1972, 1973; McGuire and Bariso, 1972; Rao *et al.*, 1973; Philibert and Raynaud, 1974; Terenius, 1974; Muechler *et al.*, 1976; Saffran *et al.*, 1976; Kokko *et al.*, 1977; Philibert *et al.*, 1977; Tamaya *et al.*, 1977) and the levels of cytoplasmic receptors decline after Day 1 of pregnancy (El-Banna and Sacher, 1977). Kokko *et al.* (1977) reported increases in nuclear progesterone receptor that preceded increases in endometrial RNA polymerase I and II activity and in chromatin template capacity after treatment of estrogen-primed rabbits with progesterone.

Endometrium was collected from rabbits on different days of early pregnancy and either frozen in liquid N_2 for RNA extraction or processed immediately for receptor assay. The procedures used for assay of mRNA activity have been described previously (Bullock *et al.*, 1976; Bullock, 1977a). A poly A-rich RNA fraction was prepared from a total nucleic acid extract by binding to oligo dT cellulose. mRNA activity was assessed by translation *in vitro* in the wheat germ system using the incorporation of [^{35}S]methionine Amersham/Searle, sp. act. 800–1200 Ci/mmol) into synthesized peptides. Total mRNA activity was determined by precipitation of the translational products with TCA, and mRNA activity by immunoprecipitation with purified anti-uteroglobin IgG prepared by affinity chromatography (Bullock, 1977b).

For assay of progesterone receptors, endometrial nuclei were purified by the method of Hardin *et al.* (1976) and receptor content was determined by an exchange assay similar to that described by Kokko *et al.* (1977) and validated in our laboratory (Young *et al.*, 1980). Optimum exchange conditions for the binding of [^{3}H]-R5020 (New England Nuclear Corp., sp. act. 86 Ci/mmol) were established as 3 hr at 15°C and the specific binding was analyzed by a Scatchard plot. Studies of the specificity of binding revealed that the relative binding affinities compared to progesterone (ratio of amount of progesterone required for 50% displacement vs. amount of competitor required for 50% displacement) were 2.8 for R5020, 0.26 for 5α-dihydroprogesterone, 0.07 for 20β-dihydroprogesterone, and <0.01 for testosterone, in agreement with other reports (Philibert *et al.*, 1977). In an estrogen-primed rabbit there were approximately 3000 binding sites per nucleus and 30 min after injecting progesterone there were approximately 12,000 sites per nucleus, demonstrating that the nuclear-bound radioactivity arose from translocation of cytoplasmic receptors.

A comparison of changes in $mRNA_{UG}$ activity and nuclear receptor levels is shown in Figure 5. There was about a sevenfold rise between nonpregnant (Day 0) and 4-day-pregnant animals in the proportion of total mRNA activity accounted for by $mRNA_{UG}$ in the poly A fraction. The pattern correlates well with the changes in rate of uteroglobin synthesis and is similar to our earlier report (Bullock *et al.*, 1976), with the addition that these data show that $mRNA_{UG}$ activity has begun to decline on Day 5 (Figure 5A).

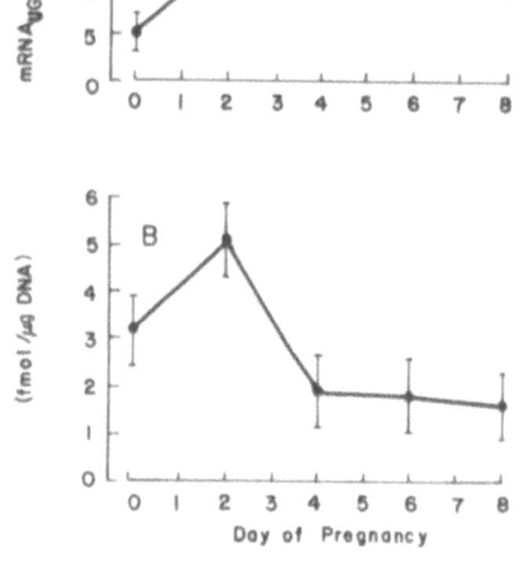

Figure 5. Correlation between uteroglobin mRNA activity and nuclear progesterone receptor concentration in rabbit endometrium during early pregnancy. (A) Uteroglobin mRNA activity as a percentage of total mRNA activity assessed by translation *in vitro* of poly A-rich RNA. (B) Progesterone receptor concentration assessed by specific binding of [^{3}H]-R5020 to purified nuclei under exchange conditions. Vertical lines show ± S.E.M.

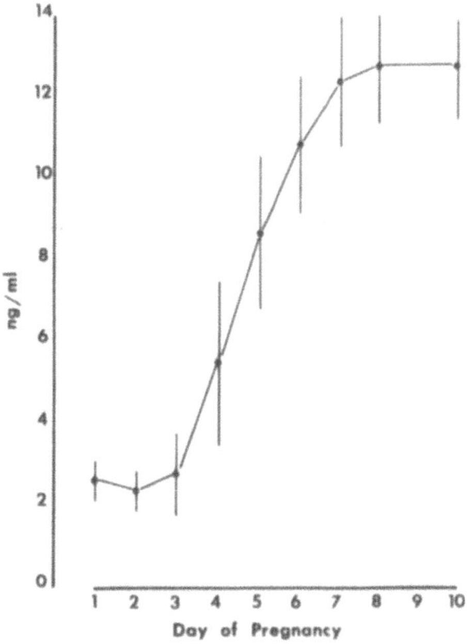

Figure 6. Concentration of peripheral plasma progesterone in rabbits during early pregnancy. Vertical lines show ± S.E.M. ($n = 4$). Drawn from the data of Fuchs and Beling (1974).

Levels of nuclear progesterone receptor showed little change throughout this stage of pregnancy (Figure 5B). There was a small, but statistically significant ($p < 0.05$), rise in the concentration of receptors on Day 2. This small rise precedes the increase in $mRNA_{UG}$ activity, but receptor levels are low when $mRNA_{UG}$ is at its peak. It is unlikely that nuclear receptors existed that were undetected by the assay. Complete exchange was shown to occur and the number of binding sites per cell (approximately 16,000 on Day 2) is similar to the number we obtained after translocation of cytoplasmic receptors by injection of progesterone. The equilibrium K_D did not change significantly between the days of pregnancy examined, suggesting that endogenous progesterone did not interfere with the exchange assay.

The transient rise in receptors may be sufficient to trigger the synthesis of long-lived mRNA or the low level of nuclear receptor may exceed the threshold required for stimulation of $mRNA_{UG}$ activity. It is equally possible that progesterone induces $mRNA_{UG}$ activity by mechanisms that are difficult to correlate with changes in level of its nuclear receptor. These correlative studies refer only to steady-state conditions and establishment of cause-and-effect relationships will require a kinetic analysis of the events following stimulation by progesterone and the binding of the receptor to nuclear chromatin. Such analysis will be facilitated by the recent synthesis of a DNA complementary to $mRNA_{UG}$ (Arnemann *et al.*, 1979; Atger *et al.*, 1980; Beato *et al.*, 1980).

The nuclear receptor data are also difficult to reconcile with our present understanding of receptor translocation. Rising steroid levels are considered to cause a shift in receptors from the cytoplasm to the nucleus, where the steroid–receptor complex exerts its biochemical actions. Thus reductions in cytoplasmic progesterone receptor during the human menstrual cycle are reflected in corresponding increases in nuclear receptor concentration (Bayard *et al.*, 1978). The data of El-Banna and Sacher (1977) show that cytoplasmic progesterone receptor in the rabbit uterus declines and remains low after Day 1 of pregnancy. Taken together with these data, our results do not show an increase in nuclear receptor when cytoplasmic receptor levels have declined.

Although we and Kokko *et al.* (1977) have demonstrated that a rapid translocation of receptors to the nucleus can be caused by progesterone, our estimates of nuclear receptor levels do not correlate with the rise in circulating progesterone that begins on Day 4 of pregnancy (Figure 6). During the first 3 days of pregnancy, when cytoplasmic receptor levels have declined and a transient increase in nuclear receptor levels occurs, there is little change in the plasma concentration of progesterone. In fact, there appears to be little increase in blood levels of progesterone at the time the activity of the mRNA and the synthesis and secretion of this progesterone-induced protein begin to rise.

A similar discrepancy between receptor level and the tissue response to progesterone was found when uteroglobin was induced by injections of steroid hormone. Whereas a single injection of progesterone causes a transient rise in nuclear progesterone receptor (Kokko *et al.*, 1977; this paper), treatment for 5 days causes a significant decline in the number of receptor sites per nucleus while increasing uteroglobin secretion 3000-fold (Isomaa *et al.*, 1979). Furthermore, when estradiol was given to inhibit the induction of uteroglobin by progesterone, there was an increase both in cytosol and in nuclear progesterone receptors (Isomaa *et al.*, 1979). Despite the maintenance of high levels of receptor, treatment with estradiol lowered the activity of $mRNA_{UG}$ (Kopu *et al.*, 1979).

Our work thus lends weight to the conclusion that recent data about the regulation of uteroglobin by progesterone and estradiol are not compatible with prevailing concepts of receptor-mediated control derived from acute-injection experiments. Hormonal regulation of a specific uterine protein under physiological conditions may be a complex process involving multiple levels of control.

ACKNOWLEDGMENTS. We thank Dr. Roy Smith for advice about receptor assay, and Drs. Edwin Milgrom and Miguel Beato for advance notice of their results. The work was supported by NIH Grant HD 09378 and by a grant from the Rockefeller Foundation.

Discussion

MACLAUGHLIN (Boston): Is the time lag between the appearance of a peak of progesterone receptor in the nucleus, say 2 days before the maximum activity of messenger RNA, accounted for by processing of enough message to be detected in an *in vitro* assay?

BULLOCK: That is possible. I tried to emphasize that we are dealing with translatable activity of steady-state message, which is very far removed from a transcriptive event, and could be related not only to processing of gene transcript but also to rates of initiation, rates of chain elongation, or even alterations in turnover and half-life of the messenger RNA.

MACLAUGHLIN: Have you ever seen pre-uteroglobin secreted *in vivo* and does colchicine or any sort of inhibitor block its secretion?

BULLOCK: We've not tried any experiments with colchicine. Not only have we not found pre-uteroglobin in uterine flushings, we have not found it in uterine tissue. Nobody has found a preprotein inside a cell that is represented by the product plus the signal sequence, which is evidently cleaved rapidly within the endoplasmic reticulum. We did make an attempt to identify the primary translation product in the tissue and we were unsuccessful.

ILAN (Cleveland): I think you would find the primary product of translation if you resorted to the reticulocyte cell-free system.

BULLOCK: We find the primary product of translation *in vitro* because the *in vitro* systems don't have membranes and don't have the postulated clipase. The problem is to demonstrate it *in vivo*.

ILAN: You do get the primary product with the wheat germ system?

BULLOCK: Yes, the translation product of the wheat germ system is larger than native uteroglobin; that is how we first demonstrated the existence of the leader sequence.

ILAN: Peeters *et al.* (*Biochim. Biophys. Acta* **561**:502, 1979) recently published that the wheat germ system does in fact have the clipping enzyme. Also, the wheat germ system often aborts short peptides, so it would be a good idea to use the reticulocyte cell-free system.

BULLOCK: Beato (in: *Development in Mammals* (M. H. Johnson, ed.), Vol. 1, pp. 361–383, North-Holland, Amsterdam, 1977) has done a comparative study of uteroglobin translation in different *in vitro* systems. In the ascites tumor system and in the wheat germ system he obtained pre-uteroglobin, as we do with the wheat germ system. In the oocyte translation system, he obtained the processed peptide. We have difficulties utilizing the reticulocyte lysate system because our product gels are difficult to interpret, since pre-uteroglobin is very similar in molecular weight to globin. Even with a suppressed lysate system, the small amount of residual endogenous globin synthesis tends to mask the identity of the uteroglobin; that is one reason why we have used the wheat germ system.

ROBERTS (Florida): How much of the uteroglobin released was newly synthesized and how much was released at the expense of prepackaged material in the cytoplasm?

BULLOCK: Numbers on that are hard to come by. There is evidently a good deal of nonradioactive uteroglobin secreted into the medium, yet storage in the cell prior to secretion does not agree with the synthesis data. On a mass basis the amount of uteroglobin in the medium is a much smaller percentage of the total protein than is the percentage of uteroglobin on the basis of incorporation of radioactivity.

STONE (Nebraska): Similar results have been published for rat liver albumin (Judah and Nicholls, 1971, *Biochem. J.* **123**:649). In terms of the mass of albumin in the liver and the amount of the newly synthesized product released, there seems to be a large intracellular pool of albumin that does not serve as a percursor to that which is released into the medium.

BEIER (Aachen): Have you done any studies to confirm the induction claimed by Muriel Feigelson with testosterone? Would you also comment on the ideas that were originated by your own studies on switching off the whole mechanism by estrogen?

BULLOCK: Those are two very interesting questions. The specificity of induction of uteroglobin by progesterone in terms of an increase in synthesis, whether or not that is due to an induction of gene expression, is fairly good. Small amounts of the protein can be stimulated by estrogen, which in my own view is a cellular phenomenon and not an intervention in the synthetic mechanism. The studies published with other steroids, such as testosterone, suffer from being done with extremely high doses. In the work you mentioned, I think 25 mg of testosterone were used. If indeed the receptor mechanism is involved in the synthesis, I interpret those results from very high doses of testosterone as being due to a kind of false binding to progesterone receptors, as Janne *et al.* (*Int. J. Androl. Suppl.* **2**:162, 1978) suggested. We have not investigated further the mechanism of inhibition by estrogens.

BARKER (Nebraska): Uteroglobin has been found in a number of tissues, as you indicated, including the lung and other tissues with secretory epithelium. Has anyone considered the possibility that uteroglobin might be a membrane component of secretory granules? In tissues which secrete proteins by this mechanism, this molecule may be spilled into the medium. Have you looked at any possible interactions between your antibody and membrane fragments?

BULLOCK: Yes, that idea has been considered by a number of people. I have difficulty with that concept because at an ultrastructural level there does not seem to be enough secretory granules in the uterine epithelium to account for the enormous-fold increase in the synthesis and secretion of uteroglobin. There is no good evidence that the secretion of uteroglobin even involves vesicular packaging; how it gets across the membrane is not understood.

RAHMAN (Cleveland): In organ culture of uterine tissue from a pseudopregnant animal, can you maintain the level of uteroglobin by steroid treatment or bring it up to the level of a 4-day-pregnant animal?

BULLOCK:. We have done no studies with pseudopregnant animals *in vitro*.

RAHMAN: If you take endometrium from an estrous animal, can you induce uteroglobin by progesterone *in vitro?*

BULLOCK: We have made a number of attempts to study steroid effects on uteroglobin secretion *in vitro*, with mixed results. We have tried with tissue taken from nonpregnant animals and also Day 4 of pregnancy. In neither case did we obtain very convincing effects of progesterone, although they were discernible. They were more discernible in the case of the tissue from the 4-day-pregnant animal when the message activity is already high, but it was hard to demonstrate a stimulatory effect in the same dish by measuring the synthesis or secretion before and after adding progesterone. In dishes which had progesterone present continuously for 48 hr, there was a higher level of incorporation of label and a higher mass secretion of uteroglobin into the medium than in control dishes incubated without progesterone. Steroid may be necessary in the medium in order to stabilize the receptors, which are known not to survive very long under *in vitro* conditions; it then becomes difficult to do before-and-after stimulation experiments in the same tissue.

CHANG (Worcester Foundation): Since Dr. Beier's and Dr. Daniel's studies, we always think about uteroglobin as very important for implantation. Did you try to put a drop of uteroglobin into the uterus to see whether the blastocyst will implant or not?

BULLOCK: We have attempted the reverse experiment, which is to take uteroglobin away and see if implantation fails. We inhibited implantation by passive immunization with specific anti-uteroglobin antibodies injected into the uterine lumen. Other people have obtained a similar effect, but in my view it is impossible to arrive at meaningful results from this technique. We showed that nonspecific antibody put into the uterine lumen will also prevent implantation; we could not show a specific effect of anti-uteroglobin (Bhatt and Bullock, 1978, *J. Reprod. Fertil.* **54:**177).

BEIER: My interpretation of all available data is that uteroglobin may not be related to implantation in a direct way. I think it is more conceivable that it is involved in the maternal environment and is acting, especially at the end of Day 3 and the beginning of Day 4, to provide good conditions for the initial phase of the uterine development of the embryo.

BULLOCK: I think that is an interesting possibility. As I said, in my view there is no good evidence for a physiological role for the protein.

References

Arnemann, J., Heins, B., and Beato, M., 1979, Synthesis and characterization of a DNA complementary to preuteroglobin mRNA, *Eur. J. Biochem.* **99:**361.

Arthur, A. T., and Daniel, J. C., Jr., 1972, Progesterone regulation of blastokinin production and maintenance of rabbit blastocysts transferred into uteri of castrate recipients, *Fertil. Steril.* **23:**115.

Atger, M., Mornon, J. P., Savouret, J. F., Loosfelt, H., Fridlansky, F., and Milgrom, E., 1980, Uteroglobin: A model for the study of the mechanism of action of steroid hormones, in: *Steroid-Induced Uterine Proteins* (M. Beato, ed.), pp. 341–350, Elsevier, Amsterdam.

Bayard, F., Damilano, S., Robel, P., and Baulieu, E.-E., 1978, Cytoplasmic and nuclear estrogen and progesterone receptors in human endometrium, *J. Clin. Endocrinol. Metab.* **46:**635.

Beato, M., 1976, Binding of steroids to uteroglobin, *J. Steroid Biochem.* **7:**327.

Beato, M., 1977, Hormonal control of uteroglobin biosynthesis, in: *Development in Mammals,* Vol. 1 (M. H. Johnson, ed.), pp. 361–383, North-Holland, Amsterdam.

Beato, M., Arnemann, J., and Voss, H.-J., 1977, Spectrophotometric study of progesterone binding to uteroglobin, *J. Steroid Biochem.* **8:**725.

Beato, M., Arnemann, J., Heins, B., Muller, H., and Nieto, A., 1980, Correlation between uteroglobin synthesis and uteroglobin mRNA content in rabbit endometrium, in: *Steroid-Induced Uterine Proteins* (M. Beato, ed.), pp. 351–368, Elsevier, Amsterdam.

Beier, H. M., 1976, Uteroglobin and related biochemical changes in the reproductive tract during early pregnancy in the rabbit, *J. Reprod. Fertil. Suppl.* **25:**53.

Bhatt, B. M., and Bullock, D. W., 1978, Non-specific effects of passive immunization on implantation in the rabbit, *J. Reprod. Fertil.* **54:**177.

Buehner, M., and Beato, M., 1978, Crystallization and preliminary crystallographic data of rabbit uteroglobin, *J. Mol. Biol.* **120:**337.

Bullock, D. W., 1977a, Progesterone induction of messenger RNA and protein synthesis in rabbit uterus, *Ann. N.Y. Acad. Sci.* **286:**260.

Bullock, D. W., 1977b, *In vitro* translation of messenger RNA for a uteroglobin-like protein from rabbit lung, *Biol. Reprod.* **17:**104.

Bullock, D. W., and Willen, G. F., 1974, Regulation of a specific uterine protein by estrogen and progesterone in ovariectomized rabbits, *Proc. Soc. Exp. Biol. Med.* **146:**294.

Bullock, D. W., Woo, S. L. C., and O'Malley, B. W., 1976, Uteroglobin messenger RNA: Translation *in vitro*, *Biol. Reprod.* **15:**435.

Daniel, J. C., Jr., 1976, Blastokinin and analogous proteins, *J. Reprod. Fertil. Suppl.* **25:**71.

El-Banna, A. A., and Sacher, B., 1977, A study on steroid hormone receptors in the rabbit oviduct and uterus during the first few days after coitus and during egg transport, *Biol. Reprod.* **17:**1.

Faber, L. E., Sandmann, M. L., and Stavely, H. E., 1972, Progesterone-binding proteins of rat and rabbit uterus, *J. Biol. Chem.* **247:**5648.

Faber, L. E., Sandmann, M. L., and Stavely, H. E., 1973, Progesterone and corticosterone binding in rabbit uterine cytosols, *Endocrinology* **93:**74.

Fridlansky, F., and Milgrom, E., 1976, Interaction of uteroglobin with progesterone, 5α-pregnane-3,20-dione and estrogens, *Endocrinology* **99:**1244.

Fuchs, A-R., and Beling, C., 1974, Evidence for early ovarian recognition of blastocysts in rabbits, *Endocrinology* **95:**1054.

Hardin, J. W., Clark, J. H., Glasser, S. R., and Peck, E. J., Jr., 1976, RNA polymerase activity and uterine growth: Differential stimulation by estradiol, estriol and nafoxidine, *Biochemistry* **15:**1370.

Isomaa, V., Isotalo, H., Orava, M., Torkkeli, T., and Janne, O., 1979, Changes in cytosol and nuclear progesterone receptor concentrations in the rabbit uterus and their relation to induction of progesterone-regulated uteroglobin, *Biochem. Biophys. Res. Commun.* **88:**1237.

Johnson, M. H., 1974, Studies using antibodies to the macromolecular secretions of the early pregnancy uterus, in: *Immunology in Obstetrics and Gynecology* (A. Centaro and N. Caretti, eds.), pp. 123–133, Int. Congr. Ser. No. 327, Excerpta Medica, Amsterdam.

Joshi, S. G., and Ebert, K. M., 1976, Effects of progesterone on labelling of soluble proteins and glycoproteins in rabbit endometrium, *Fertil. Steril.* **27:**730.

Kokko, E., Isomaa, V., and Janne, O., 1977, Progesterone-regulated changes in transcriptional events in rabbit uterus, *Biochim. Biophys. Acta* **479:**354.

Kopu, H. T., Hemminki, S. M., Torkkeli, T. K., and Janne, O., 1979, Hormonal control of uteroglobin secretion in rabbit uterus, *Biochem. J.* **180:**491.

Krishnan, R. S., 1971, Effects of passive administration of antiblastokinin on blastocyst development and maintenance of pregnancy in rabbits, *Experientia* **27:**955.

McCarthy, S. M., Foote, R. H., and Maurer, R. R., 1977, Embryo mortality and altered uterine luminal proteins in progesterone-treated rabbits, *Fertil. Steril.* **28:**101.

McGuire, J. L., and Bariso, C. D., 1972, Isolation and preliminary characterization of a progestagen specific binding macromolecule from the 273,000*g* supernatant of rat and rabbit uteri, *Endocrinology* **90:**496.

Mayol, R. F., and Longenecker, D. E., 1974, Development of a radioimmunoassay for blastokinin, *Endocrinology* **95:**1534.

Mornon, J. P., Surcouf, E., Bally, R., Fridlansky, F., and Milgrom, E., 1978, X-Ray analysis of a progesterone-binding protein (uteroglobin): Preliminary results, *J. Mol. Biol.* **122:**237.

Mornon, J. P., Bally, R., Fridlansky, F., and Milgrom, E., 1979, Characterization of two new crystal forms of uteroglobin, *J. Mol. Biol.* **127:**237.

Muechler, E. K., Flickinger, G. L., Mastroianni, L., Jr., and Mikhail, G., 1976, Progesterone binding in rabbit oviduct and uterus, *Proc. Soc. Exp. Biol. Med.* **151:**275.

Murray, F. A., and Daniel, J. C., Jr., 1973, Synthetic pattern of proteins in rabbit uterine flushings, *Fertil. Steril.* **24:**692.

O'Malley, B. W., and Means, A. R., 1974, Female steroid hormones and target cell nuclei, *Science* **183:**610.

O'Malley, B. W., Towle, H. C., and Schwartz, R. J., 1977, Regulation of gene expression in eucaryotes, *Annu. Rev. Genet.* **11:**239.

Philibert, D., and Raynaud, J.-P., 1974, Progesterone binding in the immature rabbit and guinea pig uterus, *Endocrinology* **94:**627.

Philibert, D., Ojasoo, T., and Raynaud, J.-P., 1977, Properties of the cytoplasmic progestin-binding protein in the rabbit uterus, *Endocrinology* **101:**1579.

Ponstingl, H., Nieto, A., and Beato, M., 1978, Amino acid sequence of progesterone-induced uteroglobin, *Biochemistry* **17:**3908.

Rao, B. R., Wiest, W. G., and Allen, W. M., 1973, Progesterone "receptor" in rabbit uterus. I. Characterization and estradiol-17β augmentation, *Endocrinology* **92:**1229.

Saffran, J., Loeser, B. K., Bohnett, S. A., and Faber, L. E., 1976, Binding of progesterone receptor by nuclear preparations of rabbit and guinea pig uterus, *J. Biol. Chem.* **251:**5607.

Tamaya, T., Nioka, S., Furuta, N., Shimura, T., and Takano, N., 1977, Contribution of functional groups of 19-nor-progesterone to binding of progesterone and estradiol-17β receptors in rabbit uterus, *Endocrinology* **100:**1579.

Terenius, L., 1974, Affinities of progestogen and estrogen receptors in rabbit uterus for synthetic progestins, *Steroids* **23:**909.

Young, C. E., Smith, R. G., and Bullock, D. W., 1980, Uteroglobin mRNA activity and levels of nuclear progesterone receptor in endometrium, *Mol. Cell. Endocrinol.* (in press).

16

Regulation of the Levels of mRNA for Glucose-6-phosphate Dehydrogenase and Its Rate of Translation in the Uterus by Estradiol

Kenneth L. Barker, David J. Adams, and Terrence M. Donohue, Jr.

I. Introduction

The enzyme glucose-6-phosphate dehydrogenase (G6PD) catalyzes the oxidation of D-glucose-6-phosphate to 6-phospho-D-gluconolactone with the simultaneous reduction of $NADP^+$ to NADPH. The enzyme is the first in the pentose phosphate pathway (hexose monophosphate shunt), which produces pentose phosphates used for nucleotide and nucleic acid synthesis, and reducing equivalents in the form of NADPH for use in biosynthetic reduction reactions, including fatty acid synthesis and hydroxlations. G6PD has been purified and characterized from the tissue cytosols of several mammalian species, including human erythrocytes, rat mammary gland, rat and mouse liver, and bovine adrenal cortex (Bonsignore and DeFlora, 1972; Levy, 1979). In general, the enzymes from these sources are "similar" and are composed of subunits that, by immunological and molecular weight criteria, may be identical proteins within a species. The monomeric subunits have a molecular weight of about 60,000, based on their rate of migration on SDS–polyacrylamide gels. The monomers are catalytically inactive. In the presence of $NADP^+$, the inactive monomers form a catalytically active dimer containing 1 or perhaps 2 moles of $NADP^+$ per mole of dimer; at reduced ionic strength and pH, and in the presence of Mg^{2+}, the dimers aggregate to form tetramers and hexamers.

Using standard polyacrylamide gel electrophoretic methods, one to four "isozymes" of G6PD activity can be separated. These proteins have been identified as aggregates of the

Kenneth L. Barker, David J. Adams, and Terrence M. Donohue, Jr. • Departments of Obstetrics and Gynecology and of Biochemistry, University of Nebraska College of Medicine, Omaha, Nebraska 68105. Current address of D.J.A.: Department of Medicine, University of Texas Medical School, San Antonio, Texas. Current address of T.M.D.: Biology Division, Oak Ridge National Laboratory, Oak Ridge, Tennessee 37830.

active dimer based on: (a) their relative molecular weights, (b) their immunological cross-reactivity, and (c) their ability to generate all four "isozymes" from the purified dimer by changing ionic conditions (Holton, 1972). The amino acid compositions of rat liver, bovine adrenal, human erythrocyte, and mouse liver G6PD are known and all have a similar composition. A striking feature of the molecule is the presence of pyroglutamate on the amino terminus of the monomer of the human erythrocyte enzyme, the bovine adrenal enzyme (Yoshida, 1972; Singh and Squire, 1975), and the rat liver and uterine enzymes (Donohue *et al.*, 1977). As methionine is considered to be at the initiation site of eukaryotic protein synthesis, it should be found at the amino terminus of those proteins that are not "processed" by removal of peptides from their amino terminus. The presence of pyroglutamate at the amino terminus indicates, therefore, that G6PD must exist transiently as a "pre-enzyme" in the free and/or nascent (bound to ribosomes) state and that the pre-enzyme must be processed by proteolytic cleavage of an amino acid or peptide from the animo terminus, followed by cyclization of an amino-terminal glutamate or glutamine residue to form pyroglutamate.

The activity of G6PD increases in the uterus of the ovariectomized mature rat *in vivo* in response to: (a) estrogen administration (Scott and Lisi, 1960; Barker and Warren, 1966; Molton and Barker, 1971; Baquer and McLean, 1972), (b) intrauterine administration of the cofactor $NADP^+$ (Barker, 1967), and (c) stimuli that increase the amount of glucose that is metabolized by the pentose phosphate pathway (Smith and Barker, 1976). Increased utilization of the pentose phosphate pathway can be initiated by the intrauterine administration of 50 μmol fluoride ion, which acts by blocking glycolysis at the enolase step, or by the intrauterine administration of iodoacetate, which preferentially blocks glycolysis by complexing with the very sensitive sulfhydryl group on glyceraldehyde-3-phosphate dehydrogenase. The estradiol-induced G6PD response, but not the $NADP^+$- or fluoride-induced response, is sensitive to prior treatment of the animals with actinomycin D, while the G6PD responses to all three stimuli are inhibited by prior treatment of the animals with cycloheximide. This observation suggests that the effects of $NADP^+$ and fluoride are mediated at a posttranscriptional step of protein synthesis (Barker, 1967; Smith and Barker, 1976). The physiological significance of the $NADP^+$ and fluoride responses, relative to the estrogen response, lies in the fact that estradiol causes increased *in vivo* levels of uterine $NADP^+$ (O'Dorisio and Barker, 1970, 1976). In addition, the utilization of the pentose phosphate pathway is enhanced by estradiol prior to any increase in uterine G6PD activity (Barker and Warren, 1966). Thus the effects of $NADP^+$ and pathway utilization are likely to be an integral part of the overall uterine G6PD response to estradiol.

The mechanisms of the increase in uterine G6PD activity in response to estradiol treatment have been studied using intrauterine pulse and pulse-chase injections of radiolabeled amino acids. Immunoprecipitation of G6PD from uterine cytosol preparations has been employed to assess directly the effects of the hormone on the rates of synthesis and degradation of the uterine G6PD apoenzyme. Estradiol *in vivo* causes a 10- to 20-fold increase in the rate of synthesis of this protein and, in addition, it completely inhibits the degradation of the enzyme (Smith and Barker, 1974, 1977). The estrogen-induced increase in the rate of synthesis of uterine G6PD appears to reside both at the transcriptional level and at the posttranscriptional level of protein synthesis. Evidence for the *transcriptional* level of control includes: (a) the inhibition of the estradiol-induced increase in G6PD activity by prior treatment with actinomycin D (Barker, 1967; Moulton and Barker, 1972), (b) blockage of the continuation of the estrogen-induced increase in G6PD activity, at any time during the response, by the intraperitoneal administration of 8-azaguanine, which causes the synthesis of "fraudulent mRNAs" (Moulton and Barker, 1974), and (c) the accumulation of G6PD mRNA activity during *in vitro* inhibition by cycloheximide of protein synthesis in uteri from estrogen-treated rats (Keran and Barker, 1976). Evidence for the *posttranscriptional* level of control of G6PD synthesis includes primarily the ability of estradiol to "reinduce" an increase in uterine G6PD activity. This ability is evident in "actinomycin D-blocked rats" when a second injection of estradiol is given on the second or third day following the first treatment with the hormone (Moulton and Barker, 1972). Results presented in this chapter provide further direct evidence for regulation of G6PD synthesis by estradiol: (a) at the *transcriptional level* by quantifying the G6PD mRNA activity in isolated uterine RNA, (b) at the *translational level* by measuring the ribosome transit time required to synthesize and release G6PD into the uterine cytosol, and (c) at the *protein processing level* by estimating the lag between the release of G6PD from the ribosome and the formation of pyroglutamate at its amino terminus.

II. Regulation of Uterine G6PD Synthesis by Estradiol

The effects of estradiol on the relative rates of synthesis of the apoenzyme for uterine G6PD were determined by measuring the rate of incorporation of ^{14}C amino acids into immunoprecipitable uterine G6PD, as described by Smith and Barker (1974). Highly specific antisera to rat liver G6PD were prepared by purification to homogeneity of G6PD from livers of "nutritionally induced" rats (specific activity of 180–200 units/mg protein and a single band on SDS–polyacrylamide gel electrophoresis). Rabbits were immunized with 50 to 100 units of the pure enzyme. Antisera with immunoprecipitation titers of 20 to 70 units/ml (1 unit equals the amount of antiserum required to precipitate 1 enzyme activity unit of G6PD) have been prepared and found to cross-react, and have identical enzyme activity titration points, with rat liver G6PD and uterine G6PD from control and estradiol- and $NADP^+$-induced ovariectomized mature rats. The antiserum precipitates a single protein from ^{14}C-amino acid-labeled uterine cytosol; this protein migrates on SDS–polyacrylamide gel electrophoresis to the same point as pure rat liver G6PD subunits. Additionally, the precipitation of ^{14}C-labeled G6PD from uterine cytosol by the antisera can be completely blocked by the addition of an excess of pure liver G6PD. A unique feature of the use of the ovariectomized mature rat for studies of uterine protein synthesis is the ability to administer radiolabeled compounds or inhibitors of RNA and protein synthesis directly into the uterine lumen by a simple nonsurgical technique. Twenty-five microliters of material is administered through each of the two cervical openings while the animals are under light ether anesthesia, as described by Barker (1967). This technique allows synthesis of uterine proteins of high specific activity. When corrected for specific radioactivity of the labeled precursor in the tissue, the incorporation of a labeled amino acid into a specific protein during a measured interval allows direct comparison of the relative rates of synthesis of the proteins in a variety of treatment groups.

Using these techniques, the effects of estradiol on the relative rates of synthesis of uterine G6PD have been measured and the results are given in Figure 1 (Smith and Barker, 1974). During the first 4 hr after estradiol injection both total uterine G6PD activity and G6PD synthesis rates remained at control levels. From 4 to 8 hr the rate of G6PD synthesis increased fourfold, coinciding with the end of the lag preceding the increase in total uterine G6PD activity. From 12 to 16 hr after hormone treatment the enzyme synthesis rates increased 18 times above the control rates and then were reduced to 10 to 12 times the control rates from 20 to 48 hr. Total uterine G6PD activity was maximal after 36 hr and then decreased, even though G6PD synthesis rates remained elevated and nearly constant during

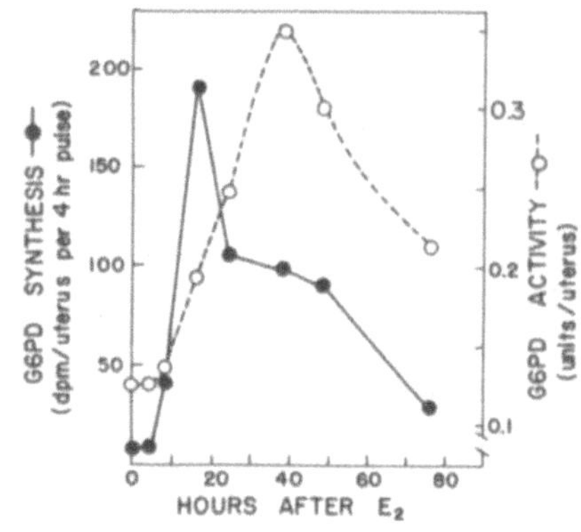

Figure 1. Effect of estradiol on the total activity and rate of synthesis of uterine glucose-6-phosphate dehydrogenase (G6PD) in the ovariectomized mature rat. ^{14}C-labeled amino acids (1.0 μCi/rat) were given at hourly intervals for 4 hr to groups of 20 identically treated rats. All animals were killed at the indicated time after estradiol administration and all pulse-labeling intervals were begun 4 hr before sacrifice. The rate of incorporation of radioactivity into immunoprecipitable uterine G6PD per 4 hr and the total G6PD activity per uterus are given as indicated (Smith and Barker, 1974).

this interval. Studies of uterine G6PD degradation rates revealed that the degration or turnover of uterine G6PD was blocked by estradiol for 36 hr, after which degradation was resumed with a half-life of 25 hr. Reinitiation of G6PD degradation accounts for the observed kinetics of the change in total uterine enzyme activity that occurred at this time (Smith and Barker, 1977). These results indicate clearly that estradiol induces *de novo* synthesis of uterine G6PD; together with the observed inhibition of estradiol induction of uterine G6PD activity by actinomycin D (Barker, 1967), they suggest that estradiol must be acting, at least in part, by inducing the synthesis of the G6PD mRNA.

III. Regulation of Uterine G6PD mRNA Levels by Estradiol

An assay for uterine G6PD mRNA has been developed in our laboratory, based on the translation of rat uterine mRNA into immunoprecipitable G6PD by a messenger-dependent rabbit reticulocyte cell-free protein synthesis system (Adams and Barker, 1979). Uterine RNA was isolated by a modification of the guanidine hydrochloride method of Stohman *et al.* (1977). Uteri from ovariectomized mature rats, given estradiol for various periods, were homogenized in 8 M guanidine–HCl, 0.1 M potassium acetate, 1 mM DTT at −10°C using a Polytron homogenizer. After centrifugation to remove cell debris, DNA, and dsRNA, the RNA in the supernatant was precipitated by addition of 0.55 vol of cold 95% ethanol. The RNA was reprecipitated three times with ethanol from a solution of 8 M guanidine–HCl, 25 mM Na_2 EDTA, 1 mM DTT. We then either fractionated the RNA on oligo dT-cellulose, according to the methods of Bantle *et al.* (1976), or assayed directly its ability to code for G6PD synthesis in the nuclease-treated reticulocyte lysate system (Pelham and Jackson, 1976). We observed a linear increase in the incorporation of [^{35}S]methionine into protein as increasing amounts of uterine RNA were added (either oligo dT-cellulose-bound

Table 1. Distribution and Activity of Uterine mRNA after Oligo dT-Cellulose Chromatography

Oligo dT RNA fraction	RNA yield (A_{260}/uterus)	mRNA activity (cpm/uterus, $\times 10^{-8}$)	G6PD mRNA activity (cpm/uterus, $\times 10^{-6}$)
Total	3.12	1.77	2.86
Unbound	2.63	0.48	0.79
Bound	0.41	0.56	0.93

RNA or total uterine RNA extracted as indicated above), to a maximum of 0.007 A_{260} units of RNA in a 25-μl translation assay. Although the specific activity of the RNA increased following oligo dT-cellulose chromatography, the total messenger activity of the bound RNA was less than one-third of that found in the unfractionated RNA (Table 1).

To minimize the effects of possible variation in procedural losses of mRNA, total unfractionated RNA was assayed to evaluate the effects of estradiol on total uterine G6PD mRNA activity. The amount of mRNA activity for G6PD was measured by immunoprecipitating the translation products with antiserum to rat liver G6PD, followed by SDS-polyacrylamide gel electrophoresis of the labeled immunoprecipitate (Figure 2). The radioactivity in three gel regions, coinciding with uterine G6PD (peak B), uterine pre-G6PD (peak A), and a predominant nascent chain of uterine G6PD, which accumulates both during the reticulocyte cell-free synthesis and during the *in vivo* uterine synthesis of G6PD (peak C), was taken as total mRNA activity. ^{35}S and ^{3}H peptide maps were made from tryptic digests of [^{35}S]-methionine-labeled protein in peaks A, B, and C, which was codigested with the G6PD subunit fraction obtained by immunoprecipitation of *in vivo* [^{3}H]-methionine-labeled uterine cytosols. The maps showed that peaks A, B, and C, produced in the reticulocyte translation assay using uterine RNA as messenger, are related to the uterine cytosol G6PD enzyme. Rat liver G6PD contains 13 methionine residues per mole of subunit protein, and peaks A, B, and C contain 7 to 9 peptides (separated by isoelectric focusing on gels over a pH range of 5 to 6.8) in common with uterine G6PD synthesized *in vivo*.

Other features we have noted are: (a) peaks A and B, but not C, are retained on NADP^{+}-Sepharose, indicating that G6PD in the first two peaks has acquired the NADP^{+}-binding site; (b) G6PD in peaks A and B, but not C, can be competed from the immunoprecipitation reaction by the addition of excess pure liver G6PD, indicating a possible preferential high-affinity binding of the nascent chain in peak C to the antibody; (c) peak C, but not peaks A and B, can be removed by centrifugation of the translation reaction mixture at 150,000*g* for 2 hr before the addition of EDTA, detergents, and the antisera, indicating that peak C protein is probably bound to the ribosome; and (d) none of the three protein peaks is precipitated when control serum is added instead of antisera to G6PD. The predominant G6PD mRNA activity both in oligo dT-cellulose-bound and in total uterine RNA sediments as a 20 S molecule on SDS–sucrose gradients.

The time course of the effects of estradiol on G6PD activity, the relative rate of G6PD synthesis, and the relative amounts of G6PD mRNA in the rat uterus are given in Figure 3. The amount of G6PD mRNA increased somewhat earlier than the increase in *de novo*

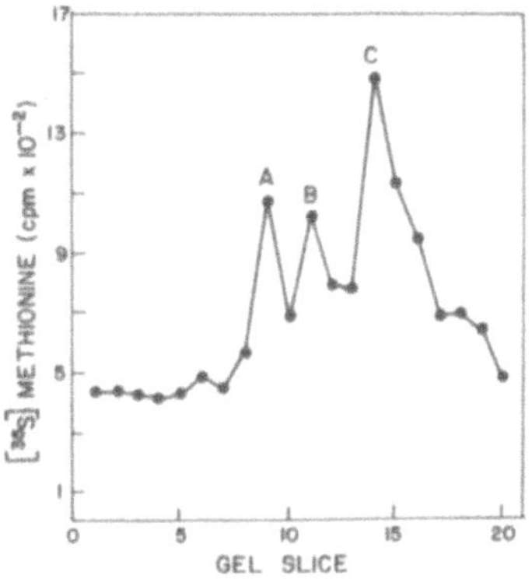

Figure 2. Separation by SDS–polyacrylamide gel electrophoresis of [^{35}S]methionine-labeled proteins in anti-rat G6PD immunoprecipitates of proteins synthesized in a nuclease-treated rabbit reticulocyte lysate cell-free translation system in response to the addition of mRNA obtained from uteri of 12-hr-estradiol-treated ovariectomized mature rats.

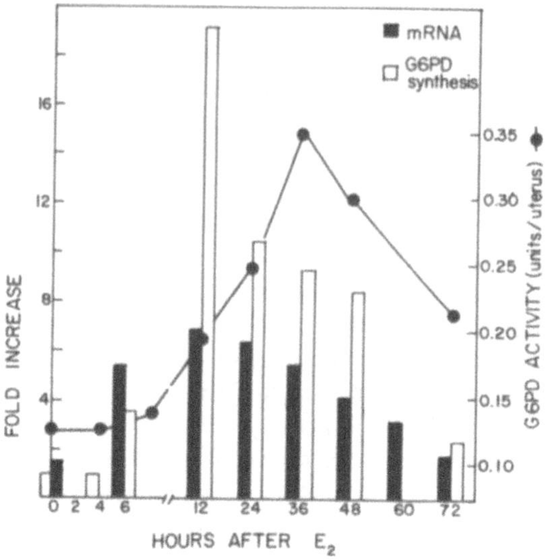

Figure 3. Effect of estradiol on the relative amounts of G6PD mRNA activity, G6PD synthesis rates, and G6PD enzyme activity in the uterus of the ovariectomized mature rat. G6PD activity and synthesis rates are replotted from Figure 1 for comparison. G6PD mRNA activity is given as the fold increase in the sum of the radioactivity in peaks A, B, and C (as noted in Figure 2) that are synthesized in a standard nuclease-treated rabbit reticulocyte lysate cell-free translation system in response to addition of 0.007 A_{260} units of uterine RNA obtained from ovariectomized mature rats given 10 μg of estradiol *in vivo* for the indicated intervals.

G6PD synthesis. Twelve hours after estrogen treatment, however, the rate of enzyme synthesis increased 18-fold, while the amount of G6PD mRNA increased only 7-fold. These results indicate that estradiol does indeed induce an increase in the amount of G6PD mRNA in the rat uterus, but suggest that other *posttranscriptional* estradiol-induced events must be occurring, which magnify the effects of the 7-fold increase in mRNA levels by a factor of two, to account for the observed increase in the overall rate of uterine G6PD synthesis.

IV. Regulation of Uterine G6PD Translation Time by Estradiol

The effect of estradiol on the time required to translate uterine G6PD under *in vivo* conditions has been assessed by measuring the average transit time. Transit time is the length of time required for a ribosome, after becoming attached to the G6PD mRNA, to complete the translation and release into the cytosol a finished molecule of G6PD. The assay technique was first described by Fan and Penman (1970) for estimating the rate of translation of total cytosol proteins in cultures of Chinese hamster ovarian cells. Palmiter (1972) and Stiles *et al.* (1976) used the method to measure the translation times for synthesis of ovalbumin in chick oviduct explants and tyrosine transaminase in hepatoma cells.

The Fan and Penman technique involves determination of the difference between the time required to observe the initial linear incorporation of a labeled amino acid into *total* protein (nascent and released protein chains) and the time required to observe initial linear incorporation of the amino acid into released proteins found in the cytosol fraction of the cell or tissues. The time required to begin linear incorporation of a labeled amino acid into nascent plus released protein chains is the time taken to equilibrate and charge tRNA with the labeled amino acid in preparation for its use in protein synthesis. The additional time to establish the initial linear rate of labeling of a specific protein in the released protein fraction represents one-half of the time required for "average ribosomes" to transit the specific mRNA. Only after 50% of the nascent chains for the specific protein are completed, in the presence of the labeled amino acid, will *all* of the positions occupied by that amino acid in the nascent chains of the specific protein (including the last chain initiated and the next to be terminated) be labeled at the same uniform rate. Linearity of the two curves establishes the extent to which the amino acyl-tRNA pool is equilibrated and maintained at a constant specific radioactivity over the interval tested.

We applied this technique *in vivo* to estimate the effect of estradiol on the ribosome transit time for uterine G6PD (Donohue *et al.*, 1979). Groups of animals were given either [^{3}H]leucine or [^{3}H]glutamate by the intrauterine route for various intervals from 2 to 16 min. The tissues were frozen quickly and postmitochondrial (15,000*g* supernatants, which contain nascent plus released protein chains) and cytosol fractions (105,000*g* supernatants, which contain only released protein chains) were prepared. G6PD was isolated from the cytosol fractions by immunoprecipitation, and the incorporation of label into each fraction was plotted as a function of pulse interval (Figure 4). Half-transit times of 2.1 and 2.8 min for

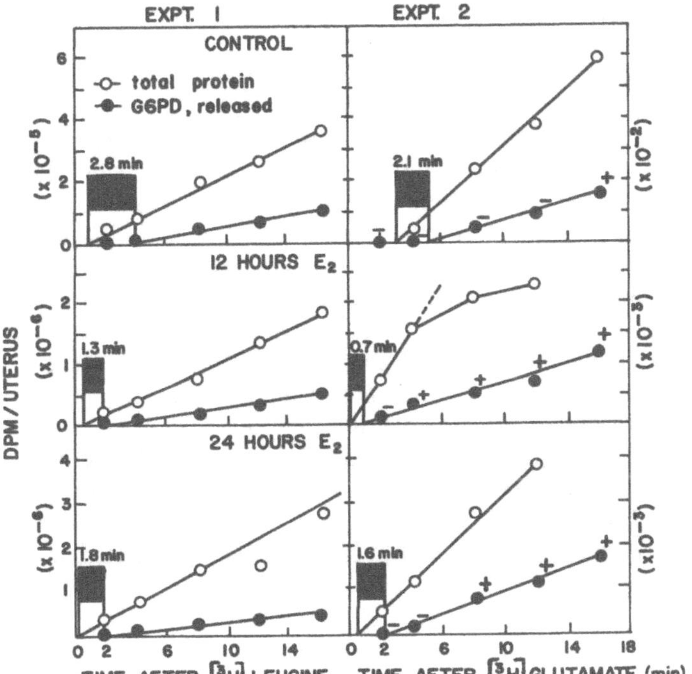

Figure 4. The effect of estradiol on the half-transit time of ribosomes on the mRNA for G6PD in the ovariectomized mature rat. In experiment 1, groups of four identically treated rats were each given 20 μCi of [^{3}Hleucine; in Experiment 2, groups of 10 estradiol-treated rats or 20 control rats were each given 95.5 μCi of [^{3}H]glutamic acid by the intrauterine route for the indicated intervals of time. Each uterus was frozen on dry ice until like tissues were pooled, homogenized, and separated into the postmitochondrial supernatant fraction (15,000*g* for 20 min) or a cytosol supernatant fraction (105,000*g* for 2 hr). Radioactivity in hot (90°C) TCA-insoluble proteins from the postmitochondrial fraction is given as total protein, and radioactivity in G6PD immunoprecipitates from the cytosol fraction is given as released G6PD. Half-transit-time estimates for each hormone treatment are given as the width of the solid black bar. In Experiment 2, the presence or absence of [^{3}H]pyroglutamate in the G6PD immunoprecipitate is noted by the + or − above each data point.

G6PD synthesis in control animals were obtained in the two independent experiments. Estimates of 0.7 and 1.3 min and 1.5 and 1.8 min were obtained in rats given estradiol for 12 and 24 hr, respectively. Thus the average times required to translate uterine G6PD mRNA are 4.9, 2.0, and 3.4 min in control, 12-hr-estradiol-treated, and 24-hr-estradiol-treated rats, respectively. These results suggest that there is a 2.5- and 1.5-fold increase in the rate of G6PD mRNA translation into the apoenzyme 12 and 24 hr after estradiol treatment, respectively. The 2.5-fold increase in the rate of translation of the G6PD mRNA, times the 7-fold greater amount of G6PD mRNA, approximates the 18-fold increase in the rate of uterine G6PD synthesis observed 12 hr after treatment of ovariectomized mature rats with estradiol.

V. Regulation of the Rate of Processing of Uterine Pre-G6PD by Estradiol

The amino terminus of rat liver and uterine G6PD is pyroglutamate. As methionine is thought to be the initiating amino acid for all eukaryotic proteins, G6PD must exist at some time either as a nascent protein chain or as a complete and released precursor molecule with an excess of one or more amino acids at its amino terminus. As part of the study to estimate the transit time for uterine G6PD, we injected [^{3}H]glutamate as the labeled precursor and estimated the amount of time required to detect [^{3}H]pyroglutamate at the amino terminus of the G6PD immunoprecipitate (Experiment 2, Figure 4). The presence or absence of labeled pyroglutamate is indicated by the + or − above the solid circles representing the amount of total radioactivity incorporated into G6PD. Pyroglutamate was detected by complete digestion of the radiolabeled immunoprecipitate with pronase. Amino acid residues containing free amino groups were separated from those without free amino groups by AG50W-X2 ion-exchange chomotography. The water eluates of these columns contained amino acid residues devoid of free amino groups; these residues were subjected to hydrolysis of the pyrrol ring in 6 N HCl at 100°C for 2 hr. The products were then chromatographed on paper in the presence of added glutamate carrier. The presence or absence of radioactivity in the glutamate spot of each chromatogram indicated the presence or absence of [^{3}H]pyroglutamate at the amino terminus of the enzyme.

The earliest time that [^{3}H]pyroglutamate was found at the amino terminus of uterine G6PD was 16, 4, and 8 min for control, 12-hr-estradiol-treated, and 24-hr-estradiol-treated animals, respectively. The differences in labeling time can be partially accounted for by the times required to initiate the incorporation of [^{3}H]glutamate into total uterine protein. How-

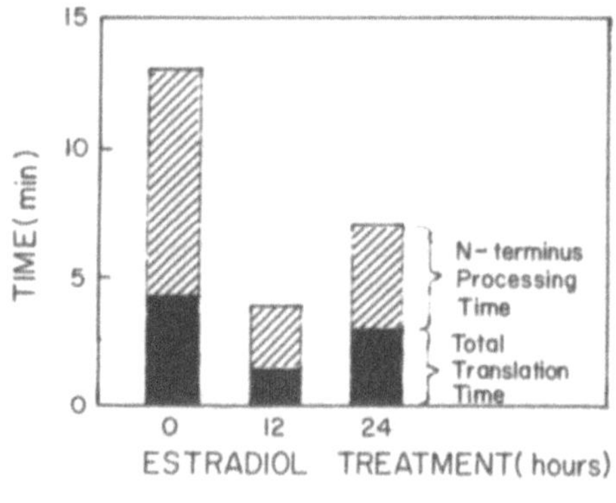

Figure 5. Estimates of the effect of estradiol on the total transit time of ribosomes on the mRNA for G6PD and the time required to process the newly synthesized G6PD to a form that contains pyroglutamate at its amino terminus in the uterus of the ovariectomized mature rat. Data taken from Experiment 2 of Figure 4.

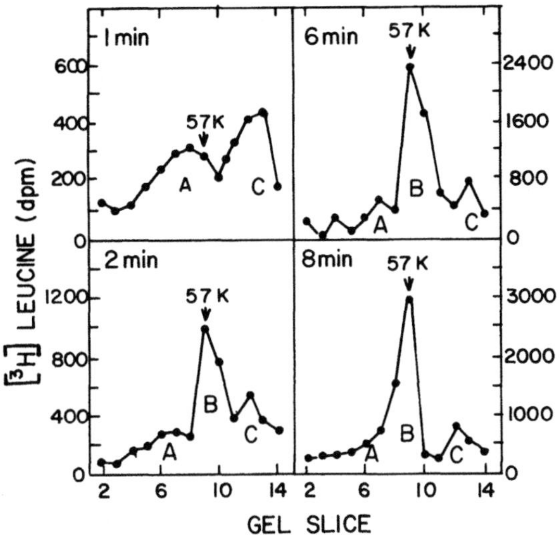

Figure 6. Effect of pulse length on the SDS–polyacrylamide gel electrophoretic profiles of *in vivo* [^{3}H]leucine-labeled G6PD immunoprecipitates from ovariectomized mature rats that had been given estradiol for 12 hr before sacrifice. The arrow designates the position in the gel of the subunit of the predominant form of G6PD that is normally found in the uterine cytosol. It has a molecular weight of 57,000.

ever, if one subtracts the time required for initial linear incorporation of label into the nascent plus released protein chains from the time required to detect labeled pyroglutamate at the amino terminus of the enzyme, it is obvious that a significant reduction in the time required to "process the amino terminus" of newly synthesized uterine G6PD occurs in response to estradiol treatment (Figure 5). Because labeled G6PD that does not contain pyroglutamate at its amino terminus is released and present in the cytosol, these results further suggest that G6PD exists in the cytosol as a released pre-G6PD molecule that is processed by a process that is itself regulated by estradiol. This possibility is further suggested by the occurrence of an immunoprecipitable form of G6PD in the cytosols of uteri labeled *in vivo* for only 1 min, at a molecular weight greater than 57,000, the size of the predominant form of the enzyme that begins to emerge after 2 min of labeling (Figure 6).

VI. Conclusions

Our studies of the mechanisms of induction of uterine G6PD clearly indicate that the amount of this enzyme protein in the uterus is controlled directly or indirectly by estradiol at at least four different molecular levels. First, at the time of maximum rate of enzyme synthesis, 12 hr after estradiol injection, there is a sevenfold increase in G6PD mRNA levels. This finding is consistent with the current concept of estrogen action on gene activation. Second, there is a twofold increase in the rate of translation of G6PD mRNA into completed and released G6PD apoenzyme. Similar activation of the translation rate of protein synthesis in the chick oviduct by estrogens has been observed by Palmiter (1972). The mechanisms of activation of the translation process are uncertain, but it is possible that the estrogen effect in the uterine G6PD system could be manifested indirectly through an increased amount of $NADP^+$. This cofactor could facilitate the final stages of translation by binding to the nascent G6PD chain, to remove the steric hindrances caused by the open configuration of the growing and very large peptide chain. Alternatively, a mechanism similar to that proposed by Sharma and Borek (1973, 1976) could exist, in which the estrogen-dependent formation of specific minor species of tRNA, present in rate-limiting amounts, could facilitate the synthesis of proteins from messengers that contain unique triplet codons requiring these tRNA molecules. These two mechanisms are consistent with our observations. The $NADP^+$-binding ability of nascent G6PD is not acquired until the chain has grown to a size of greater than 40,000. Immunoprecipitation by our antiserum indicates that specific size classes of nascent chains of the enzyme accumulate; thus the rate of mRNA translation is not uniform throughout the G6PD molecule. Third, estradiol appears to induce the activity of a process or enzyme that is rate limiting in the processing of a pre-G6PD molecule into the predominant pyroglutamate-containing form of the enzyme that is present in the uterine cytosol. Fourth, estradiol, especially during the first 36 hr following administration, appears to inhibit the turnover or degradation of all uterine cytosol proteins after which G6PD is degraded with a characteristic half-life of 25 hr. These observations suggest that estradiol, and possibly the estradiol–receptor complex, may control, directly or indirectly, multiple events in the synthesis of RNA and protein and various protein-processing events in estrogen target organs such as the uterus.

ACKNOWLEDGMENT. This research was supported by NIH Research Grant HD 02851.

Discussion

REEL (Research Triangle Institute): You see about a 2.5-fold increase in activity by 24 hr, yet the rate of synthesis is increased 18-fold, and you are blocking degradation; this does not add up in terms of activity, which should be increased more than just 2.5- to 3-fold, if all those things are happening.

BARKER: The basal level of enzyme in a tissue depends on the relative rates of synthesis and degradation. With regard to enzyme, these two processes behave kinetically as zero-order and first-order reactions, respectively. Basal enzyme activity is maintained at relatively high levels either by a low rate of degradation or by a high rate of synthesis. For this reason, estimates of the fold increase in enzyme activity relative to the fold increase in synthesis rates, taken after a time shorter than the half-life of the enzyme, will be considerably lower than expected. For G6PD the half-life in the basal state is between 50 and 100 hr and the 18-fold increase in synthesis rate is seen after an interval of only 12 hr into the response. If enzyme synthesis rates could be maintained at this elevated level by sustained hormonal stimulation, I would expect an 18-fold increase in enzyme levels to be attained.

BULLOCK (Baylor): For the half-transit-time estimates to be valid, the curves in the pre- and postribosomal fractions ought to be parallel. It seems that yours were not; could you comment on that?

BARKER: In these graphs we are plotting values for the total protein on one line, which is the nascent plus released protein pool, and for G6PD on the second line, which is only a small fraction of the total released protein. I could make these lines parallel if I changed the scale on the G6PD line. The point I would like to make is that they are linear in the range observed; they do identify the time at which the labeling and release of G6PD begins. It is more important that the lines be linear, to establish that equilibrium of the specific radioactivity of each amino acyl–tRNA pool exists during the interval used to estimate the initial labeling times.

BULLOCK: Just to clarify that point, your postribosomal count was done with immunoprecipitation and your preribosomal count was TCA?

BARKER: The postmitochondrial fraction, which included ribosomes plus cytosol, was a hot TCA precipitate that should include all proteins or peptides greater than di- or tripeptides. The G6PD was an immunoprecipitate out of the released postribosomal supernatant protein.

BULLOCK: Have you done the immunoprecipitation on the first supernatant and what would those curves look like?

BARKER: If we do a short pulse of 15 min with a lot of [^{3}H]leucine, isolate ribosomes, wash them extensively with high EDTA and detergent, to remove whatever labeled proteins adhere there, and then precipitate the nascent proteins with anti-G6PD antiserum; we obtain a group of immunoreactive peptides of molecular weight about 40,000. This molecular weight coincides with peak C shown in Figures 2 and 6 in my chapter. I believe these peptides are nascent chains of G6PD, but we have not done confirmatory studies by tryptic digestion as we did for the peak C derived from the mRNA translational products.

VAN BLERKOM (Colorado): There are a number of studies that show hormones repress as well as induce the synthesis of specific proteins. Is there any evidence that you have that sort of effect?

BARKER: We have an estradiol-induced situation with G6PD in organ culture that behaves very like that which we see *in vitro*. In that limited system, estradiol does inhibit DNA synthesis in a striking and reproducible way.

VILLEE (Harvard): After estrogen injection, G6PD was induced with about a 6-hr lag, yet the hexose monophosphate shunt increased almost immediately. What changes the shunt?

BARKER: I think the shunt responds to the rapid uptake of glucose and the activation of lipid synthesis which occur during that period. During the 6-hr lag, shunt activity goes up as the $NADP^+$ cofactor levels are elevated in response to $NADP^+$ utilization by lipid synthesis. We can also induce G6PD synthesis if we stimulate shunt utilization and $NADP^+$ levels in the absence of estrogen. If we administer $NADP^+$ into the lumen of the uterus, we can induce G6PD activity. If we introduce sodium fluoride into the lumen, the enolase enzyme in glycolysis is blocked, forcing glucose to be shunted through the pentose pathway, and G6PD synthesis is induced. The fluoride-induced response and the cofactor-induced response are inhibited by cycloheximide but not by actinomycin D. Since estradiol *in vivo* causes an increase both in shunt utilization and in $NADP^+$ levels, something related to the activity of the shunt or the availability of cofactor may be an intermediary in regulating G6PD synthesis for a posttranscriptional event. Whether or not that factor accounts for all of the so-called translational control that we see in the action of estradiol is a matter of speculation.

MANES (La Jolla): Apart from the specific situation with G6PD, regulation of the efficiency with which messages bind to ribosomes to initiate translation seems to be accounted for only in two ways. One way is a change in structure of the message itself around the initiation site, which is a little hard to envision as a hormonal effect, although it could conceivably be a result of different types of processing. The other way that has been frequently invoked but I think is currently in disfavor, would involve the presence of small molecules, either single-stranded nucleic acids or protein factors, that seem to promote specific messages and their translation. Do you have any comments regarding this aspect of the effect of estrogen?

BARKER: I believe Baglioni's group (*Biochemistry* **17**:80, 1978) reported that the presence of glucose-6-phosphate in the reticulocyte lysate system is required for maximal rate of initiation of protein synthesis. We have added the enzyme glucose-6-phosphate dehydrogenase, with a little $NADP^+$, in order to deplete G6PD in the reticulocyte system, and it suppressed the synthesis of all proteins; it suppresses most other proteins much more than it does G6PD. Another relevant point about translational control is the suggestion by Sharma and Borek [cited in the text] that isoaccepting tRNAs which might be present in rate-limiting amounts may preferentially block the completion of translation of certain messages. In our immunoprecipitation reaction we see an accumulation of a nascent chain with about 40,000 molecular weight. It is possible that a rate-limiting amount of isoaccepting tRNA may be responsible for that accumulation. The rate of accumulation of radioactivity in completed G6PD relative to that of the nascent chain is increased twofold in response to estrogen, suggesting that this may be the rate-limiting step in the translation of this particular message.

MANES: In fact, they demonstrated in the pig uterus a minor but specific isoaccepting tRNA that appeared only in the presence of estradiol.

ROBERTS (Florida): Is it possible, therefore, that the estradiol is achieving translational control by merely affecting the redox state of the cell? Secondly, is it possible that the second protein of molecular weight between 45 and 40 thousand is actin that has been brought down by your antibody?

BARKER: We have done two control experiments to identify peak C, which is the peptide to which you refer. The tryptic map of the translational products suggests that at least five or six of the peptides in normally completed G6PD are homologous with peak C, giving us some comfort that it is nascent G6PD. The other intriguing thing is that after we add albumin to the imunoprecipitation reaction with anti-albumin antiserum or control serum from the rabbit that we immunized with G6PD, we do not bring down that peak. Thus, peak C has some immunospecificity, but confirmation of identity will require a peptide map after tryptic digestion. In answer to your first question, the redox state of the cell may be an important element in the control of translation as well as other cellular events.

References

Adams, D. J., and Barker, K. L., 1979, Regulation of glucose-6-phosphate dehydrogenase mRNA levels in the uterus by estradiol, *Fed. Proc. Fed. Am. Soc. Exp. Biol.* **38**:399 (abstract).

Bantle, J. A., Maxwell, I. A., and Hahn, W. E., 1976, Specificity of oligo (dT)-cellulose chromatography in the isolation of polyadenylated RNA, *Anal. Biochem.* **72**:413.

Baquer, N. Z., and McLean, P., 1972, The effect of estradiol on the profile of constant and specific proportion groups of enzymes in the rat uterus, *Biochem. Biophys. Res. Commun.* **48**:729.

Barker, K. L., 1967, Cofactor induced synthesis of D-glucose-6-phosphate: NADP oxidoreductase in the uterus, *Endocrinology* **81**:791.

Barker, K. L., and Warren, J. C., 1966, Estrogen control of carbohydrate metabolism in the uterus: Pathways of glucose metabolism, *Endocrinology* **78**:1205.

Bonsignore, A., and DeFlora, A., 1972, Regulatory properties of glucose-6-phosphate dehydrogenase, in: *Current Topics in Cellular Regulation*, Vol. 6 (B. L. Horecker and E. R. Stadtman, eds.), p. 21, Academic Press, New York.

Donohue, T. M., Jr., Mahowald, T. A., and Barker, K. L., 1977, Identification and labeling of the NH_2-terminal residue of uterine glucose-6-phosphate dehydrogenase, *Fed. Proc. Fed. Am. Soc. Exp. Biol.* **36**:799 (abstract).

Donohue, T. M., Jr., Mahowald, T. A., and Barker, K. L., 1979, Post-transcriptional control of glucose-6-phosphate dehydrogenase synthesis in the uterus by estradiol, *Fed. Proc. Fed. Am. Soc. Exp. Biol.* **38**:329 (abstract).

Fan, H., and Penman, S., 1970, Regulation of protein synthesis in mammalian cells: Inhibition of protein synthesis at the level of initiation during mitosis, *J. Mol. Biol.* **50**:655.

Holten, D., 1972, Relationships among the multiple molecular forms of rat liver glucose-6-phosphate dehydrogenase, *Biochim. Biophys. Acta* **268**:4.

Keran, E. E., and Barker, K. L., 1976, Regulation of glucose-6-phosphate dehydrogenase activity in uterine tissue in organ culture, *Endocrinology* **99**:1386.

Levy, H. R., 1979, Glucose-6-phosphate dehydrogenases, *Adv. Enzymol.* **48**:97.

Moulton, B. C., and Barker, K. L., 1971, Synthesis and degradation of glucose-6-phosphate dehydrogenase in the rat uterus, *Endocrinology* **89**:1131.

Moulton, B. C., and Barker, K. L., 1972, Effects of RNA and protein synthesis inhibitors on preinduced levels of rat uterine glucose-6-phosphate dehydrogenase, *Endocrinology* **91**:491.

Moulton, B. C., and Barker, K. L., 1974, Effects of 8-azaguanine on the induction of uterine glucose-6-phosphate dehydrogenase, *Proc. Soc. Exp. Biol. Med.* **146**:742.

O'Dorisio, M. S., and Barker, K. L., 1970, Effects of estradiol on the levels of pyridine nucleotide coenzymes in the rat uterus, *Endocrinology* **86**:1118.

O'Dorisio, M. S., and Barker, K. L., 1976, Effects of estradiol on the biosynthesis of pyridine nucleotide coenzymes in the rat uterus, *Biol. Reprod.* **15**:504.

Palmiter, R. D., 1972, Regulation of protein synthesis in the chick oviduct. II. Modulation of polypeptide elongation and initiation roles by estrogen and progesterone, *J. Biol. Chem.* **247**:6770.

Pelham, H. R. B., and Jackson, R. J., 1976, An efficient mRNA-dependent translation system from reticulocyte lysates, *Eur. J. Biochem.* **67**:237.

Scott, D. B. M., and Lisi, A. G., 1960, Changes in enzymes of the uterus of the ovariectomized rat after treatment with estradiol, *Biochem. J.* **77**:52.

Sharma, O. K., and Borek, E., 1973, Enhancement of the synthesis of a hormone-induced protein by transfer ribonucleic acids, *J. Biol. Chem.* **248**:7622.

Sharma, O. K., and Borek, E., 1976, A mechanism of estrogen action on gene expression at the level of translation, *Cancer Res.* **36**:4320.

Singh, D., and Squire, P. G., 1975, Bovine adrenal glucose-6-phosphate dehydrogenase. V. Chemical characterization, *Int. J. Pept. Protein Res.* **7**:185.

Smith, E. R., and Barker, K. L., 1974, Effects of estradiol and NADP on the rate of synthesis of uterine glucose-6-phosphate dehydrogenase, *J. Biol. Chem.* **249**:6541.

Smith, E. R., and Barker, K. L., 1976, Effect of sodium fluoride on glucose-6-phosphate dehydrogenase activity in the rat uterus, *Biochim. Biophys. Acta* **451**:223.

Smith, E. R., and Barker, K. L., 1977, Effects of estradiol and NADP on the rate of degradation of uterine glucose-6-phosphate dehydrogenase, *J. Biol. Chem.* **252:**3709.

Stiles, C. D., Lee, K. L., and Kenney, F. T., 1976, Differential degradation of mRNA's in mammalian cells, *Proc. Natl. Acad. Sci. USA* **73:**2634.

Strohman, R. C., Moss, P. S., Micou-Eastwood, J., Spector, D., Przybyla, A., and Paterson, B., 1977, Messenger RNA for myosin polypeptides: Isolation from single myogenic cell cultures, *Cell* **10:**265.

Yoshida, A., 1972, Micro method for determination of blocked NH_2-terminal amino acids in proteins, *Anal. Biochem.* **49:**320.

17

Uterine DNA Polymerase

Acquisition of Responsiveness to Estrogen during Postnatal Development of the Rat

Alvin M. Kaye, Michael D. Walker, and Bertold R. Fridlender

I. Introduction

Although the striking growth response of the rat uterus to estrogen has been long recognized (Astwood, 1938; Mueller, 1953), our understanding of the mechanism of this process remains incomplete. Evidence from a number of other steroid responsive tissues suggests that steroids bring about their effects on specific protein synthesis by increasing the concentrations of mRNA for these particular proteins (Chan *et al.*, 1973; Schimke *et al.*, 1975; Tata, 1976; Kurtz and Feigelson, 1977), probably by transcriptional regulation (Swaneck *et al.*, 1979). Comparable analysis of estrogen action on rat uterus has been complicated by the absence of equally suitable marker proteins. However, the available evidence indicates that similar mechanisms do operate in this organ (Katzenellenbogen and Gorski, 1972).

Despite the absence of a detailed molecular understanding, some valuable information has been obtained from comparisons of the actions of strong and weak estrogens on the uterus (Anderson *et al.*, 1975). Such studies have led to the generalization that the stimulation of early effects of estrogens is unable to bring about the later effects of the hormone, such as increased DNA synthesis and cell division; rather the continued presence of the nuclear-bound steroid seems to be required for a full response. A clarification of the role of the steroid at these later times will be necessary for a full understanding of estrogen action with respect to late responses, such as DNA synthesis and cell division.

We have shown previously that the developing rat uterus attains full sensitivity to estrogen in sequential fashion (Somjen *et al.*, 1973: Kaye *et al.*, 1974). As part of this study, we characterized the estrogen response to the DNA synthesis rate (Kaye *et al.*, 1972). We sought to investigate this response at the molecular level and chose the activity of uterine DNA polymerase as a starting point.

Alvin M. Kaye, Michael D. Walker, and Bertold R. Fridlender • Department of Hormone Research, Weizmann Institute of Science, Rehovot, Israel, and Ames Yissum Ltd., Jerusalem, Israel.

Three classes of DNA polymerase have been distinguished in the mammalian cell based on their physicochemical and enzymic properties (Bollum, 1975; Weissbach, 1975, 1977): DNA polymerase α constitutes 50 to 95% of total cell polymerase activity, migrates on sucrose density gradients with a sedimentation coefficient of 6 to 8 S, and is totally inhibited in the presence of high salt (250 mM KCl) or sulfhydryl-group inhibitors such as *N*-ethylmaleimide. DNA polymerase β constitutes 5 to 50% of cell polymerase activity, exhibits a sedimentation coefficient of 3 to 4 S, is only partially inhibited by 250 mM KCl, and is insensitive to sulfhydryl-group inhibitors. DNA polymerase γ (Fridlender *et al.*, 1972) is a minor polymerase (1 to 2%) that is able to copy ribohomopolymers at a faster rate than nicked duplex DNA. Increased levels of cellular DNA synthesis, as observed in liver of partially hepatectomized rats (Chang and Bollum, 1972), developing brain neurons, cardiac muscle, and spleen (Hubscher *et al.*, 1977) and S-phase tissue culture cells (Spadari and Weissbach, 1974) are associated, in general, with significantly elevated DNA polymerase α activity. DNA polymerase β activity under such circumstances is unaffected. This observation suggests that DNA polymerase α is the enzyme involved in DNA replication, whereas DNA polymerase β may be required at all stages of the cell cycle and thus be involved in some process such as DNA repair (Hubscher *et al.*, 1979; but see Butt *et al.*, 1978). Support for this idea has come from recent work using d2TTP, an inhibitor of DNA polymerase β and γ but not α. This compound is incapable of inhibiting replicative DNA synthesis in nuclei of SV40-infected cells (Edenberg *et al.*, 1978) or HeLa cell lysates (Waqar *et al.*, 1978).

II. DNA Synthesis

Estrogen administration to the 20-day-old rat leads to an increased level of uterine DNA synthetic activity at 12 to 16 hr, reaching a peak of twofold stimulation at 24 hr (Figure 1). Similar results have been reported by Stormshak *et al.* (1976). That this stimulation represents true DNA synthesis is shown by study of the mitotic index. A clear peak at 24 to 30 hr is seen for all cell types of the uterus (Kaye *et al.*, 1972).

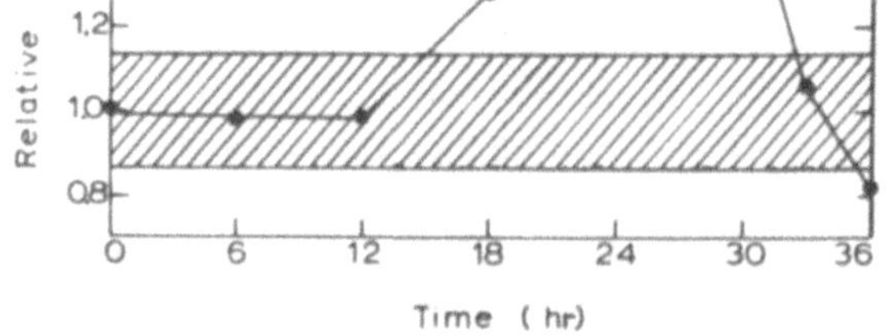

Figure 1. Increase in the rate of DNA synthesis after intraperitoneal injection of 5 μg estradiol-17β into 20-day-old rats. At the time indicated, uteri were removed and incubated for 1 hr in 1 ml medium 199 containing 3 μCi [*methyl*-^{3}H]thymidine. DNA was then isolated and its specific activity determined, expressed relative to controls (399 cpm/μg DNA/hr). The shaded band represents the 95% confidence interval for the mean control rate of DNA synthesis. E, estradiol-17β; C, control solution. From Kaye *et al.* (1972).

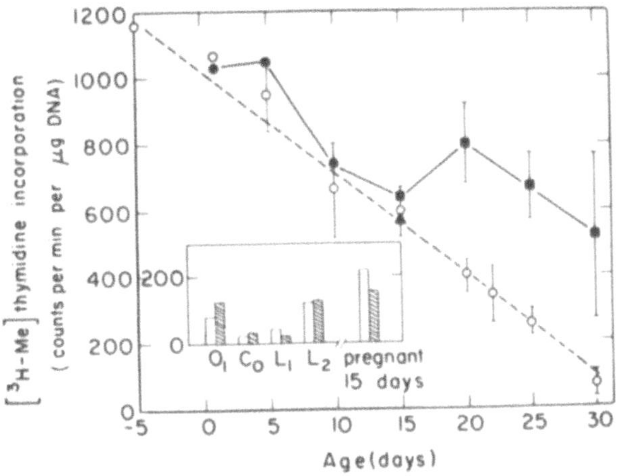

Figure 2. Responsiveness to estradiol as a function of age, measured by incorporation of [*methyl*-^{3}H]thymidine into DNA of rat uteri. Twenty-four hours after injection of estradiol-17β, uteri were removed and incubated as described in the legend to Figure 1. (○) Control rats; (●) rats given 0.05 μg estradiol; (■) rats given 0.5 μg estradiol; (▲) rats given 5 μg estradiol. Vertical lines represent 95% confidence intervals. For the inserted histogram: O_1, proestrus, C_0, estrus, L, metestrus, and L_2, diestrus. Open bars represent control rats; shaded bars = rats given 5 μg estradiol. From Kaye *et al.* (1972).

The DNA synthetic capacity of uterus decreases almost linearly from birth to 30 days of age (Figure 2). In animals aged 20 days or older, estrogen (0.05 μg) causes a significant increase in DNA synthetic capacity. However, at 15 days or younger, estrogen at the same or higher doses (5.0 μg) has no effect on this parameter.

III. DNA Polymerase

DNA polymerase activity was measured in extracts under both low- and high-salt conditions. As DNA polymerase α activity is completely inhibited in the presence of high salt, the activity under these conditions was used as a measure of DNA polymerase β, and as DNA polymerase β activity represented less than 10% of the total low-salt activity, the activity at low salt was used as a measure of DNA polymerase α. DNA polymerase β activity in the uterus of the 20-day-old rat is unaffected by estrogen administration (Figure 3). DNA polymerase α activity, on the other hand, starts to rise at 12 to 16 hr, reaching maximum values of 160% of control between 20 and 36 hr (Figure 3). Harris and Gorski (1978a) have reported similar kinetics. In animals aged 10 or 15 days, no effect of estrogen on DNA polymerase α can be detected (Figure 4), in contrast to the 50 to 60% increase seen for 20- and 25-day-old rats. At no age tested was estrogen observed to bring about an increase in DNA polymerase β activity.

To confirm this result, sucrose gradients were run of extracts of 15- and 20-day-old uteri. As expected, in no case was the DNA polymerase β (3–4 S) affected (Figure 5). However, the DNA polymerase α (6–8 S) was increased (by about 60%) following estrogen administration to 20-day-old rats but not to 15-day-old rats. To confirm their identity and to compare them with the activities found in other tissues, the two peaks were assayed under a variety of conditions (Table 1). As has been found for these enzymes from other sources, DNA polymerase α was much more susceptible than β to *N*-ethylmaleimide, to

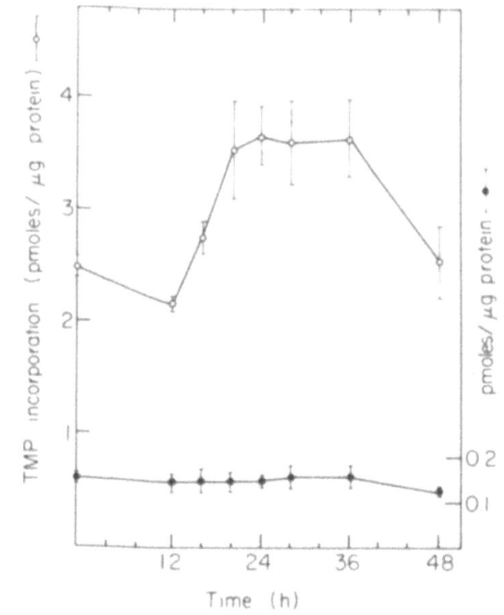

Figure 3. Time course of stimulation of uterine DNA polymerase α activity by estradiol-17β in the 20-day-old rat. At the time indicated, uteri were excised and homogenized in a cold solution containing 0.25 M sucrose, 50 mM Tris–HCl (pH 7.5), 25 mM KCl, 5 mM $MgCl_2$, and 2 mM dithiothreitol (DTT). Total cell extract was prepared from the homogenate as described by Yang *et al.* (1976). DNA polymerase activity was measured under both low- (○) and high-salt (●) conditions. The low-salt assay (taken as a measure of DNA polymerase α activity) contained in a final volume of 50µl: 100 mM Tris–HCl (pH 8.0), 2 mM $MgCl_2$, 0.5 mg/ml activated calf thymus DNA, 0.5 mg/ml bovine serum albumin (BSA), 1 mM DTT, 25 µM dATP, 25 µM dGTP, 25 µM dCTP, 25 µM [^{3}H]-TTP (308 cpm/pmol), and uterine extract, to give a final concentration of less than 10 mM KCl. The high salt assay (taken as a measure of DNA polymerase B activity) contained, in addition to the above, 250 mM KCl. After 30 min incubation, trichloroacetic-acid-insoluble material was collected on glass-fiber filters (GF/C) and radioactivity measured. Values shown are the mean ± S.E.M. of three to four independent determinations on groups of four animals per time point. From Walker *et al.* (1978).

phosphonoacetic acid, and to KCl. Both enzymes required added activated DNA, but only DNA polymerase α required all four deoxyribonucleotide triphosphates for full activity (Weissbach, 1975). Using conditions favorable for detection of DNA polymerase γ [100 mM KCl, 1 mM $MnCl_2$, and 20 µg/ml (rA)n·(dT)12] (Knopf *et al.*, 1976), no activity was observed in crude uterine extracts or in the sucrose gradient fractions of uterine extracts containing maximal activity of DNA polymerase α and β.

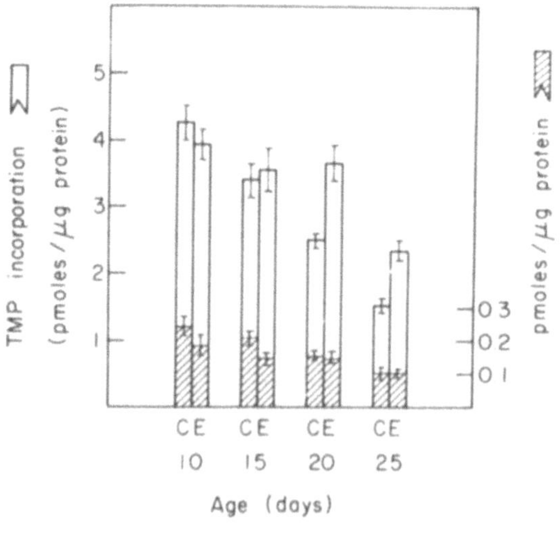

Figure 4. Age dependence of the stimulation of uterine DNA polymerase α activity by estradiol-17β at 24 hr after administration. Activity of total uterine cell extracts was measured as described in legend to Figure 3 under low-salt (open bars) and high-salt conditions (shaded bars). C, untreated rats; E., estrogen-treated rats. Values shown are the mean ± S.E.M. of three to four independent determinations on groups of four animals per age group. From Walker *et al.* (1978).

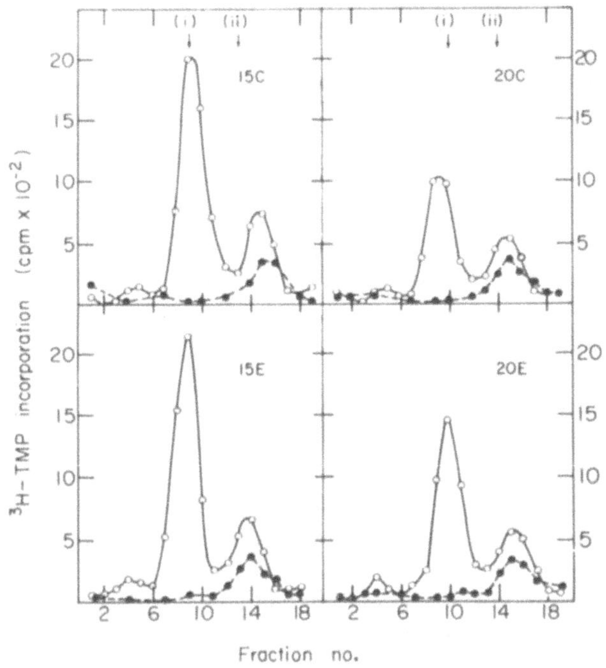

Figure 5. DNA polymerase activity in fractions of sucrose gradients of total uterine extracts from 15- and 20-day-old rats. Samples were applied to 5-ml linear sucrose gradients (10 to 20%) containing 50 mM Tris–HCl (pH 7.5), 500 mM KCl, 5 mM $MgCl_2$, 5 mM DTT, and 0.1 mg/ml BSA. Centrifugation was at 45,000 rpm for 15 hr in an SW 50.1 rotor. Fractions (14 drops) were collected and 2-μl samples were assayed under low- (x) and high-salt conditions (z). C, untreated rats; E, estrogen-treated rats. The amount of protein added to each gradient was 1.4 mg (15C), 1.3 mg (15E), 0.94 mg (20C), and 0.93 mg (20E). The arrows represent the position of (i) immunoglobulin G (6.7 S) and (ii) hemoglobin (4.3 S). The direction of sedimentation was from left to right. The decrease in the ratio of activity of DNA polymerase α to β following sucrose gradient centrifugation (cf. Figures 3 and 4) is attributable to a partial loss of activity of DNA polymerase α activity under the conditions of the centrifugation (15 hr at 2°C). From Walker *et al.* (1978).

Table 1. Properties of Separated DNA Polymerase α and β[a]

Conditions	Enzyme α	Enzyme β
	Activity (%)	
Complete system	100	100
+*N*-ethylmaleimide (5 mM)	5	48
+phosphonoacetic acid (10 μg/ml)	42	96
+KCl (100 mM)	31	107
+KCl (250 mM)	<1	46
−activated DNA	<1	2
−(dATP, dGTP, dCTP)	49	118

[a]Total cell extracts of uteri of 15-day-old rats were fractionated on sucrose gradients. Peak fractions of α and β activity were assayed under low-salt conditions. (From Walker *et al.*, 1978.)

IV. Summary and Conclusions

There is a good correlation between the effects of estrogen on DNA synthetic activity on the one hand and DNA polymerase α activity on the other, both in terms of kinetics and age dependence of the response. It is between the ages of 15 and 20 days after birth that the rat uterus first acquires the machinery necessary for a complete response to estrogen, which includes increased cell division. Harris and Gorski (1978a,b) have investigated the hormonal specificity and the effects of repeated estrogen administration on DNA polymerase α. In these cases also, a reasonable correlation could be made between DNA polymerase α activity and DNA synthesis.

These findings could imply that the activity of DNA polymerase α is in some way involved in the mediation of estrogen action on DNA synthesis. Resolution of this question will require further elucidation of the factors involved in eukaryotic DNA synthesis. In the meantime we are investigating the mechanism of increase in DNA polymerase α activity by using inhibitors of protein synthesis and by attempting to prepare antibodies to purified DNA polymerase α (Fisher and Korn, 1977) and to titrate these antibodies with extracts from uteri of control and hormone-treated rats.

ACKNOWLEDGMENTS. This work was supported in part by grants from the Ford Foundation and the Population Council, New York, to Professor H. R. Lindner, whose help and encouragement are gratefully acknowledged.

Discussion

MARTIN (ICRF, London): Do you think that the changes in polymerase α are a cause or a consequence of cells entering S phase? There are tissue culture studies with synchronized cell systems where the kinetics with which cells pass from G1 into S phase can be followed with precision. Here, it looks as though there is no real rise in polymerase levels before cells actually start to enter S; it looks as though it is a consequence of entering S.

KAYE: In our system, the time points used were not sufficiently close for us to say anything more than that both of the parameters first show a rise at 16 hr.

LEROY (Brussels): We have preliminary data showing that if we treat the animals with estrogen for 3 days, that is, from 7 to 10 days of age, we can obtain a proliferative response in uterine epithelium earlier (Day 13) than we normally observe without giving any pretreatment. Would you care to comment on this?

KAYE: I think that is very feasible. In setting up the criteria for responsiveness which I showed, we used a single injection of estrogen in the experimental protocol. We did one experiment in which we injected estrogen twice, with a delay of 48 hr between the two doses, at 13 and 15 days, and this did not give us any change in the lack of responsiveness at 15 days. It has always been in the back of our minds to go back to your old experiments and those of Price and Ortiz (*Endocrinology* **34**:215, 1944), who used repeated injections of estradiol over a period of several days and reported that there was an increase in responsiveness. I am happy to hear that you have taken up this challenge and it seems perfectly reasonable to me that the responsiveness is not an absolute function of age, but depends upon the hormonal environment. There is some variation from strain to strain of animal and development of responsiveness becomes a useful tool when one defines exactly the system in which one is working.

LEROY: My comment was about epithelial cells. Would you happen to know if the proliferative response of stromal cells to high doses of progesterone at some small dose of estrogen is also under the same kind of ontogenetic program as you have shown?

KAYE: All I can tell you about progesterone is that at 20 days of age, a point at which the stroma is also quite reactive to estrogen, the addition of progesterone slightly depresses the responsiveness to estrogen.

MANES (La Jolla): You mentioned that ornithine decarboxylase activity was inducible by estrogen and this induction could occur very early, in contrast to the DNA polymerase activity. Is this ODC activity also accompanied by ribosomal RNA production and an increased protein synthetic level in all tissues of the uterus?

KAYE: At the age of 2 days, the repertoire of protein which the uterus seems capable of making is very limited and we see the increase in ODC activity without any increase in ribosomal protein and without any increase in any other macromolecular parameter.

MANES: It has been suggested that ODC is a cofactor in the RNA polymerase I that is usually associated with increased ribosome production.

KAYE: Diane Russell and I are attempting to prove or disprove this theory and we hope to have more information soon.

KIESSLING (Oregon): Did you relate thymidine incorporation to cell division? If you express your enzyme activity as units per microgram of DNA, do you see the same effect?

KAYE: Independent of the mode of expression, we find a significant increase in the activity of the polymerase α after estrogen treatment.

KIESSLING: Presumably per cell?

KAYE: Yes, presumably per cell.

KIESSLING: Have you tried any blocking agents? Does the increase in polymerase require RNA synthesis?

KAYE: Our preliminary experiments with cycloheximide and experiments by Stormshak *et al.* (in: *Receptors and Hormone Action* (B. W. O'Malley and L. Birnbaumer, eds.), Vol. II, p. 63, Academic Press, New York, 1978) with actinomycin D are consistent with the necessity for *de novo* RNA and protein synthesis.

MOULTON (Cincinnati): Have you looked at the two different kinds of DNA polymerase in decidualizing stromal cells? I wonder if there is anything different about those enzymes in cells that are eventually going to be polyploid?

KAYE: We have done just one series of experiments: a time course during decidualization. In this study we came across nothing which made us suspect the enzymes were different, although we did not specifically look at any parameters as candidates for differences.

PSYCHOYOS (Bicêtre): In the chick oviduct there is an increase in DNA polymerase caused by estrogen, and progesterone modulates this effect. Do you have any information in your model that there is a modulatory effect of progesterone on the induction by estrogen of increased DNA polymerase?

KAYE: We ourselves have no information on progesterone effects on DNA polymerase. Hams and Gorski (*Mol. Cell. Endocrinol.* **10**:243, 1978) have shown inhibition of the full response by simultaneous treatment with progesterone and estradiol.

GLASSER (Baylor): Comparing Day 15 to Day 20, when stimulation by estrogen becomes apparent, it seems that the endogenous control level has fallen by Day 20 and the stimulation seems to be equivalent to Day 15. Would you venture a general thesis that responsiveness to the hormone is related to some type of restriction in the endogenous level of activity of any particular enzyme, in this case polymerase α?

KAYE: Yes, that is a beautiful statement of a point I should like to expand. You and other people who have investigated uterine RNA polymerase have speculated about the role of RNA synthesis. The statement has been made, with a good deal of reason, that the uterus is a deprived organ in terms of RNA synthesis. Compared to most of the other organs in the body, the endogenous level of RNA synthesis in the uterus is extremely low. When one stimulates the uterus with estrogen, one brings it back to a level which is more in keeping with the rate of synthesis seen in most other organs. In that sense, it may be that endogenous concentrations of RNA polymerase, DNA polymerase α, perhaps the IP, and many other parameters in the uterus, reflect the fact that in the absence of hormones the uterus atrophies. The word "stimulation" for the uterus, in a sense, translates into prevention of atrophy.

References

Anderson, J. N., Peck, E. J., and Clark, J. H., 1975, Estrogen-induced uterine responses and growth: Relationship to receptor estrogen binding by uterine nuclei, *Endocrinology* **96:**160–167.

Astwood, E. B., 1938, A six-hour assay for quantitative determination of estrogen, *Endocrinology* **23:**25–31.

Bollum, F. J., 1975, Mammalian DNA polymerases, *Progr. Nucleic Acids Res. Mol. Biol.* **15:**109–144.

Butt, T. R., Wood, W. M., McKay, E. L., and Adams, R. L. P., 1978, Involvement of DNA polymerase β in nuclear DNA synthesis, *Biochem. J.* **173:**309–314.

Chan, L., Means, A. R., and O'Malley, B. W., 1973, Rates of induction of specific translatable mRNA's for ovalbumin and avidin by steroid hormones, *Proc. Natl. Acad. Sci. USA* **70:**1870–1874.

Chang, L. M. S., and Bøllum, F. J., 1972, Variation in DNA polymerase activities during rat liver regeneration, *J. Bol. Chem.* **247:**7948–7950.

Edenberg, H. J., Anderson, S., and DePamphilis, M. L., 1978, Involvement of DNA polymerase α in simian virus 40 DNA replication, *J. Biol. Chem.* **253:**3273–3280.

Fisher, P. A., and Korn, D., 1977, DNA polymerase α: Purification and structural characterization of the near homogeneous enzyme from human KB cells, *J. Biol. Chem.* **252:**6528–6535.

Fridlender, B., Fry, M., Bolden, A., and Weissbach, A., 1972, A new synthetic RNA-dependent DNA polymerase from human tissue culture cells, *Proc. Natl. Acad. Sci. USA* **69:**452–455.

Harris, J., and Gorski, J., 1978a, Estrogen stimulation of DNA dependent DNA polymerase activity in immature rat uterus, *Mol. Cell. Endocrinol.* **10:**293–305.

Harris, J., and Gorski, J., 1978b, Evidence for a discontinuous requirement for estrogen in stimulation of DNA synthesis in the immature rat uterus, *Endocrinology* **103:**240–245.

Hubscher, U., Kuenzle, C. C., and Spadari, S., 1977, Variation of DNA polymerase α, β, and γ during perinatal tissue growth and differentiation, *Nucleic Acids Res.* **4:**2917–2929.

Hubscher, U., Kuenzle, C. C., and Spadari, S., 1979, Functional roles of DNA polymerases β and γ, *Proc. Natl. Acad. Sci. USA* **76:**2316–2320.

Katzenellenbogen, B. S., and Gorski, J., 1972, Estrogen action *in vitro:* Induction of the synthesis of a specific uterine protein, *J. Biol. Chem.* **247:**1299–1305.

Kaye, A. M., Sheratzky, D., and Lindner, H. R., 1972, Kinetics of DNA synthesis in immature rat uterus: Age dependence and estradiol stimulation, *Biochim. Biophys. Acta* **261:**475–486.

Kaye, A. M., Somjen, D., King, R. J. B., Somjen, G., Icekson, I., and Lindner, H. R., 1974, Sequential gene expression in response to estradiol-17β during postnatal development of rat uterus, *Adv. Exp. Med. Biol.* **44:**383–402.

Knopf, K. W., Yamada, M., and Weissbach, A., 1976, HeLa cell DNA polymerase γ: Further purification and properties of the enzyme, *Biochemistry* **15:**4540–4548.

Kurtz, D. T., and Feigelson, P., 1977, Multihormonal induction of hepatic α2u globulin mRNA as measured by hybridization to complementary DNA, *Proc. Natl. Acad. Sci. USA* **74:**4791–4795.

Mueller, G. C., 1953, Incorporation of glycine-2-^{14}C into protein by surviving uteri from α-estradiol treated rats, *J. Biol. Chem.* **204:** 77–90.

Robins, D. M., and Schimke, R. T., 1978, Differential effects of estrogen and progesterone on ovalbumin mRNA utilization, *J. Biol. Chem.* **253:**8925–8934.

Schimke, R. T., McKnight, G. S., and Shapiro, D. J., 1975, Nucleic acid probes and analysis of hormone action in the oviduct, in: *Biochemical Actions of Hormones* (G. Litwack, ed.), Vol. 3, pp. 245–269, Academic Press, New York.

Somjen, D., Somjen, G., King, R. J. B., Kaye, A. M., and Lindner, H. R., 1973, Nuclear binding of etradiol-17β and induction of protein synthesis in the rat uterus during postnatal development, *Biochem. J.* **136:**25–33.

Spadari, S., and Weissbach, A., 1974, The interrelation between DNA synthesis and various DNA polymerase activities in synchronized HeLa cells, *J. Mol. Biol.* **86:**11–20.

Stormshak, F., Leake, B., Wertz, N., and Gorski, J., 1976, Stimulatory and inhibitory effects of estrogen on uterine DNA synthesis, *Endocrinology* **99:**1501–1511.

Swaneck, G. E., Nordstrom, J. L., Kreuzaler, F., Tsai, M. J., and O'Malley, B. W., 1979, Effect of estrogen on gene expression in chicken oviduct: Evidence for transcriptional control of ovalbumin gene, *Proc. Natl. Acad. Sci USA* **76:**1049–1053.

Tata, J. R., 1976, The expression of the vitellogenin gene, *Cell* **9:**1–14.

Walker, M. D., Kaye, A. M., and Fridlender, B. R., 1978, Age-dependent stimulation by estradiol-17β of DNA polymerase α in immature rat uterus, *FEBS Lett.* **92:**25–28.

Waqar, M. A., Evans, M. J., and Huberman, J. A., 1978, Effect of 2′, 3′-dideoxythymidine-5′-triphosphate on HeLa cell *in vitro* DNA synthesis: Evidence that DNA polymerase α is the only polymerase required for cellular DNA replication, *Nucleic Acids Res.* **5:**1933–1946.

Weissbach, A., 1975, Vertebrate DNA polymerases, *Cell* **5:**101–108.

Weissbach, A., 1977, Eukaryotic DNA polymerases, *Annu. Rev. Biochem.* **46:**25–47.

Yang, W. K. Tyndall, R. L., and Daniel, J. C., 1976, DNA- and RNA-dependent DNA polymerases: Progressive changes in rabbit endometrium during preimplantation stage of pregnancy, *Biol. Reprod.* **15:**604–613.

18

The Artificially Stimulated Decidual Cell Reaction in the Mouse Uterus

Studies of RNA Polymerases and Histone Modifications

Martin J. Serra, Billy Baggett, Judith C. Rankin, and Barry E. Ledford

I. The Artificially Stimulated Decidual Cell Reaction as a Model for Studying Implantation

Implantation involves changes in all cell types of the uterine endometrium in preparation for attachment of the blastocyst to the uterine wall. The artificially stimulated decidual cell reaction (DCR) has been used as a model system for studying the uterine responses to implantation (Finn and Porter, 1975; Glasser and Clark, 1975). In the decidual cell reaction, uterine stromal cells proliferate and are transformed into decidual cells. The hormonal requirements for the oil-stimulated DCR closely approximate the hormonal preparation of the uterus at the time of natural implantation. The principal advantages offered by the artificially stimulated DCR in hormone-primed ovariectomized animals include (a) a more extensive reaction than that occurring in natural implantation; (b) a more precise time course of the reaction, particularly short times after stimulation; (c) the ability to study the roles of estrogen and progesterone in the reaction; and (d) the ability to use large numbers of animals in a single experiment.

In our model system for studying decidualization, ICR mice are bilaterally ovariectomized and rested at least one week. The mice are then primed for three consecutive days with 0.1 μg estradiol-17β (in 0.1 ml sesame oil) daily, followed by two days of no treatment (Miller and Emmens, 1969). A combination of 1 mg progesterone plus 6.7 ng estradiol-17β (in 0.1 ml sesame oil) is then given daily for the remainder of the experiment. Six hours after the third progesterone + estradiol-17β injection, the right uterine horn is stimulated by the injection of 10 μl of sesame oil intralumenally.

Martin J. Serra • Department of Chemistry, Allegheny College, Meadsville, Pennsylvania 16335. *Billy Baggett, Judith C. Rankin, and Barry E. Ledford* • Department of Biochemistry, Medical University of South Carolina, Charleston, South Carolina 29403.

II. The Biochemistry of the Decidual Cell Reaction

A number of biochemical and physiological changes have been observed in the uterus following stimulation of the decidual cell reaction (DCR). Two of the earliest biochemical events associated with the oil-stimulated DCR are increases in prostaglandin E and F levels in the stimulated horn, peaking at 5 min, and increased cyclic AMP levels in the stimulated horn, peaking at 15 min (Rankin *et al.*, 1977, 1979; Jonsson *et al.*, 1978). Growth of the uterine horns following oil stimulation occurs in a biphasic pattern. The first phase begins approximately 12 hr after stimulation and results in a doubling in stimulated horn weight by 24 hr. The second phase begins approximately 30 hr after stimulation and results in an additional eightfold increase in stimulated horn weight by 144 hr (Ledford *et al.*, 1976). Histologically, decidual cells are not apparent until the beginning of the second phase of growth.

An abrupt increase in DNA synthesis by the stimulated uterine horn begins 13 hr after oil stimulation and involves virtually all the stromal cells (Das and Martin, 1978; Ledford *et al.*, 1978). Total DNA content of the stimulated uterine horn does not change significantly until the beginning of the second phase of growth. In agreement with studies in other systems (Spalding *et al.*, 1966; Robbins and Borun, 1967), we find that histone synthesis is coupled to the rate of DNA synthesis, as shown in Figure 1 (Serra *et al.*, 1979). The rate of incorporation of [^{3}H]lysine into histone paralleled the rate of incorporation of [^{3}H]thymi-

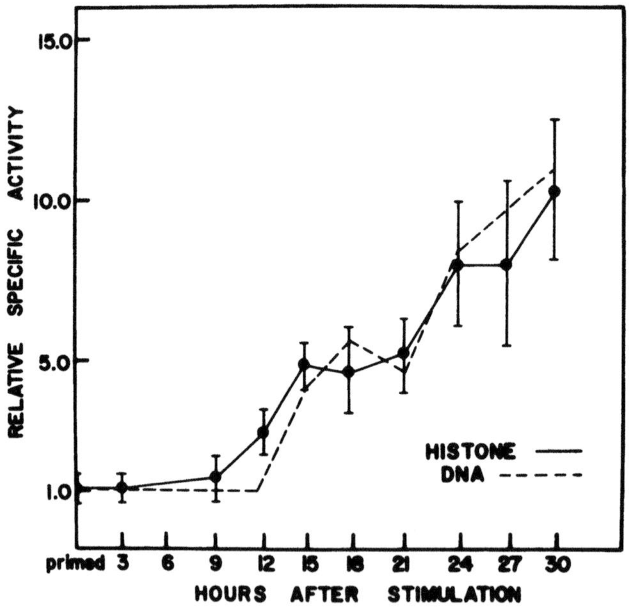

Figure 1. Synthesis of uterine histone and DNA following stimulation of the decidual cell reaction. Ovariectomized mice were hormone primed (Miller and Emmens, 1969) and the right uterine horn was stimulated intralumenally with sesame oil. For histone synthesis experiments, animals received an i.p. injection of [^{3}H]lysine 1 hr prior to sacrifice. For DNA synthesis experiments, animals received an i.p. injection of [^{3}H]thymidine 1 hr prior to sacrifice. Primed animals represent the comparison of right to left uterine horns immediately prior to the time of stimulation of the right uterine horn. The relative specific activity represents the ratio of the specific activity of the stimulated horn to the specific activity of the unstimulated horn. Bars represent S.E.M. Histone synthesis points beyond 18 hr are corrected for increased lysine pool size in the stimulated horn. From Serra *et al.* (1979); reprinted by kind permission of *Biology of Reproduction*.

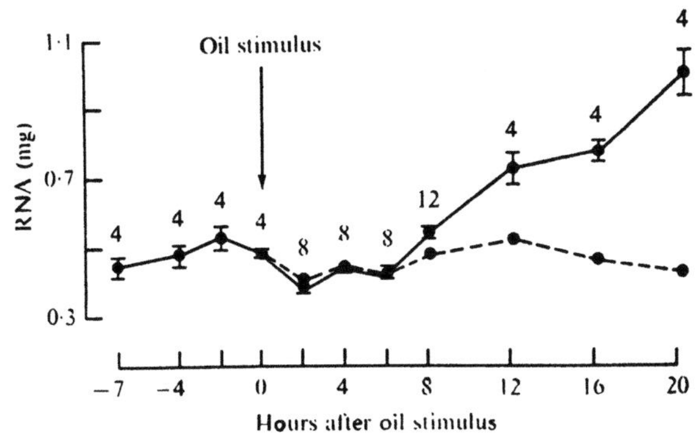

Figure 2. Changes in the amount of RNA in the mouse uterus during an interval starting 7 hr before and ending 20 hr after oil stimulation of the decidual cell reaction. Animals were primed (see text; Miller and Emmens, 1969). Results compare the mean values for the left, treated uterine horn (●——●) in mice that received an oil decidual stimulus with the right, untreated horn(●——●). Bars represent S.E.M. The number of groups of four pooled horns from which the results were determined is given above each point. From Miller (1973); reprinted by kind permission of the *Journal of Endocrinology*.

dine into DNA. The lag period of approximately 12 hr between stimulation and the onset of DNA replication is also observed in other growth-stimulated systems, such as mitogen-stimulated lymphocytes and liver regeneration following partial hepatectomy (Bender and Prescott, 1962; Grisham, 1962).

III. RNA Metabolism

The rapid growth and differentiation of uterine stromal cells into decidual cells suggest an alteration in the expression of the genetic material of these cells. Changes in the metabolism of RNA in the mouse uterus following oil stimulation of the DCR have been investigated (Miller, 1973; Shelesnyak and Tic, 1963). Increased levels of RNA are observed in the stimulated uterine horn compared to the contralateral unstimulated horn within 8 hr following stimulation. This increase in total RNA (Figure 2) is preceded by an increased rate of incorporation of radioactive uridine into RNA (Miller, 1973). A similar pattern of RNA metabolism is observed during early pregnancy in the rat (Heald *et al.*, 1975).

The rate of ribosomal RNA synthesis is increased following implantation in the rat (Heald *et al.*, 1972a,b). In addition, Heald and O'Hare (1973) have observed the presence of new species of nuclear RNA at the time of implantation. Administration of actinomycin D has been shown to inhibit the implantation of blastocysts and to delay decidualization (Finn and Martin, 1972; Finn and Bredl, 1973). Decidualization appears to be dependent, therefore, upon the synthesis of RNA at the time of implantation. This conclusion led us to examine some of the factors thought to influence the synthesis of RNA during the decidual cell reaction.

IV. RNA Polymerases

Multiple forms of DNA-dependent RNA polymerases have been observed in eukaryotes. These polymerases can be distinguished by their subunit composition, biological function, subcellular location, and sensitivity to the toxin α-amanitin (Roeder and Rutter, 1970; Weinmann and Roeder, 1974; Sklar *et al.*, 1975; Valenzuela *et al.*, 1976; Weil and Blatti, 1976). The activity of RNA polymerase can be assessed in isolated nuclei by measuring the incorporation of radioactive UTP into RNA. The activities of the three mammalian RNA polymerases can be distinguished by their sensitivity to α-amanitin and their ionic strength optima.

RNA polymerase I is responsible for the synthesis of ribosomal RNA (Roeder and Rutter, 1969; Weinmann and Roeder, 1974); it is localized in the nucleolus, is insensitive to α-amanitin, and is active at low ionic strength. Figure 3 shows a low level of RNA polymerase I activity in nuclei isolated from ovariectomized mouse uteri. Following the hormonal-priming regimen outlined above, the RNA polymerase I activity increased three- to fourfold. There is a further threefold increase by 9 hr after oil stimulation of the DCR in the stimulated horn; there is no change in the contralateral control horn. This increase in RNA polymerase I activity is consistent with the increase in ribosomal RNA synthesis observed by others (Heald *et al.*, 1972a,b). Increased total protein synthesis also occurs following stimulation of the DCR (Shelesnyak and Tic, 1963; Denari *et al.*, 1976; Ledford *et al.*, 1976; Bell, 1979), and has been shown to occur following increased synthesis of RNA (Shelesnyak and Tic, 1963). The increased RNA polymerase I activity may result in additional ribosomes, which allow for increased protein synthesis.

RNA polymerase II synthesizes heterogeneous nuclear RNA, is localized in the nucleoplasm, requires a high salt concentration, and is highly sensitive to α-amanitin (Roeder and Rutter, 1969; Weinmann and Roeder, 1974; Serra *et al.*, 1978). The nuclei from uteri

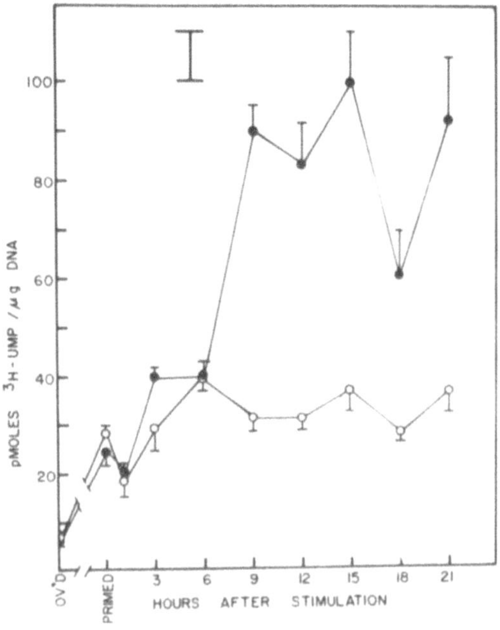

Figure 3. RNA polymerase I activity following stimulation of the decidual cell reaction. Animals were primed (see text). The point marked "OV'D" on the abscissa indicates the results of assays on ovariectomized nonprimed mice. The time marked "PRIMED" corresponds to the time of uterine oil injection for all other times. The animals used at this time point had been hormonally primed but not stimulated with sesame oil. Bars represent S.E.M. Right (stimulated) uterine horns (●); left (unstimulated) uterine horns (○). From Serra *et al.* (1978); reprinted by kind permission of *Biochimica et Biophysica Acta*.

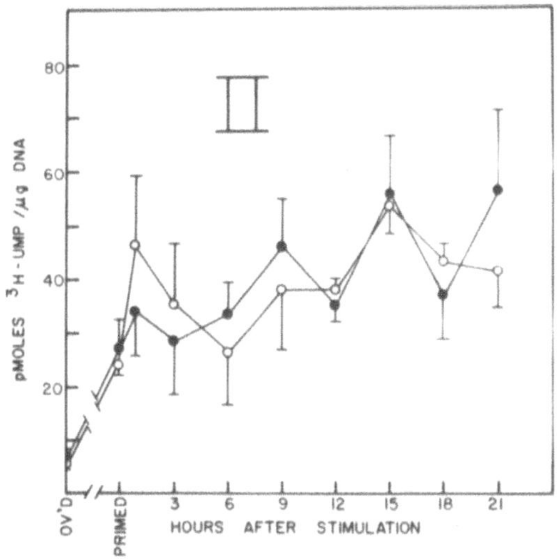

Figure 4. RNA polymerase II activity following stimulation of the decidual cell reaction. The point marked "OV'D" on the abscissa indicates the results of assays on ovariectomized nonprimed mice. The time marked "PRIMED" corresponds to the time of uterine oil injection for all other times. The animals used at this time point had been hormonally primed but not stimulated with sesame oil. Bars represent S.E.M. Right (stimulated) uterine horns (●); left (unstimulated) uterine horns (O). From Serra *et al.* (1978); reprinted by kind permission of *Biochimica et Biophysica Acta*.

of hormonally primed mice show a threefold increase in the level of RNA polymerase II compared to uterine nuclei from ovariectomized mice (Figure 4). Following oil stimulation of the decidual cell reaction, there is no additional increase in RNA polymerase II activity in nuclei from either the stimulated or the unstimulated uterine horns. This finding agrees with our inability to see any changes in the pattern of protein synthesis following oil stimulation of the DCR in the mouse (unpublished results). Denari *et al.* (1976) and Bell (1979) reported the synthesis of a "decidualization-specific protein" during pregnancy in the rat. This protein, however, may be of blastocyst rather than uterine origin; it could also be a species-specific protein such as estrogen "induced protein" appears to be in the rat (Notides and Gorski, 1966).

Transfer RNA and 5 S RNA are synthesized by RNA polymerase III (Roeder and Rutter, 1969; Weinmann and Roeder, 1974). RNA polymerase III is located in the nucleoplasm, requires a high salt concentration, and is moderately sensitive to α-amanitin. Figure 5 shows the RNA polymerase III activity in isolated mouse uterine nuclei following oil stimulation of the DCR. A five- to sixfold increase in RNA polymerase III activity is observed following the hormone-priming regimen. By 9 hr after oil stimulation, the nuclei from stimulated uterine horns had a significant increase in RNA polymerase III activity compared to nuclei from hormone-primed uteri or from the unstimulated horns. Nuclei from the unstimulated uterine horns showed no significant change in the level of RNA polymerase III activity compared to that in the hormone-primed uterus prior to stimulation. The increase in RNA polymerase III activity in stimulated uterine nuclei at 9 hr coincides with the increase in RNA polymerase I activity. Coupling of the responses of RNA polymerases I and III has been observed in other growth-stimulated model systems (Jaehning *et al.*, 1975; Viarengo *et al.*, 1975).

The synthesis of RNA is a complex biological process. Increases in nuclear RNA polymerase activity, as measured above, may represent alterations at a number of different

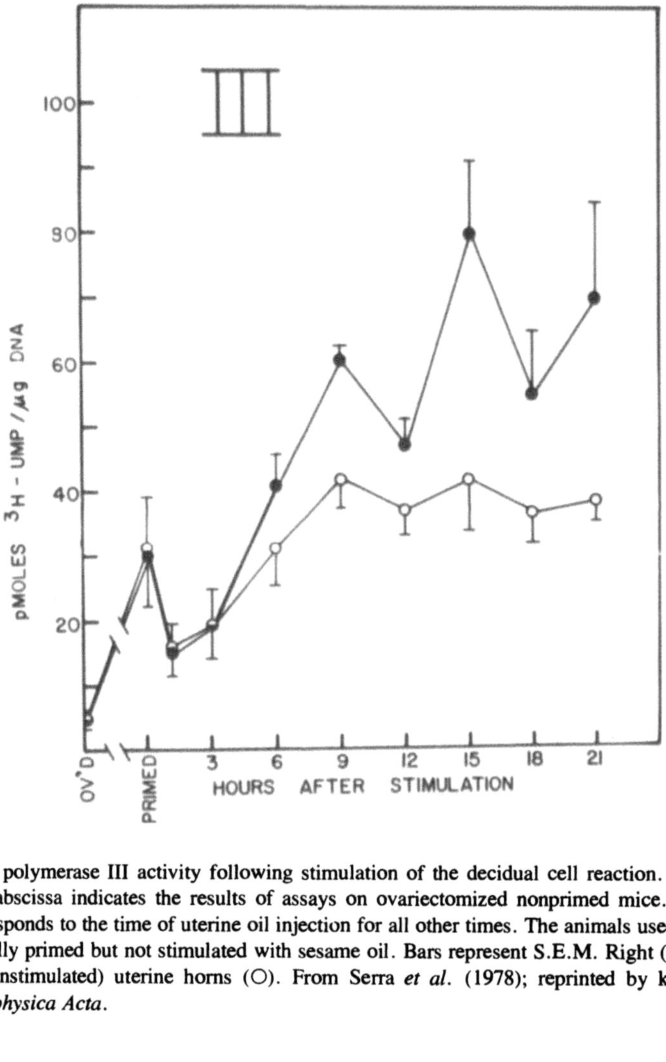

Figure 5. RNA polymerase III activity following stimulation of the decidual cell reaction. The point marked "OV'D" on the abscissa indicates the results of assays on ovariectomized nonprimed mice. The time marked "PRIMED" corresponds to the time of uterine oil injection for all other times. The animals used at this time point had been hormonally primed but not stimulated with sesame oil. Bars represent S.E.M. Right (stimulated) uterine horns (●); left (unstimulated) uterine horns (○). From Serra *et al.* (1978); reprinted by kind permission of *Biochimica et Biophysica Acta.*

levels. The increased levels of RNA polymerase activity may result from (a) increased levels of the RNA polymerase enzymes themselves; (b) changes in the proportions of RNA polymerase enzymes bound tightly and loosely to chromatin; or (c) changes in template activity.

V. Histone Modifications

The chromosomal proteins are responsible for the regulation of RNA synthesis. These proteins can be divided into two classes, the histones or basic proteins and the nonhistone or acidic proteins. Evidence has been presented for the involvement of both histone and nonhistone chromosomal proteins in genetic regulation (Wilhelm *et al.*, 1971; Spelsberg *et al.*, 1971). The precise roles of the chromosomal proteins, however, are at present unclear.

The histones are responsible for the structural integrity of the DNA in chromatin; their selective removal significantly alters the structure of the DNA (Richards and Pardon, 1970). Their evolutionary stability, while not as great as once thought, suggests a fundamental role for the histones in chromosomal packaging. Histones H2A, H2B, H3, and H4 are associated with the nucleosome, the basic unit of chromosome structure (Kornberg, 1974). The exact location of histone H1 on the chromosome is uncertain; it is believed to be associated with the linker portion of DNA (Varshavsky *et al.*, 1976).

While only a limited number of different histone types exists within somatic cells, each of the histones can display heterogeneity. This heterogeneity may be due to posttranslational modifications, such as phosphorylation or acetylation. In addition, due to the multiple copies of histone genes, nonallelic variations of the histones have been observed (Rall and Cole, 1971). These alterations in histone modification and sequence may have marked effects on the interactions of the histones with DNA. Histones undergo specific cell cycle modifications, which may culminate in the condensation of the chromosomes during mitosis. Also, increased histone modification precedes RNA and DNA synthesis in cells stimulated to divide (Gurley *et al.*, 1973, 1974, 1975).

A scan of our separation of uterine histones by gel electrophoresis is shown in Figure 6. The histones show a large degree of heterogeneity due to posttranslational modifications and nonallelic variations (Ruiz-Carrillo *et al.*, 1975, 1976; Isenberg, 1979). The most slowly migrating histone, H2A, can be resolved into four distinct bands. A similar pattern of fractionation of histone H2A was observed during embryogenesis of the sea urchin (Cohen *et al.*, 1975). Histone H1 is resolved into three subfractions. Histone H1 is evolutionarily the least conserved of the histones and has been shown to be composed of multiple subfractions with minor sequence variations (Rall and Cole, 1971). Band 7 corresponds to histone H3. This band is the most diffuse, suggesting that it is composed of subfractions not resolved with our gel system. Histone H2B appears as two bands; the more

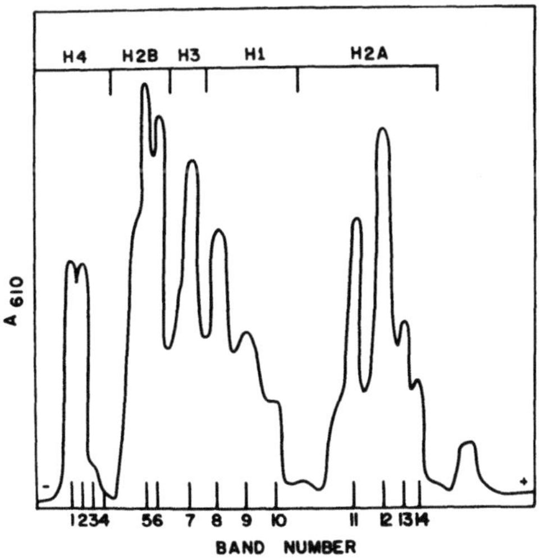

Figure 6. Triton-acid-urea gel electrophoresis of whole histone from the hormonally primed mouse uterus. The stained (numbered) histone bands were sliced from the gel and analyzed for radioactivity. The slices were dissolved in 30% H_2O_2 by heating to 110°C for 1 hr and counted in Triton–toluene. The specific activity was defined as counts per minute per square inch under the gel scan. The ratio of the specific activity of the stimulated horn to the specific activity of the unstimulated horn was defined as the relative specific activity. Each uterine horn was analyzed independently. From Serra *et al.* (1979); reprinted by kind permission of *Biology of Reproduction*.

rapidly migrating band contains a shoulder that could not consistently be resolved into a distinct band and therefore was analyzed as part of band 5. The most rapidly migrating histone, H4, displays a number of bands corresponding to various degrees of modification (Ruiz-Carrillo *et al.*, 1975).

The level of [^{32}P]phosphate incorporation into the total histone fraction during the decidual cell reaction is shown in Figure 7. Hormone-primed uterine horns have a low level of phosphate incorporation into histone. The incorporation is the same into both uterine horns. The stimulated uterine horn shows a twofold increase (vs. the corresponding unstimulated horn) in the rate of histone phosphorylation at 3 hr. This increased phosphorylation is a significant but transient effect ($p<0.05$); the levels of phosphate incorporation in the stimulated and control uterine horns are identical 6 hr after stimulation. The increase in phosphorylation in the stimulated horn at 3 hr may result from the increased cyclic AMP levels observed 15 min after oil stimulation (Rankin *et al.*, 1977). Cyclic AMP-dependent histone kinases have been extensively studied (Langan 1970, 1978; Adler *et al.*, 1972). These kinases phosphorylate specific histone serine residues. Immediately prior to increases in DNA and histone synthesis, there is a threefold increase in the degree of histone phosphorylation in the stimulated uterine horns. This increase in histone phosphorylation occurs following a second peak in cyclic AMP levels, seen 7 hr after stimulation of the DCR (Rankin *et al.*, this volume). The increased phosphorylation continues through 18 hr and then decreases.

The incorporation of phosphate into individual histone bands at various times after stimulation is shown in Figure 8. Hormone primed uteri have low levels of incorporation, predominantly into histones H2A, H1, and H3. No significant change in phosphorylation was observed in the unstimulated control horns over the period studied. The increased level of phosphorylation in the stimulated horn at 3 hr appears in histone fractions H2A, H1, and H3. Immediately before the increase in DNA and histone synthesis, the pattern of phosphorylation is altered, with increased levels of incorporation into all histone fractions. This increase in phosphorylation may be involved in the control of the initiation of DNA synthesis. At later times, the pattern reverts to that seen previously; only histones H2A, H1, and H3 have increased levels of phosphorylation. Gurley *et al.* (1975) observed changes in the pattern of histone phosphorylation during the cell cycle of Chinese hamster cells. The

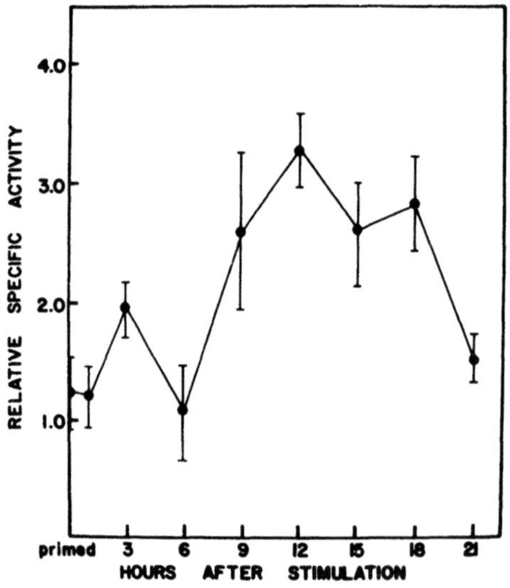

Figure 7. Phosphorylation of whole uterine histone. Animals were hormone primed prior to stimulation. Animals received an i.p. injection of [^{32}P]orthophosphoric acid 1 hr prior to sacrifice. Relative specific activity represents the ratio of the specific activity of the stimulated uterine horn to the specific activity of the corresponding unstimulated uterine horn. The time marked "primed" represents a comparison of the right to left uterine horns immediately prior to the time of stimulation of the right uterine horn. Bars represent S.E.M. From Serra *et al.* (1979); reprinted by kind permission of *Biology of Reproduction*.

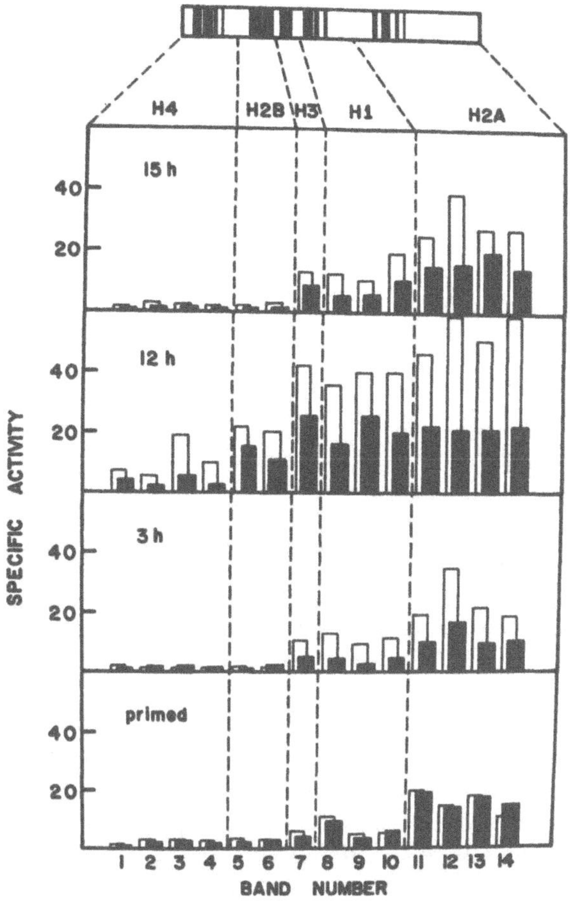

Figure 8. Phosphorylation of individual histone bands at various times after stimulation of the decidual cell reaction in the mouse uterus. Animals were hormone primed prior to stimulation. Animals received an i.p. injection of [^{32}P]orthophosphoric acid 1 hr prior to sacrifice. Specific activity is defined as the radioactivity per slice divided by the area under the gel scan for the histone fraction (cpm × 10^3/inch^2). "Primed" samples represent a comparison of right to left uterine horns immediately prior to the time of stimulation of the right uterine horn. Open bars, stimulated uterine horns; solid bars, unstimulated uterine horns. From Serra *et al.* (1979); reprinted by kind permission of *Biology of Reproduction*.

changes we observed in uterine histone phosphorylation following stimulation of the DCR may be related to changes in chromatin structure necessary for DNA replication.

Histone acetylation has been related to gene activity by a number of studies (Pogo *et al.*, 1966, 1968; Gornall and Liew, 1974). We examined the acetylation of histones by measuring the incorporation of [^{3}H]acetate into histone proteins. Figure 9 shows the incorporation of acetate into total histone. No change is observed in the level of histone acetylation in the unstimulated uterine horn. At 9 hr, there is a twofold increase in the level of acetate incorporation into total histone in the stimulated horn. This increased level of acetylation in the decidualizing uterine horn is observed at all later times studied through 18 hr.

The same levels of acetylation of the individual histone bands are observed in both uterine horns prior to stimulation (Figure 10). While the relative specific activity of total histone acetylation remains essentially unchanged 1 hr after stimulation (Figure 9), there

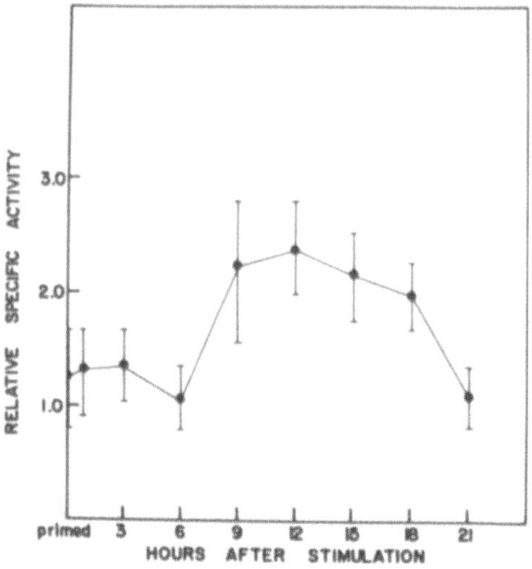

Figure 9. Acetylation of whole uterine histone. Animals were hormone primed prior to stimulation. Animals received an i.p. injection of sodium [^{3}H]acetate 1 hr prior to sacrifice. Relative specific activity represents the ratio of the specific activity of the stimulated uterine horn to the specific activity of the corresponding unstimulated uterine horn. The "primed" sample represents the comparison of right to left uterine horns immediately prior to the time of stimulation of the right uterine horn. Bars represent S.E.M. From Serra *et al.* (1979); reprinted by kind permission of *Biology of Reproduction*.

are increases in acetylation of histones H2B and H4 in stimulated horns compared to unstimulated horns (Figure 10). We observed a wide variation in the amount of radioactivity incorporated into the histones between animals after intraperitoneal injection of labeled precursor. However, when the relative specific activities are based on the contralateral unstimulated horns, the ratios between animals are in agreement. The pattern of increased acetylation seen 12 hr after stimulation is similar to that at 1 hr. These changes in histone acetylation following stimulation of decidualization may result in an increase in the template activity of uterine chromatin.

VI. Template Activity

The increases in RNA polymerase activity, together with the histone modifications, are suggestive of changes in the expression of genetic material during the decidual cell reaction. To pursue this possibility, we have studied the chromatin template activity by measuring the ability of *E. coli* RNA polymerase to synthesize RNA, using chromatin prepared at various times after stimulation of the DCR. Figure 11 shows the results. Uterine chromatin from ovariectomized mice has a low level of template activity. This result would be expected based on the low levels of RNA metabolism in such uteri. Our hormone-priming regimen leads to a slight increase in template activity. When chromatin is examined at various times after stimulation, no significant change in the template activity is seen (Figure 11). There is also no significant difference between the activities of the stimulated and unstimulated uterine horns at any time studied. O'Grady *et al.* (1975) showed increases in chromatin template activity at the implantation site compared to the interimplantation site in the rat. The inability to correlate our results with those of O'Grady *et al.* could be explained in several ways. There may be a species difference between the rat and the mouse, as noted earlier. We measured the template activity at earlier times in the decidualization process than did O'Grady *et al.* Our study utilized chromatin prepared from whole decidualizing uteri, and not specific implantation sites in natural pregnancy.

VII. Summary

Utilizing oil-induced decidualization in estradiol- and progesterone-primed ovariectomized mice, we have studied several factors that may relate to and perhaps control the increases that occur in DNA-dependent transcription and replication associated with this process. The activities of RNA polymerase I, which synthesizes rRNA, and III, which synthesizes tRNA and 5 S RNA, as measured in nuclei without added template, are signifi-

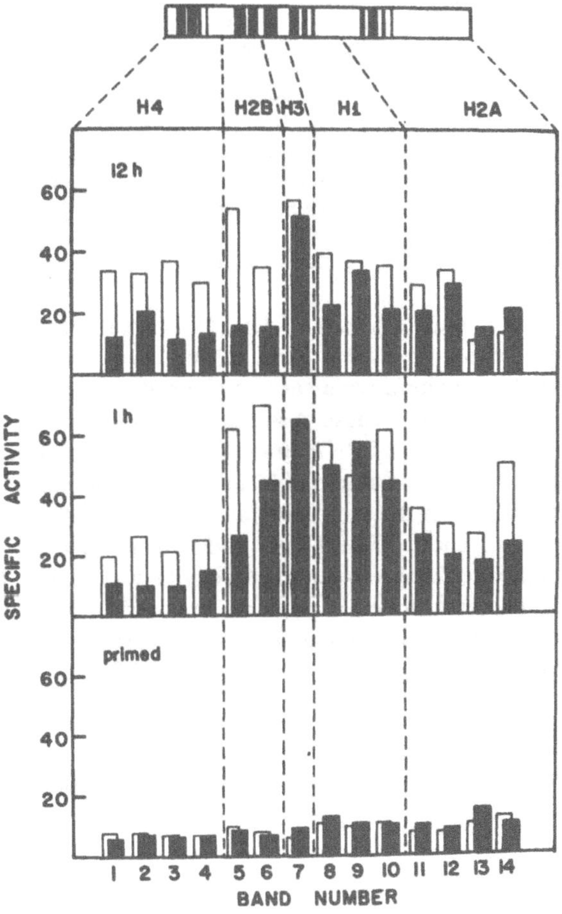

Figure 10. Acetylation of individual histone bands at various times after stimulation of the decidual cell reaction in the mouse uterus. Animals were hormone primed prior to decidualization. Animals received an i.p. injection of sodium [^{3}H]acetate 1 hr prior to sacrifice. Specific activity is defined as the radioactivity per slice divided by the area under the gel scan for the histone fraction (cpm × 10^3/inch^2). "Primed" samples represent the comparison of right to left uterine horns immediately prior to the time of stimulation of the right uterine horn. Open bars, stimulated uterine horn; solid bars, unstimulated uterine horns. From Serra *et al.* (1979); reprinted by kind permission of *Biology of Reproduction*.

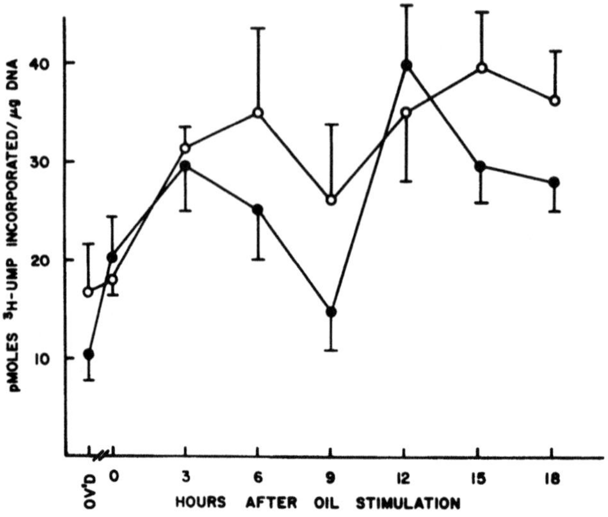

Figure 11. Template activity of chromatin isolated from mouse uterine nuclei at various times after oil stimulation of the decidual cell reaction. Animals were hormone primed prior to decidualization. The time marked "OV'D" on the abscissa indicates the results of assays on ovariectomized nonprimed mice. The time marked "0" corresponds to the time of uterine oil stimulation for all other times. The animals used at this time point had been been hormonally primed but not stimulated with sesame oil. Bars represent S.E.M. Right (stimulated) uterine horns (●); left (unstimulated) uterine horns (○).

cantly increased by 9 hr after stimulation of the DCR. No increase in the activity of RNA polymerase II, which synthesizes mRNA, was found during the period studied. Increases in histone phosphorylation and acetylation were seen at about the same time, particularly in certain specific fractions of the histone. These changes nearly all coincide with the first significant change in the rate of synthesis of RNA in the stimulated uterine horn, and precede the burst of DNA synthesis that begins at about 15 hr. Significant increases in the rate of total histone synthesis coincide with the increased rate of DNA synthesis. In studies over the first 18 hr following stimulation of the DCR, with excess bacterial DNA-dependent RNA polymerase, no significant change in template activity was found. This observation suggests that the modifications of histone are probably involved in controlling replication, not transcription, in this rapidly growing tissue. Actual changes in the levels of the RNA polymerases may regulate transcription.

ACKNOWLEDGMENT. This work was supported by Grant HD 06362 from the National Institute of Child Health and Human Development of the National Institutes of Health.

Discussion

MANES (La Jolla): How did you isolate and identify the various RNA polymerases? Were these all *in vitro* assays?

SERRA: Yes, they were all done with isolated nuclei and were determined by sensitivity to α-amanitin.

MANES: So the assays were done on isolated nuclei rather than the isolated polymerases?

SERRA: Yes, with different concentrations of the toxin.

MARSTON (Birmingham): There is lumenal continuity between the right and left uterine horn in most mouse strains. You were lucky that the strain you used did not have a functional lumenal continuity, so you could use one horn as control and the other as treated. In many mouse strains, you would not get away with that.

SHELESNYAK (Washington): A tremendous amount of work has been done in this study and I personally find it tragic. You want to study decidualization and it is understandable that you don't want to work with a blastocyst interfering with your system. On the other hand, you set up a model to study biochemical parameters and you spoil that system by trying artificially to create an environment in the uterus that exists already in a pseudopregnant animal. A pseudopregnant animal has the natural hormonal and other systemic requirements for successful implantation. Whatever biochemical parameters you may measure then are not confounded by additional progesterone or additional estrogen. I say this because there has been tremendous work trying to understand what decidualization is, that has not advanced much.

GLASSER (Baylor): I have just discovered today why in Texas we export oil, we don't put it down uteri. The work is very interesting and I simply want to go on record as saying that in rat systems the early stimulation of decidualization is radically different. We have used both the pseudopregnant rat and, in an effort to understand and differentiate between the effect of progesterone and/or estrogen, we have used the castrate rat supported with various combinations of hormones. We have unpublished work on stimulation of the decidual cell reaction by intralumenal instillation of isotonic Hanks solution. Under those conditions, carefully adjusting for the possibility of wound healing, we find an immediate and precipitous increase in polymerase II, monitored in terms of the species of RNA which eventually follows. At about 9 to 12 hr, we can identify the appearance of either selectively synthesized or induced proteins and we find an enhanced chromatin template activity, although we have not done initiation-site assays. The species of protein produced and some of the RNA patterns have been confirmed by O'Grady and Bell [in: *Development in Mammals,* Vol. 1 (M. H. Johnson, ed.), p. 165, North-Holland, Amsterdam, 1977], again using the rat. I think decidualization is replication and differentiation of cells with new information. Our data are consistent with that idea. Your results show that you have caused transformation of stromal cells to decidual cells and you show some evidence of cell replication. Yet you present little, if any, data that during this process of differentiation any new information is being generated in this different population of cells. My confusion cannot be resolved on the basis of species differences.

SERRA: The decidual cell reaction between the rat and mouse may be different, because we have looked at protein synthesis in a number of different ways and have been unable to detect synthesis of any new protein species. Rather we have observed increases in total protein synthesis.

MARTIN (ICRF, London): I disagree with John Marston; at least three strains of mice that we have worked with, the Sydney White, the QS, and the TO, do not have complete continuity between horns and you can induce unilateral decidual responses in them. In regard to the pseudopregnant animal, you use either cervical stimulation, which I believe Dr. Shelesnyak was one of the first to describe, or else you wield the knife on the male. In either case, the system is artificial. There are two other points I would like to make about the artificial system. Professor Finn, who first used oil in the mouse, showed clearly that the resulting decidual reaction is morphologically identical to that produced by the blastocyst. Second is the question of getting a lot of animals at the same time. I have worked with pregnant mice and with castrate mice put on the same hormonal regime used by Dr. Serra here. Pregnant mice vary greatly in the time at which implantation and progestation start. If we want to do precise experiments of this sort, and to relate biochemical events precisely to changes in the proliferation rate or production of new cell types in the organ, we need to use the castrate system, which brings me finally to your data. Using autoradiography, we find exactly the same time course of DNA synthesis, but we start to see the primary decidual cells differentiating morphologically at about 18 hr. All the changes are very rapid and we also see a lot of cells in mitosis in the outer regions at about the same time. I notice you found fluctuations in thymidine uptake, a drop in RNA polymerase, and changes in histone phosphorylation, which correlate very nicely with the changes in cell proliferation and differentiation that we observe.

SERRA: We notice that all stromal cells are going into DNA synthesis in a synchronous manner.

MANES: We are probably seeing a stem cell response. Probably following stimulation there is asymmetric cell division going on, so that one retains the stem cell population plus a new population of cells that is now making new types of proteins and has new information. Does this fit the data? Is there a basal population of stem cells in the decidua and stroma?

SERRA: From our autoradiographic data of DNA synthesis, it appears that all stromal cells are going through DNA synthesis synchronously. We expected to see a small population of stromal cells at the point where the deciduoma would develop, but that is not what we saw through the first cell division. We are still not sure what happens between the first cell division and the formation of the decidual cell mass.

MANES: A period of DNA synthesis followed by at least one cell division would be required to segregate cell types if that is what is happening. I am not sure if the time course leads to that conclusion.

FINN (Liverpool): I think you said that because DNA synthesis did not start, the oil was put in and then you got a reaction. Something must be happening during those 9 hr and you looked for it but could not find it. As far as I can see, you were looking for changes in the stromal cells. You were applying the stimulus to the epithelial cell surface, however, so one assumes that there has got to be some change in those cells before the message can reach the stroma. Although you were taking epithelial cells, you probably would not detect a change because the epithelium is such a small proportion of the total. This factor might explain the difference with Stan Glasser's results, because the rat seems to be much more sensitive to trauma, and trauma seems to affect the stroma directly. So, it is possible there are changes over part of those 9 hr that are simply confined to epithelial cells, whereby the stimulus is transduced and transmitted to the stroma.

SERRA: We come to the point of asking where are these changes taking place? I think that is where Stan Glasser's system for separating cell types may be of great assistance. Is the RNA polymerase activity increasing in a certain population of cells?

CHANG (Worcester Foundation): I would like to ask you, or Dr. Shelesnyak or Dr. Finn, did you every try to put in oil or air during pregnancy? Does that affect implantation or not?

FINN: I did try this a very long time ago, and all that happens is that the oil seems to take over and you see a normal response to the oil. I don't know what happens to the blastocyst.

CHANG: Did the mice stay pregnant or not?

FINN: No, not if you do it in both horns.

SHELESNYAK: The insertion of any fluid into the uterus will interfere with pregnancy. In the rat, we have used a technique for inducting decidualization by an antihistamine, pyrathiazine (Kraicer and Shelesnyak, 1958, *J. Endocrinol.* **17**:324). Administered systemically, we do not have to touch the uterus at all; we can do precise timing. When this drug is administered to a pregnant rat, there is an overriding decidualization, but mostly around the decidual sites associated with pregnancy. In fact, there is a very interesting phenomenon about the decidualization process. By the use of the pyrathiazine, within 6 to 10 hr, we find localized areas that decidualize almost like sites of nidation. We have never explored precisely why this happens but it's not uniformly distributed. It is at an early enough stage so that there is no problem of tubular constriction.

CHANG: I always feel you are not working on implantation, you are working on decidua formation. I make this remark because in other animals, especially pigs or sheep or cows, there is no such thing as a decidua as you describe in mice and rats. Talk to a medical doctor and ask where is the decidua in the human placenta, you never get a clear answer. That is the rude remark I wanted to make!

MANES: This business of the localized decidual response sounds intriguing and I have often wondered whether there really are sites where implantation can occur, that allow for the spacing of embryos in the uterus, or whether the embryos create their own spots. It reminds me of what Dr. Biggers said yesterday about Elephantulus, the African elephant shrew. This little animal, which has something like 300 fertilized eggs with each ovulation, only allows one to implant in a very special localized spot in the uterus, which is a remarkable strategy for reproduction. That is just an off-the-cuff comment, but in terms of decidual response, there are probably preferred sites most of the time.

ORSINI (Wisconsin): Some years ago I published photographs of decidualized areas in a number of animals (Orsini, 1962, *J. Reprod. Fertil.* **3**:288, Orsini, 1963, in: *Delayed Implantation* (A. C. Enders, ed.), pp. 156–167, Univ. of Chicago Press, Chicago). I found out that it was possible to decidualize with air alone injected into the uterus of the pseudopregnant hamster, mouse, or rat. During the sensitive period, air usually produces distinct nodule development in one horn. At the wrong moment, air gives no swelling, and with too much pressure, it gives a solid uterus. I published two pictures (Orsini, 1963, *J. Endocrinol.* **28:**119) of a rat that showed a cleared specimen with the nodule caused by air. I have some of these specimens in bottles in my lab that I like to bring out when I have an anatomist around, especially an embryologist. One can't tell which are embryonic and which are DCRs induced by air. The interesting thing is that at early periods they appear at the antimesometrical border of the lumen, the way the first attachment of an embryo begins. A little later swelling begins, but this is long before any decidua appear. If you follow the development, the same pattern of decidua forms as in a normal conceptus, and then the same cavity forms but there isn't any embryo there (Orsini *et al.*, 1970, *Am. J. Obstet. Gynecol.* **106**:14). These deciduomata in pseudopregnant hamsters will degenerate. Air-induced deciduomata in the sterile horn of a unilaterally pregnant hamster, however, can persist until term and the vascular changes which occur in the mesometrium are like those at an embryonic site (Orsini, 1957, *J. Morphol.* **100**:565).

References

Adler, A. J., Langan, T. A., and Fasman, G. D., 1972, Complexes of deoxyribonucleic acid with lysine-rich (f1) histone phosphorylated at two separate sites: Circular dichroism studies, *Arch. Biochem. Biophys.* **153**:769.

Bell, S. C., 1979, Protein synthesis during deciduoma morphogenesis in the rat, *Biol. Reprod.* **20**:811.

Bender, M. S., and Prescott, D. M., 1962, DNA synthesis and mitosis in cultures of human peripheral leukocytes, *Exp. Cell Res.* **27**:221.

Cohen, L. H., Newrock, K. M., and Zweidler, A., 1975, Stage-specific switches in histone synthesis during embryogenesis of the sea urchin, *Science* **190**:994.

Das, R. M., and Martin, L., 1978, Uterine DNA synthesis and cell proliferation during early decidualization induced by oil in mice, *J. Reprod. Fertil.* **53**:125.

Denari, J. H., Nestor, I. G., and Rosner, J. M., 1976, Early synthesis of uterine proteins after a decidual stimulus in the pseudopregnant rat, *Biol. Reprod.* **15**:1.

Finn, C. A., and Bredl, J. C. S., 1973, Studies on the development of the implantation reaction in the mouse uterus: Influence of actinomycin D, *J. Reprod. Fertil.* **34**:247.

Finn, C. A., and Martin, L., 1972, Temporary interruption of the morphogenesis of deciduomata in the mouse uterus by actinomycin D, *J. Reprod. Fertil.* **31**:353.

Finn, C. A., and Porter, D. G., 1975, *The Uterus*, Publishing Sciences Corp., Acton, Mass.

Glasser, S. R., and Clark, J. H., 1975, A determinant role for progesterone in the development of uterine sensitivity to decidualization and ovoimplantation, in: *The Developmental Biology of Reproduction* (C. L. Markert and J. Papaconstantinou, eds.), pp. 311–345, Academic Press, New York.

Gornall, A. G., and Liew, C. C., 1974, Covalent modification of proteins at times of gene activation and protein synthesis, *Adv. Enzyme Regul.* **12**:267.

Grisham, J. W., 1962, A morphologic study of deoxyribonucleic acid synthesis and cell proliferation in regenerating rat liver: Autoradiography with thymidine-H^3, *Cancer Res.* **22**:842.

Gurley, L. R., Walters, R. A., and Tobey, R. A., 1973, The metabolism of histone fractions. VI. Differences in the phosphorylation of histone fractions during the cell cycle, *Arch. Biochem. Biophys.* **154**:212.

Gurley, L. R., Walters, R. A., and Tobey, R. A., 1974, Cell cycle-specific changes in histone phosphorylation associated with cell proliferation and chromosome condensation, *J. Cell Biol.* **60**:1356.

Gurley, L. R., Walters, R. A., and Tobey, R. A., 1975, Sequential phosphorylation of histone subfractions in the Chinese hamster cell cycle, *J. Biol. Chem.* **250**:3936.

Heald, P. J., and O'Hare, A., 1973, Changes in rat uterine RNA during early pregnancy, *Biochim. Biophys. Acta* **324**:86.

Heald, P. J., O'Grady, J. E., and Moffat, G. E., 1972a, The incorporation of [^{3}H]uridine into nuclear RNA in the uterus of the rat during early pregnancy, *Biochim. Biophys. Acta* **281**:347.

Heald, P. J., O'Grady, J. E., O'Hare, A., and Vass, M., 1972b, Changes in uterine RNA during early pregnancy in the rat, *Biochem. Biophys. Acta* **262**:66.

Heald, P. J., O'Grady, J. E., O'Hare, A., and Vass, M., 1975, Nucleic acid metabolism of cells of the luminal epithelium and stroma of the rat uterus during early pregnancy, *J. Reprod. Fertil.* **45**:129.

Isenberg, I., 1979, Histones, *Annu. Rev. Biochem.* **48**:159.

Jaehning, J. A., Stewart, C. C., and Roeder, R. G., 1975, DNA-dependent RNA polymerase levels during the response of human peripheral lymphocytes to phytohaemagglutinin, *Cell* **4**:51.

Jonsson, H. T., Jr., Rankin, J. C., Ledford, B. E., and Baggett, B., 1978, Prostaglandin levels following stimulation of the decidual cell reaction in the mouse uterus, Program of the 60th Meeting of the Endocrine Society, Miami, Florida, p. 326.

Kornberg, R. D., 1974, Chromatin structure: A repeating unit of histones and DNA, *Science* **184**:868.

Langan, T. A., 1970, Phosphorylation of histones *in vivo* under the control of cyclic AMP and hormones, *Biochem. Psychopharmacol.* **3**:307.

Langan, T. A., 1978, Isolation of histone kinases, *Methods Cell Biol.* **19**:143.

Ledford, B. E., Rankin, J. C., Markwald, R. R., and Baggett, B., 1976, Biochemical and morphological changes following artificially stimulated decidualization in the mouse uterus, *Biol. Reprod.* **15**:529.

Ledford, B. E., Rankin, J. C., Froble, V. L., Serra, M. J., Markwald, R. R., and Baggett, B., 1978, The decidual cell reaction in the mouse uterus: DNA synthesis and autoradiographic analysis of responsive cells, *Biol. Reprod.* **18:**506.

Miller, B. G., 1973, Metabolism of RNA and pyrimidine nucleotides in the uterus during the early decidual cell reaction, *J. Endocrinol.* **59:**275.

Miller, B. G., and Emmens, C. W., 1969, The effects of oestradiol and progesterone on the incorporation of tritiated uridine into the genital tract of the mouse, *J. Endocrinol.* **43:**427.

Notides, A., and Gorski, J., 1966, Estrogen-induced synthesis of a specific uterine protein, *Proc. Natl. Acad. Sci. USA* **56:**230.

O'Grady, J. E., Moffat, G. E., McMinn, L., Vass, M., O'Hare, A., and Heald, P. J., 1975, Uterine chromatin template activity during the early stages of pregnancy in the rat, *Biochim. Biophys. Acta* **407:**125.

Pogo, B. G. T., Allfrey, V. G., and Mirsky, A. E., 1966, RNA synthesis and histone acetylation during the course of gene activation in lymphocytes, *Proc. Natl. Acad. Sci. USA* **55:**805.

Pogo, B. G. T., Pogo, A. O., Allfrey, V. G., and Mirsky, A. E., 1968, Changing patterns of histone acetylation and RNA synthesis in regeneration of the liver, *Proc. Natl. Acad. Sci. USA* **59:**1337.

Rall, S. C., and Cole, R. D., 1971, Amino acid sequence and sequence variability of the amino-terminal regions of lysine-rich histones, *J. Biol. Chem.* **246:**7175.

Rankin, J. C., Ledford, B. E., and Baggett, B., 1977, Early involvement of cyclic nucleotides in the artificially stimulated decidual cell reaction in the mouse uterus, *Biol. Reprod.* **17:**549.

Rankin, J. C., Ledford, B. E., Jonsson, H. T., Jr., and Baggett, B., 1979, Prostaglandins, indomethacin and the decidual cell reaction in the mouse uterus, *Biol. Reprod.* **20:**399.

Richards, B. M., and Pardon, J. F., 1970, The molecular structure of nucleohistone (DNH), *Exp. Cell Res.* **62:**184.

Robbins, E., and Borun, T. W., 1967, The cytoplasmic synthesis of histones in HeLa cells and its temporal relationship to DNA replication, *Proc. Natl. Acad. Sci. USA* **57:**409.

Roeder, R. G., and Rutter, W. J., 1969, Multiple forms of DNA-dependent RNA polymerase in eukaryotic organisms, *Nature (London)* **224:**234.

Roeder, R. G., and Rutter, W. J., 1970, Specific nucleolar and nucleoplasmic RNA polymerases, *Proc. Natl. Acad. Sci. USA* **65:**675.

Ruiz-Carrillo, A., Wangh, L. J., and Allfrey, V. G., 1975, Processing of newly synthesized histone molecules, *Science* **190:**117.

Ruiz-Carrillo, A., Wangh, L. J., and Allfrey, V. G., 1976, Selective synthesis and modification of nuclear proteins during maturation of avian erythroid cells, *Arch. Biochem. Biophys.* **174:**273.

Serra, M. J., Ledford, B. E., Rankin, J. C., and Baggett, B., 1978, Changes in RNA polymerase activity in isolated mouse uterine nuclei during the decidual cell reaction, *Biochim. Biophys. Acta* **521:**267.

Serra, M. J., Ledford, B. E., and Baggett, B., 1979, Synthesis and modification of the histones during the decidual cell reaction in the mouse uterus, *Biol. Reprod.* **20:**214.

Shelesnyak, M. C., and Tic, L., 1963, Studies on the mechanism of decidualization. IV. Synthetic processes in the decidualizing uterus, *Acta Endocrinol.* **42:**465.

Sklar, V. E. F., Schwartz, L. B., and Roeder, R. G., 1975, Distinct molecular structures of nuclear class I, II and III DNA-dependent RNA polymerases, *Proc. Natl. Acad. Sci. USA* **72:**348.

Spalding, J., Kagiwara, K., and Mueller, G. C., 1966, The metabolism of basic proteins in HeLa cell nuclei, *Proc. Natl. Acad. Sci. USA* **56:**1535.

Spelsberg, T. C., Wilhelm, J. A., and Hnilica, L. S., 1971, Nuclear proteins in genetic restriction. II. The nonhistone proteins in chromatin, *Subcell. Biochem.* **1:**107.

Valenzuela, P., Hager, G. L., Weinberg, F., and Rutter, W. J., 1976, Molecular structure of yeast RNA polymerase III: Demonstration of the tripartite transcriptive system in lower eukaryotes, *Proc. Natl. Acad. Sci. USA* **73:**1024.

Varshavsky, A. J., Bakayer, V. V., and Georgiev, G. P., 1976, Heterogeneity of chromatin subunits *in vitro* and location of histone H1, *Nucleic Acids Res.* **3:**477.

Viarengo, A., Zoncheddu, A., Taningher, M., and Orunesu, M., 1975, Sequential stimulation of nuclear RNA polymerase activities in livers from thyroidectomized rats treated with triiodothyronine, *Endocrinology* **97:**955.

Weil, P. A., and Blatti, S. P., 1976, HeLa cell deoxyribonucleic acid dependent RNA polymerases: Function and properties of the class III enzymes, *Biochemistry* **15:**1500.

Weinmann, R., and Roeder, R. G., 1974, Role of DNA-dependent RNA polymerase III in the transcription of the tRNA and 5 S RNA genes, *Proc. Natl. Acad. Sci. USA* **71:**1790.

Wilhelm, J. A., Spelsberg, T. C., and Hnilica, L. S., 1971, Nuclear proteins in genetic restriction. I. The histones, *Subcell. Biochem.* **1:**39.

VI

Blastocyst–Uterine Interactions

Introduction
Blastocyst–Uterine Interactions

Henning M. Beier

The title of this part presupposes that the blastocyst and uterus actively influence each other. While the existence of such interactions is generally accepted, their nature, timing, and extent are obscure. Formerly, the mammalian uterus was thought to be merely a fertile soil from which the embryo emerged like plants in a field after the seed has been successfully deposited. Since Von Baer (1827) detected the mammalian oocyte, our views have changed profoundly and today we are able to study the morphology, physiology, and biochemistry of the two systems involved, namely, the blastocyst and genital tract, using sophisticated techniques.

The developing mammalian embryo, particularly during the cleavage and blastocyst stages, is no longer thought to be autonomous within the maternal reproductive tract. Maternal endocrine control, follicular fluid, and oviductal and uterine secretions (which contain distinct substrates of low and high molecular weight) all contribute to the limitations of the embryo's independence before attachment or implanation takes place. Considerable attention has been paid to the synthetic and secretory capacity of the endometrium and its cellular transformations, but the metabolic and synthetic capacity of the blastocyst has not been ignored.

Mother and embryo are genetically independent, each acting to accomplish the goal of normal pregnancy and development. Mammalian embryonic development, however, depends increasingly on maternal support for the delivery of sufficient offspring to preserve the species. While the embryonic differentiation is unlikely to be influenced by maternal factors, embryonic growth is well established to be a nutrition-dependent process, and is thus influenced by maternal metabolism and physiology (Naftolin, 1979). Endocrine modulation of the mother controls the proliferation and transformation of the genital tract, which is essential for the establishment of pregnancy (Segal *et al.*, 1973).

These preparations for pregnancy, in turn, are a prerequisite for embryonic development. The crucial point, however, is whether or not embryonic influences in the earliest

Henning M. Beier • Department of Anatomy and Reproductive Biology, Medical Theoretical Institute, Rheinisch-Westfälische Technische Hochschule, Aachen, Federal Republic of Germany.

phases of pregnancy, particularly during the preimplantation stage, actively modify the maternal system. Whether the decisive actions brought about by the blastocyst are realized by stimulatory factors or by inhibitors of some physiological components is still an open question. In this section, blastocyst signals are discussed, which may condition the genital tract and its intralumenal environment in order to meet the demands of the blastocyst (Wolstenholme and O'Connor, 1965; Whelan and Heap, 1979). These signals can be hormones, metabolites, or mechanical stimuli mediated by cellular contact. Contributions to Section VI thus deal with the presence and role of blastocyst–uterus interactions as prerequisites for the establishment of the pre- and post implantation stages of normal pregnancy.

References

Baer, K. E. von, 1827, De ovi mammalium et hominis genesi, epistolam ad Academiam Imperialem Scientiarum Petropolitanam, Lipsiae Sumptibus Leopoldi Vossii (Leopold Voss), Leipzig.

Naftolin, F. (ed.), 1979, *Abnormal Fetal Growth: Biological Bases and Consequences,* Dahlem Workshop, Life Sciences Research Report 10, Dahlem Konferenzen, Berlin.

Segal, S. J., Crozier, R., Corfman, P. A., and Condliffe, P. G. (eds.), 1973, *The Regulation of Mammalian Reproduction,* Thomas, Springfield, Ill.

Whelan, J., and Heap, R. B. (eds.), 1979, *Maternal Recognition of Pregnancy,* Ciba Foundation Symposium 64 (new series), Excerpta Medica, Amsterdam.

Wolstenholme, G. E. W., and O'Connor, M. (eds.), 1965, *Preimplantation Stages of Pregnancy,* Ciba Foundation Symposium, Churchill, London.

19

Embryonic Signals and Maternal Recognition

R. B. Heap, A. P. F. Flint, and J. E. Gadsby

I. Introduction

In the first paper of this volume Professor Amoroso argues that viviparity is a reproductive stratagem that has evolved many times in widely separated taxonomic groups of the animal kingdom. Retention of embryos within the body and the birth of living young occurs among representatives of all classes of vertebrates except birds, and in many groups of invertebrates. In the light of this wide distribution it is not surprising that solutions to problems posed by this evolutionary innovation are numerous (Amoroso *et al.*, 1979). In eutherian and marsupial mammals, in which viviparity has become the preferred mode of reproduction, a prominent role has emerged for the corpus luteum. Progesterone secretion by the corpus luteum is essential for the establishment of gestation in all eutherians so far studied, and mechanisms have evolved that result in converting the female reproductive system from a cyclic pattern, in which oscillations of progesterone and estrogen secretion occur, to a noncyclic pattern, in which progesterone secretion is dominant. The time in gestation when this transformation occurs is an early indication of the maternal recognition of pregnancy (Short, 1969), and recent studies have focused on the nature of embryonic factors that signal this event (Heap and Perry, 1977).

II. Embryonic Signals

Pregnancy depends in the first instance on the prolonged functional activity of the corpus luteum. Among certain eutherian mammals, as in sheep, cow, and pig, the prolongation of luteal function is one of the earliest indications of maternal recognition, for it precedes the events of attachment and implantation. The embryo signals its presence by

R. B. Heap, A. P. F. Flint, and J. E. Gadsby • A.R.C. Institute of Animal Physiology, Babraham, Cambridge, England. Current address of J.E.G.: Reproductive Endocrinology Program, Department of Pathology, University of Michigan, Ann Arbor, Michigan.

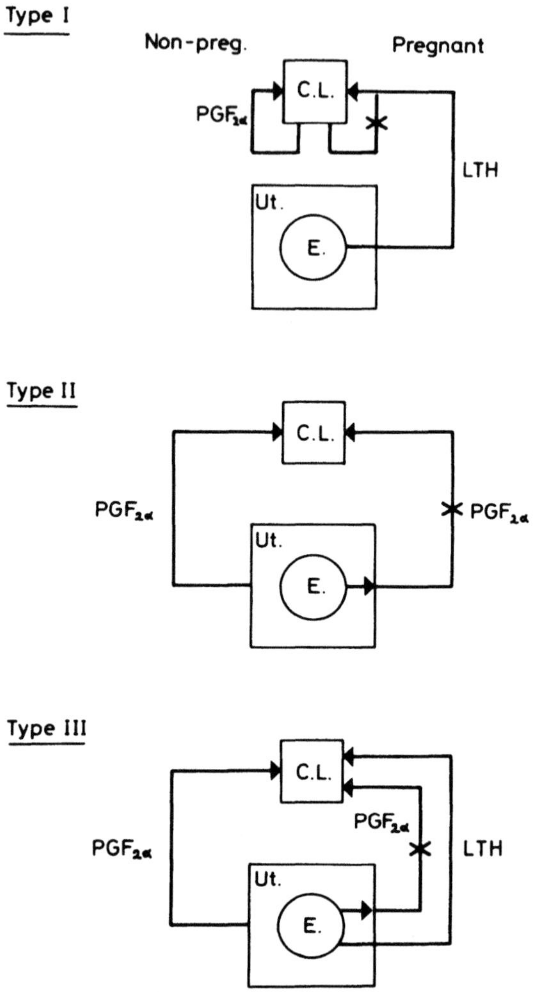

Figure 1. Putative embryonic signals in early pregnancy. E., embryo; C.L., corpus luteum; Ut., uterus; LTH, luteotropic hormone complex; $PGF_{2\alpha}$, prostaglandin $F_{2\alpha}$. A cross on a pathway indicates that the effect of the luteolytic $PGF_{2\alpha}$ may be neutralized by an embryonic luteotropin (Type I), that the release of $PGF_{2\alpha}$ may be blocked by the presence of a conceptus (Type II), or that an embryonic luteotropin may override the effect of $PGF_{2\alpha}$ secretion (Type III). Reprinted by kind permission of the *Journal of Reproduction and Fertility*.

secreting compounds that ensure the survival of the corpus luteum (Heap *et al.*, 1979). Among the variety of mechanisms by which luteal survival may be achieved, three hypothetical patterns have been described (Figure 1; Perry *et al.*, 1976): a chorionic gonadotropin that overrides luteolysis by neutralizing the effects of compounds probably ovarian in origin (e.g., CG in women and rhesus monkeys); antiluteolytic substances that prevent the release of prostaglandin $F_{2\alpha}$ and protect the corpus luteum from lysis (trophoblastin in sheep; Martal *et al.*, 1979); and luteotrophic–antiluteolytic effects (e.g., estrogens in pigs; Flint *et al.*, 1979). We have studied the latter example and in this paper review the evidence for estrogen synthesis by the early embryo, the temporal relation between estrogen synthesis and maternal recognition, the pathways of estrogen synthesis, and possible modes of action of estrogens.

In the pig, plasma progesterone concentrations reach maximum values at about Day 12 postcoitum (p.c.). Thereafter they are maintained, though at a lower level, for the duration of pregnancy, and the corpora lutea formed after ovulation remain and are indispensible throughout gestation. In the event of a sterile mating the corpora lutea regress rapidly, and by Day 16 p.c. progesterone levels have declined to very low values. In the pregnant animal blastocysts reach a diameter of about 5 mm by Day 10 p.c., and after Day 12 p.c. they elongate rapidly to a length of up to 1 m (Anderson, 1978). Localized points of attachment to the endometrium can be found after that time, and definitive attachment by interlocking microvilli develops by about Day 18 p.c. Placentation is noninvasive and the maternal epithelium retains its integrity, unlike that of many eutherian mammals in which destruction of maternal tissue occurs to a varying degree. Thus, in the pig, although trophoblast cells express an inherent invasiveness when transplanted to ectopic sites (Samuel, 1971; Samuel and Perry, 1972), this property is attenuated inside the uterus. It is also notable that the striking morphological changes of stromal tissue associated with decidualization and the development of the maternal component of the placenta, as found in rodents, do not take place in the pig.

The abundance of preimplantation trophectoderm, the availability of readily separated embryonic and maternal tissues, and the protracted attachment phase make the pig a good model for the study of preimplantation embryonic signals. Production of such a signal occurs by Day 12 p.c., as removal of embryos on or after this time results in prolonged luteal function in some, although not all, animals, whereas embryo removal before Day 12 results in luteal regression at the time expected in nonpregnant animals (Dhindsa and Dziuk, 1968).

III. Onset of Trophoblast Estrogen Synthesis

The notion has emerged that estrogens produced by the embryo constitute part of the early interaction between embryo and mother. Reasons for considering estrogens as possible candidates for the role of an embryonic signal have been discussed previously (Flint *et al.*, 1979); they derive from earlier findings that estrogens are luteotrophic in this species (Kidder *et al.*, 1955; Gardner *et al.*, 1963; du Mesnil du Buisson, 1967), from more recent studies that show that preimplantation embryos contain aromatase by conversion of [^{3}H]androstenedione, [^{3}H]testosterone (Figure 2), and [^{3}H]dehydroepiandrosterone into estrogens *in vitro* (Perry *et al.*, 1973, 1976), and from the work of Bazer and his colleagues in

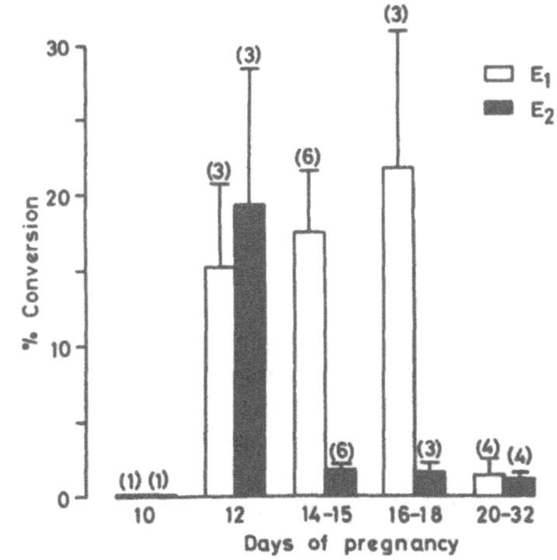

Figure 2. Aromatase in pig blastocyst and early embryonic tissue. Results show the mean conversion (±S.E.) of [^{3}H]androstenedione or [^{3}H]testosterone into estrone (E_1) and estradiol-17β (E_2) when incubated *in vitro* with 300 mg tissue for 3 hr in 5% CO_2 in oxygen. Figures in parentheses refer to number of animals. Results in Day 12 group obtained with incubations containing up to seven blastocysts per flask; those in Day 20–32 group with pooled chorionic and allantochorionic tissue.

Florida who have shown that estrogens suppress the uterine release of $PGF_{2\alpha}$ into venous blood in nonpregnant animals (Frank *et al.*, 1977; Moeljono *et al.*, 1977). If estrogens are to function as signals from embryo to mother, however, aromatase activity should be present before the time of maternal recognition. Activity is clearly demonstrable by Day 12 p.c. and appears closely correlated with the time of rapid blastocyst elongation (Figure 3). Endogenous estrogens have been detected in pre-elongation blastocysts; in spherical blastocysts, 5–10 mm in diameter, the concentration of endogenous estrone and estradiol-17β is 260 and 80 pg/mg protein, respectively (Gadsby and Heap, 1978). These results indicate that aromatase induction precedes the process of elongation but that the activity is normally below the sensitivity of the assay method. Further evidence for aromatase induction preceding the onset of blastocyst elongation and the maternal recognition of pregnancy comes from experiments in ovariectomized animals treated with medroxyprogesterone acetate (MPA; Depo-Provera, Upjohn Ltd.) to maintain pregnancy; aromatization was enhanced and clearly detectable on Day 10 when blastocysts were still spherical (Flint *et al.*, 1979).

IV. Endogenous Precursors of Trophoblast Estrogens

The finding of Gadsby *et al.* (1976) that ^{3}H-labeled progesterone was converted to estrone and estradiol-17β by preimplantation trophoblast incubated *in vitro* with added cofactors raised the question of whether maternal progesterone of luteal origin was an endogenous precursor of early embryo estrogen production. [^{3}H]Progesterone was infused close-arterially into the gravid uterine horn of three anesthetized sows on Day 22 of gestation, a time when estrogen production in early pregnancy is rising to high values as reflected by circulating levels of estrone sulfate (Robertson and King, 1974). Labeled progesterone concentration in uterine vein blood reached a steady value within 100 min of a continuous infusion over 3 hr. Uterine venous blood collected during the last hour of in-

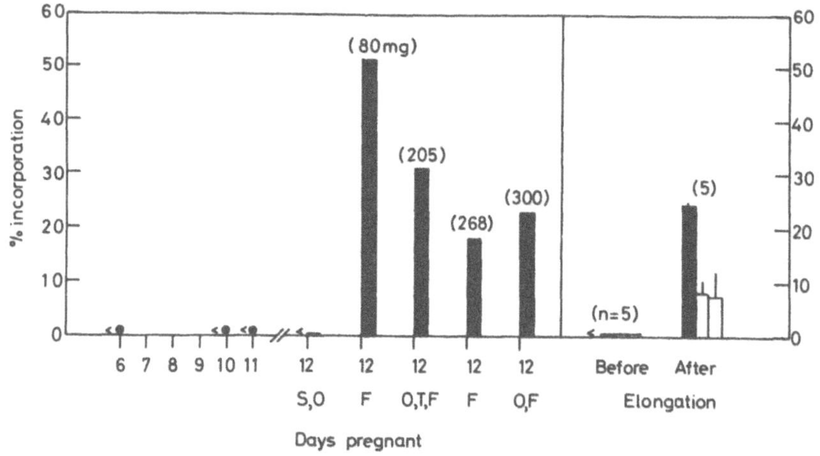

Figure 3. Onset of aromatase activity in pig blastocysts determined *in vitro*. Results in left panel show conversion of [^{3}H]androstenedione by spherical (S), ovoid (O), tubular (T), and filamentous (F) blastocysts. For classification of blastocysts see Anderson (1978). Wet weight of tissue given in parentheses; 6 to 25 spherical blastocysts were used in incubations before Day 12. Results in right panel show mean conversion ($\pm$ S.E.) in all experiments of [^{3}H]androstenedione to phenolic compounds (solid bar), estrone (middle bar), and estradiol-17β (right bar) before and after elongation.

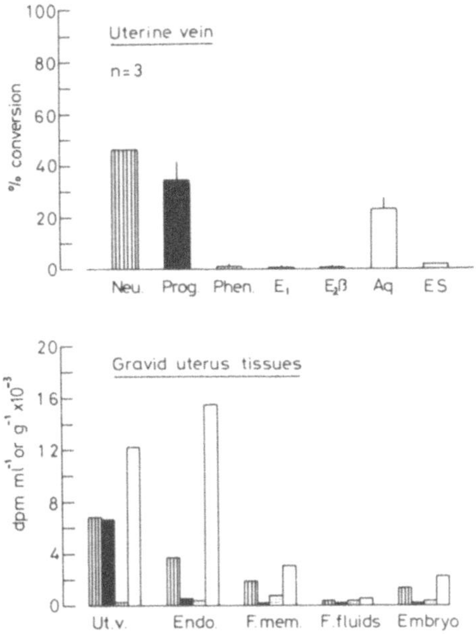

Figure 4. Conversion of circulating progesterone to estrogens in the gravid uterus of the pig on Day 22 pc. In three animals [^{3}H]progesterone was infused close-arterially at approximately 1 μCi/min for up to 180 min. Top panel shows the distribution of radioactivity in uterine venous blood between 135 and 170 min after start of infusion; neutral fraction (Neu., 1 experiment), progesterone (Prog.), phenolic fraction (Phen.), estrone (E_1), estradiol-17β ($E_2\beta$), aqueous fraction (Aq.), and estrone sulfate (ES). Bottom panel shows the distribution of radioactivity in uterine vein (Ut.v.), endometrium (Endo.), fetal membrane (F.mem.), fetal fluids (F.fluids), and embryo (for each tissue, bars from left to right refer to neutral, progesterone, phenolic, and aqueous radioactivity).

fusion showed a small conversion of progesterone to labeled estrogens (Figure 4; estrone, 0.1% of infusate; estradiol-17β, 0.01%; conjugated estrogens, 1.2%). Fetal membranes and endometrial tissue removed at the end of the experiment contained labeled unconjugated phenolic compounds (18,558 and 15,289 dpm/g wet wt; infusion rate, 1.9 μCi/min) from which estrone and estradiol-17β were identified by recrystallization.

Although these results show conversion of circulating labeled progesterone to estrogens by the gravid uterus in early pregnancy, further studies have indicated that progesterone can also be synthesized *de novo*. Day 16 p.c. trophoblast explants were cultured in medium 199 for 96 hr by the method of Wyatt (1976). Increasing amounts of progesterone were detected in the medium, reaching maximum values after 96 hr of culture (Figure 5). On average, a total of over 40 ng progesterone was produced in each culture dish containing trophoblast tissue of approximately 50 mg wet wt. When tissue was cultured in medium

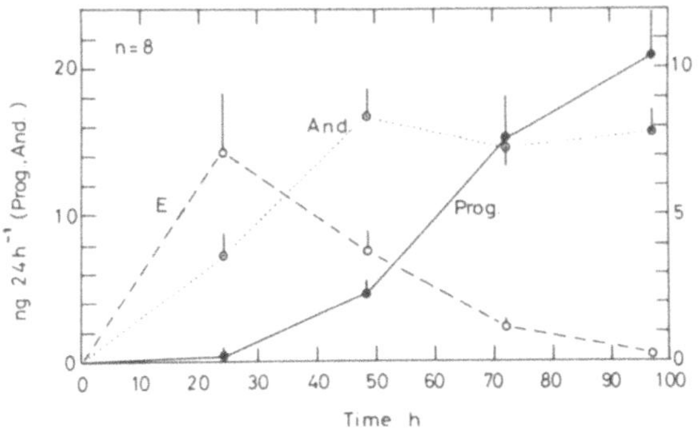

Figure 5. Production of total unconjugated estrogens (E), progesterone (Prog.), and androstenedione (And.) by pig trophoblast (Day 16) cultured for 96 hr in medium 199 by the method of Wyatt (1976). Medium was removed every 24 hr and steroids were determined by radioimmunoassay (means ± S.E.).

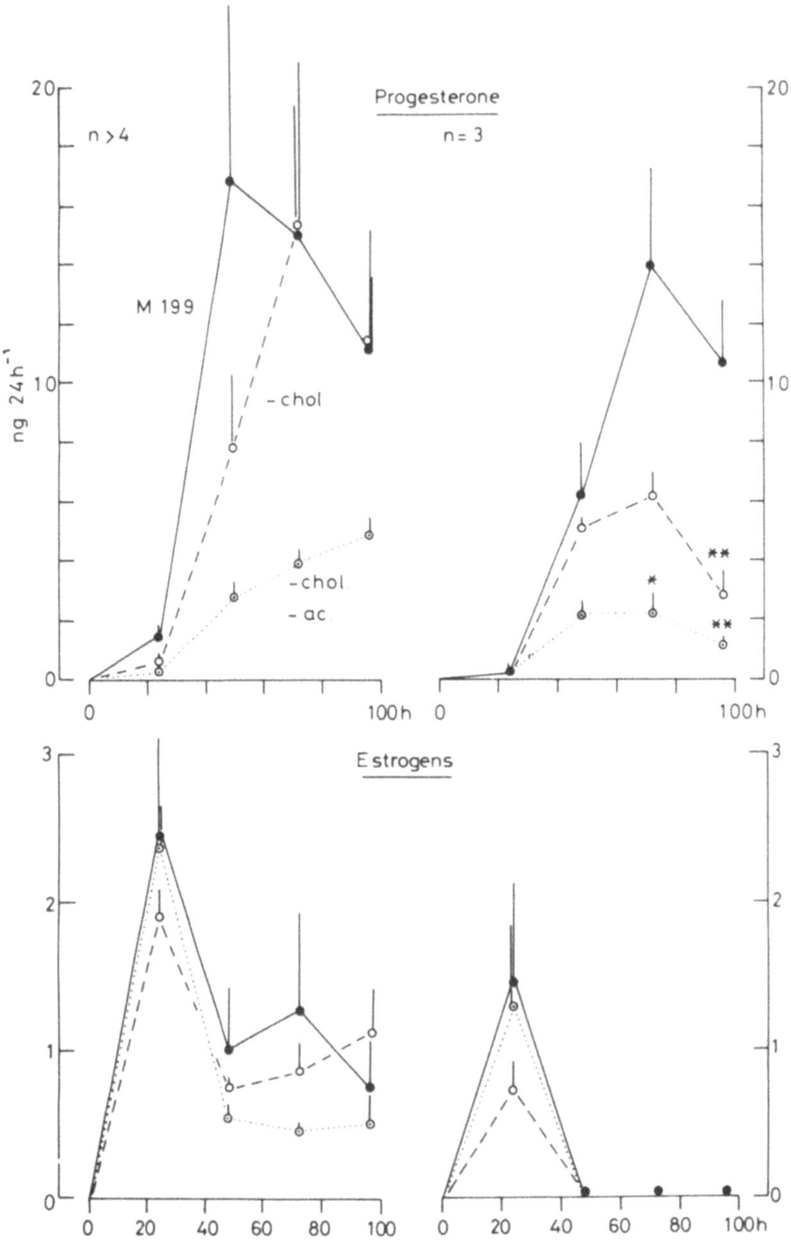

Figure 6. Influence of composition of culture medium on production of total unconjugated estrogens and progesterone (means ± S.E.). Pig trophoblast (Day 16) was cultured for 96 hr in medium 199 (M199, ——), M199 minus cholesterol (−chol., ----), M199 minus cholesterol and acetate (−chol., −ac., · · · ·). Medium was removed every 24 hr and steroids were determined by radioimmunoassay. Level of statistical significance compared with results using complete M199: $^{*}p<0.05$, $^{**}p<0.01$.

199 deficient in acetate and cholesterol, progesterone production was reduced, but not eliminated. Unconjugated estrogens were synthesized during the first 24 hr of culture in medium 199 but their production then declined. This pattern of estrogen synthesis occurred irrespective of medium composition (Figure 6). These results show that 16-day trophoblast tissue has the enzymatic capacity to synthesize progesterone *de novo* from acetate and cholesterol, and that estrogens are probably synthesized from endogenous C-21 precursors. During culture for 96 hr aromatase activity declined, resulting in an accumulation of progesterone and androstenedione and a decrease in estrogens in the medium (Figure 5). This finding may reflect changes in trophoblast function *in vivo,* as aromatase activity appears to decline by Days 20–32 p.c. compared with that observed between Days 12 and 18 p.c. (Figure 2).

The findings so far do not resolve the question of the relative importance of circulating progesterone and *de novo* synthesis in the production of trophoblast estrogens, but experiments on pigs ovariectomized during the first 9 days p.c. and treated with MPA to maintain gestation throw further light on this problem. Plasma concentrations of estrone sulfate were measured in peripheral blood taken from indwelling vascular catheters. Estrone sulfate, the dominant form of estrogens secreted by the gravid uterus due to sulfoconjugation in uterine tissue (Pack and Brooks, 1974; Robertson and King, 1974; Perry *et al.*, 1976), was present in similar concentration in peripheral plasma of ovariectomized, MPA-treated and control animals (Figure 7). Plasma progesterone values were less than 1 ng/ml in treated animals compared with normal values of 10–20 ng/ml. Although plasma progesterone concentration was reduced to about 10% of the normal values, endometrial progesterone concentrations at Days 21–22 p.c. were similar to those in control animals. These results suggest that trophoblast estrogens are produced by *de novo* synthesis rather than from circulating progesterone, and that endometrial progesterone may be derived partially from embryonic

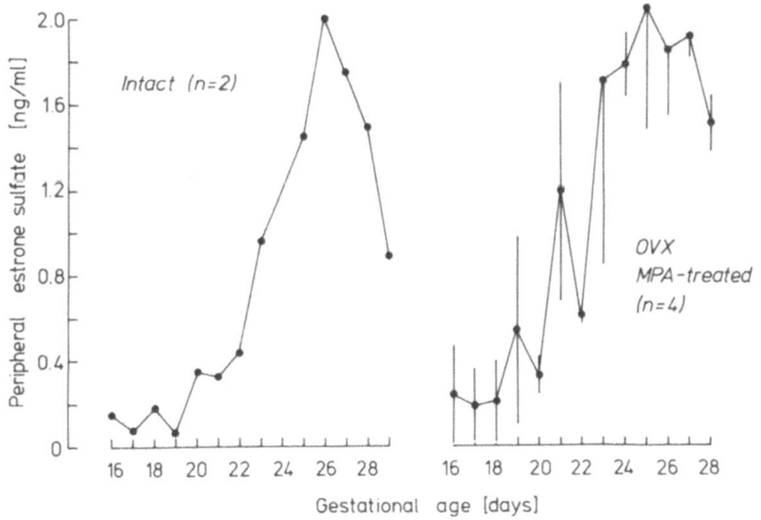

Figure 7. Removal of ovarian progesterone in early pregnancy and its effect on circulating levels of estrone sulfate. Ovariectomy (OVX) was performed in the first 10 days of gestation and animals were treated with medroxyprogesterone acetate (MPA) to maintain gestation (100 mg MPA on day before surgery and 10 mg per day or every alternate day starting on day of surgery). Peripheral blood samples were taken daily from indwelling vascular catheters. Values in intact animals were similar to normal values previously published (Robertson and King, 1974).

synthesis. The results do not exclude, however, the possibility of endometrial synthesis of estrogens and their precursors though it seems unlikely that circulating progesterone is quantitatively a major precursor.

Preimplantation trophoblast in the pig shows many of the features required of a classical endocrine organ in that it possesses a wide range of enzymes involved in steroid biosynthesis (Figure 8), it releases biologically active estrogens into the surrounding medium *in vitro,* and plasma estradiol concentrations in the utero-ovarian vein are higher on Days 12 to 17 in pregnant than in nonpregnant animals (Moeljono *et al.,* 1977).

V. Mechanism of Action of Trophoblast Estrogens

If trophoblast estrogens act as a preimplantation signal in the maternal recognition of pregnancy, we should next consider the mechanisms that prolong the life span of the corpus luteum and ensure that the endometrium remains under progesterone domination. Thus trophoblast estrogens may exert a short-range antiluteolytic effect by reducing endometrial $PGF_{2\alpha}$ release into uterine venous blood, possibly by a redirection into the uterine lumen as proposed by Bazer and colleagues (Frank *et al.,* 1978). This hypothesis is

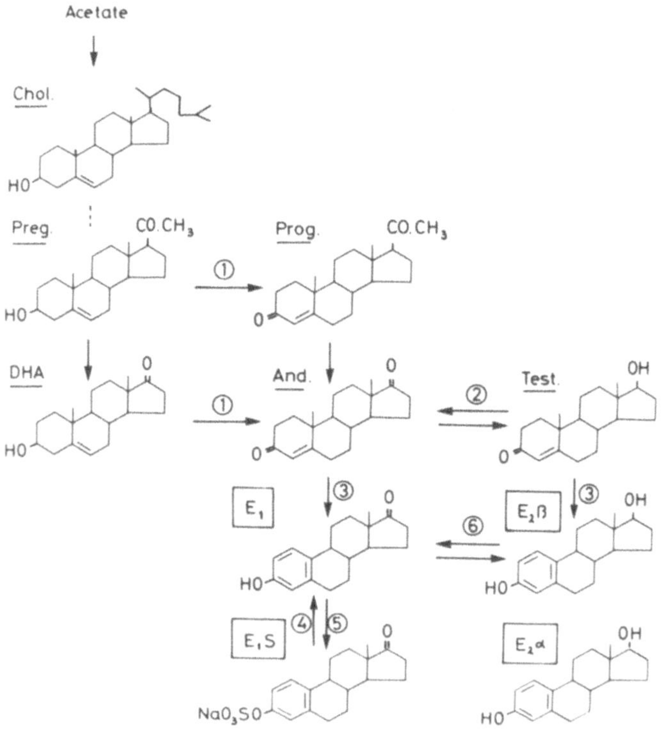

Figure 8. Steroid synthesis in the pig blastocyst. Definitive identification of estrone (E_1), estradiol-17β ($E_2\beta$), estradiol-17α ($E_2\alpha$) has been obtained by recrystallization of the products to constant specific radioactivity. Enzymes: 1, 3β-hydroxy-Δ^5-steroid dehydrogenase (EC 1.1.1.145); 2, testosterone 17β-dehydrogenase and 17β-oxidoreductase (EC 1.1.1.63, 1.1.1.64); 3, aromatase; 4, arylsulfatase (EC 3.1.6.1); 5, estrogen sulfotransferase. Chol., cholesterol; Preg., pregnenolone; Prog., progesterone; DHA, dehydroepiandrosterone; And., androstenedione; Test., testosterone; E_1S, estrone sulfate. Dashed line indicates that pathway has not been identified.

supported by the findings referred to above that $PGF_{2\alpha}$ concentration in utero-ovarian venous blood is reduced in early pregnancy (Moeljono *et al.*, 1977), that estradiol valerate treatment during the estrous cycle reduces the uterine release of $PGF_{2\alpha}$ into utero-ovarian venous blood, and that the same estrogen treatment results in raised $PGF_{2\alpha}$ levels in the uterine lumen (Frank *et al.*, 1977). The biochemical mechanisms involved in these changes have not been elucidated.

In addition to a local, antiluteolytic effect, trophoblast estrogens in the form of sulfoconjugated compounds may have a long-range influence either on the hypothalamo-pituitary system (Guthrie *et al.*, 1972), causing LH secretion that is luteotrophic after Day 14 p.c. in this species, or on the corpus luteum directly. These tissues contain aryl sulfatase and 17β-oxidoreductase (Perry *et al.*, 1976; Heap *et al.*, 1977) required for the conversion of estrone sulfate to biologically active estrogens. Maternal recognition of pregnancy, therefore, may derive from the multiple effects of trophoblast estrogens leading to prolongation of the life of the corpus luteum.

An additional requirement of this hypothesis is that local concentrations of biologically active estrogens are prevented from neutralizing a secretory endometrium, which is promoted by progesterone and is a prerequisite of implantation. Pack *et al.* (1979) have recently shown that endometrial sulfotransferase (and 17β-oxidoreductase) remain high in activity over the period of implantation and early pregnancy. This finding confirms the idea that sulfotransferase activity is directly related to plasma progesterone level. Moreover, this high level of enzyme activity is associated with a low level of estradiol nuclear receptors. The authors interpret these observations to mean that trophoblast estrogen is deactivated (by oxidation and conjugation) and the estradiol–receptor complex prevented from forming, thereby ensuring a secretory rather than a proliferative endometrium.

VI. Species Differences

Evidence for steroid synthesis and metabolism in preimplantation embryos of various species has been published recently and this work has been discussed elsewhere (Dickmann *et al.*, 1976; Heap *et al.*, 1979; Singh and Booth, 1979). Comparative measurements of aromatase activity in preimplantation trophoblast or implantation sites indicate that the values in the pig are markedly higher than in the other species so far studied (Figure 9). The reason for this high activity in the pig is not known. It is notable that among the species described above, the pig is the only one in which implantation is noninvasive, though as yet it is not possible to assess whether trophoblast estrogen synthesis and the form of placentation are causally related.

We have already argued that the estrogens may comprise a signal by which the mother recognizes the presence of an embryo, but there are other species, such as the cow and sheep, in which maternal recognition also precedes implantation, as in the pig, yet trophoblast estrogen synthesis is undetectable at this critical stage. In the cow, estrogens (including estrone sulfate) are luteolytic rather than luteotrophic at the time of maternal recognition, about Day 17 p.c. (Lemon, 1975; Eley *et al.*, 1979a). In this species, embryonic synthesis of estrogens appears to develop later, that is, by about the time of definitive attachment, and significant quantities of estrone sulfate are detectable in plasma, allantoic fluid (Robertson *et al.*, 1978; Eley *et al.*, 1979b), and milk (Figure 10; Heap and Hamon, 1979) by Day 40 p.c. These findings therefore support the view that trophoblast estrogens may also have short-range effects associated with the process of implantation, another sign of maternal recognition.

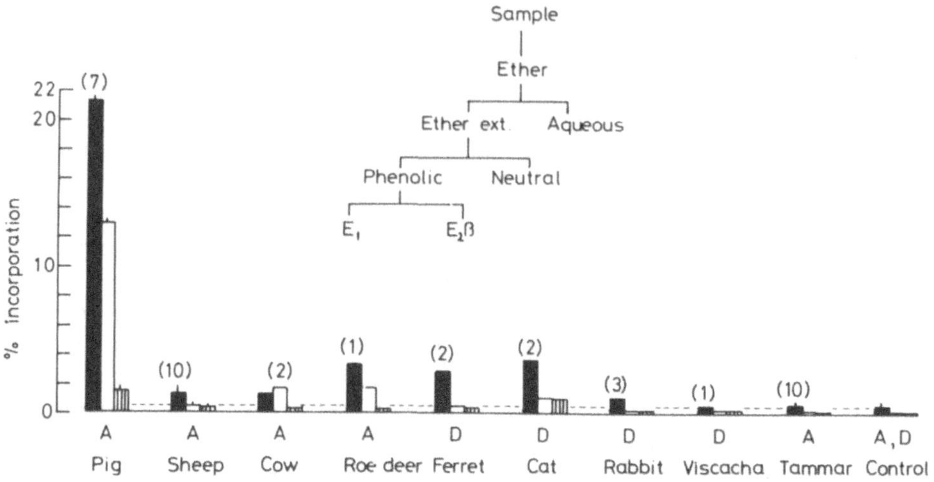

Figure 9. Aromatase activity in trophoblast tissue from various animals about the time of implantation or maternal recognition of pregnancy (pig, 14–18 days; sheep, 16–18 days; cow, 16 and 22 days; roe deer, allantochorionic tissue after attachment; ferret, 13 and 15 days; cat, 11 and 13 days; rabbit, 6 days; plains viscacha, 18 days; tammar wallaby, 19 to 25 days; control, no tissue). Precursors used were androstenedione, A, or dehydroepiandrosterone, D. Separation of phenolic fraction (solid bars), estrone (open bars), and estradiol-17β (lined bars) indicated; details of method have been published previously (Perry *et al.*, 1976). Horizontal dashed lines indicate amount of radioactivity recovered in a phenolic fraction when precursors were incubated in absence of tissue. Amount of wet weight tissue approximately 300 mg except in ferret (650 mg implantation site), rabbit (60 mg), and viscacha (more than 100 blastocysts, wet weight not known).

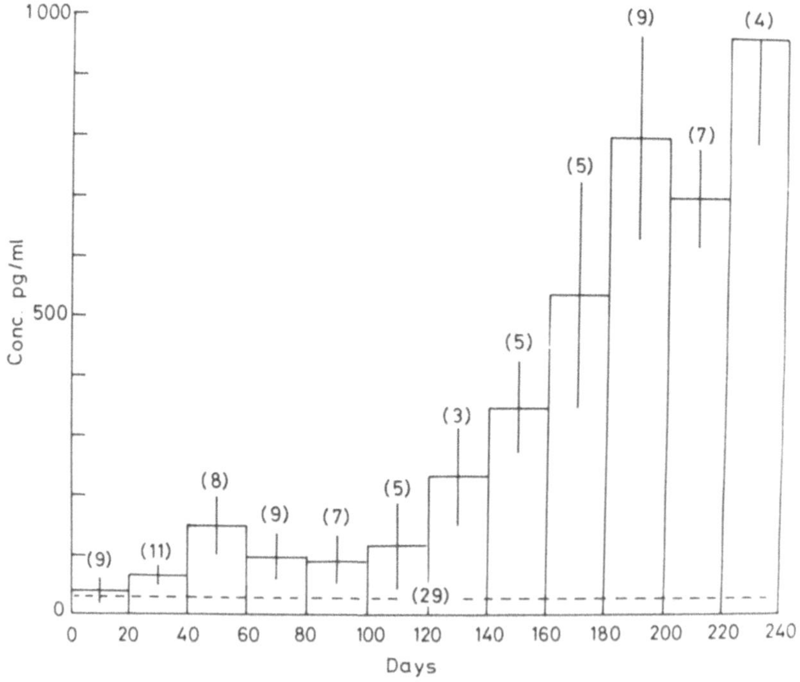

Figure 10. Estrone sulfate concentration (in estrone equivalents, mean ± S.E.) in aqueous phase of milk of Jersey cows during gestation. Values in nonpregnant animals shown by horizontal dashed line. Reprinted by kind permission of the *British Veterinary Journal*.

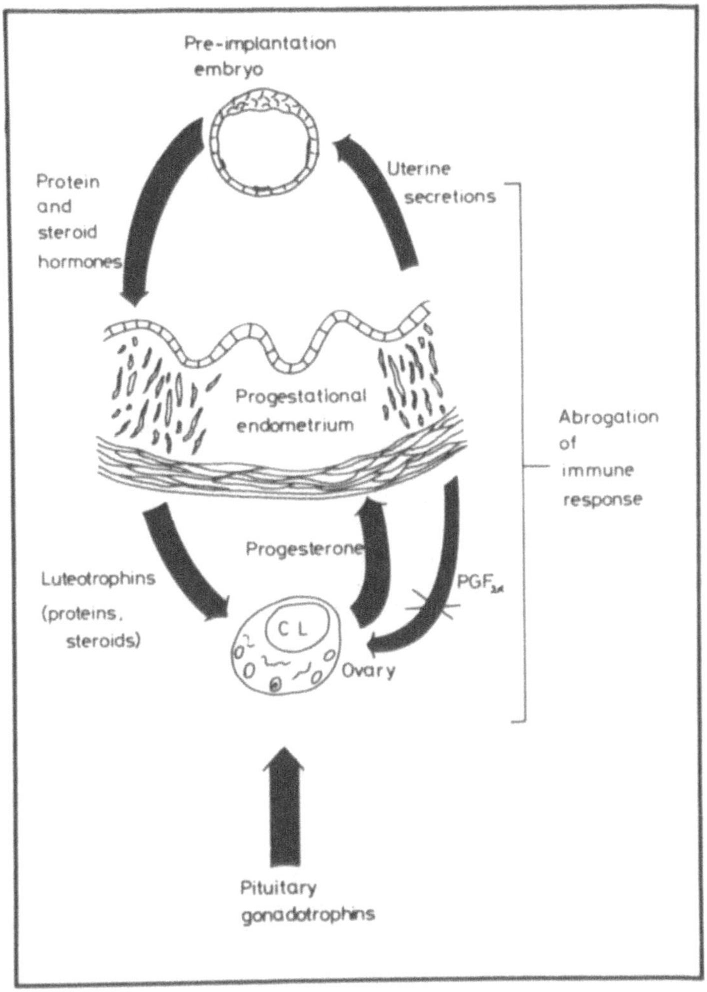

Figure 11. Summary of two-way interaction between embryo and mother during the establishment of pregnancy. CL, corpus luteum, $PGF_{2\alpha}$, prostaglandin $F_{2\alpha}$. Reprinted by kind permission of the *British Medical Bulletin*.

VII. Two-Way Interaction

We have so far considered signals of embryonic origin that are involved in transforming the female reproductive system from a cyclic to a noncyclic, progesterone-dominated pattern by prolonging corpus luteum function (Figure 11). Preimplantation embryos in domestic animals undergo considerable elongation prior to attachment (e.g., pig) and it is conceivable that this event may depend on the removal of inhibitory substances or the production of stimulatory factors. Recent studies suggest that the pig endometrium may enhance embryo growth. Trophoblast protein synthesis was stimulated *in vitro* when explants were incubated in coculture with endometrium but not with other maternal tissues, nor in the presence of added bovine serum albumin, fetal calf serum, or uteroferrin (Wyatt, 1976; Rice and Heap, unpublished observations). The elaboration of embryonic signals associated with maternal recognition, of whatever form, depends on the normal survival and proliferation of trophoblast, and work is in progress to investigate the nature of putative regulators of trophoblast growth.

ACKNOWLEDGMENTS. We thank Mrs. N. Ackland, Mr. R. D. Burton, and Miss E. L. Sheldrick for excellent technical assistance. Part of this work received financial support from the World Health Organization, the Meat and Livestock Commission, and the Royal College of Veterinary Surgeons Trust Fund.

Discussion

DICKMANN (Kansas): The rabbit needs at least four blastocysts to maintain the corpora lutea. Has the problem of how many embryos are required to maintain the corpora lutea been studied in the pig?

HEAP: Yes. Polge *et al.* (*J. Reprod. Fertil.* **12**:395, 1966) investigated this problem and demonstrated that the pig required more than four embryos in order to maintain the corpus luteum and to establish gestation.

DICKMANN: Around 1972 or 1973, when we were investigating whether blastocysts can make steroid and polypeptide hormones, we were very much in the dark, but I think today it is well established that the blastocyst has the capacity to make hormones. I think we will see a diversity of functions for these hormones, perhaps related to the size of the blastocyst and hence the amount of hormone produced. The pig has a number of 1-m-long blastocysts that may produce enough hormone to act systematically. In the rat I do not think there will be a systemic effect or even an effect through the local circulation between the uterus and the ovary. In species where the blastocyst is relatively small, as in the rat, mouse, and human, the effect of these hormones is likely to be limited to a direct, local effect on the endometrial tissue.

FORD (Iowa State): Maternal recognition of pregnancy occurs in the pig at about Day 12 to 13, and in the cow at about Day 14 to 18. In the cow, we see a two- to threefold increase in uterine arterial blood flow going only to the gravid uterine horn during this 4-day period (Ford *et al.*, 1979, *J. Reprod. Fertil.* **56**:63). In the pig, a four- to fivefold increase in uterine blood flow occurs on Days 12 and 13, corresponding to a decrease in $PGF_{2\alpha}$ secretion from the gravid uterine horn (Ford and Christenson, 1979, *Biol. Reprod.* **21**:617). We have measured estrogen levels in the cow and in the pig and found that estrogen increased both in uterine luminal fluid and uterine venous blood, at the time of maternal recognition of pregnancy and the tremendous elongation of the blastocysts in these species. I think estrogen secreted by the blastocyst can have many effects, one of which is an increase in uterine arterial blood flow, which may have an effect on luteal function or uterine secretions and could have other effects on the utero-ovarian system.

HEAP: Your results are certainly very interesting and that was why I qualified my remarks on the reduction of $PGF_{2\alpha}$ concentrations in uterine vein blood of pregnant animals as a reason for estrogens acting as an embryonic signal in the pig. Calculations from your figures of the increase in blood flow may show that the reduction in $PGF_{2\alpha}$ secretion in a uterine vein in pregnancy is not as great as was first supposed when compared to that in nonpregnant animals. If this is the case, then it raises problems about how estrogens of embryonic origin protect the corpus luteum and exert their luteotrophic effect.

Harrison (*J. Physiol.* **290**:36P, 1979) transplanted the ovary to the neck in the nonpregnant pig and found that the corpus luteum regresses at the normal time and cycle length is about 21 days, apparently because pulmonary metabolism of $PGF_{2\alpha}$ is considerably less than that of other species so far studied (Davies *et al.*, 1980, J. *Physiol.* **301**:86P), and systemic levels of $PGF_{2\alpha}$ are therefore relatively high. It is possible that other endogenous components exist that cause regression of the corpus luteum. Watson and Maule-Walker (*J. Reprod. Fertil.* **51**:393, 1977) have presented evidence for another endometrial luteolytic component, which raises the question of whether this factor(s) is also regulated by estrogens.

BAZER (Florida): It is difficult to defend the prostaglandin dilution thesis because we did not measure blood flow in those studies. Terqui *et al.* (unpublished), however, measured the prostaglandin metabolite, 13,14-dihydro-15-keto-$PGF_{2\alpha}$ in peripheral blood, in which the pool size should be uniform. Shille *et al.* (*Zentralbl. Veteringermed. Reihe A* **26**:169, 1979) have also measured this metabolite in peripheral blood. Both groups find low levels in pregnant animals, but high levels between Days 12 and 17 of the estrous cycle. The data we published for the pregnant and nonpregnant pig or pseudopregnant and nonpseudopregnant stage (Moeljono *et al.*, 1977, *Pros-*

taglandins **14**:543; Frank *et al.*, 1977, *Prostaglandins* **14**:183) show that the number of peaks and the average concentration of the peaks are markedly reduced in the pregnant and pseudopregnant animal. These data continue to support our theory for the maternal recognition of pregnancy (Bazer and Thatcher, *Prostaglandins* **14**:397, 1977). Defining a peak as a mean plus two standard deviations, using the overall mean of both treatment groups, shows no peaks in the pregnant or pseudopregnant animals, but there are peaks if the data are analyzed on a within-group basis. I wanted to ask you, did you isolate and measure conversion to androgens and conjugated estrogens as well as free estrogen?

HEAP: We looked at the conversion to conjugated estrogens and that conversion again was extremely low, less than 2% in the majority of cases. We did not look at the conversion to androgens.

MARTIN (ICRF, London): The signal in the pig appears to originate at the same time as the blastocyst elongates. Has anyone done the experiment to see whether dilation of the uterus or insertion of foreign bodies would in any way prolong luteal function, and does estrogen applied directly into the pig uterus have any effect on luteal function?

HEAP: Insertion of polyethylene spirals (IUDs) on the first day of estrus into one or both uterine horns of cyclic gilts did not prolong luteal function. In mated gilts the IUDs resulted in a high incidence of embryonic mortality between 8 and 14 days (Gerrits *et al.*, *J. Reprod. Fertil.* **17**:501, 1968). So far as your second question is concerned, I am not aware that estrogen has been applied intralumenally in the sow. The critical period in nonpregnant animals when estrogen is required to prolong the life span of the corpus luteum extends from Day 11 to 15 (Frank *et al.*, *Prostaglandins* **14**:1183, 1977).

SHERMAN (Roche Institute): You indicated that aromatase activity falls off in culture whereas production of progesterone and androstenedione continues. Do you have any speculations about why that might be? What's happening to the aromatase?

HEAP: We do not yet know the explanation. It is possible that aromatase activity declines normally since the percentage conversion of [^{3}H]androstenedione or [^{3}H]testosterone into estrogens declines by Day 20 to 32 p.c. when expressed relative to unit weight of tissue (see our Figure 2). Evidence *in vivo* shows that levels of estrone sulfate in peripheral blood decline between between Day 40 and 60 p.c. suggesting an alteration in trophoblast estrogen synthesis (Robertson and King, 1974, *J. Reprod. Fertil.* **40**:133). Moreover, Ainsworth and Ryan (*Endocrinology* **79**:875, 1966) failed to demonstrate the conversion of C-21 steroids to estrogens by placental tissue obtained in late pregnancy, in contrast to our findings in early pregnancy (Gadsby *et al.*, 1976, *J. Endocrinol.* **71**:45P). I think these results imply controls of trophoblast steroid synthesis that we do not yet understand.

SHERMAN: What if you take the embryos out earlier and place them into culture? Can you get the peak in estrone sulfate?

HEAP: Production of estrone sulfate depends on the interaction between blastocyst tissue and endometrium, and we have not examined this phenomenon *in vitro* after Day 20.

SHERMAN: You also indicated that if you remove acetate and cholesterol from your medium, you see a drop in progesterone and androstenedione production but not in estrogen. Do you attribute this to the presence of an adequate amount of progesterone and androstenedione in the cells to act as substrate for the aromatase?

HEAP: Yes, I think that is an obvious interpretation. However, the concentration of progesterone and of estrogens in these tissues at the start of culture is low compared to nanogram concentrations in these cultures, so I am not sure it is an adequate explanation.

SHERMAN: Do you have an adequate explanation?

HEAP: No. We have more experiments to do.

SHERMAN: In some ways I think you are reaching a little too hard to try to generalize in terms of the function and role of estrogens in implantation. Notwithstanding Dr. Dickmann's comment, there is no good evidence of estrogen synthesis or indeed steroid biosynthesis in preimplantation rat embryos and we are pretty certain that there is none in preimplantation mouse embryos (Sherman and Altinza, 1977, *Biol. Reprod.* **16**:190). Embryologists have become quite accustomed to the fact that embryogenesis is very different amongst different species. To try to generalize across species can be risky.

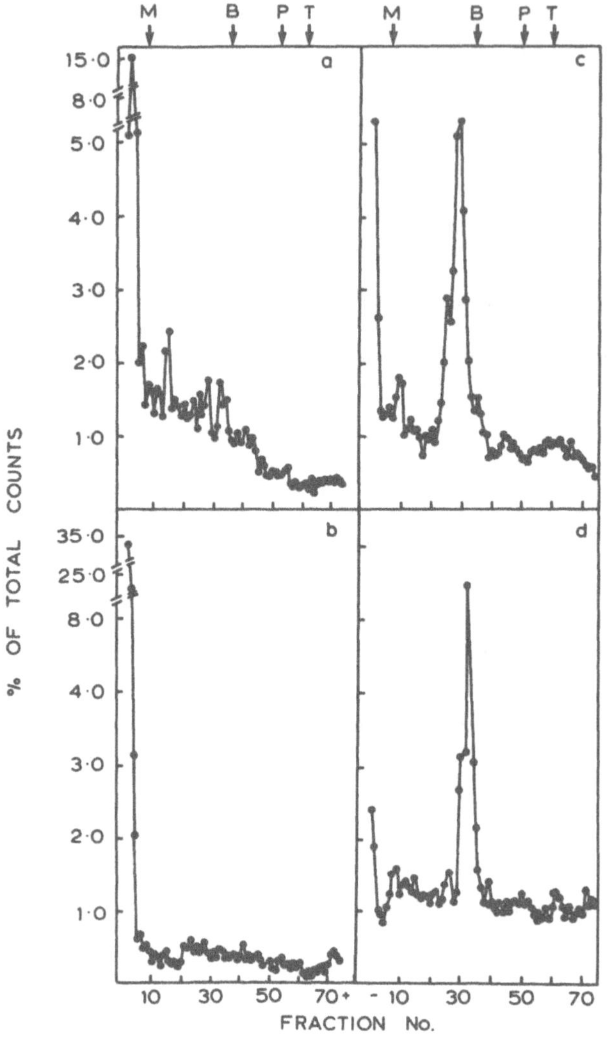

Figure D1. Qualitative analysis of the incorporation of [^{3}H]glucosamine into (a) control blastocysts, (b) experimental blastocysts treated with 1 μg tunicamycin/ml, (c) control extracellular medium, and (d) experimental extracellular medium from (b). The samples were analyzed on 6% polyacrylamide-sodium dodecyl sulfate gels. From Fishel, S. B., and Surani, M. A. H., *J. Reprod. Fert.* **59:**181–185, 1980.

HEAP: In defense, I am not trying to generalize for all mammalian species. I have described some evidence that we have obtained for blastocyst estrogen synthesis in the pig and prefaced my remarks by drawing attention to the fact that this has a very different type of implantation compared to laboratory rodents or primates, for example.

BULLOCK (Baylor): I think the point raised by Dr. Ford about the presence of steroids in the uterine lumen is pertinent. This observation has always made it difficult for me to see how the hypothesis of the local effect of the very small amounts of estrogen that are evidently produced by blastocysts of species other than the pig, attractive though that hypothesis is, could possibly function against this large background of lumenal steroid. The maternal recognition system in the human, as we know, is considered to be production of chorionic gonadotropin; Saxena and Asch have presented evidence at this meeting for its existence in the rabbit. Have you anything to say about the presence of peptide signals in the species that you have examined?

HEAP: There is increasing evidence for chorionic gonadotropins in the sheep (Lacroix and Martal, 1979, *C.R. Acad. Sci. Ser. D.* **288:**771) and pig (Saunders *et al., J. Endocrinol.,* **85:**25P) produced from an early stage of gestation, and possibly from about the time of implantation.

SURANI (Cambridge): We have recently looked at synthesis and secretion of glycoproteins by mouse blastocysts. The Figure (D1) illustrates the glycopeptides detected in whole blastocysts, and the glycopeptides that are released into the medium. There is one group of glycopeptides released into the medium that has an approximate molecular weight of 87,000. At present we are not aware what function it serves, but it could have either a local effect on the uterus or perhaps indirectly on the ovary.

KENNEDY (W. Ontario): I was intrigued by your comment in passing that there apparently was not a pontamine blue reaction in the pig uterus. I wonder if you would care to elaborate further on why the pig might be so different from other species in this regard?

HEAP: We have failed to note localization of pontamine blue at attachment sites. There is a general distribution of dye throughout the endometrium.

KENNEDY: So it is possible that there is a general increase in the vascular permeability under these conditions?

HEAP: I think that may well be a possibility. But there is an absence of the classical local pontamine blue reaction.

KENNEDY: Have you considered comparing a gravid horn with a nongravid horn within the same animal?

HEAP: I think that certainly would be a valid comparison, but we have not done that experiment.

References

Amoroso, E. C., Heap, R. B., and Renfree, M. B., 1979, Hormones and the evolution of viviparity, in: *Hormones and Evolution* (E. J. Barrington, ed.), pp. 925–989, Academic Press, New York.

Anderson, L. L., 1978, Growth, protein content and distribution of early pig embryos, *Anat. Rec.* **190:**143.

Dhindsa, D. S., and Dziuk, P. J., 1968, Effect on pregnancy in the pig after killing embryos or fetuses in one uterine horn in early gestation, *J. Anim. Sci.* **27:**122.

Dickmann, Z., Dey, S. K., and Sen Gupta, J., 1976, A new concept: Control of early pregnancy by steroid hormones originating in the preimplantation embryo, *Vitam. Horm.* (*N.Y.*) **34:**215.

du Mesnil du Buisson, F., 1967, Contrôle du maintien du corps jaune de la truie, *Arch. Anat. Microsc. Morphol. Exp.* **56:**358.

Eley, R. M., Thatcher, W. W., and Bazer, F. W., 1979a, Luteolytic effect of oestrone sulphate on cyclic beef heifers, *J. Reprod. Fertil.* **55:**191.

Eley, R. M., Thatcher, W. W., and Bazer, F. W., 1979b, Hormonal and physical changes associated with bovine conceptus development, *J. Reprod. Fertil.* **55:**181.

Flint, A. P. F., Heap, R. B., Gadsby, J. E., and Saunders, P. T. K., 1979, Blastocyst oestrogen synthesis and the maternal recognition of pregnancy, in: *Maternal Recognition of Pregnancy* (J. Whelan, ed.), Ciba Foundation Colloquium No. 64 (new series), pp. 209–228, Excerpta Medica, Amsterdam.

Frank, M., Bazer, F. W., Thatcher, W. W., and Wilcox, C. J. 1977, A study of prostaglandin $F_{2\alpha}$ as the luteolysin in swine. III. Effects of estradiol valerate on prostaglandin F, progestins, estrone and estradiol concentrations in the utero-ovarian vein of nonpregnant gilts, *Prostaglandins* **14:**1183.

Frank, M., Bazer, F. W., Thatcher, W. W., and Wilcox, C. J., 1978, A study of prostaglandin $F_{2\alpha}$ as the luteolysin in swine. IV. An explanation for the luteotrophic effect of estradiol, *Prostaglandins* **15:**151.

Gadsby, J. E., and Heap, R. B., 1978, Steroid hormones and their synthesis in the early embryo, in: *Novel Aspects of Reproductive Physiology* (C. H. Spilman and J. W. Wilks, eds.), 7th Brook Lodge Workshop on Problems of Reproductive Biology, 1977, Spectrum Publications, New York.

Gadsby, J. E., Burton, R. D., Heap, R. B., and Perry, J. S., 1976, Steroid metabolism and synthesis in early embryonic tissue of the pig, *J. Endocrinol.* **71:**45.

Gardner, M. L., First, N. L., and Casida, L. E., 1963, Effect of exogenous estrogens on corpus luteum maintenance in gilts, *J. Anim. Sci.* **22:**132.

Guthrie, H. D., Henricks, D. M., and Handlin, D. L., 1972, Plasma estrogen, progesterone and luteinizing hormone prior to estrus and during early pregnancy in pigs, *Endocrinology* **91:**675.

Heap, R. B., and Hamon, M., 1979, Oestrone sulphate in milk as an indicator of a viable conceptus in cows, *Br. Vet. J.* **125:**355.

Heap, R. B., and Perry, J. S., 1977, Maternal recognition of pregnancy, in: *Contemporary Obstetrics and Gynaecology* (G. V. P. Chamberlain, ed.), pp. 3–7, Northwood Publications, London.

Heap, R. B., Perry, J. S., Burton, R. D., Gadsby, J. E., Wyatt, C., and Jenkin, G., 1977, Blastocyst steroidogenesis and embryo–maternal interactions in the establishment of pregnancy, in: *Reproduction and*

Evolution (J. H. Calaby and C. H. Tyndale-Biscoe, eds.), pp. 341–347, Australian Academy of Science, Canberra.

Heap, R. B., Flint, A. P. F., and Gadsby, J. E., 1979, Role of embryonic signals in the establishment of pregnancy, *Br. Med. Bull.* **35:**129.

Kidder, H. E., Casida, L. E., and Grummer, R. H., 1955, Some effects of estrogen injections on the estrual cycle of gilts, *J. Anim. Sci.* **14:**470.

Lemon, M., 1975, Effect of oestrogens alone or in association with progestagens on the formation and regression of the corpus luteum of the cyclic cow, *Ann. Biol. Anim. Biochim. Biophys.* **15:**243.

Martal, J., Lacroix, M.-C., Loudes, C., Saunier, M., and Wintenberger-Torrès, S., 1979, Trophoblastin, an antiluteolytic protein present in early pregnancy in sheep, *J. Reprod. Fertil.* **56:**63.

Moeljono, M. P. E., Thatcher, W. W., Bazer, F. W., Frank, M., Owens, L. J., and Wilcox, C. J., 1977, A study of prostaglandin $F_{2\alpha}$ as the luteolysin in swine. II. Characterization and comparison of prostaglandin F, estrogen and progestin concentrations in utero-ovarian vein plasma of non-pregnant and pregnant gilts, *Prostaglandins* **14:**543.

Pack, B. A., and Brooks, S. C., 1974, Cyclic activity of estrogen sulfotransferase in the gilt uterus, *Endocrinology,* **95:**1680.

Pack, B. A., Brooks, C. L., Dukelow, W. R., and Brooks, S. C., 1979, The metabolism and nuclear migration of estrogen in porcine uterus throughout the implantation process, *Biol. Reprod.* **20:**545.

Perry, J. S., Heap, R. B., and Amoroso, E. C., 1973, Steroid hormone production by pig blastocysts, *Nature (London)* **245:**45.

Perry, J. S., Heap, R. B., Burton, R. D., and Gadsby, J. E., 1976, Endocrinology of the blastocyst and its role in the establishment of pregnancy, *J. Reprod. Fertil. Suppl.* **25:**85.

Robertson, H. A., and King, G. J., 1974, Plasma concentration of progesterone, oestrone and oestradiol-17β and of oestrone sulphate in the pig at implantation, during pregnancy and at parturition, *J. Reprod. Fertil.* **40:**133.

Robertson, H. A., King, G. J., and Carnegie, J. A., 1978, Appearance of oestrone sulphate in the allantoic fluid of the cow, *J. Endocrinol.* **79:**243.

Samuel, C. A., 1971, The development of pig trophoblast in ectopic sites, *J. Reprod. Fertil.* **27:**494.

Samuel, C. A., and Perry, J. S., 1972, The ultrastructure of pig trophoblast transplanted to an ectopic site in the uterine wall, *J. Anat.* **113:**139.

Short, R. V., 1969, Implantation and the maternal recognition of pregnancy, in: *Foetal Autonomy* (G. E. W. Wolstenholme and M. O'Connor, eds.), Ciba Foundation Symposium, pp. 2–26, Churchill, London.

Singh, M. M., and Booth, W. D., 1979, Origin of oestrogen in pre-implantation rabbit blastocysts, *J. Steroid Biochem.* **11:**723.

Wyatt, C., 1976, Endometrial components involved in protein synthesis by 16-day pig blastocyst tissue in culture, *J. Physiol. London* **260:**73.

20

Uterine Blastotoxic Factors

Alexandre Psychoyos and Viviane Casimiri

I. Introduction

The classical study of Chang (1950) showed that the intrauterine survival of a fertilized egg and its implantation in the rabbit require a synchrony between embryonic and uterine development. The importance of such a synchronization has been established in all species studied, although in many the chronological limits for a successful transfer appear extended to several days. As a general rule intrauterine survival of transferred eggs of any stage becomes impossible beyond the time at which nidation would normally occur.

In the rat, by the end of the fifth day of pregnancy or pseudopregnancy, the endometrium is no longer capable of a decidual response and the uterine environment becomes detrimental to unimplanted embryos. The appearance of a transient phase of receptivity on this fifth day leads to a uterine state of "nonreceptivity," which lasts until the end of pregnancy or pseudopregnancy. The hormonal conditioning of these uterine changes has been the subject of numerous studies, in particular in the rat and the mouse. A model (which will be discussed below) has been established in these two species that may be valid for a large variety of mammals. Several ultrastructural and biochemical correlates of the receptive uterus have been defined in the rat and mouse, as well as in other species. However, serious gaps exist in our present knowledge of the "nonreceptive" state of the uterus. What makes the endometrium insensitive to decidual stimuli? Why and how does the uterine milieu become hostile to blastocysts and ova of the preblastocyst stages?

In an attempt to answer the second of these questions, we report here the evidence we have concerning the involvement in uterine hostility of a blastotoxic substance present in uterine fluid.

II. The Uterine State of "Nonreceptivity"

Dickmann and Noyes (1960) observed in the rat that when 3-day ova were recovered after 24 hr of residence in a 4-day uterus, or when 2-day ova were recovered after 48 hr in

---

Alexandre Psychoyos and Viviane Casimiri • Laboratoire de Physiologie de la Reproduction, C.N.R.S.–ER 203, Hopital de Bicêtre (INSERM), Bicêtre, France.

a 3-day uterus, these ova appeared normal. However, when 4-day ova were recovered after more than 9 hr residence in a 5-day uterus, they were severely damaged. Thus, in this species, the endometrial environment prior to implantation (Days 3 and 4) allows survival and development of underdeveloped ova, whereas a sudden change coinciding with the time of implantation (Day 5) makes this environment unfavorable for egg survival. The same conclusion has been drawn from experiments in which 5-day blastocysts implanted normally after transfer to 4- or 5-day uteri. If the transfer of 5-day blastocysts was delayed beyond the fifth day of pregnancy, the result was always the same, no blastocysts ever implanted; they degenerated and were expelled within 10 hr (Psychoyos, 1966). One should note, however, that in this same species an ovariectomy performed early in pregnancy keeps the uterus in a "neutral" state for a long period, during which the blastocyst, though in diapause, maintains its viability unimpaired.

It is now well established that the same basic hormonal sequence, i.e., a small amount of estrogen in a 48-hr progesterone-dominated uterus, initiates a receptive phase and exerts a biphasic effect leading to the establishment of a "nonreceptive" state (Psychoyos, 1973; Finn, 1977; Leroy, 1978). Donor blastocysts can implant in the uteri of virgin ovariectomized rats treated with such a progesterone–estrogen regimen. If they are transferred within 24 hr after the injection of estrogen, which completes the hormonal sequence initiating the receptive phase. If the blastocysts are transferred later than this time, implantation fails to occur; the uterus has already entered the "nonreceptive" or refractory state (Psychoyos, 1963a). Meyers (1970), using the endometrial response to decidual stimuli as an index, found that this refactory uterine state is then maintained without any other hormonal change as long as progesterone administration continues. Recovery of the endometrial progestational potential requires the withdrawal of progesterone for a minimal interval of some 48 hr (Meyers, 1970; Psychoyos, 1973; Glasser and McCormack, 1979).

According to the above model, when the hormonal sequence establishing the state of "nonreceptivity" is assured earlier than the normal timing, it interferes with egg survival and blocks implantation by inducing an advanced uterine hostility. In fact, a preovulatory treatment with small amounts of progesterone desynchronizes the ovo–endometrial relationships, and the fertilized ova, when they enter the uterine cavity, face a hostile environment. Such a phenomenon was shown in the rat (Psychoyos, 1963b), as well as in the rabbit (Chang, 1969; McCarthy *et al.*, 1977), and there is evidence that it may be the case also in the human (Psychoyos, 1976).

III. Evidence for the Existence of a Blastotoxic Substance

Considering the possible causes of the hostility shown by the "nonreceptive" uterus toward the unimplanted embryo, one could adopt as a working hypothesis either the absence of a factor essential for egg survival or the presence of an inhibitory (toxic) substance. It is noteworthy that blastocyst degeneracy occurs in the "nonreceptive" uterus with striking rapidity, whereas the embryo at this stage exhibits a remarkable resistance when incubated *in vitro*, even in highly simplified media.

Direct evidence for the existence in uterine fluid of an inhibitor of blastocyst metabolic activity was offered by our results in the rat (Psychoyos, 1973; Psychoyos *et al.*, 1975) and those of Weitlauf (1976, 1978) in the mouse. In our experiments, we have studied the inhibitory activity of uterine flushings on the incorporation *in vitro* of [^{3}H]uridine into blastocyst RNA. Blastocyst RNA synthesis was estimated 22 hr after placing the blastocyst in culture, by pulse labeling with uridine for 45 min and extracting the TCA-soluble and

TCA-insoluble radioactivity. The uterine flushings were obtained just prior to the *in vitro* assay by flushing 1 ml of culture medium sequentially through three or four uteri, and the protein concentration was adjusted in all samples to 160 μg/ml. Two hours before pulse labeling with uridine, the cultured blastocysts were transferred to uterine flushing-culture medium.

Flushings were obtained from animals in various hormonal conditions. Similar results were obtained whether the flushings came from uteri of ovariectomized animals, either untreated or treated with progesterone, or from animals sacrificed on one of the first 4 days of pregnancy. All culture media containing uterine flushings inhibited by about 50% both the total uptake of uridine (TCA-soluble label) and the incorporation of label into the TCA-insoluble fraction compared to the controls. The inhibitory activity of flushings obtained from the uteri of animals on the second day of pregnancy was lower than the activity of flushings obtained on Days 3 and 4 of pregnancy. Likewise, the inhibitory activity of uterine flushings from untreated ovariectomized animals was lower than that of the flushings obtained from the uteri of ovariectomized progesterone-treated animals.

Flushings from the uteri of 5-day pseudopregnant rats had a higher inhibitory activity than those of Day 4, whereas the flushings from Days 6 and 7 of pseudopregnancy inhibited completely the uridine uptake by blastocysts, indicating the presence in these flushings of a toxic factor.

Aitken (1977), studying the capacity of uterine flushings to affect the *in vitro* development of blastocysts in the mouse, did not notice any effect of flushings obtained from delayed-implanting (neutral) animals. When he used flushings from 4-day (receptive) mouse uteri or "nonreceptive" uteri (ovariectomized mice treated with hormones), he observed that the percentage of blastocysts hatching from their zona pellucida was significantly greater than in unsupplemented medium. These results do not favor the existence of some inhibitory activity in flushings from "nonreceptive" mouse uteri.

However, in the rabbit, McCarthy *et al.* (1977), attempting to elucidate the reasons for the embryonic mortality induced by preovulatory progesterone administration, concluded from their results that the uterine fluid can be embryotoxic in this case. Guilbert-Blanchette and Lambert (1978) found that rabbit fluid, collected at different periods of pseudopregnancy, allowed up to 50% of morulae to become expanded blastocysts after 2 days of *in vitro* culture in this fluid. Nevertheless, under the same conditions, none of the pronuclei embryos developed normally and the authors consider this effect as evidence for an embryotoxic component.

Human uterine flushings collected at various stages of the menstrual cycle have also been tested for their effect on mouse blastocysts cultured *in vitro*. With one exception (uterine flushings collected on the last day of the menstrual cycle), blastocyst hatching and attachment were not impaired by flushings collected before or after ovulation (Aitken and Maathuis, 1978).

We believe that these controversial reports may be explained by differences concerning the concentration of the collected uterine fluid. Our data show clearly, at least in the rat, that uterine flushings collected on Day 6 of pseudopregnancy exhibit a strong blastotoxic effect *in vitro*. As shown in Table 1, within 24 hr of culture in such flushings all blastocysts have degenerated, whereas under the same conditions all blastocysts cultured in flushings collected on Day 2 of pseudopregnancy exhibited the normal hatching, attachment, and outgrowth *in vitro*. In both cases the culture medium was prepared by flushing the excised uterus with 200 μl Eagle's medium. Flushings from 15 animals were pooled, centrifuged at 5000 rpm for 10 min, and the supernatant was filtered through a 0.4-μm Millipore filter, stored at −20°C, and supplemented before use with 5% FCS.

Table 1. In Vitro Development of Rat Blastocysts Cultured for 24 hr in Medium Containing Uterine Flushings Collected on Day 2 or Day 6 of Pseudopregnancy

Supplement	Number of blastocysts	Number normal	Percentage degenerated
None	40	40	0
Day 2 flushings	38	38	0
Day 6 flushings	45	0	100

Considering these results and those mentioned above, where flushings from Day 6 or 7 of pseudopregnancy were able to block the uptake of uridine by blastocysts *in vitro* within 2 hr of exposure, the existence in the fluid of "nonreceptive" uteri of a component affecting embryonic metabolic activity and viability appears obvious to us.

IV. The Effective Substance

We thought initially that the inhibitory (toxic) component in uterine secretion was a protein. However, fractionation of uterine flushings from "nonreceptive" uteri on Sephadex G-200 columns gave at least four protein fractions, all of which *stimulated* the metabolic activity of blastocysts *in vitro*. The inhibitory activity of unfractionated uterine fluid was found to be dialyzable. In fact, with the exception of those who consider that uteroglobin could be embryotoxic in the rabbit if present at the wrong time, all investigators who have studied uterine embryotoxic factors or inhibitors of blastocyst metabolic activity agree that the uterine fluid loses its toxic or inhibitory properties after dialysis.

The different effects of the high- and low- (<10,000) molecular-weight components of 6-day uterine flushings upon blastocyst viability are shown in Table 2. Dialysis completely abolishes the toxic effect of 6-day flushings. The blastotoxic component is restricted to the dialysate. In these experiments each horn was flushed with 0.5 ml glass-distilled water. The flushings from 20 animals were pooled, placed in a dialysis bag immersed in 500 ml cold glass-distilled water, and changed three times at 6-hr intervals. The nondialyzable material and the total dialysate were then lyophilized and stored at −20°C until use. They were then rehydrated with Eagle's medium supplemented with 5% FCS, at a concentration equivalent to the material recovered per uterine horn, diluted to a microdrop (50 μl) of culture medium.

We studied the effect of varying the concentration of low-molecular-weight material

Table 2. In Vitro Development of Rat Blastocysts Cultured for 48 hr in Medium Containing 6-Day Uterine flushings Treated by Dialysis

Supplement	Number of blastocysts	Number normal	Percentage degenerated
None	35	34	3
Nondialyzed	50	0	100
Dialyzed (A)	22	22	0
Dialysate (B)	30	0	100
A + B	20	0	100

Table 3. In Vitro Development of Blastocysts Cultured for 24 hr in Medium Containing 6-Day Uterine Flushings Treated by Fractionation on Sephadex G-25/10 Column

Supplement	Number of blastocysts	Number normal	Percentage degenerated
None	42	42	0
Fraction I	37	37	0
Fraction II	30	30	0
Fraction III	36	30	0
Fraction IV	34	0	100
Fraction V	30	27	0
Fraction VI	30	30	0

(dialysate) of 6-day flushings in the culture medium. With a concentration equivalent to the amount collected per horn and diluted to 50 μl of medium, all blastocysts degenerated within 24 hr. Decreased concentrations (one-half or one-quarter) exhibited a delayed toxicity, whereas a concentration equivalent to one-eighth was found not to affect blastocyst viability for at least 6 days of culture.

In further experiments, the dialysate of flushings from 6-day uteri was fractionated on a Sephadex G-25/10 column with glass-distilled water as eluant. All fractions absorbing at 220–260 nm, designated fractions I (void volume), II, IV, and V, as well as those showing no significant absorption, fractions III and VI, were lyophilized and tested for their activity after being rehydrated with Eagle's medium supplemented with 5% FCS, at a concentration equivalent to the material recovered per horn and diluted to 50 μl of medium. As shown in Table 3, the effective substance appears to elute in fraction IV. All blastocysts cultured in medium containing this fraction degenerated within 24 hr of culture. None of the other fractions exhibited such an effect within this period. However, fractions III and V, which elute just before and just after the active fraction, exhibited a slight toxicity on subsequent days of culture, as if they were contaminated by the active component of fraction IV.

Dilutions of fraction IV gave similar results to those obtained by the dilution of the unfractionated dialysate. With a concentration equivalent to the material collected per horn and diluted to 50 μl of medium, all blastocysts degenerated within 24 hr. Decreased concentrations (one-half or one-quarter) exhibited a delayed toxicity, whereas the viability of the blastocysts that were cultured at a one-eighth dilution was not impaired for at least the first 6 days of culture.

Further characterization of the blastotoxic component of fraction IV is under way. However, based on the absorbance profile of this fraction and its time of elution, we believe that it could be a polypeptide.

Weitlauf (1978), studying the factors in mouse uterine fluid that inhibit the incorporation of [^{3}H]uridine by blastocysts *in vitro,* found that gel filtration on a G-25 column of fluid flushed from the uteri of delayed-implanting and implanting mice gave several fractions that reduced this incorporation. However, he noticed more inhibitory activity in the void volume fraction of flushings from delayed-implanting animals. He suggested that this fraction may contain a factor responsible for the metabolic dormancy of embryos during the diapause associated with delayed implantation. Whether this factor is related to the blastotoxic one we isolate in the fluid flushed from "nonreceptive" uteri in the rat remains to be clarified. Until more information is obtained about the chemical structure and biological properties of this toxic substance, we propose to call it *blastocidin*.

The way in which blastocidin affects blastocyst viability *in vitro* and its possible implication in rendering the "nonreceptive" uterus hostile to unimplanted embryos *in vivo* remain to be determined.

Discussion

A. KAYE (Weizmann Institute): Can you give us an approximate molecular weight from the Sephadex column for the blastocidin?

PSYCHOYOS: It behaves as a molecule of about 1000 molecular weight.

LEROY (Brussels): We have also been interested in the refractory phase induced by nidatory estrogen, but from the point of view of the mechanism of production of this refractory phase. We think that this is a good criterion of nidatory estrogen action because it is the end effect of this biphasic phenomenon of sensitization and desensitization of the uterus. We use a different criterion than toxicity to the blastocyst, the ability to produce the decidual reaction, which, as you know, it also abolished in the refractory phase. We found we could abolish the refractory phase by giving a high dose of cycloheximide but not by giving very high doses of actinomycin D. This finding suggests posttranscriptional control for nidatory estrogen action and fits in well with what Dr. Glasser showed regarding the reduction of transcriptional activity in the uterus. We could curtail the refractory phase and still get a decidual reaction by giving the inhibitor as late as 12 to 16 hr after the estrogen. This is about the time when Surani (*J. Reprod. Fertil.* **5**:289, 1977) found production of *de novo*-synthesized high-molecular-weight protein after estrogen action in the progestational uterus. Would you comment on this?

PSYCHOYOS: I cannot at the moment, before knowing the chemical identity of the toxic factor.

LEROY: I suggest you incubate blastocysts and test their metabolic activity in media containing flushings of the uteri from animals treated with different inhibitors. One snag would be to get rid of the contaminating inhibitor in the flushing so as not to intoxicate the embryos directly by the inhibitor.

PSYCHOYOS: That would be a good experiment.

SURANI (Cambridge): We do find *de novo* synthesis of several macromolecules about 18 hr after injection of estrogen in ovariectomized rats given progesterone. Several years ago, you proposed that delay of implantation was due to production of an inhibitor by the uterus and, as you know, I disagree with that view. I would like to ask if you now consider what you call blastocidin to be equivalent to the inhibitor which you originally thought kept embryos in delay?

PSYCHOYOS: We need more experiments to see what is behind either the toxic effect or the inhibition of blastocyst RNA synthesis which are seen with uterine flushings, in order to know whether blastocidin is equivalent to the inhibitor which, as I believe, keeps the embryos *in utero* in delay. We tried to see if the difference between these two factors was quantitative rather than qualitative and we attempted to create *in vitro* an intermediate situation equivalent to delay by using diluted material. There was either a toxic effect or no effect. If the eggs were incubated in a one-fourth dilution for instance, they started to outgrow at the proper time for the *in vitro* conditions and then died. We failed to obtain any indication that blastocysts, subjected to a sublethal concentration of the toxic fraction, remain in diapause *in vitro*.

DICKMANN (Kansas): Some mammals can go into delayed implantation whereas others, in contrast, cannot. Dormancy can occur only at the blastocyst stage. In insects, dormancy occurs at all stages of the life cycle except in preblastocyst embryos (Lees, 1955, *The Physiology of Diapause in Arthropods*, Cambridge Monographs in Exp. Biol. No. 4, Cambridge Univ. Press, London). When we put rat morulae in 5-day uteri, Day 1 being the day of finding a plug, they degenerate (Dickmann 1970, *Fertil. Steril.* **21**:541). What happens in the morula-to-blastocyst transition, which gives the embryo protection against this toxic effect?

PSYCHOYOS: The tolerance of the embryo to blastocidin seems to increase as the egg develops towards the blastocyst stage. An increased toxicity parallel with the tolerance of the egg may be the key for the need for synchrony. In other words, an egg less than 5 days old could not tolerate the concentration of the toxic factor present in a 5-day-old uterus and would degenerate, whereas a 5-day-old blastocyst could tolerate this concentration and would survive.

NILSSON (Uppsala): There are problems with the use of uterine flushings, especially on Day 6 in the rat, since there is hardly any uterine secretion present. There is even a closure of the uterine lumen. You should not use the word uterine fluid or uterine secretion but just uterine flushings; this also implies that perhaps the factors are not present in the lumen and are not available for the blastocyst.

MARTIN (ICRF, London): I would like to sound the same cautionary note, because we find a great deal of damage in the mouse. The flushing not only leaches materials from the lumen, it also breaks through into the stroma, always, oddly enough, in the antimesometrial area, rupturing blood vessels and leaching stromal cells. We see differences in damage depending on the stage of pregnancy or the hormonal regime. Damage seems to be associated with lumenal closure. If the rat lumen is closed on Day 6, as Dr. Nilsson said, do you find a difference in the total amount of material that you flush out on Day 6 as opposed to Day 2? Have you looked for damage?

PSYCHOYOS: The protein content is about the same on Day 2 and Day 6 of pseudopregnancy. Only on the fifth day in the rat is there an increase in the protein content of uterine flushings.

C. WARNER (Iowa State): I want to pursue the biochemistry of fraction IV a little. Do you know if it is heat stable or labile, and sensitive to RNAse or DNAse?

PSYCHOYOS: No, we have a lot of things left to do.

YOSHINAGA (NICHD): You reported (Psychoyos, 1966, in: *Egg Implantation* (G. E. W. Wolstenholme and M. O'Connor, eds.), Ciba Foundation Study Group No. 23, pp. 4–15, Churchill, London) that rat blastocysts transferred into 6-day pseudopregnant uteri were expelled to the vagina. So this substance may not only have a toxic effect on the blastocyst but may also stimulate contraction of the uterus. Do you have any data on the contractility of the uterus with the substance?

PSYCHOYOS: That is an interesting point. At the moment I do not have any evidence to support that possibility. In the experiments you mentioned, I ligated the horn and found blastocysts had degenerated. In the nonligated horn I could not find blastocysts by flushing 10 hr after egg transfer, indicating that they were expelled by that time.

GLASSER (Baylor): You present us only two numbers, 0 and 100. Or 100 and 0. Has this substance any effect on morula-to-blastocyst transformation, or on trophoblast outgrowth? Have you looked at the microvilli or the membranes when you found inhibition of uptake of uridine from the fluid?

PSYCHOYOS: I am sorry for this 100 or 0, but that is what happened. The effect on uptake of uridine is very early. By placing the eggs into this toxic medium, i.e., containing 6-day flushings, for only 2 hr there was a complete block of uridine uptake. To simplify testing for toxicity we determined survival by observing eggs under a light microscope after 24 hr of culture.

SHERMAN (Roche Institute): Have you tried this factor on cell types other than blastocyst or embryos?

PSYCHOYOS: Of course.

SHERMAN: What happens?

PSYCHOYOS: At concentrations toxic to embryos, it is not toxic to epithelium and kidney.

S. JOSHI (Albany): I have a strong suspicion that the cytotoxic effect you find is derived from a substance which resides intracellularly. It could be a product of the cell damage produced during the flushing of the uterine horn.

PSYCHOYOS: It is interesting that the uterine epithelium contains this factor on Day 6 but not on Day 2 of pseudopregnancy, i.e., during the refractory period for nidation, whether or not it is a component of uterine secretions. This substance may be a product of cell damage caused by flushing, but it exists at the appropriate time and has a toxic effect.

References

Aitken, R. J., 1977, The culture of mouse blastocysts in the presence of uterine flushings collected during normal pregnancy, delayed implantation and pro-oestrus, *J. Embryol. Exp. Morphol.* **41:**295.

Aitken, R. J., and Maathuis, J. B., 1978, Effect of human uterine flushings collected at various states of the menstrual cycle on mouse blastocysts *in vitro*, *J. Reprod. Fertil.* **53:**137.

Chang, M. C., 1950, Development and fate of transferred rabbit ova or blastocysts in relation to the ovulation time of recipients, *J. Exp. Zool.* **114:**197.

Chang. M. C., 1969, Fertilization, transportation and degeneration of the egg in pseudopregnant or progesterone-treated rabbits, *Endocrinology* **84:**356.

Dickmann, Z., and Noyes, R. W., 1960, The fate of ova transferred into the uterus of the rat, *J. Reprod. Fertil.* **12:**197.

Finn, C. A., 1977, The implantation reaction, in: *Cellular Biology of the Uterus* (R. E. Wynn, ed.), pp. 246–308, Plenum Press, New York.

Glasser, S. R., and McCormack, S. A., 1979, Estrogen-modulated uterine gene transcription in relation to decidualization, *Endocrinology* **104:**1112.

Guilbert-Blanchette, L., and Lambert, R. D., 1978, Analysis of the rabbit uterine fluid collected by a continuous collecting technique: Evidence for an embryotoxic component, *Biol. Reprod.* **19:**1125.

Leroy, F., 1978, Aspects moléculaires de la nidation, in: *L'implantation de l'oeuf* (F. DuMesnil DuBuisson, A. Psychoyos, and K. Thomas, eds.), pp. 81–92, Masson, Paris.

McCarthy, S. M., Foote, R. H., and Maurer, R. R., 1977, Embryo mortality and altered uterine luminal proteins in progesterone-treated rabbits, *Fertil. Steril.* **28:**101.

Meyers, K., 1970, Hormonal requirements for the maintenance of oestradiol-induced inhibition of uterine sensitivity in the ovariectomized rat, *J. Endocrinol.* **46:**341.

Psychoyos, A., 1963a, Précisions sur l'état de "non-réceptivité" de l'uterus, *C. R. Acad. Sci.* **257:**1153.

Psychoyos, A., 1963b, Nouvelles remarques sur l'état uterin de "non-réceptivité," *C. R. Acad. Sci.* **257:**1367.

Psychoyos, A., 1966, Recent research on egg-implantation, in: *Ciba Foundation Study Group on Egg Implantation* (G. E. W. Wolstenholme and M. O'Connor, eds.), pp. 4–28, Churchill, London.

Psychoyos, A., 1973, Hormonal control of ovo-implantation, *Vitam. Horm. (N.Y.)* **32:**201.

Psychoyos, A., 1976, Hormonal control of uterine receptivity for nidation, *J. Reprod. Fertil. Suppl.* 25, pp. 17–18.

Psychoyos, A., Bitton-Casimiri, V., and Brun, J. L., 1975, Repression and activation of the mammalian blastocyst, in: *Regulation of Growth and Differentiated Function in Eukaryotic Cells* (G. P. Talwar, ed.), pp. 509–514, Raven Press, New York.

Weitlauf, H. M., 1976, Effect of uterine flushings on RNA synthesis by "implanting" and "delayed implanting" mouse blastocysts *in vitro*, *Biol. Reprod.* **14:**566.

Weitlauf, H. M., 1978, Factors in mouse uterine fluid that inhibit the incorporation of ^{3}H-uridine by blastocysts *in vitro*, *J. Reprod. Fertil.* **52:**321.

21

Lysosomal Mechanisms in Blastocyst Implantation and Early Decidualization

Bruce C. Moulton and Sudha Elangovan

I. Introduction

The lysosome with its diverse content of hydrolytic enzymes has the capacity to degrade virtually every macromolecule of biological origin. Precise control of this self-destructive potential provides eukaryotic cells with the means for intracellular digestion and the means for adaptation to changes in nutritional and hormonal variables. Intracellular digestion of exogenous material taken into cells by pinocytosis or phagocytosis provides substrate for the synthesis of new cellular organelles and enzymes. Endogenous macromolecules and pieces of cytoplasmic material sequestered and digested by lysosomes during autophagia enable redirected physiological function as cellular components are degraded and resynthesized. These lysosomal mechanisms appear to be involved in the pinocytosis of the uterine lumenal epithelium during the preimplantation period of pregnancy, the penetration of the lumenal epithelium by the blastocyst, and the extensive remodeling of the endometrial stroma during decidualization.

Other physiological activities of the uterus, including cyclic changes in the uterine en dometrium during the reproductive cycle and the massive resorption of various uterine tissues during postpartum involution, also appear to require precise hormonal control of lysosomal function and enzyme content. Lysosomal involvement in these physiological functions of the uterus has been reviewed previously as has the subject of lysosome-mediated enhancement of nucleocytoplasmic communication in the hormone-activated target cell (Woessner, 1969; Wood, 1973; Szego, 1975).

II. Lysosomal Activity during Early Pregnancy

During the preimplantation period of early pregnancy, unknown biochemical changes in the uterine endometrium controlled by progesterone and estrogen result in the develop-

Bruce C. Moulton and Sudha Elangovan • Departments of Obstetrics and Gynecology and of Biological Chemistry, University of Cincinnati College of Medicine, Cincinnati, Ohio 45267.

ment of endometrial sensitivity to the blastocyst. Development of endometrial sensitivity could depend upon the accumulation of lysosomal enzymes in the endometrium or changes in the intracellular activity of endometrial lysosomes. Many histochemical studies of the uterus during the estrous cycle have identified concentrations of presumed lysosomal enzymes in endometrial epithelial cells that increased to maximal levels toward the end of early estrus (Woessner, 1969). Similar patterns of change in endometrial lysosomes were observed during the human menstrual cycle although there was more evidence for the autophagic process in epithelial cells during the rodent estrous cycle (Woessner, 1969; Henzl *et al.*, 1972). Both estrogen and progesterone treatment of ovariectomized rats restored uterine lysosomal structure and enzyme content.

Effects of estrogen and progesterone on levels of enzyme activity in lumenal epithelial cells and on lysosomal activity within these specific cells could have potential significance for the blastocyst implantation. During early pregnancy, both multivesicular bodies and more typical lysosomes were found in lumenal and glandular epithelial cells (Enders and Given, 1977). Light microscopic histochemical studies identified lysosomal phosphatase activity in the lumenal and glandular epithelium during early pregnancy (Christie, 1966; Abraham *et al.*, 1970; Smith and Wilson, 1971). On the day of implantation lysosomelike

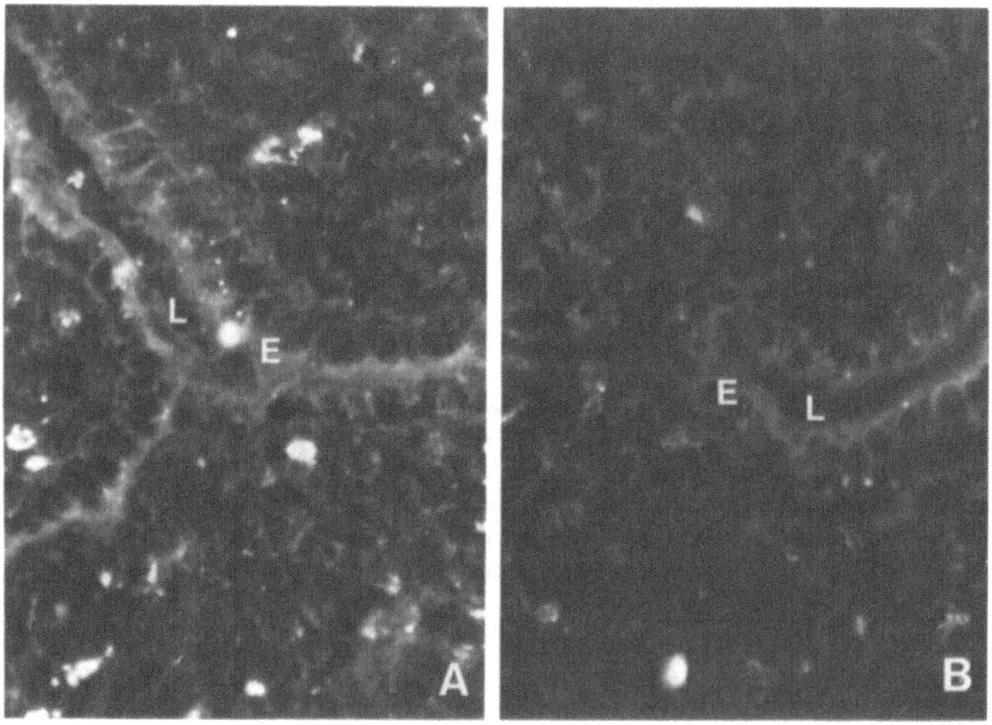

Figure 1. Immunofluorescent localization of cathepsin D in cross sections of implantation and interimplantation uterine segments on Day 5 of pregnancy. (A) Interimplantation site, (B) implantation site. Uterine tissues were excised, frozen in a hexane–dry ice mixture, and stored at −70°C before 10-μm sections were cut with a cryostat at −20°C. Sections on slides were incubated in acetone at −10°C for 6–8 min, air-dried, and then incubated with either goat antiserum to cathepsin D or preimmune serum at room temperature for 1 hr. After washing, the slides were incubated with fluorescein-conjugated rabbit anti-goat 7 S globulin. Slides were then examined and photographed at constant exposure times. E, lumenal epithelium; L, uterine lumen.

Table 1. Effect of Progesterone and Estradiol on Rates of Cathepsin D Synthesis in the Ovariectomized Rat Uterus [a]

Treatment	cpm cathepsin D / cpm sol fraction	cpm cathepsin D / cpm TCA-ppt. ($\times 10^{-4}$)
Control	2760 [b] (2586, 2934)	5.15 (3.71, 6.60)
Progesterone 2 mg, 6 hr	8257 (5290, 11,225)	9.54 (9.18, 9.90)
Control	558 (169, 947)	1.39 (0.794, 1.98)
Estradiol 5 μg, 6 hr	297 (55.4, 538)	1.52 (0.272, 2.78)

[a] Rats ovariectomized for 3 weeks were injected with progesterone (2 mg, s.c.) or estradiol (5 μg, s.c.). Groups of 12 control or 6 hormone-treated rats were given [^{3}H]leucine (32 μCi/40 μl) by transcervical administration 2 and 1 hr before sacrifice. Cathepsin D in homogenates of pooled uteri was partially purified by acid pH precipitation and by ammonium sulfate fractionation before addition of anti-cathepsin D antiserum. The immunoprecipitate was solubilized and dissociated into subunits by SDS digestion. A single peak of radioactivity coincidental with the molecular weight of cathepsin D enzyme protein was observed, and the radioactivity in this peak was taken as the radioactivity incorporated into uterine cathepsin D protein. Cathepsin D radioactivity was expressed per cpm in the TCA-soluble fraction (an estimate of the available pool of amino acids) or per cpm in the TCA-precipitable protein (an estimate of the rate of general protein synthesis).
[b] Mean of parenthetically indicated values obtained from two experiments.

structures of lumenal epithelial cells appeared to fuse with endocytic vacuoles formed from irregular protrusions at the apical surfaces of these cells (Parr and Parr, 1974). Fusion of these endocytic vesicles with lysosomes in the epithelial cells would allow the lysosomal degradation of macromolecules from the lumenal fluid and the eventual release of low-molecular-weight compounds for blastocyst nutrition. Observations providing some support for this mechanism include the maximal ingestion of intrauterine ferritin by lumenal epithelial cells on Day 4 of pregnancy* and the subsequent digestion of this protein within the epithelial lysosomal system (Parr and Parr, 1978).

Very few biochemical studies have examined levels of lysosomal enzymes in the uterine endometrium during early pregnancy. Wood and Psychoyos (1967) found that the activity of β-glucuronidase in the endometrium of the pseudopregnant rat remained relatively constant, whereas acid cathepsin D activity increased from Day 1 of pseudopregnancy to reach a maximal specific activity on Day 4 as the uterus attained maximal sensitivity to decidualizing stimuli. This increase in cathepsin D activity was inhibited by the presence of an intrauterine thread (Wood, 1969). Increases in endometrial cathepsin D activity were also observed in ovariectomized rats treated with appropriate hormonal replacement to induce uterine sensitization to decidualizing stimuli (Wood and Psychoyos, 1967). Progesterone treatment of ovariectomized rats increased the rate of synthesis of uterine cathepsin D as measured by incorporation of labeled amino acid into immunoprecipitable cathepsin D protein (Table 1), but the uterine cells involved in the increased rate of enzyme synthesis have not been identified. As shown in Figure 1, major concentrations of cathepsin D protein identified by immunohistochemical staining were located in the lumenal

* The presence of sperm in the vaginal smear indicated Day 0 of pregnancy.

epithelium of cross sections of pregnant rat uteri. The accumulation of cathepsin D activity in these cells during early pregnancy would contribute to the means for epithelial intracellular degradation of protein.

During the preimplantation period, ovarian secretion of progesterone and estrogen controls the development and duration of uterine sensitivity to decidualizing stimuli (Finn, 1977). In ovariectomized rats maintained with progesterone, a decidual reaction could be induced by a traumatic stimulus such as a knife scratch (DeFeo, 1963a,b). Intrauterine administration of oil was not an effective stimulus, however, unless estrogen was injected in addition to progesterone (Finn, 1965). These observations suggested that estrogen altered the capacity of epithelial cells to receive and transduce a decidualizing stimulus. Pronounced differentiative changes observed in epithelial cells during the preimplantation period may include increases in epithelial concentrations of lysosomal enzymes and intracellular lysosomal activity. Concentrations of lysosomal enzymes localized in uterine epithelial cells appeared to be controlled by progesterone and estrogen, and greatest pinocytotic and lysosomal activity occurred during the period of greatest uterine sensitivity. The acute effect of estrogen on the stability of lysosomal membranes in target tissues of ovariectomized rats (Szego *et al.*, 1977) may indicate a possible mechanism by which endometrial lysosomes could be activated for the brief period of uterine sensitization. While the epithelial location of lysosomal enzyme concentrations and their control by progesterone and estrogen suggest a lysosomal function in endometrial sensitization, establishment of this mechanism will require careful study of both the enzyme content of endometrial lysosomes and their intracellular activity during early pregnancy.

III. Lysosomes and Epithelial Penetration

The morphological diversity of interactions between blastocyst and uterine epithelium in different animal species dictates a diversity of cellular mechanisms of lysosomal involvement in implantation. Differences between the three types of interaction of trophoblast with uterine epithelium during penetration of the epithelium (displacement, fusion, and intrusion penetration) could depend upon differences in the proportion of autophagic activity in trophoblast and epithelial cells, the extracellular release of hydrolytic enzymes, and other differences in cellular activities such as pinocytosis or phagocytosis that also involve lysosomal activity. Blastocyst implantation in rats and mice has received considerable research attention and will be examined in greatest detail in this review.

Disappearance of the uterine lumenal epithelium in response to the implanting blastocyst appeared to be a significant feature of blastocyst implantation in the rat and mouse (Schlafke and Enders, 1975). During the first 24 hr after initial contact of the blastocyst with the lumenal epithelium, epithelial cells showed increasing frequency of necrosis and degeneration (El-Shershaby and Hinchliffe, 1975). As implantation proceeded, almost all of the epithelial cells disappeared and the remaining cells, with or without contact with the basement membrane, appeared about to be engulfed by the trophoblast (Enders and Schlafke, 1967; Tachi *et al.*, 1970). Although these observations suggested that lumenal epithelial cells are eventually phagocytosed by the trophoblast, earlier deterioration of the epithelium could have resulted from autolytic activity within the cells. The death of limited numbers of cells during earliest blastocyst implantation in the mouse did not appear to involve lysosomal activity, but the later accumulation of portions of epithelial cell cytoplasm within epithelial lysosomes indicated that deterioration of these cells involved self-digestion rather than trophoblast attack or secretion (El-Shershaby and Hinchliffe, 1975).

Other evidence indicated that destruction of the epithelial cells was initiated from within the epithelial cells rather than by the presence of the blastocyst (Abraham *et al.*, 1970). Uterine lumenal epithelial cells disappeared in response to the intrauterine administration of oil without the presence of a blastocyst (Finn, 1977). Differences between the endometrial responses to intrauterine oil and to the implanting blastocyst appeared to be the result only of the increased rate of epithelial deterioration after oil administration (Hinchliffe and El-Shershaby, 1975; Lundkvist *et al.*, 1977). Many other features of the endometrial response appeared similar. Further evidence for the control of epithelial cell death resulted from experiments using actinomycin D (Finn and Bredl, 1973, 1977). Administration of this compound did not affect blastocyst activation, attachment, or stromal edema, but degeneration of the uterine epithelial cells was inhibited. Although unknown nonspecific effects of actinomycin D on blastocyst or epithelial cells cannot be excluded, the implication of this experiment was that epithelial cell death during implantation was an active intracellular process.

Deterioration and eventual removal of the uterine epithelial cells might reasonably depend upon the autophagic activity of epithelial lysosomes and their content of hydrolytic enzymes, particularly cathepsin D, which appears to be involved in intracellular protein degradation (Barrett, 1977). During the preimplantation period, lysosomes of the lumenal epithelial cells degraded protein taken into the epithelial cells by endocytosis (Parr and Parr, 1978). The accumulation of cathepsin D within the endometrium during early pregnancy (Wood and Psychoyos, 1967) in response to progesterone secretion would provide uterine epithelial cells with the potential for increased autophagic activity and increased intracellular protein degradation, facilitating the deterioration of these cells and their eventual removal. Mechanisms by which the blastocyst might initiate localized changes in endometrial lysosomal function remain to be elucidated.

The deterioration of the lumenal epithelial cells with their elevated content of cathepsin D during blastocyst implantation would explain the decrease in cathepsin D activity measured at the implantation site (Figures 1 and 2) (Moulton, 1974). Of five lysosomal enzymes measured in implantation and interimplantation uterine segments, only cathepsin D activity decreased at the implantation site (Moulton, 1974; Moulton *et al.*, 1978). Cathepsin B_1 and β-glucuronidase activities did not change and arylsulfatase B and acid phosphatase increased, but only some 48 hr after the first contact of the blastocyst with the endometrium. Cathepsin D enzyme lost at the blastocyst implantation site could be degraded within the lumenal epithelial cells or released to the lumenal fluids and possibly taken up by the trophoblast by micropinocytosis. While the fate of the cathepsin D is unknown, an extracellular action of the enzyme seems unlikely because of its inactivity above pH 5. Various proteolytic activities identified in uterine secretions and in trophoblast and uterine tissues by gelatin digestion experiments have not had sufficient biochemical characterization to establish a lysosomal origin (Denker, 1977). Peptidase activities in uterine secretions, characterized in more detail, have shown neutral or alkaline pH optima, which would seem to exclude a lysosomal origin (Rosenfeld and Joshi, 1977; Hoversland and Weitlauf, 1978; Mahaboob Basha *et al.*, 1978).

Deterioration of the lumenal epithelium during blastocyst implantation and reduction of the cytoplasm of glandular epithelial cells during mammary gland involution (Woessner, 1969) appear to involve precisely controlled lysosomal mechanisms. Degeneration of cells and tissues is also an important component of normal embryonic morphogenesis, yet the mechanisms initiating cell death and controlling their degeneration are not understood (Saunders, 1966). The appearance of lysosomes within prenecrotic cells characterizes one type of cell death associated with the removal of organ anlagen or large tissue units such as

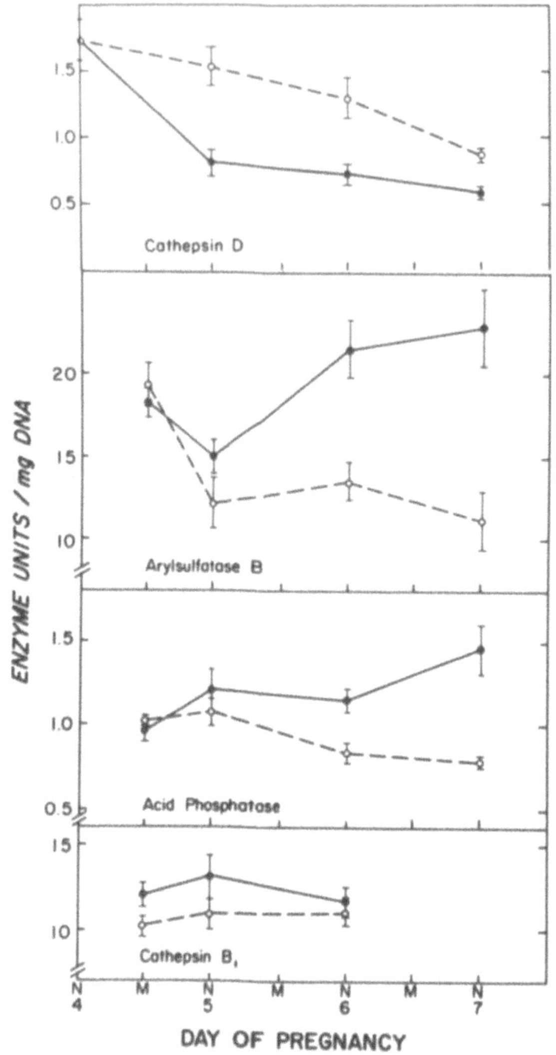

Figure 2. Effect of blastocyst implantation on uterine lysosomal enzyme activities. Enzyme activities were measured in interimplantation uterine segments (○) and implantation segments (●) as previously described (Moulton, 1974; Moulton *et al.*, 1978). This figure is redrawn from those references. Data are the mean ± S.E.M. of at least five animals.

the epithelial plates after palatal closure during development of the mammalian secondary palate (Schweichel and Merker, 1973). Palatal fusion involves initial contact between epithelial surfaces followed by adhesion of these epithelia, autolysis of epithelial cells at the midline seam, and fusion of the mesenchyme with phagocytic removal of the epithelial cells. Epithelial cells cease DNA synthesis 24 hr prior to contact, lysosomal enzyme synthesis increases, and levels of cAMP in the whole palate increase near the time of palate fusion (Mato *et al.*, 1972; Pratt and Martin, 1975). During the day before blastocyst implantation, mitotic activity in the uterine endometrial epithelium ceases as stromal mitosis continues, endometrial cathepsin D activity increases, and levels of cAMP increase tenfold after deciduogenic stimulus (Wood, 1969; Finn, 1977; Rankin *et al.*, 1977). In these two examples of epithelial removal and subsequent fusion of tissues, cellular deterioration and autolytic activity appear to depend upon lysosomes and their content of hydrolytic enzymes.

IV. Lysosomes and Early Decidualization

Implantation of the blastocyst initiates complex growth and differentiation processes characterized by increased rates of DNA, RNA, and protein synthesis at the implantation site (Finn, 1977). Increased cell growth and metabolism of decidualization are accompanied by cellular differentiation that implicitly involves the tearing down of old cellular structure as new components are synthesized. Decidual lysosomes could accomplish this intracellular turnover of cellular structure and enable eventual decidual regression later in pregnancy.

During decidualization, collagen bundle disintegration, fraying, and disaggregation indicated lysosomal activity at the implantation site (Fainstat, 1963). Lysosomes first appeared in decidual cells on the day after the initiation of decidualization and became increasingly numerous during the next 3 days of decidual growth (Jollie and Bencosme, 1965). Lobel *et al.* (1965) demonstrated that primary decidual cells showed a strong acid phosphatase stain that spread outwards as more endometrial stromal cells developed into decidual cells. In the rabbit, increases in both acid phosphatase and arylsulfatase were observed at the implantation site (Abraham *et al.*, 1970).

Diverse changes in various lysosomal enzyme activities have been measured biochemically during decidualization. Cathepsin D, cathepsin B_1, and β-glucuronidase activities remained low during the growth of the deciduoma, in contrast to levels measured during decidual regression (Wood and Barley, 1970; Moulton, 1974; Moulton *et al.*, 1978). The low activities of these enzymes could be associated with the accumulation of mucopolysaccharides and protein during early decidualization. On the other hand, acid phosphatase and arylsulfatase activities increased during early decidualization coincident with the appearance of lysosomes in the decidual tissue (Manning *et al.*, 1967; Wood and Barley, 1970; Moulton *et al.*, 1978). The diversity of changes in lysosomal enzyme activities indicated a heterogeneity of lysosomal activity in specific decidualizing cells and/or a heterogeneity of lysosomes themselves in the decidual cells. The specific changes in stromal lysosomal enzyme activities probably are related to the synthetic or degradative processes required for decidual differentiation.

V. Summary

Increased endometrial synthetic activity during blastocyst implantation coexists with the intracellular degradation of exogenous material taken into uterine epithelial cells by pinocytosis and phagocytosis, with the autophagic activity of epithelial cells as they deteriorate and disappear, and with the extensive remodeling of stromal cells as they differentiate into decidual cells. These lysosomal activities could be accomplished by concentrations of lysosomal enzymes localized in lumenal epithelial cells and by the accumulation of lysosomes with specific enzyme activities in stromal cells as decidualization proceeds. Progesterone and estrogen secretion controls the enzyme content of endometrial lysosomes and presumably their lability, providing the endometrium with the capacity for controlled self-destruction and for responses to stimuli initiating decidualization. While the location of lysosomal enzyme concentrations and their control by progesterone and estrogen suggest a lysosomal function in blastocyst implantation and decidualization, establishment of these lysosomal mechanisms will require careful study of both the enzyme content of endometrial lysosomes and their intracellular activity during early pregnancy.

ACKNOWLEDGMENTS. We thank Dr. James Lessard for his assistance with immunohistochemical staining techniques and fluorescence photography, and Dr. Roger Ganschow for his advice in the preparation of the antibody. Carol Bates and Beth Koenig provided excellent technical assistance. This research was supported by Grants HD-07255 and HD-10721 from the National Institutes of Health.

Discussion

FINN (Liverpool): Stan Glasser was worried about making a verb out of "vortex." There is a verb "to oil," and it means to lubricate, so I think it sounds strange to be oiling the uterus! I am interested in your feeling that these enzymes might have been released in the process of programmed cell death. You might get some information on this by using actinomycin D. We have shown (Finn and Bredl, 1973, *J. Reprod. Fertil.* **34**:247) that actinomycin D can stop these cells breaking down, or at least delay the process. In fact there appears to be a buildup of cells and I would be interested to see what happens to your enzyme during this process.

MOULTON: We are aware of that work and would like to study the effect of various inhibitors on the rate of synthesis of cathepsin D. We have done some experiments with pepstatin, a specific inhibitor of cathepsin D, but pepstatin must be dissolved in DMSO to deliver an effective dose to the uterus. As a result, we have not used pepstatin to try to inhibit implantation, because DMSO has some effects itself.

GLASSER (Baylor): In terms of the actual measurements of the cathespin D protein, you showed data only with progesterone or estrogen alone. You showed only enzyme activity with a combination of the hormones. If you establish a progesterone dominance and then give a single dose of estradiol, what happens to synthesis of cathespin D protein?

MOULTON: We have not yet done that experiment. We also want to look at the immunohistochemical localization of the enzyme in response to progesterone and estradiol.

DICKMANN (Kansas): You mentioned that estrogen is a labilizer of lysosomes and raised a question as to what triggers this activity. It seems obvious to me that the blastocyst is doing this. The way I see it, the blastocyst releases estrogen, which releases the enzymes from the lysosomes and these enzymes are responsible for the degeneration of the epithelium.

MOULTON: One of the difficulties is that there are a number of other compounds that have the same effect. Prostaglandins labilize lysosomes and I would not be surprised if histamine does so as well. So, I would not want to ascribe lysosomal labilization to a specific effect of estradiol.

HEAP (Cambridge): Following up from Zeev Dickmann, it might be thought that his idea seems an obvious hypothesis, but there are obvious problems. In studies using artificial membranes (liposomes), estrogens can be shown to be labilizers and progesterone can be shown to be a stabilizer, but the two steroids together have some quite striking interactions (Heap *et al.*, 1971, *Biochim. Biophys. Acta* **233**:307). A whole range of stabilizing and labilizing effects that you are discussing can be reproduced by altering the relative amounts of the two steroids. There is a lot of emphasis on the possibility of trophectoderm producing estrogen, but we should not ignore the possibility that it is also producing progesterone locally. We ought to consider, if these steroids exert a local action by their membrane effects, that there may be an important interaction between a number of steroids.

MARTIN (ICRF, London): Following up those two points of local production of estrogen, do you see a very narrow localization around the implantation site? In other words, do you see disappearance of the enzyme histochemically only in the antimesometrial cleft, where the blastocyst is, and have you done serial sections to see how far the blastocyst influence extends?

MOULTON: It appears that the epithelial activity disappears not only in the antimesometrial portion of the epithelium but in all parts of it. We have not done serial sections to see how far it extends on either side.

MARTIN: Was it only in the lumenal epithelium and not in the glands?

MOULTON: It seems to be only in epithelial cells.

BULLOCK (Baylor): I take it you measured total cellular enzyme activity and I wonder if you have any information about enzyme activity in isolated lysosomes?

MOULTON: No, we have not. The uterus is a difficult tissue from which to isolate lysosomes. The deciduoma would be much better for those studies, but we have not attempted to isolate decidual lysosomes.

References

Abraham, R., Hendy, R., Dougherty, W. J., Fulfs, J. C., and Golberg, L., 1970, Participation of lysosomes in early implantation in the rabbit, *Exp. Mol. Pathol.* **13:**329–345.

Barrett, A. J., 1977, Cathepsin D and other carboxyl proteinases, in: *Proteinases in Mammalian Cells and Tissues* (A. J. Barrett, ed.), pp. 209–248, North-Holland, Amsterdam.

Christie, G. A., 1966, Implantation in the rat embryo: Glycogen and alkaline phosphatases, *J. Reprod. Fertil.* **12:**279.

DeFeo, V. J., 1963a, Temporal aspects of uterine sensitivity in the pseudopregnant or pregnant rat, *Endocrinology* **72:**305.

DeFeo, V. J., 1963b, Determination of the sensitive period in the rat by different inducing procedures, *Endocrinology* 73:488.

Denker, H.-W., 1977, Implantation: The role of proteinases and blockage of implantation by proteinase inhibitors, *Adv. Anat. Embryol. Cell Biol.* **53,** part 5.

El-Shershaby, A. M., and Hinchliffe, J. R., 1975, Epithelial autolysis during implantation of the mouse blastocyst: An ultrastructural study, *J. Embryol. Exp. Morphol.* **33:**1067–1080.

Enders, A. C., and Given, R. L., 1977, The endometrium of delayed and early implantation, in: *Biology of the Uterus* (R. M. Wynn, ed.), pp. 203–243, Plenum Press, New York.

Enders, A. C., and Schlafke, S., 1967, A morphological analysis of the early implantation states in the rat, *Am. J. Anat.* **120:**185–226.

Fainstat, T., 1963, Extracellular studies of the uterus. I. Disappearance of the discrete collagen bundles in endometrial stroma during various reproductive states in the rat, *Am. J. Anat.* **112:**337–369.

Finn, C. A., 1965, Oestrogen and the decidual cell reaction of implantation in mice, *J. Endocrinol.* **32:**223.

Finn, C. A., 1977, The implantation reaction, in: *Biology of the Uterus* (R. M. Wynn, ed.), pp. 245–308, Plenum Press, New York.

Finn, C. A., and Bredl, J. C. S., 1973, Studies on the development of the implantation reaction in the mouse uterus: Influence of actinomycin D, *J. Reprod. Fertil.* **34:**247.

Finn, C. A., and Bredl, J. C. S., 1977, Autoradiographic study of the effect of actinomycin D on decidual differentiation of stromal cells in the mouse uterus, *J. Reprod. Fertil.* **50:**109–111.

Henzl, M. R., Smith, R. E., Boost, G., and Tyler, E. T., 1972, Lysosomal concept of menstrual bleeding in humans, *J. Clin. Endocrinol.* **34:**860.

Hinchliffe, J. R., and El-Shershaby, A. M., 1975, Epithelial cell death in the oil-induced decidual reaction of the pseudopregnant mouse: An ultrastructural study, *J. Reprod. Fertil.* **45:**463–468.

Hoversland, R. C., and Weitlauf, H. M., 1978, The effect of estrogen and progesterone on the level of amidase activity in fluid flushed from the uteri of ovariectomized mice, *Biol. Reprod.* **19:**908.

Jollie, W. P., and Bencosme, S. A., 1965, Electron microscopic observations on primary decidua formation in the rat, *Am. J. Anat.* **116:**217.

Lobel, B. L., Tic, L., and Shelesnyak, M. C., 1965, Studies on the mechanism of nidation: Histochemical analysis of decidualization in the rat, *Acta Endocrinol.* **50:**469.

Lundkvist, Ö., Ljungkvist, J., and Nilsson, O., 1977, Early effects of oil on rat uterine epithelium sensitized for decidual induction, *J. Reprod. Fertil.* **51:**507–509.

Mahaboob Basha, S. M., Horst, M. N., Bazer, F. W., and Roberts, R. M., 1978, Peptidases from pig reproductive tract: Purification and properties of aminopeptidases from uterine secretions, allantoic fluid, and amniotic fluid, *Arch. Biochem. Biophys.* **185:**174.

Manning, J. P., Steinetz, B. G., Giannina, T., and Meli, A., 1967, Effect of estrogen and progesterone on uterine acid and alkaline phosphatase and β-glucuronidase activity in mated ovariectomized rats, *Proc. Soc. Exp. Biol. Med.* **125**:508–512.

Mato, M., Smiley, G. R., and Dixon, A. D., 1972, Epithelial changes in the presumptive regions of fusion during secondary palate formation, *J. Dent. Res.* **51**:1451–1456.

Moulton, B. C., 1974, Ovum implantation and uterine lysosomal enzyme activity, *Biol. Reprod.* **10**:543–548.

Moulton, B. C., Koenig, B. B., and Borkan, S. C., 1978, Uterine lysosomal enzyme activity during ovum implantation and early decidualization, *Biol. Reprod.* **19**:167–170.

Parr, M. B., and Parr, E. L., 1974, Uterine luminal epithelium: Protrusions mediate endocytosis, not apocrine secretion, in the rat, *Biol. Reprod.* **11**:220.

Parr, M. B., and Parr, E. L., 1978, Uptake and fate of ferritin in the uterine epithelium of the rat during early pregnancy, *J. Reprod. Fertil.* **52**:183–188.

Pratt, R. M., and Martin, G. R., 1975, Epithelial cell death and cyclic AMP increase during palatal development, *Proc. Natl. Acad. Sci. USA* **72**:874–877.

Rankin, J. C., Ledford, B. E., and Baggett, B., 1977, Early involvement of cyclic nucleotides in the artificially stimulated decidual cell reaction in the mouse uterus, *Biol. Reprod.* **17**:549–554.

Rosenfeld, M. G., and Joshi, M. S., 1977, A possible role of a specific uterine fluid peptidase in implantation in the rat, *J. Reprod. Fertil.* **51**:137.

Saunders, J. W., Jr., 1966, Death in embryonic systems, *Science* **154**:604–612.

Schlafke, S., and Enders, A. C., 1975, Cellular basis of interaction between trophoblast and uterus at implantation, *Biol. Reprod.* **12**:41–65.

Schweichel, J.-U., and Merker, H.-J., 1973, The morphology of various types of cell death in prenatal tissues, *Teratology* **7**:253–266.

Smith, M. S. R., and Wilson, I. B., 1971, Histochemical observations on early implantation in the mouse, *J. Embryol Exp. Morphol.* **25**:165.

Szego, C. M., 1975, The lysosome in nucleocytoplasmic communication, in: *Lysosomes in Biology and Pathology* (J. T. Dingle and R. T. Dean, eds.), Vol. 4, pp. 385–477, North-Holland, Amsterdam.

Szego, C. M., Nazareno, M. B., and Porter, D. D., 1977, Estradiol-induced redistribution of lysosomal proteins in rat preputial gland, *J. Cell Biol.* **73**:354.

Tachi, S., Tachi, C., and Lindner, H. R., 1970, Ultrastructural features of blastocyst attachment and trophoblastic invasion in the rat, *J. Reprod. Fertil.* **21**:37–56.

Woessner, J. F., Jr., 1969, The physiology of the uterus and mammary gland, in *Lysosomes in Biology and Pathology* (J. T. Dingle and H. B. Fell, eds.), Vol. 1, pp. 299–329, Wiley, New York.

Wood, C., and Psychoyos, A., 1967, Activité de certaines enzymes hydrolytiques dans états de réceptivité utérine chez la ratte, *C.R. Acad. Sci.* **265**:141–144.

Wood, J. C., 1969, Effect of an intrauterine thread on acid cathepsin activity in the endometrium of the pseudopregnant rat, *Biochem. J.* **114**:665–666.

Wood, J. C., 1973, Lysosomes of the uterus, *Adv. Reprod. Physiol.* **6**:221–230.

Wood, J. C., and Barley, V. L., 1970, Biochemical changes in forming and regressing deciduoma in the rat uterus, *J. Reprod. Fertil.* **23**:469–475.

VII

Mechanisms of Implantation

Introduction

Mechanisms of Implantation

Fuller W. Bazer

In the evolution of reproductive processes, mechanisms have developed for improved efficiency of nutrient exchange between the material system and the conceptus. In many fishes, estrogen, of ovarian origin, induces production by the maternal liver of vitellogenin that is incorporated into ova regardless of whether they will be fertilized. The avian species represent a stage in which yolk material from the maternal liver and secretions of the reproductive tract are produced in response to ovarian steroids and are incorporated into the egg. It is remarkable that the appropriate quantities of nitrogenous substances, water, carbohydrates, and minerals are present to allow development of the chick. Again, however, the maternal system expends energy to secrete these substances into an egg that may or may not be fertilized.

Mammalian species appear to place more restrictions on the reproductive process. In the absence of fertilized ova, or of embryos developing in synchrony with the endometrium, pregnancy fails and the endometrium undergoes only limited development before involuting. If endometrium–conceptus interactions occur, the conceptus signals its presence, which results in maintenance of the corpora lutea and continued endometrial development. Implantation occurs early in pregnancy and involves establishment of specialized membranes for nutrient and endocrine communication between embryonic placental and maternal systems.

Steroid hormones of ovarian and/or placental origin and protein hormones of placental and/or pituitary origin may act on the endometrium to induce secretions for nourishment of the conceptus during various periods of gestation. The mammalian uterine endometrium, therefore, assumes the role played by the maternal liver in lower species. Using the terminology of Schlafke and Enders (1975), fusion implantation, as found in the pig, sheep, horse, and cow, is associated with endometrial histotroph secretion for the majority of pregnancy. In these species, energy requirements appear relatively high, as the endometrium must be provided with various substrates that are in turn converted to other substrates to be utilized by the conceptus.

Fuller W. Bazer • Department of Animal Science, University of Florida, Gainesville, Florida 32611.

Mammals having displacement implantation (Schlafke and Enders, 1975) appear to utilize endometrial secretions only until implantation is established. After that time, nutrient transfer appears to involve the energy-efficient direct exchange of nutrient substrates between the maternal capillary bed and that of the conceptus.

The chapters in this section thus deal with: (1) the need for synchrony between the uterine endometrium and the embryo for establishment of pregnancy; (2) morphological events associated with implantation; (3) direct or indirect chemical signals, e.g., estrogens and prostaglandins, which enhance vascular exchange across the placenta; and (4) early embryogenesis following implantation.

Reference

Schlafke, S., and Enders, A. C., 1975, Cellular basis of interaction between trophoblast and uterus at implantation, *Biol. Reprod.* **12**:41.

22

The Role of Prostaglandins in Endometrial Vascular Changes at Implantation

T. G. Kennedy and D. T. Armstrong

I. Introduction

One of the earliest signs of blastocyst implantation in all species investigated is a localized increase in endometrial vascular permeability (Psychoyos, 1973). This increase in permeability can be visualized readily by the intravenous injection of a macromolecular dye such as Evans blue, which leaves the vascular system only in areas where the permeability is abnormally high (Psychoyos, 1971). Examination of the uterus about 15 min after injection reveals colored sites (dye sites), each one of which in polytocous species indicates the presence of a blastocyst. Increased endometrial vascular permeability also precedes decidualization, which can be induced by the application of a wide variety of stimuli to the properly sensitized uterus, and is thought to be essential if decidualization is to occur (Psychoyos, 1973). While it is well established that these endometrial changes depend upon the proper steroid milieu, which may vary among species, the identity of the factor(s) responsible for the increase in endometrial vascular permeability and the subsequent decidualization of appropriately sensitized uteri are at present uncertain. In this communication we will summarize the evidence that indicates that prostaglandins are involved in mediating the changes in permeability in response to blastocysts and artificial stimuli.

II. Changes in Endometrial Vascular Permeability in Response to Blastocysts

The earliest indications that prostaglandins are involved in implantation came from studies in which inhibitors of prostaglandin biosynthesis, such as indomethacin (Vane,

T. G. Kennedy and D. T. Armstrong • Departments of Obstetrics and Gynecology and of Physiology, The University of Western Ontario, London, Ontario, Canada.

1971), were used. Gavin *et al.* (1974) and Saksena *et al.* (1976) reported that indomethacin inhibited implantation in the rat and mouse, respectively, as indicated by the absence of implantation swellings. However, as this inhibitor of prostaglandin synthesis also inhibits the decidual cell reaction in response to artificial stimuli (Castracane *et al.*, 1974; Sananes *et al.*, 1976; Tobert, 1976; Rankin *et al.*, 1979a), it was not clear from these studies whether it was the initiation of implantation or the subsequent decidualization that had been affected.

There are similarities between the early stages of implantation and the inflammatory response, with increased vascular permeability being a common feature. Because prostaglandins had been proposed as mediators of the increased permeability during the early inflammatory response (Crunkhorn and Willis, 1971; Kaley and Weiner, 1971; Kaley *et al.*, 1972; Williams and Morley, 1973), we tested the hypothesis that the increase in endometrial vascular permeability at the initiation of implantation is mediated by prostaglandins. As reported by Kennedy (1977), indomethacin given to rats on Day 5 of pregnancy (Day 1 = first day of vaginal sperm) inhibited the appearance of uterine dye sites on the evening of Day 5 (Table 1). This inhibition was transitory, and on Day 6 increased vascular permeability could be demonstrated in both treated and control rats, although the dye sites are smaller in indomethacin-treated animals, suggesting a delay in implantation (Table 1). Indomethacin also effectively inhibited the uterine dye site response in ovariectomized rats to which exogenous steroids, adequate for the initiation of implantation, were given (Kennedy, 1977). This finding indicates that the inhibitor of prostaglandin synthesis affected the initiation of implantation by a mechanism that was independent of an effect on ovarian steroidogenesis. Also, the concentrations of prostaglandins of both the E and the F series were greater in the uterine dye sites than elsewhere in the uterus of the rat (Table 2; Kennedy, 1977), providing additional suggestive evidence for a role for prostaglandins in the initiation of implantation.

The ability of indomethacin to inhibit the initiation of implantation has also been reported for the hamster (Evans and Kennedy, 1978a) and the rabbit (Hoffman *et al.*, 1978). In addition, for the hamster, Evans and Kennedy (1978a) found that the concentration of prostaglandins of the E, but not of the F, series was elevated within the uterus at the site of implantation. This observation is of interest as it is the E series prostaglandins that have been implicated in the vascular changes during the inflammatory process.

Table 1. Incidence of Uterine Dye Sites on the Evening of Day 5 (7 P.M.) and the Morning of Day 6 (6–7 A.M.) in Pregnant Rats with or without Indomethacin Treatment on Day 5

	Day 5	Day 6		
Treatment	Proportion of rats with uterine dye sites	Proportion of rats with uterine dye sites	Number of uterine dye sites[a]	Weight of uterine dye sites (mg)[a]
Sesame oil	10/10	10/12	13.4 ± 0.3	8.52 ± 0.35
Indomethacin[b]	1/10[c]	12/13	11.5 ± 0.9	7.15 ± 0.34[d]

[a] Mean ± S.E.M. for rats with uterine dye sites.
[b] 1 mg given s.c. in oil at 8 A.M. and again at 1 P.M. on Day 5.
[c] $p < 0.001$, compared with control.
[d] $p < 0.02$, compared with control.

Table 2. Prostaglandin (PG) Concentrations in Rat Uteri on the Evening of Day 5 of Pregnancy

Area of uterus	Number of rats	PGE concentration[a] (pg/mg)	PGF concentration[a] (pg/mg)
Nondye site	14	10.1 ± 1.2	7.1 ± 0.7
Dye site	14	20.7 ± 2.2[b]	11.8 ± 1.0[b]

[a] Mean ± S.E.M.
[b] $p < 0.001$, compared with control.

III. Changes in Endometrial Vascular Permeability in Response to Artificial Stimuli

There are several problems with the interpretation of the data obtained from studies with pregnant animals. It is not known, for example, if the uterine concentration of prostaglandins E was elevated at the uterine dye sites as a cause or as a consequence of the initiation of implantation. In addition, the source of the prostaglandins was unknown; they may have been produced by the blastocysts or by the endometrium. If the latter is the case, what then is the nature of the signal that indicates the presence of a blastocyst and results in localized increased levels of prostaglandins? To address some of these problems, we have conducted studies in rats pretreated with hormones so that, at the time of the experiment, their uteri were sensitized for the decidual cell reaction.

Initially, experiments were conducted to determine if prostaglandins mediate the changes in endometrial vascular permeability induced by artificial stimuli. This question was of interest because of the possibility that blastocysts, by virtue of their close contact with the endometrial epithelium (Psychoyos, 1973), and artificial stimuli may have the common property of damaging the endometrium, thereby stimulating prostaglandin production. There is evidence that artificial deciduogenic stimuli cause tissue damage (Finn, 1977; Lundkvist *et al.*, 1977), and in other tissues, injury is known to stimulate prostaglandin biosynthesis (Ramwell and Shaw, 1970; Piper and Vane, 1971).

The artificial stimulus used in our studies was the unilateral injection into the uterine lumen of 50 μl phosphate-buffered saline containing gelatin (PBS-G). Changes in endometrial vascular permeability were quantified using ^{125}I-labeled bovine serum albumin ([^{125}I]-BSA) (Psychoyos, 1961). The data obtained suggest that prostaglandins, probably of the E series, mediate the increase in endometrial vascular permeability in response to intraluminal PBS-G (Kennedy, 1979a). Evidence for this conclusion includes: first, indomethacin, an inhibitor of prostaglandin biosynthesis, inhibited the permeability response to intrauterine PBS-G (Table 3); second, uterine concentrations of prostaglandins E, as well as prostaglandins F, were elevated following the intrauterine injection (preceding detectable changes in endometrial vascular permeability); and third, the injection into the uterine lumen of small amounts of prostaglandin E_2 resulted in increased endometrial vascular permeability in rats in which endogenous prostaglandin production had been inhibited (Table 4).

The proposal that prostaglandins mediate the changes in endometrial vascular permeability following an artificial stimulus is supported by the work of others. First, the uterine content of prostaglandins E and F in mice is elevated by deciduogenic stimuli (Jonsson *et*

Table 3. Effect of Indomethacin on the Endometrial Vascular Permeability Indices of Rats Given a Unilateral Intrauterine Injection of 50 μl PBS-G[a]

Treatment	Endometrial vascular permeability index[b]	
	Autopsy at 4 hr[c]	Autopsy at 8 hr[c]
Sesame oil	1.19±0.07[d]	1.64±0.18
Indomethacin[e]	0.96±0.07	1.06±0.11

[a]Phophate-buffered saline containing gelatin.
[b]Ratio of concentrations of radioactivity in the injected and noninjected uterine horns 15 min after i.v. [^{125}I]-BSA.
[c]Relative to intrauterine injection.
[d]All values are mean ± S.E.M.; six rats per group.
[e]2 × 1 mg given s.c. in oil.

Table 4. Effect of a Unilateral Intrauterine Injection of 10 μg Prostaglandin E_2 on the Endometrial Vascular Permeability Indices of Indomethacin-Treated[a] *Rats*

Treatment	Endometrial vascular permeability index[b]	
	Autopsy at 4 hr[c]	Autopsy at 8 hr[c]
Vehicle[d]	1.23±0.14[e]	1.03±0.08
Prostaglandin E_2	1.46±0.08	1.54±0.10

[a]2 × 1 mg given s.c. in oil.
[b]Ratio of the concentrations of radioactivity in the injected and noninjected uterine horns 15 min after i.v. [^{125}I]-BSA.
[c]Relative to the time of intrauterine injection of PBS-G.
[d]50 μl phosphate-buffered saline containing gelatin (PBS-G).
[e]All values are mean ± S.E.M.; six to seven rats per group.

al., 1978; Rankin *et al.*, 1979a). Second, indomethacin treatment inhibits decidualization induced by artificial stimuli in rats (Castracane *et al.*, 1974; Sananes *et al.*, 1976; Tobert, 1976) and mice (Rankin *et al.*, 1979a). As increased endometrial vascular permeability is thought to be prerequisite for the decidual cell reaction (Psychoyos, 1973), it is possible that indomethacin inhibits decidualization by preventing the normal increase in endometrial vascular permeability. However, as Tobert (1976) observed inhibition of decidualization even when indomethacin administration was delayed until 8 hr after the stimulus, by which time the increase in permeability has presumably already occurred (Psychoyos, 1973; Kennedy, 1979a), it seems likely that indomethacin affects not only endometrial vascular permeability, but also the later stages of the decidual cell reaction. Finally, prostaglandins administered into the uterine lumen of the rat (Sananes *et al.*, 1976) and rabbit (Hoffman *et al.*, 1977) induce decidualization.

IV. Timing of Uterine Sensitivity

Implantation of the blastocyst requires strict synchronization between the development of the embryo and that of the endometrium (Psychoyos, 1973). In addition, decidualization in response to artificial stimuli can only be obtained during a limited period of pregnancy or pseudopregnancy, or when the uterus has been sensitized by giving the animals an appropriate regimen of hormone treatments (Psychoyos, 1973; Finn and Porter, 1975). As the production of prostaglandins by uterine homogenates from pseudopregnant rats is maximal on Day 5 (Fenwick *et al.*, 1977), corresponding to the time at which the uterus is most sensitive to deciduogenic stimuli (De Feo, 1963), we investigated the possibility that this critical timing was related to the ability of the uterus to respond to deciduogenic stimuli with prostaglandin production (Kennedy, 1979b).

Immature rats were treated with hormones and ovariectomized so that they were at the equivalent of Day 4, 5, or 6 of pseudopregnancy. By varying the time of the deciduogenic stimulus in this way, differential sensitization of the uterus for the decidual cell reaction resulted; as indicated by uterine weights 5 days after the unilateral intrauterine injection of PBS-G, sensitivity was greatest in rats on the equivalent of Day 5 of pseudopregnancy. The increase in endometrial vascular permeability 8 hr after stimulation of the uterus was also greatest in animals treated on Day 5, as indicated by radioactivity levels in the stimulated horn 15 min after the intravenous injection of [^{125}I]-BSA. Compared to the noninjected uterine horns, the uterine concentrations of prostaglandins E and F were elevated 15 min and 2 hr following intrauterine PBS-G; however, there were no significant differences in prostaglandin concentrations between Days 4, 5, and 6, suggesting that the differential endometrial vascular permeability responses previously obtained were not due to differences in uterine prostaglandin production. The absence of a difference between the days was unexpected, because of the previously mentioned report of Fenwick *et al.* (1977) that prostaglandin production by uterine homogenates was greatest on Day 5. A possible explanation for the differences may be that homogenates may not reflect *in vivo* events accurately. Alternatively, more prostaglandins may have been synthesized *in vivo* on Day 5 in our studies, but were not retained within the uterus. (If it is assumed that changes in vascular permeability within the endometrium depend on local prostaglandin concentrations, then uterine levels of prostaglandins are presumably of greater importance than overall uterine production.)

As uterine production of prostaglandins in response to artificial stimuli did not readily

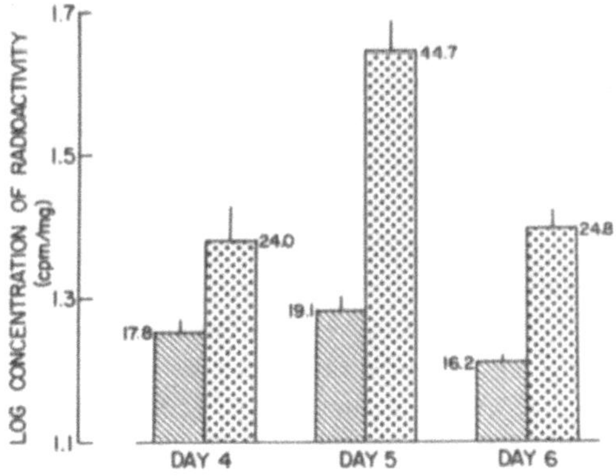

Figure 1. Effects of a unilateral intrauterine injection of 10 μg prostaglandin E_2 in 50 μl phosphate-buffered saline containing gelatin, given to indomethacin-treated rats on the equivalent of Day 4, 5, or 6 of pseudopregnancy, on the concentrations of radioactivity in the uterine horns 15 min after an i.v. injection of [^{125}I]-BSA. The rats were sacrificed 8 hr after the intrauterine injection. Each bar represents the mean ($\pm$ S.E.M.) of the logarithmically transformed data of eight animals. The geometric mean is given beside each bar. Striped bar, noninjected horn; dotted bar, injected horn.

explain the changes in uterine sensitivity, the ability of the endometrium to respond to prostaglandin E_2 with increased vascular permeability was examined. Rats that had been treated with indomethacin to inhibit endogenous prostaglandin production were used; in these animals, intrauterine PBS-G does not increase endometrial vascular permeability unless it contains prostaglandin E_2. When prostaglandin E_2 was administered into the uterine lumen on the equivalent of Day 4, 5, or 6 of pseudopregnancy, the greatest change in endometrial vascular permeability (as indicated by uterine radioactivity levels after intravenous $[^{125}I]$-BSA) was in animals treated on Day 5 (Figure 1). One possible explanation for this finding is that the timing of uterine sensitivity is related entirely to the ability of the uterus to respond to prostaglandins with increased endometrial vascular permeability. This ability may in turn be related to the properties of endometrial receptors for prostaglandin E_2, which may, for example, be present in the greatest concentrations on Day 5. Alternatively, changes in endometrial vascular permeability may require mediators in addition to prostaglandins, and the production, release, or action of these other mediators may determine the timing of uterine sensitivity. These possibilities are discussed further in Section VII of this chapter.

V. Which Prostaglandin(s)?

The question of which prostaglandins may be involved in mediating the vascular permeability changes leading to implantation has been approached in several ways. Based on the measurements of prostaglandins at the site of implantation, prostaglandins of the E and I series are possible mediators of the changes in endometrial vascular permeability. Levels of E series prostaglandins are elevated in uterine dye sites, relative to the other areas of the uterus, in rats (Kennedy, 1977) and hamsters (Evans and Kennedy, 1978a). As indicated by measurements of 6-ketoprostaglandin $F_{1\alpha}$, the stable breakdown product of PGI_2, the levels of prostaglandin I_2 are also elevated in uterine dye sites of rats (Kennedy and Zamecnik, 1978). By contrast, levels of prostaglandins F are elevated in implantation sites in rats (Kennedy, 1977) but not in hamsters (Evans and Kennedy, 1978a).

After the application of an artificial deciduogenic stimulus to the sensitized rat uterus, the levels of prostaglandins E, F, and I_2 (measured as 6-ketoprostaglandin $F_{1\alpha}$) are elevated within the uterus (Kennedy, 1979a,c). Intrauterine administration of prostaglandin E_2, but not prostaglandin $F_{2\alpha}$ (Kennedy, 1979a) or prostaglandin I_2 (Kennedy, 1979c), to rats in which endogenous prostaglandin production had been inhibited by indomethacin, increased endometrial vascular permeability. However, because of the instability of prostaglandin I_2 at neutral pH, the absence of a permeability change in response to this prostaglandin does not exclude the possibility that it has a role as a mediator.

The induction of decidualization in response to intrauterine prostaglandin $F_{2\alpha}$ has been reported by Sananes *et al.* (1976) for the rat and by Hoffman *et al.* (1977) for the rabbit. In the latter study, prostaglandin E_2 was considerably more effective than prostaglandin $F_{2\alpha}$. The results of Sananes *et al.* (1976) and Kennedy (1979a) are difficult to reconcile, although both the preparation of the test animals and the endpoints measured were different. Kennedy (1979a) found that prostaglandin $F_{2\alpha}$ inhibited the endometrial vascular permeability response to prostaglandin E_2. As increased endometrial vascular permeability is a prerequisite for decidualization (Psychoyos, 1973), it seems likely that prostaglandin $F_{2\alpha}$ should inhibit, rather than induce, decidualization, as reported by Sananes *et al.* (1976).

The suggestion that prostaglandin I_2 may be the mediator of decidualization has come from Rankin *et al.* (1979b), who found that tranylcypromine, a selective inhibitor of prostaglandin I_2 synthesis (Gryglewski *et al.*, 1976), inhibits decidualization in mice. However, the selectivity of this inhibitory action of tranylcypromine has been questioned (Rajtar and de Gaetano, 1979), and the inhibition has not been overridden with prostaglandin I_2.

Thus there is considerable evidence for, and none against, a role for prostaglandin E_2 as a mediator, some suggestive evidence for prostaglandin I_2, and evidence for and against a role for prostaglandin $F_{2\alpha}$. Given the similarities between the early stages of implantation and the inflammatory response, it is of interest to note that it has been prostaglandins of the E, and more recently the I, series that have been implicated in the changes in vascular permeability during the inflammatory response (Williams and Morley, 1973; Kuehl *et al.*, 1977; Williams and Peck, 1977; Murota and Morita, 1978; Williams, 1979).

VI. Source of Prostaglandins

The two most likely sources of the prostaglandins that are involved in the initiation of blastocyst implantation are the blastocysts themselves and the endometrium.

The observation that rabbit blastocysts contain prostaglandins (Dickmann and Spilman, 1975) raised the possibility that the blastocyst is the source of prostaglandins (Kennedy, 1977); these could be secreted by the blastocyst to act locally on the endometrium to bring about the localized increase in endometrial vascular permeability. An attempt has been made to determine if rat blastocysts are capable of prostaglandin biosynthesis *in vitro*. To obtain large numbers of blastocysts, the system of Nuti *et al.* (1975) was used to synchronize estrus, ovulation, and mating in immature rats. As indicated by the presence of sperm in the vaginal smears, a high proportion of the rats, treated when 30 days old with 8 IU of pregnant mare serum gonadotropin, mated at the expected time when placed with males. Nuti *et al.* (1975) have reported that rats, treated in this manner, can conceive, maintain viable fetuses to term, and deliver live pups in the normal manner. Thus blastocysts collected from these treated animals presumably do not differ from normal blastocysts.

On the morning or afternoon of Day 5 of pregnancy, blastocysts were flushed from excised uteri with a stream of culture medium (Brinster and Thomson, 1966), using a needle on a hypodermic syringe. Blastocysts from up to 30 animals were collected and pooled. Sterile techniques were used throughout.

The blastocysts were washed three times with culture medium as described by Brinster (1967) and, in most experiments, divided into two pools containing equal numbers of blastocysts, one pool being incubated and the other being used for the determination of preincubation prostaglandin levels. The incubations were conducted as described by Torbit and Weitlauf (1974), except that paraffin oil was omitted, at 37°C in a water-saturated atmosphere of 5% CO_2 in air. The incubations ranged in duration from 2 to 24 hr. In some experiments, blastocysts were examined at the end of the incubation period by phase-contrast microscopy; in virtually all cases, the blastocysts appeared healthy and the incubations were not contaminated.

The incubations were terminated by the addition of ethanol to the incubation tubes, which were then stored at −20°C until assayed for prostaglandins E and F by radioimmunoassay. Culture medium without blastocysts was subjected to the same procedures, including incubation, and served as blanks in the prostaglandin assays. The incubations containing blastocysts were considered to have detectable prostaglandins only when the assay values were greater than the sum of the mean, plus two standard deviations, of the blanks.

Despite the fact that up to 150 blastocysts per incubation tube were used, it was not possible, under any of the conditions used, to demonstrate either that blastocysts contain prostaglandins E or F or that they synthesized prostaglandins E or F during culture. Conditions used included:

a. Incubation for up to 24 hr in Brinster's medium without serum.
b. As for (a), but with 5% fetal calf serum added to the medium; the addition of the serum elevated the blank values in the assays, presumably because the serum contained prostaglandins
c. Incubation for up to 24 hr in Brinster's medium containing 10 ng/ml arachidonic acid, with or without fetal calf serum. In these experiments, prostaglandins E and F were separated from each other and from arachidonic acid as described by Kennedy (1978).

There are a number of possible reasons for the failure to detect prostaglandin biosynthesis by rat blastocysts, including:

a. Blastocysts are incapable of prostaglandin biosynthesis.
b. The incubation conditions were inappropriate.
c. The sensitivity of the prostaglandin radioimmunoassays was inadequate.
d. Blastocysts produce prostaglandins other than prostaglandins E and F, or rapidly metabolize prostaglandins E and F to compounds that do not cross-react with the antibodies used in the assays.

The failure to find detectable amounts of prostaglandins in rat blastocysts contrasts with the report of Dickmann and Spilman (1975) for rabbit blastocysts; this presumably reflects the difference in development of these mammalian embryos. Mature rabbit blastocysts contain thousands of cells, whereas rat blastocysts contain less than a hundred (McLaren, 1972; Sherman and Atienza, 1977).

Having failed to find evidence for significant prostaglandin production by rat blastocysts, we then examined endometrial tissue as an alternative source of the prostaglandins involved in implantation. As reviewed above, there is much evidence that artificial stimuli result in elevated uterine levels of prostaglandins, presumably as a result of increased endometrial synthesis in response to tissue injury. If the hypothesis of an endometrial source is true for blastocyst implantation as well, then the nature of the signal by which the blastocyst induces localized increased endometrial levels of prostaglandins is of great interest. The signal may be mechanical, as a result of the close contact between the blastocysts and the endometrial epithelium that is established just prior to the increased permeability (Psychoyos, 1973; Lundkvist *et al.*, 1979). Alternatively, the blastocyst signal may be chemical in nature, possibly estrogen, as proposed by Dickmann and his colleagues (Dickmann *et al.*, 1976, 1977; Dickmann, 1979). However, there is little direct evidence for blastocyst estrogen production, especially in species such as rodents in which the blastocyst remains small prior to implantation (Antila *et al.*, 1977; Sherman and Atienza, 1977). In addition, Evans and Kennedy (1978b) reported the initiation of implantation and elevated levels of prostaglandins E in uterine dye sites in ovariectomized, adrenalectomized, medroxyprogesterone acetate-treated hamsters receiving an inhibitor of steroidogenesis. If it can be assumed that the inhibitor reached the blastocysts in effective concentrations, then these data suggest that the initiation of implantation and localized elevated levels of prostaglandins E were induced by a blastocyst signal that was independent of possible blastocyst steroidogenesis.

That endometrial cells are a source of prostaglandins is an attractive hypothesis, as it is capable of explaining the increases in endometrial vascular permeability brought about by both blastocysts and artificial stimuli.

VII. Mode of Action of Prostaglandins

Little is known about the mechanisms by which prostaglandins bring about increased endometrial vascular permeability and the subsequent decidualization. The effects of prostaglandins on vascular permeability have been most studied in investigations of the inflammatory response; however, the mechanisms by which prostaglandins increase permeability in this response are poorly understood and controversial. There is evidence for direct effects of prostaglandins on the microvasculature (Arora *et al.*, 1970; Kaley and Weiner, 1971; Kaley *et al.*, 1972; Grennan *et al.*, 1977), for indirect effects mediated by prostaglandin-induced release of histamine (Horton, 1969; Crunkhorn and Willis, 1971), and for prostaglandins potentiating the effects of histamine (Williams and Morley, 1973; Johnston *et al.*, 1975; Chahl, 1976; Grennan *et al.*, 1977) and bradykinin (Williams and Morley, 1973; Ikeda *et al.*, 1975; Johnston *et al.*, 1975; Katori *et al.*, 1975; Vane and Ferreira, 1975; Chahl, 1976; Grennan *et al.*, 1977).

Recently, Williams (1977) and Williams and Peck (1977) have emphasized that during the inflammatory response, there is not only an increase in vascular permeability but also vasodilation, and that these two components of the response may have different chemical mediators. According to this hypothesis, expression of an increase in vascular permeability requires not only an increase in permeability but also vasodilation. Thus inflammatory edema can be suppressed by inhibiting the production, or action, of either type of mediator. There is general agreement that prostaglandins, probably of the E or I series, mediate the vasodilation (Kuehl *et al.*, 1977; Williams, 1977, 1979; Williams and Peck, 1977), but disagreement about the mediators of the changes in vascular permeability, with histamine and bradykinin being proposed by Williams and Peck (1977), and the endoperoxide, prostaglandin G_2, or some nonprostaglandin product of it, being proposed by Kuehl *et al.* (1977).

There are indications that this hypothesis of two mediators may be applicable to the changes in endometrial vascular permeability. The data of Bitton *et al.* (1965) indicate that, in pseudopregnant rats, vasodilation accompanies the increased endometrial vascular permeability induced by an artificial stimulus. This vasodilation may be mediated by prostaglandins of the E or I series. The involvement of histamine in the endometrial vascular permeability response is suggested by the demonstration of Brandon and Wallis (1977) that histamine H_1- and H_2-receptor antagonists reduce the number and intensity of uterine dye sites.

At the cellular level, the effects of prostaglandins may be mediated by alterations in intracellular cyclic AMP levels. Prostaglandins of the E and I series are stimulators of cyclic AMP synthesis in a variety of cell types (Kuehl *et al.*, 1976; Singhal *et al.*, 1976; Goff *et al.*, 1978; Omini *et al.*, 1979; Toth *et al.*, 1979) and artificial deciduogenic stimuli bring about a rapid increase in uterine cyclic AMP levels (Leroy *et al.*, 1974; Rankin *et al.*, 1977, 1979a). The increase in cyclic AMP levels in response to intrauterine oil is inhibited by indomethacin, indicating that the response is prostaglandin mediated (Rankin *et al.*, 1979a).

The cells within the endometrium that respond to prostaglandins with increased cyclic AMP synthesis are unknown. In the rat mesentery, under the influence of prostaglandins, increases in permeability and cyclic AMP levels are correlated, suggesting a causal relationship (Kahn and Brachet, 1978). Thus it would be of interest to know if prostaglandins stimulate cyclic AMP synthesis in endothelial cells within the endometrium, as this may alter their function. However, if, as postulated in the two-mediator hypothesis of vascular permeability, prostaglandins are primarily vasodilators, then they would act presumably to reduce the tone of arteriolar smooth muscle; this effect may be mediated by cyclic AMP (Dunham *et al.*, 1974; Kadowitz *et al.*, 1975). Alternatively, cyclic AMP synthesis may be stimulated by prostaglandins in stromal cells. This stimulation may be of importance for the differentiation of these cells into decidual cells, although this seems unlikely as the intrauterine application of cyclic AMP or dibutyryl cyclic AMP does not induce decidualization (Leroy *et al.*, 1974; Webb, 1975; Hoffman *et al.*, 1977; Rankin *et al.*, 1977). However, cholera toxin, a potent stimulator of cyclic AMP synthesis, induces decidualization when injected into the uterine lumen of mice (Rankin *et al.*, 1977). Perhaps cyclic AMP and its dibutyrl analog gave negative results because cyclic AMP levels were not elevated sufficiently within responsive cells to trigger a response.

VIII. Summary and Conclusions

We have reviewed evidence suggesting an involvement of prostaglandins in the endometrial vascular permeability changes associated with the initiation of implantation and the decidual cell reaction. The evidence includes the observation that inhibitors of prostaglandin synthesis delay or prevent the increase in endometrial vascular permeability in response to blastocysts or artificial stimuli. The concentrations of prostaglandins are higher in areas of increased endometrial vascular permeability than elsewhere in the uterus. When introduced into the uterine lumen, prostaglandin E_2 increases endometrial vascular permeability.

We suggest that blastocysts and artificial deciduogenic stimuli may share the common property of causing injury to the endometrium, thereby stimulating prostaglandin biosynthesis, particularly of the E and/or I series. The prostaglandins may then have two functions. First, they may act on arteriolar smooth muscle, thereby causing the vasodilation necessary for changes in endometrial vascular permeability. Second, they may stimulate cyclic AMP production by stromal cells, thereby initiating the differentiation of decidual cells.

ACKNOWLEDGMENTS. The research reported in this review was supported by grants from the Medical Research Council (Canada) and the World Health Organization. T.G.K. is a Scholar, and D.T.A. a Career Investigator, of the Medical Research Council. We thank Dr. B. G. Miller for numerous valuable discussions.

Discussion

LEROY (Brussels): You were careful not to implicate prostaglandins in the direct production of decidual cells. There is no doubt that prostaglandins play a role in vascular permeability during the blue response. It all boils down to how these reactions are related to triggering decidualization. Sananes *et al.* (*Mol. Cell. Endocrinol.* **6**:153, 1976) showed that they could induce decidualization with $PGF_{2\alpha}$ almost as much as with scratching in immature animals treated with progesterone. By giving indomethacin they could prevent the effect of scratching. When we tried this, the only difference being that we were dealing with ovariectomized animals treated with progesterone, we did not succeed. Judith Rankin told me that in South Carolina they have tried to obtain deciduomas by injecting prostaglandins in mice and they could not succeed either. Have you tried, and what do you think of this discrepancy?

KENNEDY: In preliminary experiments I have tried to induce decidualization with intralumenal applications of prostaglandins without success. I think the absence of an effect may well be due to the fact that the prostaglandins may not stay around for a sufficiently long time to bring about decidualization. It is important to bear in mind Tobert's data (*J. Reprod. Fertil.* **47**:39, 1976), which indicated that delaying the administration of indomethacin until 8 hr after the deciduogenic stimulus will still inhibit the decidual cell reaction. By 8 hr there was a full-blown vascular permeability response in my hands. It is possible that one requires prostaglandin production for a prolonged period after the deciduogenic stimulus in order to get a full decidual cell reaction. Unless one has some technique for slow release of prostaglandins within the uterine lumen, one is not going to get reasonable decidualization. Hoffman *et al.* (*J. Reprod. Fertil.* **50**:231, 1977) gave implants of prostaglandins into the uterine lumen of rabbits and obtained decidualization in response to prostaglandin E_2. They obtained some decidualization also in response to prostaglandin $F_{2\alpha}$, but it was not nearly as extensive as with prostaglandin E_2.

NILSSON (Uppsala): One of the many interesting problems of this issue is, where is prostaglandin produced? We have investigated with electron microscopy what happens when we block implantation by adding indomethacin. We found that the trophoblast cell looks quite normal. Attachment occurs quite normally, but the uterine epithelium has some large vesicles. Thus the only novel change from an electron microscopic point of view is in the uterine epithelium. Could it be that the uterine epithelium produces the prostaglandin?

KENNEDY: I have no information about where the prostaglandins are being produced within the uterus. If it is within the endometrium, it could be within the lumenal epithelium or within the stromal cells. I think the only way one is going to get at this is to separate these cell types and even then I am not sure that this is going to address the problem. Perhaps immunohistochemical studies would be useful.

M. JOSHI (N. Dakota): Another important factor in this matter is the role of estrogen in the increased vascular permeability. Schlough and Meyer (*Fertil. Steril.* **20**:439, 1969) showed that antiestrogens suppress the blue dye reaction. Estrogen has been shown in mice to be involved in changing the permeability of uterine capillaries (Martin *et al.*, 1973, *J. Endocrinol.* **56**:309). Also, estrogen induces release of prostaglandins from the uterus (Sherman and Lau, 1973, *Prostaglandins* **3**:317; Blatchley and Poyser, 1974, *J. Reprod. Fertil.* **40**:205). Have you any comments on the involvement of estrogen in this matter of prostaglandin and vascular permeability?

KENNEDY: Our data indicate that prostaglandins are involved in some way in increasing permeability. In pregnant animals, estrogens would have to be produced locally to get this local effect, and estrogens decrease rather than elevate the levels of prostaglandin E_2. If a local production of estrogens is involved, I would have expected the prostaglandins of the E series to have been decreased at the uterine dye site rather than increased. Over and above that, we have data that indicate that inhibitors of prostaglandin synthesis will inhibit the response, which argues against a direct role of estrogens.

DICKMANN (Kansas): I am a proponent of the hypothesis that blastocyst estrogen plays a role in implantation. McCormack *et al.* (*Biol. Reprod.* **20**:98A, 1979) have shown that the rat blastocyst produces estrogen in culture. We know also that estrogen has the capacity to release prostaglandin (Ryan *et al.*, 1974, *Prostaglandins* **5**:257). In a paper we published 2 years ago (*Science* **195**:687, 1978), we described experiments in which eggs at the four-cell stage were incubated for 2 hr in a salt solution with or without estrogen. We transferred the eggs into a 5-day pseudopregnant uterus and found that those that were estrogen treated gave the blue reaction whereas the other eggs, which were treated identically except for estrogen, did not give the blue reaction. This evidence suggests that estrogen from the blastocysts is involved in the initiation of implantation.

KENNEDY: We have attempted to investigate this hypothesis by treating hamsters, which do not require a maternal source of estrogen for implantation, with inhibitors of steroidogenesis. We were unable to demonstrate any effect of inhibitors of steroidogenesis on the uterine dye site reaction. There are problems with interpretation of those data, because we cannot demonstrate that the inhibitors of steroidogenesis inhibited the putative blastocyst estrogen, since no one has been able to demonstrate directly estrogen production by the hamster blastocyst. One of the things which appeals to me about the prostaglandin hypothesis is that it provides an explanation of the changes in vascular permeability, which occur not only at implantation but also in response to deciduogenic stimuli. There is no source of blastocyst estrogen in the artificial deciduogenic reaction.

MOULTON (Cincinnati): Phospholipase is a lysosomal enzyme, so perturbing the lysosomes in the epithelial cells might be a means for initiating synthesis of prostaglandins.

BIGGERS (Harvard): A year ago we published a paper showing that meclofenamic acid and 7-oxa-13-prostynoic acid would inhibit hatching of mouse blastocysts in culture (Biggers *et al.*, 1978, *Biol. Reprod.* **19**:519). This paper was related to other work on water transport across the trophectoderm, which causes the swelling of the blastocyst. Movement of water across epithelia, such as the frog skin and the toad bladder, and the kidney and intestine, involves PGE_2. According to Ramwell, the substance may work by controlling sodium permeability and coupled water transport. We have now shown the effectiveness of other inhibitors of prostaglandin function, and the question arises, can the blastocyst make its own endogenous prostaglandins? Dr. Catherine Rice incubated rabbit blastocysts in medium containing [^{14}C]arachidonic acid and tried to isolate the products of the arachidonic acid cascade, with no success. Dr. Needelman suggested that we carry out the isolations after preloading the phospholipids in the blastocyst memberanes with radioactive arachidonic acid, and then use a calcium ionophore to activate phospholipase A_2 to release endogenous arachidonic acid. Following this procedure, Rice found PGE_2, $PGF_{2\alpha}$, and thromboxane B, showing that the arachidonic acid cascade can occur in the rabbit blastocyst which could thus be its own source of prostaglandins. Dickmann and Spilman reported that they could detect prostaglandins in 6-day rabbit blastocysts by radioimmunoassay.

Inhibitors like meclofenamic acid, and a new one, 18,18,20-trimethyl-PGE_2, injected locally into the uterus in nanogram amounts are very effective in preventing implantation. The question arises, do prostaglandins produced by the blastocyst control maternal functions, such as the local vascular response, or are they acting independently in mother and embryo? PGE_2 in the blastocyst, for example, may be necessary to maintain the trophoblast epithelium for water transport. In the uterus, prostaglandins may be involved in the vascular and decidual responses. An inhibitor may exert multiple independent effects to produce infertility.

C. Tachi (Tokyo): We have succeeded in systemic induction of decidualization by injecting prostaglandins $F_{2\alpha}$ and E_2 (Tachi and Tachi, 1974, in: *Physiology and Genetics of Reproduction* (E. M. Coutinho and F. Fuchs, eds.), Part B, pp. 263–286, Plenum Press, New York). The experiment was done at very high levels of prostaglandins and we had some problems in getting reproducible results. In certain batches of animals we had a low rate of induction. Nevertheless, we were much interested in this effect because prostaglandin is the first natural substance shown to cause decidualization by systemic means. The first substance was pyrathiazine (Shelesnyak and Kraicer, 1963, *J. Reprod. Fertil.* **2**:438), but pyrathiazine is not a natural compound.

If this induction of decidualization by systemic administration of prostaglandins can be taken as a valid experiment, then the source of prostaglandins should be maternal and not the trophoblast. I believe the action of prostaglandins is not limited to changes in the vascular permeability, but might affect directly the proliferation of stromal cells.

Psychoyos (Bicêtre): Considering the estrogen effect on capillary permeability, I can say that there is no change in the permeability of the endometrial capillaries after injection of 1 μg of estradiol in the rat, although the blood volume of the uterus changes. By using only iodine-labeled protein you measure essentially the blood volume but not the amount of protein which comes out of the vessels. To estimate this amount you have to calculate separately the blood volume. I suggest that you estimate in parallel the blood volume and permeability changes. You may then see a much stronger effect of prostaglandins on vasodilation or on capillary permeability.

Kennedy: Thank you very much for your suggestion. We plan to use dual-isotope techniques to look at changes in blood volume within the uterus and the extravasation of proteins.

Nilsson: I should like to follow the line that prostaglandins are involved in decidual cell transformation. In the human there are predecidual cells at the end of the menstrual cycle. I give some women indomethacin to cure premenstrual tension. I should very much like to see biopsies from those women to find out if indomethacin blocks the appearance of the predecidual cells. Do you happen to have any information on that?

Kennedy: I have some anecdotal information, based on the dating of endometrial biopsies from women who were treated with indomethacin for rheumatoid arthritis, that development of a secretory-phase endometrium was retarded. This story suggests that prostaglandins are involved in the differentiation of the human endometrium.

Martin (ICRF, London): You fail to induce decidualization, but you do induce the vascular permeability change with prostaglandin. Is that a very short-lived response or does it go on as long as in the early stages of decidualization?

Kennedy: I have not looked after 8 hr.

Martin: It would be interesting to know if there is an abortive response.

Kennedy: Yes, I agree.

References

Antila, E., Koskinen, J., Niemelä, P., and Saure, A., 1977, Steroid metabolism by mouse preimplantation embryos *in vitro*, *Experientia* **33:**1374.

Arora, S., Lahiti, P. D., and Samyal, R. K., 1970, The role of prostaglandin E_1 in the inflammatory process in the rat, *Int. Arch. Allergy Appl. Immunol.* **39:**186.

Bitton, V., Vassent, G., and Psychoyos, A., 1965, Réponse vasculaire de l'utérus au traumatisme, au cours de la pseudogestation chez la ratte, *C.R. Acad. Sci.* **261:**3474.

Brandon, J. M., and Wallis, R. M., 1977, Effect of mepyramine, a histamine H_1-, and burimamide, a histamine H_2-receptor antagonist, on ovum implantation in the rat, *J. Reprod. Fertil.* **50:**251.

Brinster, R. L., 1967, Protein content of the mouse embryo during the first five days of development, *J. Reprod. Fertil.* **13:**413.

Brinster, R. L., and Thomson, J. L., 1966, Development of eight-cell mouse embryos *in vitro*, *Exp. Cell Res.* **42:**308.

Castracane, V. D., Saksena, S. K., and Shaikh, A. A., 1974, Effect of IUDs, prostaglandins and indomethacin on decidual cell reaction in the rat, *Prostaglandins* **6:**397.

Chahl, L. A., 1976, Interactions of bradykinin, prostaglandin E_1, 5-hydroxytryptamine, histamine and adenosine-5′-triphosphate on the dye leakage response in rat skin, *J. Pharm. Pharmacol.* **28:**753.

Crunkhorn, P., and Willis, A. L., 1971, Cutaneous reactions to intradermal prostaglandins, *Br. J. Pharmacol.* **41:**49.

De Feo, V. J., 1963, Determination of the sensitive period for the induction of deciduomata in the rat by different inducing procedures, *Endocrinology* **73:**488.

Dickmann, Z., 1979, Systemic versus local hormonal requirements for blastocyst implantation:A hypothesis, *Perspect. Biol. Med.* **22:**390.

Dickmann, Z., and Spilman, C. H., 1975, Prostaglandins in rabbit blastocysts, *Science* **190:**997.

Dickmann, Z., Dey, S. K., and Sen Gupta, J., 1976, A new concept: Control of early pregnancy by steroid hormones originating in the preimplantation embryo, *Vitam. Horm. (N.Y.)* **34:**215.

Dickmann, Z., Sen Gupta, J., and Dey, S. K., 1977, Does "blastocyst estrogen" initiate implantation?, *Science* **195:**687.

Dunham, E. W., Haddox, J. K., and Goldberg, N. D., 1974, Alteration of vein cyclic 3′:5′ nucleotide concentrations during changes in contractility, *Proc. Natl. Acad. Sci. (USA)* **71:**815.

Evans, C. A., and Kennedy, T. G., 1978a, The importance of prostaglandin synthesis for the initiation of blastocyst implantation in the hamster, *J. Reprod. Fertil.* **54:**255.

Evans, C. A., and Kennedy, T. G., 1978b, Evidence against a role for blastocyst steroidogenesis in the initiation of implantation in the hamster, *Soc. Study Reprod., 11th Annual Meeting,* Abstract No. 135.

Fenwick, L., Jones, R. L., Naylor, B., Poyser, N. L., and Wilson, N. H., 1977, Production of prostaglandins by the pseudopregnant rat uterus, *in vitro*, and the effect of tamoxifen with the identification of 6-keto-prostaglandin $F_{1\alpha}$ as a major product, *Br. J. Pharmacol.* **59:**191.

Finn, C. A., 1977, The implantation reaction, in: *Biology of the Uterus* (R. M. Wynn, ed.), pp. 245–308, Plenum Press, New York.

Finn, C. A.., and Porter, D. G., 1975, *The Uterus,* Publishing Sciences, Acton, Mass.

Gavin, M. A., Dominguez Fernandez-Tejerina, J. C., Montañes de las Heras, M. F., and Vijil Maeso, E., 1974, Efectos de un inhibidor de la biosíntesis de las prostaglandinas (indometacina) sobre la implantación en la rata, *Reproduccion* **1:**177.

Goff, A. K., Zamecnik, J., Ali, M., and Armstrong, D. T., 1978, Prostaglandin I_2 stimulation of granulosa cell cyclic AMP production, *Prostaglandins* **15:**875.

Grennan, D. M., Mitchell, W., Miller, W., and Zeitlin, I. J., 1977, The effects of prostaglandin E_1, bradykinin and histamine on canine synovial vascular permeability, *Br. J. Pharmacol.* **60:**251.

Gryglewski, R. J., Bunting, S., Moncada, S., Flower, R. J., and Vane, J. R., 1976, Arterial walls are protected against deposition of platelet thrombi by a substance (prostaglandin X) which they make from prostaglandin endoperoxides, *Prostaglandins* **12:**685.

Hoffman, L. H., Strong, G. B., Davenport, G. R., and Frölich, J. C., 1977, Deciduogenic effect of prostaglandins in the pseudopregnant rabbit, *J. Reprod. Fertil.* **50:**231.

Hoffman, L. H., DiPietro, D. L, and McKenna, T. J., 1978, Effects of indomethacin on uterine capillary permeability and blastocyst development in rabbits, *Prostaglandins* **15:**823.

Horton, E. W., 1969, Hypothesis on physiological roles of prostaglandin, *Phyisol. Rev.* **49:**122.

Ikeda, K., Tanaka, K., and Katori, M., 1975, Potentiation of bradykinin-induced vascular permeability increase by prostaglandin E_2 and arachidonic acid in skin, *Prostaglandins* **10:**747.

Johnston, M. G., Movat, H. Z., and Ranadive, N. S., 1975, Enhanced vascular permeability induced by chemical mediators, mediator-releasing agents and the effect of prostaglandins, *Fed. Proc. Fed. Am. Soc. Exp. Biol.* **34,** Abstract No. 4691.

Jonsson, H. T., Rankin, J. C., Ledford, B. E., and Baggett, B., 1978, Prostaglandin levels following stimulation of the decidual cell reaction in the mouse uterus, *Endocrine Soc., 60th Annual Meeting,* Abstract No. 502.

Kadowitz, P. J., Joiner, P. D., Hyman, A. L., and George, W. J., 1975, Influence of prostaglandins E_1 and $F_{2\alpha}$ on pulmonary vascular resistance, isolated lobar vessels and cyclic nucleotide levels, *J. Pharmacol. Exp. Ther.* **192:**677.

Kahn, A., and Brachet, E., 1978, The role of various prostaglandins on the correlation between permeability to albumin and cAMP levels in the isolated mesentery, *Prostaglandins* **16:**939.

Kaley, G., and Weiner, R., 1971, Prostaglandin E_1: A potential mediator of the inflammatory response, *Ann. N.Y. Acad. Sci.* **180:**338.

Kaley, G., Messina, E. J., and Weiner, R., 1972, The role of prostaglandins in microcirculatory regulation and inflammation, in: *Prostaglandins in Cellular Biology* (P. W. Ramwell and B. B. Pharriss, eds.), pp. 309–327, Plenum Press, New York.

Katori, M., Isoda, K., and Tanaka, K., 1975, Potentiation of increased vascular permeability of bradykinin by prostaglandin E_2 and arachidonic acid, and effects of indomethacin, *Life Sci.* **16:**806 (Abstract).

Kennedy, T. G., 1977, Evidence for a role for prostaglandins in the initiation of blastocyst implantation in the rat, *Biol. Reprod.* **16:**286.

Kennedy, T. G., 1978, Effect of oxytocin on estrogen-induced uterine luminal fluid accumulation in ovariectomized rats and the role of prostaglandins, *Can. J. Physiol. Pharmacol.* **56:**908.

Kennedy, T. G., 1979a, Prostaglandins and increased endometrial vascular permeability resulting from the appplication of an artificial stimulus to the uterus of the rat sensitized for the decidual cell reaction, *Biol Reprod.* **20:**560.

Kennedy, T. G., 1979b, Timing of uterine sensitivity for the decidual cell reaction (DCR) in rats: Role of prostaglandins (PGs), *Endocrine Soc., 61st Annual Meeting,* Abstract No. 577.

Kennedy, T. G., 1979c, Does prostaglandin I_2 (PGI_2) mediate the increased endometrial vascular permeability which results from the application of a deciduogenic stimulus to the sensitized rat uterus?, *Soc. Study Reprod., 12th Annual Meeting,* Abstract No. 175.

Kennedy, T. G., and Zamecnik, J., 1978, The concentration of 6-keto-prostaglandin $F_{1\alpha}$ is markedly elevated at the site of blastocyst implantation in the rat, *Prostaglandins* **16:**599.

Kuehl, F. A., Cirillo, V. J., Zanetti, M. E., Beveridge, G. C., and Ham, E. A., 1976, The effect of estrogen upon cyclic nucleotide and prostaglandin levels in the rat uterus, *Adv. Prostaglandin Thromboxane Res.* **1:**313.

Kuehl, F. A., Humes, J. L., Egan, R. W., Ham, E. A., Beveridge, G. C., and Van Arman, C. G., 1977, Role of prostaglandin endoperoxide PGG_2 in inflammatory processes, *Nature (London)* **265:**170.

Leroy, F., Vansande, J., Shetgen, G., and Brasseur, D., 1974, Cyclic AMP and the triggering of the decidual reaction, *J. Reprod. Fertil.* **39:**207.

Lundkvist, Ö, Ljungkvist, I., and Nilsson, O., 1977, Early effects of oil on rat uterine epithelium sensitized for decidual induction, *J. Reprod. Fertil.* **51:**507.

Lundkvist, Ö, Nilsson, O., and Bergström, S., 1979, Studies on the trophoblast–epithelial complex during decidual induction in rats, *Am. J. Anat.* **154:**211.

McLaren, A., 1972, The embryo, in: *Reproduction in Mammals: Embryonic and Fetal Development* (C. R. Austin and R. V. Short, eds.), pp. 1–42, Cambridge University Press, London.

Murota, S., and Morita, I., 1978, Effect of prostaglandin I_2 and related compounds on vascular permeability response in granuloma tissues, *Prostaglandins* **15:**297.

Nuti, K. M., Sridharan, B. N., and Meyer, R. K., 1975, Reproductive biology of PMSG-primed immature female rats, *Biol. Reprod.* **13:**38.

Omini, C., Folco, G. C., Pasargiklian, R., Fano, M., and Berti, F., 1979, Prostacylin (PGI_2) in pregnant human uterus, *Prostaglandins* **17:**113.

Piper, P., and Vane, J., 1971, The release of prostaglandins from lung and other tissues, *Ann. N.Y. Acad. Sci.* **180:**363.

Psychoyos, A., 1961, Perméabilité capillaire et decidualisation utérine, *C.R. Acad. Sci.* **252:**1515.

Psychoÿos, A., 1971, Methods for studying changes in capillary permeability of the rat endometrium, in: *Methods in Mammalian Embryology* (J. C. Daniel, Jr., ed.), pp. 334–338, Freeman, San Francisco.

Psychoyos, A., 1973, Endocrine control of egg implantation, in: *Handbook of Physiology,* Section 7, Volume II, Part 2, (R. O. Greep, E. G. Astwood, and S. R. Geiger, eds.), pp. 187–215, Am. Physiol. Soc., Washington, D.C.

Rajtar, G., and de Getano, G., 1979, Tranylcypromine is not a selective inhibitor of prostacyclin in rats, *Thromb. Res.* **14:**245.

Ramwell, P. W., and Shaw, J. E., 1970, Biological significance of the prostaglandins, *Recent Prog. Horm. Res.* **26:**139.

Rankin, J. C., Ledford, B. E., and Baggett, B., 1977, Early involvement of cyclic nucleotides in the artificially stimulated decidual cell reaction of the mouse uterus, *Biol. Reprod.* **17:**549.

Rankin, J. C., Ledford, B. E., Jonsson, H. T., and Baggett, B., 1979a, Prostaglandins, indomethacin and the decidual cell reaction in the mouse uterus, *Biol. Reprod.* **20:**399.

Rankin, J. C., Ledford, B. E., and Baggett, B., 1979b, The effect of transylcypromine on the artificially stimulated decidual reaction in the mouse uterus, *Soc. Study Reprod., 12th Annual Meeting,* Abstract No. 174.

Saksena, S. K., Lau, I. F., and Chang, M. C., 1976, Relationship between oestrogen, prostaglandin $F_{2\alpha}$ and histamine in delayed implantation in the mouse, *Acta Endocrinol.* (*Copenhagen*) **81:**801.

Sananes, N., Baulieu, E. E., and Le Goascogne, C., 1976, Prostaglandin(s) as inductive factor of decidualization in the rat uterus, *Mol. Cell. Endocrinol* **6:**153.

Sherman, M. I., and Atienza, S. B., 1977, Production and metabolism of progesterone and androstenedione by cultured mouse blastocysts, *Biol. Reprod.* **16:**190.

Singhal, R. L., Tsang, B. K., and Sutherland, D. J. B., 1976, Regulation of cyclic nucleotide and prostaglandin metabolism in sex steroid-dependent cells, in: *Cellular Mechanisms Modulating Gonadal Hormone Action* (R. L. Singhal and J. A. Thomas, eds.), pp. 324–424, Univ. Park Press, Baltimore.

Tobert, J. A., 1976, A study of the possible role of prostaglandins in decidualization using a nonsurgical method for the instillation of fluids into the rat uterine lumen, *J. Reprod. Fertil.* **47:**391.

Torbit, C. A., and Weitlauf, H. M., 1974, The effect of oestrogen and progesterone on CO_2 production by "delayed implanting" mouse embryos, *J. Reprod. Fertil.* **39:**379.

Toth, M., Todd, H., and Hertelendy, F., 1979, A comparison of the effects of PGI_2, PGE_2 and PGH_2 on the cyclic nucleotide levels in rat anterior pituitary glands *in vitro*, *Prostaglandins* **17:**105.

Vane, J. R., 1971, Inhibition of prostaglandin synthesis as a mechanism of action of aspirin-like drugs, *Nature* (*London*) *New Biol.* **231:**232.

Vane, J. R., and Ferreira, S. H., 1975, Interactions between bradykinin and prostaglandins, *Life Sci.* **16:**804.

Webb, F. T. G., 1975, The inability of dibutyryl adenosine 3′, 5′-monophosphate to induce the decidual reaction in intact pseudopregnant mice, *J. Reprod. Fertil.* **42:**187.

Williams, T. J., 1977, Chemical mediators of vascular responses in inflammation: A two mediator hypothesis, *Br. J. Pharmacol.* **61:**447P.

Williams, T. J., 1979, Prostaglandin E_2, prostaglandin I_2 and the vascular changes of inflammation, *Br. J. Pharmacol.* **65:**517.

Williams, T. J., and Morley, J., 1973, Prostaglandins as potentiators of increased vascular permeability in inflammation, *Nature* (*London*) **246:**215.

Williams, T. J., and Peck, M. J., 1977, Role of prostaglandin-mediated vasodilatation in inflammation, *Nature* (*London*) **270:**530.

23

Comparison of Implantation in Utero and in Vitro

Allen C. Enders, Daniel J. Chávez, and Sandra Schlafke

I. Introduction

Because implantation is the increasingly intimate interaction of a blastocyst with the endometrium of an intact uterus, there is really no such thing as *in vitro* implantation. However, trophoblast *in vitro* continues to differentiate beyond the preimplantation stage, and double organ culture can be used for cell associations. Therefore, it is possible to make models, if not of normal implantation, at least of ectopic implantation. The precise morphological events of implantation are sufficiently different in different groups of mammals to preclude all but the broadest generalizations concerning implantation (Wimsatt, 1975; Schlafke and Enders, 1975). In considering those species that have been studied by *in vitro* implantation models, attention has been confined to examples in which there is considerable invasion of the endometrium. Many domestic animals and most of the marsupials, in which implantation is more superficial, have been excluded from these studies.

These forms in which the blastocyst becomes either partially or completely "interstitial" (within the endometrium as opposed to within the uterine lumen) can be subdivided into two groups. In one group, the nature of the trophoblast initially involved in adhesion to the uterine epithelium and penetration of that epithelium is syncytial. This group includes carnivores, some rodents (e.g., guinea pig, ground squirrel), rabbits, and probably most higher primates. The other group, in which the first adhesion and epithelial penetration are accomplished by cellular trophoblast, includes many bats and the laboratory rat and mouse.

In the case of the human, we can only surmise from the histological appearance of the early implantation stages in the Carnegie collections (Figure 1) that the preceding stages may have involved syncytial trophoblast (Enders, 1976b). In the case of the rhesus monkey, a similar situation would exist except that we have recently obtained an epithelial

Allen C. Enders, Daniel J. Chávez, and Sandra Schlafke • Department of Human Anatomy, University of California School of Medicine, Davis, California 95616.

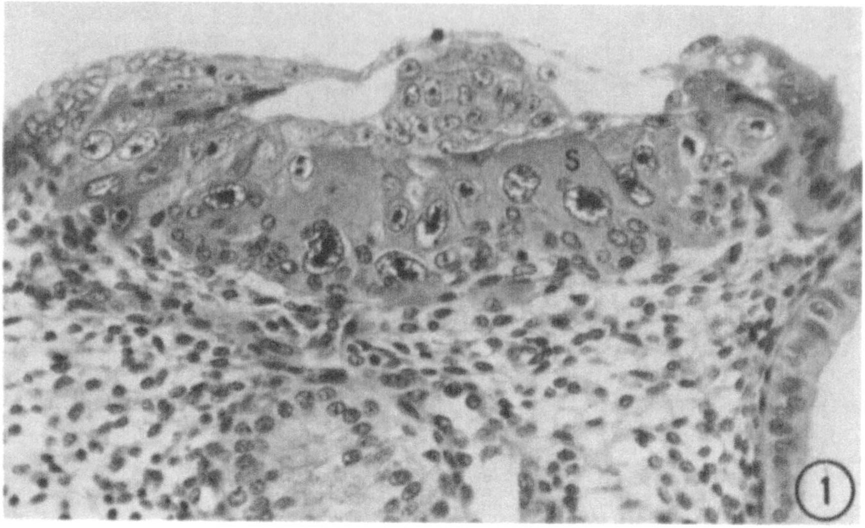

Figure 1. An early human implantation site, illustrating apparent syncytial trophoblast (S) at the invasive front of the blastocyst on the endometrium. Stage 5a, No. 8020, in the collections of the Carnegie Institution, Department of Embryology, Davis Division.

penetration stage for electron microscopic studies (Figure 2). Interestingly the rabbit, which has syncytial knobs, was one of the first animals studied *in vitro*. However, Glenister's (1962, 1963) pioneering joint organ culture method proved a difficult and not entirely adequate model, for the blastocyst would attach to clots, stroma, and areas of degeneration as well as uterine epithelium. The invasive tissue was cellular rather than syncytial trophoblast, and necrosis within the endometrium and disorganization were widespread. Blandau (1971) has used short-term culture of the guinea pig blastocyst and of the dissected postimplantation stages to study the relationships between growth of syncytium and its proteolytic enzyme activity. Unfortunately there has been little other *in vitro* work with the "syncytial trophoblast" group at the earliest implantation stages.

By far the greatest activity has concerned the laboratory rat and mouse, species in the "cytotrophoblast" group. The 1970s may well be considered the decade of blastocyst outgrowth, for although the method had its origin in the middle 1960s (Mintz, 1964; Cole and Paul, 1965; Gwatkin, 1966), the widespread use of the method is largely a product of the 1970s. The process has now been used to analyze a wide variety of events that occur in the peri-implantation period. The conditions for outgrowth have been studied extensively (Spindle and Pedersen, 1973; McLaren and Hensleigh, 1975) and appropriate substrates for attachment and outgrowth have been determined (Jenkinson and Wilson, 1973; Sherman, 1978).

Many of the studies on metabolic requirements can be considered to be elucidating possible uterine control of blastocysts (i.e., delay vs. activation). The experimental depletion of glucose, which does not appear to be lacking in normal delay of implantation, has been used to prevent outgrowth; it suggests the possibility of metabolic deficiencies as controls (Wordinger and Brinster, 1976; Sherman *et al.*, 1979). Other studies include the effects of rat uterine fluid on blastocyst development using *in vitro* outgrowth (Surani, 1977) and the chronology of steroid hormone synthesis or its absence during blastocyst development (Salomon and Sherman, 1975; Sherman and Salomon, 1975). The outgrowth method has also been used to study aspects of the trophoblast surface such as antigenicity (Heyner,

1973) or the nature of the glycocalyx (Jenkinson and Searle, 1977), as well as the effects of enzymatic digestion of surface materials (Glass *et al.*, 1979).

In vitro outgrowth of mouse blastocysts has also been used to study the effects of inhibitors of RNA and protein synthesis of blastocysts (Rowinski *et al.*, 1975) and the X-ray sensitivity of preimplantation stages (Goldstein, 1975). Prolonged *in vitro* cultivation of blastocysts has been employed in studies of differentiation (Pedersen and Spindle, 1976; Wiley and Pedersen, 1978). Developmental problems of mice with mutations at the T locus (Erickson and Pedersen, 1975; Sherman and Wudl, 1977) and the agouti locus (Pedersen, 1974) have been studied with cultured embryos. The interaction with cells from different endometrial and nonendometrial sources has been studied using monolayers *in vitro* (Sherman and Salomon, 1975). In addition, both trophoblast giant cell formation and the proliferation of trophoblast have been investigated using outgrowths (Gwatkin, 1966; Sherman and Salomon, 1975), as has the production of lytic enzymes, such as plasminogen activator, by the blastocyst (Strickland *et al.*, 1976).

This arbitrary and incomplete list of some of the uses that have been made of the trophoblast outgrowth technique illustrates its importance in recent studies of the rat and mouse blastocyst during the peri-implantation period. The separation of blastocyst from maternal components is clearly such an important investigational tool that it will continue to be used. We should try therefore to keep clear the similarities (model aspects) and dissimilarities (potential activities) between *in vivo* and *in vitro* conditions.

Most of the investigators cited and the major reviews of this procedure (Sherman and Wudl, 1976; Jenkinson, 1978; Sherman *et al.*, 1979) stress the importance of comparing

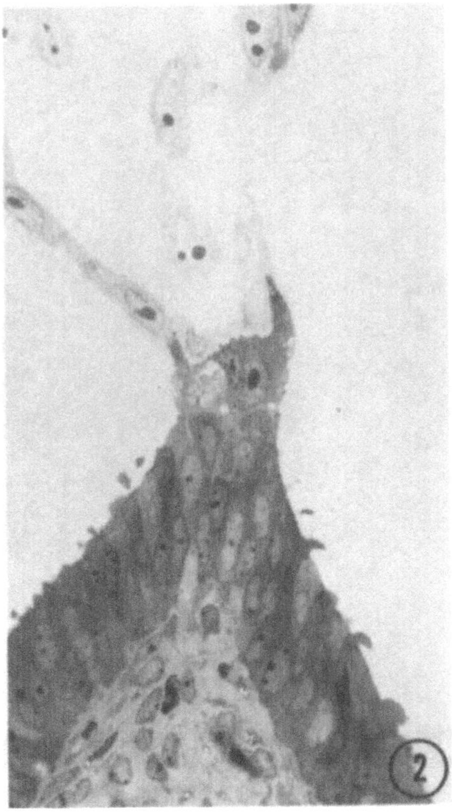

Figure 2. An early stage in implantation in the rhesus monkey. A small region of apparent syncytial trophoblast forms the adherent portion of the blastocyst and has penetrated into the uterine epithelium. Day 9.5 (Enders and Hendrickx, in preparation).

the *in vitro* events with *in vivo* development. This report will compare cell development through process of attachment and outgrowth, concentrating on the trophoblast with limited observations on the endoderm.

II. Comparison of Development

The events of *in vitro* outgrowth in the rat and mouse seem similar from the descriptions thus far available. This description is based on rat blastocyst outgrowths in our laboratory over a period of years, and on recent experiments by one of the authors (D.C.) with mouse blastocysts.

A. Hatching

Blastocysts, on being placed *in vitro* within the zona pellucida, first undergo a hatching phenomenon. In the absence of uterine proteolytic factors, hatching appears to be a more pronounced splitting than is seen *in vivo,* and occurs later, even allowing for the time lag characteristic of *in vitro* culture (Sherman and Wudl, 1976). Blastocysts *in vitro* achieve a greater size before hatching than *in vivo.* Rat blastocysts in particular are distinctly larger than *in vivo,* spherical rather than pear shaped, and the inner cell mass tends to occupy a crescent rather than a third segment of a sphere just prior to hatching.

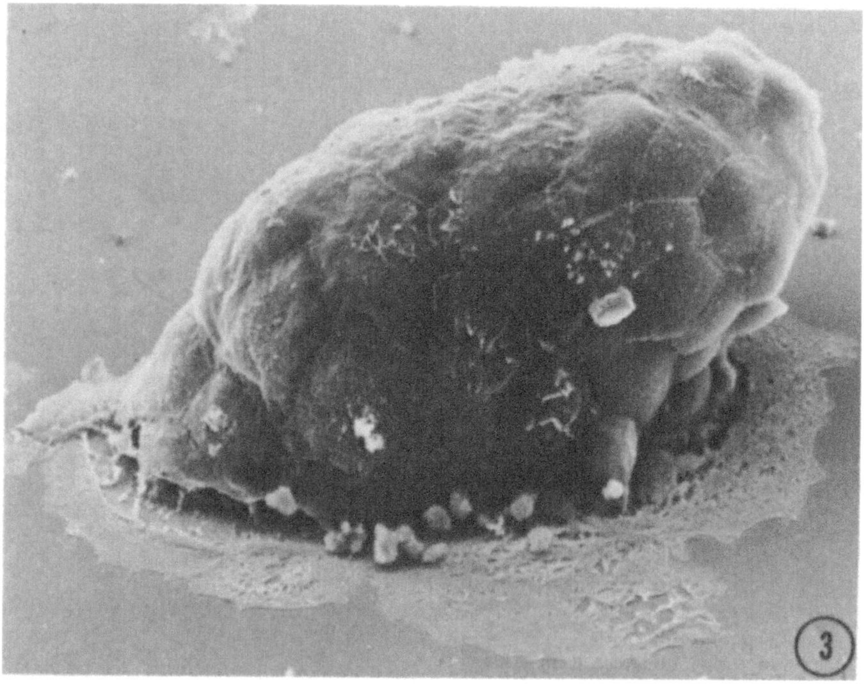

Figure 3. Mouse blastocyst adhering to the surface of a culture dish. The trophoblast layer is intact over the surface; there is variation in length and number of microvilli, and trophoblast outgrowth onto the surface has begun. Ridges at the junctions between cells are prominent.

B. Adhesion Stage

After hatching, the blastocysts attach to the substrate (Figure 3). That is, they become sufficiently adhesive that a gentle swirling does not displace them (Sherman, 1978). At this stage rat and mouse blastocysts *in utero* would be clasped by closure of the lumen (Hedlund *et al.*, 1972; Enders, 1976a) and thus would have much more surface in contact with the uterus than gravitational contact provides with the *in vitro* substrate. Furthermore, the numerous microvilli of the uterus would slowly be undergoing the transformation that allows more extensive contact with the trophoblast (Enders, 1975). A major deviation in the two processes, then, is that *in vitro* there is a gravitational contact with an inert substrate at the time of adhesion and *in vivo* the blastocyst is held in place over most of its surface against a living, changing, and probably transporting surface.

Embryonically at this time, the inner cell mass of both the rat and the mouse undergoes a loosening; that is, there is more intercellular space between the cells of the ICM than was present in the previous stage (Enders *et al.*, 1978; Figures 4, 5). At the close of the adhesion stage *in vivo,* two types of endoderm are apparent: parietal endoderm extending along the mural trophoblast, and visceral endoderm associated with the small but again compacted ICM. It is not clear that these events always take place similarly *in vitro,* as this is the stage at which the majority of the blastocysts collapse prior to outgrowth. Sherman (1978) and Sherman *et al.* (1979) suggest that even when attachment is prevented by growing blastocysts on agarose or interfered with by omitting fetuin from the culture medium, some developmental processes continue. In some circumstances many of the blastocysts remain expanded during initial outgrowth (Chávez, 1979; and Van Blerkom, this volume). It would be interesting to examine the cytology of the inner cell mass, especially the developing endoderm, in such specimens.

C. Outgrowth

Rat blastocysts, in our hands, and mouse blastocysts under many conditions collapse about the time that migration and cell processes from the blastocysts onto the substrate can be visualized (Figure 6). Individual blastocysts may not follow this pattern; Jenkinson and Wilson (1973) have shown that with a more three-dimensional substrate it is easier to maintain the orientation of blastocyst cells.

The first outgrowths of trophoblast onto the substrate show many of the features of tissue culture cells. Filopodia, lobopodia, and lamellipodia may all be present as flanges of cytoplasm extend along the substrate (Figures 7–10). These structures are associated with "ruffling" and the conditions for their formation have been investigated using several cell lines (Revel *et al.*, 1978). All of these modifications of the cell surface are ectoplasmic and devoid of the larger organelles. As spreading continues, processes extend around the blastocyst, usually one cell thick at the periphery but several cells thick close to the ICM. Both microtubules and bundles of filaments are characteristic of the cell processes involved in this migration (Figure 11). Individual cell surfaces show a great deal of variation; cells with lamellae may be adjacent to microvillous cells (Figure 9), and the heterogeneity is generally much greater than seen in trophoblast cells *in vivo*. Individual cells become detached from the blastocyst and wander off along the substrate. Cytological modifications that are not readily explained as ruffling also occur *in vitro*. Annulate lamellae are more abundant (Figure 12) and the quadruple membranes of adjacent cisternae of back-to-back endoplasmic reticulum, most commonly seen in prometaphase, are seen in interphase cells (Figure 13).

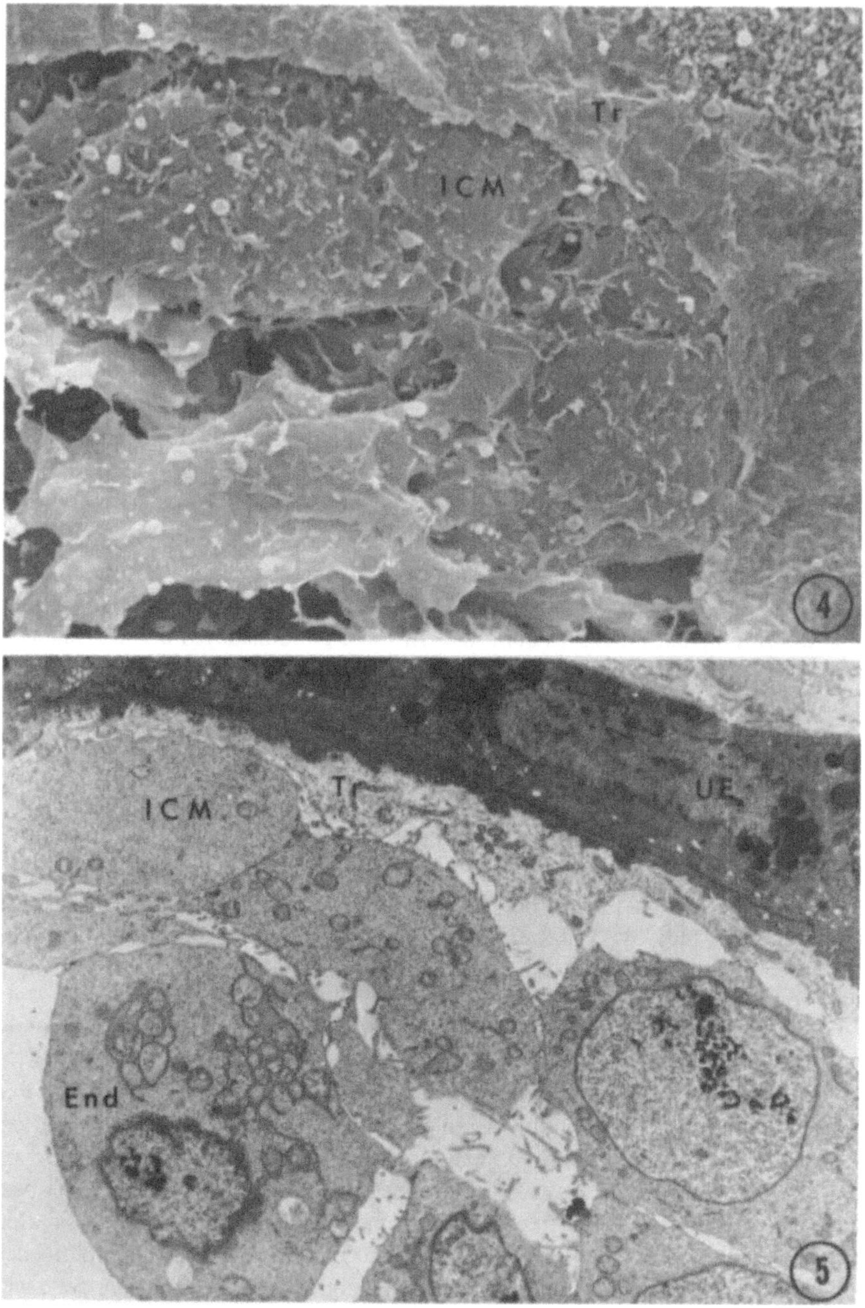

Figures 4, 5. Scanning and transmission electron micrographs of the ICM of the rat blastocyst during the adhesion stage of implantation. The ICM cells are separated by more intercellular space than at the previous developmental stage and microvilli are prominent. Uterine epithelium (UE); trophoblast (Tr); inner cell mass (ICM); and endoderm (End.). Note the prominent cisternae of endoplasmic reticulum in the cross section of the endoderm cells (bottom, Figure 5).

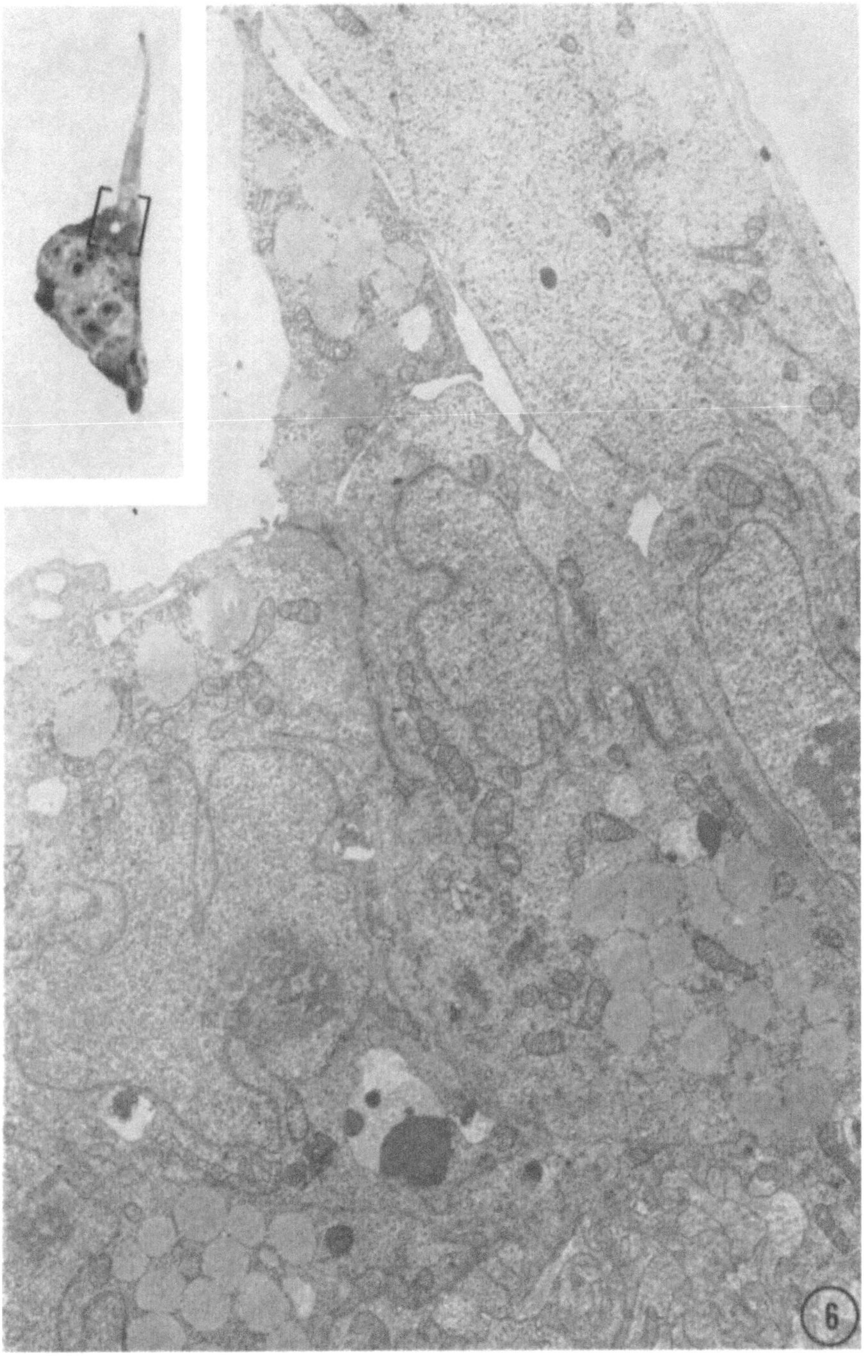

Figure 6. An early stage of outgrowth of a rat blastocyst. The blastocyst has collapsed and cells are extending around the periphery of the embryo (see inset). The cells covering the ICM contain numerous lipid droplets.

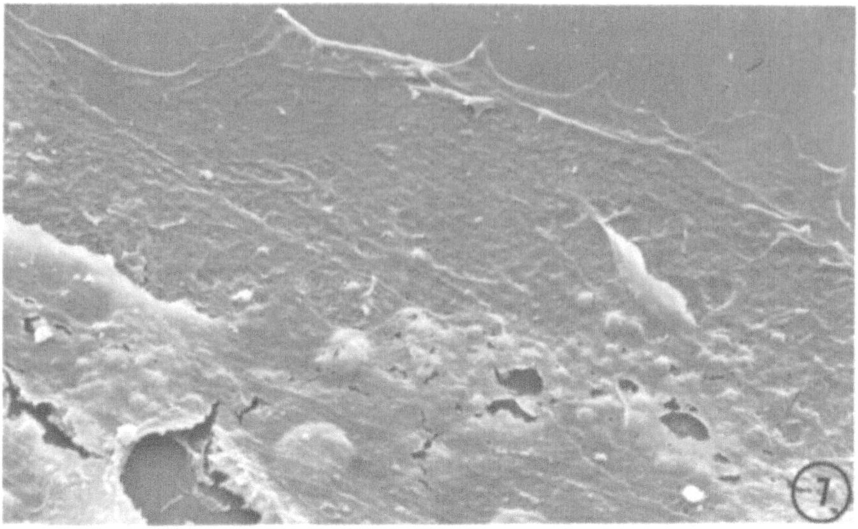

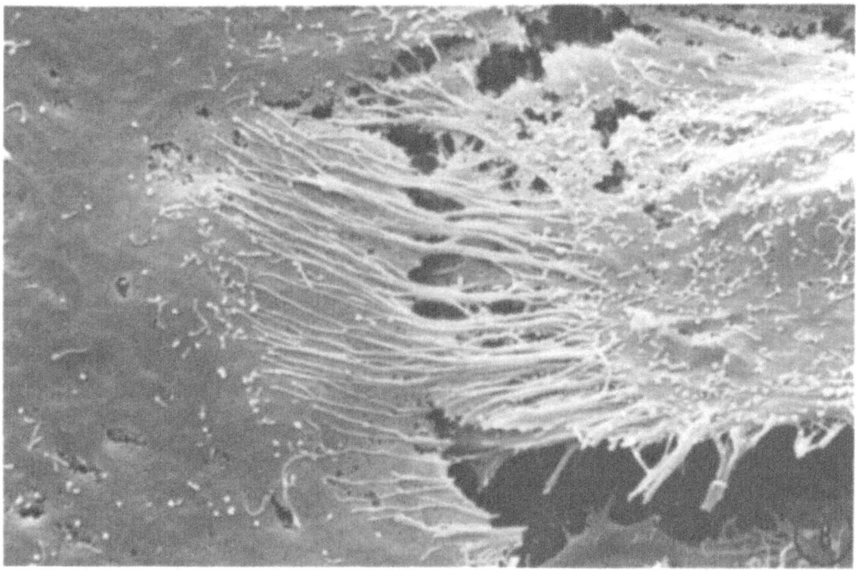

Figure 7. Scanning electron micrograph of the peripheral edge of trophoblast onto substrate. Long thin flanges such as these are common in cultured cells. The breaks in the cells toward the bottom of the micrograph are probably preparation artifacts due to shrinkage.
Figure 8. In addition to flanges of cells growing onto other cells or onto substrate, thin "microspikes" or extremely long filopodia may adhere either to other cells or to the substrate.

At some time during outgrowth, it is common for the layer of trophoblast overlying the ICM to become incomplete (Figure 14), exposing the underlying endoderm cells that are separated *in vivo* from the trophoblast by a thick basement membrane, Reichert's membrane, which is policed by parietal endoderm cells and is continuous with the endodermal basal lamina of the egg cylinder. There tends to be a relatively finite period of outgrowth from a given blastocyst, although length of trophoblast survival is more indeterminate (Sherman and Salomon, 1975).

D. Outgrowth and Monolayers

When blastocysts are grown on a cellular substrate, a clear area develops around the blastocyst that is devoid of substrate cells; this has been interpreted as an indication of proteolytic activity (Sherman and Wudl, 1976; Jenkinson, 1978). More recently it has been shown that there is an initial contact stage (Chávez, 1979; Glass *et al.*, 1979; Chávez and Van Blerkom, this volume). Glass *et al.* (1979) suggested that the pattern is that of contact inhibition, with cells withdrawing after contact rather than undergoing proteolysis. How-

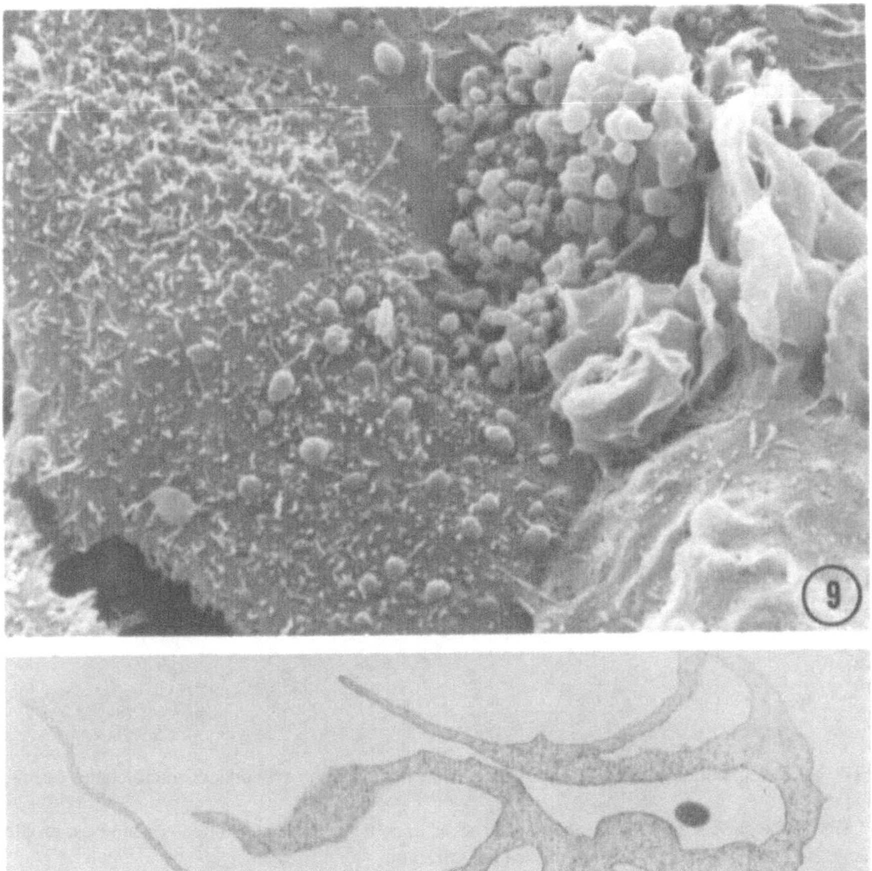

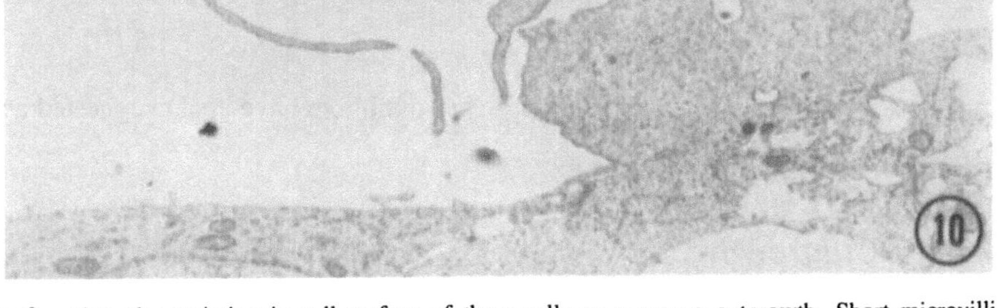

Figure 9. Note the variation in cell surface of these cells on a mouse outgrowth. Short microvilli, stubby lobopodia, and extensive lamellipodia can be seen, as well as a cell with few surface modifications (lower right).
Figure 10. Lamellipodia on the surface of a rat outgrowth. Note the ectoplasmic nature of the thin flanges.

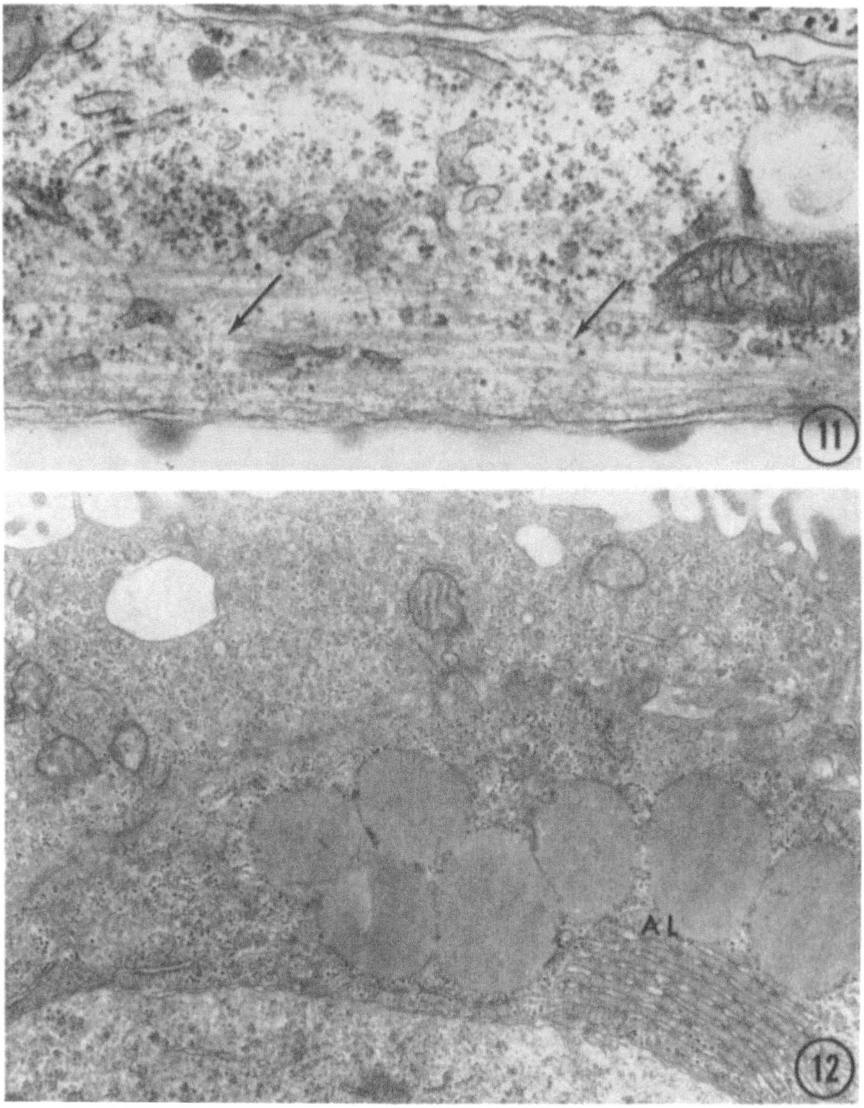

Figure 11. The basal trophoblast adjacent to the culture dish often contains networks of microtubules (arrows) and microfilaments.

Figure 12. Cells on the surface of a rat blastocyst after 48 hr in culture. These cells often contain lipid droplets and annulate lamellae (AL), as well as various surface modifications.

ever, further studies by Pedersen's group using protease inhibitors have again suggested active proteases (Kudo *et al.*, 1979).

III. Analysis of Outgrowth

Outgrowth *in vitro* is sufficiently different from development *in vivo* that it is even difficult to know what stage is appropriate for comparison. If we use elapsed time, then 24 hr

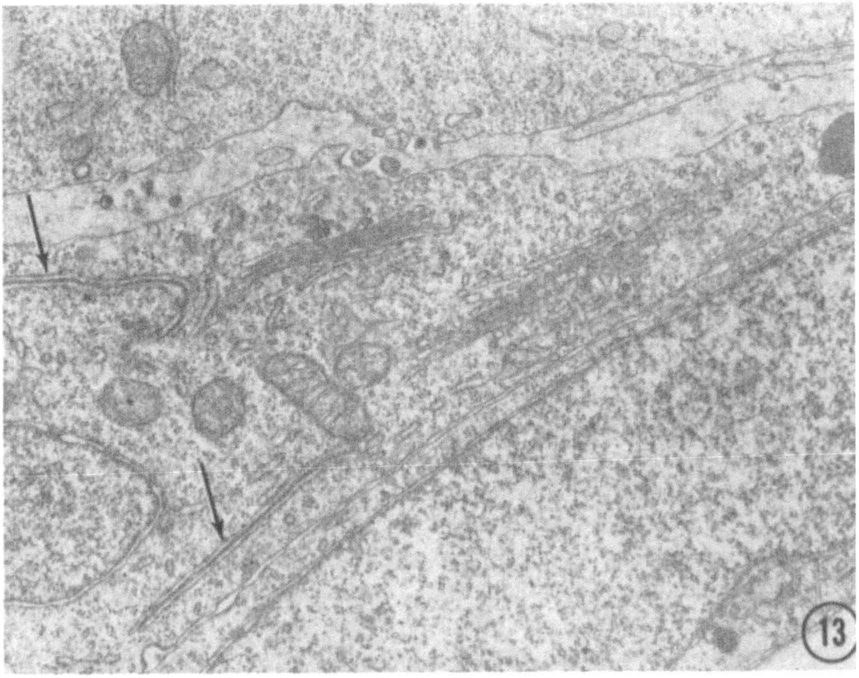

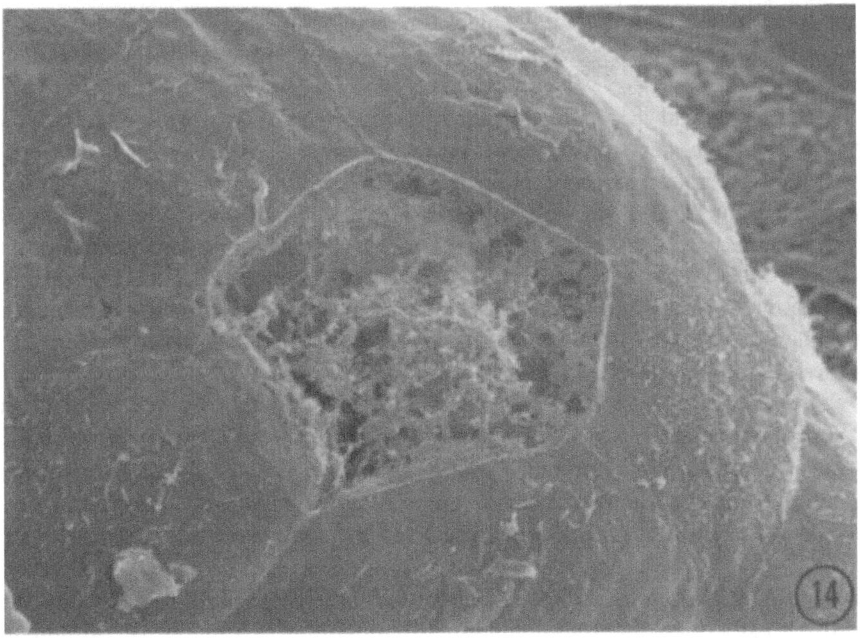

Figure 13. Cells in the basal area of a rat outgrowth, including a cell adjacent to the culture dish (lower right) and a cell containing quadruple membranes (arrows). The intercellular substance above these cells is probably the rough equivalent of Reichert's membrane and is interspersed between a presumptive endoderm cell (above) and the trophoblast cells (below).

Figure 14. The trophoblast layer that is initially on the surface of adhering blastocyst is often lost. This phenomenon may be the result of migration of trophoblast, or some trophoblast cells may degenerate as appears to have occurred here.

after adhesion *in vivo* we find a polarized seemingly passive stage. The implantation chamber has become more tubular and the blastocyst, rather than flattening, has become more elongated from inner cell mass to abembryonic trophoblast, while becoming more spherical in section through the mural trophoblast (lateral trophoblast). In the central zone of the implantation chamber, the lumenal epithelial cells and fragments of these have been largely phagocytosed by the lateral trophoblast cells, which maintain their position in the wall of the blastocyst and also their junctions with adjacent trophoblast cells (Figure 15). The majority of the trophoblast cells of the lateral walls of the blastocyst have thus come to lie in contact with the basal lamina that had underlain the uterine lumenal epithelium. The trophoblast cells at the abembryonic extreme in the rat have undergone enlargement, are more irregular, and are presumably giant cells corresponding to those in the mouse (Zybine, quoted by Sherman *et al.,* 1972).

Only at the margin of the ICM near the mesometrial end of the implantation chamber do extensive processes from trophoblast cells extend under the uterine epithelial cells (Figure 16). Here again, they overlie the basal lamina of the lumenal epithelium and maintain contact with adjacent trophoblast cells. At this location flanges of cytoplasm are seen that resemble only roughly those found in outgrowths, as only the finest of these processes are strictly ectoplasmic although more have filaments and microtubules.

The first projections through the basal lamina of the uterine lumenal epithelium appear to be from decidual cells (Figure 16), not from trophoblast (Enders and Schlafke, 1979), making the trophoblast of the rat appear less invasive than that of some other species. Only later do these cells, together with the ectoplacental cone cells, invade the mesometrial side of the uterus to form the chorioallantoic placenta, which together with the visceral endoderm cells of the yolk sac forms the significant tissue for exchange of materials from maternal to fetal organism.

Aside from the physical shape of the outgrowth compared to the corresponding *in vivo* implantation stage, the greatest contrast between the two is the loss of integrity of trophoblast overlying the visceral endoderm that occurs *in vitro,* resulting in what might be considered a premature inversion of the yolk sac. It is interesting that this distortion of the developing fetal membranes does not necessarily interfere with embryonic development, allowing further study of the ICM derivatives (Wiley and Pedersen, 1978; Hsu, this volume). Even apparent similarities, such as the flanges of intruding trophoblast at the margin of the polar zone that are seen *in vivo* and the flanges of trophoblast outgrowing *in vitro,* may have little in common other than shape.

Although superficially there is some resemblance between the residual basal lamina and the protein coat of an *in vitro* substrate, the probability of transfer of substances through the basal lamina *in vivo* and improbability of transfer through the *in vitro* substrate make them rather different. The uterine basal lamina is at the apical end of trophoblast cells *in vivo,* but polarity *in vitro* is poorly established and the substrate coat is "basal" to the trophoblast cells.

Even though outgrowth may not be equivalent to the corresponding stage of implantation *in vivo,* it probably does represent a stage in differentiation of trophoblast, for it cannot be initiated early by removal of the zona pellucida (Sherman, 1978).

IV. Prospects and Problems

The *in vitro* culture of the isolated blastocyst as a tool for studying implantation would appear to have promise for studying factors involved in adhesiveness of the blastocyst sur-

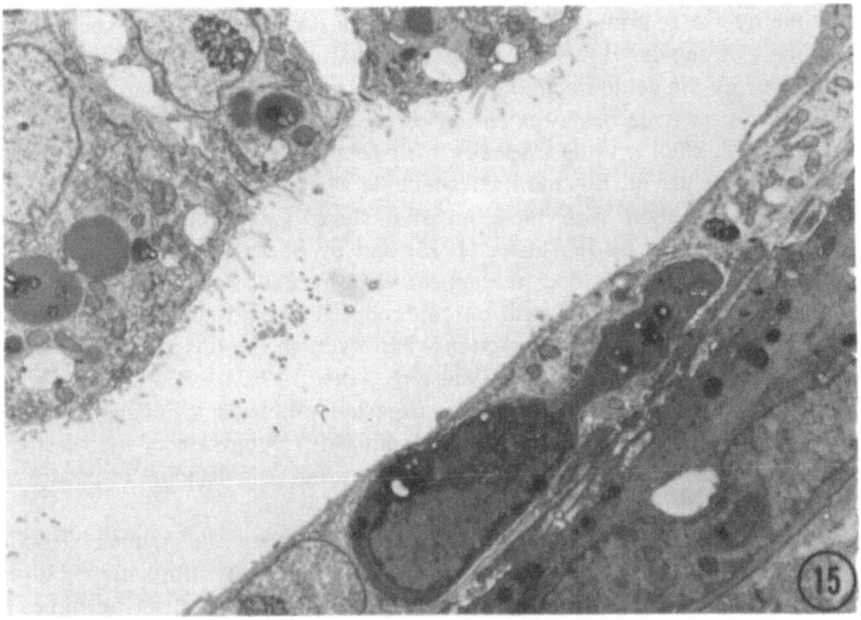

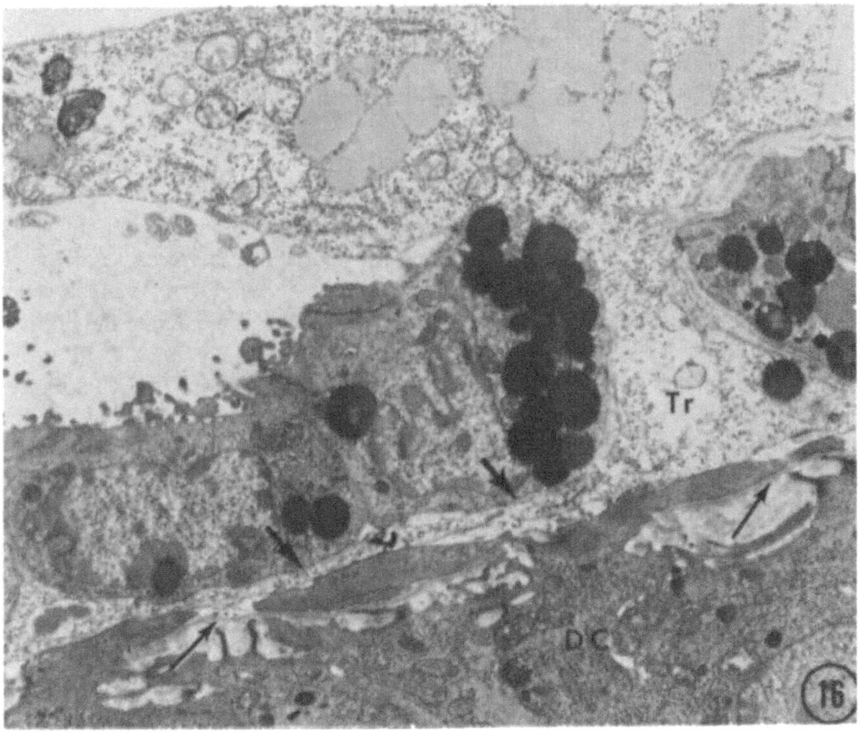

Figure 15. During normal implantation *in vivo*, trophoblast cells phagocytose uterine lumenal epithelial cells. The parietal yolk sac layer is composed of a few isolated cells (a process of one appears at the extreme upper right) and the visceral yolk sac consists of columnar cells with a distinctive pattern of apical organization.

Figure 16. The nearest the rat blastocyst comes to invasion at 7 days of gestation is the intrusion of processes from trophoblast cells (Tr) under the uterine epithelium (large arrows) in the region adjacent to the ectoplacental cone. Note also that processes from the underlying decidual cell (DC) are interposed between the trophoblast and the residual basal lamina (small arrows).

face. Certainly the mouse and rat, as the cheapest and most readily available species, will often be the animals of choice. However, a better internal control of whether the changes observed are induced by the culture conditions would be to use a species that adheres only on a limited region of its surface. An extension of the studies of Blandau (1971) on the guinea pig, or adhesion studies using a species with polarized implantation but without a zona pellucida, would be useful. It is more questionable whether outgrowth per se has anything significantly in common with the "invasive stage" at implantation. The time sequence, which was stressed by Jenkinson (1978) and by Sherman (Sherman and Salomon, 1975; Sherman *et al.*, 1979), does not appear valid insofar as the rat is concerned.

Combined organ culture or outgrowth on cultured cells would appear to be a useful means of studying blastocyst–uterine interactions. However, the results to date have been disappointing. As pointed out previously, Glenister's (1962, 1963) series of experiments with the rabbit endometrium revealed many unsuspected problems. Similarly, although Grant *et al.* (1975) showed that progesterone is essential for outgrowth of the blastocyst into cultured immature mouse uterus, the uterus did not produce a decidual response, and in other respects proved an incomplete model.

The use of cell monolayers (Cole and Paul, 1965; Sherman and Salmon, 1975) is useful with regard to specific cell adhesion or potential lytic enzyme formation, but certainly does not resemble invasion. On Day 5 of late gestation in the rat, for example, the uterine lumenal epithelium is tall columnar and can be shown to transport material from the apical surface to the lateral compartment and underlying stroma (Enders and Nelson, 1973). Similarly, it is presumed that *in vivo* there is transfer of material to the area around the blastocysts through the uterine lumenal epithelium from the vascular system. This system cannot function *in vitro*.

Even in using *in vitro* methods to indicate proteolysis, care must be taken with interpretation of data. In a study of *in vitro* dissolution of a gelatin membrane, Owers and Blandau (1971) reported little proteolytic activity of postimplantation trophoblast, ectoplacental cone, or uterine epithelium, almost none for decidua, and a great deal for yolk sac endoderm. This finding, of course, does not mean that the endoderm is invasive, nor would it lead to the observed possibility that decidual cell processes penetrate the epithelial basement membrane.

Another pitfall is cell identification after loss of initial organization. It is customary to identify cells by position, shape, and staining at the light microscopic level, and by shape and pattern of organelles at the electron microscopic level. In particular, it is important to assess whether a pattern is random, has intermediates (hence variation in one population), or represents distinctly separate entities. The problem is difficult not only because "new" features such as ruffling may develop *in vitro,* but more importantly the cytological distinction between cell types is often reduced *in vitro*. For example, the endoderm cells adjacent to other ICM components in outgrowths, which should correspond to visceral endoderm, have a marked paucity of apical absorption canaliculi. When cell position is no longer a valid criterion, identification is weakened. Pedersen and Spindle (1977), in a study of the effects of BuDR on mouse outgrowth, reported as a minor conclusion that the effect on endoderm (identified by light microscopy) was a lack of vacuolation in the treated embryos. However, from their micrographs, an interpretation of retention of trophoblast resulting in internal endoderm appears likely.

In conclusion, *in vitro* culture of blastocysts is a highly useful method capable of providing many indications of the potential function of trophoblast and embryonic cells. However, suggestions that the outgrowth mimics or gives specific information concerning invasion of the uterus must be viewed with caution. We hope the method will be used carefully and improved so that our words of caution will prove too conservative. Methods should be developed that are appropriate for study of invasion by blastocysts that develop only a limited area of adhesion and by blastocysts in which the invasive tissue is syncytium. In addition, detailed and carefully timed studies of the cytology of implantation both *in vivo* and *in vitro* are necessary to allow us to follow the changes occurring in specific cell populations a day or more after implantation *in vivo* or attachment *in vitro*.

Discussion

LEROY (Brussels): I got the impression that there were quite a few mitotic figures in the stroma beneath an attached rhesus monkey blastocyst.

ENDERS: I think you were probably looking at the leukocyte population of that stroma rather than mitotic figures. There are a lot of mitotic figures in the monkey, but not on Day 9; there are many more by late Day 10. The effects of the blastocyst begin first as pronounced edema, a large number of neutrophilic leukocytes and macrophages, and, following that, a plaque reaction which is an epithelial proliferation and epithelial polyploidy.

LEROY: This concerns only epithelial cells?

ENDERS: Oh, no! After implantation we see division in the endothelial cells of the blood vessels in the area and in some of the stromal cells, but the specialized plaque is composed of epithelial cells. That is only one small part of the response to the blastocyst in the rhesus monkey.

LEROY: Have you seen any signs which would suggest some passage of information from epithelium to stroma to basement membrane before there is any damage to the epithelium in species like mice and rats?

ENDERS: I would have to say not at the primary implantation site. Wislocki and Streeter (*Contrib. Embryol.* **160,** 1938) found plaques where they did not find the blastocyst; I was fortunate enough to see that once. Remember the monkey implants first on one side and then on the second side, so it has two areas. I found two additional areas opposite one another, with some epithelial damage on one of them. I do not know what caused the alteration, but it is a possibility, as Wislocki and Streeter suggested, that the blastocyst had lodged at this site first. That is the only indication I know that there is a preattachment effect; if so that could as easily have been due to damage as to message.

SHERMAN (Roche Institute): I want to bring up the matter of decidual cells appearing to penetrate the basement membrane. Do you see this often enough to eliminate the possibility that trophoblast cells have first penetrated the basement membrane and, after subsequent withdrawal, that the space then becomes occupied by decidual cells?

ENDERS: We see these cells where the trophoblast shows few surface projections. In other words, it is not just at the margins of the ectoplacental cone but all along the lateral aspect of the implantation chamber. I do not see breaks in the first stages without decidual processes extending through them. I do see decidual processes on the epithelial side of the basal lamina, where there are no breaks, but that observation is problably a three-dimensional problem of examining individual sections. So, the impression I received is that there is no penetration by the trophoblast. I would point out, if you want to speculate, that the basal lamina is presumably type IV collagen and there is probably a lot of reorganization of collagen by the decidual cells in that implantation chamber.

SHERMAN: The fact that the decidual cells penetrate through the basement membrane does not settle the issue of which cells produce the collagenase or protease activity. It is possible, for example, that trophoblast cells, which produce plasminogen activator, may activate a zymogen of collagenase and thus facilitate penetration of the basement membrane by decidual cells. The system is complex and sooner or later we have to try to isolate the proteolytic components involved.

SURANI (Cambridge): The blastocyst is capable of developing in extrauterine sites, but when it is present in the uterus, the uterus has to be sensitized before implantation can occur. We have to consider the changes in the lumenal environment, both in terms of macromolecules and ions, as well as the changes in the surface properties of the trophectoderm and the epithelium. I wonder if you would like to comment on that?

ENDERS: Study of the surface coats has been frustrating. Grinnell's group showed that the uterine surface binds less cationized ferritin at about the time of implantation (Hewitt *et al.*, 1979, *Biol. Reprod.* **21**:691). Our own results indicate that there is not a great deal bound by the uterine surface. When we have the two tissues together we find little binding by the surface epithelium but more by the trophoblast. There seems to be a secreted material between these adhering processes that also will bind cationized ferritin. Those of us who use concanavalin Awere probably working with a sugar too proximal to the polypeptide chain. We do not have sufficient evidence to say anything about the terminals of the oligosaccharides or what molecular changes are necessary for adhesion. Work with uterine fluid, such as yours, and work by other people on that surface should provide some of the molecular information.

BULLOCK (Baylor): You made a statement to the effect that if you wanted to get things through the surface coat material, you would rather do it *in vitro* than *in vivo*. Could you comment on the relative thickness of the surface coat material both on the trophoblast and on the epithelium *in vitro* and *in vivo* and tell us why you would rather do it *in vitro*?

ENDERS: You have the liberty of a greater variety of markers and methods *in vitro* than *in vivo*. For example, there are too many artifacts that develop from penetration in some of the cytochemical methods *in vivo*. You can't do a concanavalin A–peroxidase stain on apposed surfaces with any reliability; it fades out when two come close together. You can't really do a good ruthenium red stain when they are apposed. This is why we need the combined methods. One of the reasons we tried cytochemistry in the rat and mouse is that these species do not have a thick uterine glycocalyx as does the ferret, a species in which we can see the disappearance of that glycocalyx just be transmission electron microscopy, before apposition of the two surfaces.

CHANG (Worcester Foundation): I would like to hear your opinion about ectopic pregnancy, why it occurs in humans and not in other species.

ENDERS: Ectopic implantation has been a difficult subject. People have argued that there is endometriosis in the human and possibly, therefore, an endometriumlike area. I am a little skeptical of that idea, because damaged oviducts increase the frequency of ectopic implantation. Perhaps the trophoblast of the human is more aggressive in that it has to do more of the job of insinuating through epithelial cells than is necessary in the rat and mouse, which have a maternal system geared to making an implantation chamber. That is not saying a lot, but it does go back to my plea of trying to work more with species that have a synctial attachment to see if that is different from those that have a cellular attachment.

MANES (La Jolla): To add more fuel to that fire, we see, about once or twice a year, rabbit embryos in the peritoneal cavity, implanted either on the kidney capsule or in the mesentery. Inevitably we find a little scar in the uterus which was presumably the point at which the expanding blastocyst ruptured through before full attachment. I don't know what this says about invasiveness or the properties of the rabbit trophoblast as opposed to other species but there are obviously other sites for implantation or attachment.

ENDERS: The magic of the uterus is that it performs all its functions, not that any one of them takes a mystical substance. It allows sperm passage, spaces the blastocysts, allows nidation, allows growth without loss of organization, continues the appropriate environment for quiescence of the uterine muscle, and then it kicks the little buggers out at the end. It is marvelous that it does all these things, but I do not think there is only one way to do any of them. Studies in different areas are telling us some of the potential things that trophoblast can do and some of the potential things that can be done by other interactions. I do not think there are any magical substances and people who attached names to substances a few years ago were acting prematurely.

DENKER (Aachen): One question to the electron microscopist, which is pertinent to the plasminogen activator story. In the rabbit, we see material with the electron microscope which appears to be fibrin. We find this in the blood vessels which are going to be eroded by the trophoblast, and sometimes between the trophoblast and uterine epithelium, but also in the blastocyst cavity after attachment occurs. We found it puzzling to see deposition of material which appears to be fibrin, rather than fibrinolysis in these early stages of implantation. Is there any such material in other species and is there any morphological evidence for lysis of this material which could point to an active plasminogen activator/plasmin system?

ENDERS: I am afraid your question is too extensive for me, because you have to consider time sequences here. After implantation comes organization of the placenta and a whole series of events involving leakage from blood vessels. The rabbit is an excellent example where, although penetration of the epithelium is by fusion, penetration into the endothelium of the blood vessel is not. Those vessels leak before the trophoblast shares junctional complexes with either side of the endothelial cell and sends through processes that are not at all like the ruffling that occurs *In vitro*. I cannot this quickly go through in my mind all the areas where fibrinlike material is found, such as in the basal plate of the human, to consider evidence of plasminogen activity.

PSYCHOYOS (Bicêtre): If we throw out the magic substances, can you explain the hostility of the uterine environment on the basis of morphological criteria and the impossiblity of obtaining the decidual reaction after a precise moment?

ENDERS: I have no evidence that would explain why the epithelium is capable of eliciting the decidual response from the underlying stroma at any particular time. Some of the more striking things that the epithelium does, such as pinocytosis (as opposed to micropinocytosis), which most epithelial cells do not do except in culture, occur over a broad period, not just the most receptive period.

References

Blandau, R. J., 1971, Culture of guinea pig blastocyst, in: *Biology of the Blastocyst* (R. J. Blandau, ed.), pp. 59–69, Univ. of Chicago Press, Chicago.

Chavez, D. J., 1979, *Delayed Implantation: Molecular and Cellular Basis*, Ph.D. dissertation, Colorado State University, Fort Collins, Color.

Cole, R. J., and Paul, J., 1965, Properties of cultured preimplantation mouse and rabbit embryos, and cell strains derived from them, in: *Preimplantation Stages of Pregnancy* (G. E. W. Wolstenholme and M. O'Conner, eds.), pp. 82–112, Academic Press, New York.

Enders, A. C., 1975, The implantation chamber, blastocyst, and blastocyst imprint of the rat: A scanning electron microscope study, *Anat. Rec.* **182:**137.

Enders, A. C., 1976a, Anatomical aspects of implantation, *J. Reprod. Fertil. Suppl.* **25:**1.

Enders, A. C., 1976b, Cytology of human early implantation, *Res. Reprod.* **8:**1.

Enders, A. C., and Nelson, D. M., 1973, Pinocytotic activity of the uterus of the rat, *Am. J. Anat.* **138:**277.

Enders, A. C., and Schlafke, S., 1979, Comparative aspects of blastocyst–endometrial interactions at implantation, in: *Maternal Recognition of Pregnancy*, Ciba Foundation Symposium 64, pp. 3–32, Elsevier/North-Holland, Amsterdam.

Enders, A. C., Given, R. L., and Schlafke, S., 1978, Differentiation and migration of endoderm in the rat and mouse at implantation, *Anat. Rec.* **190:**65.

Erickson, R., and Pedersen, R. A., 1975, *In vitro* development of t^6/t^6 embryos, *J. Exp. Zool.* **193:**377.

Glass, R. H., Spindle, A. I., and Pedersen, R. A., 1979, Mouse embryo attachment to substratum and interaction of trophoblast with cultured cells, *J. Exp. Zool.* **208:**327.

Glenister, T. W., 1962, Embryo–endometrial relationships during nidation organ culture, *J. Obstet, Gynaecol. Br. Commonw.* **69:**809.

Glenister, T. W., 1963, Observations on mammalian blastocysts implanting in organ culture, in: *Delayed Implantation* (A. C. Enders, ed.), pp. 171–182, Univ. of Chicago Press, Chicago.

Goldstein, L. S., Spindle, A. I., and Pedersen, R. A., 1975, X-Ray sensitivity of the preimplantation mouse embryo *in vitro, Radiat. Res.* **62:**276.

Grant, P. S., Ljungkvist, I., and Nilsson, O., 1975, The hormonal control and morphology of blastocyst invasion in the mouse uterus *in vitro, J. Embryol. Exp. Morphol.* **34:**310.

Gwatkin, R. B. L., 1966, Amino acid requirements for attachment and outgrowth of the mouse blastocyst *in vitro, J. Cell. Physiol.* **68:**335.

Hedlund, K., Nilsson, O., Reinius, S., and Aman, G., 1972, Attachment reaction of the uterine luminal epithelium at implantation: Light and electron microscopy of the hamster, guinea pig, rabbit and mink, *J. Reprod. Fertil.* **29:**131.

Heyner, S., 1973, Detection of H-2 antigens on the cells of the early mouse embryo, *Transplantation* **16:**675.

Jenkinson, E. J., 1978, The *in vitro* blastocyst outgrowth system as a model for the analysis of peri-implantation development, in: *Development in Mammals*, Vol. 2 (M. Johnson, ed.), p. 157, North-Holland, Amsterdam.

Jenkinson, E. J., and Searle, R. F., 1977, Cell surface changes on the mouse blastocyst at implantation, *Exp. Cell Res.* **106:**386.

Jenkinson, E. J., and Wilson, I. B., 1973, *In vitro* studies on the control of trophoblast outgrowth in the mouse, *J. Embryol. Exp. Morphol.* **30:**21.

Kubo, H., Spindle, A. I., and Pederson, R. A., 1979, Possible involvement of protease in mouse blastocyst implantation *in vitro, Biol. Reprod. Suppl.* **20:**50A.

McLaren, A., and Hensleigh, H. C., 1975, Culture of mammalian embryos over the implantation period, in: *The Early Development of Mammals* (M. Balls and A. E. Wild, eds.), pp. 45–60, Cambridge Univ. Press, London.

Mintz, B., 1964, Formation of genetically mosaic embryos and early development of lethal (t^{12}/t^{12})–normal mosaics, *J. Exp. Zool.* **157**:273.

Owers, N. O., and Blandau, R. J., 1971, Proteolytic activity of the rat and guinea pig blastocyst *in vitro*, in: *Biology of the Blastocyst* (R. J. Blandau, ed.), pp. 207–223, Univ. of Chicago Press, Chicago.

Pedersen, R. A., 1974, Development of lethal yellow (A^y/A^y) mouse embryos *in vitro*, *J. Exp. Zool.* **188**:307.

Pederson, R. A., and Spindle, A. I., 1976, Genetic effects of mammalian development during and after implantation, in *Embryogenesis in Mammals*, Ciba Foundation Symposium 40, pp. 133–154, Elsevier, Amsterdam.

Pedersen, R. A., and Spindle, A. I., 1977, Interference with *in vitro* development of mouse inner cell mass by 24-hr treatment with 5-bromodeoxyuridine, *Differentiation* **9**:43.

Revel, J. P., Darr, G., Griepp, E. B., Johnson, R., and Miller, M. M., 1978, Cell movement and intercellular contact, in: *The Molecular Basis of Cell–Cell Interaction* (R. A. Lerner and D. Bergsma, eds.), pp. 67–81, Liss, New York.

Rowinski, J., Solter, D., and Koprowski, H., 1975, Mouse embryo development *in vitro:* Effects of inhibitors of RNA and protein synthesis on blastocyst and post-blastocyst embryos, *J. Exp. Zool.* **192**:133.

Salomon, D. S., and Sherman, M. I., 1975, Implantation and invasiveness of mouse blastocysts on uterine monolayers, *Exp. Cell Res.* **90**:261.

Schlafke, S., and Enders, A. C., 1975, Cellular basis of interaction between trophoblast and uterus at implantation, *Biol. Reprod.* **12**:41.

Sherman, M. I., 1978, Implantation of mouse blastocysts *in vitro*, in: *Methods in Mammalian Reproduction* (J. C. Daniel, Jr., ed.), pp. 247–257, Academic Press, New York.

Sherman, M. I., and Salomon, D. S., 1975, The relationships between the early mouse embryo and its environment, in: *The Developmental Biology of Reproduction* (C. L. Markert and J. Papaconstantinou, eds.), pp. 277–309, Academic Press, New York.

Sherman, M. I., and Wudl, L. R., 1976, The implanting mouse blastocyst, in: *The Cell Surface in Animal Embryogenesis and Development* (G. Poste and G. L. Nicolson, eds.), pp. 81–125, Elsevier/North-Holland, Amsterdam.

Sherman, M. I., and Wudl, L., 1977, T-complex mutations and their effects, in: *Concepts in Mammalian Embryogenesis* (M. I. Sherman, ed.), pp. 136–234, MIT Press, Cambridge, Mass.

Sherman, M. I., McLaren, A., and Walker, P. M. B., 1972, Mechanism of accumulation of DNA in giant cells of mouse trophoblast, *Nature (London) New Biol.* **238**:175.

Sherman, M. I., Shalgi, R., Rizzino, A., Sellens, M. H., Gay, S., and Gay, R., 1979, Changes in the surface of the mouse blastocyst at implantation, in: *Maternal Recognition of Pregnancy*, Ciba Foundation Symposium 64, pp. 33–47, Elsevier/North-Holland, Amsterdam.

Spindle, A. I., and Pedersen, R. A., 1973, Hatching, attachment, and outgrowth of mouse blastocysts *in vitro:* Fixed nitrogen requirements, *J. Exp. Zool.* **186**:305.

Strickland, S., Reich, E., and Sherman, M. I., 1976, Plasminogen activator in early embryogenesis: Enzyme production by trophoblast and parietal endoderm, *Cell* **9**:231.

Surani, M. A. H., 1977, Response of preimplantation rat blastocysts *in vitro* to extracellular uterine luminal components, serum and hormones, *J. Cell Sci.* **25**:265.

Wiley, L. M., and Pedersen, R. A., 1978, Morphology of mouse egg cylinder development *in vitro:* A light and electron microscopic study, *J. Exp. Zool.* **200**:389.

Wimsatt, W. A., 1975, Some comparative aspects of implantation, *Biol. Reprod.* **12**:1.

Wordinger, R. J., and Brinster, R. L., 1976, Influence of reduced glucose levels on the *in vitro* hatching, attachment, and trophoblast outgrowth of the mouse blastocyst, *Dev. Biol.* **53**:294.

24

Time-Lapse Cinematography of Mouse Embryo Development from Blastocysts to Early Somite Stage

Yu-Chih Hsu

I. In Vitro Development of Mouse Embryos from Blastocyst to Early Somite Stage

Mouse embryos develop in culture from the two-cell egg stage to the early somite stage (Hsu *et al.*, 1974) and are morphologically indistinguishable from embryos developed *in utero* (Hsu, 1971, 1972, 1973; Hsu *et al.*, 1974; Wiley and Pedersen, 1977; Chen and Hsu, 1979; Gonda and Hsu, 1980; Libbus and Hsu, 1980a,b).

The sequence of *in vitro* development was classified by stages according to the *in vivo* criteria of Theiler (1972) and Witschi (1972). By continuous modification and improvement of the culture method, about 50 to 70% of individually cultured blastocysts (stage 6) were able to differentiate to stage 15, if there were frequent medium changes to remove embryonic waste (Hsu, 1979). With the present method of culturing mouse embryos, it takes 7 days in culture to develop blastocysts (stage 6) from 3½ days of gestation to the early somite stage (stage 15), equivalent to 8½ days of gestation. The majority of embryos developed synchronously in culture.

This success in culturing mouse embryos through the stages of early organogenesis with a high efficiency makes it possible, for the first time, to record postimplantation developments and to study morphogenetic movement using time-lapse cinematography.

Mouse blastocysts (stage 6) were cultured in 35-mm plastic culture dishes with 2 ml medium CMRL 1066 plus 10% fetal calf serum for the first 2 days and human placental cord serum thereafter. The method has been described in detail (Hsu, 1979). The culture dishes were enclosed in plexiglass chambers that were flushed constantly with 5% CO_2 in air; the ambient temperature was 38°C. The cultures were observed under a Nikon inverted phase microscope (100×, 60×) and a Wild dissecting microscope (50×, 25×, and

Yu-Chih Hsu • Laboratory of Mammalian Development, Department of Pathobiology, School of Hygiene and Public Health, The Johns Hopkins University, Baltimore, Maryland 21205.

12.5×). Embryo development was filmed with Ektachrome Film EF, 7242, with 1-min intervals timed by a Nikon Cine Autotimer.

Mouse blastocysts underwent contraction (Figure 1A) and relaxation (Figure 1B) several times before shedding the zona pellucida. The contraction persisted for 1 to 5 hr, depending upon the degree of contraction. Blastocysts rotated in the culture medium, most of the time in a counterclockwise direction, while the blastocoele expanded to form the inner cell mass (ICM). Contracted blastocysts also rotated within the zona pellucida before returning to the relaxed state. The movement of cell components, mitchondria, and nucleus was remarkably active within mural trophoblasts. Denuded blastocysts attached to the culture dishes with random location of mural trophoblast cells. Trophoblast cells migrated outward rapidly with many microvilli, blebs, and ruffles (Figure 2A,B). As the blastocoele collapsed, the inner cell mass remained as a cell aggregate.

The inner cell mass became covered by primary endoderm, then rapidly increased its diameter 1.5 times within 12 hr (Figure 3A,B). As the egg cylinders grew, the embryonic region with its proamniotic cavity was distinguishable from the extraembryonic region (Figure 4A,B). There was a constant flow of culture medium toward the embryos, suggesting microvillous movement to accelerate nutrient uptake and waste removal.

As the ectoplacental cones proliferated on the plastic substrate, the egg cylinder rotated and waved in the medium. The whole egg cylinder contracted abruptly but rather rhythmically. At the end of 6 days of cultivation, a notochord formed in the embryonic shield (Figure 5A). As the neural fold formed, the foregut invaginated and five pairs of somites aggregated along the notochord. A strong cardiac contraction (Figure 5B) continued for about 1 day before the whole embryo became atrophic.

II. In Vitro Monozygotic Twin Formation in Mouse Embryos

In a series of experiments in developing mouse embryos *in vitro,* we observed that about one in one hundred cultured blastocysts formed monozygotic twins.

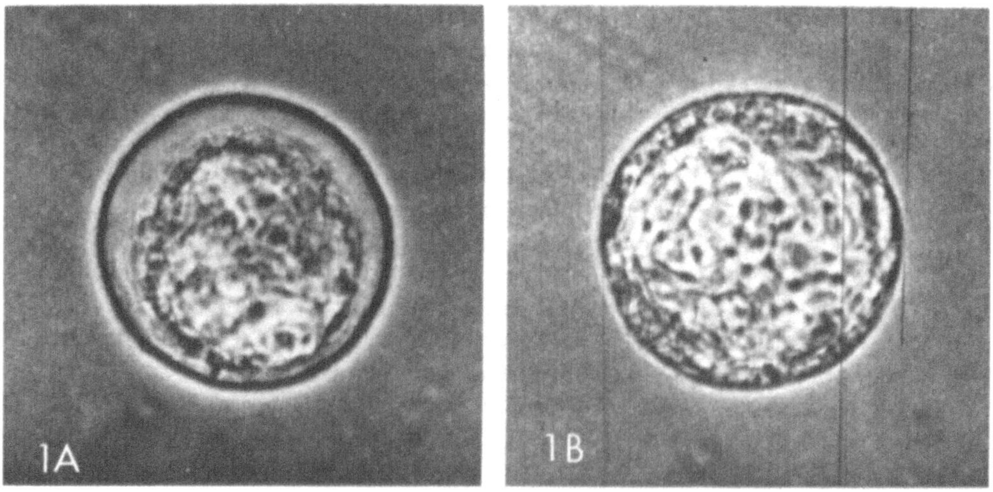

Figure 1. (A) Contraction and (B) relaxation of the blastocyst. Major contractions last 4 to 5 hr, minor contractions up to 1 hr.

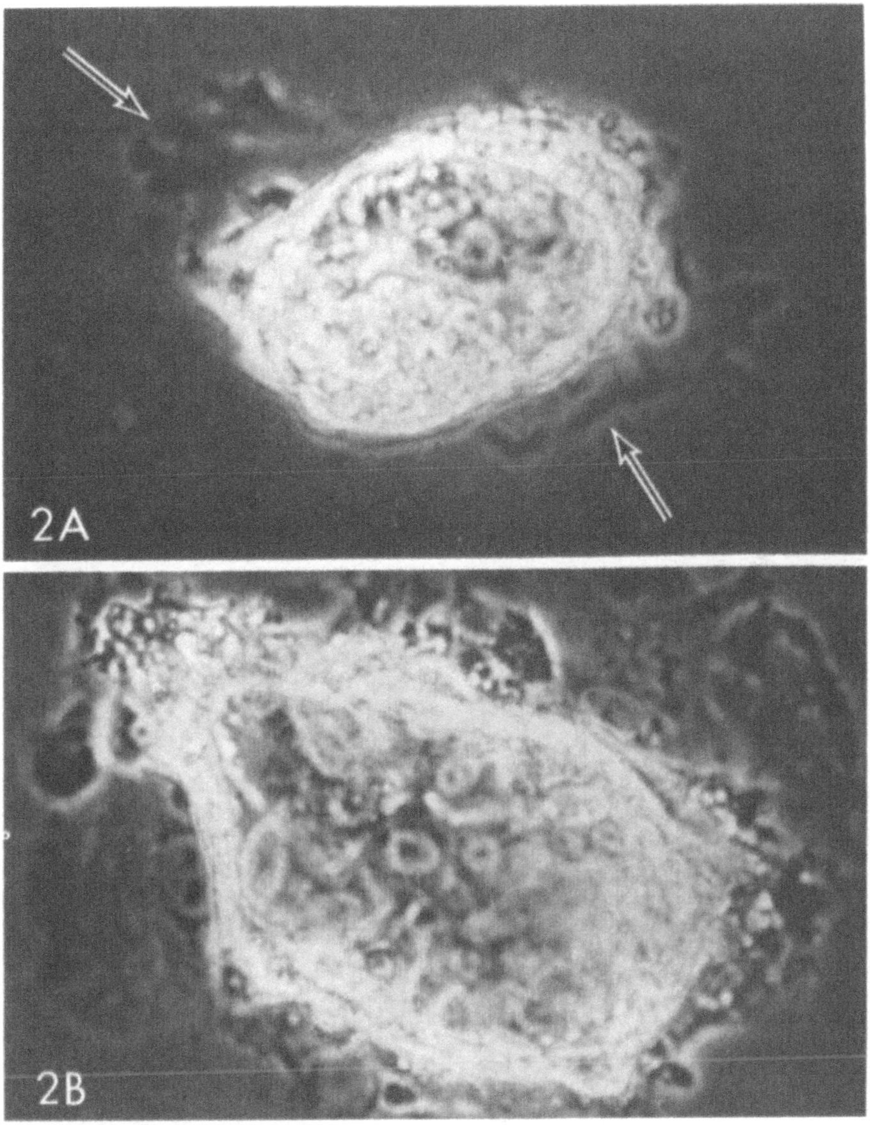

Figure 2. (A,B) The denuded blastocyst contracted soon after attachment to the plastic substrate. The mural trophoblast cells migrate outward with many microvilli, blebs, and ruffles (arrows).

After hatching, blastocysts attached to the surface of the plastic culture dishes. Blastocysts never attached at the polar trophoblasts, which are apposed to the ICM, but usually attached at a random location of the mural trophoblast. Therefore, attached blastocysts were more asymmetric at the point of attachment with regard to the ICM (Figure 6A, left), except for a few that attached at the antipolar mural trophoblast cells (symmetric attachment, Figure 6A, right).

The blastocoelic cavity collapsed due to the radial migration of spreading mural trophoblasts and enlargement of the ICM. At this stage, the ICM was covered with primary endoderm on the blastocoelic cavity side and by polar trophoblast on the medium side (Fig-

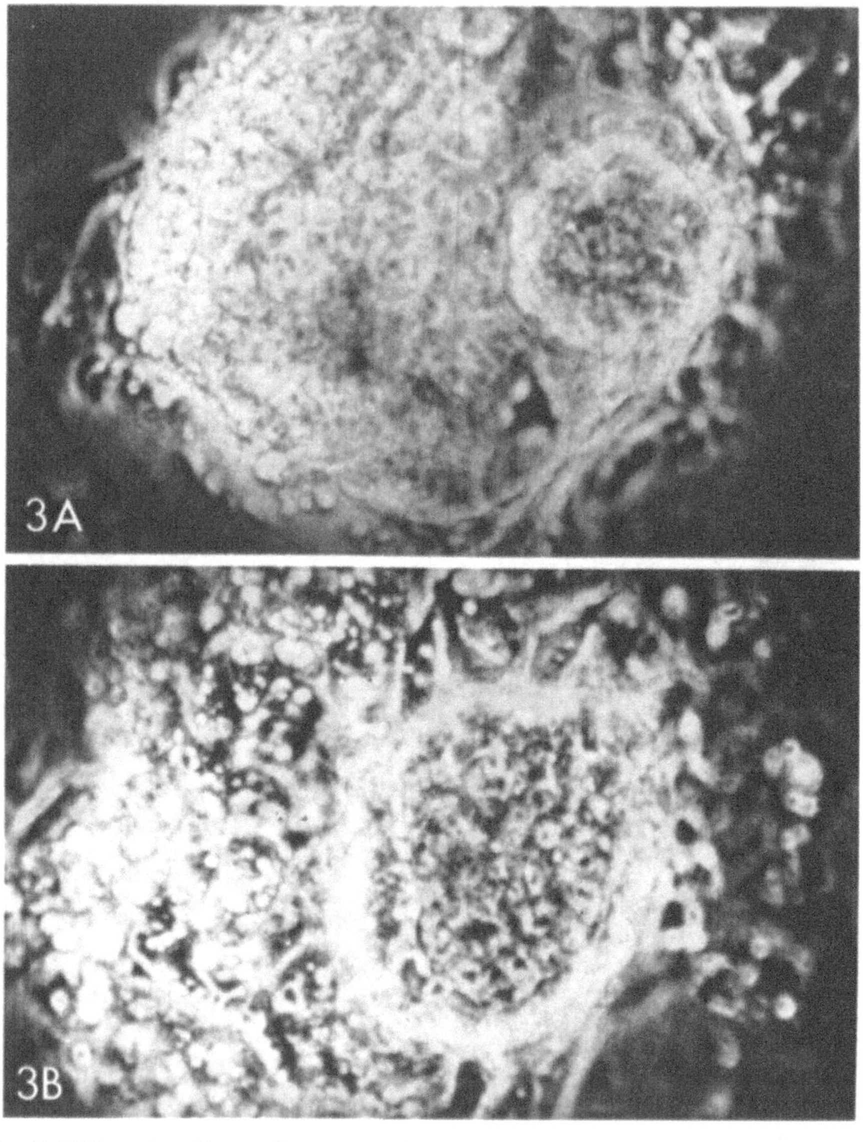

Figure 3. (A,B) Expansion of inner cell mass covered by primary endoderm, one of the most critical stages for mouse embryos to develop *in vitro*.

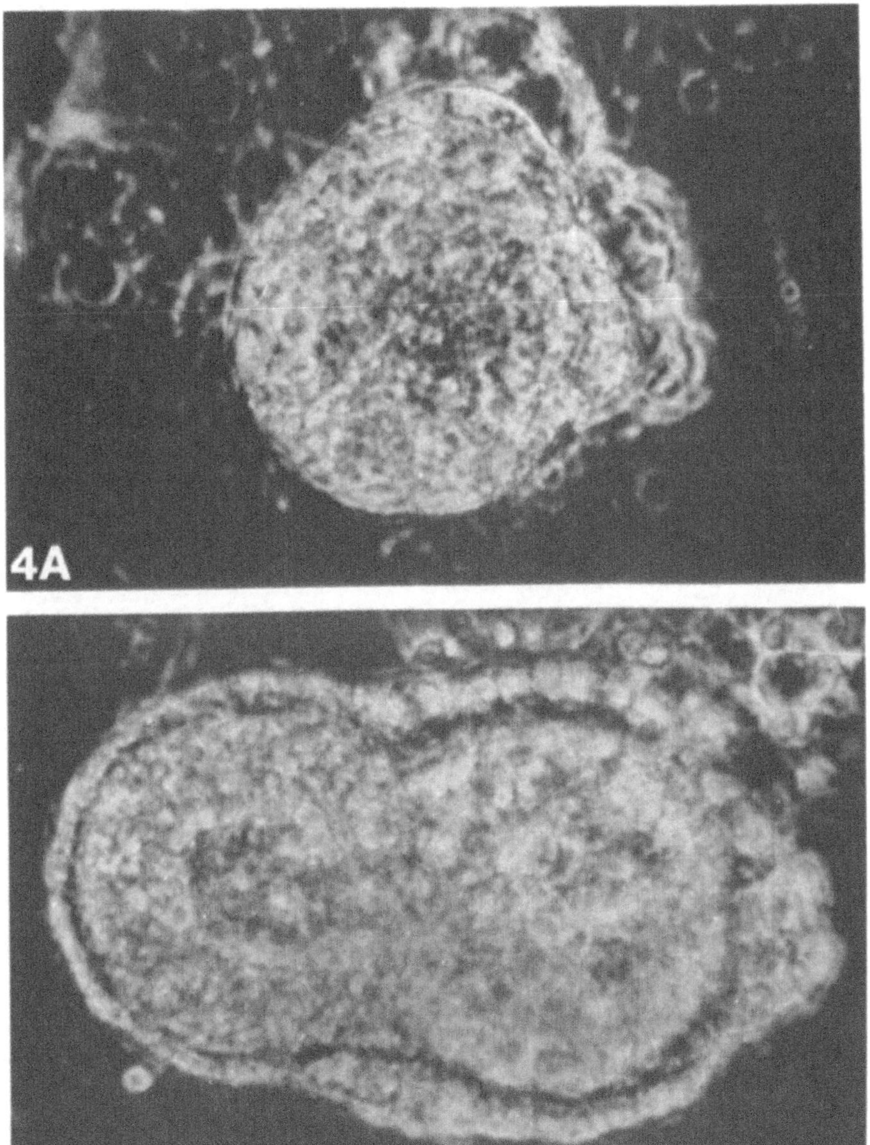

Figure 4. (A,B) Elongation and expansion of the egg cylinder.

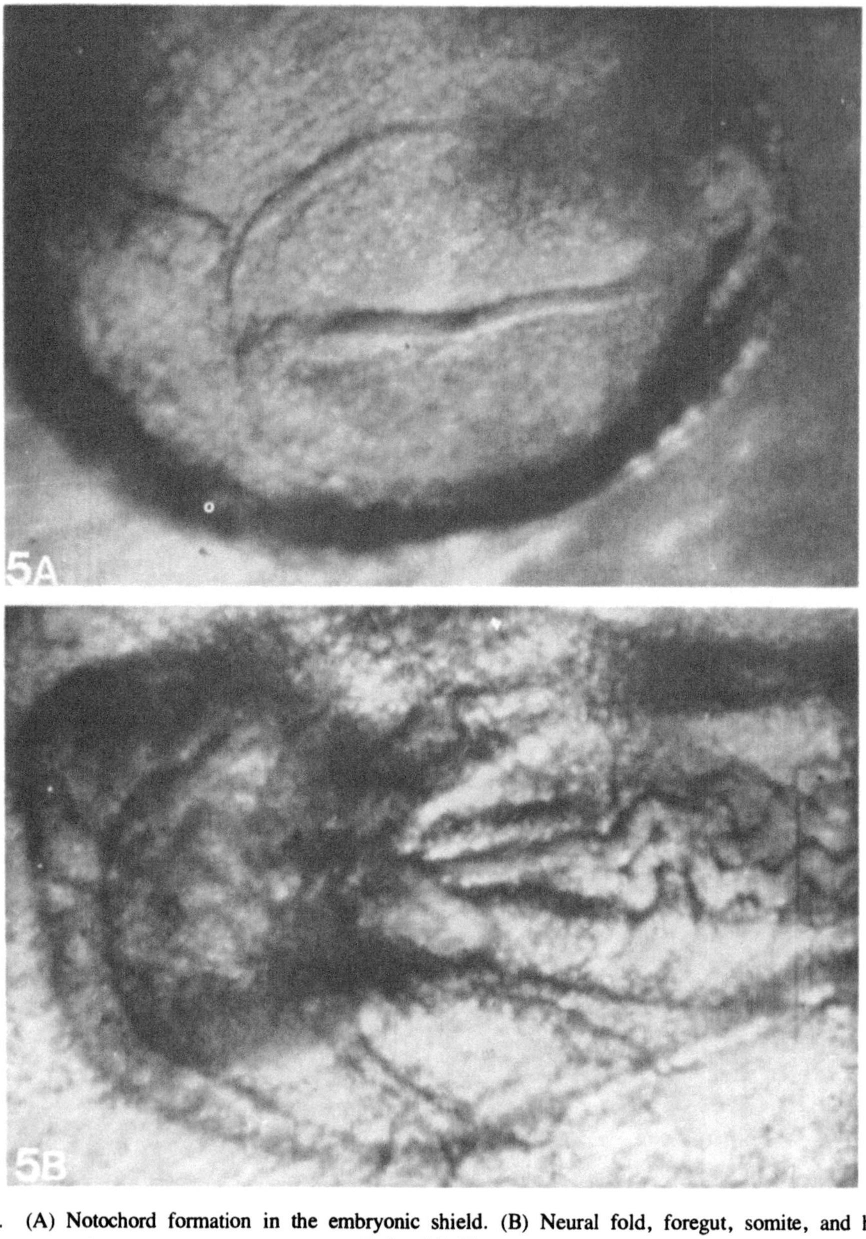

Figure 5. (A) Notochord formation in the embryonic shield. (B) Neural fold, foregut, somite, and heart are distinguishable.

ure 6B, left). In asymmetric attachment, the polar trophoblast cells are positioned to one side of the embryo. With the embryo attached in this manner, the enlarged ICM is physically restricted from further downward expansion by the plastic substrate and by lack of space in the collapsed blastocoelic cavity. The ICM elongated sideways and then upwards, protruding into the culture medium in a break between the polar and the mural trophoblast cells.

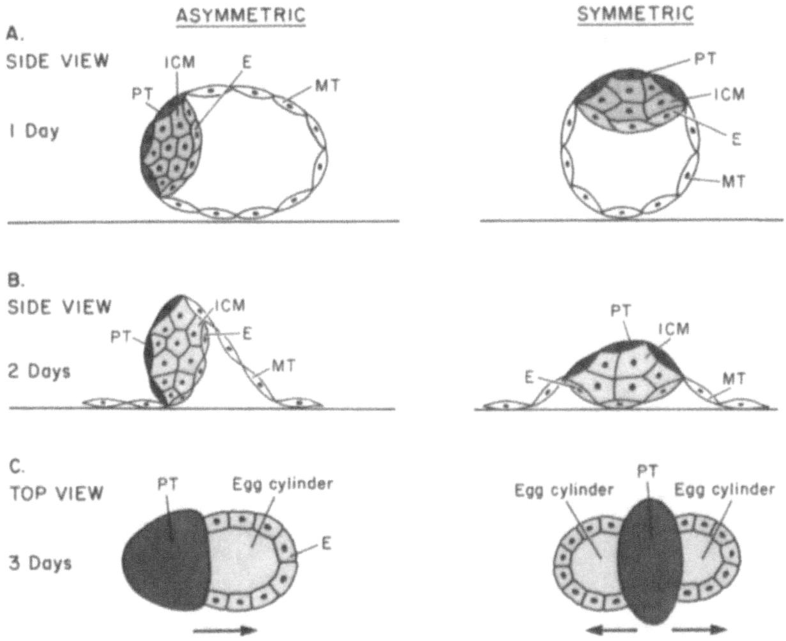

Figure 6. Schematic drawing of blastocyst attachment (Day 1 in culture), radial outgrowth of mural trophoblasts (Day 2 in culture), and subsequent egg cylinder elongation (Day 3 in culture). (A) On Day 1 of culture, a side view of a denuded blastocyst attached to the plastic substrate shows mural trophoblast (MT) cells (left, asymmetric attachment) and antipolar trophoblast cells (right, symmetric attachment). (B) On Day 2 of culture, the mural trophoblast cells migrate outward radially, resulting in collapse of the blastocoelic cavity. (C) On Day 3 of culture, a top view of mouse embryo development and egg cylinder elongation in asymmetrically attached mouse embryos (left) and symmetrically attached embryos (E) (right) is shown. Left: While polar trophoblast (PT) proliferated to the left becoming the ectoplacental cone, the egg cylinder elongated toward the right. Right: Polar trophoblast cells sit on top of the ICM, and the egg cylinders grow bilaterally away from polar trophoblast cells.

The polar trophoblast cells of asymmetrically attached embryos continued to proliferate to form the ectoplacental cone in one direction, and the ICM, covered with primary endoderm, grew away from the polar trophoblast cells, forming the early egg cylinders (Figure 6C, left).

In contrast, the ICM of more symmetrically attached blastocysts separated into two parts due to restrictions placed on its growth by the apically attached polar trophoblast and the plastic substrate. The ICM is forced to migrate bilaterally away from the polar trophoblast cells (Figure 6C, right), presumably through breaks that have occurred between mural and polar trophoblasts. Thus monozygotic twins are formed on the plastic culture dish through the physical separation of totipotent cells into each subdivided ICM.

With the lateral migration of each growing ICM, the ectoplacental cone became positioned basally (in relation to the direction of growth) and proliferated to form the ectoplacental cone. As the egg cylinder elongated, the direction of growth of the embryo changed from horizontal to more upright as in the asymmetrically attached embryos.

Although one embryo of twins is usually larger and its growth stage is slightly more advanced than its counterpart, this is not always the case if the attachment is truly symmetric with regard to the ICM.

However, equal size of the twin embryos at the same developmental stage has also

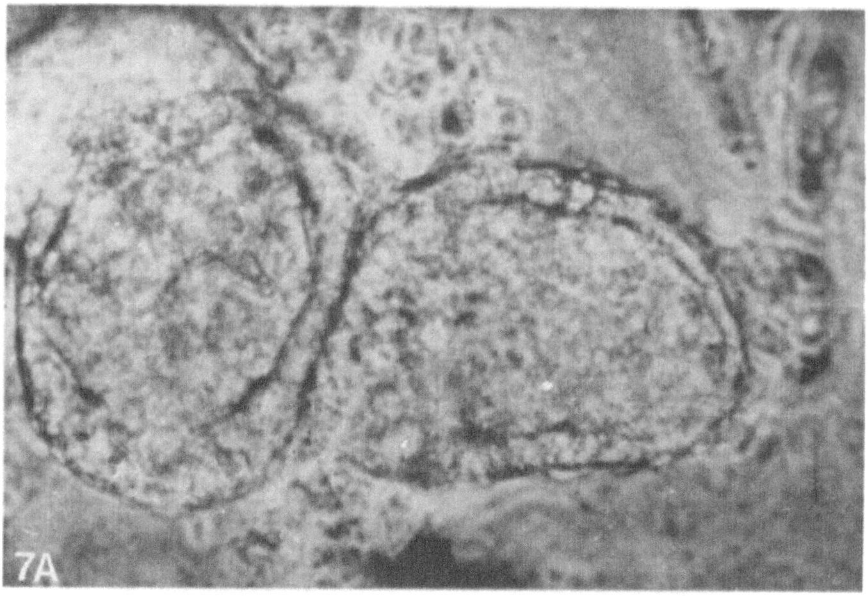

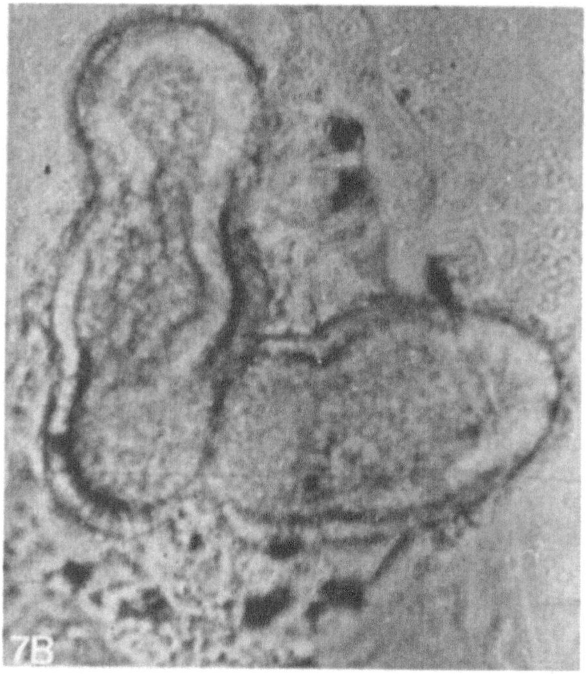

Figure 7. Monozygotic twins developed *in vitro* at the early egg cylinder stage (A) and at the late presomite neurula stage (B).

been observed. In all cases, both embryos of twins developed to the early somite stage, including cardiac contraction.

The development of twinning from the early egg cylinder stage to the late egg cylinder stage is shown in frames from a film (Figure 7A,B). The genesis of monozygotic twins will be described in detail elsewhere (Hsu and Gonda, 1980).

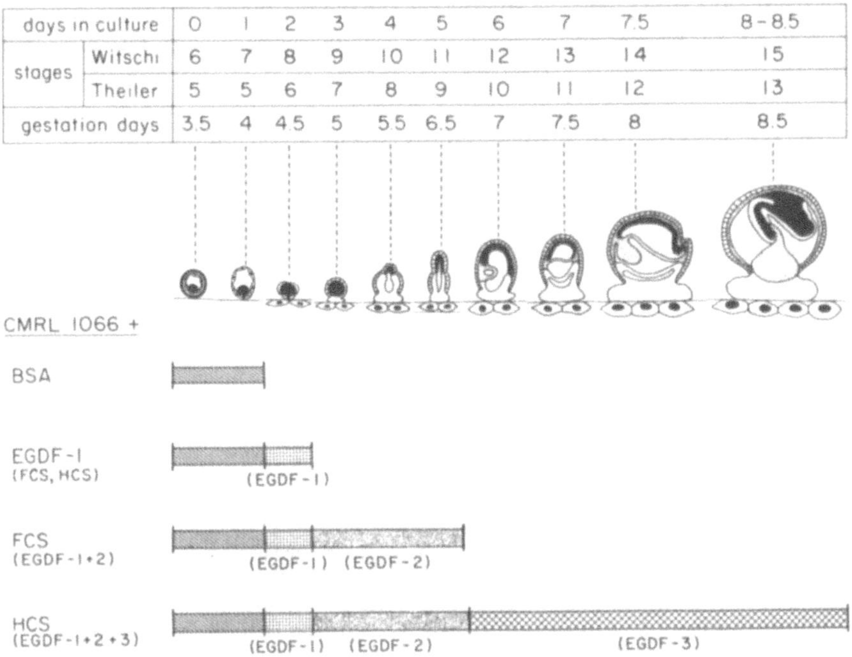

Figure 8. Stages of mouse embryos reached with various embryo growth and differentiation factors (EGDF). Blastocysts [stage 6 of Witschi (1972), stage 5 of Theiler (1972)] in medium CMRL 1066 plus 0.25% BSA developed to stage 7 after 2 days of culture. Trophoblasts did not spread out on culture dishes, and embryos eventually died at stage 7. Blastocysts (stage 6 of Witschi) developed to stage 11 in medium CMRLK 1066 plus 10% FCS after 5 days of culture. Fetal calf serum was separated into two fractions, a larger molecular weight fraction (EGDF-1) and a smaller molecular weight fraction (EGDF-2) by molecular sieving with DIAFLO membranes XM300, XM100, and PM30 (Amicon). EGDF-1 is required for embryos to develop from stage 7 to stage 8, and EGDF-2 promotes embryonic growth from stage 8 to stage 11. Although some degree of embryonic growth continued beyond stage 11 in medium using FCS as the sole source of macromolecules, the structure of the embryos had become increasingly disproportional compared to that developed *in utero*. Neural plates became atrophic and eventually disappeared. Embryos resumed their normal growth to stage 15 *in vitro* if FCS was replaced by HCS beyond stage 11. Therefore, HCS contains a factor (EGDF-3) that is indispensable for mouse embryos to grow from stage 11 to stage 15. Blastocysts (stage 6) were also able to develop to stage 15 in medium containing HCS as the sole source of macromolecules; therefore, HCS also contains EGDF-1 and EGDF-2.

III. Embryo Growth and Differentiation Factors in Embryonic Sera

For mouse embryos to grow *in vitro* continuously from stage 6 to stage 15, which takes 8 days in culture, we have found that fetal calf serum (FCS) is required from stage 6 to stage 11 and human placental cord serum (HCS) from stage 11 to stage 15. Mouse embryos develop well *in vitro* from stage 6 to stage 11 with FCS, but then begin to lose their organization. If FCS is replaced by HCS beyond stage 11, the embryos resume their growth from stage 11 to 15. Apparently then, FCS and HCS contain different factors that are required for the growth and differentiation of mouse embryos. These factors have been designated tentatively as embryo growth and differentiation factors (EGDF) (Figure 8).

Our preliminary results show that FCS (important from stage 6 to 11) can be separated into at least two fractions, one of larger molecular weight (EGDF-1) and one of smaller molecular weight (EGDF-2), using DIAFLO membrane filters (Amicon XM300, XM100, and PM30). EGDF-1 is required for mouse embryos to develop from stage 6 to stage 8 (implantation stage), and is required for embryos to develop *in vitro* from stage 8 to stage 11. EGDF-3 from HCS, required for stages 11 to 15, is not permeable to DIAFLO membrane XM100 (Hsu, 1980).

ADDENDUM. Dr. Hsu has copies of a 16-mm movie available for use upon request.

ACKNOWLEDGMENTS. I thank Dr. A. Cohen and Dr. B. Libbus for instruction in time-lapse cinematography.

References

Chen, L. T., and Hsu, Y., 1979, Hemopoiesis of the cultured whole mouse embryo, *Exp. Hematol.* **7**:231–244.

Gonda, M., and Hsu, Y., 1980, Correlative scanning electron, transmission electron, and light microscopic studies of the *in vitro* development of mouse embryos on a plastic substrate at the implantation stage, *J. Embryol. Exp. Morphol.* **56**:23–39.

Hsu, Y., 1971, Post-blastocyst differentiation *in vitro, Nature (London)* **232**:100–102.

Hsu, Y., 1972, Differentiation *in vitro* of mouse embryos beyond the implantation stage, *Nature (London)* **239**:200–202.

Hsu, Y., 1973, Differentiation *in vitro* of mouse embryos to the stage of early somite, *Dev. Biol.* **33**:403–411.

Hsu, Y., 1979, *In vitro* development of individually cultured whole mouse embryos from blastocyst to early somite stage *Dev. Biol.* **68**:453–461.

Hsu, Y., 1980, Embryo growth and differentiation factors in embryonic sera of mammals, *Dev. Biol.* **76**:465–474.

Hsu, Y., and Gonda, M., 1980, Monozygotic twin formation in mouse embryos *in vitro, Science* **209**:605–606.

Hsu, Y., Baskar, J., Stevens, L., and Rash, J., 1974, Development *in vitro* of mouse embryos from the two-cell egg stage to the early somite stage, *J. Embryol. Exp. Morphol.* **31**:235–245.

Libbus, B., and Hsu, Y., 1980a, Sequential development and tissue organization in whole mouse embryos cultured from blastocyst to early somite stage, *Anat. Rec.* **197**:317–329.

Libbus, B., and Hsu, Y., 1980b, Changes in S-phase associated with differentiation of mouse embryos in culture from blastocyst to early somite stage, *Anat. Embryol.* **159**:235–244.

Theiler, K., 1972, *The House Mouse,* Springer-Verlag, Berlin/New York.

Wiley, L. M., and Pedersen, R. A., 1977, Morphology of mouse egg cylinder development *in vitro:* A light and electron microscope study, *J. Exp. Zool.* **200**:389–402.

Witschi, E., 1972, Characterization of developmental stages, in: *Biology Data Book* (L. Altman and D. S. Dittmer, eds.), 2nd ed., Part II, Vol. 1, pp. 178–180, Fed. Am. Soc. Exp. Biol., Bethesda, Md.

VIII

Short Communications

1 The Effects of Estradiol-17β and Progesterone on the Volume of Uterine Fluid in Ovariectomized Mice

Roger C. Hoversland and H. M. Weitlauf

Numerous reports have appeared dealing with various molecular components of uterine fluid during implantation or delayed implantation (Pinsker *et al.*, 1974; Aitken, 1977; Surani, 1977; Hoversland and Weitlauf, 1978) and the findings have led to the suggestion that components of uterine fluid influence or even regulate metabolism in preimplantation embryos. In those experiments, uterine fluid was usually collected by flushing buffer through the uterus. The resulting fluids were analyzed for the total amount of various components, or the ratio of those components to protein. One difficulty in interpreting such data is that the concentration of the various substances *in situ* cannot be determined without an estimate of the original volume of uterine fluid. To resolve this difficulty, the present experiments were undertaken to determine the volume of uterine fluid in mice under various endocrine regimens. The method used was a modification of the technique described by Kulangara (1972).

Virgin white Swiss mice were bilaterally ovariectomized and allowed to recover for 10 days; they were divided into three groups and treated as follows: Group 1, 0.1 ml sesame seed oil for 2 days followed by 25.0 ng estradiol-17β for 2 days; Group 2, 2.0 mg progesterone for 4 days (i.e., conditions for delayed implantation); Group 3, 2.0 mg progesterone for 2 days followed by 25.0 ng estradiol-17β in combination with 2.0 mg progesterone (i.e., conditions for implantation). All hormones were dissolved in 0.1 ml sesame seed oil and were injected s.c. at noon on each day of treatment. Animals were killed at noon on the third day of treatment, 6 A.M. on the fourth day, noon on the fourth day, or noon on the fifth day, and their uteri were removed.

Three to four microliters of Kreb's Ringer bicarbonate buffer (pH 7.4) containing 1 mg/ml [*Me*-^{14}C]-BSA (1.0–1.2 μCi/mg, New England Nuclear Corp.) was flushed through each uterine horn followed by a 45-μl bolus of air to purge the lumen. Buffer and air were injected into the lumen of the uterine horn near the uterotubal junction with a 50-μl Hamilton syringe and a 25-gauge needle. The effluent (i.e., buffer diluted with uterine fluid) was collected via a cannula (made from a blunt needle) inserted into the cervical end of the uterine horn and held in place by a modified hemostat and bibulous paper cushion. The effluents from both horns of one animal were pooled. The change in concentration of [*Me*-^{14}C]-BSA caused by dilution of the original buffer with uterine fluid was determined by estimating the radioactivity in a known volume before and after its passage through the uterus.

Four hundred microliters of cold PBS (pH 7.4, 1 mg/ml BSA) was flushed through

Roger C. Hoversland and H. M. Weitlauf • Department of Anatomy, School of Medicine, University of Oregon Health Sciences Center, Portland, Oregon 97201.

each horn to recover residual [*Me*-^{14}C]-BSA and was collected in a separate scintillation vial. To determine the amount of [*Me*-^{14}C]-BSA "bound" (i.e., lost due to either specific or nonspecific binding, phagocytosis, or diffusion), uterine horns were dissolved in NCS (Amersham) and the amount of radioactivity in each pair of uterine horns was determined. For purposes of calculation, the volume of Kreb's Ringer bicarbonate buffer containing [*Me*-^{14}C]-BSA flushed through each pair of uterine horns was assumed to be equal to the sum of [*Me*-^{14}C]-BSA in the effluent, the residual, and that "bound" to the uterine horns divided by the original concentration of [*Me*-^{14}C]-BSA.

The volume of uterine fluid in each pair of uterine horns was calculated using the formula $(C_iV_i - C_eV_i)/C_e = V_u$, where C_i is the concentration of labeled BSA in the buffer flushed into the uterine horns, C_e is the concentration of labeled BSA in the effluent from the uterus, V_i is the volume of buffer flushed into the uterus, and V_u is the volume of uterine fluid. [*Me*-^{14}C]-BSA "bound" to the uterus was substracted from the quantity $(C_iV_i - C_eV_i)$.

The experiment was repeated several times with 5 to 14 animals at each point. Data were ln-transformed and analyzed for statistical differences by a two-way analysis of variance followed by a Duncan's multiple range test.

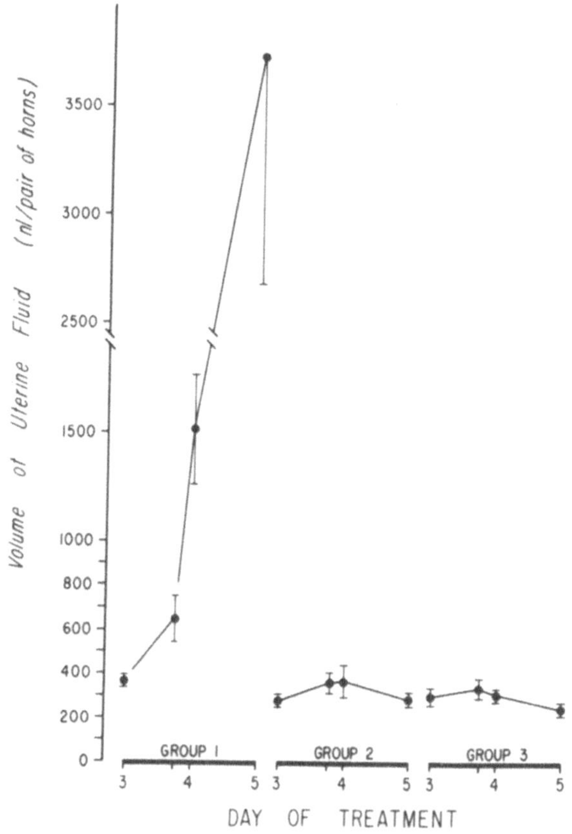

Figure 1. The effect of estradiol-17β and progesterone on the volume of uterine fluid in ovariectomized mice. Animals were treated as follows: Group 1, sesame seed oil for 2 days followed by estradiol-17β for 2 days; Group 2, progesterone alone for 4 days (i.e., conditions for delayed implantation); Group 3, progesterone in combination with estradiol-17β (i.e., conditions for implantation). The mean and S.E.M. of 5 to 14 animals are given at each point. The increases in volume in Group 1 are significant ($p < 0.05$).

Volume of the uterine fluid increased following treatment with estradiol-17β alone (i.e., from 360 nl to 3735 nl within 48 hr, Figure 1, Group 1). The increase in volume was statistically significant ($p < 0.05$) within 18 hr. By contrast, the volume of uterine fluid did not change ($p > 0.05$) following treatment with either progesterone alone (i.e., mean volumes were between 277 and 369 nl, Group 2) or progesterone in combination with estradiol-17β (i.e., mean volumes were between 246 and 335 nl, Group 3). Data collected on the fifth day of treatment are shown in Figure 1 but were not included in the statistical analysis due to heteroscedasticity that could not be corrected by ln transformation.

The results demonstrate that the volume of uterine fluid in ovariectomized mice treated with progesterone alone or progesterone in combination with estradiol-17β does not differ. Therefore, it appears that differences in the total amounts of various substances found in the uterine fluid of either "delayed-implanting" or "implanting" mice do actually reflect differences in concentration *in situ*. By contrast, the increase in volume occurring after treatment with estradiol-17β alone demonstrates that significant changes in the volume of uterine fluid do occur under some conditions and could lead to erroneous conclusions.

ACKNOWLEDGMENT. This work represents a portion of the work submitted by R.C.H. to the Graduate Council, University of Oregon Health Sciences Center, as partial fulfillment of the requirements for the Ph.D. degree and was supported by NICHD Grant HD 07133-01, NICHD Grant HD 00020, and PHS Grant HD 08496.

References

Aitken, R. J., 1977, Embryonic diapause, in: *Development in Mammals* (M. H. Johnson, ed.), Vol. 1, pp. 307–359, North-Holland, Amsterdam.

Hoversland, R. C., and Weitlauf, H. M., 1978, The effect of estrogen and progesterone on the level of amidase activity in fluid flushed from the uteri of ovariectomized mice, *Biol. Reprod.* **19**:908–912.

Kulangara, A. C., 1972, Volume and protein concentration of rabbit uterine fluid, *J. Reprod. Fertil.* **28**:419–425.

Pinsker, M. C., Sacco, A. G., and Mintz, B., 1974, Implantation-associated proteinase in mouse uterine fluid, *Dev. Biol.* **38**:285–290.

Surani, M. A. H., 1977, Cellular and molecular approaches to blastocyst uterine interactions at implantation, in: *Development in Mammals* (M. H. Johnson, ed.), Vol. 1, pp. 245–305, North-Holland, Amsterdam.

2 Presence of hCG-like Material in the Preimplantation Rabbit Blastocyst

Brij B. Saxena and Ricardo H. Asch

I. Introduction

That the blastocyst and the uterus exchange information prior to implantation is suggested by the greater increase in secretion of progesterone in pregnant than in pseudopregnant rabbits on Days 4 to 6 and by the secondary rise of CG–LH-like material in the serum of pregnant rabbits on Days 3 and 4 postcoitum (Singh and Adams, 1978; Varma *et al.*, 1979). The maintenance of the corpus luteum of pregnancy in the presence of blastocysts in the oviduct and after transplantation of the blastocyst under the kidney capsule in the rabbit suggests that the blastocyst may be the source of luteotropic stimuli prior to implantation (Zielmaker and Verhamme, 1978). Dickmann and Dey (1974) suggested that steroidogenesis in the rabbit blastocyst may be activated by a gonadotropic material of blastocyst origin.

The detection of gonadotropic material on the surface of the rabbit morula and blastocyst has been documented (Asch *et al.*, 1978; Varma *et al.*, 1979). The presence of a luteotropic substance in preimplantation rabbit blastocyst fluid has also been demonstrated by radioimmunoassay (RIA), by radioreceptor assay (RRA), and by the ability of the blastocyst fluid to stimulate morphological luteinization and progesterone synthesis in porcine and simian granulosa cell cultures (Haour and Saxena, 1974; Fujimoto *et al.*, 1975; Channing *et al.*, 1975; Channing *et al.*, 1978).

In view of reports in the literature that are in conflict with the evidence for the presence of a luteotropic material in the preimplantation rabbit blastocyst (Sundaram *et al.*, 1975; Holt *et al.*, 1976), further investigations have been undertaken independently in our laboratories.

II. Partial Purification and Characterization of the hCG-Like Material from Rabbit Blastocyst Fluid

The hCG-like material in preimplantation rabbit blastocyst fluid is not dialyzable and is susceptible to proteolytic degradation, suggesting that the luteotropic activity is associated with a high-molecular-weight protein moiety (Haour and Saxena, 1974; Channing *et al.*, 1978). Blastocysts have been collected by laparotomy from rabbits on Day 6 after mating as described earlier (Varma *et al.*, 1979). An average of 16.6 ± 3 μl fluid (mean $\pm$ S.E.M.) was recovered from each blastocyst. Four batches of 2.5, 3.2, 4.5, and

Brij B. Saxena • Departments of Obstetrics and Gynecology and of Medicine, Cornell University Medical College, New York, New York 10021. *Ricardo H. Asch* • Department of Obstetrics and Gynecology, University of Texas Health Science Center, San Antonio, Texas 78284.

14.4 ml blastocyst fluid, representing 155, 193, 271, and 843 blastocysts, respectively, have been subjected to chromatography on concanavalin A–Sepharose and gel filtration on Sephadex G-100 (Saxena, 1979). The fraction containing hCG-like activity has been assayed by RRA and RIA, as well as *in vitro* bioassays measuring progesterone synthesis in monkey granulosa cells and testosterone production by rat Leydig cell homogenate (Catt *et al.*, 1972; Haour and Saxena, 1974; Channing *et al.*, 1978; Asch *et al.*, 1979a).

Serial dilutions of the rabbit blastocyst fluid yield dose–response curves similar to the hCG standard in both RIAs and RRAs (Haour and Saxena, 1974; Asch *et al.*, 1979a).

In the gel filtration studies, a radioimmunoassay for hCG, using antiserum against hormone-specific β subunit, showed an immunoreactive material from the rabbit blastocyst fluid eluting in the region corresponding to native hCG; no activity was detected in extracts from unfertilized ova or nonpregnant uteri (Asch *et al.*, 1979a). The K_{av} of hCG-like material eluted from the Sephadex G-100 column is similar to that of hCG. The hCG-like material of the blastocyst fluid is bound by concanavalin A and can be eluted with 0.1 M α-methylmannoside, indicating its glycoprotein nature (Asch *et al.*, 1979; Saxena, 1979).

Each milliliter of blastocyst fluid contained 1234 μg protein and 87 ng of hCG-like material, as determined by RRA. In the absence of estimates of the secretion and clearance rates, estimates of the hCG-like material produced prior to or at the time of implantation are at present only tentative. Fractionation by chromatography on concanavalin A–*Sepharose* and gel filtration on Sephadex G-100 and G-10 increased the specific activity in the blastocyst fluid from 0.7 ng hCG-like material/μg protein to 2 ng hCG-like material/μg protein, representing a 28-fold purification and a 60% recovery. The hCG-like activity of the purified blastocyst luteotropin in RRA and RIA as well as in rat Leydig cell testosterone production and monkey granulosa cell progesterone assay, respectively, is 2.0, 0.78, 1.4, and 2.5 ng hCG-like material/μg protein (Saxena, 1979). Hence, the material needs at least 500-fold purification to achieve homogeneity. With a yield of 26 μg protein from 4.5 ml blastocyst fluid from approximately 280 blastocysts, purification is going to be a difficult feat.

These results indicate that a protein moiety associated with hCG-like activity can be partially purified from rabbit blastocyst fluid. A preliminary amino acid analysis of the purified material indicates a preponderance of Asp and Gln, as in hCG, but approximately 50% lower quantities of Pro, ½Cys, Lys, and Arg than in hCG (Saxena, 1979). Carbohydrate analysis of the purified blastocyst luteotropin reveals the presence of monosaccharide units similar to hCG, except significantly lower quantities of sialic acid; this may enhance the clearance rate of the blastocyst luteotropic material and thus pose problems in its detection (Saxena, 1979).

III. Discussion

The facts that the human blastocyst can implant at sites different than endometrium, for example in ectopic pregnancies, and can produce hCG provide evidence that it is the blastocyst that carries the message and is equipped to synthesize hCG. If the hCG-like substance is produced in the rabbit blastocyst, its presence in the blood suggests an active transport of material through the uterine wall before implantation and prior to the establishment of vascular connections. It is interesting to note that exogenous hCG introduced in rabbit and human uteri appears in the peripheral circulation within 30 min and 6 hr, respectively (Saxena *et al.*, 1977). The presence of transitory hCG-like material in women bearing an IUD, though a controversial issue at present, also raises the interesting question of the activity of the blastocyst and the occurrence of implantation in the presence of an IUD (Saxena and Landesman, 1978).

A possible role of hCG-like material as a barrier to immunological rejection of the blastocyst prior to implantation has been implicated (Beer and Billingham, 1978). Following the establishment of the corpora lutea, a luteotropic substance from the blastocyst may bind to the receptor in the corpus luteum to sustain the secretion of progesterone. The presence of hCG-like material in sperm has been implicated as a precursor of the morula and blastocyst luteotropin, which may also stimulate local steroidogenesis in the blastocyst (Asch *et al.*, 1979b). The exact chemical nature, source, and mode of action of hCG-like material of the preimplantation blastocyst, however, need to be further investigated. Knowledge in this area is of great significance for the development of contraceptive vaccines (Talwar, 1978) in particular, and in the search for new avenues of fertility regulation in general.

References

Asch, R. H., Fernandez, E. O., Magnasco, L. A., and Pauerstein, C. J., 1978, Demonstration of a chorionic gonadotropin-like substance in rabbit morulae, *Fertil. Steril.* **29:**123.

Asch, R. H., Fernandez, E. O., Siler-Khodr, T. M., and Pauerstein, C. J., 1979a, Evidence for a human chorionic gonadotropin-like material in the rabbit blastocyst, *Fertil. Steril.* **32:**697.

Asch, R. H., Fernandez, E. O., Siler-Khodr, T. M., and Pauerstein, C. J., 1979b, Presence of an hCG-like substance in human sperm, *Am. J. Obstet. Gynecol.* **135:**1041.

Beer, A. E., and Billingham, R. E., 1978, Immunoregulatory aspects of pregnancy, *Fed. Proc. Fed. Am. Soc. Exp. Biol.* **37:**2374.

Catt, K. J., Dufau, M. L., and Tsuruhara, T., 1972, Radioligand receptor assay of LH and chorionic gonadotropin, *J. Clin. Endocrinol. Metab.* **34:**123.

Channing, C. P., Stone, S. L., Sakai, C. N., Haour, F., and Saxena, B. B., 1978, A stimulatory effect of the fluid from preimplantation rabbit blastocysts upon luteinization of monkey granulosa cell cultures, *J. Reprod. Fertil.* **54:**477.

Dickmann, Z., and Dey, S. K., 1974, Steroidogenesis in the preimplantation rat embryo and its possible influence on the morula–blastocyst transformation and implantation, *J. Reprod. Fertil.* **37:**91.

Fujimoto, S., Euker, J. S., Riegle, G. D., and Dukelow, W. R., 1975, On a substance cross-reacting with luteinizing hormone in the preimplantation blastocyst fluid of the rabbit, *Proc. Jpn. Acad. Sci.* **51:**123.

Haour, F., and Saxena, B. B., 1974, Detection of a gonadotropin in rabbit blastocyst before implantation, *Science* **185:**444.

Holt, J. A., Heise, W. F., Wilson, S. M., and Keyes, P. L., 1976, Lack of gonadotropin activity in the rabbit blastocyst prior to implantation, *Endocrinology* **98:**904.

Saxena, B. B., 1979, Current studies of a gonadotropin-like substance in preimplanted rabbit blastocyst, in: *Recent Advances in Reproduction and Regulation of Fertility* (G. P. Talwar, ed., pp. 319–332, Elsevier/North-Holland, Amsterdam.

Saxena, B. B., and Landesman, R., 1978, Does implantation occur in the presence of an IUD?, *Res. Reprod.* **10:**(3):1.

Saxena, B. B., Kaali, S., and Landesman, R., 1977, The transport of chorionic gonadotropin through the reproductive tract, *Eur. J. Obstet. Gynecol.* **7:**1.

Singh, M. M., and Adams, C. E., 1978, Luteotropic effect of the rabbit blastocyst, *J. Reprod. Fertil.* **53:**331.

Sundaram, K., Connel, K. G., and Passatino, T., 1975, Implication of absence of hCG-like gonadotropin in the blastocyst for control of corpus luteum function in pregnant rabbit, *Nature (London)* **256:**739.

Talwar, G. P., 1978, Anti-hCG immunization, *Contraception* **18:**19.

Varma, S. K., Dawood, M. Y., Haour, F., Channing, C. P., and Saxena, B. B., 1979, Gonadotropin-like substance in the preimplanted rabbit blastocyst, *Fertil. Steril.* **31:**68.

Zielmaker, G. H., and Verhamme, C. M. P. M., 1978, Luteotropic activity of ectopically developing rat blastocysts, *Acta Endocrinol. (Copenhagen)* **88:**589.

3 Electron Microscopic Study of Rat ↔ Mouse Chimeric Blastocysts Produced by Embryo Aggregation

S. Tachi and C. Tachi

Chimeric embryos artificially produced by conjoining whole or partial embryos from two different species serve as a unique tool for analyzing the mechanisms underlying the mode of cellular interactions during embryogenesis and ovum implantation.

In mammals, chimeras between rats and mice have been produced by aggregation (Mulnard, 1973; Stern, 1973; Zeilmaker, 1973) and by microinjection (Gardner and Johnson, 1973). However, as no electron microscopic studies of such interspecific chimeric embryos have been published, we summarize herein the results of our investigation.

Eight- to twelve-cell-stage embryos were collected from oviducts of rats (Wistar–Imamichi strain) and mice (ICR strain). The embryos were placed in phosphate-buffered saline containing Pronase (Seikagaku-Kogyo & Co. Ltd., Tokyo, Japan) at a concentration of 0.5% to remove the zona pellucida (Mintz, 1962).

Aggregation of mouse and rat embryos was carried out in a small drop of standard egg culture medium supplemented with fetal calf serum (1:1) under paraffin oil (Merck & Co. Ltd., West Germany). The aggregated embryos were cultured for 20½–30 hr at 37°C under an atmosphere of 5% CO_2 in air. When they reached the blastocyst stage, they were fixed with 3% glutaraldehyde solution in cacodylate buffer (pH 7.2), postfixed with OsO_4, and embedded in Araldite resin. After staining with uranyl acetate and lead citrate, thin sections were observed in a Hitachi Model 500 electron microscope.

We have so far made 114 combinations and obtained 26 chimeric embryos that successfully developed into blastocysts. When they were examined electron microscopically, the cells of rat and mouse origin could easily be identified by the appearance of cytoplasmic inclusions. While the cytoplasm of rat-derived cells contain "plaques," those of mouse-derived cells have fibrillar and crystalloid structures (Figure 1); the species-specific cytoplasmic inclusions of rat and mouse embryonic cells were first described by Enders and Schlafke (1965).

Using the cytoplasmic inclusions as an unambiguous marker to identify the origin of each cell in the composite embryos, it was possible to analyze the pattern of differentiation of blastomeres in the chimeric embryos (Figure 2). So far six rat ↔ mouse chimeric blastocysts have been examined in serially cut thin sections.

Whereas both rat- and mouse-derived cells differentiated equally well into either inner cell mass cells or trophoblast cells, there was a slight but noticeable tendency for more rat-derived cells to differentiate into trophoblast cells than mouse-derived cells. Cells of the

S. Tachi • Department of Anatomy, Institute of Basic Medical Sciences, University of Tsukuba, Niihari-gun, Ibaraki-ken, Japan. *C. Tachi* • Zoological Institute, Faculty of Science, University of Tokyo, Tokyo, Japan.

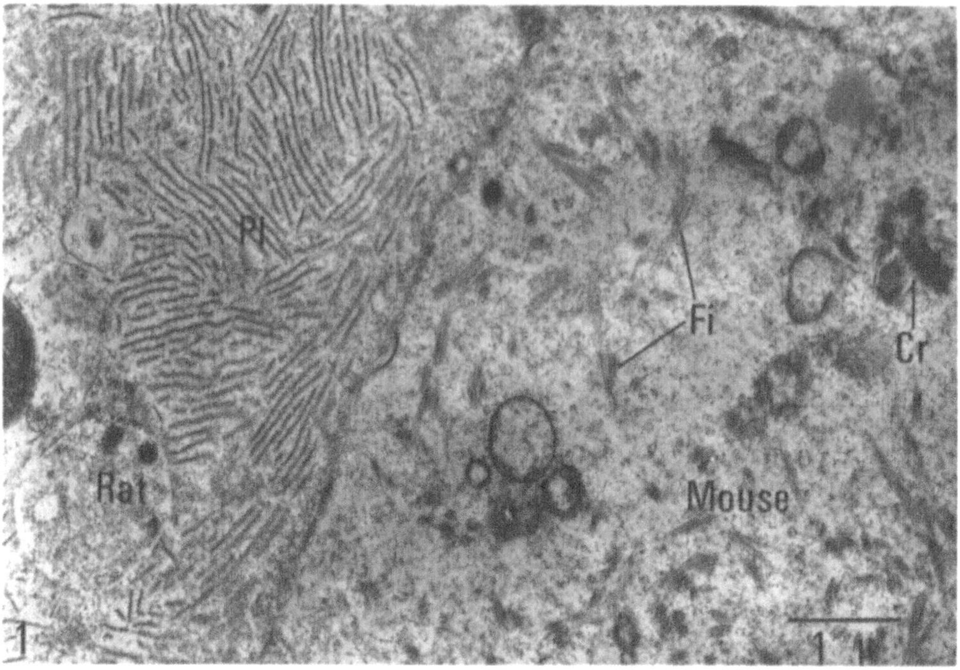

Figure 1. Rat- and mouse-derived cells are seen in close contact in the inner cell mass of the composite embryo. The cytoplasm of the rat-derived cell is filled with "plaques" (Pl), whereas the mouse-derived cell contains fibrillar (Fi) and crystalloid (Cr) structures.

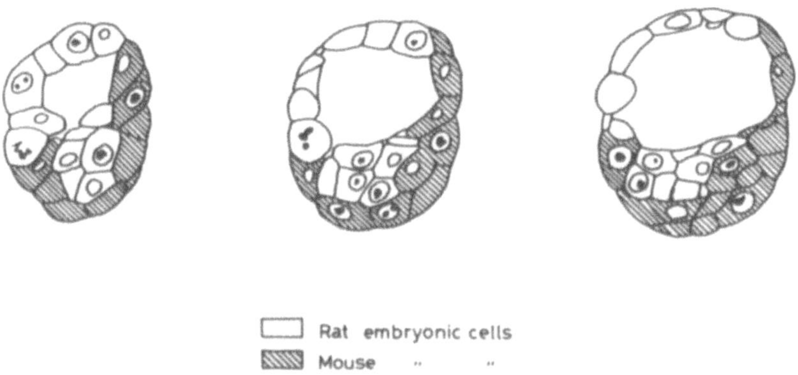

Figure 2. Tracings of electron micrographs showing the distribution of rat- and mouse-derived cells at three different planes along a longitudinal axis of the chimeric blastocyst.

same species remained as a group, and no random mixing of blastomeres took place in the chimeric blastocysts.

ACKNOWLEDGMENTS. This work was supported in part by grants from the Ministry of Education of Japan (267002-1977 and 210709-1977) and from the Ford Foundation, New York.

References

Enders, A. C., and Schlafke, S. J., 1965, The fine structure of the blastocyst: Some comparative studies, in: *Preimplantation Stages of Pregnancy* (G. E. W. Wolstenholme and M. O'Conner, eds.), pp. 29–59, Churchill, London.

Gardner, R. L., and Johnson, M. H., 1973, Investigation of early mammalian development using interspecific chimaeras between rat and mouse, *Nature (London) New Biol.* **246:**86.

Mintz, B., 1962, Experimental study of the developing mammalian egg: Removal of the zona pellucida, *Science* **138:**594.

Mulnard, J. G., 1973, Formation de blastocystes chimériques par fusion d'embryons de rat et de souris au stage VIII, *C.R. Acad. Sci.* **276:**379.

Stern, M. S., 1973, Chimaeras obtained by aggregation of mouse eggs with rat eggs, *Nature (London)* **243:**472.

Zeilmaker, G., 1973, Fusion of rat and mouse morulae and formation of chimaeric blastocysts, *Nature (London)* **242:**115.

4 Characteristics of an Endogenous Inhibitor of Progesterone Binding in Rat Trophoblast

Thomas F. Ogle

I. Introduction

This study investigates the existence of an endogenous inhibitory substance (I) in the cytosol of rat trophoblast that acts to decrease the affinity of progesterone (P) for the progesterone receptor (PR). The kinetic behavior of I at several reproductive stages and its separation from the cytosol PR are reported.

II. Materials and Methods

Trophoblast tissue from Long–Evans rats was prepared on Days 9, 12, 14, and 18 of pregnancy. Appearance of vaginal sperm marked Day 1. Histologic examination showed the preparation to contain principally spongiotrophoblast and very little labyrinth tissue. Uteri from nonpregnant rats were also prepared in some experiments. Tissues were homogenized, incubated briefly with dextran-coated charcoal to remove endogenous steroids, and then centrifuged for 50 min at 165,000*g* (4°C). The supernatant (cytosol) was incubated with various concentrations of [^{3}H]progesterone with and without a 140-fold excess of unlabeled progesterone in 10 mM Tris, 30% glycerol (v/v), and 1 mM dithiothreitol at pH 7.8 (4°C). After a 20 to 24-hr equilibration period at 4°C, free [^{3}H]progesterone was separated from bound by dextran-coated charcoal and counted. Data were evaluated by analysis of variance followed by the Student–Newman–Keuls multiple range test.

III. Results and Discussion

A. Presence of an Endogenous Inhibitor in Trophoblast Cytosol

Dilution experiments were performed to determine the effect of cytosol concentration on the kinetics of P–PR binding at several reproductive stages. Figure 1 depicts representative findings. It is clear that the negative slope of each line increased with each decrement in [cytosol] without a corresponding change in B_{max}. This pattern was observed in nonpreg-

Thomas F. Ogle • Department of Physiology, School of Medicine, Medical College of Georgia, Augusta, Georgia 30912.

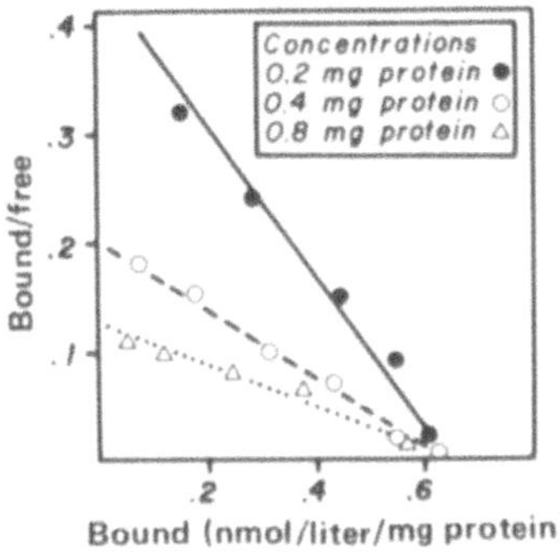

Figure 1. Scatchard analysis of the effects of cytosol concentration on the kinetics of P–PR binding. A representative plot of Day 12 cytosol at three concentrations is presented.

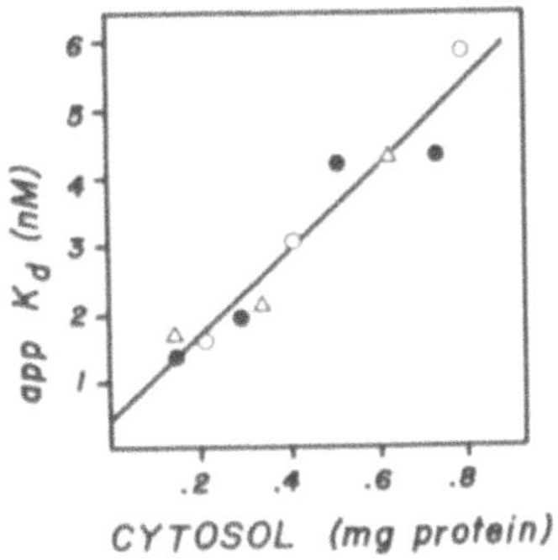

Figure 2. Effects of cytosol concentration on app K_d. The concentration of cytosol protein per tube is plotted with the corresponding app K_d. Each symbol represents a different Day 12 cytosol preparation.

nancy and on Days 9, 12, and 14 of pregnancy. These results suggest that an endogenous inhibitor may be present in trophoblast cytosol that interacts with P and PR in such a fashion as to decrease the slope of a Scatchard plot (-1/app K_d). These effects can be more readily examined by replotting app K_d vs. the corresponding [cytosol]. Such a secondary plot exhibited a linear relationship similar to Figure 2 for all reproductive stages. The slope (Δ app K_d/mg protein) provides a measure of I potency. The intercept estimates the theoretical limit for K_d, i.e., [I] = 0. Similarly, a secondary plot of B_{max} vs. [cytosol] reveals the influence of cytosol concentration on the availability of binding sites. The intercept estimates B_{max} when [I] is zero. Figure 3 summarizes these findings. The upper panel shows that the potency of I did not change during nonpregnancy and early pregnancy but declined by Day 14. The theoretical limit for K_d was equal at all reproductive stages until Day 18 (middle panel). Thus, the form of PR probably remained constant through Day 14 and the alteration in I potency may reflect changes in inhibitor dynamics rather than in PR per se. The number of binding sites also remained constant until Day 14 when it declined (lower panel). By Day 18 all evidence of I was absent and the K_d was greatly enhanced whereas B_{max} was much reduced. This may reflect an alteration in the form of PR or perhaps the presence of another binder.

B. Receptor and Inhibitor Fractions of Trophoblast Cytosol

Cytosol from Day 12 trophoblast was fractionated by differential precipitation with a saturated solution of $(NH_4)_2SO_4$. Each fraction was examined for PR binding and inhibitory activity. Only the 0–50% fraction exhibited binding activity. Mixing this PR-rich fraction with the 50–70% fraction altered the app K_d without affecting the total number of binding sites, as exemplified by the results of the following experiment (values were determined by Scatchard and double reciprocal plot analysis): 0–40% alone, app $K_d = 2.57$ nM, $B_{max} = 1.13$ nmol/liter/mg; 0–40% + 40–50%, app $K_d = 1.87$ nM, $B_{max} = 1.66$ nmol/liter/mg; 0–40% + 50–70%, app $K_d = 6.11$ nM, $B_{max} = 1.33$ nmol/liter/mg.

The kinetic characteristics of other cytosol preparations are summarized below. The 0–40% $(NH_4)_2SO_4$ fraction contained essentially all of the PR and exhibited the same app K_d as whole cytosol (app $K_d = 2.2 \pm 0.1$ and 3.5 ± 0.3 nM, respectively, $n = 10$), but specific binding activity was enhanced about 25% ($B_{max} = 1.1 \pm 0.1$ and 0.8 ± 0.1 nmol/li-

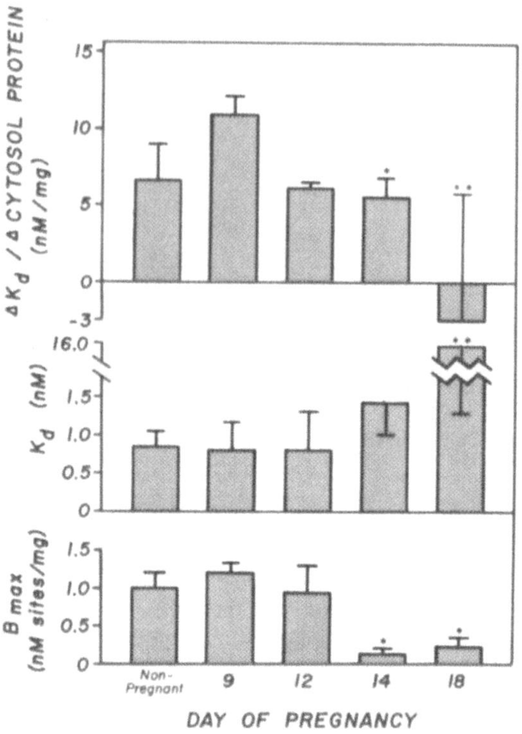

Figure 3. Summary of kinetic behavior of the endogenous inhibitor at several reproductive stages. The top panel shows rate of change in app K_d per milligram change in trophoblast cytosol (slope of replot). The middle panel shows the theoretical limit for K_d (intercept of replot). The lower panel shows B_{max} when [I] = 0. Values are means ±S.E.M of four to five cytosol preparations. *Mean differs from Day 9, $p < 0.05$. **Mean differs from all other stages, $p < 0.01$.

ter/mg, respectively, $n = 10$). The 40–50% fraction contained PR but not I and exhibited the same kinetic behavior as the 0–40% fraction (app $K_d = 2.3 \pm 0.1$ and 2.2 ± 0.1 nM and $B_{max} = 1.0$ and 1.1 nmol/liter/mg, respectively, $n = 3$). The 50–70% fraction exhibited no specific binding but an enhanced app K_d when added to the 0–40% fraction or to whole cytosol (inhibitory potency = 64.0 ± 12.9 nM Δ app K_d/mg, $n = 5$). Subfractions were active but less potent. Warming whole cytosol to 35°C for 30 min essentially eliminated PR but inhibitor potency (Δ app K_d/mg) was enhanced about twofold above that of the control cytosol aliquot kept at 4°C (13.0 ± 3.4 vs. 6.3 ± 0.4 nM, respectively, $n = 4$). Inhibitory activity was destroyed by heating whole cytosol or the 50–70% fraction to 70°C for 30 min.

An inhibitory substance is present in rat trophoblast that decreases the affinity of P for PR but has no influence on the availability of binding sites. The chemical nature of this substance is as yet unknown. However, preliminary experiments have shown that it is not adsorbed by dextran-coated charcoal, is nondialyzable, is denatured by $HgCl_2$, and is heat labile. The inhibitor elutes from 10-ml columns of Sephadex G-25 immediately following the void-volume peak. Based on these preliminary findings the inhibitor may be a protein of greater than 12,000 molecular weight.

ACKNOWLEDGMENTS. The excellent technical assistance of Mrs. Marybeth Kidd is gratefully acknowledged. Financial support of the NICHD (Grant 1-RO-1-HD12139-01) is also acknowledged.

5 Histone Synthesis in Preimplantation Mouse Embryos

Peter L. Kaye and R. G. Wales

Recent advances in biochemical technology allow more detailed examination of gene expression during cleavage development of the mammalian embryo. As nuclear proteins are so obviously involved in embryonic activity at this time, a study of the expression and metabolism of these proteins (particularly the histones) should prove valuable in elucidating genetic regulatory mechanisms in mammalian development.

One of the main questions in preimplantation development is the genetic source of the embryonic proteins. It is generally accepted that the large stores of RNA and protein in the egg serve as a source of material for use during cleavage: this is the case for histone mRNA in the sea urchin (Ruderman and Gross, 1974) and for histones in *Xenopus laevis* (Adamson and Woodland, 1974). Furthermore there is now evidence for two sets of histone genes in the sea urchin that are expressed differentially as development proceeds (Kunkel and Weinberg, 1978). Possibly a similar system may operate in the mammal. In order to investigate these possibilities, we began a study of histone synthesis in the preimplantation mouse embryo.

Embryos at various developmental stages were collected from superovulated mice, washed, and then cultured for 4 hr in medium containing [^{3}H]lysine (approx 0.5 mCi/ml, 1 mCi/mmol). The embryos were then collected, washed twice in the same medium containing cold lysine, and stored frozen at $-80°C$.

Following the addition of carrier calf thymus histones, the embryos were fractured by freeze-thawing; histones were isolated from whole embryos and separated by electrophoresis essentially as described by Newrock *et al.* (1977). The stained gels were sketched, scanned, sliced, and counted.

Embryos of all stages of development examined incorporated [^{3}H]lysine into acid soluble protein (Figure 1). Examination of this material by the sensitive Triton–acetic acid–urea system showed significant peaks of radioactivity that migrated with the calf thymus histone markers. Early experiments revealed only slight differences in the mobilities of the major histones (except for H1) of mouse liver and calf thymus. For convenience the commercial preparation of calf thymus histones was used as a marker.

A large amount of radioactivity is associated with heterogeneous material that predominantly has a mobility less than H2A and enters only the top of the gel. The amount of this material appears to increase with development. There are also significant peaks of radioactivity migrating between H2A and H1. These proteins have mobilities in the range of some of the minor histone variants and HMG proteins (Franklin and Zweidler, 1977). However, this material may be contaminating ribosomal proteins; it is unlikely to be due to

Peter L. Kaye • Division of Medical Biochemistry, Faculty of Medicine, University of Calgary, Alberta, Canada, and School of Veterinary Studies, Murdoch University, Western Australia. *R. G. Wales* • School of Veterinary Studies, Murdoch University, Western Australia.

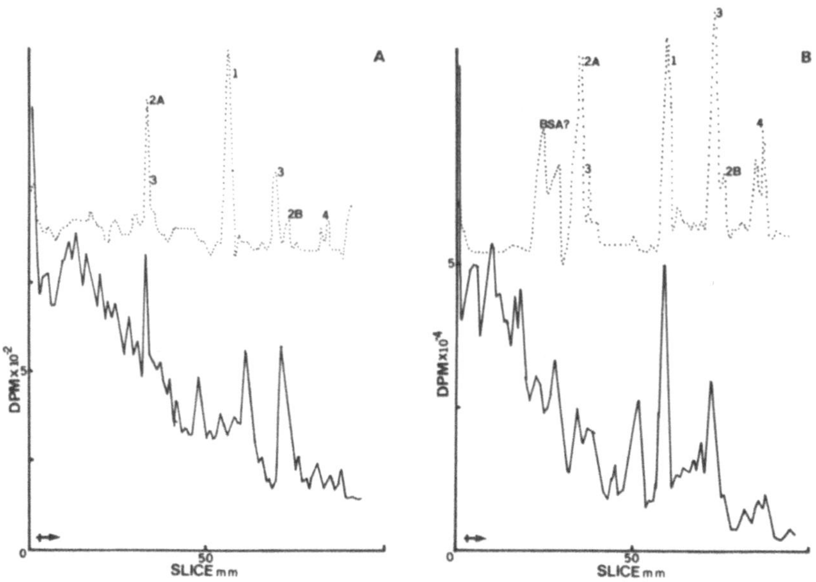

Figure 1. Electrophoretic profiles of acid-soluble proteins labeled with [^{3}H]lysine of mouse embryos at two developmental stages and analyzed as described in the text. (A) Two-cell embryos (197); (B) morulae (506). (····) Scan of gel showing position of marker calf thymus histones; (——) radioactivity.

degradation because bisulfite was used as a protease inhibitor. In addition, a parallel extraction of tryptophan-labeled material revealed very little contamination of the histones.

There are a number of radioactive peaks with mobilities slightly less than those of histone markers; they may represent modified forms of the embryo histones or histone variants as reported by Franklin and Zweidler (1977). A further possibility is that some of these may be histone species that are specific for particular developmental stages as observed in the sea urchin (Newrock *et al.*, 1977); their identification awaits further experiments.

ACKNOWLEDGMENTS. This work was supported by a grant from the Australian Research Grants Committee. P.L.K. is the recipient of a C. J. Martin Travelling Fellowship from the National Health and Medical Research Council of Australia.

References

Adamson, E. D., and Woodland, H. R., 1974, Histone synthesis in early amphibian development: Histone and DNA synthesis are not coordinated, *J. Mol. Biol.* **88**:263.

Franklin, S. G., and Zweidler, A., 1977, Non-allelic variants of histones 2a, 2b and 3 in mammals, *Science* **266**:273.

Kunkel, N. S., and Weinberg, E. S., 1978, Histone gene transcripts in the cleavage and mesenchyme blastula embryo of the sea urchin, *S. Pupuratus*, *Cell* **14**:313.

Newrock, K. M., Alfageme, C. R., Nardi, R. V., and Cohen, L. H., 1977, Histone changes during chromatin remodelling in embryogenesis, *Cold Spring Harbor Symp. Quant. Biol.* **42**:421.

Ruderman, J. V., and Gross, P. R., 1974, Histones and histone synthesis in sea urchin development, *Dev. Biol.* **36**:386.

6 Use of Concanavalin A to Monitor Changes in Glycoprotein Synthesis during Early Mouse Development

Terry Magnuson and Charles J. Epstein

One- and two-dimensional gel electrophoresis have been used to examine proteins synthesized by preimplantation mouse embryos and, for each stage analyzed, about 400–600 distinct polypeptides have been resolved (Epstein and Smith, 1974, Van Blerkom and Brockway, 1975, Levinson *et al.*, 1978). In an attempt to characterize further these proteins, we have used concanavalin A (Con A), a plant lectin that binds to mannose residues, to precipitate both total and cell surface glycoproteins synthesized during the preimplantation period. The precipitates were subsequently analyzed by two-dimensional gel electrophoresis and autoradiography.

For analysis of total glycoprotein synthesis, embryos were labeled for 4 hr with [^{35}S]methionine (2 mCi/ml). Total cellular protein was extracted with 0.5% Nonidet P40, reacted with Con A (2 mg/ml), and then exposed to rabbit antiserum directed against Con A. The resulting immune complexes were precipitated with a 10% suspension of heat-killed, formalin-fixed *Staphylococcus aureus* (Cowan 1) in 0.5% Nonidet P40. After washing, Con A-bound glycoproteins were eluted with 0.25 M α-D-methylmannoside. About 6–12% of the total acid-precipitable radioactivity was released from unfertilized eggs in this manner, while 4–8% of the total radioactivity from 2-cell, 8- to 16-cell, and late blastocyst embryos was eluted with mannoside. In control experiments, which included addition of mannoside prior to the addition of Con A, only 0.1–0.6% of the total radioactivity was eluted, indicating little release of proteins nonspecifically bound to the *Staphylococcus aureus* anti-Con A–Con A complex. The binding of Con A to labeled glycoproteins was not affected by pretreatment of the protein extract with 0.25 M galactose. Furthermore, galactose did not release any Con A-bound glycoproteins.

Analysis of autoradiograms of two-dimensional gel patterns revealed 29 glycoproteins that were precipitated from unfertilized eggs by Con A and approximately 50–70 from 2-cell, 8- to 16-cell, or late blastocyst embryos (Table 1). These glycoproteins represent about 5–15% of the 400–600 polypeptides normally resolved by two-dimensional gel electrophoresis of total proteins synthesized by the embryo. Most of the polypeptides present on gel patterns of precipitated glycoproteins could also be identified on autoradiograms of total proteins. When two-dimensional gel patterns of ova and embryos were compared, stage-specific glycoproteins could be identified (Table 1). However, glycoproteins present in more than one stage were also found (Table 1), and for some of these peptides, quantitative differences were detected.

In an attempt to determine which, if any, of the Con A-precipitated glycoproteins

Terry Magnuson and Charles J. Epstein • Departments of Pediatrics and of Biochemistry and Biophysics, University of California, San Francisco, California 94143.

Table 1. Developmental Program of Con A-Precipitated Glycoproteins

Glycoprotein No.	Unfertilized eggs	2-Cell	8-to 16-cell	Late blastocyst
1–2, 4–6, 9–21	×××××××××			
7	×××××××××	×××××××××		
3, 8, 22–29	×××××××××	×××××××××	×××××××××	×××××××××
52, 54, 55, 57, 58, 67–72		×××××××××		
30, 32, 40, 46–50, 56		×××××××××	×××××××××	
31, 33–39, 41–45, 51, 53, 59–66, 73–75		×××××××××	×××××××××	×××××××××
79, 80, 83, 95			×××××××××	
76–78, 81, 82, 84–94, 96–100, 110			×××××××××	×××××××××
101–109, 111				×××××××××

were components of the cell surface, intact ^{35}S-labeled embryos were incubated in Con A and then anti-Con A *prior* to extraction with Nonidet P40. Less than 1% of the total radioactivity was eluted from embryos treated in this manner. When two-dimensional gel patterns of these precipitated cell surface glycoproteins were examined, a complex of three pairs (doublets) of basic peptides (numbers 46–48 in Table 1) was enriched in experimental gels of 2-cell and 8- to 16-cell embryos when compared to gels of control samples (prepared as described above). This complex was not present at the late blastocyst stage.

The appearance and disappearance of glycoproteins binding to Con A in two-dimensional gels of labeled proteins of developing embryos were easily followed and thus provided a pattern of synthesis during normal development for a specific class of proteins. Such a developmental program may be useful for analysis of preimplantation lethal mutations. Levinson *et al.* (1978) have observed that the major qualitative changes in two-dimensional gel patterns of total proteins synthesized by early embryos occurred by the 8- to 16-cell stage. Our results with Con A-precipitated glycoproteins were generally similar, although we have observed some qualitative differences between early cleavage and late blastocyst stage embryos. Those glycoproteins present in unfertilized eggs and also in early cleavage stage embryos may be the result of stored maternal message.

Interestingly, the cell surface doublet precipitated from early cleavage stage embryos appears to be the same 69,000- to 71,000-molecular-weight complex that Johnson and Calarco (1980) precipitated with a rabbit antiserum (A-BL2) prepared against mouse blastocysts. Consistent with our data, these investigators found A-BL2 to react maximally with 8- to 12-cell embryos. The nature and function of this complex remain to be determined.

Acknowledgment. This work was supported by NIH Grant HD-03132. T.M. was a postdoctoral fellow of the National Science Foundation, and C.J.E. is an investigator of the Howard Hughes Medical Institute.

References

Epstein, C. J., and Smith, S. A., 1974, Electrophoretic analysis of proteins synthesized by preimplantation mouse embryos, *Dev. Biol.* **40**:233–244.

Johnson, L. V., and Calarco, P. G., 1980, Mammalian preimplantation development: The cell surface, *Anat Rec.* **196**:201–219.

Levinson, J., Goodfellow, P., Vadeboncoeur, M., and McDevitt, H., 1978, Identification of stage-specific polypeptides synthesized during murine preimplantation development, *Proc. Natl. Acad. Sci. USA* **75**:3332–3336.

Van Blerkom, J., and Brockway, G. O., 1975, Qualitative patterns of protein synthesis in the preimplantation mouse embryo. I. Normal pregnancy, *Dev. Biol.* **44**:148–157.

7 The Role of Glycoproteins in the Development of Preimplantation Mouse Embryos

M. A. H. Surani and S. J. Kimber

Cell surface glycoproteins apparently play a crucial role in preimplantation development of mouse embryos (Bennett *et al.*, 1971). Preimplantation mouse embryos are able to synthesize a variety of glycoconjugates including sialoglycoproteins, fibronectin, fucosyl glycopeptides, and nonsulfated polysaccharides (Surani, 1979). Modifications in the cell surface glycoproteins were observed during preimplantation development of mouse embryos (Pinsker and Mintz, 1973), and changes in the binding of plant lectins to the cell surface also have been found (Rowinski *et al.*, 1976). Developmentally regulated glycoproteins may thus play an important role in early mammalian embryogenesis. We have studied the synthesis of glycoproteins and the effects of inhibition of protein glycosylation by tunicamycin on preimplantation development of mouse embryos.

Embryos were retrieved from the genital tracts of MF1 albino outbred mice at the 2-cell, 8-cell, or blastocyst stage. The 2- and 8-cell embryos were cultured in BMOC-3 medium; the blastocysts were cultured in a complex medium with 10% fetal calf serum. The majority of 2- and 8-cell embryos in the control group developed to the blastocyst stage within 72 hr in culture. However, 2- and 8-cell embryos cultured in the presence of 1.0 μg tunicamycin/ml medium underwent cleavage up to approximately the 32-cell stage without compacting. Early cleavage divisions up to the 8-cell stage also were not affected by as much as 5.0 μg tunicamycin/ml. The embryos were sectioned and examined by light and transmission electron microscopy. In the 8-cell embryos treated with tunicamycin for 30 hr, the contact between adjacent blastomeres varied, but was generally reduced compared to the control group of compacted embryos. The blastomeres were either spherical or irregular in shape and the outermost blastomeres especially were rounded up with furrows between them. After 44 hr in culture, when embryos in the control group had reached the blastocyst stage, most of the blastomeres in the experimental group contained a large number of vacuoles. These blastomeres also revealed the presence of adherens as well as gap junctions between them. Gap junctions were especially prominent at the outermost regions of adjacent blastomeres (Kimber and Surani, unpublished). Blastocysts in the control group also attached to the petri dish with extensive giant cell outgrowths. Those cultured in the presence of tunicamycin failed to show trophoblast adhesion and giant cell outgrowths.

Incorporation of L-[4, 5-^{3}H]leucine, D-[6-^{3}H]glucosamine, and D-[2-^{3}H]mannose was examined in embryos in the control group and compared with those cultured in 1.0 μg tunicamycin/ml. A modified glucose-free culture medium was used to allow measurable incorporation of labeled sugars (Surani, 1979). In the eight-cell embryos cultured for 24 hr in

M. A. H. Surani and S. J. Kimber • A.R.C. Institute of Animal Physiology, Cambridge, England.

the presence of the precursors, [^{3}H]leucine incorporation was inhibited by only 14% while that of [^{3}H]mannose and [^{3}H]glucosamine was inhibited by about 60%. When blastocysts were cultured for up to 24 hr, the inhibition of [^{3}H]leucine incorporation was 17% and that of [^{3}H]glucosamine and [^{3}H]mannose 28 and 80%, respectively. These studies therefore suggest that protein synthesis is maintained at a high level but that the incorporation of the sugar precursors is inhibited in the presence of tunicamycin, which is known to specifically block glycosylation of asparaginyl residues of *N*-glycosidically linked glycoproteins.

Labeled proteins and glycoproteins were qualitatively analyzed by polyacrylamide disc gel electrophoresis. No large qualitative changes in [^{3}H]leucine-labeled polypeptides were observed, but the majority of the [^{3}H]glucosamine- and [^{3}H]mannose-labeled glycopeptides were substantially inhibited. Glycosylation of the majority of the proteins is therefore inhibited by tunicamycin except for the protein–polysaccharide fractions migrating near the top of the gels, which contained about 40–50% of incorporated [^{3}H]glucosamine and were insensitive to inhibition by tunicamycin. This finding may account for the relatively low inhibition of [^{3}H]glucosamine (28%) compared to the inhibition of [^{3}H]mannose (80%) at the blastocyst stage. Analysis of the glycoproteins labeled with [^{3}H]glucosamine and [^{3}H]mannose by two-dimensional gel electrophoresis revealed 20 to 30 glycopeptides, all of which were diminished in blastocysts cultured in the presence of tunicamycin. However, polypeptides labeled with [^{35}S]methionine showed no detectable difference when compared with the labeled polypeptides of blastocysts in the control group (Surani and Braude, unpublished). This finding is surprising as the unglycosylated polypeptides would be expected to migrate differently compared to the normal glycosylated polypeptides owing to their charge differences.

Differences in cell surface properties also were found in the control and experimental groups. The binding of concanavalin A was substantially reduced in blastocysts treated with tunicamycin. Cell surface components were labeled with ^{125}I by the lactoperoxidase method and were analyzed by disc gel electrophoresis. No differences were detected in the cell surface components of blastocysts in the control and experimental groups cultured in the presence of tunicamycin. However, eight-cell embryos cultured in 1.0 μg tunicamycin/ml showed the absence of a component of approximate molecular weight 60,000 to 80,000. The majority of the unglycosylated components probably migrate normally to the cell surface.

These studies show that cell surface glycoproteins and their oligosaccharide moieties play an important role in the development of preimplantation mouse embryos during compaction and trophoblast adhesion. Detailed analysis of the developmentally regulated cell surface glycoproteins is thus essential for elucidating their precise role in early mammalian development.

ACKNOWLEDGMENTS. This work was supported in part by a Ford Foundation Grant. S.J.K. was supported by an MRC postdoctoral fellowship. We thank Simon Fishel for comments, Sheila Barton for technical assistance, and Dr. R. L. Hamill of Eli Lilly, USA, for a gift of tunicamycin.

References

Bennett, D., Boyse, E. A., and Old, L. J., 1971, Cell surface immunogenetics in the study of morphogenesis, in: *Cell Interactions* (L. G. Silvestri, ed.), pp. 247–263, North-Holland, Amsterdam.

Pinsker, M. C., and Mintz, B., 1973, Changes in cell-surface glycoproteins of mouse embryos before implantation, *Proc. Natl. Acad. Sci. USA* **70:**1645.

Rowinski, J., Solter, D., and Koprowski, H., 1976, Changes in Concanavalin A induced agglutinability during preimplantation mouse development. *Exp. Cell Res.* **100:**404.

Surani, M. A. H., 1979, Glycoprotein synthesis and inhibition of glycosylation by tunicamycin in preimplantation mouse embryos: Compaction and trophoblast adhesion, *Cell* **18:**217.

8 The Role of Blastolemmase in Implantation Initiation in the Rabbit

Hans-Werner Denker

I. Introduction

Implantation of the mammalian embryo in the uterus involves the formation of cellular contact between embryo and maternal tissues. The molecular–biological mechanisms of this process are poorly understood. Systematic investigations in the rabbit have provided evidence that certain enzymes, particularly a peculiar trophoblast-dependent proteinase, play an interesting role.

II. Results

Implantation in the rabbit follows the central type. At the time of attachment, i.e., 7 days postcoitum, the blastocyst is considerably expanded (approximately 5 mm in diameter); it is oriented, by an unknown mechanism, with the embryonic disk facing the mesometrial endometrium and the abembryonic trophoblast facing the antimesometrial endometrium. Up to this stage, extracellular blastocyst coverings are still interposed between trophoblast and uterine epithelium. These blastocyst coverings are equivalent to the zona pellucida that has been dissolved in earlier stages; they are derived from secretions of the tube, the uterus, and the trophoblast (Denker and Gerdes, 1979). According to histochemical investigations, their composition is comparable to that of epithelial mucins: protein backbone and carbohydrate side chains with sulfate ester groups and terminal sialic acid (Denker, 1970). A particularly well-developed glycocalyx is present at the surface of the uterine epithelium and shows a somewhat similar composition.

Contact between trophoblast and uterine epithelium is first established in the abembryonic–antimesometrial region. It is preceded by an increase in stickiness of blastocyst coverings and uterine epithelial surface, followed by dissolution of the barrier formed by the blastocyst coverings. This process may involve enzymatic changes of the glycoprotein substances. In fact, considerable activity of various glycosidases as well as exo- and endopeptidases is demonstrable histochemically at these sites during initiation of implantation. Of particular interest is a peculiar endopeptidase (proteinase) activity which is trophoblast-dependent and which appears exactly at the time (and the site) of antimesometrial implantation and which disappears immediately thereafter (Denker, 1971). This enzyme is called blastolemmase. As shown by experiments with various proteinase inhibitors *in vitro,* blastolemmase is closely related to trypsin. Substrate specificity, however, is more re-

Hans-Werner Denker • Abteilung Anatomie der RWTH, D-5100 Aachen, West Germany.

stricted: arginyl bonds are much more easily hydrolyzed than lysyl bonds, and hydrolysis rates are strongly influenced by the type of amino acid present in subsite positions P_2 and P_3 adjacent to the arginyl residue (P_1) (Denker and Fritz, 1979).

Proteinase inhibitors that inhibited blastolemmase effectively *in vitro* (such as aprotinin = Trasylol, antipain, NPGB) were administered intrauterally *in vivo* at 6 days 12 hr pc, i.e., 12 hr before the time of implantation initiation (Denker, 1977). This treatment resulted in blockage of dissolution of blastocyst coverings, which remained interposed between trophoblast and uterine epithelium so that attachment was impossible. The unattached blastocysts continued to expand, and as a result the undissolved blastocyst coverings ruptured. In such places a delayed and locally restricted attachment can occur. These experiments illustrate that blastolemmase plays an important role in dissolution of the blastocyst coverings in a manner reminiscent of a hatching enzyme. The possible roles of uterine-secretion proteinases and of other enzymes of the trophoblast and the endometrium (like glycosidases) remain to be defined. Whether or not these enzymes are also directly involved in the process of formation of cellular contact between trophoblast and uterine epithelium needs further investigation.

References

Denker, H.-W., 1970, Topochemie hochmolekularer Kohlenhydratsubstanzen in Frühentwicklung und Implantation des Kaninchens, I and II, *Zool. Jahrb. Abt. Allg. Zool. Physiol. Tiere* **75:**141, 246.

Denker, H.-W., 1971, Enzym-Topochemie von Frühentwicklung und Implantation des Kaninchens, III. Proteasen, *Histochemie* **25:**344.

Denker, H.-W., 1977, "Implantation: The Role of Proteinases, and Blockage of Implantation by Proteinase Inhibitors," *Adv. Anat. Embryol. Cell. Biol.* **53:**Part 5, Springer-Verlag, Berlin.

Denker, H.-W., and Fritz, H., 1979, Enzymic characterization of rabbit blastocyst proteinase with synthetic substrates of trypsin-like enzymes, *Hoppe-Seyler's Z. Physiol. Chem.* **360:**107.

Denker, H.-W., and Gerdes, H.-J., 1979, The dynamic structure of rabbit blastocyst coverings. I. Transformation during regular preimplantation development, *Anat. Embryol.* **157:**15.

9 Estradiol-17β Dehydrogenase in Rat Endometrium at the Time of Implantation

O. Kreitmann and F. Bayard

In order to determine the possible role of estrone (E_1) formation during the implantation process, estradiol-17β dehydrogenase (E_2DH) activity was investigated in rat endometrium during the estrous cycle, the first days of pregnancy, and in castrated rats treated with estradiol (E_2) alone or in combination with progesterone (P). Endometrium of mature Sprague–Dawley rats was removed from the myometrium by gentle scraping of longitudinally opened uterine horns. E_2DH assays were performed on endometrial homogenate according to the method of Tseng and Gurpide (1974) (E_2DH in pmol/mg protein ·hr).

During the estrous cycle, E_2DH activity is maximal at estrus (E) and proestrus (PO) and decreases during metestrus and diestrus ($p<0.001$) (Figure 1). During pregnancy, the activity decreases progressively on Days 1–2 (P_1–P_2, Day 1 = spermatozoa present in vaginal smears) from the values of PO, to minimal values on P_3–P_4, then increases slightly on P_5 ($p<0.05$) and dramatically on P_6 (even higher than for PO, $p<0.005$), and remains higher than P_3–P_4 at least until P_{10}. No diurnal variation was observed on P_4–P_5 (Figure 1). The ligation of one oviduct on P_1 does not prevent the appearance of normal E_2DH values; thus blastocysts and deciduoma are apparently not necessary for the development of E_2DH activity.

The hormonal milieu of early pregnancy (Psychoyos, 1976) was mimicked in castrated rats (> 10 days) by daily injections of E_2 and P (Leroy 1978) as described in Table 1. E_2 injection on Day 7 seems to determine E_2DH activity on the following days: without this E_2 treatment, no E_2DH activity appears; with 0.1 μg E_2 (to mimic the "nidatory E_2" surge), a lag is observed between E_2 administration and the induction of E_2DH activity (Figure 2); for higher doses (2 μg) this lag is shortened. Thus the "nidatory E_2" seems to be responsible for the appearance of E_2DH activity on P_6. Progesterone priming is necessary as three daily injections of 0.5 μg E_2 are required to induce a significant increase in enzyme activity in castrated rats (data not shown). In these experiments, high E_2DH values can be maintained after Day 10 by daily injections of 2.5 mg P.

Similarly treated animals were used to investigate the uterine sensitivity for decidualization [0.5 μg E_2 on Day 10 followed 17 hr later by uterine intralumenal injection of 50 μl arachid oil, and subsequent E_2 (0.2 μg) and P (2 mg) administration daily until Day 15]. When "nidatory E_2" injection is not done on Day 7, E_2DH value is low on Day 10 (156 ± 36 pmol/mg protein × hr, n = 5), and similarly treated animals have a very thick deciduoma on Day 15 (uterine weight 3086 ± 123 mg, n = 5). After "nidatory E_2" injection on Day 7, E_2DH activity is high on Day 10 (3213 ± 1159 pmol/mg protein × hr, n = 4), and no deciduoma formation occurs (444 ± 23 mg, n = 5) (refractory state).

O. Kreitmann and F. Bayard • INSERM U 168, Department of Endocrinology CHU Toulouse Rangueil, Universite Paul Sabatier, Toulouse, France.

Table 1. Protocol of E_2 and P Administration to Castrated Rats (>10 Days)[a]

Day of treatment[b]									
1	2	3	4	5	6	7	8	9	10
E_2	E_2	—	—	P	P	PE_2	P	P	P

[a]After Leroy (1978).
[b]0.5 μg E_2 on Days 1 and 2 corresponds to proestrous E_2 increase. 0.1 μg E_2 on Day 7 mimics "nidatory E_2" surge of pregnancy (Day P_4). P, 2.5 mg throughout.

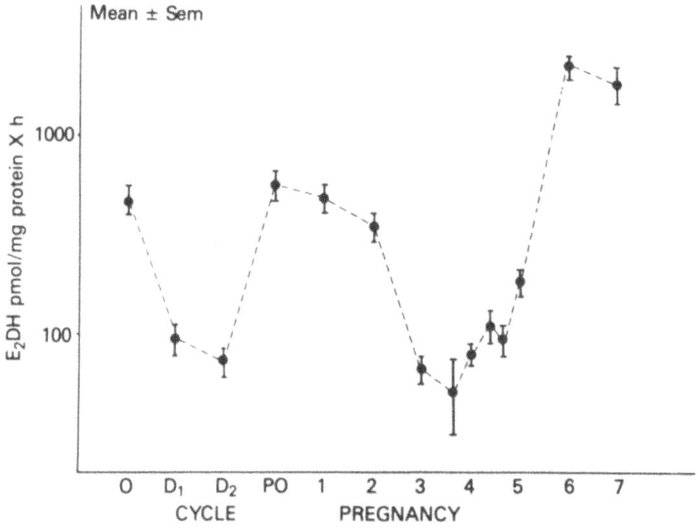

Figure 1. E_2DH activity in rat endometrium at various phases of the estrous cycle (mean ± S.E.). The time of sampling was usually 2 P.M. but additional samples were taken at 8 A.M. and 10 P.M. on Day 4 and at 1 A.M. on Day 5 of pregnancy.

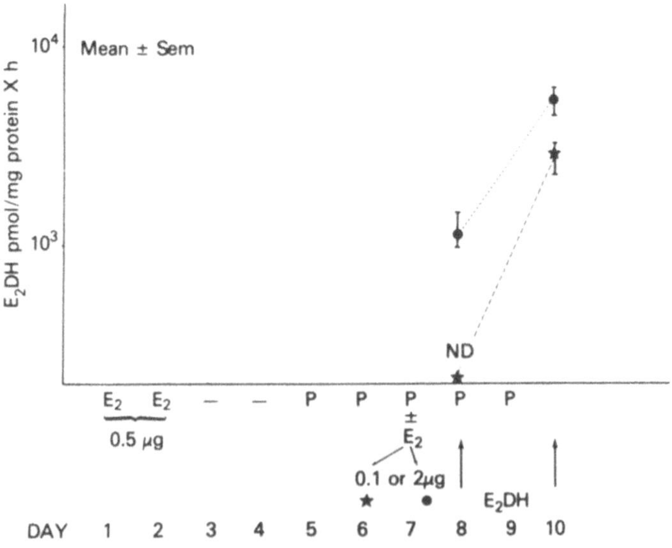

Figure 2. Endometrial E_2DH activity measured on Days 8 and 10 after the sequence of treatment described in Table 1 (n = 10; ordinate in log scale). E_2 (●, 2μg; ★, 0.1 μg) given on Day 7.

The sensitivity of endometrium for decidualization disappears when hormonal conditions induce a high E_2DH activity.

The results indicate that E_2 metabolism, with E_1 production *in situ*, under the influence of E_2DH, may be involved in the receptivity/refractoriness of endometrium for implantation in the rat.

References

Leroy, F., 1978, Aspects moleculaires de la nidation, in: *L'implantation de l'oeuf* (F. du Mesnil du Buisson, A. Psychoyos, and K. Thomas, eds.), pp. 81–92, Masson, Paris.

Psychoyos, A., 1976, Hormonal control of uterine receptivity for nidation, *J. Reprod. Fertil. Suppl.* **25:**17–28.

Tseng, L., and Gurpide, F., 1974, Estradiol and 20α-dihydroprogesterone dehydrogenase activities in human endometrium during the menstrual cycle, *Endocrinology* **94:**419–423.

10 The Pig Uterus Secretes a Progesterone-Induced Inhibitor of Plasminogen Activator

Deborra E. Mullins, Fuller W. Bazer, and R. Michael Roberts

I. Introduction

Pigs undergo central implantation in which embryonic development occurs entirely within the lumen of the uterus. The trophoblast does not invade into the stroma, and the epithelium remains intact throughout pregnancy. If, however, the embryo is transplanted to some ectopic site, such as the kidney capsule or into the stroma of the uterus, it becomes highly invasive (Samuel, 1971; Samuel and Perry, 1972). In animals that undergo true implantation, proteolytic enzymes have been detected in the uterus during the invasive period (Kirchner, 1972; Denker, 1974; Pinsker *et al.*, 1974). The most noteworthy of these uterine proteases is plasminogen activator (PA), an enzyme that is defined operationally by its ability to convert the zymogen plasminogen to plasmin. Such an enzyme is produced by mouse trophoblast cells and has been suggested as being the agent responsible for allowing the trophoblast to penetrate the uterine epithelium and invade the underlying stroma (Strickland *et al.*, 1976). Because the pig trophoblast is not invasive within the uterus, we were led to question whether the trophoblast produced PA or, alternatively, whether the uterus might secrete an inhibitor of this enzyme.

II. Does the Porcine Trophoblast Produce PA?

Attachment of the conceptus to the uterine wall begins around Day 12 after the onset of previous estrus and is complete by about Day 20. This process also involves expansion of a spherical blastocyst (3-mm diameter at Day 10) to an elongated thread up to a meter in length at Day 18. Expanding Day 12 blastocysts were flushed from the uterus under sterile conditions and incubated *in vitro* for 48 hr. Aliquots of the medium were assayed for PA by means of the indirect [^{125}I]fibrin plate method (Unkeless *et al.*, 1973). Enzyme was released by these conceptuses in a time-dependent manner, with a final recovery of about 800 units per embryo (Figure 1).

Deborra E. Mullins and R. Michael Roberts • Department of Biochemistry and Molecular Biology, University of Florida, Gainesville, Florida 32610. *Fuller W. Bazer* • Department of Animal Science, University of Florida, Gainesville, Florida 32611.

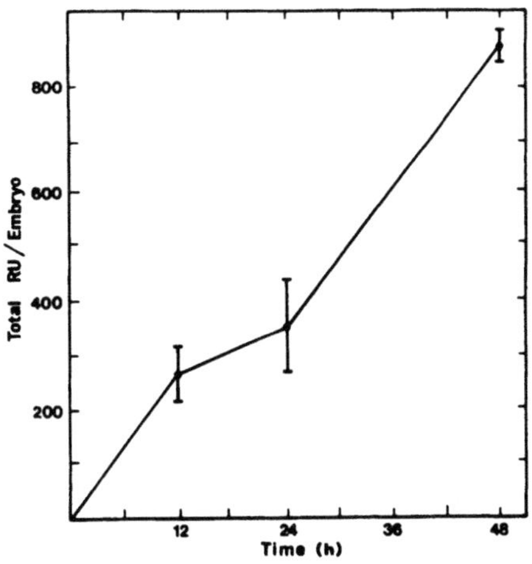

Figure 1. Secretion of plasminogen activator by Day 12 pig conceptuses cultured *in vitro*. Conceptuses were cultured in MEM medium and aliquots assayed for PA activity by the [^{125}I] fibrin plate assay. Data are expressed as the total amount of PA activity in the medium per conceptus ± S.E. One reference unit is defined as 5% of the counts removed by 0.25% (w/v) trypsin in the 3-hr assay.

III. Is PA Detectable in the Uterus?

The uterine lumen of pregnant and nonpregnant pigs was flushed with saline to remove soluble proteins and the flushings were assayed for PA activity. Activities were high (averaging 11 units/150 μg protein) at the beginning (Day 3) and toward the end (Day 18) of the cycle, but low in midcycle when serum progesterone levels are high. Furthermore, PA activity in the flushings from pregnant uteri also was low from Day 12 onwards even though the conceptuses are known to be capable of producing large amounts of PA *in vitro*.

IV. Does the Uterus Secrete a PA Inhibitor under the Influence of Progesterone?

Samples (150μg protein) of midcycle flushings (Day 15) were mixed with 150μg of late-cycle flushings of high PA activity and then assayed by the fibrin plate method. The combined activity of the mixed sample was 65.8% lower than would have been expected if the PA activity of the two samples had been additive, suggesting that luteal-phase flushings contain a protease inhibitor. In order to determine if this inhibitor is hormonally induced, flushings were obtained from ovariectomized animals maintained on either progesterone, progesterone plus estrogen, estrogen, or corn oil, the vehicle for the steroids. Whereas the latter two samples had high PA specific activity, flushings from animals administered progesterone or progesterone plus estrogen had very low activities. Mixing experiments with equivalent amounts of protein verified that the progesterone-induced samples were able to inhibit the PA activity in the flushings from estrogen- or corn-oil-treated animals, as well as the PA activity secreted by Day 12 conceptuses *in vitro*. Equivalent amounts of pig serum had no detectable inhibitory activity. The progesterone-induced proteins (750 μg) were also able to inhibit 48.1% of the activity of 30 units of urokinase using a direct fluorometric assay with Cbz-Gly-Gly-Arg-methoxy-β-naphthylamide as substrate, thus verifying the presence of a PA inhibitor in the flushings. However, progesterone-induced flushings are also active against trypsin, chymotrypsin, and plasmin, but not elastase. Thus it is not clear whether several protease inhibitors or one inhibitor of broad specificity are present in the flushings.

V. Summary and Conclusions

We have shown that a PA inhibitor, which is apparently progesterone induced, is secreted by the porcine uterus at the same time that PA-producing conceptuses are beginning to attach. This may explain how the uterine epithelium is able to resist invasion by a trophoblast that is highly invasive when implanted elsewhere.

References

Denker, H. W., 1974, Trophoblastic factors involved in lysis of the blastocyst coverings and in implantation in the rabbit, *J. Embryol. Exp. Morphol.* **32:**739.

Kirchner, C., 1972, Uterine protease activities and lysis of the blastocyst covering in the rabbit, *J. Embryol. Exp. Morphol.* **28:**177.

Pinsker, M. C., Sacco, A. G., and Mintz, B., 1974, Implantation-associated proteinase in mouse uterine fluid, *Dev. Biol.* **38:**285.

Samuel, C. A., 1971, The development of the pig trophoblast in ectopic sites, *J. Reprod. Fertil.* **27:**494.

Samuel, C. A., and Perry, J. S., 1972, The ultrastructure of the pig trophoblast transplanted to an ectopic site in the uterine wall, *J. Anat.* **113:**139.

Strickland, S., Reich, E., and Sherman, M. I., 1976, Plasminogen activator in early embryogenesis: Enzyme production by trophoblast and parietal endoderm, *Cell* **9:**231.

Unkeless, J. C., Tobia, A., Ossowski, L., Quigley, J. P., Rifkin, D. B., and Reich, E., 1973, An enzymatic function associated with transformation of fibroblasts by oncogenic viruses: Chick embryo fibroblast cultures transformed by avian RNA tumor viruses, *J. Exp. Med.* **137:**85.

11 Role of a Uterine Endopeptidase in the Implantation Process of the Rat

M. G. Rosenfeld and M. S. Joshi

I. Introduction

The genital tract fluids of mammalian species provide the milieu in which oocyte maturation, fertilization, early development of cleaving embryos, and blastocyst implantation take place. Proteins secreted into the lumen of the female genital tract, especially those regulated by reproductive hormones, may participate crucially in these reproductive events. Studies on the uterine lumenal proteins of mice (Gore-Langton and Surani, 1976) and rats (Surani, 1975, 1976) suggested a hormone-controlled sequence of protein changes that coincide with the time of implantation. Joshi *et al.* (1970) have isolated and characterized an endopeptidase from estrous fluid obtained from proestrous rats and from immature or overiectomized rats treated with estradiol. The uterine peptidase demonstrated strong affinity to sperm membrane and zona pellucida. Immunological studies (Joshi and Murray, 1976) showed that the enzyme was unique to uterus and oviduct and was not detected in serum. Appearance of the enzyme on Day 5 of pregnancy (Joshi and Murray, 1976) suggested a role for this protein in the implantation process (Rosenfeld and Joshi, 1977).

II. Materials and Methods

Wistar rats from Carworth Farms were used. The collection of uterine secretion and the purification of the endopeptidase were as described in our earlier work (Joshi *et al.*, 1970; Joshi and Murray, 1976). Disc gel electrophoresis of native protein and denatured protein was performed using the technique of Davis (1964) and Weber and Osborn (1969), respectively. Zona lytic effect of the rat uterine peptidase on unfertilized and fertilized eggs was studied *in vitro*. Immature rats were superovulated and eggs with cumulus mass were dissected out from ampullae and suspended in 0.1% hyaluronidase containing 1% BSA to remove cumulus cells. To obtain fertilized eggs, the superovulated rats were caged with males of proven fertility. The cumulus-free eggs were divided into four groups and placed in depression slides containing 0.2 ml incubation medium (Krebs–Ringer bicarbonate with 4 mg/ml BSA and antibiotics) and (1) 30 μg of purified rat uterine peptidase, (2) 30μg boiled purified endopeptidase, (3) incubation medium alone, or (4) 0.1% trypsin. The depression slides were placed in a CO_2 incubator at 37°C. Eggs were observed at hourly intervals with a phase contrast microscope. Boromycin and 1,3-bis(4-chlorocinnamylidereamino)guanidine, the drugs used to release cortical granules, were gifts from Dr. R. B. L. Gwatkin, Merck Institute.

M. G. Rosenfeld • Department of Anatomy, Downstate Medical Center, Brooklyn, New York 11203.
M. S. Joshi • Department of Anatomy, University of North Dakota, Grand Forks, North Dakota 58201.

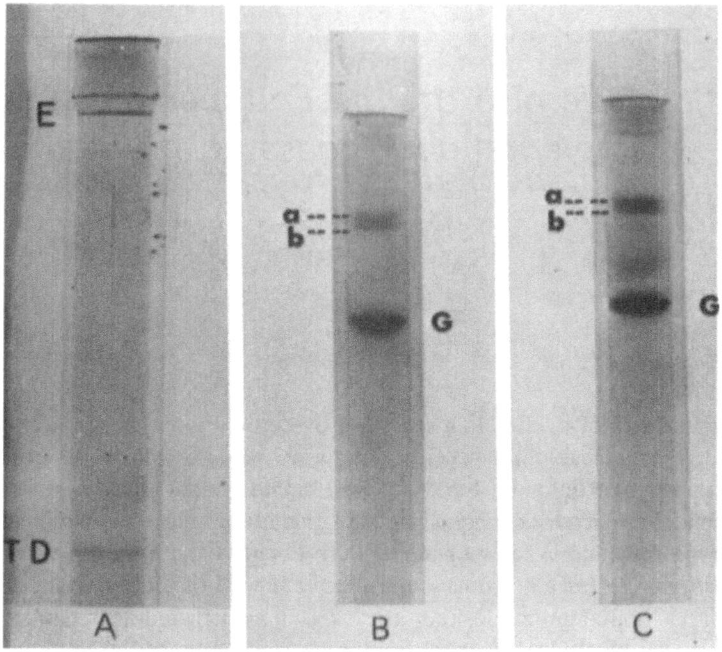

Figure 1. (A) Polyacrylamide gel electrophoresis (PAGE) of pure fraction (PF) from G-200 column under nondenaturing conditions in 5% running gel containing 80–100 μg protein. E, endopeptidase; TD, tracking dye. (B) SDS–PAGE (10%) of excised agarose precipitin line from Ouchterlony analysis. (C) SDS–PAGE of PF + IgG used in Ouchterlony analysis. a, Heavy component of molecular weight 110,000; b, lighter component of molecular weight 100,000; G, heavy (γ) chain of IgG.

III. Results

The preliminary results of the SDS–polyacrylamide gel electrophoresis experiment indicate that the endopeptidase consists of two nonidentical components with molecular weights of approximately 110,000 and 100,000 (Figure 1). Both proteins are PAS positive on the gel. The carbohydrate content of the native protein was estimated to be between 30 and 35%.

The lysis of zonae of unfertilized rat eggs began approximately 3 hr after incubation with uterine endopeptidase and complete lysis was achieved by 5 hr (Figure 2). The fertilized eggs were unaffected by the rat endopeptidase (Figure 3) through 16 hr of incubation. The results suggest that the zona pellucida of the rat egg has been modified during the fertilization process making it resistant to proteolytic digestion by the estrogen-dependent uterine peptidase. The zonae of fertilized eggs were readily lysed by trypsin.

The zona lytic effect of rat endopeptidase was tested on unfertilized eggs treated with drugs that induce cortical granule discharge. Treatment of unfertilized eggs with either boromycin or a guanidine derivative mimics the results obtained with fertilized eggs (Figure 4). The zonae of these eggs were resistant to proteolytic digestion by uterine endopeptidase but not by trypsin. It appears that the release of cortical granule material, either by fertilization or by drugs, modified the structure of rat zona making it resistant to the action of endogenous peptidase. This physiological event could be important for preventing pre-

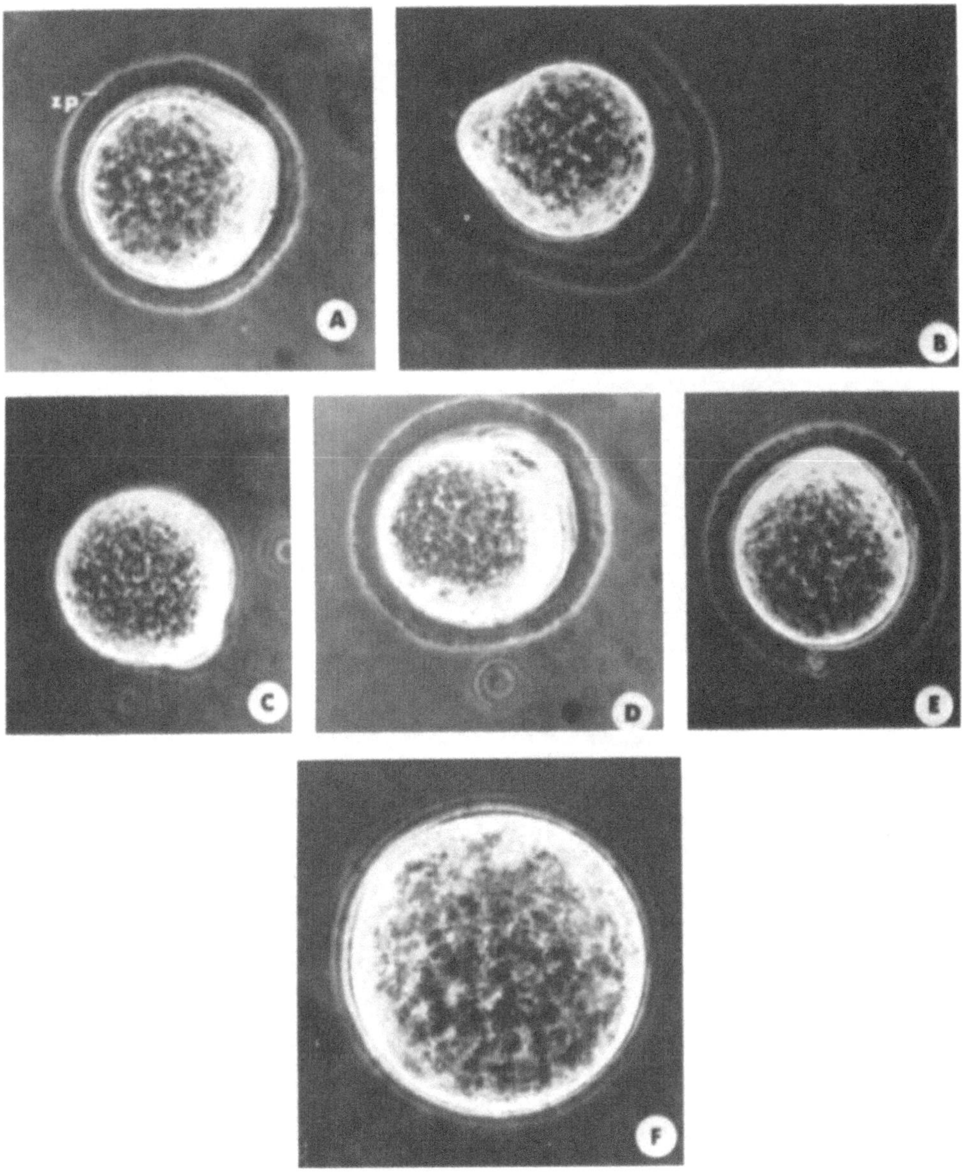

Figure 2. Unfertilized rat eggs after incubation with PF (150 μg/ml) in a modified Krebs–Ringer medium. (A) 3 hr; zona pellucida (zp) intact. (B) 4 hr; zp beginning to thin. (C) 5 hr; zp lysed. (D) 16 hr, with PF inactivated by boiling. (E) Incubation medium alone. (F) 0.1% trypsin.

mature dissolution of zona pellucida by the enzyme in the oviduct and uterus, which could lead to disruption of normal implantation. Though the results do not provide evidence to indicate that uterine peptidase alone is responsible for zona lysis of the blastocyst, the endopeptidase may be involved in zona lysis by acting in conjunction with other proteins secreted by the uterus or blastocyst at the time of implantation.

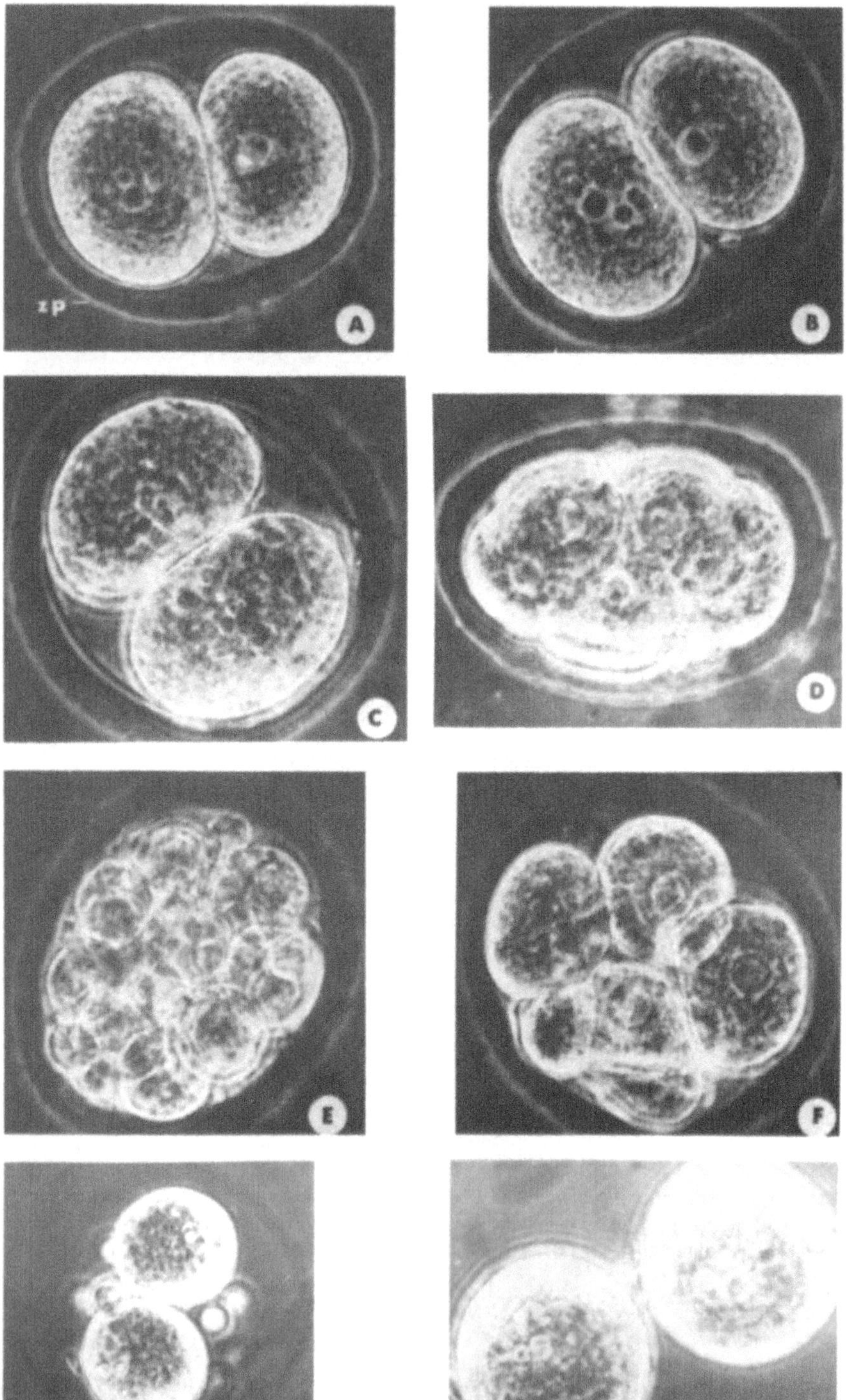

Figure 3. Fertilized eggs incubated with PF (150 μg/ml). (A) 4 hr; zona pellucida (zp) intact. (B) 8 hr. (C) 16 hr. (D) Eight-cell egg, 16 hr. (E) Morula, 16 hr. (F) Eight-cell egg, 24 hr. (G) 0.1% trypsin, 10 min. (H) 0.1% trypsin, 20 min.

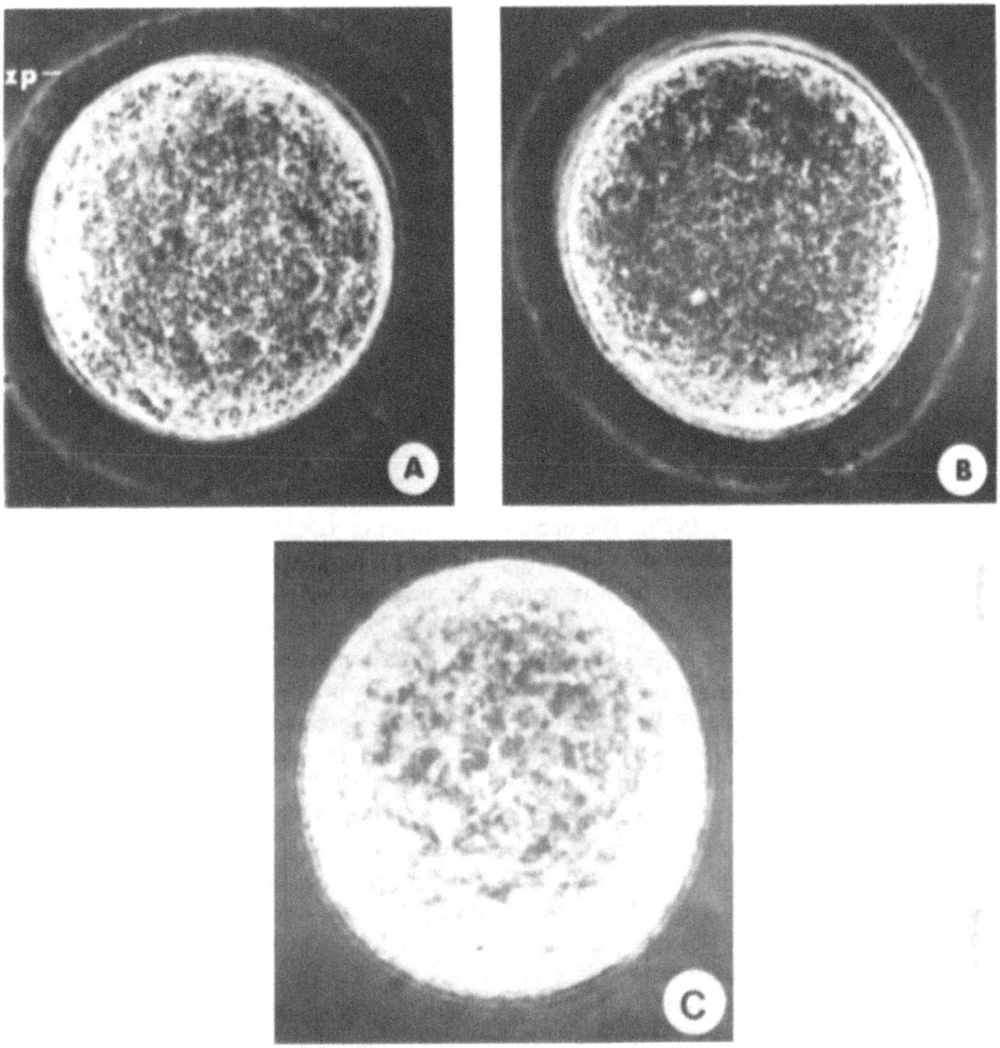

Figure 4. Unfertilized eggs incubated for 30 min with boromycin (10 μg/ml) before exposure to PF. (A) 16 hr. (B) 24 hr. (C) Treated with boromycin and trypsin.

References

Davis, B. J., 1964, Method and application to human serum proteins, *Ann. N.Y. Acad. Sci.* **121:**404.

Gore-Langton, R. E., and Surani, M. A. H., 1976, Uterine luminal proteins of mice, *J. Reprod. Fertil.* **46:**271.

Joshi, M. S., and Murray, I. M., 1976, Immunological studies of the rat uterine fluid peptidase, *J. Reprod. Fertil.* **37:**361.

Joshi, M. S., Yaron, A., and Lindner, H. R., 1970, An endopeptidase in the uterine secretion of the proestrous rat and its relation to a sperm decapacitating factor, *Biochem. Biophys. Res. Commun.* **38:**52.

Rosenfeld, M. G., and Joshi, M. S., 1977, A possible role of a specific uterine fluid peptidase in implantation in the rat, *J. Reprod. Fertil.* **51:**137.

Surani, M. A. H., 1975, Hormonal regulation of proteins in the uterine secretion of ovariectomized rats and the implications for implantation and embryonic diapause, *J. Reprod. Fertil.* **43:**411.

Surani, M. A. H., 1976, Uterine luminal proteins at the time of implantation in rats, *J. Reprod. Fertil.* **48:**141.

Weber, K., and Osborn, M., 1969, The reliability of molecular weight determinations by dodecyl sulfate–*polyacrylamide* gel electrophoresis, *J. Biol. Chem.* **244:**4406.

12 The Role of Prostaglandins and Cyclic Nucleotides in the Artifically Stimulated Decidual Cell Reaction in the Mouse Uterus

Judith C. Rankin, Barry E. Ledford, and Billy Baggett

The transformation of uterine stromal cells into decidual cells is an integral part of normal implantation (Finn and Porter, 1975). The decidual cell reaction (DCR) can be induced in a uterine horn of a hormone-primed ovariectomized mouse by intralumenal injection of sesame oil or by crushing the uterine horn (trauma). Traumatic stimulation does not require the small "nidatory" dose of estradiol a few hours prior to the stimulus; this is required for oil induction of the response just as it is required for natural implantation. Either stimulus leads to the transformation of uterine stromal cells into rapidly proliferating decidual cells. Decidual cells are not histologically identifiable until 24 hr after stimulation. However, various biochemical and physiological changes precede this transformation and can be quantitated at early times following artificial stimulation of the DCR.

In our model system (Ledford *et al.*, 1976), 5 min after either oil or traumatic stimulation, prostaglandin (PG) E_2 and $F_{2\alpha}$ levels have increased 10-fold in the stimulated horn (Jonsson *et al.*, 1978; Rankin *et al.*, 1979). A second early event in the decidualization process is a 10-fold increase in cyclic AMP (cAMP) levels in the stimulated horn 15 min after oil stimulation (Figure 1). When trauma is the stimulus, a 3- to 4-fold increase in cAMP levels in the stimulated horn is measured at 15 min (Rankin *et al.*, 1977). Cyclic AMP returns to control levels within 2 hr following either stimulus. A second peak in the level of cAMP in the stimulated horn is seen at 7 hr following oil stimulation (Figure 1).

Pretreatment of animals with indomethacin, an inhibitor of prostaglandin synthetase, prevents increases in PGE_2 and $PGF_{2\alpha}$ levels following oil or traumatic stimulation (Rankin *et al.*, 1979). This indomethacin pretreatment also prevents the increase in cAMP and the DCR when oil is the stimulus (Rankin *et al.*, 1979). When trauma is the stimulus, indomethacin-pretreated animals show a significant increase in the level of cAMP in the stimulated horn and decidualization still occurs (Rankin *et al.*, 1979).

Measurement of 6-keto-$PGF_{1\alpha}$, the stable metabolite of prostacyclin (PGI_2), 15 min after oil stimulation of the DCR reveals a two- to threefold increase in its level in the stimulated horn (Table 1). Pretreatment of the animals with tranylcypromine, an inhibitor of prostacyclin biosynthesis, prevented this increase but did not prevent the increases in PGE^2 and PGF^2_α levels seen 5 min after oil stimulation (Table 1). This confirms the selectivity of the inhibitory action of tranylcypromine. Tranylcypromine pretreatment

Judith C. Rankin, Barry E. Ledford, and Billy Baggett • Department of Biochemistry, Medical University of South Carolina, Charleston, South Carolina 29403.

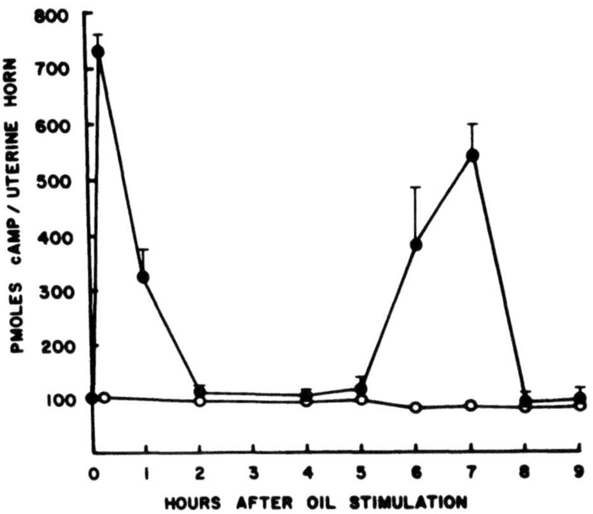

Figure 1. Uterine cAMP levels following oil stimulation of the DCR. Ovariectomized mice were hormone primed (Ledford *et al.*, 1976); the right uterine horn was stimulated intralumenally with sesame oil. Mice were killed at various times after oil stimulation and uterine horns assayed individually for cAMP content. Points represent the means from at least six animals per group; bars represent the standard errors of the means. (•) Stimulated horn; (o) unstimulated horn.

had no significant effect on the trauma-stimulated increase in cAMP levels and subsequent decidualization (Table 2). This pretreatment did, however, significantly reduce the oil-stimulated increase in the level of cAMP and greatly reduced the subsequent DCR (Table 2).

These results are indicative of a role of a prostaglandin, probably prostacyclin, as a mediator of the decidualization induced by oil injection into the uterine horn of a properly hormonally primed mouse. The inhibitory action of indomethacin and tranylcypromine on the oil-stimulated DCR implicates the transformation of arachidonate to the endoperoxide, PGG_2, as an early, and perhaps rate-controlling, reaction in the process of oil-stimulated decidualization, with prostacyclin as the active mediator of further events. Trauma, however, seems to initiate cAMP increases leading ultimately to decidualization without the necessary involvement of the prostaglandin system at this early stage.

Table 1. The Effect of Tranylcypromine Pretreatment on Uterine Prostaglandin Levels following Stimulation of the DCR [a]

Treatment	PGE_2 levels (pg/horn)		$PGF_{2\alpha}$ levels (pg/horn)		6-Keto-PGF_1 levels (pg/horn)	
	Stimulated	Unstimulated	Stimulated	Unstimulated	Stimulated	Unstimulated
Sesame oil	3328 ± 701	333 ± 60	3320 ± 439	312 ± 43	812 ± 143	304 ± 23
Tranylcypromine + sesame oil	2428 ± 432	367 ± 32	3050 ± 332	262 ± 32	197 ± 49	173 ± 38

[a] Ovariectomized mice were hormone primed (Ledford *et al.*, 1976). Tranylcypromine (20 mg/kg body wt) was administered i.p. 30 min prior to stimulation. PG levels were measured 5 min after stimulation (Rankin *et al.*, 1979). 6-Keto-$PGF_{1\alpha}$ levels were measured 15 min after stimulation. Values are the means from six animals per group ± S.E.M.

Table 2. The Effect of Tranylcypromine Pretreatment on Uterine cAMP Levels and on Uterine Horn Weights following Oil or Traumatic Stimulation of the DCR[a]

Treatment	cAMP (pmol/horn)		Horn weight (mg)	
	Stimulated	Unstimulated	Stimulated	Unstimulated
Sesame oil	748 ± 32	95 ± 5	279 ± 14	29 ± 3
Tranylcypromine + sesame oil	354 ± 87	93 ± 3	58 ± 10	29 ± 1
Trauma	311 ± 45	91 ± 5	89 ± 13	28 ± 2
Tranylcypromine + trauma	296 ± 81	87 ± 11	91 ± 9	31 ± 3

[a] Ovariectomized mice were hormone primed (Ledford *et al.*, 1976). Cyclic AMP levels were measured 15 min after stimulation (Rankin *et al.*, 1977). Horn weights were determined 72 hr after stimulation. Values are the means from nine animals per group ± S.E.M.

ACKNOWLEDGMENT. This work was supported by Grant HD 06362 from the National Institute of Child Health and Human Development of the National Institutes of Health.

References

Finn, C. A., and Porter, D. G., 1975, *The Uterus*, Publishing Sciences Corp., Acton, Mass.

Jonsson, H. T., Jr., Rankin, J. C., Ledford, B. E., and Baggett, B., 1978, Prostaglandin levels following stimulation of the decidual cell reaction in the mouse uterus, Program of the 60th Meeting of the Endocrine Society, Miami, p. 326.

Ledford, B. E., Rankin, J. C., Markwald, R. R., and Baggett, B., 1976, Biochemical and morphological changes following artificially stimulated decidualization in the mouse uterus, *Biol. Reprod.* **15**:529.

Rankin, J. C., Ledford, B. E., and Baggett, B., 1977, Early involvement of cyclic nucleotides in the artificially stimulated decidual cell reaction in the mouse uterus, *Biol. Reprod.* **17**:549.

Rankin, J. C., Ledford, B. E., Jonsson, H. T., Jr., and Baggett, B., 1979, Prostaglandins, indomethacin and the decidual cell reaction in the mouse uterus, *Biol. Reprod.* **20**:399.

13 Inhibitory Effect of Indomethacin on Nuclear Binding of Progesterone in the Uterus and on Differentiation of Decidual Cells in Culture

Sara Peleg and Hans R. Lindner

Indomethacin, an inhibitor of prostaglandin synthesis, was reported to inhibit the decidual cell reaction (DCR) in the uterus of the rat *in vivo* (Sananes *et al.*, 1976). Our interest was in determining if prostaglandins are essential for initiating or for maintaining the differentiation of decidual cells.

We studied the action of indomethacin on the DCR in a culture system. Endometrial cells, explanted from the uterus of the pseudopregnant rat on the fourth day of leukocytic smear (L_4), were cultured (as described by Vladimirsky *et al.*, 1977) for 4 days. At this time differentiation into decidual cells was observed, characterized by an increased nuclear

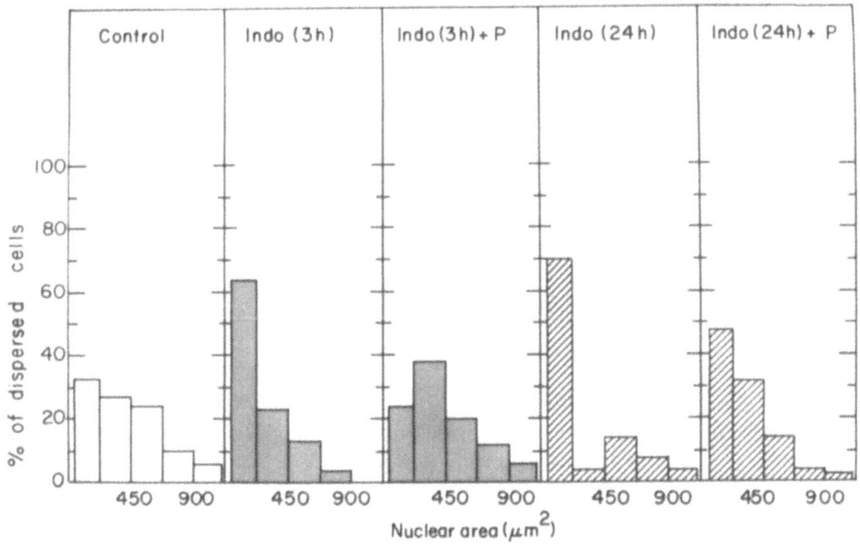

Figure 1. Effect of indomethacin administration *in vivo* on behavior of endometrial cells in culture, and reversal of the effect by progesterone. Indomethacin (1 mg per rat) was injected 3 or 24 hr before sacrifice on Day L_4. Progesterone (2×10^{-8} M) was added to the medium where indicated. Distribution of nuclear area was determined after 96 hr culture. Note partial suppression of shift to polyploidy by indomethacin pretreatment and restoration of decidual differentiation by progesterone *in vitro*. Indo, indomethacin; P, progesterone.

Sara Peleg and Hans R. Lindner • Department of Hormone Research, The Weizmann Institute of Science, Rehovot, Israel.

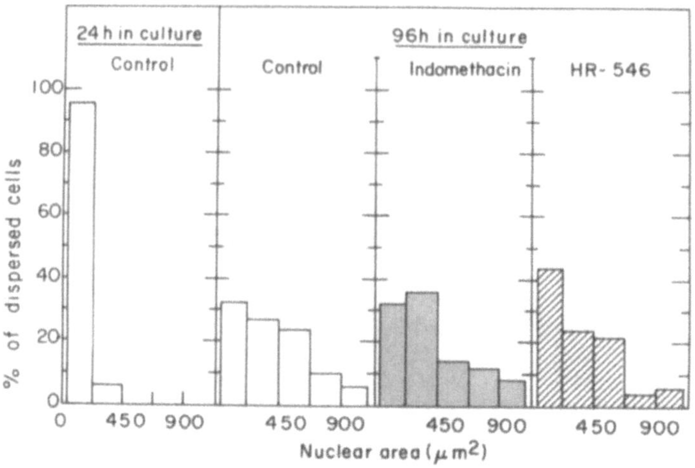

Figure 2. Differentiation of decidual cells *in vitro* in the presence of a PG-synthetase inhibitor or a PG-antagonist (HR-546). Distribution of nuclear area was determined before differentiation (at 24 hr) and after 96 hr culture. Endometrial scrapings from pseudopregnant (L_4) rats were trypsinized and cultured in medium containing indomethacin (5 μg/ml) or HR-546 (5 μg/ml) or in control medium. Nuclear area was estimated as $[\pi/4 (D_1 \times D_2)]$, where D_1 and D_2 are the smallest and largest diameters, respectively, measured in 100 cells.

area, indicative of increased cell ploidy (Barlow and Sherman, 1972). Pretreatment of the animals with indomethacin (1 mg per animal, i.p.) 3 or 24 hr before preparation of the cells for culture reduced the number of polyploid cells in the 4-day-old cultures (Figure 1). When progesterone (2×10^{-8} M) was added to cells from indomethacin-treated animals, the distribution profile of the nuclear area on the fourth day was similar to that observed in control cultures (Figure 1). Addition of 5 μg/ml HR-546, an antagonist of prostaglandins (Adaikan and Karim, 1977), or 5 μg/ml indomethacin, an inhibitor of prostaglandin synthesis, had no effect on overt differentiation (Figure 2).

These results suggest that (1) once triggered, the differentiation of endometrial cells into polyploid decidual cells *in vitro* is independent of prostaglandins; and (2) indomethacin interfered with progesterone action *in vivo*.

We measured the distribution of progestin receptors between nuclear and cytoplasmic

Table 1. The Effect of Indomethacin Administration on Nuclear and Cytoplasmic Binding of Progesterone Analog (R-5020) in Uteri of Pseudopregnant (L_4) Rats

	Specific binding of R-5020 (fmol/μg DNA)[a]			
	Cytosol 37,000g supernatant	Nuclei 800g pellet	Total binding	Nuclear binding/ total binding
Control	5.2±0.8	3.4±0.6	8.6±1.1	0.38±0.04
Indomethacin[b]	6.5±1.1	1.7±0.8	8.3±1.6	*0.17±0.05

[a]Triplicate samples of cytoplasmic and nuclear fractions from each uterus were exposed to [^{3}H]-R-5020 and binding was determined according to Peleg *et al.* (1979). Results are expressed as mean ± S.E. (seven uteri per treatment group).
[b]Indomethacin (1 mg per animal, i.p.) was injected on Day L_3.
*Significantly different from control values ($p < 0.05$).

fractions of uterine preparations, and found that on Day L_4, nuclear binding of progestin (R-5020), expressed as a fraction of total binding, was reduced in the uteri of indomethacin-treated rats to half the level seen in control rats (Table 1).

The possibility that indomethacin inhibits the decidual cell reaction through interference with the translocation of progesterone receptors into the nuclei of uterine cells will be examined further.

References

Adaikan, P. G., and Karim, S. M. M., 1977, Nonspecific prostaglandin antagonism by 8-ethoxycarbonyl-5-methyl-10,11-dihydro-PGA_1 ethyl ester (HR-546) on some smooth muscle preparations *in vitro, Prostaglandins* **14**:653.

Barlow, P. W., and Sherman, M. I., 1972, The biochemistry of differentiation of mouse trophoblast: Studies on polyploidy, *J. Embryol. Exp. Morphol.* **27**:447.

Peleg, S., Bauminger, S., and Lindner, H. R., 1979, Oestrogen and progestin receptors in deciduoma of the rat, *J. Steroid Biochem.* **10**:139.

Sananes, N., Baulieu, E. E., and Le Goascogne, C., 1976, Prostaglandin(s) as inductive factor of decidualization in the rat uterus, *Mol. Cell. Endocrinol.* **6**:153.

Vladimirsky, F., Chen, L., Amsterdam, A., Zor, U., and Lindner, H. R., 1977, Differentiation of decidual cells in cultures of rat endometrium, *J. Reprod. Fertil.* **49**:61.

14 Maternal Recognition of Pregnancy in Swine

M. T. Zavy, R. H. Renegar, W. W. Thatcher, and Fuller W. Bazer

I. Introduction

Sexually mature female pigs have recurring estrous cycles of 18 to 21 days unless interrupted by pregnancy. Mating occurs during estrus (24 to 72 hr) and pregnancy lasts about 114 days.

II. Uterine Endometrium–Corpora Lutea (CL) Relationship

Hysterectomy or destruction of uterine endometrial epithelium results in prolonged CL function (Anderson *et al.*, 1969). Prostaglandin $F_{2\alpha}$ is produced by pig endometrium (Patek and Watson, 1976) and exerts a luteolytic effect in swine (Moeljono *et al.*, 1976). Establishment of pregnancy or injection of estrogens, e.g., 5 mg estradiol valerate (E_2V) per day, on Days 11 through 15 of the estrous cycle leads to prolonged CL maintenance, "pseudopregnancy" (Frank *et al.*, 1977).

III. Theory of Maternal Recognition of Pregnancy in Swine

Two assumptions must be made to put the theory of maternal recognition of pregnancy in swine (Bazer and Thatcher, 1977) into perspective. First, estrogen, either injected (Frank *et al.*, 1977, 1978) or of conceptus origin (Heap *et al.*, 1979), is assumed to be the luteotropic signal. Second, $PGF_{2\alpha}$ is assumed to be the uterine luteolysin (Moeljono *et al.*, 1976).

As uterine endometrium is the source of the porcine luteolysin, immunoreactive prostaglandin F (PGF) concentrations in utero-ovarian vein plasma were compared between Day 12 and onset of the next estrus in nonpregnant gilts and Days 12 and 24 of pregnancy (Moeljono *et al.*, 1977). Average concentration, number of peaks ($>\bar{X} + 2$ S.D.), and mean peak concentrations of PGF were significantly reduced in pregnant versus nonpregnant gilts. Using a similar approach, utero-ovarian vein plasma PGF concentrations were compared in E_2 V-treated and control gilts (Frank *et al.*, 1977). E_2 V-treated gilts had significantly lower average concentrations, fewer peaks, and lower peak concentrations of PGF.

M. T. Zavy, R. H. Renegar, and Fuller W. Bazer • Department of Animal Science, University of Florida, Gainesville, Florida. *W. W. Thatcher* • Department of Dairy Science, University of Florida, Gainesville, Florida 32611.

In pregnant and E_2 V-treated gilts, CL maintenance was associated with markedly reduced utero-ovarian vein PGF concentrations.

To determine if PGF production was inhibited during pregnancy or E_2 V-induced pseudopregnancy, PGF content in uterine flushings was determined. Total recoverable PGF ($\bar{X}$ ± S.E.M., ng) per uterine horn for control and E_2V-treated gilts, respectively, was ($p < 0.01$): 2 ± 0.09 and 2 ± 0.8; 27 ± 8 and 19± 3; 31 ± 10 and 2504 ± 676; 210 ± 59 and 4506 ± 1137; 66 ± 30 and 5113 ± 1735 on Days 11, 13, 15, 17, and 19 after onset of estrus (Frank *et al.*, 1978). Similarly, total recoverable PGF per two uterine horns ($\bar{X}$ ± S.E.M., ng) for pregnant and nonpregnant gilts, respectively, was ($p < 0.01$): 14 ± 3 and 68 ± 27; 19 ± 3 and 28 ± 4; 17 ± 5 and 23 ± 6; 111 ± 54 and 39 ± 8; 2427 ± 1548 and 204 ± 68; 1620 ± 868 and 658 ± 573; 15,866 ± 7588 and 828 ± 348; 22,466 ± 1761 and 821 ± 127 for Days 6, 8, 10, 12, 14, 15, 16, and 18 after onset of estrus. These data indicate that endometrial PGF production is not inhibited during pregnancy or pseudopregnancy, but uterine lumenal content is significantly elevated.

The theory of maternal recognition of pregnancy in swine (Bazer and Thatcher, 1977) can now be summarized. In nonpregnant pigs, endometrial PGF is released in an endocrine fashion and elevated utero-ovarian vein PGF concentrations lead to CL regression and a new estrous cycle. In pregnant pigs, the endometrium also produces PGF, but estrogens of blastocyst origin maintain secretion of PGF into the uterine lumen (exocrine direction). The PGF is therefore unavailable to exert a luteolytic effect on the CL that is essential for pregnancy.

References

Anderson, L. L., Bland, K. P., and Melampy, R. M., 1969, Comparative aspects of uterine luteal relationships, *Rec. Prog. Horm. Res.* **25**:27.

Bazer, F. W., and Thatcher, W. W., 1977, Theory of maternal recognition of pregnancy in swine based on estrogen controlled endocrine versus exocrine secretion of prostaglandin $F_{2\alpha}$ by the uterine endometrium. *Prostaglandins* **14**:397.

Frank, M., Bazer, F. W., Thatcher, W. W., and Wilcox, C. J., 1977. A study of prostaglandin $F_{2\alpha}$ as the luteolysin in swine. III. Effects of estradiol valerate on prostaglandin F, progestins, estrone and estradiol concentrations in the utero-ovarian vein of nonpregnant gilts, *Prostaglandins* **14**:1183.

Frank, M., Bazer, F. W., Thatcher, W. W., and Wilcox, C. J., 1978. A study of prostaglandin $F_{2\alpha}$ as the luteolysin in swine IV. An explanation for the luteotrophic effect of estradiol, *Prostaglandins* **15**:151.

Heap, R. B., Flint, A. P. F., Gadsby, J. E., and Rice, C., 1979. Hormones of the early embryo and the uterine environment, *J. Reprod. Fertil.* **55**:267.

Moeljono, M. P. E., Bazer, F. W., and Thatcher, W. W., 1976, A study of prostaglandin $F_{2\alpha}$ as the luteolysin in swine I. Effect of prostaglandin $F_{2\alpha}$ in hysterectomized gilts, *Prostaglandins* **11**:737.

Moeljono, M. P. E., Thatcher, W. W., Bazer, F. W., Frank, M., Owens, L. J., and Wilcox, C. J., 1977, A study of prostaglandin $F_{2\alpha}$ as the luteolysin in swine II. Characterization and comparison of prostaglandin F, estrogens and progestin concentrations in utero-ovarian vein plasma of nonpregnant and pregnant gilts, *Prostaglandins* **14**:543.

Patek, C. E., and Watson, J., 1976, Prostaglandin F and progesterone secretion by porcine endometrium and corpus luteum *in vitro*, *Prostaglandins* **12**:97.

15 Effects of Preimplantation Bovine and Porcine Conceptuses on Blood Flow and Steroid Content of the Uterus

S. P. Ford, J. R. Chenault, R. K. Christenson, S. E. Echternkamp, and J. J. Ford

I. Introduction

The embryonic signal that initiates luteal maintenance occurs by about Day 12 in the sow (Dhindsa and Dzuik, 1968), Day 13 in the ewe (Moor and Rowson, 1966), and Day 16 in the cow (Betteridge *et al.*, 1978). The mechanism responsible for maintenance of a corpus luteum (CL) during early pregnancy, and in particular the way in which the embryo influences this process, is not clearly understood but may involve a local affect of the conceptus on utero-ovarian blood flow. Data presented by Ford *et al.* (1976) demonstrated a preferential effect of the ovine and bovine conceptus for reducing *in vitro* constriction of the uterine artery supplying the gravid horn on Day 15 and Day 17 postcoitus, respectively. Transient increases in blood flow to gravid uterine horns of ewes have been observed on Days 13 to 15 of pregnancy (Greiss and Anderson, 1970).

It is known that the pig blastocyst develops the capacity to synthesize estrogen *in vitro* by Day 12 of pregnancy (Perry *et al.*, 1976). Increased uterine blood flow following estrogen administration has been observed in the cow (Roman-Ponce *et al.*, 1978), sow (Dickson *et al.*, 1969), and ewe (Huckabee *et al.*, 1970), and injection of estrogen into the lumen of an isolated uterine horn in ewes causes a rapid unilateral increase in blood flow (Greiss and Miller, 1971). Thus estrogens of embryonic origin may act locally and directly to increase uterine and/or ovarian blood flow in these species, establishing optimal conditions for continuation of pregnancy.

The present studies were conducted to determine patterns of blood flow to the uterus and estrogen secretion by uteri of cows and sows during the estrous cycle and early pregnancy.

II. Results and Discussion

Blood flow to the uterine horns of cows and sows was determined using electromagnetic blood flow probes placed around left and right middle uterine arteries. The pattern of blood flow to the uteri of pregnant and nonpregnant cows was similar until Day 14 (first day of estrus = Day 0), while the pattern of blood flow to the uteri of pregnant and nonpregnant sows was similar until Day 11. Between Days 14 and 18 postcoitus in cows,

S. P. Ford, J. R. Chenault, R. K. Christenson, S. E. Echternkamp, and J. J. Ford. • Roman L. Heuska U.S. Meat Animal Research Center, Clay Center, Nebraska 68933. Present address of S.P.F.: Department of Animal Science, Iowa State University, Ames, Iowa 50011. Present address of J.R.C.: The Upjohn Company, Kalamazoo, Michigan 49001.

and Days 11 and 14 postcoitus in sows, blood flow to the uterine horns containing conceptuses increased ($p < 0.01$) severalfold. No corresponding increase in blood flow to the uterine horns was observed in nonpregnant cows and sows. The increase in blood flow to the gravid uterine horns of cows and sows occurred on days when the presence of embryos had been shown to extend the lifespan of the CL.

In other studies, uteri of cows (Days 14, 16, and 18 after mating or estrus) and sows (Days 11, 13, and 15 after mating or estrus) were exposed through a midventral incision and a branch of an artery and vein supplying and draining each uterine horn was cannulated. Following blood collection, uterine horns ipsilateral to ovaries containing CL were flushed with phosphate-buffered saline to obtain embryo(s) and/or uterine flushings. Embryos (bovine only), uterine flushings, and blood plasma were stored separately at −20°C until assay of estrone (E_1) and estradiol-17β (E_2) by radioimmunoassay. Conceptuses were not separated from uterine flushings of pregnant sows; thus "uterine flushings," when referring to sows, contained embryonic tissue. On Days 14, 16, and 18 of pregnancy in the cow, concentrations of E_2 were greater ($p < 0.05$) in uterine venous blood draining a gravid uterine horn than those in the uterine arterial blood supplying the horn (9.2 ± 2.3 vs. 4.9 ± 1.3 pg/ml). Detectable E_2 (>6.7 pg) was found in uterine flushings and embryos of some cows as early as Day 14 of pregnancy. By Day 18 of pregnancy all uterine flushings and embryos contained detectable E_2 (26.7 ± 4.7 and 10.9 ± 1.5 pg, respectively). No E_2 was detected in uterine flushings from nonpregnant cows. On Days 11 and 13 of pregnancy in sows, only E_2 concentrations were greater ($p < 0.05$) in uterine venous blood than uterine arterial blood (7.2 ± 2.0 vs. 3.9 ± 0.5 and 7.6 ± 1.4 vs. 4.4 ± 1.1 pg/ml, respectively). On Day 15 pregnancy, E_1 concentrations were greater ($p < 0.05$) in uterine venous than uterine arterial blood (13.8 ± 0.6 vs. 9.7 ± 0.6 pg). No arterial–venous differences in E_1 or E_2 were observed in nonpregnant sows. On Day 11 of pregnancy, only sows with blastocysts greater than 8 mm in diameter had elevated concentrations of estrogens in uterine flushings compared to flushings from nonpregnant controls. On Days 13 and 15 of pregnancy, uterine flushings from all sows contained markedly elevated ($p < 0.001$) levels of E_1 and E_2 compared to flushings from nonpregnant controls. In addition, uterine flushings from sows on day 13 of pregnancy had a greater ($p < 0.05$) content of E_1 and E_2 than uterine flushings from sows on Day 15 of pregnancy (7204 ± 3753 and 2029 ± 646 vs. 1327 ± 488 and 819 ± 290 pg, respectively).

Results of these studies clearly suggest a local influence of the bovine and porcine conceptus on uterine blood flow before any definitive attachment between trophoblast and uterine wall. Increased levels of estrogens in uterine flushings and uterine blood are temporally associated with the increased blood flow to gravid uterine horns observed on Day 15–17 in the cow and Days 12–13 in the sow. It is conceivable that the early conceptus in these species produces or stimulates synthesis of estrogen that dilates the ipsilateral utero-ovarian vasculature at a time when the presence of the embryos has been determined to extend the lifespan of the CL.

ACKNOWLEDGMENT. The cooperation of the Nebraska Agricultural Experiment Station, University of Nebraska, Lincoln, is acknowledged.

References

Betteridge, K. J., Eaglesome, M. D., Randall, G. C. B., Mitchell, D., and Sugden, E. A., 1978, Maternal progesterone levels as evidence of luteotrophic or antiluteolytic effects of embryos transferred to heifers 12–17 days after estrus, *Theriogenology* **9**:86 (abstract).

Dhindsa, D. S., and Dziuk, P. J., 1968, Effect on pregnancy in the pig after killing embryos or fetuses in one uterine horn in early gestation, *J. Anim. Sci.* **27:**122.

Dickson, W. M., Bosc, M. J., and Locatelli, A., 1969, Effect of estrogen and progesterone on uterine blood flow in castrate sows, *Am. J. Physiol.* **217:**1431.

Ford, S. P., Weber, L. J., and Stormshak, F., 1976, *In vitro* response of ovine and bovine uterine arteries to prostaglandin $F_{2\alpha}$ and periarterial sympathetic nerve stimulation *Biol. Reprod.* **15:**58.

Greiss, F. C., Jr., and Anderson, S. G., 1970, Uterine blood flow during early ovine pregnancy, *Am. J. Obstet. Gynecol.* **106:**30.

Greiss, F. C., Jr., and Miller, H. B., 1971, Unilateral control of uterine blood flow in the ewe, *Am. J. Obstet. Gynecol.* **111:**299.

Huckabee, W. E., Crenshaw, C., Curet, L. B., Mann, L., and Barron, D. H., 1970, The effect of exogenous oestrogen on the blood flow and oxygen consumption of the uterus of the nonpregnant ewe, *Q. J. Exp. Physiol* **55:**16.

Moor, R. M., and Rowson, L. E. A., 1966, The corpus luteum of the sheep: Effect of the removal of embryos on luteal function, *J. Endocrinol.* **34:**497.

Perry, J. S., Heap, R. B., Burton, R. D., and Gadsby, J. E., 1976, Endocrinology of the blastocyst and its role in the establishment of pregnancy, *J. Reprod. Fertil. Suppl.* **25:**85.

Roman-Ponce, H., Thatcher, W. W., Caton, D., Barron, D. H., and Wilcox, C. J., 1978, Thermal stress effects of uterine blood flow in dairy cows, *J. Anim. Sci.* **46:**175.

16 Effects of Nicotine on Conceptus Cell Proliferation and Oviductal/Uterine Blood Flow in the Rat

Robert E. Hammer, Jerald A. Mitchell, and Harold Goldman

I. Introduction

Daily administration of nicotine to rats during Days 0–5 of pregnancy delays loss of the zona pellucida, obliteration of the blastocyst cavity, and implantation (Card and Mitchell, 1979). Prior to implantation the unattached conceptus depends on oxygen and other essential metabolic substrates available within the lumen of the reproductive tract to sustain its continued growth and development. The availability of oxygen within the uterine lumen of the rat increases prior to implantation (Yochim and Mitchell, 1968) and declines rapidly upon vasoconstriction of the uterine vasculature (Mitchell and Yochim, 1968). Because conceptus development requires optimal oxygen tension (Brinster, 1972) and nicotine is a potent vasoactive substance, it was suggested that the nicotine-induced alterations in conceptus development and implantation may result, in part, from changes in reproductive tract blood flow (Hammer and Mitchell, 1979). The following study was undertaken to establish whether nicotine alters cell proliferation in embryos prior to implantation and if so whether such alterations are associated with reduced oviductal and/or uterine blood flow.

II. Materials and Methods

Adult Sprague–Dawley rats maintained under controlled environmental conditions were bred (positive sperm = Day 0 of pregnancy). Beginning on Day 0, rats received two daily injections (10 A.M. and 3 P.M.) of either saline or nicotine (5 mg/kg body wt) through Day 5 of pregnancy. Embryos retrieved at selected times from the oviduct or uterus were examined, the cells dispersed, and nuclei counted (Tarkowski, 1966). Oviductal and uterine blood flow were measured by ^{86}Rb fractionation (Sapirstein, 1958) on Day 4 in pseudopregnant rats at selected times following a single injection of saline or nicotine solution (5 mg/kg body wt).

III. Results

Nicotine treatment (Days 0–5) slows intraoviductal conceptus cell proliferation as indicated by a reduction in mean number of nuclei per conceptus at all retrieval times (Figure

Robert E. Hammer, Jerald A. Mitchell, and Harold Goldman • Departments of Anatomy and Pharmacology, Wayne State University, School of Medicine, Detroit, Michigan 48201.

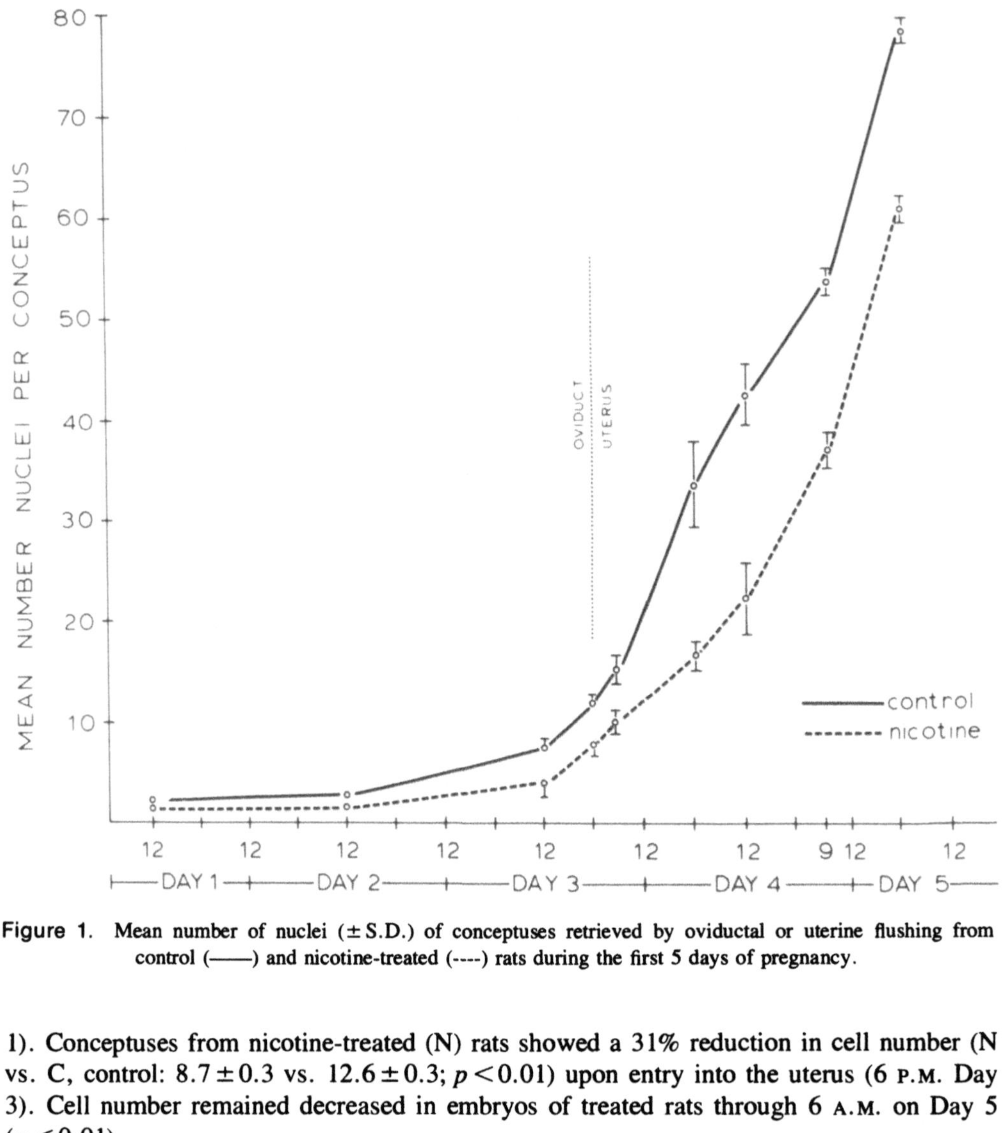

Figure 1. Mean number of nuclei (± S.D.) of conceptuses retrieved by oviductal or uterine flushing from control (——) and nicotine-treated (----) rats during the first 5 days of pregnancy.

1). Conceptuses from nicotine-treated (N) rats showed a 31% reduction in cell number (N vs. C, control: 8.7 ± 0.3 vs. 12.6 ± 0.3; $p < 0.01$) upon entry into the uterus (6 P.M. Day 3). Cell number remained decreased in embryos of treated rats through 6 A.M. on Day 5 ($p < 0.01$).

Ten minutes following nicotine injection, oviductal blood flow decreased from a control value of 0.61 ± 0.06 ml/min/g to 0.45 ± 0.03, a reduction of 40%, and remained suppressed through 2 hr ($p < 0.005$) (Figure 2). Similarly, uterine blood flow was decreased 40% at 10 min following nicotine and remained reduced for 2 hr ($p < 0.005$).

IV. Discussion

As embryos of nicotine-treated rats consist of markedly fewer cells upon entry into the uterus, the growth-suppressing effects of the alkaloid are exerted on the conceptus while it is in transit through the oviduct. That the difference in cell number between embryos of C vs. N rats persists after entry into the uterus suggests that growth-retarding effects continue to be exerted on conceptuses *in utero*. The fact that a single injection of nicotine results in a rapid, marked, and sustained reduction in both oviductal and uterine blood flow suggests

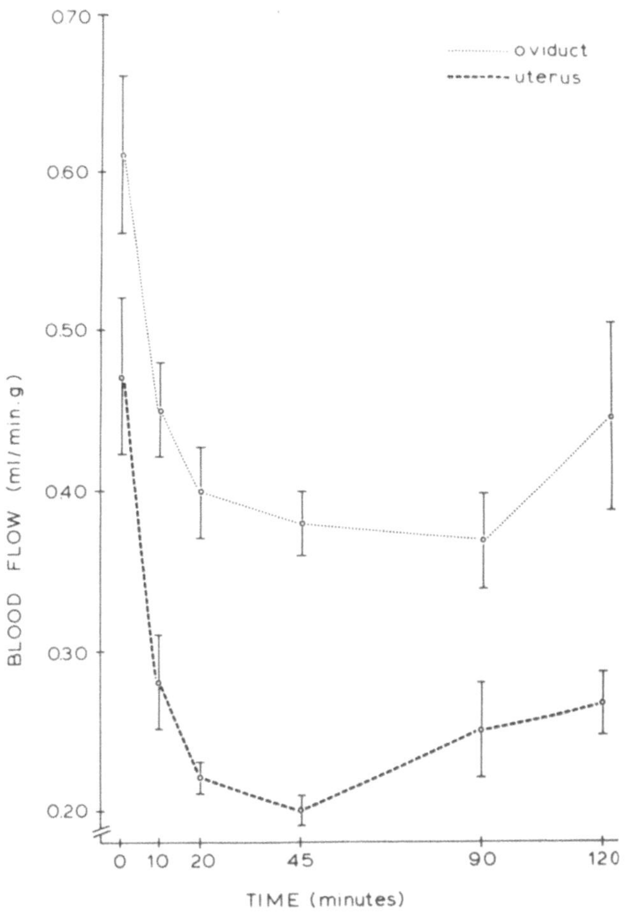

Figure 2. Oviductal (····) and uterine (----) blood flow (± S.E.M.; $n \geq 5$ at each point) following nicotine injection to 4-day pseudopregnant rats. Oviductal and uterine blood flows were reduced in treated animals through 2 hr ($p < 0.005$).

that alterations in conceptus growth result in part from reduced substrate availability within the lumina of the reproductive tract. How nicotine reduces blood flow remains to be determined. However, the observations that the uterine vasculature of the rat receives sympathetic innervation (Adham and Schenk, 1969), that nicotine evokes systemic epinephrine release (Hazard *et al.*, 1957), and that intravenous or topically applied epinephrine markedly reduces intrauterine oxygen tension (Mitchell and Yochim, 1968) suggest nicotine-evoked activation of the sympathetic nervous system.

Acknowledgment. This work was supported by Grant 1012 from the Council for Tobacco Research, USA.

References

Adham, N., and Schenk, E., 1969, Autonomic innervation of the rat vagina, cervix and uterus and its cyclic variation, *Am. J. Obstet. Gynecol.* **104:**508–516.

Brinster, R., 1972, Developing zygote, in: *Reproductive Biology* (H. Balin and S. Glasser, eds.), pp. 748–775, Excerpta Medica, Amsterdam.
Card, J. P., and Mitchell, J. A., 1979, The effects of nicotine on implantation in the rat, *Biol. Reprod.* **20:**532–539.
Hammer, R. E., and Mitchell, J. A., 1979, Nicotine reduces embryo growth, delays implantation and retards parturition in rats, *Proc. Soc. Exp. Biol. Med.* **162:**333–336.
Hazard, R., Beauvallet, M., and Larno, S., 1957, Adrenaline and noradrenaline content of adrenal vein in the dog subjected to the action of nicotine. *C. R. Soc. Biol.* **151:**210–212.
Mitchell, J. A., and Yochim, J. M., 1968, Measurement of intrauterine oxygen tension in the rat and its regulation by ovarian steroid hormones, *Endocrinology* **83:**691–700.
Sapirstein, L. A., 1958, Regional blood flow by fractional distribution of indicators, *Am. J. Physiol.* **193:**161–163.
Tarkowski, A. K., 1966, An air drying method for chromosome preparations from mouse eggs, *Cytogenetics* **5:**394–400.
Yochim, J. M., and Mitchell, J. A., 1968, Intrauterine oxygen tension in the rat during progestation: Its possible relation to carbohydrate metabolism and the regulation of nidation, *Endocrinology* **83:**706–712.

17 *The Uterine Epithelium as a Transducer for the Triggering of Decidualization in the Rat*

F. Leroy and B. Lejeune

In rats and mice, the uterine surface epithelium on Day 5 of pregnancy appears still intact while the decidual reaction is already spreading (Finn, 1977). Therefore the physiological signal from the blastocyst that initiates decidualization must be conveyed in some way through the epithelial layer. Uterine trauma, such as scratching or crushing, can elicit a decidual cell reaction (DCR) in pseudopregnant or adequately sensitized spayed animals, but whereas a "normal" DCR requires previous intervention of small amounts of estrogen (Finn, 1977), traumatic deciduomas can also be obtained with progesterone alone (De Feo, 1967). It has therefore been argued that luteal estrogen primarily acts upon epithelial cells by enabling them to transmit the embryonic message, whereas trauma would bypass this step and act on the stroma directly (Finn and Martin, 1974). Another opinion holds that the epithelium is a compulsory intermediate in any type of DCR, for traumatic procedures would be efficient only if they also damage epithelial cells (Psychoyos, 1974; Thibault, 1978).

Our aim was to determine which of the foregoing interpretations for the role of the epithelium is correct and to investigate how this tissue acts as a transducer in the onset of normal decidualization.

Seven days after ovariectomy, rats were given hormonal treatment mimicking progestational ovarian secretions; that is, 2 days at 500 ng estradiol-17β per day followed by 2 days of rest, after which a daily dose of 5 mg progesterone was given for 3 days. On the last day, 50 ng estradiol-17β was also injected. At maximal uterine sensitivity (i.e., 15 hr after this last estrogen administration), animals underwent laparotomy and both uterine horns were transected at the isthmus. This allowed the lumenal epithelium to be removed by gentle longitudinal squeezing of the cornu. This method for *in vivo* ablation was derived from that used by Bitton-Casimiri *et al.* (1977) on dissected horns. Histological controls showed that in the majority of cases, the removed tissue ribbon was pure epithelium, the *in situ* stroma remaining intact and still covered by the basal membrane.

Different treatments (scratching and/or intralumenal injections) were applied to the uterine horns immediately after uni- or bilateral removal of the epithelium. In each animal one side was treated in such a manner as to serve as an internal control. In some groups the epithelium was squeezed out or merely dissociated *in situ* without its removal from the lumen, in only the upper or lower half of the experimental horn. After 4 days the animals were killed and the DCR assessed, during which time they had received 2 mg progesterone

F. Leroy and B. Lejeune • Laboratory of Gynecology and Research on Human Reproduction, St. Pierre Hospital, Free University, Brussels, Belgium. F.L. and B.L. are Chercheur qualifié and Aspirant at the Belgium F.N.R.S., respectively.

plus 200 ng estradiol-17β. One group was left to rest for 12 days after unilateral ablation of the epithelium before being hormonally resensitized and submitted to scratching of both uterine cornua.

The results indicated that (1) although submitted to the same DCR induction (scratching + PBS injection) as the reactive control side, deepitheliated horns had become incapable of producing a DCR; (2) none of several substances that are possibly involved in the triggering of decidualization, i.e., histamine (Shelesnyak *et al.*, 1970)) and prostaglandins E_2 and $F_{2\alpha}$ (Sananes *et al.*, 1976), could elicit a DCR when injected into cornua deprived of their epithelium; (3) even intralumenal injection of concentrated 60,000*g* supernatant or sediment of ground epithelium isolated from identically sensitized donor rats did not restore the capacity for decidualization of deepitheliated uterine horns; (4) mere dissociation *in situ* of the epithelium without pulling it out of the uterine lumen was enough to prevent the DCR from occurring; (5) detachment of the epithelium in the upper or lower half of a horn precluded DCR in this treated region but at the same time a massive response was induced in the adjacent untouched segment in 66% of the animals; (6) 8 to 15 days after its ablation the epithelium had been regenerated from the glands (Leroy *et al.*, unpublished observations), allowing decidualization to occur again at 65% of maximal weight.

Because squeezing of the uterus might have caused histologically unapparent damage to stromal cells and thereby inhibited DCR, the effect of total longitudinal crushing of one horn per rat versus that of two transverse pinches was tested in adequately sensitized animals. In spite of this trauma, which entails extensive tissue damage, the crushed horns were still capable of reacting by a DCR amounting to 75% of the massive heterolateral response. It therefore appears that unresponsiveness of deepitheliated horns is related to the absence of the lumenal epithelium and not to stromal cell destruction.

Our results thus indicate that in the triggering of decidualization the endouterine epithelium plays the role of an obligatory transducer that cannot be bypassed. This implies that although different types of stimuli can induce the DCR (De Feo, 1967), the message ultimately conveyed to the stroma depends on epithelial specificity. The inability of epithelial extracts to restore the capacity to decidualize and the observation that mere disruption of topographical relationships between both tissues would prevent DCR, suggest that membrane changes such as permeability modifications and/or electrical depolarization might be involved. It is possible, however, that deciduogenic substances emanating from the epithelium would only be produced or synthesized with some delay after the application of an inductive stimulus; mere traumatic disruption of some epithelial cells may not suffice to make these substances immediately available.

References

Bitton-Casimiri, V., Rath, N. C., and Psychoyos, A., 1977, A simple method for separation and culture of rat uterine epithelial cells, *J. Endocrinol.* **73:**537.

De Feo, V. J., 1967, Decidualization, in: *Cellular Biology of the Uterus* (R. M. Wynn, ed.), pp. 191–290. Appleton–Century–Crofts, New York.

Finn, C. A., 1977. The implantation reaction, in: *Biology of the Uterus* (R. M. Wynn, ed.), pp. 246–306, Plenum Press, New York.

Finn, C. A., and Martin, L., 1974, The control of implantation, *J. Reprod. Fertil.* **39:**795.

Psychoyos, A., 1973, Hormonal control of implantation, *Vitam. Horm.* **32:**201.

Sananes, N., Baulieu, E. E., and Le Goascogne, C., 1976, Prostaglandin(s) as inductive factor of decidualization in the rat uterus, *Mol. Cell. Endocrinol.* **6:**153.

Shelesnyak, M. C., Marcus, G. J., and Lindner, H. R., 1970, Determinants of the decidual reaction, in: *Ovoimplantation, Human Gonadotropins and Prolactin* (P. O. Hubinont, F. Leroy, C. Robyn, and P. Leleux, eds.), Karger, Basel.

Thibault, C., 1978, L'implantation: sa programmation, in: *L'implantation de l'oeuf* (F. du Mesnil du Buisson, A. Psychoyos, and K. Thomas, eds.), pp. 1–19, Masson, Paris.

18 Cellular Contact between the Trophoblast and the Endometrium at Implantation in the Ewe

M. Guillomot, J. E. Fléchon, and S. Wintenberger-Torrès

I. Introduction

Implantation in the ewe takes place well after blastocyst differentiation, and the trophoblast and the uterine epithelium first establish contact between Days 15 and 20 (Boshier, 1969).

We have tried to determine the morphological and ultrastructural changes occurring from Day 12 on the interfaces of the trophoblast/endometrium, and we have studied the nature and development of the coat of the endometrial and trophoblast cell surfaces which intervenes in the processes of cellular interaction.

Differences between cyclic and pregnant ewes first appeared on Day 14. There was a depression in the center of the pregnant ewe caruncle; folds extended from the depression (Figure 1). The cyclic ewe caruncle had a smooth surface. The trophoblast adhered to the endometrial surface in several places and might be caught in the caruncular folds.

At Day 15, the embryo was seen adhering to the endometrium at many points and around the glandular openings; in following the epithelial folds, the trophoblast covered several caruncles.

Between Days 16 and 20, the trophoblast covered an increasing number of caruncles. The uterine epithelium was uneven; some voluminous cells were prominent among the epithelial cells.

The contacts between the trophoblast and the uterine epithelium became narrower. The uterine microvilli and the folds of the trophoblast plasma membrane interpenetrated.

Villous proliferations about 100 μm long appeared in the areas of the trophoblast covering the edges of the caruncles and the intercaruncular spaces. Electron microscopy showed that these proliferations penetrated the uterine gland lumina (Figure 2); they constituted a real, temporary histotrophic system by which the embryo absorbed uterine secretions.

At Day 18, the unlysed uterine epithelium flattened and became syncytial.

II. Cytochemical Studies

The cell coat observed on both the uterine and the trophoblast cells was composed of glycoproteins and glycolipids. On the uterine epithelium, this coat was not modified during implantation. But on the trophoblast cells, concanavalin A receptors and Ruthenium red

M. Guillomot, J. E. Fléchon, and S. Wintenberger-Torrès • Station de Physiologie Animale–I.N.R.A., Jouy-en-Josas, France.

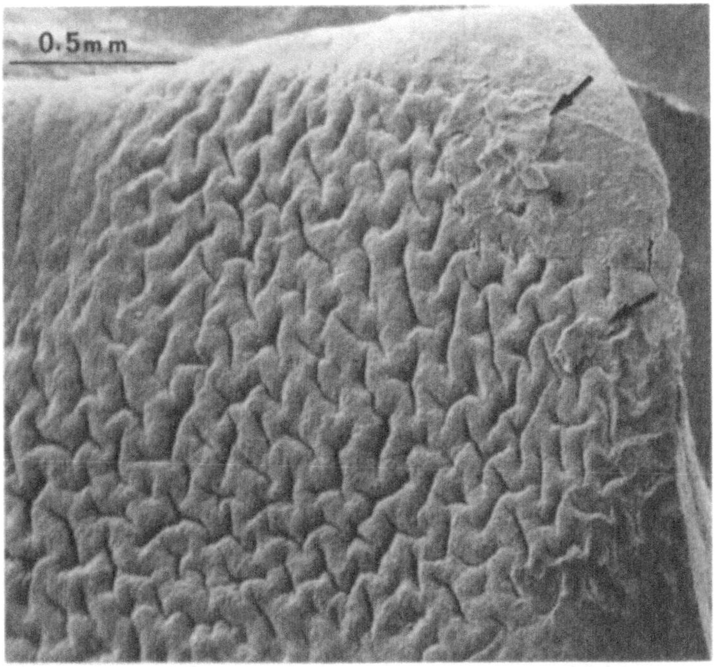

Figure 1. Uterine caruncle of a pregnant ewe (Day 16). Note characteristic wrinkling of epithelial surface. Pieces of trophoblast are still adhering (arrows). Scanning electron microscopy.

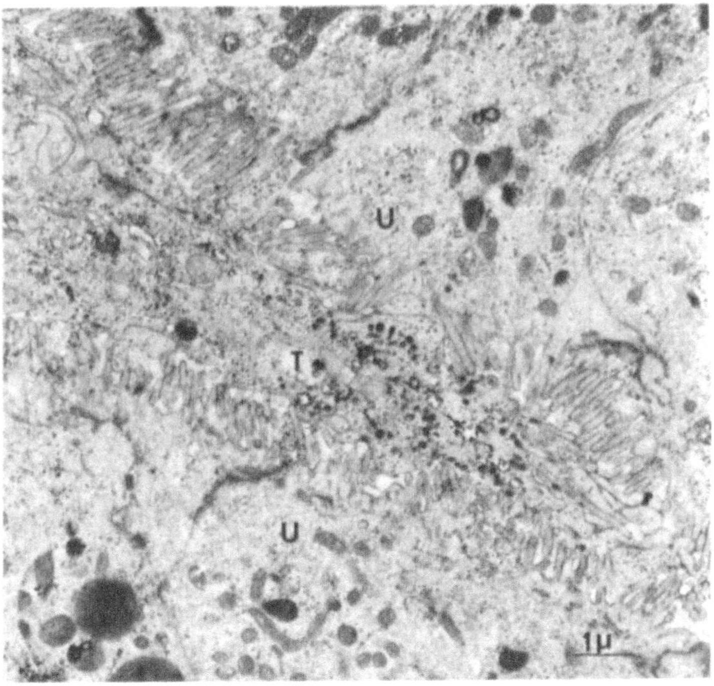

Figure 2. Pregnant ewe (Day 15). A trophoblast villous proliferation (T) is seen invading the lumen of a uterine gland. Glandular epithelium cells (U) are provided with microvilli. Note endocytotic vesicles (arrow) within the cytoplasm of trophoblast cell. Transmission electron microscopy. Uranylacetate–lead citrate staining.

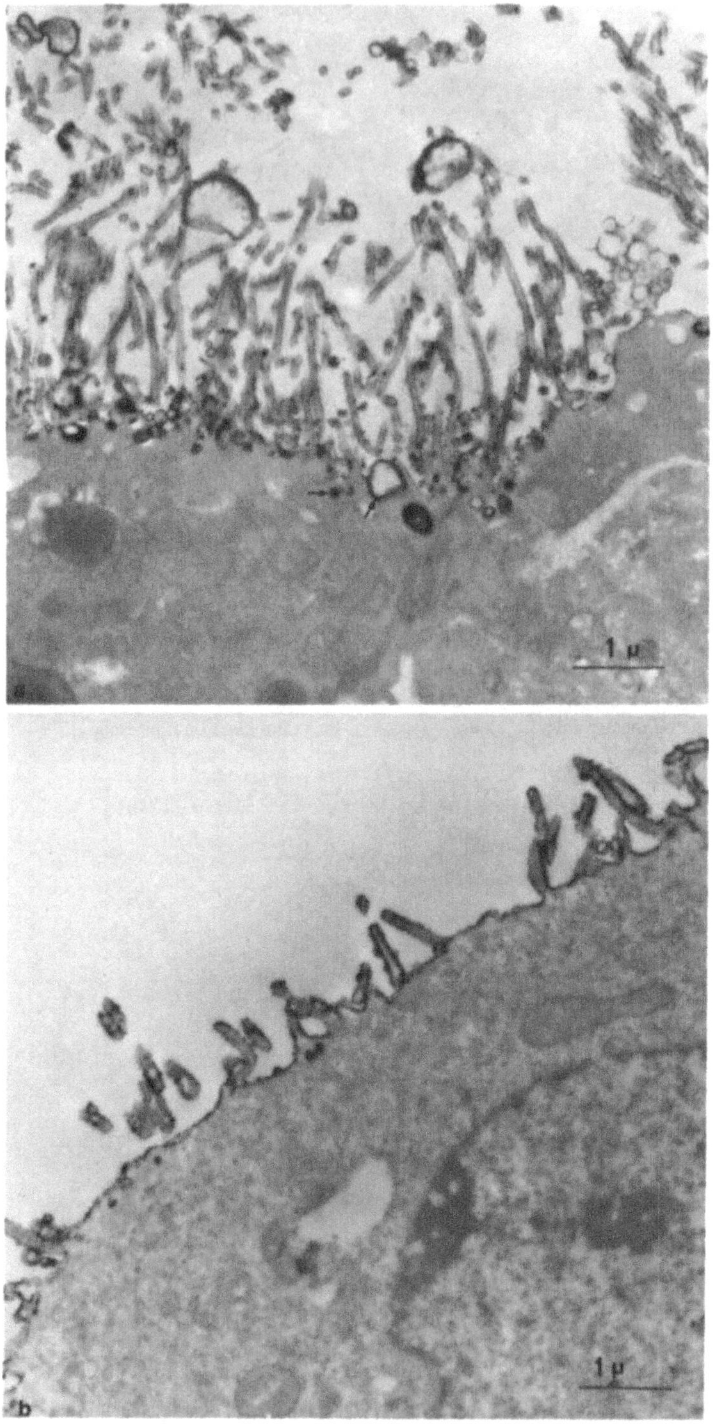

Figure 3. Cytochemical study of surface coats. (a) Trophoblast cell (Day 13) after concanavalin A–peroxidase procedure. Peroxidase reaction product is only seen coating invaginations of the plasma membrane. Surface of microvilli is nearly devoid of reaction product. Unstained section. (B) Trophoblast cell (Day 15). Same procedure as (a). Peroxidase reaction product covers entire cell surface. Apical microvilli have nearly disappeared at this stage. Compare with (a). Unstained section.

staining increased at the beginning of implantation at Day 15 (Figure 3). Negative charges, probably due to acid groups, were also distributed more homogeneously at that stage. Nevertheless, the presence of negative charges on both the uterine and the trophoblast cells did not confirm the concept that adhesion is a simple matter of electrostatic attraction (Enders and Schlarfke, 1974).

The presence of PTA-positive material between maternal and embryonic tissues might correspond to fibrinoids acting as immunosuppressive agents already described in the mouse (Bradbury *et al.*, 1965).

Endocytosis of exogenous material was observed in the uterine epithelium at all stages studied. But at Day 15 endocytosis was more extensive, and absorbed material crossing the basal membrane was found 30 min later in the connective tissue.

This absorption of exogenous material from the uterine lumen might facilitate adhesion of the conceptus, but might also be a way of transmitting "information" from the conceptus to the maternal organism for maintenance of pregnancy.

III. Conclusions

We found extensive changes at the beginning of implantation (Day 15) in the surface structure of the epithelium and in uterine cell activity as seen in endocytosis. The factors responsible for these modifications and their role during implantation remain to be determined.

References

Boshier, D. P., 1969, A histological and histochemical examination of implantation and early placentome formation in sheep, *J. Reprod. Fertil.* **19:**51–61.

Bradbury, S., Billington, W. D., and Kirby, D. R. S., 1965, A histochemical and electron microscopical study on the fibrinoid of the mouse placenta, *J. R. Microsc. Soc.* **84:**199–211.

Enders, A. C., and Schlafke, S., 1974, Surface coats of the mouse blastocyst and uterus during the preimplantation period, *Anat. Rec.* **180:**31–46.

19 *Cell Motility and the Distribution of Actin and Myosin during Mouse Trophoblast Development in Vitro*

J. S. Sobel, R. H. Glass, R. Cooke, A. I. Spindle, and Roger A. Pedersen

Mouse trophoblast development *in vitro* can be used as a model system for the study of preimplantation events (Sherman and Wudl, 1976; Glass *et al.*, 1979). We have utilized this system to examine the relationship between cell motility and the organization of contractile proteins during trophoblast development.

Mouse blastocysts were cultured in modified Eagle's medium supplemented with 5% each of fetal and newborn calf serum (Glass *et al.*, 1979). The blastocysts attached to the substrate on the second day of culture and trophoblast spread out as a sheet by active movement of the marginal cells (Figure 1a). Trophoblast outgrowth was measured on sequential photographs and on tracings of time-lapse cinematographic recordings with a planimeter. The mean area of blastocysts at the time of attachment was 0.02 ± 0.01 mm^2 (n = 10) and the area increased to a maximum outgrowth size of 0.28 ± 0.05 mm^2 (n = 10) on the sixth or seventh culture day. A representative culture increased its area at a maximum rate of 0.13 μm^2/min, with a linear increase of 0.17 μm/min, during the third and fourth culture days. Cultures exhibited a slow and variable decline in rate of area increase during the subsequent 2 days of spreading. Within a day of achieving maximal outgrowth size, cell shape became more elongate as cells began to pull apart from each other and a few small spaces appeared between cells in the previously intact trophoblast sheet. These spaces were formed by active contractile movements of at least one of a pair of separating cells. The first interstices were quickly resealed, but those that formed on the following day (Figure 1b) persisted and continued to enlarge during the seventh to tenth culture day. In this way the trophoblast sheet was gradually converted into a networklike structure consisting of open spaces enclosed by interconnecting cells. Although trophoblast exhibited only small local alterations in its periphery during this time, the area actually occupied by cells declined as spaces between cells enlarged and the cells themselves contracted. The few cells that succeeded in completely detaching themselves from neighboring cells assumed a rounded configuration and did not translocate although they continued to exhibit active surface movement.

When trophoblast was cocultured with either L cells or macrophages, the advancing edge of the trophoblast outgrowth displaced these cells, demonstrating the "invasiveness"

J. S. Sobel, R. H. Glass, R. Cooke, A. I. Spindle, and Roger A. Pedersen • Laboratory of Radiobiology, and Department of Obstetrics, Gynecology, and Reproductive Sciences, and Department of Biochemistry and Biophysics, University of California, San Francisco, California 94143. *Roger A. Pedersen* • Laboratory of Radiobiology, and Department of Anatomy, University of California, San Francisco, California 94143. Present address of J.S.S.: Department of Anatomical Sciences, State University of New York at Buffalo, Buffalo, New York, 14214.

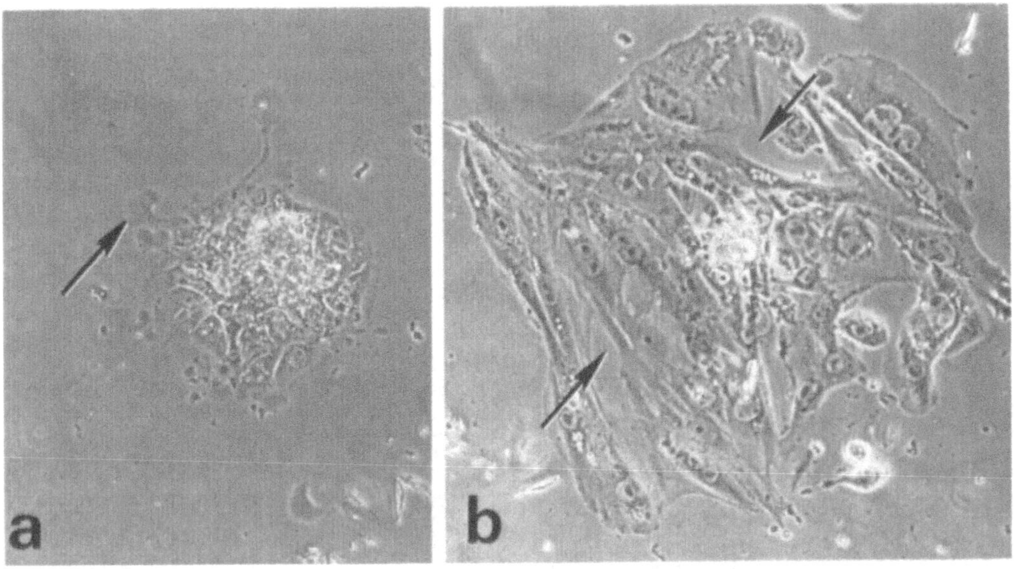

Figure 1. Development of trophoblast in culture, phase-contrast photomicrographs. (a) Day 2 culture. Trophoblast is starting to spread out on the substrate and marginal cells extend fan-shaped lamella (arrow). (b) Day 7 culture. Trophoblast has attained maximal outgrowth size. The cells are elongated and spaces are present between some cells (arrows).

of the trophoblast. Similar behavior has been described elsewhere (Cole and Paul, 1965; Salomon and Sherman, 1975; Glass *et al.*, 1979). Trophoblast was also cocultured with L cells and macrophages through the seventh to tenth culture days when the tissue ceased to expand and was thus no longer "invasive." Even during this period, trophoblast cells continued to be motile and capable of displacing cocultured cells.

Actin and myosin distribution were examined in blastocyst cultures that were fixed at various times during development with paraformaldehyde and made permeable with Triton X-100. Actin was detected with fluoresceinated subfragment 1 of myosin (F-S-1) (Cooke, 1972; Schloss *et al.*, 1977) and myosin was identified by indirect immunofluorescence using rabbit antimyosin to human platelet myosin (a gift of K. Fujiwara) and rhodamine-conjugated goat anti-rabbit IgG. Specificity of staining was indicated by the inhibition of the F-S-1 staining reaction with 1 mM ATP and of antimyosin by the absence of staining with preimmune serum. Double immunodiffusion tests showed no reaction between goat anti-rabbit IgG and subfragment 1 (prepared from rabbit myofibrils), and no reaction between subfragment 1 and antimyosin.

During trophoblast spreading (second to seventh culture days), the fan-shaped lamellae of the marginal cells contained actomyosin fibers in various stages of development. In the earliest stages the proximal areas of these lamellae contained a few delicate, radially oriented actin fibers that stained discontinuously for myosin. The lamellae exhibited a cortical actin layer but no cortical myosin. Lamellae in more advanced stages of development contained longer and thicker actomyosin fibers that originated from a single point near the nucleus and splayed out into the lamella (Figure 2a,b). Cortical actin was usually absent at this stage. Finally, when trophoblast reached its maximal outgrowth size, on the sixth or seventh culture day, the marginal cells stopped forming lamellar projections and their peripheral borders became densely packed with thick bundles of radially oriented actomyosin fibers.

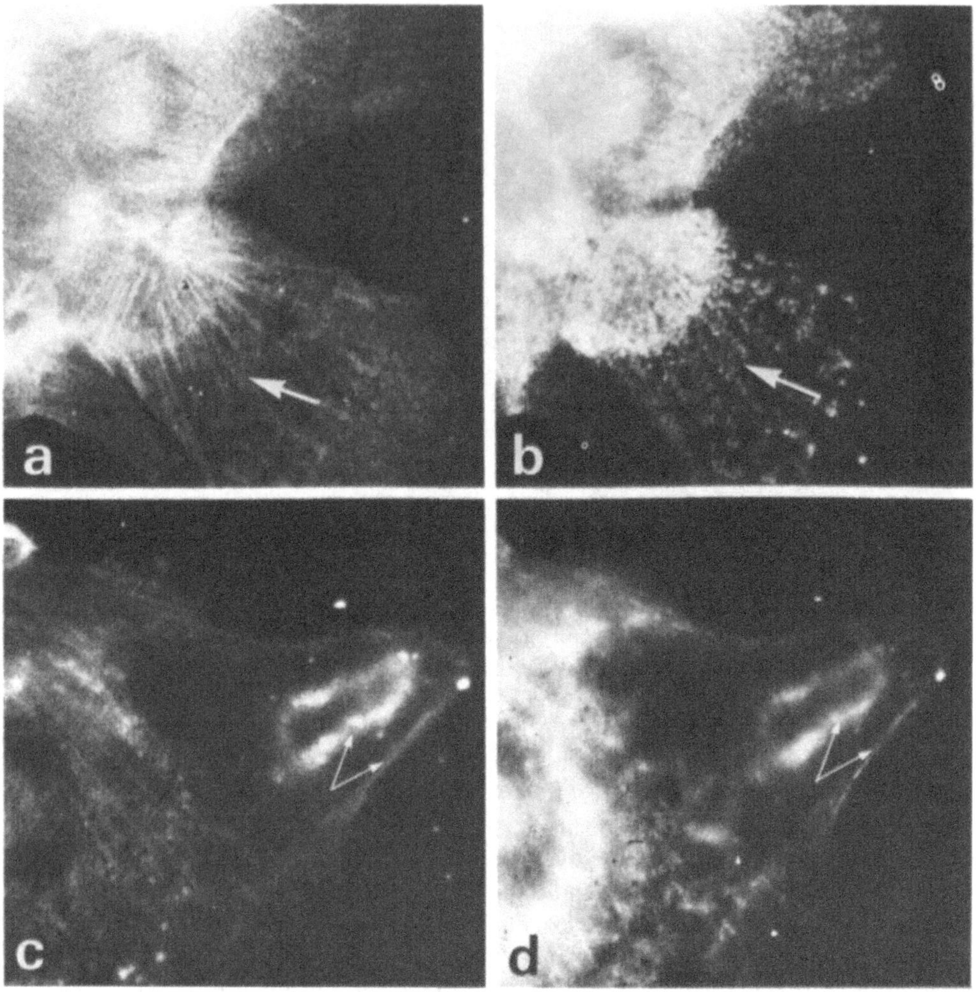

Figure 2. Actin and myosin distribution in trophoblast cultures. (a, b) Fan-shaped lamella in a Day 2 culture. Fibers extend from a proximal point near the nucleus into the lamella and stain continuously for actin (a) and discontinuously for myosin (b) (arrows). There is diffuse actin-staining material in the distal part of the lamella but no distinct cortical layer of actin or myosin. From Sobel *et al.* (1980); reprinted by kind permission of *Developmental Biology*. (c, d) Lamella in a Day 9 culture contains a ring-shaped condensation and cortical layer that stain for actin (c) and myosin (d) (arrows).

Conversion of the trophoblast sheet into a network during the seventh to tenth culture days (Figure 1b) was associated with changes in the distribution of actin and myosin. The cells of the network showed extensive development of microspikes, clawlike projections containing actin and myosin in diffuse form or organized into fibers, and simplified lamellar structures that contained distinctive actomyosin condensations (Figure 2c,d).

Cessation of trophoblast spreading is thus characterized by tissue reorganization, modified cell motility, and a different distribution of actin and myosin. Comparison of trophoblast development in culture and *in utero* reveals that cessation of spreading *in vitro* and loss of invasiveness *in vivo* occur at about the same equivalent gestation time (Kirby, 1965; Sherman and Salomon, 1975). Furthermore, loss of invasiveness *in vivo* involves conversion of the trophoblast into a loose meshwork of giant cells with long interdigitating cytoplasmic processes (Everett, 1935), which resembles the cell network formed by trophoblast under *in vitro* conditions. It seems possible that the observed changes in cultured trophoblast are analogous to the alterations that occur during the loss of trophoblast invasiveness *in vivo*.

ACKNOWLEDGMENTS. This work was supported by the U.S. Department of Energy. We thank Dr. K. Fujiwara for the gift of antimyosin.

References

Cole, R. J., and Paul, J., 1965, Properties of cultured preimplantation mouse and rabbit embryos, and cell strains derived from them, in: *Preimplantation Stages of Pregnancy* (G. E. W. Wolstenholme and M. O'Connor, eds.), pp. 82–112, Little, Brown, Boston.

Cooke, R., 1972, A new method for producing myosin subfragment-1, *Biochem. Biophys. Res. Commun.* **49:**1021–1028.

Everett, J. W., 1935, Morphological and physiological studies of the placenta in the albino rat, *J. Exp. Zool.* **70:**243–287.

Glass, R. H., Spindle, A. I., and Pedersen, R. A., 1979, Mouse embryo attachment to substratum and interaction of trophoblast with cultured cells, *J. Exp. Zool.* **208:**327–335.

Kirby, D. R. S., 1965, The "invasiveness" of the trophoblast, in: *The Early Conceptus, Normal and Abnormal* (W. W. Park, ed.), pp. 63–78, Univ. of St. Andrews Press, Edinburgh.

Salomon, D. S., and Sherman, M. I., 1975, Implantation and invasiveness of mouse blastocysts on uterine monolayers, *Exp. Cell Res.* **90:**261–268.

Schloss, J. A., Milstead, A., and Goldman, R. D., 1977, Myosin subfragment binding for the localization of actin-like microfilaments in cultured cells: A light and electron microscope study, *J. Cell Biol.* **74:**794–815.

Sherman, M. I., and Salomon, D. S., 1975, The relationships between the early mouse embryo and its environment, in: *The Developmental Biology of Reproduction* (C. L. Markert and J. Papaconstantinou, eds.), pp. 277–309, Academic Press, New York.

Sherman, M. I., and Wudl, L. R., 1976, The implanting mouse blastocyst, in: *The Cell Surface in Animal Embryogenesis and Development* (G. Poste and G. L. Nicolson, eds.), pp. 81–125, North-Holland, Amsterdam.

Sobel, J. S., Cooke, R., and Pedersen, R. A., 1980, Distribution of actin and myosin in mouse trophoblast: Correlation with changes in invasiveness during development *in vitro*, *Dev. Biol.* **78:**365–379.

20 Autoradiographic Studies of Mouse Uterine Glands during the Periimplantation Period

Randall L. Given and Allen C. Enders

Significant changes in the uterine fluid composition of rats and mice have been noted during the periimplantation period. These changes may play a role in the interaction of the uterus and blastocyst at implantation. Biochemical analysis has shown an increase in the amount of uterine fluid protein at implantation from levels seen prior to implantation or during the delayed-implantation period (Aitken, 1977; Surani, 1977b). Many of these proteins are uterine specific and of endometrial origin (Surani, 1977a).

Several sources can contribute to uterine fluid, including endometrial structures such as the glands. Finn (1977) has noted that the glands in the mouse appear active and accumulate PAS-positive material. Cytological changes indicative of increased secretory activity have also been observed during the periimplantation period under normal and induced implantation conditions (Given and Enders, 1978, 1980).

However, the synthetic and secretory capability of mouse uterine glands during the peri-implantation period has not been determined. This study was initiated to examine the functional capabilities of uterine glands during this period using both light and electron microscopic autoradiography.

On Days 4, 5, and 6 of pregnancy Swiss–Webster female mice were given an intravenous injection of 1 mCi of [^{3}H]leucine in Earle's balanced salt solution (presence of a vaginal plug is designated Day 1 of pregnancy): 20, 45, and 90 min after injection the animals were anesthetized and infused with fixative, the uteri removed, and interimplantation site areas were processed using routine electron microscopic methods. For light microscopic autoradiography 0.5-μm sections were coated with Kodak NTB-2 emulsion and then exposed for 1 to 2 weeks before development. For electron microscopic autoradiography thin sections of [^{3}H]leucine-treated tissue from Days 5 and 6 of pregnancy were coated with Ilford L-4 emulsion and exposed 3 to 6 months before development. Developed sections were photographed and the micrographs analyzed using the method of Feeney and Wissig (1972). Grain counts were expressed as a percent of total grains counted and as relative grain concentration (percent total grains vs. percent area occupied by that organelle). Relative grain concentration gives an estimate of activity associated with the various cell organelles by correlating labeling intensity with area occupied by the organelle.

In vitro [^{3}H]galactose pulse–chase studies were conducted using mice on Days 5 and 6 of pregnancy. The mice were anesthetized, the uteri removed, and the interimplantation site tissue cut into small pieces and placed in pulse medium for 20 min (250 μCi [^{3}H]galac-

Randall L. Given • Department of Anatomy, School of Medicine, University of Oregon Health Sciences Center, Portland, Oregon 97201. *Allen C. Enders* • Department of Human Anatomy, University of California School of Medicine, Davis, California 95616.

tose/ml MEM). After pulse incubation tissue pieces were washed and placed in chase medium (MEM with 5 mM [^{1}H]galactose) for 10, 20, 45, and 80 min. At the end of the pulse and chase incubations the tissue pieces were fixed and prepared as above for light microscopic autoradiography.

Light microscopic autoradiography of uterine glands after administration of leucine showed similar labeling on Days 4, 5, and 6 of pregnancy. Both tracers were incorporated into both glandular epithelium and the surrounding stroma at 20 min after administration. By 45 min after administration some label was also seen over the glandular lumen; at 90 min after administration the concentration of grains was even higher.

[^{3}H]Galactose pulse–chase studies revealed that the tracer was incorporated into the gland epithelia during the 20-min pulse incubation on both Day 5 and Day 6 of pregnancy (Figure 1). By the end of the subsequent 10-min chase some label was found over the glandular lumen (Figure 2). Even higher concentrations of label were found over the lumina after 20, 45, and 80 min of chase incubation (Figure 3). In these autoradiographs silver grains were confined primarily to the apical portion of the glandular cells while almost no labeling was present over the stromal cells.

Analysis of electron microscopic autoradiographs of the glandular epithelium on Days 5 and 6 of pregnancy showed that the relative concentration of grains was initially high over the RER and Golgi complex at 20 min after [^{3}H]leucine injection and declined at the

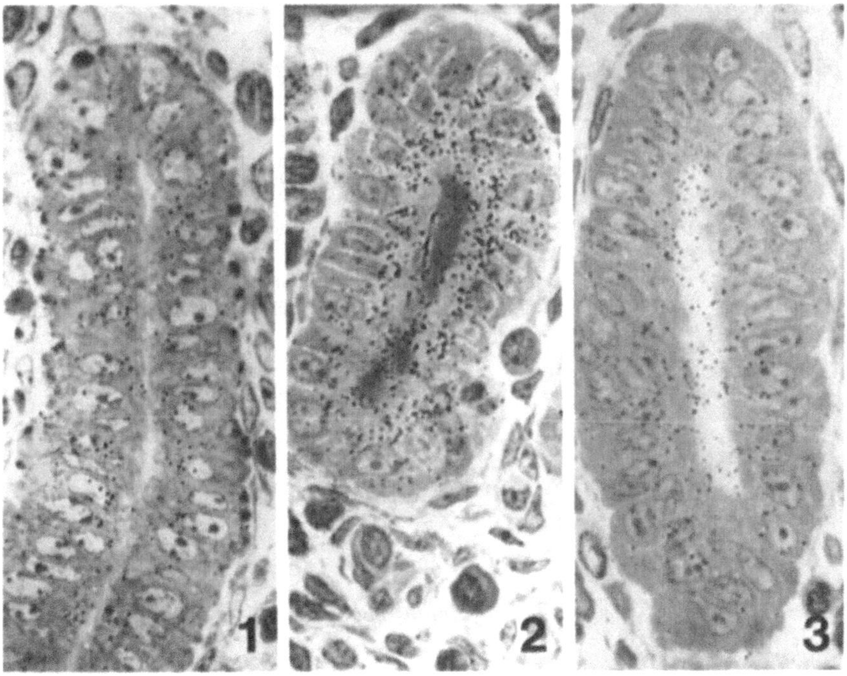

Figures 1–3. Light microscopic autoradiographs of mouse uterine glands after *in vitro* administration of [^{3}H]galactose.

Figure 1. After a 20-min pulse incubation labeling is present over the apical portion of these glandular cells. Day 5 of pregnancy. ×710.

Figure 2. Grains are located over the glandular epithelium and lumen of endometrium following a 20-min pulse–10-min chase incubation. Note the lack of labeling over the stromal cells. Day 6 of pregnancy. ×740.

Figure 3. In this tissue, fixed after 80 min of chase incubation, labeling is again seen over the glandular cells and lumen. Day 5 of pregnancy. ×740.

subsequent intervals (45 and 90 min). Activity over the glandular lumen increased during the time points observed. Small cytoplasmic vesicles appeared to provide a pathway for movement of label from the Golgi complex to the glandular lumen on both Day 5 and Day 6 of pregnancy. The large vesicles present in the glandular cells on Day 5 of pregnancy also appear to provide an additional transport pathway as do the granules present on Day 6 of pregnancy.

These studies show that during the periimplantation period the uterine glands have a constant rate of synthesis and secretion of proteins and/or glycoproteins. These glands are active secretory structures that demonstrate intracellular pathways of transport similar to other exocrine cells. The uterine glands of the mouse must be considered a source of uterine fluid protein and/or glycoprotein that may contribute to qualitative changes in total uterine fluid protein at the time of implantation.

ACKNOWLEDGMENT. This work was supported by NIH Grants HD 10342 and GM 02025.

References

Aitken, R. J., 1977, Embryonic diapause, in: *Development in Mammals,* Vol. 1 (M. H. Johnson, ed.), pp. 307–359, Elsevier/North-Holland Biomedical Press, Amsterdam.

Feeney, L., and Wissig, S. L., 1972, A biochemical and radiographic analysis of protein secretion by thyroid lobes incubated *in vitro, J. Cell Biol.* **53:**510.

Finn, C. A., 1977, The implantation reaction, in: *Biology of the Uterus* (R. M. Wynn, ed.), pp. 245–308, Plenum Press, New York.

Given, R. L., and Enders, A. C., 1978, Mouse uterine glands during the delayed and induced implantation periods, *Anat. Rec.* **190:**271.

Given, R. L., and Enders, A. C., 1980, Mouse uterine glands during the periimplantation period: Fine structure, *Am. J. Anat.* **157:**169.

Surani, M. A. H., 1977a, Radiolabeled rat uterine luminal proteins and their regulation by oestradiol and progesterone, *J. Reprod. Fertil.* **50:**289.

Surani, M. A. H., 1977b, Cellular and molecular approaches to blastocyst uterine interactions at implantation, in: *Development in Mammals* (M. H. Johnson, ed.), Vol. 1, pp. 245–306, Elsevier/North-Holland Biomedical Press, Amsterdam.

21 *In Vitro* Attachment and Outgrowth of Mouse Trophectoderm

Daniel J. Chávez and Jonathan Van Blerkom

Implantation of a mammalian embryo into the uterine lining is a morphodynamic process that is accompanied by marked changes in cellular architecture and subcellular organization. In order to determine the extent to which such changes are mandated by factors external to the embryo, we have examined surface and intracellular changes in fine structure during the progressive outgrowth of mouse trophectoderm in the presence and absence of a cellular monolayer derived from primary explants of endometrium (''quasi-implantation'').

Subcellular alterations associated with implantation also occur *in vitro* by embryos outgrowing in the presence or absence of a cellular substratum (Figures 1–4). By contrast, extensive and complex projections of the cell membrane (lamellipodia) are quite evident

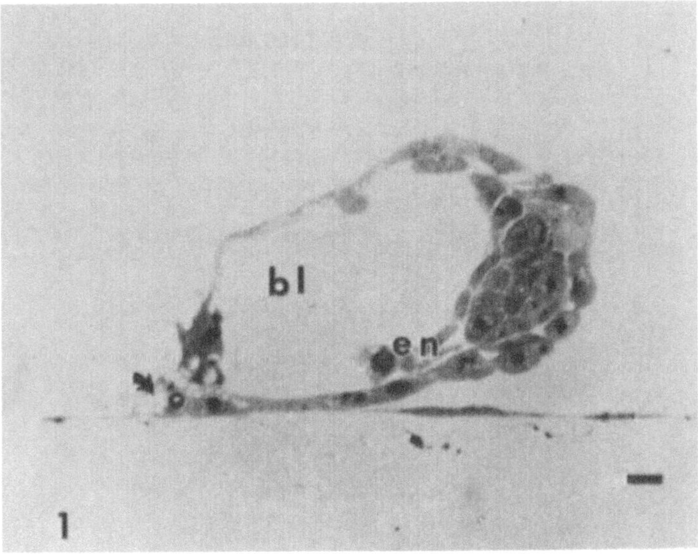

Figure 1. Light micrograph of an embryo during the initial stages of attachment and outgrowth onto a cellular substrate. This blastocyst had been cultured overnight prior to fixation. The vacuolization of trophectoderm cells (arrow) is typical of early outgrowth. The vacuolated trophoblast cell has displaced the uterine cell and is in direct contact with the surface of the culture dish. Note the persistence of the blastocoele (bl) and the formation of endoderm (en). (Bar corresponds to 10 μm.)

Daniel J. Chávez • Department of Human Anatomy, School of Medicine, University of California, Davis, California 95616. *Jonathan Van Blerkom* • Department of Molecular, Cellular, and Developmental Biology, University of Colorado, Boulder, Colorado 80309.

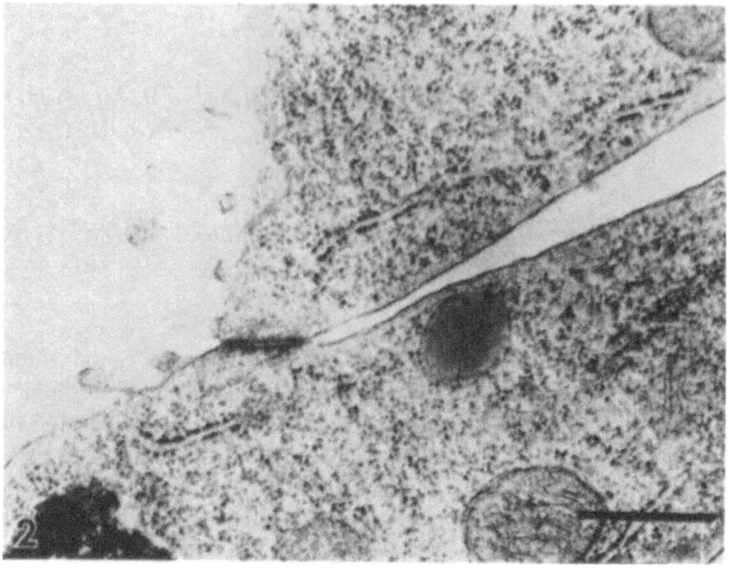

Figure 2. Transmission electron micrograph of a junctional complex formed between a trophoblast cell (above) and a uterine cell in monolayer culture. There are numerous microvilli on the surface of the trophoblast cell and the uterine cell is relatively free of microvilli. The surface of trophoblast cells that contact the substrate during adhesion becomes smooth in appearance. (Bar corresponds to 0.5 μm.)

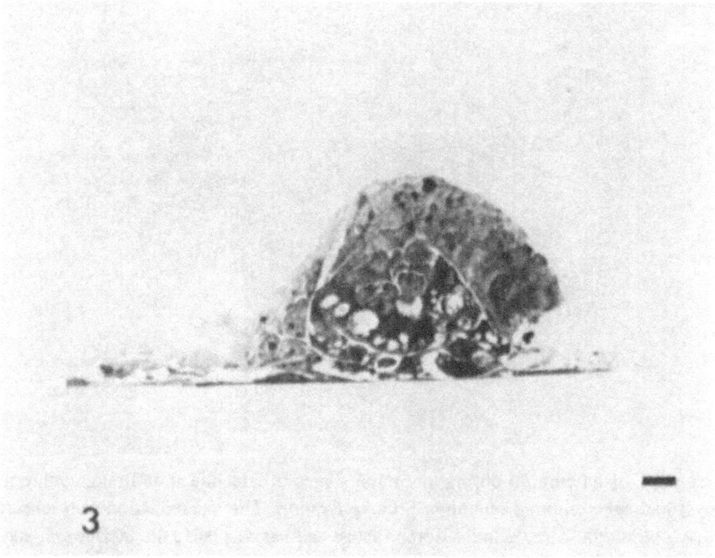

Figure 3. Light micrograph of an embryo during later stages of outgrowth. Approximately 20–40% of cultured embryos develop egg cylinder structures that resemble those *in utero*. The outgrown trophoblast cells remain vacuolated and have completely displaced the uterine cells previously in monolayer. It has not been determined whether displacement is accomplished by physical intrusion of trophoblast cells or by synthesis of lytic enzymes. The ability of blastocysts to displaced cultured cells may be useful in defining the property of invasiveness but is not a suitable model for studying the invasion that occurs during implantation. (Bar corresponds to 10 μm.)

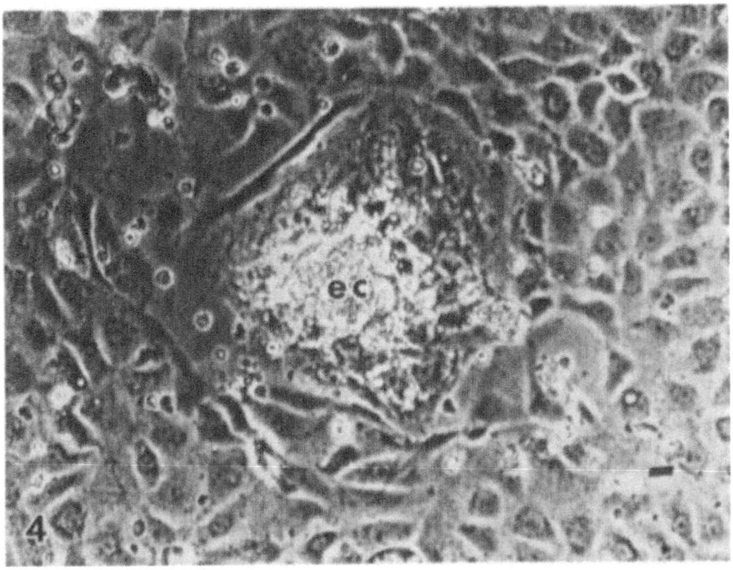

Figure 4. Phase contrast micrograph of an outgrowing embryo that has displaced uterine cells and is now entirely in contact with the surface of the culture dish. Evident in this photograph are intercellular spaces between outgrowing trophoblast and uterine cells. ec, egg cylinder. (Bar corresponds to 10 μm.)

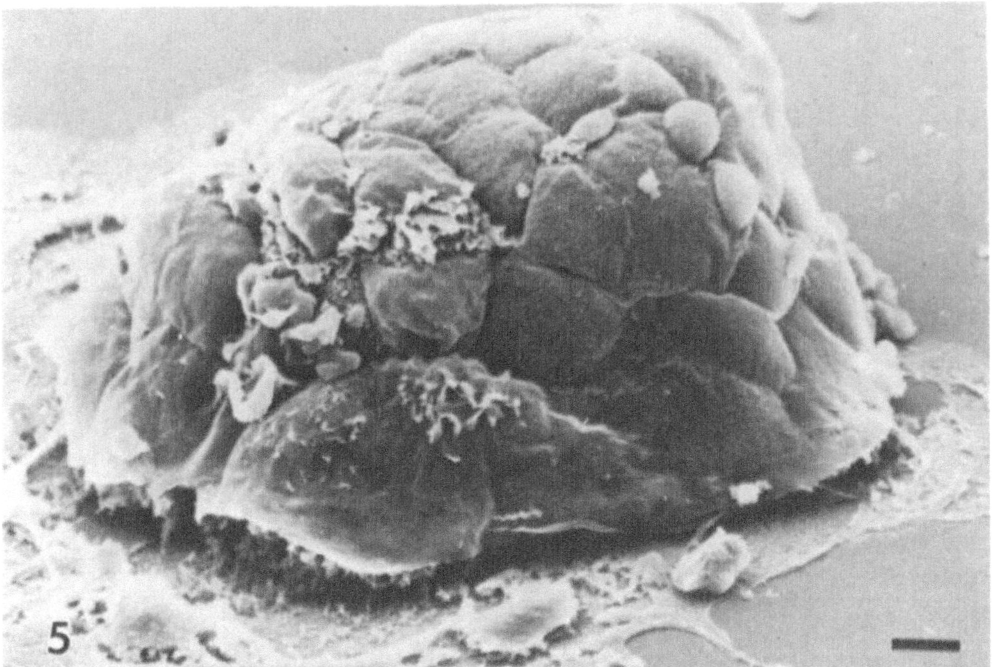

Figure 5. Scanning electron micrograph of an outgrowing embryo demonstrating numerous lamellipodia on the surface of the trophoblast. Lamellipodia formation is typical of rapidly growing cells in tissue culture and probably represents embryonic adaptation to *in vitro* conditions. (Bar corresponds to 10 μm.)

during trophectoderm outgrowth but not during implantation *in utero* (Figure 5). The elaboration of lamellipodia most probably represents a cellular response to *in vitro* culture conditions rather than a fundamental difference in the initial stages of implantation.

During the progressive outgrowth of trophectoderm onto a uterine monolayer, zones of fusion between embryonic and uterine cells become apparent (Figure 2). On the other hand, uterine cells adjacent to but not in direct contact with trophectoderm appear to be "pushed away" from the embryo, forming a pronounced "halo" that encircles the outgrowing trophectoderm (Figures 3 and 4). In conjunction with high-resolution two-dimensional polyacrylamide gel electrophoretic analysis of embryonic protein synthesis, the results suggest that cellular and molecular differentiation during the initial stages of implantation more likely reflect an intrinsic, embryonic program than a differentiative program that requires external cues from the uterus.

Only a small percentage of embryos cultured through the implantation period develop embryonic structures that resemble those *in utero* (Figure 3). On the other hand, greater than 90% experience attachment and outgrowth. Attachment and monolayer formation are typical of many types of cultured cells. Therefore, one must interpret the analogies with caution. However, *in vitro* attachment and outgrowth do provide a unique model for the study of adhesiveness. The ability of trophoblast to displace cultured cells may provide an assay for invasiveness though probably not a model for studying the mechanisms of invasion. In other experiments we have used attachment and outgrowth as a test for *in vitro* activation from delayed implantation and as a criterion for whether or not we have caused delayed implantation by manipulating the components of culture media.

22 Hormone Production by Rat Blastocysts and Midpregnancy Trophoblasts in Vitro

Shirley A. McCormack and Stanley R. Glasser

The production of hormones by the trophoblast as well as its susceptibility to regulation by them are important factors in the elucidation of the mechanisms by which the synchrony of the fetal and maternal organisms is accomplished during pregnancy. In the past, clues to the presence in trophoblast tissue of enzyme systems for steroidogenesis have been found by histochemical means (Deane *et al.*, 1962; Ferguson and Christie, 1967; Botte *et al.*, 1968). In spite of the intrinsic limitations of data obtained from *in vitro* systems, the separation of maternal from fetal contributions is essential if hormone production by the trophoblast is to be detected.

We have incubated rat blastocysts and midpregnancy trophoblast *in vitro* and measured hormone production into the medium. Blastocysts were collected on Day 4 of pregnancy, washed once in Hanks' balanced salt solution without calcium or magnesium (Hanks' BSS – Ca – Mg), and incubated (10–20 blastocysts per dish) in 3 ml NCTC-135 medium (Gibco) containing 10% fetal calf serum (heat inactivated), 100 units penicillin/ml, and 100 μg streptomycin/ml in 35-mm plastic culture dishes at 37°C in air:CO_2 (95:5) for up to 10 days. The medium was replaced every 24 hr. The blastocysts were observed under phase microscopy and attachment and outgrowth were noted. Midpregnancy trophoblast was obtained by dissecting the entire trophoblast shell from Day 10 embryos. The trophoblasts were washed once with Hanks' BSS – Ca – Mg, dissociated with trypsin–EDTA or with the neutral protease Dispase (Boehringer–Mannheim Biochemicals, Indianapolis), and cultured in 3 ml medium at approximately 50,000 cells/ml. Every set of blastocysts or trophoblasts included dishes of media without cells as controls.

The medium collected daily from experimental and control dishes was frozen for subsequent hormone assay. The highest value obtained for the controls was subtracted from the values obtained with cells. Entire experiments were assayed as sets and compared with their own internal controls.

Steroids were extracted from the medium with ether, fractionated into estradiol-17β (E), testosterone (T), and progesterone (P) on Sephadex LH-20 columns, assayed by standard radioimmunoassay (RIA) procedures using antisera of known cross-reactivity and corrected for percent recovery.

Rat placental lactogen (rPL) was assayed by a radioreceptor assay (RRA) (Shiu *et al.*, 1973) performed by Dr. Michael Blank in the laboratory of Dr. Henry Friesen.

The results of these experiments are summarized in Tables 1 and 2.

Although the production of E by blastocysts on equivalent gestation days (ECD) 7–8 was considerable (Table 1), this phenomenon probably does not represent secre-

Shirley A. McCormack and Stanley R. Glasser • Department of Cell Biology, Baylor College of Medicine, Houston, Texas 77030.

Table 1. Hormone Production by Blastocysts Placed in Culture on Day 4 of Pregnancy

Hormone	pg/blastocyst/24 hr[a]	EGD[b]	n[c]
E	44.3 ± 15.0	7–8	4
T	6.2 ± 2.1	6–14	5
P	28.4 ± 18.0	10–13	5
rPL	1820 ± 600	11–12	4

[a] Maximal concentration of hormone produced after subtraction of the highest concentration found in the medium control. Mean ± S.E.
[b] The range of equivalent gestation days on which the maximal concentration occurred in different dishes.
[c] The number of experimental dishes.

Table 2. Hormone Production by Trophoblasts Placed in Culture on Day 10 of Pregnancy[a]

Hormone	pg/trophoblast/24 hr	EGD	n
E	3.5 ± 1.2	12–15	4
T	22.3 ± 7.8	12–15	6
P	250 ± 128	12–15	5
rPL	143,500 ± 615	12	2

[a] Symbols are defined in footnotes of Table 1.

tion because Marcal *et al.* (1975) could not detect Δ^5-3β-hydroxysteroid dehydrogenase activity in the rat trophoblast before EGD 11 (EGD = day blastocyst or trophoblast is removed from rat + number of days in culture). Furthermore, the E present in rabbit blastocysts has been shown to be primarily the result of uptake of maternal E (Singh and Booth, 1978). Midpregnancy trophoblast produced very little, if any, E (Table 2).

The production of T by blastocysts was negligible (Table 1). Later, midpregnancy trophoblast did show low levels of T production (EGD 12–15) (Table 2). Information on the rat trophoblast is lacking, but mouse blastocysts at EGD 5–6 could not be shown to synthesize androstenedione, an immediate precursor of T, from dehydroepiandrosterone (DHEA), while blastocyst cultures at EGD 10–11 were able to do so as well as to metabolize the androstenedione further (Sherman and Atienza, 1977). However, Okker-Reitsma (1976), who identified metabolites of DHEA in the mouse placenta, found androstenedione and five α-reduced metabolites and was unable to detect T, dihydrotestosterone, or E.

The production of P by rat blastocysts was variable and low (Table 1). However, midpregnancy trophoblast produced levels 10 times that of the blastocysts during the period EGD 12–15 (Table 2). The rat trophoblast produced 0.8 pmol P/trophoblast/24 hr, comparable to the maximum amounts of P produced by cultured mouse blastocysts (0.5 to 5.0 pmol/blastocyst/24 hr; Sherman and Atienza, 1977).

We measured peripheral levels of E and P and found concentrations of 9.8 ± 3.3 pg E/ml and 64.2 ± 18.9 pg P/ml ($n = 15$) on Day 4 of pregnancy and 7.8 pg E/ml and 54.7 pg P/ml ($n = 3$) on Day 10. The P levels are similar to the values of 64 and 76 pg/ml on Days 4 and 10, respectively, found by Morishige *et al.* (1973). Levels in ovarian vein blood at these times may be of greater significance to the blastocyst or trophoblast in the uterus.

Concentrations of 200 ± 43 pg E/ml and 1.6 ± 0.4 ng T/ml have been reported on Day 4 of pregnancy in 10-min collections of rat ovarian vein blood (Brodie *et al.*, 1977).

Therefore, a maternal origin must be suspected for steroids released by blastocysts or trophoblasts, particularly where others have not been able to show conversion of labeled immediate precursors as in the case of E and T. In the case of P production, however, the release of bound steroid seems an unlikely explanation of our results, especially for midpregnancy trophoblast. In addition to an initial wash, the trophoblasts had undergone a subsequent dissociation step and did not reach their maximal concentration of P until 2 to 5 days after being placed in culture.

The rPL production by blastocysts on EGD 12 was ~1% (Table 1) of that produced by trophoblasts on the same day (Table 2). The peak levels occurred on the day on which others (Matthies, 1967; Kelly *et al.*, 1973) have found maximal levels of rPL in rat serum.

Both P and rPL were produced by midpregnancy trophoblast in amounts much greater (10 and 100 times, respectively) than those of blastocysts at the same EGD. The fact that there may have been more cells in the dissected trophoblasts than in the cultured blastocysts may be sufficient to explain the differences, but maternal regulation between Days 5 and 10 could also occur.

In conclusion, our results indicate the production of E and rPL by blastocysts in the culture on EGD 7–8 and 11–12, respectively, but not of T or P. Midpregnancy trophoblast in culture produced considerable P and rPL on EGD 12–15 and 12, respectively, but no E or T. The production of E by blastocysts could have been due to the slow release of bound maternal E or to secretion by the blastocysts themselves, while the production of P and rPL by midpregnancy trophoblast was probably due to secretion by the trophoblast cells.

References

Botte, V., Tramontana, S., and Chieff, G., 1968, Histochemical distribution of some hydroxysteroid dehydrogenases in the placenta, foetal membranes and uterine mucosa of the mouse, *J. Endocrinol.* **40:**189.

Brodie, A. M. H., Schwarzel, W. C., Shaikh, A. A., and Brodie, H. J., 1977, The effect of an aromatase inhibitor, 4-hydroxy-4-androstene-3,17-dione, on estrogen dependent processes in reproduction and breast cancer, *Endocrinology* **100:**1684.

Deane, H. W., Rubin, B. L., Driks, E. C., Lobel, B. L., and Leipsner, G., 1962, Trophoblast giant cells in placentas of rats and mice and their probable role in steroid-hormone production, *Endocrinology* **70:**407.

Ferguson, M. M., and Christie, G. A., 1967, Distribution of hydroxysteroid dehydrogenases in the placentae and foetal membranes of various mammals, *J. Endocrinol.* **38:**291.

Kelly, D. A., Shiu, R. P. C., Robertson, M. C., and Friesen, H. G., 1973, Studies of rat chorionic mammotropin by radioreceptor assay, *Fed. Proc. Fed. Am. Soc. Exp. Biol.* **32:**213.

Marcal, J. M., Chew, N.J., Salomon, D. S., and Sherman, M. I., 1975, $\Delta^5 - 3\beta$-hydroxysteroid dehydrogenase activities in rat trophoblast and ovary during pregnancy, *Endocrinology* **96:**1270.

Matthies, D. L., 1967, Studies of the luteotropic and mammotropic factor found in trophoblast and maternal peripheral blood of the rat at midpregnancy, *Anat. Rec.* **159:**55.

Morishige, W. K., Pepe, G. J., and Rothchild, I., 1973, Serum luteinizing hormone, prolactin and progesterone levels during pregnancy in the rat, *Endocrinology* **92:**1527.

Okker-Reitsma, G. H., 1976, Metabolism of (^{3}H)-dehydroepiandrosterone by mouse placental tissue *in vitro, K. Ned. Akad. Wet. Amsterdam Proc. Ser. C* No. 3, **79:**290.

Sherman, M. I., and Atienza, S. B., 1977, Production and metabolism of progesterone and androstenedione by cultured mouse blastocysts, *Biol. Reprod.* **16:**190.

Shiu, R. P. C., Kelly, P. A., and Friesen, H. G., 1973, Radioreceptor assay for prolactin and other lactogenic hormones, *Science* **180:**968.

Singh, M. M., and Booth, W. D., 1978, Studies on the metabolism of neutral steroids by preimplantation rabbit blastocysts *in vitro* and the origin of blastocyst oestrogen, *J. Reprod. Fertil.* **53:**197.

23 Separation and Characterization of Endocervical Cells

Beverly S. Chilton, Santo V. Nicosia, and Don P. Wolf

The influence of steroid hormones on the cytodifferentiation of reproductive tract tissues is well documented for oviductal and endometrial cells. By using an *in vitro* approach that utilizes isolated endocervical cell populations, we now present evidence consistent with a hormone-related modulation of endocervical cell morphology and population size.

Endocervical cell suspensions from estrous (E), 5-day pseudopregnant (P), and ovariectomized (OVX) New Zealand white rabbits were obtained by tissue dissociation in pronase followed by mechanical disaggregation (Figure 1). Trypan blue dye exclusion by 95–98% of the cells and [^{3}H]leucine incorporation confirmed viability, while ultrastructural analysis of dispersed cells demonstrated cytostructural integrity. Endocervical cell suspensions from E rabbits consisted of 25% histochemically (periodic acid–Schiff, PAS) distinct mucous cells (21% vacuolated and 79% granular). This percentage was reduced to only 12% mucous cells (8.5% vacuolated and 91.5% granular) in P rabbits and to approximately 4% mucous cells (50% vacuolated and 50% granular) in OVX animals. Dispersed cells ($1.6–2.0 \times 10^7$) were separated at unit gravity on a 0.49–2.35% linear gradient of bovine serum albumin (BSA) into four distinct populations that were designated as vacuolated, PAS-negative; vacuolated and granular, PAS-positive; and ciliated cells. Some physical characteristics of these cell populations from E animals are shown in Table 1 and distribution profiles are shown in Figure 1. The distribution of vacuolated, PAS-negative cells was similar in gradients of E and P populations while this cell type predominated throughout the gradient for OVX animals. The two mucous-containing cell types isolated from P animals, in contrast to E animals, were found in the lower BSA concentrations, suggesting a progesterone-dependent reduction in sedimentation rate and cell diameter, concomitant with a reduction in the number of intracellular mucous granules.

To characterize biosynthetic activity, isolated vacuolated and granular cells from E and P animals and heterogeneous cell populations from OVX animals were incubated for 36 hr with *N*-[^{14}C]acetyl-D-glucosamine. Its incorporation, expressed as mean cpm per presumptive mucous cell $\times 10^{-2} \pm$ S.E., into intracellular (I) and secreted (S) glycoproteins was taken as an index of mucin synthesis and secretion, respectively. For E animals mucin synthesis and secretion were similar for vacuolated ($I = 2.49 \pm 0.46$; $S = 1.34 \pm 0.34$) and granular ($I = 2.43 \pm 0.25$; $S = 1.22 \pm 0.29$) PAS-positive cells. These values were also similar to the relative incorporation and secretion levels observed with PAS-positive cells from P animals. Endocervical cells from OVX animals did not incorporate significant amounts of the radiolabel. However, with the subcutaneous administration of estradiol-17β (5 μg/

Beverly S. Chilton, Santo V. Nicosia, and Don P. Wolf • Department of Obstetrics and Gynecology, University of Pennsylvania Medical School, Philadelphia, Pennsylvania 19104.

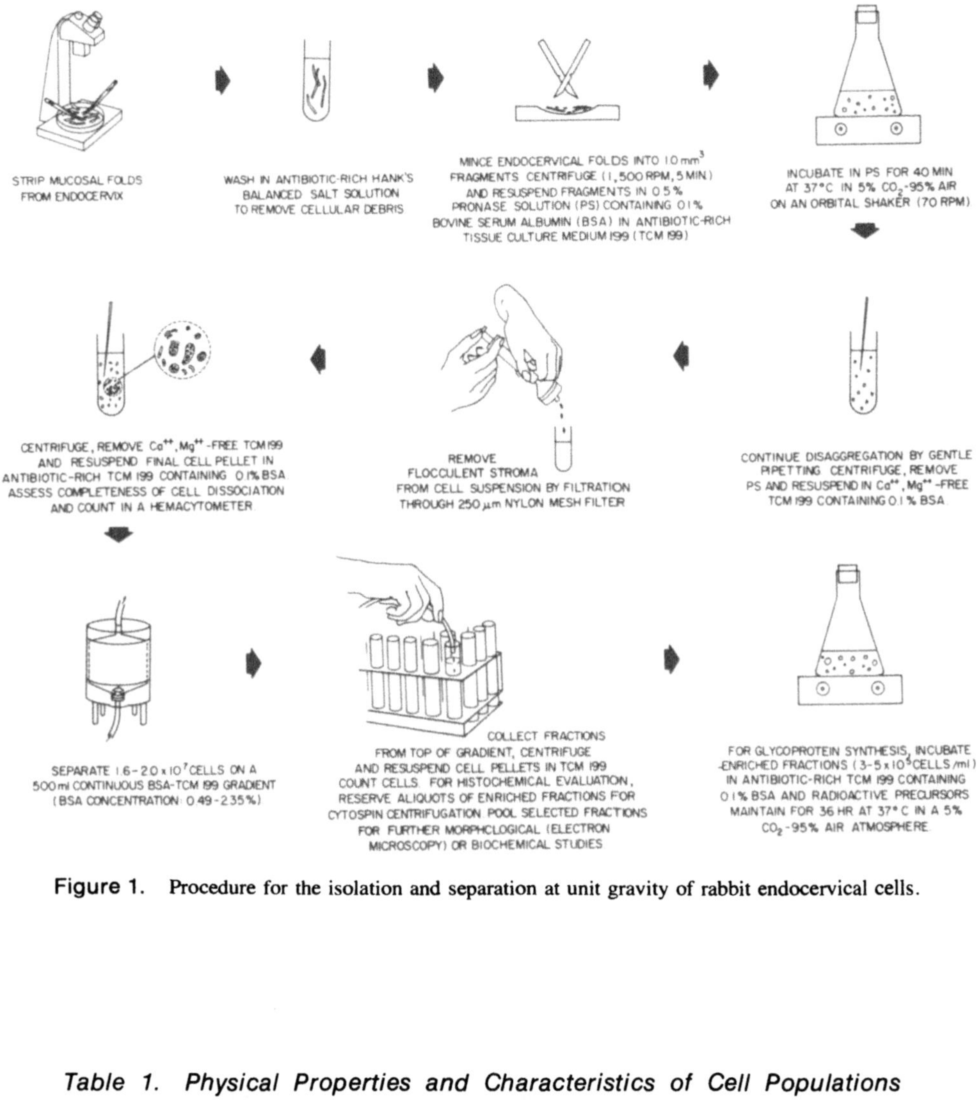

Figure 1. Procedure for the isolation and separation at unit gravity of rabbit endocervical cells.

Table 1. Physical Properties and Characteristics of Cell Populations Separated at Unit Gravity from Estrous Rabbits

Cell type	Density[a] (g/ml)	Diameter (μm)	Enrichment[b]
Vacuolated, PAS-negative	1.011–1.012	12–24	5- to 6-fold
Vacuolated, PAS-positive	1.013–1.015	11.5–17.2	4- to 5-fold
Granular, PAS-positive	1.015–1.017	8.5–12.2	1.5- to 2.0-fold
Ciliated	1.017	14–20	3- to 5-fold

[a] BSA density at which cells banded.
[b] Relative to original inoculum.

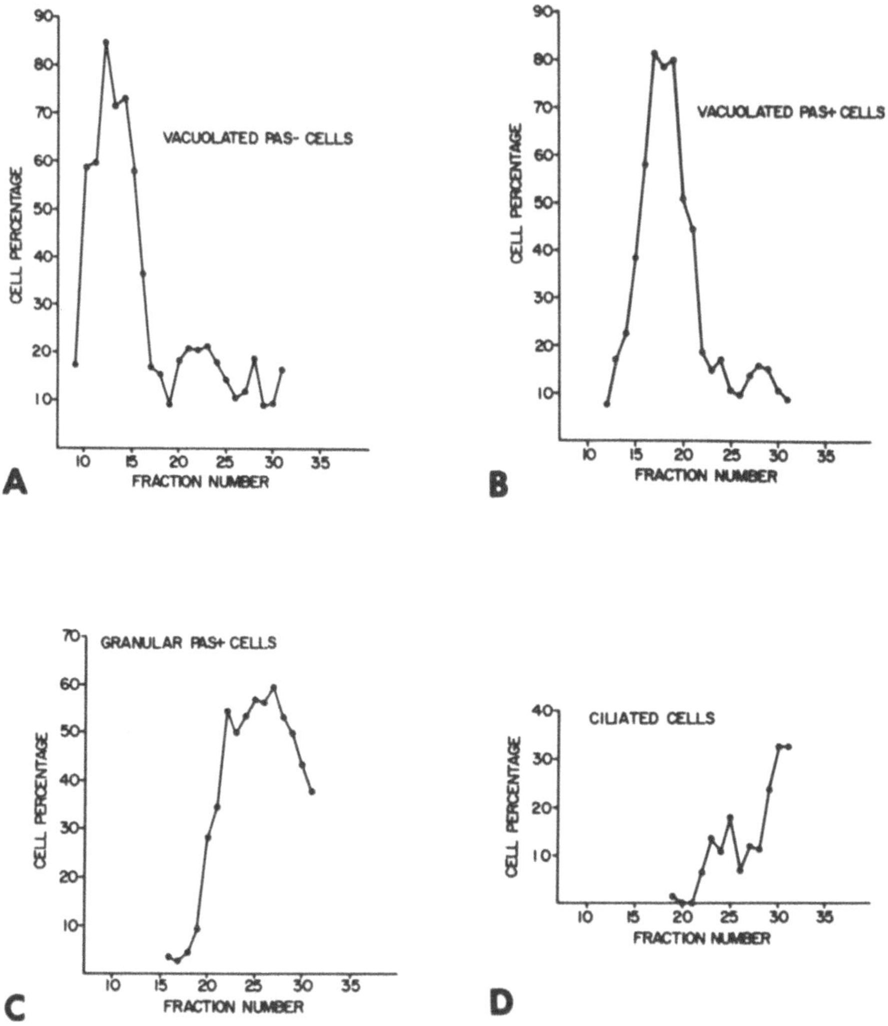

Figure 2. Enrichment profiles for (A) vacuolated, PAS-negative cells; (B) vacuolated, PAS-positive cells; (C) granular, PAS-positive cells; and (D) ciliated cells expressed as percentage of the total cell number in individual fractions. Each gradient was collected in 15-ml fractions for a total of 36 fractions (approximately 500 ml).

12 hr/animal) for 10 days to OVX animals, precursor incorporation and mucous cell populations resembled E levels. These results suggest that steroid hormones may exert their effects indirectly by modulating the type rather than the biosynthetic activity of mucous cells. Support for an indirect role of steroids also comes from another series of experiments from this laboratory, where the addition of exogenous steroids was without effect on the synthesis and secretion of mucin by isolated endocervical cells.

Acknowledgments. This work was supported by USPHS Grant HD-06274-Sub 4 and by the Population Council.

24 Role of the Major Histocompatibility Complex in the Timing of Early Mammalian Development

Kathryn M. Verbanac and Carol M. Warner

The timing of preimplantation mammalian development is crucial for successful implantation. At the time of implantation, both the embryo and the uterus must be at the correct developmental stage. In the mouse, several studies have suggested that there are "fast" and "slow" developing mouse strains, and that genetic factors influence the timing of development (Whitten and Dagg, 1962; McLaren and Bowman, 1973; Titenko, 1977). Both CBA and C3H mice are "slow" developing strains, and both strains are of the $H\text{-}2^k$ haplotype. This led us to postulate that the control of the timing of preimplantation mouse development might reside in genes of the *H-2* (major histocompatibility) complex.

To test this hypothesis we examined a number of inbred and congenic mouse strains of different *H-2* haplotypes. The mice were superovulated with 5 IU PMS (Organon) at 4 P.M. followed 48 hr later by 10 IU hCG (ICN Nutritional Biochemicals). The mice were mated and the embryos collected at set intervals post-hCG injection. The number of cells per embryo was assessed by the method of Tarkowski (1966) for all embryos beyond the four-cell stage of development. To determine the time of ovulation, superovulated unfertilized eggs were collected at 2-hr intervals post-hCG injection. The number of eggs per clutch was determined after treatment with hyaluronidase.

Table 1 shows the data collected for three inbred strains, two congenic strains, and two F_1 hybrids for the "eight-cell" stage (65 hr post-hCG) and the blastocyst stage (89 hr post-hCG). The congenic strain B10.BR is similar to the CBA strain with which it shares its *H-2* haplotype, and different from the C57BL/10Sn strain with which it shares its background genes. (The SJL and B10.S strains are included as a control.) The interpretation is that there is at least one gene in the *H-2* complex that influences the timing of early mouse embryo development. We have called this *H-2*-linked gene *Ped: preimplantation embryo development*. The F_1 data in Table 1 show that the expression of the "fast" allele is dominant or codominant, and there is no apparent influence of the maternal cytoplasm on gene expression.

Figure 1 shows a plot of $\log_2$ of the number of cells per embryo vs. time for the C57BL/10Sn, CBA, and B10.BR strains. The slopes of the CBA and B10.BR lines are the same and are significantly ($P<0.01$) less than the slope of the C57BL/10Sn line; This suggests that the *Ped* gene influences the *rate* of development. The displacement of the CBA and B10.BR lines in Figure 1 is partially explained by the data shown in Figure 2. These data show that the B10.BR strain ovulates earlier than the CBA strain, and the C57BL/10Sn strain ovulates earlier than either of the other strains. However, difference in

Kathryn M. Verbanac and Carol M. Warner • Department of Biochemistry and Biophysics, Iowa State University, Ames, Iowa 50010.

Table 1. Number of Cells in Preimplantation Mouse Embryos from Inbred, Congenic, and F_1 Hybrid Mice at 65 and 89 hr after Superovulation

		65 hr post-hCG			89 hr post-hCG		
Mouse strain	*H-2* haplotype	No. expts.	No. embryos	Mean cell No./ embryo (S.E.)	No. expts.	No. embryos	Mean cell No./ embryo (S.E.)
C57BL/10Sn	b	3	63	7.6 (0.2)	8	48	33.1 (2.1)
CBA	k	5	45	4.4 (0.2)[b]	7	45	18.9 (1.0)[b]
B10.BR	k	2	59	6.0 (0.3)[a]	6	50	22.8 (1.3)[b]
(C57BL ♀ × B10.BR ♂) F_1	b/k	—	—	—	5	41	29.7 (1.6)
(B10.BR ♀ × C57BL ♂) F_1	k/b	—	—	—	5	60	28.6 (1.2)
SJL	s	7	46	7.1 (0.3)	5	48	31.9 (2.0)
B10.S	s	3	51	8.0 (0.4)	3	28	33.8 (2.4)

[a] Significantly different from the C57BL strain at the $p < 0.05$ level.
[b] Significantly different from the C57BL strain at the $p < 0.001$ level.

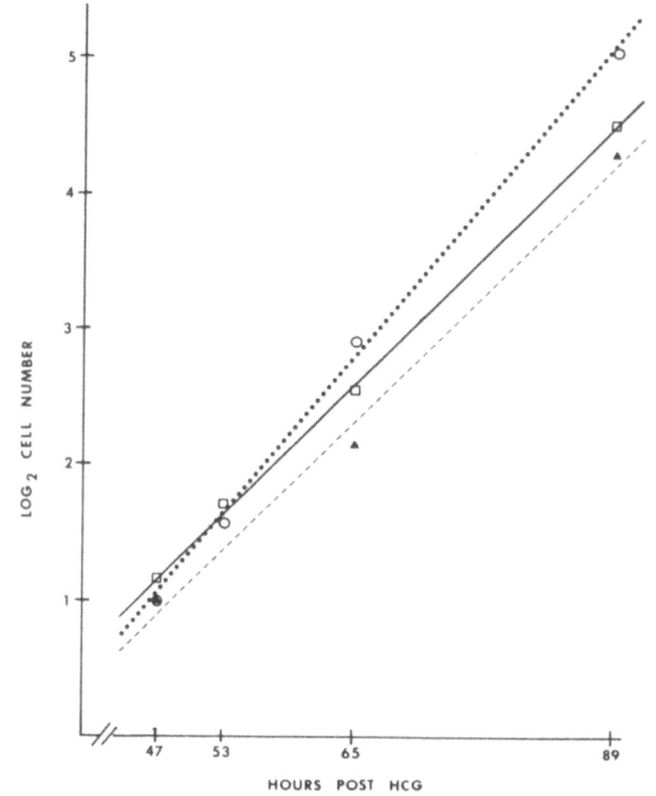

Figure 1. Log_2 cell number of C57BL/10Sn (○···○), B10.BR (□–□), and CBA (▲---▲) embryos as a function of time post-hCG injection. Each line is the least-squares linear regression through the points. The C57BL/10Sn line differs significantly in slope from the other two lines ($P < 0.01$). All three lines have a correlation coefficient $r = 0.99$. The equations for each line are C57BL/10Sn: $y = 0.097x - 3.50$; B10.BR: $y = 0.080x - 2.58$; CBA: $y = 0.078x - 2.75$.

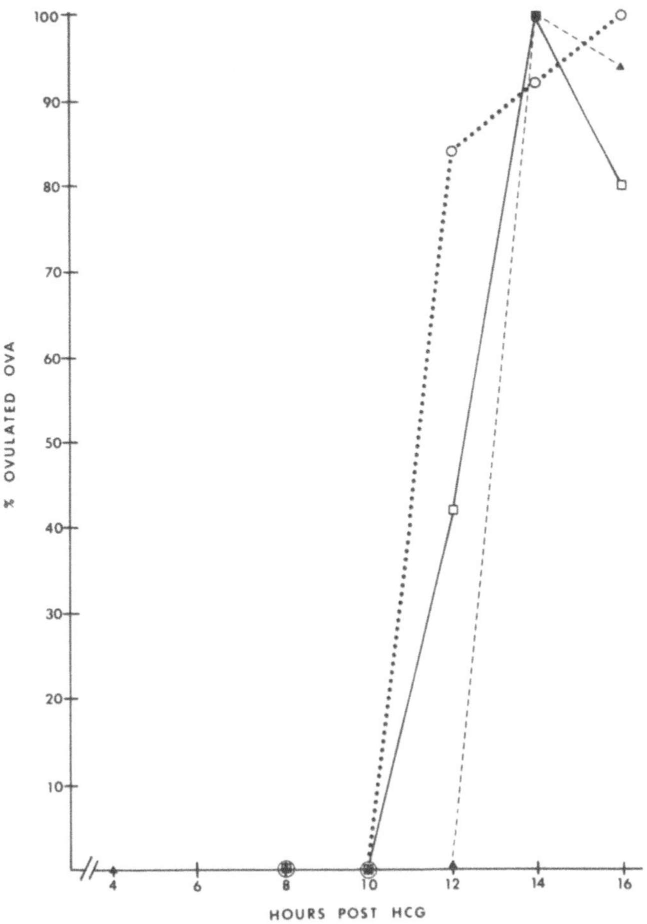

Figure 2. Time of superovulation of C57BL/10Sn (O···O), B10.BR (□—□), and CBA (▲---▲) mice. Number of ova ovulated is graphed as a percent of the total vs. hours post-hCG injection.

the time of ovulation alone cannot account for the marked differences in the number of cells per embryo seen at the blastocyst stage.

In conclusion, there is at least one gene, called *Ped,* in the *H-2* complex that affects the timing of preimplantation mouse embryo development. Over 60 traits have been reported to be associated with the *H-2* complex (Klein, 1978, 1979). It is not clear if these represent different genes or pleiotropic effects of a few genes. The *H-2*-linked *Ped* gene described here apparently influences both the rate of cleavage and the time of ovulation. This is the first report of a role of the major histocompatibility complex in early mammalian development.

References

Klein, J., 1978, The *H-2* mutations: Their genetics and effect on immune functions, *Adv. Immunol.* **26**:55.

Klein, J., 1979, The major histocompatibility complex of the mouse, *Science* **203**:516.

McLaren, A., and Bowman, P., 1973, Genetic effects on the timing of early development in the mouse, *J. Embryol. Exp. Morphol.* **30**:491.

Tarkowski, A. K., 1966, An air-drying method for chromosome preparations from mouse eggs, *Cytogenetics* **5**:394.

Titenko, N. V., 1977, [Translation] Preimplantation development of mouse embryos in homo- and heterogeneous crosses, *Ontogenez* **8**:27.

Whitten, W. K., and Dagg, C. P., 1962, Influence of spermatozoa on the cleavage rate of mouse eggs, *J. Exp. Zool.* **148**:173.

25 Pregnancy-Associated Endometrial Protein in Women

Sharad G. Joshi, Richard A. Smith, Donald H. Szarowski, and Edgar S. Henriques

Several investigators have attempted to identify progestagen-dependent endometrial proteins (PEPs) that may regulate certain key events in pregnancy. We reported that human endometrium synthesizes an antigenic PEP (Joshi, 1977; Joshi *et al.*, 1980a,b). Analysis of extracts of endometria in different developmental stages by Ouchterlony two-dimensional immunodiffusion (ID) test and by immunoelectrophoresis (IEP) revealed the presence of the PEP in all 42 specimens of decidua of early pregnancy (5–12 weeks) and in 18 of 38 samples of secretory-phase endometria, but not in any of 32 samples of proliferative-phase endometria. *In vitro* labeling of tissues with radioactive leucine followed by immunoprecipitation demonstrated that proliferative-phase endometria also synthesizes PEP but in amounts far below the sensitivity limits of ID or IEP. Furthermore, the tissue concentration of PEP, determined by rocket immunoelectrophoresis (RIEP), was about three times greater in decidua of early pregnancy than in mid-secretory-phase endometria. In individual cycling women the occurrence of PEP in the endometrium is related to high serum progesterone levels ($>$5 ng/ml).

Biochemical studies strongly suggest that the PEP is a glycoprotein that has an isoelectric point of approximately 4.9; it can be reductively dissociated into two subunits, each having a molecular weight of approximately 27,000. The PEP is not related to several of the known pregnancy-associated proteins, including transferrin, ceruloplasmin, α_1-antitrypsin, ferritin, uteroglobin, α-fetoprotein, the β subunit of human chorionic gonadotropin, pregnancy zone protein (Schoultze, 1974), and pregnancy-associated plasma proteins (Lin *et al.*, 1974). However, the PEP is similar to α uterine protein detected in amniotic fluid by Sutcliffe *et al.* (1978).

PEP could not be detected by IEP or ID in adult male sera, pregnancy and nonpregnancy sera, extracts of immature or mature fetal placenta, umbilical cord, ovary, Fallopian tube, myometrium, or cervix that were dissected from the total abdominal hysterectomy specimens of women in the proliferative or secretory phase of the menstrual cycle. However, PEP was detected in amniotic fluid, and its levels, as determined by RIEP, were maximal at 16–18 weeks and thereafter declined rapidly.

We conclude, therefore, that the human endometrium contains a glycoprotein the synthesis of which is markedly stimulated during pregnancy and that it is transported to the amniotic fluid.

ACKNOWLEDGMENTS. This work was supported by Ford Foundation Grant 680-0900 and NICHD Grant HD09622.

Sharad G. Joshi, Richard A. Smith, Donald H. Szarowski, and Edgar S. Henriques • Department of Obstetrics and Gynecology, Albany Medical College, Albany, New York 12208.

References

Joshi, S. G., 1977, In discussion (pp. 246–247) of the paper entitled "Protein Composition of Human Endometrium and Its Secretions at Different Stages of the Menstrual Cycle," Hirsch, P. J., Fergusson, I. L. C., and King, R. J. B., *Ann. N.Y. Acad. Sci.* **286**:233–295.

Joshi, S. G., Ebert, K., and Swartz, D., 1980a, Detection and synthesis of a progestagen-dependent protein in human endometrium, *J. Reprod. Fertil.* **59**:273–285.

Joshi, S. G., Ebert, K. M., and Smith, R. A., 1980b, Properties of the progestagen-dependent protein of the human endometrium, *J. Reprod. Fertil.* **59**:287–296.

Lin, T., Halbert, S. P., Kiefer, D., and Spellacy, W. N., 1974, Characterization of the four pregnancy-associated plasma proteins, *Am. J. Obstet. Gynecol.* **118**:223–236.

Schoultz, B., 1974, A quantitative study of the pregnancy zone protein in the sera of pregnant and puerperal women, *Am. J. Obstet. Gynecol.* **119**:792–797.

Sutcliffe, R. G., Brock, D. J. H., Nicholson, L. V. B., and Dunn, E., 1978, Fetal- and uterine-specific antigens in human amniotic fluid, *J. Reprod. Fertil.* **54**:84–90.

26 Survival of Rat Blastocysts in the Uterus of the Mouse Following Xenogeneic Implantation

C. Tachi, S. Tachi, M. Yokoyama, and N. Osawa

Xenogeneic ovum implantation offers a unique opportunity to analyze the complex mechanisms that underlie the natural process of ovum implantation in mammals.

Rat blastocysts transferred to the uterus of pseudopregnant mice on Day 3 successfully undergo, as revealed by electron microscopy, the stages of ovum implantation equivalent to the early and late attachment stages of normal implantation in rats and mice (Tachi and Tachi, 1979). The foreign implants, however, are invariably destroyed shortly after the penenetration of the basement membrane of the lumenal epithelial cells (Tachi and Tachi, 1979). While disparity existed among the results reported by different investigators concerning the fate of rat blastocysts transferred to the uterus of the mouse, none of the xenogeneic implants have been observed surviving beyond 96 hr after transfer or about 170 hr p.c. (Tarkowski, 1962; Potts *et al.*, 1970; Tachi and Tachi, 1979). This period, therefore, appears to represent a temporal limit so far unsurmounted for the survival of rat blastocysts in the uterus of the mouse (Tachi and Tachi, 1979). Various endocrine, immune, or cellular factors can conceivably cause the premature death of the foreign implants.

Electron microscopic observation of the degenerative process of rat blastocysts following implantation in the endometrium of the mouse (Tachi and Tachi, 1979) indicated to us that complex immunological responses of the host might be implicated in the destruction of the foreign implants. We therefore examined the fate of rat blastocysts transferred to the uterus of nude mice, which congenitally lack the thymus and are defunct in T-cell-mediated immunity.

Nude mice (nu/nu) of BALB/c background were bred and maintained under SPF conditions at the Central Institute of Experimental Animals, Kawasaki, Japan. They were approximately 70 days old at the time of the experiments. Pseudopregnancy was induced by mating with vasectomized males of strain BALB/c.

Wistar rats (100 days old) served as the blastocyst donors. Trophoblast tissues were obtained from ectoplacental cones on Day 10 of gestation.

Blastocysts were collected in standard egg culture medium (Biggers *et al.*, 1971) and transferred to the uterus of nude mice under a sterile hood.

When blastocysts of ordinary BALB/c mice were transferred to the uterus of the nude mouse as a control experiment, they gave birth to healthy pups after a normal gestation period, and nursed them to weaning.

C. Tachi • Zoological Institute, Faculty of Science, University of Tokyo, Tokyo, Japan. *S. Tachi* • Department of Anatomy, Institute of Basic Medical Sciences, University of Tsukuba, Niihari-gun, Ibaraki-ken, Japan. *M. Yokoyama* • Central Institute for Experimental Animals, Kawasaki, Japan. *N. Osawa* • The 3rd Department of Internal Medicine, Faculty of Medicine, University of Tokyo, Tokyo, Japan.

Table 1. Implantation of Rat Blastocysts in the Uterus of the Nude Mouse

Time of sacrifice (hr)	Approximate age of blastocysts at sacrifice (hr pc)	No. of blastocysts transferred	No. of successful implantation sites
48	85	32	3
72	180	206	12
92	200	57	0
Total		295	15

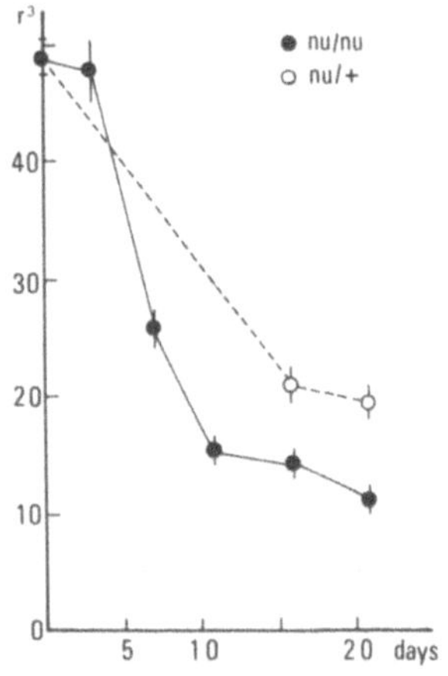

Figure 1. Changes in the size of rat trophoblast tissues transplanted subcutaneously to the nude mouse. Ordinate represents arbitrary units proportional to the actual volume of the grafts. Abscissa shows days after transplantation.

We have transferred 295 rat blastocysts to the uterus of pseudopregnant nude mice and recovered 15 xenogeneic implants (Table 1). No implants, however, have been observed in the host uterus 96 hr after transfer (Table 1).

We then examined the fate of the trophoblast tissues obtained from rat placentae and transplanted subcutaneously to the nude mouse. None of the grafts succeeded, and they regressed approximately at the same rate as in the control mice (nu/+) (Figure 1). However, a striking difference was observed in the histological appearance of the grafts recovered 14 days after the transfer between those from nude mice and those from nu/+ mice. Whereas the grafts from nu/+ mice contained many leukocytes and lymphocytes (Figure 2a), those from nude mice were entirely free of such cells (Figure 2b).

Our results, taken together, strongly indicate that cellular immunity mediated by T cells plays little role in the destruction of rat blastocysts in the uterus of the mouse following xenogeneic implantation.

ACKNOWLEDGMENTS. This work was supported in part by grants from the Ministry of Education of Japan (267002-1977 and 210709-1977) and from the Ford Foundation, New York.

References

Biggers, J. D., Whitten, W. K., and Whittingham, D. G., 1971, The culture of mouse embryos *in vitro*, in: *Methods in Mammalian Embryology* (J. C. Daniel, ed.), pp. 86–116, Freeman, San Francisco.

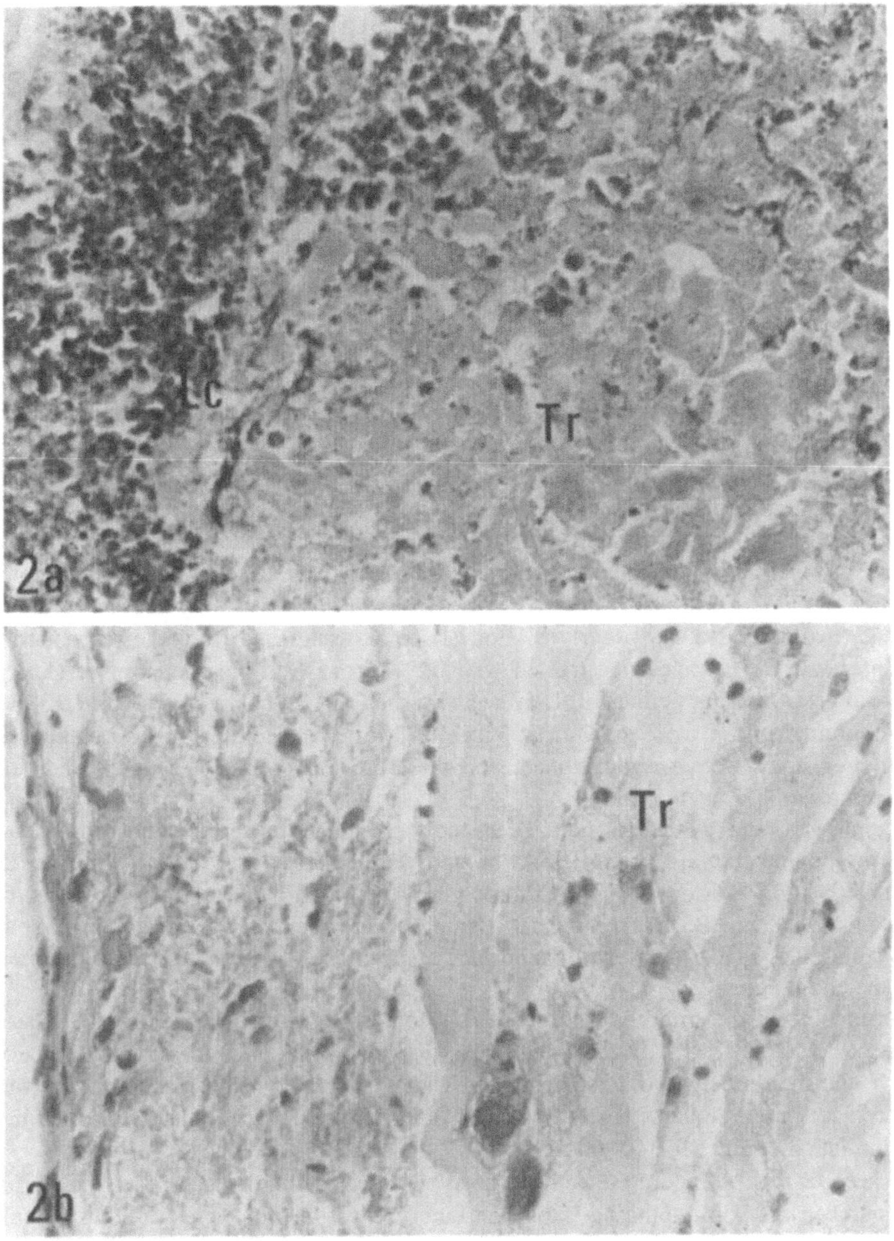

Figure 2. (a) Rat trophoblast tissue transplanted subcutaneously to nu/ + mouse, and recovered 14 days after surgery. LC, leukocytes and lymphocytes; Tr, trophoblast cells. (b) Rat trophoblast tissue transplanted subcutaneously to the nude mouse (nu/nu), and recovered 14 days after surgery. Tr, trophoblast cells.

Potts, D. M., Wilson, I. B., and Smith, M. S. R., 1970, Observations on rat eggs transplanted to the mouse uterus, *J. Reprod. Fertil.* **22:**425.

Tachi, S., and Tachi, C., 1979, Ultrastructural studies on maternal–embryonic cell interaction during experimentally induced implantation of rat blastocysts to the endometrium of the mouse, *Dev. Biol.* **68:**203.

Tarkowski, A. K., 1962, Interspecific transfers of eggs between rat and mouse, *J. Embryol. Exp. Morphol.* **10:**476.

27 Concanavalin-A-Binding Capacity of Preimplantation Mouse Embryos

J. T. Wu

Cell surface changes during embryonic development and differentiation may be reflected in differential cell agglutinability induced by lectins (Moscona, 1971; Oppenheimer, 1977) or in differential lectin binding capacity (O'Dell *et al.*, 1973). Changes in agglutinability of mouse embryos during cleavage have been shown by several studies (Pienkowski, 1974; Rowinski *et al.*, 1976; Sobel and Nebel, 1976; Rector and Granholm, 1978). The present study was designed to determine whether or not there is a change in the number of concanavalin A (Con A) binding sites during the preimplantation development of mouse embryos. Con A binds specifically to α-D-mannose and α-D-glucose residues (Sharon and Lis, 1972).

Day 2 (two-cell) to Day 4 (early blastocyst) embryos flushed from the reproductive tracts of pregnant CD-1 mice were treated with 0.5% Pronase in bicarbonate-free Hanks' solution to digest the zona pellucida. They were washed in two changes of Whitten's medium and cultured in the same medium at 37°C for 3 hr. The late blastocysts obtained on Day 5 of pregnancy were without zonae, and therefore the Pronase treatment and subsequent culture at 37°C were omitted.

All embryos were washed in cold Delbecco's phosphate-buffered saline (PBS) containing 4 mg bovine serum albumin (BSA)/ml and were exposed at 4°C for 30 min to 25, 50,100, or 200 μg [^{3}H]-Con A (38.8 Ci/mmol, New England Nuclear Corp., Boston)/ml PBS containing 1.8 mg BSA/ml. Duplicate samples were incubated in [^{3}H]-Con A solutions containing 0.06 M α-methyl-D-mannoside, a specific Con-A-binding sugar, to measure nonspecific binding.

After incubation the embryos were washed in four changes of PBSA and 1 to 3 Day 5 blastocysts or 3 to 12 Day 2, 3, or 4 embryos were placed in a 20-ml glass scintillation vial. Each vial then received 0.2 ml deionized water, and after being frozen and thawed twice, received 7.3 ml Aquasol (New England Nuclear Corp.). Each vial was counted in a liquid scintillation counter for 40–60 min.

The specific-Con-A-binding data were analyzed by Scatchard plot (Rosenthal, 1967) and the number of Con-A-binding sites per embryo was estimated as follows: 2-cell, 12×10^7; 4-cell, 13×10^7; 8- to 10-cell, 17×10^7; early blastocyst, 19×10^7; and implanting late blastocyst, 150×10^7. Thus the Con-A-binding capacity of mouse embryos increased with development, especially during implantation. However, the association constants decreased from 3×10^6 M^{-1} in 2-cell embryos to 0.83×10^6 M^{-1} in late blastocysts. This suggests that the Con-A-binding sites (probably glycoproteins) on the surface of the embryo may vary, during development, in the ratio of different binding groups (or sugars) and/or in the groups surrounding the binding sites.

J. T. Wu • Worcester Foundation for Experimental Biology, Shrewsbury, Massachusetts 01545.

ACKNOWLEDGMENTS. This study was supported by NIH Grant HD-12047 and by American Cancer Society Grant IN-100-D.

References

Moscona, A. A., 1971, Embryonic and neoplastic cell surfaces: Availability of receptors for concanavalin-A and wheat germ agglutinin, *Science* **171**:905.

O'Dell, D. S., Ortolani, G., and Monroy, A., 1973, Increased binding of radioactive concanavalin-A during maturation of *Ascidia* eggs, *Exp. Cell Res.* **83**:488.

Oppenheimer, S. B., 1977, Interactions of lectins with embryonic cell surfaces, *Curr. Top. Dev. Biol.* **11**:1.

Pienkowski, M., 1974, Study of the growth regulation of preimplantation mouse embryos using concanavalin-A, *Proc. Soc. Exp. Biol. Med.* **145**:464.

Rector, J. T., and Granholm, N. H., 1978, Differential concanavalin-A-induced agglutination of eight-cell preimplantation mouse embryos before and after compaction, *J. Exp. Zool.* **203**:497.

Rosenthal, H. E., 1967, A graphic method for the determination and presentation of binding parameters in a complex system, *Anal. Biochem.* **20**:525.

Rowinski, J., Solter, D., and Koprowski, H., 1976, Change of concanavalin-A induced agglutinability during preimplantation mouse development, *Exp. Cell Res.* **100**:404.

Sharon, M., and Lis, H., 1972, Lectins: Cell-agglutinating and sugar specific proteins, *Science* **177**:949.

Sobel, J. S., and Nebel, L., 1976, Concanavalin-A agglutinability of developing mouse trophoblast, *J. Reprod. Fertil.* **47**:399.

28 The Use of Pharmacological Agents to Study Implantation

D. W. Hahn and J. L. McGuire

The diverse aspects of implantation can be studied by many different approaches, including biochemical, morphological, physiological, and pharmacological. Although often not considered to be basic research, studies of drug effects and mechanism of drug action studies have historically provided much of the information we have today about basic physiology.

During the last 10 years, a number of groups (Emmens, 1970; Duncan and Wheeler, 1975; Aref and Hafez, 1977), including our own, have investigated a variety of pharmacological agents that have effects on implantation, and which could offer investigators new experimental tools for studying the process of implantation. During the course of those studies, we learned that various drugs affect implantation by different mechanisms of action, only some of which we believe will be of value to reproductive biologists.

ORF 3858 (2-methyl-3-ethyl-4-phenyl-Δ^4-cyclohexene carboxylic acid), coined the "morning-after pill" or "postcoital antifertility agent," has been studied extensively ever since its synthesis by Ortho in 1961 (Morris and Van Wagenen, 1966; Yard *et al.*, 1969, Blye, 1970, McGuire *et al.*, 1971; Greenslade and Hahn, 1975). Our studies indicated that (a) ORF 3858 stimulates tubal transport of embryos in rats as does diethylstilbestrol; (b) ORF 3858 does not bind to uterine cytosol estrogen receptors whereas the *p*-hydroxylated metabolites do bind with high affinity; (c) ORF 3858 induces delayed uterine RNA synthesis whereas the *p*-hydroxylated derivative and estrogens induce an immediate increase in uterine RNA synthesis; and (d) the anti-implantation activity of ORF 3858 is lost when biotransformation of this compound to the *p*-hydroxylated form is inhibited. We concluded that ORF 3858 merely acts as a proestrogen, and we do not believe it is a unique tool for the study of implantation.

ORF 8511 [1-diphenylmethylenyl-2-methyl-3-ethyl-4-acetoxycyclohexane] is a weak estrogenic compound and is structurally similar to ORF 3858 and F 6103 [bis(*p*-acetoxyphenyl)-2-methyl-cyclohexylidene methane]. At the time of our studies, many investigators believed that, with these compounds and other nonsteroidal estrogenic compounds, a separation of estrogenicity and antifertility activities had been achieved (Bacic *et al.*, 1970; Blye, 1970; Emmens, 1970). We hypothesized, however, that a relationship existed between the hormonal activity of these compounds and their estrogenicity. Our studies in four species suggested that ORF 8511 and similar nonsteroidal compounds may owe their postcoital antifertility activity to their estrogenicity and that there is a species difference in sensitivity to the estrogenicity of these compounds (Hahn *et al.*, 1974). If our conclusions are correct, neither ORF 8511 nor F 6103 could be considered to have unique effects on implantation.

Other drugs we have studied, however, may be unique and therefore add to our under-

D. W. Hahn and J. L. McGuire • Research Laboratories, Ortho Pharmaceutical Corporation, Raritan, New Jersey 08869.

standing of the dynamic aspects of the implantation mechanism. ORF 5513 [3,5-bis(dimethylamino)-1,2,4-dithiazolium chloride] is a nonsteroidal anti-implantation agent that lacks endocrine activity. This compound is unique among all the compounds we have studied in that it has pronounced effects throughout the entire reproductive process; it inhibits ovulation and implantation and interrupts established pregnancy, in both early and late stages of gestation (Hahn *et al.*, 1977b). Most compounds affect specific stages of the reproductive process, so there must be some common denominator with ORF 5513, which is essential for the normal development of the embryo, including implantation. If its action can be identified, it may well be useful in better understanding the process of implantation.

ORF 9371 (17α-pregna-20-yne-4-en-3-one-17β-acetoxyoxine) and ORF 9326 (17β-acetoxy-2α-chloro-3-(*p*-nitrophenoxy)imino-5-androstane) are both steroidal agents that inhibit implantation. ORF 9371 emerged from a program designed to identify progesterone antagonists to progestin receptor binding. It inhibits implantation in several species and prevents uterine proliferation induced by progesterone in rabbits (Hahn *et al.*, 1975). ORF 9326 is a derivative of dihydrotestosterone. If it is administered orally prior to implantation in rats, it does not prevent implantation but resorption of embryos subsequently occurs. ORF 9326 also interrupts the postimplantation stage of gestation (Hahn *et al.*, 1977a). Although the mode of action of these latter pharmacological agents is yet to be clearly identified, we believe that knowledge of how these compounds work will lead to a better understanding of the process of implantation.

References

Aref, I., and Hafez, E. S. E., 1977, Postcoital contraception: Physiological and clinical parameters, *Obstet. Gynecol. Surv.* **32**:417–437.

Bacic, M., Engstrom, L., Johannisson, E., Leideman, T., and Diczfalusy, E., 1970, Effect of F 6103 on implantation and early gestation in women, *Acta Endocrinol. (Copenhagen)* **64**:705–717.

Blye, R. P., 1970, The effect of estrogens and related substances on embryonic viability, *Adv. Biosci.* **4**:326–343.

Blye, R. P., and Homm, R., 1967, Antizygotic activity of 2-methyl-3-ethyl-4-cyclohexene carboxylic acid in the rat, *Fed. Proc. Fed. Am. Soc. Exp. Biol.* **26**:486.

Duncan, E. W., and Wheeler, R. G., 1975, Pharmacological and mechanical control of implantation, *Biol. Reprod.* **12**:143–175.

Emmens, C. W., 1970, Postcoital contraception, *Br. Med. Bull.* **26**:45–51.

Greenslade, F. C., and Hahn, D. W., 1975, Stimulation of uterine RNA synthesis in mice by the postcoital antifertility agent, ORF 3858, *J. Reprod. Fertil.* **45**:401–403.

Hahn, D. W., Allen, G., McGuire, J. L., and Da Vanzo, J. P., 1974, A relationship between estrogenicity and antifertility activity of ORF 8511 and similar nonsteroidal antiimplantive agents, *Contraception* **9**:393–401.

Hahn, D. W., Allen, G. O., Polidoro, J. P., and McGuire, J. L., 1975, ORF 9371: A new contragestational agent which possesses antiprogestational activity, *Proceedings, 57th Annual Meeting of the Endocrine Society*, p. 245.

Hahn, D. W., Allen, G. O., McConnell, R. F., and McGuire, J. L., 1977a, ORF 5513: Prototype for a new class of antifertility agents, *Proceedings, 10th Annual Meeting of the Society for the Study of Reproduction*, p. 64.

Hahn, D. W., Allen, G. O., and Greenslade, F. C., 1977b, The post-implantive contragestational activity of 17β-acetoxy-2α-chloro-3-(*p*-nitrophenoxy)imino-5α-androstane (ORF 9326) in the rat, *Fed. Proc. Fed. Am. Soc. Exp. Biol.* **36**:342.

McGuire, J. L., Turner, G. D., and Greenslade, F. C., 1971, Studies on the effect of 2-methyl-3-ethyl-4-phenyl-Δ^4-cyclohexene carboxylic acid (ORF 3858) and its *p*-hydroxylated metabolites on uptake of ^{3}H-estradiol-17β by the uterus, *Proc. Soc. Exp. Biol. Med.* **136**:146–149.

Morris, J. M., and Van Wagenen, G., 1966, Compounds interfering with ovum implantation and development. III. The role of estrogens, *Am. J. Obstet. Gynecol.* **96**:804–813.

Yard, A. S., Juhasz, L., and Grimes, R. M., 1969, Studies on the anti-fertility effects and metabolism of a new, postcoital oral contraceptive, 2-methyl-3-ethyl-4-phenyl-Δ4-cyclohexene carboxylic acid, sodium salt (ORF 3858), *J. Pharmacol. Exp. Ther.* **167**:105–116.

Participants

E. C. Amoroso
A.R.C. Institute of Animal Physiology
Babraham
Cambridge
England

D. T. Armstrong
Departments of Obstetrics and Gynecology and of Physiology
University of Western Ontario
London, Ontario
Canada

Ricardo H. Asch
Department of Obstetrics and Gynecology
University of Texas Health Science Center
San Antonio, Texas 78284

H. J. Baeten
Division of Natural Science
St. Norbert College
De Pere, Wisconsin 54115

Billy Baggett
Department of Biochemistry
Medical University of South Carolina
Charleston, South Carolina 29403

Uriel Barkai
Department of Zoology
University of Tel Aviv
Ramat Aviv
Tel Aviv
Israel

Kenneth L. Barker
Departments of Obstetrics and Gynecology and of Biochemistry
University of Nebraska College of Medicine
Omaha, Nebraska 68105

Fuller W. Bazer
Department of Animal Science
University of Florida
Gainesville, Florida 32611

Henning M. Beier
Department of Anatomy and Reproductive Biology
Medical Theoretical Institute
Rheinisch-Westfälische Techniche Hochschule
Aachen, Federal Republic of Germany

John D. Biggers
Laboratory of Human Reproduction and Reproductive Biology
Harvard Medical School
Boston, Massachusetts 02115

Larry R. Boots
Department of Obstetrics and Gynecology
University of Alabama Medical Center
Birmingham, Alabama 35294

David W. Bullock
Departments of Cell Biology and of Obstetrics and Gynecology
Baylor College of Medicine
Houston, Texas 77030

James R. Carollo
Department of Anatomy
University of Oregon Health Science Center
Portland, Oregon 97201

M. C. Chang
Worcester Foundation for Experimental Biology
Shrewsbury, Massachusetts 01545

Daniel J. Chávez
Department of Human Anatomy
School of Medicine
University of California
Davis, California 95616

J. R. Chenault
Animal Health–Reproduction Unit
The Upjohn Company
Kalamazoo, Michigan 49001

Beverly S. Chilton
Department of Obstetrics and Gynecology
University of Pennsylvania Medical School
Philadelphia, Pennsylvania 19104

Hans-Werner Denker
Abteilung Anatomie
der RWTH
D-5100 Aachen
West Germany

Dharam S. Dhindsa
Reproductive Biology Study Section
Dviision of Research Grants
NIH
Bethesda, Maryland 20205

Zeev Dickmann
Department of Gynecology and Obstetrics
University of Kansas Medical Center
Kansas City, Kansas 66103

Alan B. Dudkiewicz
Department of Biology
University of Houston
Houston, Texas 77004

Allen C. Enders
Department of Human Anatomy
University of California School of Medicine
Davis, California 95616

Charles J. Epstein
Departments of Pediatrics and of Biochemistry and Biophysics
University of California
San Francisco, California 94143

C. A. Finn
Departments of Veterinary Physiology and Pharmacology
University of Liverpool
Liverpool
England

S. P. Ford
Department of Animal Sciences
Iowa State University
Ames, Iowa 50010

Randall Given
Department of Anatomy
School of Medicine
University of Oregon Health Sciences Center
Portland, Oregon 97201

Robert Glass
Laboratory of Radiobiology
Department of Obstetrics, Gynecology and Reproductive Science, and Department of Biochemistry and Biophysics
University of California
San Francisco, California 94143

Stanley R. Glasser
Department of Cell Biology
Baylor College of Medicine
Houston, Texas 77030

D. W. Hahn
Research Laboratories
Ortho Pharmaceutical Corporation
Raritan, New Jersey 08869

Linda Hall
Department of Cell Biology
Baylor College of Medicine
Houston, Texas 77030

Robert E. Hammer
Departments of Anatomy and Pharmacology
Wayne State University, School of Medicine
Detroit, Michigan 48201

James Hawkins
Department of Cell Biology
Baylor College of Medicine
Houston, Texas 77030

R. B. Heap
A.R.C. Institute of Animal Physiology
Babraham
Cambridge
England

Loren H. Hoffman
Department of Anatomy
Vanderbilt Medical School
Nashville, Tennessee 37232

Paula C. Hoos
Department of Anatomy
Vanderbilt University Medical School
Nashville, Tennessee 37232

Roger C. Hoversland
Department of Anatomy
School of Medicine
University of Oregon Health Sciences Center
Portland, Oregon 97201

Yu-Chih Hsu
Laboratory of Mammalian Development
Department of Pathobiology
School of Hygiene and Public Health
The Johns Hopkins University
Baltimore, Maryland 21205

Judith Ilan
Department of Reproductive Biology
Case Western Reserve University School of Medicine
Cleveland, Ohio 44106

M. H. Johnson
Department of Anatomy
University of Cambridge
Cambridge
England

O'Neal Johnston
Department of Biochemistry
Merrell Research Center
Cincinnati, Ohio 45215

M. S. Joshi
Department of Anatomy
University of North Dakota
Grand Forks, North Dakota 58201

Sharad G. Joshi
Department of Obstetrics and Gynecology
Albany Medical College
Albany, New York 12208

Joanne Julian
Department of Cell Biology
Baylor College of Medicine
Houston, Texas 77030

Alvin M. Kaye
Department of Hormone Research
Weizmann Institute of Science
Rehovot
Israel

Peter L. Kaye
Division of Medical Biochemistry
Faculty of Medicine
University of Calgary
Calgary, Alberta
Canada

T. G. Kennedy
Departments of Obstetrics and Gynecology and of Physiology
The University of Western Ontario
London, Ontario
Canada

Gerald M. Kidder
Laboratory of Radiobiology
University of California
San Francisco, California 94122

A. A. Kiessling
Department of Anatomy
School of Medicine
University of Oregon Health Sciences Center
Portland, Oregon 97201

Gordon King
Department of Animal and Poultry Science
University of Guelph
Guelph, Ontario N1G 2W1
Canada

Peretz Kraicer
Department of Zoology
University of Tel Aviv
Ramat Aviv
Tel Aviv, Israel

O. Kreitmann
Department of Endocrinology
CHU Toulouse Rangueil
Universite Paul Sabatier
Toulouse
France

William B. Langan
Department of Biology
Villanova University
Villanova, Pennsylvania 19085

Fernand O. G. Leroy
Laboratory of Gynecology and Research on Human Reproduction
St. Pierre Hospital
Free University
Brussels
Belgium

David T. MacLaughlin
Vincent Research Laboratories
Massachusetts General Hospital
Boston, Massachusetts 02114

Shirley A. McCormack
Department of Cell Biology
Baylor College of Medicine
Houston, Texas 77030

Terry Magnuson
Departments of Pediatrics and of Biochemistry and Biophysics
University of California
San Francisco, California 94143

Cole Manes
La Jolla Cancer Research Foundation
La Jolla, California 92037

John Marston
Department of Anatomy
Medical School
Birmingham B15 2TJ
United Kingdom

Len Martin
Hormone Physiology Department
Imperial Cancer Research Fund
Lincoln's Inn Fields
London WC2A 3PX
United Kingdom

Tony May
Department of Cell Biology
Baylor College of Medicine
Houston, Texas 77030

Rodney A. Mead
Department of Biological Sciences
University of Idaho
Moscow, Idaho 83843

Bevan G. Miller
Department of Obstetrics/Gynecology
University of Western Ontario Medical School
London, Ontario N6A 5A5
Canada

Jerald A. Mitchell
Department of Anatomy and Pharmacology
Wayne State University, School of Medicine
Detroit, Michigan 48201

Marilyn Monk
MRS Mammalian Development Unit
4 Stephenson Way
London NW1 2HE
United Kingdom

Bruce C. Moulton
Departments of Obstetrics and Gynecology and of Biological Chemistry
University of Cincinnati College of Medicine
Cincinnati, Ohio 45267

Deborra E. Mullins
Department of Biochemistry and Molecular Biology
University of Florida
Gainesville, Florida 32610

S. R. Munshi
Division of Biology
Institute for Research in Reproduction
Bombay 400 012
India

Rudolph Neri
Department of Physiology
Schering Plough Research Division
Bloomfield, New Jersey 07003

Ove Nilsson
Department of Anatomy
Biomedical Centre
Uppsala S-751-23
Sweden

Thomas F. Ogle
Department of Physiology
School of Medicine
Medical College of Georgia
Augusta, Georgia 30912

Margaret Ward Orsini
Department of Anatomy
University of Wisconsin Medical School
Madison, Wisconsin 53706

Helen A. Padykula
Department of Anatomy
University of Massachusetts Medical School
Worcester, Massachusetts 01605

Roger A. Pedersen
Laboratory of Radiobiology and Department of Anatomy
University of California
San Francisco, California 94143

Sara Peleg
Department of Hormone Research
The Weizmann Institute of Science
Rehovot
Israel

Alexandre Psychoyos
Laboratoire de Physiologie de la Reproduction
C.N.R.S.-ER 203
Hopital de Bicêtre (INSERM)
Bicêtre
France

S. Rahman
Department of Reproductive Biology
Case Western Reserve University
University Hospital of Cleveland
Cleveland, Ohio 44106

Judith C. Rankin
Department of Biochemistry
Medical University of South Carolina
Charleston, South Carolina 29403

Jerry R. Reel
Chemistry and Life Sciences Group
Research Triangle Institute
Research Triangle Park
North Carolina 27709

Catherine Rice
Laboratory of Human Reproduction and Reproductive Biology
Harvard Medical School
Boston, Massachusetts 02115

George S. Richardson
Department of Gynecology (Vincent I)
Massachusetts General Hospital
Boston, Massachusetts 02114

R. Michael Roberts
Department of Biochemistry and Molecular Biology
University of Florida
Gainesville, Florida 32610

J. Rossant
Department of Biological Sciences
Brock University
St. Catherines, Ontario
Canada

William A. Sadler
Population and Reproduction Branch
Center for Population Research (NICHD)
NIH (Landow Bldg. C-733)
Bethesda, Maryland 20205

P. G. Satyaswaroop
Department of Obstetrics and Gynecology
Mount Sinai School of Medicine
New York, New York 10029

Brij B. Saxena
Departments of Obstetrics and Gynecology and of Medicine
Cornell University Medical College
New York, New York 10021

Pepper Schedin
Department MCD Biology
University of Colorado
Boulder, Colorado 80320

Joel Schindler
Roche Institute of Molecular Biology
Nutley, New Jersey 07110

Gilbert A. Schultz
Division of Medical Biochemistry
University of Calgary
Calgary, Alberta
Canada

J. Elliot Scott
Department of Anatomy
University of Manitoba
Winnipeg, Manitoba
Canada

Sheldon Segal
The Rockefeller Foundation
New York, New York 10036

Martin J. Serra
Department of Chemistry
Allegheny College
Meadsville, Pennsylvania 16335

M. C. Shelesnyak
River House
Cherryfield Road
Drayden, Maryland 20630

Michael I. Sherman
Roche Institute of Molecular Biology
Nutley, New Jersey 07110

J. Sabina Sobel
Laboratory of Radiobiology
University of California
San Francisco, California 94143

Pierre Soupart
Department of Obstetrics and Gynecology
Vanderbilt University Medical School
Nashville, Tennessee 37232

A. Spindle
Laboratory of Radiobiology
University of California
San Francisco, California 94143

George Stancel
Department of Pharmacology
University of Texas Health Science Center
Houston, Texas 77025

Roger T. Stone
USDA—MARC (Box 166)
Clay Center, Nebraska 68933

Sidney Strickland
Rockefeller University
New York, New York 10021

M. A. H. Surani
A.R.C. Institute of Animal Physiology
Cambridge
England

Daniel Szollosi
Station de Physiologie Animale
INRA
Jouy-en Josas 78350
France

C. Tachi
Zoological Institute
Faculty of Science
University of Tokyo
Tokyo
Japan

S. Tachi
Department of Anatomy
Institute of Basic Medical Sciences
University of Tsukuba
Niihari-gun, Ibaraki-ken
Japan

Richard J. Tasca
School of Life and Health Sciences
University of Delaware
Newark, Delaware 19711

Charles A. Turbil
Department of Obstetrics and Gynecology
Vanderbilt University School of Medicine
Nashville, Tennessee 37232

Henry G. Trapido-Rosenthal
Department of Biochemistry and Biophysics
University of Hawaii
Honolulu, Hawaii 96822

Jonathan Van Blerkom
Department of Molecular, Cellular, and Developmental Biology
University of Colorado
Boulder, Colorado 80309

Claude A. Villee
Department of Biological Chemistry and Laboratory of Human Reproduction and Reproductive Biology
Harvard Medical School
Boston, Massachusetts 02115

Carol M. Warner
Department of Biochemistry and Biophysics
Iowa State University
Ames, Iowa 50010

Marlene Warner
Department of Obstetrics and Gynecology
Baylor College of Medicine
Houston, Texas 77030

H. M. Weitlauf
School of Medicine
Department of Anatomy
University of Oregon Health Sciences Center
Portland, Oregon 97201

S. Wintenberger-Torrès
Station de Physiologie Animale–I.N.R.A.
Jouy-en-Josas
France

J. T. Wu
Worcester Foundation for Experimental Biology
Shrewsbury, Massachusetts 05145

Linda Wudl
Department of Genetic Toxicology
Allied Chemical Co.
Morristown, New Jersey 07960

Jerome M. Yochim
Department of Physiology and Cell Biology
University of Kansas
Lawrence, Kansas 66045

Koji Yoshinaga
Population and Reproduction Branch
Center for Population Research (NICHD)
NIH (Landow Bldg. C-733)
Bethesda, Maryland 20205

Index

GPSR Compliance

The European Union's (EU) General Product Safety Regulation (GPSR) is a set of rules that requires consumer products to be safe and our obligations to ensure this.

If you have any concerns about our products, you can contact us on ProductSafety@springernature.com

In case Publisher is established outside the EU, the EU authorized representative is:

Springer Nature Customer Service Center GmbH
Europaplatz 3
69115 Heidelberg, Germany

Batch number: 10838062

Printed by Printforce, the Netherlands